McGRAW-HILL YEARBOOK OF
Science &
Technology

2002

McGRAW-HILL YEARBOOK OF
Science &
Technology

2002

**Comprehensive coverage of recent events and research as compiled by
the staff of the McGraw-Hill Encyclopedia of Science & Technology**

McGraw-Hill
New York   Chicago   San Francisco   Lisbon   London   Madrid   Mexico City   Milan
New Delhi   San Juan   Seoul   Singapore   Sydney   Toronto

Library of Congress Cataloging in Publication data

McGraw-Hill yearbook of science and technology.
1962–   . New York, McGraw-Hill.

  v. illus. 26 cm.
  Vols. for 1962–   compiled by the staff of the
McGraw-Hill encyclopedia of science and technology.
    1. Science—Yearbooks.   2. Technology—
Yearbooks.   1. McGraw-Hill encyclopedia of
science and technology.
Q1.M13   505.8   62-12028

ISBN 0-07-137416-7
ISSN 0076-2016

McGraw-Hill

A Division of The McGraw-Hill Companies

1 2 3 4 5 6 7 8 9 0   DOW/DOW   0 7 6 5 4 3 2 1

This book was printed on acid-free paper.

*It was set in Garamond Book and Neue Helvetica Black Condensed by
TechBooks, Fairfax, Virginia. The art was prepared by TechBooks.
The book was printed and bound by R. R. Donnelley & Sons Company,
The Lakeside Press.*

Contents

Editing, Design, & Production Staff

Roger Kasunic, Director of Editing, Design, and Production

Joe Faulk, Editing Manager

Frank Kotowski, Jr., Senior Editing Supervisor

Ron Lane, Art Director

Thomas G. Kowalczyk, Production Manager

Consulting Editors

Dr. Milton B. Adesnik. *Department of Cell Biology, New York University School of Medicine, New York.* CELL BIOLOGY.

Prof. Eugene A. Avallone. *Consulting Engineer; Professor Emeritus of Mechanical Engineering, City College of the City University of New York.* MECHANICAL AND POWER ENGINEERING.

A. E. Bailey. *Formerly, Superintendent of Electrical Science, National Physical Laboratory, London, England.* ELECTRICITY AND ELECTROMAGNETISM.

Prof. William P. Banks. *Chairman, Department of Psychology, Pomona College, Claremont, California.* PHYSIOLOGICAL AND EXPERIMENTAL PSYCHOLOGY.

Dr. Eugene W. Bierly. *American Geophysical Union, Washington, DC.* METEOROLOGY AND CLIMATOLOGY.

Prof. Carrol Bingham. *Department of Physics, University of Tennessee, Knoxville.* NUCLEAR AND ELEMENTARY PARTICLE PHYSICS.

Dr. Chaim Braun. *Altos Management Consultants Inc., Los Altos, California.* NUCLEAR ENGINEERING.

Dr. Melbourne Briscoe. *Ocean, Atmosphere, and Space and Technology, Office of Naval Research, Arlington, Virginia.* OCEANOGRAPHY.

Robert D. Briskman. *Executive Vice President, Engineering Department, Sirius Satellite Radio, New York.* TELECOMMUNICATIONS.

Prof. Wai-Fah Chen. *Dean, College of Engineering, University of Hawaii.* CIVIL ENGINEERING.

Dr. John F. Clark. *Director, Graduate Studies, and Professor, Space Systems, Spaceport Graduate Center, Florida Institute of Technology, Satellite Beach.* SPACE TECHNOLOGY.

Prof. David L. Cowan. *Chairman, Department of Physics and Astronomy, University of Missouri, Columbia.* CLASSICAL MECHANICS AND HEAT.

Dr. Carol Creutz. *Department of Chemistry, Brookhaven National Laboratory.* INORGANIC CHEMISTRY.

Prof. Ron Darby. *Department of Chemical Engineering, Texas A&M University, College Station.* CHEMICAL ENGINEERING.

Dr. Michael Descour. *Optical Sciences Center, University of Arizona.* ELECTROMAGNETIC RADIATION AND OPTICS.

Prof. Turgay Ertekin. *Chairman, Department of Petroleum and Natural Gas Engineering, Pennsylvania State University, University Park.* PETROLEUM ENGINEERING.

Barry A. J. Fisher. *Director, Scientific Services Bureau, Los Angeles County Sheriff's Department, Los Angeles, California.* FORESIC SCIENCE AND TECHNOLOGY.

Dr. John Gordon. *School of Forestry and the Environment, Yale University, New Haven, Connecticut.* FORESTRY.

Dr. Richard L. Greenspan. *The Charles Stark Draper Laboratory, Cambridge, Massachusetts.* NAVIGATION.

Article Titles and Authors

The 2002 *McGraw-Hill Yearbook of Science & Technology* provides the reader with a wide overview of important recent developments in science, technology, and engineering, as selected by our distinguished board of consulting editors. At the same time, it satisfies the reader's need to stay informed about important trends in research and development that will fundamentally influence future understanding and practical applications of knowledge in fields ranging from astronomy to zoology. Readers of the *McGraw-Hill Encyclopedia of Science & Technology* will find the *Yearbook* to be a valuable companion publication, enhancing the timeliness and depth of the *Encyclopedia*.

In the 2002 edition, we continue to document the rapid advances in such areas as cell biology and molecular medicine, cosmology, chemistry, environmental science, forensics, and theoretical and experimental physics, as well as the profound influence of advances in computing and information technology on all areas of engineering and science.

Each contribution to the *Yearbook* is a concise yet authoritative article authored by one or more specialists in the field. We are pleased that noted researchers have been supporting the *Yearbook* since its first edition in 1962 by taking time to share their knowledge with our readers. The topics are selected by our consulting editors in conjunction with our editorial staff based on present significance and potential applications. McGraw-Hill strives to make each article as readily understandable as possible for the nonspecialist reader through careful editing and the extensive use of graphics, much of which is prepared specially for the *Yearbook*.

Librarians, students, teachers, the scientific community, journalists and writers, and the general public continue to find in the *McGraw-Hill Yearbook of Science & Technology* the information they need in order to follow the rapid pace of advances in science and technology and to understand the developments in these fields that will shape the world of the twenty-first century.

Mark D. Licker
PUBLISHER

McGRAW-HILL YEARBOOK OF
Science &
Technology

2002

African superplume

The largest discrete structure in the Earth's interior, known as the African superplume, has been found beneath southern Africa within the outer layer known as the mantle. The mantle, comprising solid but flexible rock about 3000 km (1800 mi) thick, is where the flow associated with plate tectonics occurs. While it is not certain what the African superplume is exactly, it may be a giant upwelling of mantle material. Southern Africa's high topography and unusual land surface are indicative of uplift that may be dynamically supported by this rising superplume. This interpretation is qualitatively consistent with a range of observations showing that Africa is experiencing active intraplate volcanism and rifting, as well as with seismic and other quantitative evidence of anomalous mantle features below the continent.

One interpretation of the African superplume is that it represents a long-lived upwelling of hot mantle material. It is likely, however, that at least part of the superplume is denser than the surrounding mantle, which slows its rise upward.

Seismic evidence. Geophysicists use seismic waves generated by earthquakes and recorded by global networks of seismometers to image features within the Earth. One class, called primary or P waves, is identical to familiar acoustic or pressure waves by which sound travels. Another class of waves, called shear or S waves, occurs only in solids. Seismic waves are very sensitive to the temperature and composition of the material through which they pass. In recent years, the resolution of seismological techniques has improved considerably, and global seismologists are now using a method called tomography, akin to CAT scans used in medicine, to image the interior of the Earth.

Velocity variations. Using tomography, seismologists have discovered that the most significant seismic structure in the deep mantle lies beneath southern Africa and the South Atlantic Ocean. In the structure, referred to as the African anomaly, seismic wave velocities are lower than normal; this may indicate that the structure is hotter than the surrounding mantle. Studies of S waves show that the low-velocity anomaly extends at least 1500 km (900 mi) above

the core-mantle boundary and has an average velocity a few percent lower than normal mantle material. Tomography shows that the anomaly has a complex three-dimensional shape (see **illus.**). This anomaly covers an extensive region in the lowermost mantle, about the size of Africa in map view. At a depth of 2000 km (1200 mi), the anomaly is centered beneath southern Africa. The African mantle anomaly appears to be tilted in an easterly-northeasterly direction and may be connected to a low-velocity structure in the upper mantle beneath the East African Rift. This rift is a region where the continental crust is in the early stages of being split apart by horizontal motions, much like the way it is thought that a supercontinent splits apart. All of this evidence supports the concept that the anomaly reflects a hot superplume of mantle rock rising and spreading beneath Africa. *See also* ULTRALOW-VELOCITY ZONE (SEISMOLOGY).

Through most of the Earth's mantle, down to at least 2000 km depth, the velocities of the two wave types, P and S, are well correlated. This is compelling evidence that much of the variation in seismic velocity observed from place to place in the upper two-thirds of the mantle is due to changes in temperature, as expected in a mantle that is undergoing thermal convection. However, seismic images indicate that the velocities of the two wave types are poorly correlated below 2000 km depth. This may mean that there is widespread variation in the composition of rocks in the lowest portions of the mantle, perhaps due to the presence of the superplume.

Density variations. If we could measure the density of rock from place to place, we would have a firmer understanding of mantle dynamics and the origin of the African superplume. Unfortunately, P and S waves are relatively insensitive to density. However, the planet-wide oscillations generated by a large earthquake are more sensitive to density. Recently, although with significant uncertainty, the first seismic images of lateral variations in density have been made. The expectation was that the African superplume would have a density lower than normal due to thermal expansion of the presumably hot rock. However, it was found that the African anomaly may actually be slightly denser than the surrounding rocks

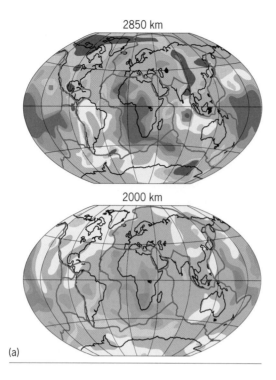

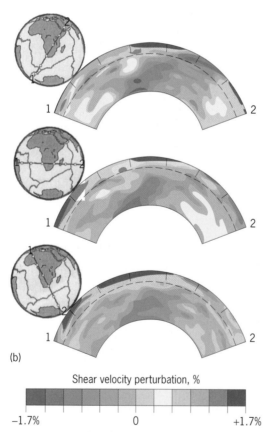

Shear velocity perturbation, %

−1.7%   0   +1.7%

Seismic image of African superplume. (a) Two maps
showing S-wave velocity zones at 2000 km and 2850 km
(1200 mi and 1750 mi) into the mantle. The later map is just
above the core-mantle boundary. (b) Cross sections of the
superplume showing how the huge anomaly extends from
the core-mantle boundary to the shallower mantle just
below the Red Sea. (*Reproduced by permission from
M. Gurnis et al., Geochemistry, Geophysics, Geosystems 1,
Pap. 1999-GC000035; copyright 2000, American
Geophysical Union*)

in a zone that extends up to about 500 km (300 mi)
above the core-mantle boundary. The measurements
suggest that high-density mantle rock may be surrounded by a giant low-density region, which could
be a thermal halo (warmer material surrounding
cooler material).

Surface evidence. The African superplume is believed to influence the Earth's surface in three ways:
through lifting the topography upward, through volcanism, and through rifting (or splitting the Earth's
surface apart). However, directly above the superplume itself, in southern Africa, there is no volcanism at all, and this is quite mysterious because the
topography is elevated to an unusual height.

Africa not only has the highest elevation for a
continent which has not undergone any recent collision with another continent, but also has a distinctive shape reflecting broad-scale uplift. A spectacular feature indicative of the uplift is the Great
Escarpment, which extends nearly unbroken for almost 3000 km (1800 mi) around the southern margin of Africa. The Great Escarpment separates a vast
interior plateau, elevated nearly a kilometer above
sea level, from a narrow coastal strip. Moreover, the
elevated topography extends from southern Africa
through east Africa and the East African Rift Valley
where the Earth's surface is actively splitting apart.
This rift valley intersects the Red Sea, a place where
new oceanic crust has formed by sea-floor spreading. In addition, the elevated African topography is
continuous with unusually shallow ocean basins surrounding Africa. The entire region from the South
Atlantic to eastern Africa has been referred to as the
African superswell.

The historical development of the southern
African topography corresponds quite well to the
predictions of some computer models of mantle dynamics, which simulate the slow motions within the
Earth's interior and shed light on long-term dynamics behind mantle and surface features. Such models
estimate that uplift of southern Africa has been occurring since the breakup of Gondwanaland, with a
significant amount occurring during the last 65 million years of Earth history. The African superplume is
found below what was once the center of the ancient
Gondwanaland supercontinent. Improved understanding of the superplume might thus help to solve
the puzzle of why supercontinents assemble and
break apart.

Gravity anomalies. The African superplume appears to be the largest structure in the mantle with
a low seismic velocity. However, beneath the central Pacific Ocean another large low-seismic-velocity
structure called the Pacific superplume is found that,
like the African, is associated with the swelling upward of the Earth's surface. This bathymetric (water
depth) swelling, called the Pacific superswell, may
have formed in the Cretaceous Period (before about
65 million years ago), and so may be older than the
African superplume. Both the African and the
Pacific superplumes are located in the center of
giant anomalies in the Earth's gravity field. These

gravity anomalies indicate that the Earth's surface is being swelled upward, supporting the inferences made from geology. Within the boundaries of these gravity anomalies are found nearly all of the world's hot spots. Hot spots, such as Hawaii, are isolated sources of volcanism which occur within the interior of plates and do not seem to be related to plate tectonics, but may be related to superplumes.

Origin. When the surface expression of the superplume—high topography in southern Africa and high topography, volcanism, and rifting in eastern Africa—is combined with the seismic features of the Earth's deep interior and evidence from computer modeling, a tentative picture of the origin of the superplume can be assembled.

Sophisticated computer software has yielded dynamic models of the Earth's interior which are consistent with both seismic images and surface geology. These models show that if the African superplume were caused only by hot mantle, the topographic uplift would be kilometers higher than it is today. If the anomaly is only partially thermal, however, the models match all of the observations.

The African anomaly may be a thermal (that is, warmer than the surrounding material) structure with a high-density core. The high-density core would counteract the thermal buoyancy so that the feature rises more slowly than expected. Essentially, a dense structure with an anomalous chemical composition is surrounded by a thermal halo. In fact, computer models of the evolution of the mantle have long suggested that if dense material exists at the core-mantle boundary, so-called mantle dregs, then such material would tend to accumulate at the base of giant mantle upwellings. There are various hypotheses regarding the origin of this dense material, including that it may be primordial mantle left over from the origin of the Earth; it may be a product of the reaction between the mantle and the molten iron core; or it may be a result of the settling of ocean crust after billions of years of plate tectonic cycling.

For background information *see* AFRICA; EARTH, GRAVITY FIELD OF; EARTH INTERIOR; HOT SPOTS (GEOLOGY); PLATE TECTONICS; SEISMOLOGY in the McGraw-Hill Encyclopedia of Science & Technology.
                                    Michael Gurnis

Bibliography. M. Gurnis, Sculpting the earth from inside out, *Sci. Amer.*, 284(3)40–47, 2001; M. Gurnis et al., Constraining mantle density structure using geological evidence of surface uplift rates: The case of the African superplume, *Geochemistry, Geophysics, Geosystems 1*, Pap. 1999GC000035, 2000; M. Ishii and J. Tromp, Normal-mode and free-air gravity constraints on lateral variations in velocity and density of Earth's mantle, *Science*, 285:1231–1236, 1999; C. Lithgow-Bertelloni and P. G. Silver, Dynamic topography, plate driving forces and the African superswell, *Nature*, 395:269–272, 1998; J. Ritsema, H. J. van Heijst, and J. H. Woodhouse, Complex shear wave velocity structure related to mantle upwellings beneath Africa and Iceland, *Science*, 286:1925–1928, 1999.

Aging (genetics)

Aging is a complex, multifactorial process that is influenced by both genes and the environment. Aging occurs in almost all animals and is a major risk factor for the onset of a range of human diseases. The molecular and physiological causes of aging remain unknown. However, a number of genetic factors that help determine animal life span have been identified, and it is likely that some mechanisms may be elucidated in the near future.

Defining aging is a challenge. Measurements of many biochemical and physiological parameters show that aging is associated with profound changes. These changes occur at the molecular, cellular, and tissue levels and affect the function of the entire animal. Most changes are detrimental and are sometimes characterized by an accumulation of molecular and cellular damage. Thus, aging is usually described as the sum of a series of time-dependent changes that increase the probability of death with increasing chronological age of an organism. This increase in age-dependent mortality is the hallmark of biological aging, and the rate of this increase is strongly influenced by genes leading to distinctive species-specific life spans. Such genes may be termed gerontogenes.

Evolutionary origin. The evolutionary origin of aging and gerontogenes is the subject of well-developed theories. Why are animals not immortal? Extrinsic hazards, such as disease and predation, make indefinite survival highly unlikely, even for a potentially immortal animal. As a consequence of these hazards, natural populations exhibit an age structure in which young organisms outnumber old organisms. This structure results in the decline of the force of natural selection with age because the cohort of young animals will contribute more progeny to the next generation than the cohort of old animals. As a consequence, optimum Darwinian fitness is heavily weighted on the function of young animals.

The effect of genes on aging is, therefore, not due to direct selection on aging characters; rather, aging is a nonadaptive process in which gerontogenes either have no influence on fitness or have been selected for beneficial effects early in life. No genes have been selected to actively promote aging; rather, aging is an outcome of a lack of selective pressure on late-life characteristics. One prediction of this view is that genes affecting aging are likely to affect other characteristics as well. Indeed, invertebrate genetic mutants with extended life spans usually exhibit changes in the rate or type of development and reduced or delayed fertility.

Molecular mechanisms. Genetic effects on life span have revealed a great deal about the molecular mechanisms of aging. In some species, mutation of just one gene can extend life span by as much as 100%. Over 50 mutations of this type have been identified in the microscopic nematode roundworm *Caenorhabditis elegans*, in which simple genetic breeding experiments have led to roundworms with 300%

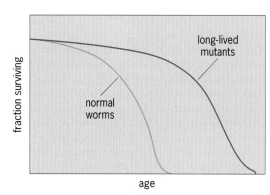

Fig. 1. Survival curve for a typical *Caenorhabditis elegans* longevity mutant. Life span is determined in *C. elegans* by following the survival of a cohort of approximately 100 worms that are synchronously aging. The normal mean life span of *C. elegans* is 20 days, whereas the long-lived mutants have a mean life span of 36 days.

increases in life span—the normal mean life span of this species is approximately 20 days (**Fig. 1**). *Caenorhabditis elegans* is widely used as a model for aging studies because of its short life span and the wealth of genetic and molecular information available—its genome was the first from a multicellular organism to be sequenced entirely. Its normal habitat is a hazardous, changeable soil environment; and it is clearly adapted for rapid growth, high reproductive rates (each individual adult hermaphrodite can produce over 300 progeny), and survival during periodic starvation. Large increases in the life span of the fruit fly *Drosophila melanogaster* have also resulted from selective breeding. More recently, genetic engineering techniques have been used to increase fruit fly life span.

Such simple genetic alterations provide a way to uncover the causes of aging. In many cases, the genes determining life span have been identified, and the protein product encoded by the genes has been inferred from the DNA sequence. In *C. elegans*, a number of gerontogenes are known to be similar to

human genes, and remarkable new ways of thinking about aging have emerged. One genetic pathway that determines the life span of *C. elegans* is very similar to a pathway that is implicated in the insulin response in humans. When insulin binds to its target tissues in humans, a signal is relayed from the cell surface to the interior, where a succession of metabolic and gene expression modifications take place. Some gerontogenes in *C. elegans* encode for protein components of an insulinlike signaling pathway. Reducing signaling through this pathway in the roundworm promotes an increased life span without slowing other traits, such as development. Current experiments are investigating the cellular or hormonal processes controlled by this pathway in the roundworm as well as the influence of this pathway on the aging of other species.

A second class of worm gerontogenes also shares similarity with genes in mammals. The *clk*, or "clock," genes appear to control the timing of a whole range of biological processes, including the rate at which the roundworms grow. Mutation of these genes promotes longevity, and at least one of these *clk* genes encodes a protein involved in regulating metabolism in the mitochondria, the main manufacturer of cellular chemical energy. This may be a genetic clue to one of the major causes of aging—the toxic by-products of metabolism, oxygen radicals.

Oxygen radicals. Oxygen radicals are produced in cells, primarily in the mitochondria, as a normal consequence of oxidative metabolism. Oxygen radicals are highly reactive molecules that contain oxygen and are known to cause damage to proteins, DNA, and the lipid bilayer that makes up the cell membranes. Some experiments provide direct evidence for the theory that, over a life span, the accumulation of such oxidative damage compromises cell and tissue function and that the rate of aging is determined by the rate of damage accumulation.

All organisms have natural defense systems that counteract the action of oxygen radicals. Without these systems, life in the presence of oxygen would be very difficult. The defenses comprise a series of small molecules, such as antioxidant vitamins, and enzymes that catalyze the conversion of oxygen radicals to nontoxic molecular species. If aging is caused by reactive oxygen radicals, the genes that encode the antioxidant enzymes may affect the rate of aging. This hypothesis has been tested by genetically engineering fruit flies to carry extra copies of these genes and hence make more antioxidant enzymes. In some experiments these flies were indeed long-lived. Nematode worm life span was also significantly increased by feeding the roundworms small-molecule drugs that mimic the action of these antioxidant enzymes. In addition, the roundworm gerontogene mutations in the insulinlike pathway made the worms resistant to oxygen radicals.

This all suggests that oxygen radicals may be important in the aging of a number of diverse animals. But are oxygen radicals important for mammals? A genetically mutant mouse, the p66^shc mouse,

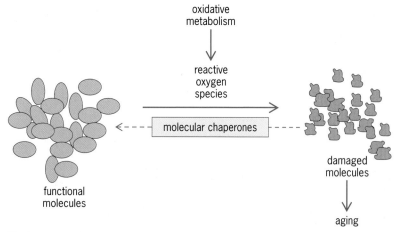

Fig. 2. Factors affecting aging. Aging may be due to the accumulation of molecular damage resulting from the action of reactive oxygen radicals. Evidence for this mechanism comes from genetic alterations that increase the levels of antioxidant enzymes and appear to promote longevity. Molecular chaperones also promote longevity, by restoring damaged proteins to their healthy folded state.

indicates that they might be. This mouse strain is 30% longer-lived than the similar mouse lacking the mutation. The mutant mice are also resistant to oxygen radicals, and even when cells from this mouse strain are grown in culture, the cells are resistant to chemicals that produce oxygen radicals.

Other factors. Although it appears that resistance to oxygen radicals is a major determinant of longevity, long-lived mutants are also resistant to a range of other stresses. For example, mutation in the nematode worm *age-1* gene, which encodes for a protein in the insulinlike signaling pathway, confers resistance to oxygen radicals, heavy metals, heat, and ultraviolet radiation. For this to happen, the signaling pathway is probably affecting the levels of not only antioxidant enzymes but also other stress-resistance proteins. This includes proteins known as molecular chaperones that are generally associated with maintaining other cellular proteins in their correctly folded and active state. Heat stress causes a protein to unfold, and molecular chaperones facilitate refolding (**Fig. 2**) or target the damaged protein for degradation. Worms and flies engineered to carry extra copies of genes encoding molecular chaperones also exhibit increased life spans, suggesting that increasing an organism's resistance to various stresses can generally confer longevity.

Human aging. Human aging has also been the subject of genetic analysis, but very little is known of the human gerontogenes compared with the invertebrate counterparts. Longevity studies of twins have found that about 25% of the life-span differences in human populations can be accounted for by genes. Efforts to find these genes are under way, and some success has come from studies of specific genes in centenarian populations. If a particular form of a gene is associated with longevity, that form should be found more frequently in the centenarian population compared with the general population. Some studies have shown that one particular variant of a gene called APOE was overrepresented in the centenarian population, suggesting that it contributed to the longevity of this group.

In addition to gene association studies, the human aging genes have been discovered by investigations of rare premature aging syndromes. An example is Werner's syndrome. This disease occurs in adults and is associated with loss of muscle, premature graying and loss of hair, heart valve calcification, and atherosclerosis. Although these characteristics together suggest that aging has been accelerated, the disease does not cause accelerated brain aging. The gene that is altered in Werner's syndrome has been identified (it is involved in DNA replication), and it is hoped that therapies for the condition may emerge from this work.

The main concern in studies on human aging is not life span so much as the impact that aging has on health. Aging is the most important factor for the onset of disease in developed countries, but the link between aging and diseases such as Alzheimer's is unknown. However, genetic studies are at the heart of the current understanding of some age-related diseases. For example, some forms of Alzheimer's disease are familial and are caused by inheriting certain forms of genes. The discovery of Alzheimer's disease genes has led to the identification of potential causal factors such as the protein β-amyloid that is found in deposits in the brains of such patients.

For background information *see* AGING; ALZHEIMER'S DISEASE; GENE; HUMAN GENETICS; MITOCHONDRIA in the McGraw-Hill Encyclopedia of Science & Technology.   Gordon J. Lithgow

Bibliography. S. N. Austad, *Why We Age*, John Wiley, New York, 1997; M. R. Rose, *Evolutionary Biology of Aging*, Oxford University Press, New York, 1991; E. L. Schneider and J. Rowe, *Handbook of the Biology of Aging*, 5th ed., Academic Press, San Diego, 1996; O. Toussaint et al., Molecular and Cellular Gerontology, *Ann. N. Y. Acad. Sci.*, vol. 908, 2000.

Agricultural ecology

Agricultural ecology (or agroecology) uses ecological knowledge about organism distribution and abundance to enhance crop yields and protect crops from insect or disease attack. In turn, agricultural ecology is used as a model system for exploring further applied and basic ecological phenomena. Climate, soils, crop physiology and chemistry, and plant and insect population biology are critical elements of agricultural ecology.

Predator-prey interactions and biological control. Lessons learned from ecological studies of predator-prey interactions, especially those involving insects and other agroecosystem arthropods, are readily applied to many agricultural systems. The use of natural enemies, such as predators or parasitoids, to reduce crop pest populations is a time-honored approach that has received vigorous attention from agricultural ecologists. In particular, the use of imported predators to curb exotic pest populations, referred to as classical biological control, has been a staple of both large-scale commercial farming and smaller-scale organic farming for centuries. Other techniques aimed at bolstering resident populations of natural enemies, including the direct augmentation of predator abundances and the manipulation of noncrop plants (for example, flowering plants, which may serve as an appealing nectar source for predators), make up much of the scientific literature concerning biological pest control.

Apart from simply facilitating encounters between predators and prey, agricultural ecology research focuses special attention on the details of both predator and prey behavior. Predator foraging behavior is well studied for many systems (for example, aggregation behavior of ladybird beetles toward prey) and is incorporated in the planning stages of many agricultural systems. Likewise, escape and defensive behaviors of prey insects are crucial areas of study. Recently, innovative studies combining both predator and prey behavior have contributed to greater understanding of the holistic nature of ecological interactions in agroecosystems. For instance, in some settings, disturbance of pea aphid pests by ladybird beetles triggers an escape response, in which the aphids drop from plants only to be eaten by ground-dwelling carabid beetles. Neither ladybird beetles nor carabid beetles by themselves control aphid pests, but together these natural enemies make up an effective natural-enemy complex. The discovery of such multispecies synergisms has inspired further research into how multiple predator species interact with prey. To date, research results indicate that multiple predator species acting in concert may disproportionately reduce prey, may reduce prey in a strictly additive sense, or may in fact impact prey populations less than they would if each predator acted independently.

Evolutionary relationships between predators, prey, and their host plants often also yield valuable insight into agricultural issues. For example, some predators and parasitoids have developed a keen sensitivity to chemical signals (volatiles) emitted by crop plants under attack, indicating that predators and parasitoids are attracted to damaged crop plants rather than actual prey densities. In such cases, researchers are attempting to facilitate this interaction by better understanding crop plant chemistry. Many instances in which plants increase their own defensive chemical compounds in response to insect attack ("induced defenses") have already been documented. This phenomenon in conjunction with the action of natural enemies can greatly reduce the need for synthetic chemical pesticides. Apart from being more environmentally friendly, combinations of natural phytochemistry and natural enemies can reduce costs to growers by slowing insect pest resistance to chemical pesticides.

A special case of biological control is the use of herbivorous insects to suppress weeds. Weed control constitutes a major problem in sustainable agriculture because weed outbreaks can result in major reductions in crop yield. In some cases, herbivorous biological control agents have been introduced to keep weed populations in check. Since herbicide usage far outweighs insecticide usage, biological control of weeds may offer more environmental benefits than biological control of insects.

Despite years of ecological research into predator-prey dynamics, establishing and maintaining long-term biological control over pest insects is often frustrating and elusive. Successful examples of sustained biological control often involve specialized insects (for example, braconid wasps) that are able to track prey populations closely and persist on alternate prey during lean times. In many cases, however, predators and parasitoids have to be reintroduced on a regular basis in order to maintain control over pest insect outbreaks. Difficulties in establishing and maintaining biological control often stem from differences between predator and pest reproductive, dispersal, and mortality rates.

Landscape manipulation. Ecologists have long recognized the importance of habitat heterogeneity (the diversity and spatial distribution of different habitat types) on resident organisms such as plants and animals. Some of these principles have been incorporated into common applications in agricultural ecology, including crop rotation, hedgerows, intercropping, and trap cropping. Crop rotation (rotating the crops planted in a particular field through time) can help enhance soil fertility as well as reduce weed and insect pest problems. In addition, the regular rotation of nitrogen-fixing crops or cover crops (for example, alfalfa and clover) can reduce the need for synthetic fertilizers, especially if cover crops are plowed into the soil at the end of the rotation to produce "green manure." Hedgerows are frequently incorporated into crop areas to provide windbreaks that help reduce erosion and may also interfere with herbivore colonization. An added benefit of hedgerows is that they often support other desirable wildlife (for example, birds and rodents) in agricultural areas. Intercropping involves planting two or more crops together in the same field in alternating rows or other spatial configurations (often called mixed planting or companion planting). This technique is commonly used to help reduce insect pest damage, either by inhibiting colonization of insect pests or by bolstering natural enemy populations. In both cases, weedy plants or ornamentals (which may emit olfactory/ chemical signals that confuse herbivores or offer rich nectar awards to predators) may be used to enhance heterogeneity in crop fields. Trap cropping, in which small fractions of crop fields are devoted to the production of a sacrificial crop that is deemed especially palatable to economically damaging pests, is another technique commonly used to minimize pest problems. After pests preferentially colonize the trap crop, growers may use a pesticide or physically remove (for example, mow or plow) the trap crop, thereby reducing the pest population.

In all landscape manipulations, an important consideration is the trade-off between arable land devoted to the production of a grower's principal crop(s) and the amount of land devoted to alternate or trap vegetation. Apart from the direct economic consequences of producing less acreage of the principal crop, competition among plants may play an important role. In some cases, weeds or alternate crops may reduce yields of principal crops through direct (for example, allelopathy) or indirect means (for example, competitive dominance in procuring nutrients). In some cases, the effectiveness of mixed planting in conferring protection to principal crops depends on the particular plant species involved; thus some combinations of plants (for example, corn and tomatoes) are used more often than others. In low-input schemes such as no-till agriculture, in which soil disturbance is kept to a minimum, there may be a trade-off between supporting soil-dwelling fauna (for example, natural enemies) and providing habitat for overwintering of pests (when pests spend the winter in or around crop vegetation). In general, growers must carefully weigh the biological and economic benefits and disadvantages of applying ecological principles.

Biotechnology. With the increasingly widespread deployment of genetically modified organisms (GMO) in agricultural settings, some novel issues have come to the forefront of agricultural ecology in recent years. One major concern, from a production perspective, is the risk of squandering new technologies aimed at reducing growers' reliance on synthetic pesticides and fertilizers. It is anticipated that resistance problems will arise from the use of genetically modified insect-resistant crops, just as decades of inadvertent artificial selection of insects from extensive use of chemical insecticides has resulted in widespread insect resistance to many pesticides. This problem is further exacerbated by the fact that some of the antiherbivore defenses being genetically spliced into high-acreage crop plants such as soybeans, corn, and cotton are toxins to which insects have already had years of exposure. For example, researchers have already documented insect populations that are resistant to the pesticide Bt, a toxin derived from the bacterium *Bacillus thuringiensis* that has been used in spray form by organic farmers for several decades. Bt insecticide production is a primary trait incorporated into insect-resistant GMOs. In anticipation of such problems, ecologists have been engaged in extensive field experiments and computer modeling projects for the past two decades in an attempt to develop protocols for slowing resistance. One proposed solution is the employment of mixtures of GMO and non-GMO crops in order to dilute resistant populations of herbivorous pests locally. Recent technological efforts have focused on genetically modifying crops that would produce antiherbivore toxins only in response to an otherwise harmless activator spray, thus reducing the exposure of the toxin to insect populations unless absolutely necessary.

From a broader environmental perspective, the widespread deployment of genetically modified crop plants has raised the specter of a range of potential ecological problems. Concerns over the possibility of transgenes escaping from target plants to other populations, especially weedy relatives of target plants, has stimulated decades of risk assessment research. Because gene flow is in many cases a function of pollinator behavior, which is random at best, efforts to quantify the risk of gene escape posed by deploying genetically engineered crops have yielded only advisory guidelines (for example, buffer zones of recommended sizes/scales). Rather than completely eliminating such risks, researchers emphasize focusing on reducing risk as much as possible while learning about the potential shifts in population and community ecology due to gene escape. *See also* GENETICALLY ENGINEERED CROPS.

For background information *see* AGRICULTURAL SCIENCE (PLANT); AGROECOSYSTEM; ALLELOPATHY; BIOTECHNOLOGY; ECOLOGY; GENETIC ENGINEERING; INSECT CONTROL (BIOLOGICAL); PESTICIDE;

POPULATION ECOLOGY in the McGraw-Hill Encyclopedia of Science & Technology.   John Banks

Bibliography. M. Altieri, *Agroecology: The Scientific Basis of Sustainable Agriculture*, Westview Press, 1996; C. R. Carroll, J. H. Vandermeer, and P. Rosset (eds.), *Agroecology*, McGraw-Hill, 1990; S. Gliessman, *Agroecology: Ecological Processes in Sustainable Agriculture*, Lewis Publisher, 1997.

Allergy

Allergies result from immune responses to environmental compounds (allergens), which are normally harmless agents on their own, derived from sources such as pollen, animal dander, and house mites. Allergens trigger disease through activation of inflammatory pathways in sensitized individuals, initiating immune defense mechanisms that are out of proportion to the allergen threat to the individual. The inflammation pathways are thought to involve the production of the IgE class of antibodies, which develop specificity for particular allergens. The IgE antibodies on their own are incapable of activating an immune response, which requires the interaction of these antibodies with specific cells carrying a high-affinity receptor for the antibody. IgE bound to the surface of cells acts as a sensitive sentry for the presence of small amounts of allergens. The recognition of allergens triggers the cellular receptors and the activation of these cells, which release histamines, leukotrienes, and other mediators of the allergic response. Recent studies have elucidated the atomic interactions between the antibodies and their high-affinity receptors that are at the heart of this

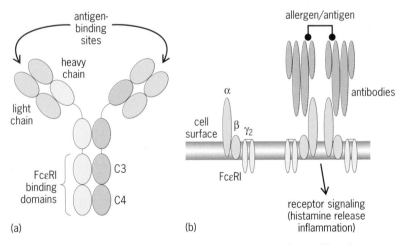

(a)   (b)

Fig. 1. IgE and FcεRI structures. (*a*) The IgE domain structure is shown with each immunoglobulin domain represented by an oval. The two light chains are formed by two domains. The heavy chains are formed by a total of five domains each, with the two heavy chains forming the overall dimeric structure of the antibody. The two antigen-binding sites are located at the interface of the heavy and light chains, and the receptor-binding domains (C3 and C4) are located at the ends of the two heavy chains. (*b*) The full IgE receptor structure is shown on the left and is composed of four protein chains: an α chain, a β chain, and two γ chains anchored in the cell membrane. On the right, a simple signaling complex consists of two antibodies crosslinked by an antigen or allergen, bound to two receptors at the cell surface. Co-localization of two receptors induces cellular activation, which leads to activation of the inflammation pathways associated with the allergic response.

inflammatory pathway, providing new tools to develop therapeutic treatments for allergies and asthma.

IgE antibodies. IgE antibodies, like other antibodies of the immune system, are produced by specialized cells called B cells. Each B cell carries out a sophisticated shuffling and rearrangement of its antibody genes to produce a unique antibody protein. The genetic shuffling allows the immune system to generate antibodies that can recognize almost any molecular shape, even for substances never encountered by the immune system. Thus, one B cell is committed to making one type of antibody, and the immune system produces a very large number of such cells to provide defense against pathogens. Upon encountering foreign substances, such as allergens or antigens, B cells can become activated to secrete antibodies that are specific for intruding molecules. Secreted antibodies can then bind specifically to the invaders and eliminate them by a variety of cellular and biochemical pathways.

The antibody is composed of two heavy and two light polypeptide chains, which assemble into a Y-like structure (**Fig. 1a**). Each chain is constructed of modular immunoglobulin domains. The regions of the antibody that are variable between antibodies and bind to allergens or antigens are known as antigen-binding sites. They are located at the top of the two arms of the Y structure, at the interface of the heavy and light chains. At the base of the Y, the two heavy chains form the effector, or Fc, region that is conserved between all IgE antibodies. The Fc region binds to high-affinity IgE receptors expressed on cells, coupling the antigen specificity of the antibody to cells, that can carry out valuable functions of the immune response. However, in the allergic response, activation of these Fc-mediated pathways is associated with a hypersensitivity to environmental agents. The modular, domain-based structure of antibodies allows many antigens to interact with a variety of conserved cellular pathways in the immune system. High-affinity interactions with cellular receptors have been mapped primarily to the two terminal domains, C3 and C4, of the IgE heavy chain.

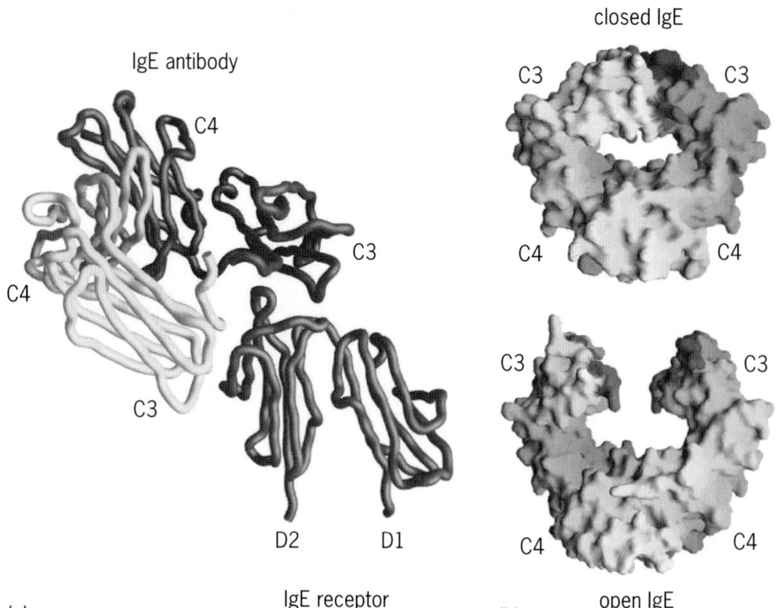

High-affinity IgE receptor. The high-affinity IgE receptor (FcεRI) is found on a variety of cells associated with the immune system, such as mast cells and basophils. The mast cells are resident in peripheral epithelial tissues and are thought to play a major role in the initial stages of the allergic response. The high-affinity IgE receptor is expressed on the surface of mast cells as a complex of four polypeptide chains: the α, β, and two γ chains (**Fig. 1b**). IgE antibodies coat the surface of these cells, poised to encounter antigens. Antigens or allergens that can cluster two or more receptors together at the cell surface initiate a signal transduction cascade inside the cells. Cellular activation leads to the immediate release of preformed inflammatory agents, such as histamine, as well as the activation of biosynthetic and transcription mechanisms to produce additional compounds and proteins that trigger inflammation.

The interaction between IgE and FcεRI is mediated by the extracellular domains of the α subunit of the receptor complex. Expression of soluble fragments of the α subunit can be achieved by introduction of a stop codon prior to the transmembrane-spanning region of the α-chain gene, and such fragments retain similar binding affinity for the antibody. The separation of the antibody-binding and transmembrane-signaling components of the receptor has allowed the visualization of the receptor IgE interaction.

Structure of IgE-FcεRI complex. The atomic structure of a complex of the soluble FcεRI α subunit bound to the IgE-Fc region has been revealed using x-ray crystallographic techniques. The structure of the complex shows how the dimeric antibody Fc region interacts with two distinct binding surfaces of the receptor (**Fig. 2a**). The receptor structure is formed by two nearly antiparallel immunoglobulin domains, labeled D1 and D2, which are arranged as an inverted-V structure relative to the cell membrane. The receptor thus presents a broad, convex surface of interaction for the antibody, rather than a localized, active binding pocket. The two identical C3 domains of the IgE-Fc region interact with two distinct sites on the receptor: the back side of the D2 domain and at the top of the D1D2 domain interface. Both interaction sites contribute to the high affinity of the interaction ($\sim$1 nM), as mutations in both regions affect the binding affinity.

Although the crystal structure of this complex does not include the full structure of the antibody, it

Fig. 2. Crystal structures of IgE bound to FcεRIα and unbound. (*a*) Ribbon model (tracing the position of each amino acid in the protein chains) of the IgE:FcεRIα complex. The four antibody domains (C3 and C4) are labeled, as are the two domains of the receptor α chain (D1 and D2). The two chains of the antibody fragment are bound to the IgE receptor through interactions of the C3 domains with two distinct surfaces of the receptor. (*b*) Molecular surface models (using all atoms of the IgE-Fc structure) in the closed and open forms. The regions of the IgE C3 domains that contact the receptor are indicated in darker gray. The closed form is observed in the absence of FcεRIα, and the open form is observed in the complex with FcεRIα.

provides some clues as to the orientation of the antibody relative to the cell membrane. The linker regions that connect the IgE-Fc region to the remainder of the antibody structure form an arch across the top, convex surface of the receptor, with both ends of the two heavy chains pointing up and away from the receptor and the cell surface. Binding of the antibody to the receptor would induce a significant bend in the antibody structure, allowing the antigen-binding arms of the antibody to freely engage allergens and antigens. Asymmetry and flexible orientation of the antigen recognition domains of the antibody relative to the receptor are likely to be important in allowing signal transduction to be initiated by many different antibody:antigen complexes. [The Protein Data Bank contains three-dimensional structures of FcεRIα (PDB ID: 1F2Q), the IgE-Fc region (PDB ID: 1FP5), and the IgE:FcεRIα complex (PDB ID: 1F6A).]

Structural flexibility in IgE-Fc region. In addition to the structure of the IgE:FcεRI complex, the structure of the IgE-Fc region alone has revealed novel conformational flexibility in this class of antibodies. The structure of the IgE-Fc region changes in the absence of the receptor, forming a closed conformation, compared with a more open conformation when bound to FcεRIα (Fig. 2b). The two receptor-binding domains (C3) move closer to the lower C4 domains, restricting the space between the two chains that form the binding site for FcεRI. This conformational change is not observed in the major class of antibodies, IgG (which are structurally similar to the IgE antibodies), and is potentially controlled by significant sequence and structural differences between the two antibody classes. Since IgE antibodies also interact with a second, lower-affinity receptor, CD23, it may be that this conformational change allows both FcεRI and CD23 to form complexes with the antibody.

The conformational change in the IgE-Fc region suggests new strategies for the development of therapeutic agents to treat allergy and asthma. Recent clinical trials of anti-IgE therapy, injection of an engineered IgG antibody that blocks IgE binding to FcεRI, have shown therapeutic efficacy in the treatment of severe allergic asthma. Based on the recent structural studies of the IgE antibody and its receptor complex, the design of inhibitors that stabilize the closed conformation of the IgE antibody might prove to be an effective therapeutic strategy. By disconnecting all of the IgE antibodies that recognize different allergens from the receptors that trigger the inflammatory response, such an inhibitor could have widespread therapeutic benefits in allergic reactions of all types.

Activation of mast cells. The process by which allergens and antigens activate mast cells remains to be fully understood. As with many other receptors of the immune system, such as T-cell receptors, B-cell receptors, and natural killer cell receptors, activation occurs after crosslinking of multiple receptors brings receptors into proximity. The co-localization of receptors at the cell surface then triggers phosphorylation of adjacent intracellular tails of the receptor γ chains by lyn kinase. Lyn kinase belongs to a family of enyzmes involved in cellular signal transduction cascades that operate by attaching phosphate groups to targeted tyrosine residues in proteins, which then act as signals for the next step in the signal transduction cascade. For example, lyn kinase phosphorylates the IgE receptor on tyrosines in the cytoplasmic tails of the γ chains, which then become docking sites for the syk kinase, leading to its activation. Subsequent phosphorylation events trigger the immediate release of histamine from these cells and the initiation of other inflammation pathways.

There are two models for the initiation of these phosphorylation events. In one model, weak interactions of lyn kinase with the receptor allow cross-phosphorylation of adjacent receptors upon antigen-induced co-localization. In a second model, crosslinking of the receptors by antigens may induce movement of receptor complexes into specialized membrane regions, known as rafts, which are already enriched in lyn kinase. Movement into the membrane rafts would therefore lead to enhanced phosphorylation of the receptor γ chains due to the co-localization with lyn kinase. Evidence for both models has been obtained, and these models may not be mutually exclusive in explaining the initiation of receptor signal transduction. Further elucidation of the mechanisms underlying IgE receptor signal transduction may provide new therapeutic routes to the alleviation of allergies and asthma.

For background information *see* ALLERGY; ANTIBODY; ANTIGEN; ANTIGEN-ANTIBODY REACTION; ASTHMA; IMMUNOGLOBULIN in the McGraw-Hill Encyclopedia of Science & Technology.

Theodore Jardetzky

Bibliography. T. W. Chang, The pharmacological basis of anti-IgE therapy, *Nat. Biotechnol.*, 18(2):157–162, 2000; [PDB ID: 1F6A] S. C. Garman et al., Structure of the Fc fragment of human IgE bound to its high-affinity receptor Fc(Epsilon)RI(Alpha), *Nature*, 406:259, 2000; [PBD ID: 1F2Q] S. C. Garman, J. P. Kinet, and T. S. Jardetzky, Crystal structure of the human high-affinity IgE receptor, *Cell*, 95:951, 1998; H. Turner and J. P. Kinet, Signalling through the high-affinity IgE receptor Fc epsilonRI, *Nature*, 402(6760 Suppl.):B24–B30, 1999; [PDB ID: 1FP5] B. A. Wurzburg, S. G. Garman, and T. S. Jardetzky, Structure of the human IgE-Fc Cepsilon3-Cepsilon4 reveals conformational flexibility in the antibody effector domains, *Immunity*, 13:375, 2000.

Ancient DNA

Within the past decade, biochemical methods have advanced so much that deoxyribonucleic acid (DNA) can be copied and sequenced from individuals representing extinct populations, or even from individuals representing entire species now extinct. Such DNA, from long-gone sources, is called ancient DNA (aDNA). The introduction of the polymerase chain reaction (PCR) into the molecular biologist's tool box has made this extraordinary retrieval possible. The

PCR is a biochemical technique in which free nucleotides, DNA primer sequences, and a polymerase (*Taq*) are employed to amplify DNA in vitro. Theoretically, it is possible to begin with a single target DNA molecule and end with millions of copies that can easily be sequenced with routine methods.

Almost immediately after the introduction of the PCR method, a multitude of investigators began exploiting the technique for amplifying DNA from ancient sources. In the early 1990s, there was an explosion of studies reporting DNA sequences from extinct organisms, some of which were of extraordinary antiquity, such as the Tyrolean Ice Man, mum-

mies, frozen mammoths, a 7000-year-old human brain, Miocene leaf fossils, and insects entombed in amber for millions of years. These studies generated intense interest in the method of ancient DNA amplification as well as in the science to which it was applied.

Research challenges. The challenges of aDNA studies can be summarized as (1) poor DNA quality and quantity, (2) contamination from exogenous DNA, and (3) PCR inhibition. DNA from extinct organisms is typically fragmented, usually in sizes of 100–300 base pairs, and in low quantity. Once the mechanism of DNA repair in a living organism is

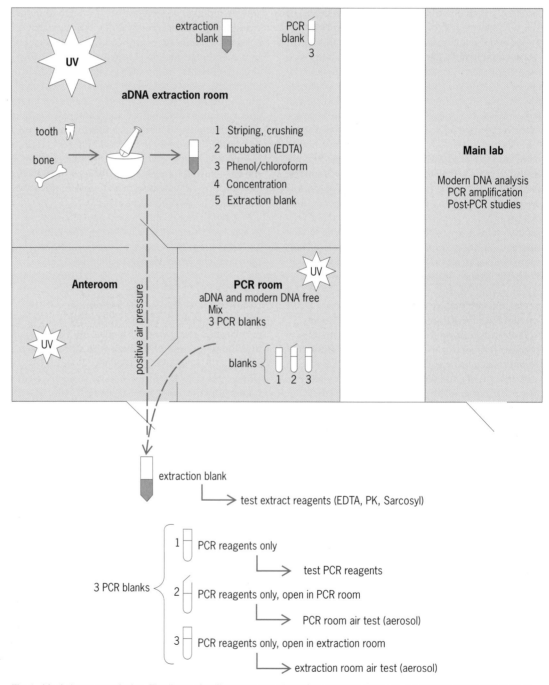

Fig. 1. Ideal clean-room design. The figure also illustrates use of multiple negative controls, exposed to different stages of ancient DNA preparation, for monitoring contamination. EDTA = ethylenediamine tetraacetic acid; PK = proteinase K.

turned off with death, DNA quickly degenerates and is modified by breakage due to loss of bases and may be rendered inaccessible to enzymatic amplification due to cross-linkage. These characteristics conspire to introduce contamination as a continual challenge for aDNA researchers. In any PCR amplification, the DNA polymerase will always favor intact contaminating DNA to the desired aDNA. Contamination usually stems from two possible sources: (1) human genomic DNA from handling and preparation of ancient samples and (2) PCR amplicons (small, replicating DNA fragments) from labs in which modern DNA is being investigated.

Low-quality DNA can also result in amplification of artifact sequences. If aDNA is poorly represented in the PCR reaction, modern DNA of any abundance can compete for primer annealing. Mispriming of modern DNA, combined with weak priming of target DNA, can result in the generation of artifact sequences early in the reaction that can then be exponentially amplified in subsequent PCR cycles. The result is a chimeric sequence that can mimic what would be expected of an authentically ancient sequence.

These problems can be compounded or, even worse, amplification made impossible by the copurification of PCR inhibitors. Extraction modifications that are favorable to DNA retention are also favorable to the copurification of PCR inhibitors. Soil-derived degradation products (collectively known as humus) often become associated with a subfossil specimen during the organic decay process. These products can act as strong inhibitors of *Taq* polymerase. Another cause of inhibition results when cross links form between reducing sugars and amino groups (the Maillard reaction), resulting in chemical bonds that prevent amplification. Thus, the investigator may not know if the lack of amplification is due to a lack of target DNA or simply to the presence of inhibitors.

Research techniques. Contamination control is absolutely necessary for good aDNA technique. **Figure 1** shows an ideal clean-room design. This laboratory must be physically separated from the main laboratory and independently accessible. There should be three chambers: an anteroom for donning clean-room garb, a small PCR preparation room, and an aDNA extraction room. The chambers must maintain positive pressure with respect to outside air, and as an additional contamination-control measure, they should be routinely exposed to short-wave UV light. By isolating the various phases of aDNA preparation and amplification, the investigator can also identify the precise step in which contamination might be introduced (Fig. 1).

Currently, inhibition is the only cause of unsuccessful PCR amplification that can be unambiguously identified. If adding an aliquot of aDNA extract to a positive control ("spiking") prevents amplification of the control, polymerase inhibition is immediately confirmed. There are also techniques for identifying the intensity of PCR inhibitors in a given aDNA extract (**Fig. 2**). By adding progressively greater

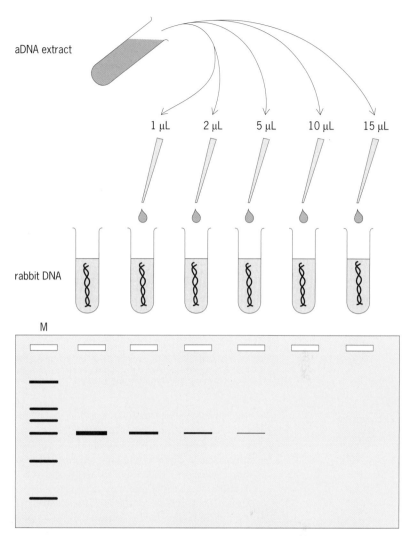

Fig. 2. Progressive spiking of a positive polymerase chain reaction control for assessing the degree of PCR inhibition. At 10 μL, amplification of control is completely inhibited.

amounts of an ancient extract to a positive PCR control (for example, if attempting to amplify fossil human DNA, modern rabbit DNA might be chosen as the positive control), the investigator can determine the precise amount of extract required to prevent amplification. Therefore, the investigator knows that less than that amount must be used in a PCR reaction to have successful aDNA amplification.

Primer design is critical for addressing the challenge of poor DNA quality. Although modern DNA can be amplified in long sections, with primers conveniently designed to anneal to conserved regions of the genome, aDNA must be amplified in short segments due to its fragmentary nature (**Fig. 3**). Because primers must be designed more frequently, perhaps every 50–100 base pairs versus every 500–1000 base pairs with modern DNA, the investigator always confronts the possibility that the aDNA primers will be too specific (if they are designed from the sequences of a single organism) or they will be too general (amplifying numerous organisms, but poorly). Moreover, due to the demand for designing many primers, it is unavoidable that relative efficiency of individual primer pairs will vary.

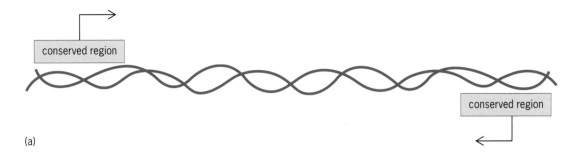

(a)

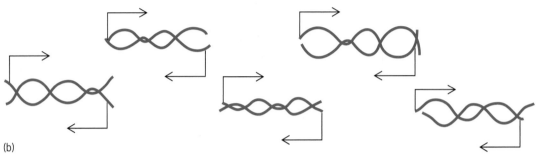

(b)

Fig. 3. Comparison of PCR amplification strategy for contemporary versus ancient DNA. (*a*) In a long strand of contemporary DNA, only two primers (bent arrows) are necessary to amplify the entire region of interest. (*b*) Because ancient DNA is so fragmented, many more primers are required to amplify an equivalent region.

Virtually every successful aDNA project to date has focused on mitochondrial DNA (mtDNA) because most cells possess multiple mitochondria but only a single nucleus. For every single-copy nuclear gene within a given cell, there will be approximately a thousandfold excess of mitochondrial genes. Since only very small amounts of total DNA can usually be recovered from ancient specimens, there is a much greater chance of recovering a mitochondrial gene via PCR, due to the high quantity, than a nuclear gene.

Ancient human DNA. The negative publicity that resulted from a number of flawed studies led some aDNA investigators to propose explicit measures for assuring the repeatability of their results. These methods for verifying aDNA authenticity are as follows:

1. Prepare independent extracts from the same sample in two laboratories, geographically separated.
2. Prepare multiple PCR reactions for each extract.
3. Require several positive tubes for each PCR reaction.
4. Test multiple clones per positive tube.
5. Employ species-specific primers, if possible.
6. Design an amplification and sequencing strategy that allows for contiguous assembly from multiple overlapping fragments.
7. Always include negative extraction and PCR controls.
8. Continuously apply stringent contamination-control methods.

As the ultimate test of authenticity, aDNA researchers

agreed that sequences (especially ancient human sequences) should be confirmed via independent procedures in two laboratories that are geographically separated.

Using these new guidelines, investigators have begun to examine a controversial theory relating to the origins of anatomically modern humans. It is well established that *Homo erectus* emerged from Africa, approximately 1.5 million years ago, to establish populations in virtually every part of the globe. The controversy relates to subsequent events. In one model, it is hypothesized that modern human features evolved by slow parallel evolution in *Homo erectus* populations, with genetic continuity maintained by individual migration and occasional interbreeding. This model of human origins is generally referred to as the multiregional hypothesis. In the opposing, "out-of-Africa" hypothesis, geneticists have concluded that a wave of anatomically modern humans emerged from Africa, between 200,000 and 100,000 years ago, to replace their nonmodern predecessors. The out-of-Africa model implies that Neanderthals, although temporarily coincident in time and space with these modern humans, would not (and probably could not) have interbred with them.

Until recently, the choice between these two models depended entirely upon interpretations of the fossil record and on DNA comparisons from living humans. However, in 1997, Svante Pääbo, Mark Stoneking, and colleagues reported that they had characterized a small region of the mitochondrion directly from the Neanderthal type specimen, originally recovered in western Germany in 1856. The study found that the Neanderthal sequence falls well outside the range of modern-human mtDNA variation. The results were confirmed with phylogenetic analysis, revealing that the Neanderthal sequence is the clear evolutionary outlier to the clade containing modern human mtDNA sequences from all reaches of the planet. Therefore, the researchers concluded that Neanderthals are not the ancestors of modern humans and did not interbreed with them, thus supporting the out-of-Africa model of human origins.

However, another recent study of human fossil DNA reached an opposing conclusion. G. Adcock and colleagues determined that mtDNA recovered from a 60,000-year-old anatomically modern human found in Australia might tell a different story of human origins. These investigators found that DNA from this specimen is quite distinct from extant (and a few fossil) mtDNA lineages found in both Australia and other parts of the world, including Africa. The mtDNA from this fossil (called LM3) was found to group with a nuclear insert sequence that is generally believed to have diverged prior to the diversification of modern-human mtDNA. Peacock and colleagues therefore reasoned that their results cast doubt on both the geographic and temporal aspects of the out-of-Africa hypothesis, concluding that if replacement did occur, part of the replacement must have occurred in Australia and some of those replaced must have been anatomically modern.

It is possible to envisage a model whereby the results of these seemingly opposing studies are compatible with a unified replacement theory. In **Fig. 4**, the 1856 Neanderthal specimen and the Australian LM3 specimen are roughly contemporaneous but are drawn from two different human lineages. The 1856 specimen is drawn from the terminal phase of Neanderthal evolution, shortly before replacement by early modern humans, and the LM3 specimen is drawn from the anatomically modern lineage but, by chance, represents an mtDNA lineage that is now extinct via the process of lineage sorting. Lineage sorting occurs because mtDNA is maternally inherited. When a woman has only sons, she does not pass her mtDNA on to subsequent generations. If she has no maternal relatives, or has maternal relatives who also have only sons, the entire mtDNA lineage can become extinct even though her descendants live on.

One of the inescapable handicaps of aDNA studies is that the investigator is entirely the slave of happenstance with regard to choice of samples for analysis. Many studies have demonstrated the numerous ways in which a gene tree (the small lines within the cylinders in Fig. 4) and a species tree (the cylinders) may or may not agree. This can be particularly problematic for aDNA researchers because they are constrained to examine relatively recent events wherein lineage sorting may still be occurring. Due

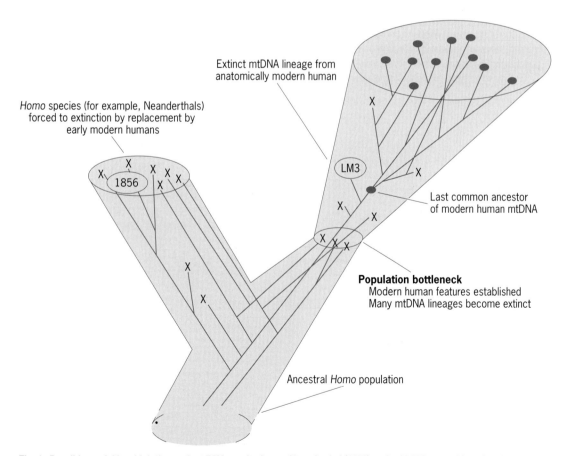

Extinct mtDNA lineage from anatomically modern human

Homo species (for example, Neanderthals) forced to extinction by replacement by early modern humans

1856

LM3

Last common ancestor of modern human mtDNA

Population bottleneck
Modern human features established
Many mtDNA lineages become extinct

Ancestral *Homo* population

Fig. 4. Possible model in which the ancient DNA results from a Neanderthal (1856) and a 60,000-year-old modern human from Australia (LM3) are both consistent with a unified replacement hypothesis. X represents mtDNA lineage extinction, either through natural process of lineage sorting or through species extinction.

to the effects of differential mtDNA lineage extinction, the investigator may or may not find the mtDNA alleles that precisely track the history of the species or population under investigation.

Despite the manifold methodological and theoretical limitations of aDNA studies, the technique will continue to reveal historical events that are beyond the scope of contemporary comparative studies. These events range over ancient pathologies (for example, malaria's role in the fall of Rome; the existence of tuberculosis in the New World prior to the arrival of Europeans; the precise culprit in the Black Death), the reconstruction of population origins and subsequent migrations, and the determination of relationships among extinct and extant organisms.

For background information *see* DEOXYRIBONU-CLEIC ACID (DNA); EARLY MODERN HUMANS; FOSSIL HUMAN; GENE; GENE AMPLIFICATION; NEANDERTALS in the McGraw-Hill Encyclopedia of Science & Technology.   Thomas Delefosse; Anne D. Yoder

Bibliography. G. J. Adcock et al., Mitochondrial DNA sequences in ancient Australians: Implications for modern human origins, *Proc. Nat. Acad. Sci. USA*, 98:537–542, 2001; M. Krings et al., Neanderthal DNA sequences and the origin of modern humans, *Cell*, 90:19–30, 1997; S. R. Woodward, N. J. Weyand, and M. Bunnel, DNA sequence from Cretaceous period bone fragments, *Science*, 266:1229–1232, 1994; H. Zischler et al., Detecting dinosaur DNA, *Science*, 268:1191–1193, 1995.

Anthropoid origins

A wide variety of biological evidence, ranging from fossils to DNA, indicates that living monkeys, apes, and humans share a common ancestry that excludes all other organisms, including other primates. Known collectively as anthropoids, these "higher" primates differ from their living prosimian relatives (lemurs, lorises, bushbabies, and tarsiers) in many key features of anatomy, physiology, and behavior. Until recently, the fossil record has only weakly illuminated the wide gulf separating living anthropoids and prosimians. As a result, one of the largest remaining gaps in knowledge of primate evolution concerns the origin and early diversification of anthropoids. Novel paleontological discoveries are finally beginning to resolve the long-standing issue of exactly how anthropoids fit on the primate evolutionary tree. Newly discovered fossils are also leading to unexpected insights regarding the paleobiology of the earliest anthropoids. These new findings are threatening to overturn an earlier consensus about when and where anthropoids originated.

Eocene prosimians. Classical models hold that anthropoids originated sometime near the Eocene-Oligocene boundary (about 34 million years ago), having evolved from one of two major groups of Eocene prosimians, the adapiforms and omomyids. Adapiforms share many anatomical features with living lemurs, and most paleontologists agree that lemurs evolved from some unknown adapiform ancestor. Similarly, omomyids share some important traits with living tarsiers, although in most respects omomyids are far more primitive. Because tarsiers appear to be more closely related to anthropoids than lemurs are, the hypothesis that anthropoids evolved from an omomyid (rather than adapiform) ancestor is more consistent with both the fossil record and the biology of living animals.

Both of these traditional models of anthropoid origins have been challenged by fossils unearthed during the 1990s. For example, the well-preserved skulls and limb bones of the omomyid *Shoshonius cooperi* recovered from early Eocene rocks (about 50 million years old) in Wyoming's Wind River Basin revealed numerous tarsierlike features, suggesting that the tarsier lineage was already established by this early date. This inference was subsequently validated by discoveries of two different fossil tarsiers, *Xanthorhysis tabrumi* and *Tarsius eocaenus*, at somewhat younger sites in China dating to the middle Eocene (40–45 million years ago). Documenting such early dates for the tarsier lineage has significant ramifications for anthropoid origins. That is, if tarsiers are the nearest living relatives of anthropoids, and if the tarsier lineage dates at least to the early Eocene, the anthropoid lineage must be equally ancient. Over the past decade or so, fossils of anatomically primitive anthropoids dating to the middle Eocene have been recovered for the first time.

Early African anthropoids. For many years, the oldest uncontested anthropoids documented in the fossil record came from early Oligocene strata of the Jebel Qatrani Formation in the Fayum region of northern Egypt. Beginning in 1989, anthropoid fossils from a new site located lower in the local section began to be described. This locality, called L-41, has yielded relatively complete and abundant fossils of early anthropoids, now known to date to the late Eocene (about 35–36 million years ago). Crushed skulls of *Proteopithecus sylviae* and *Catopithecus browni* from L-41 show that such diagnostic characters as the postorbital septum, which forms the back part of the bony eye socket that is typical of anthropoids, were already present in these taxa. The high taxonomic diversity of anthropoids at L-41, which includes at least the families Proteopithecidae and Oligopithecidae, indicates that anthropoids had experienced an interval of adaptive radiation significantly prior to the late Eocene.

Elsewhere on the African continent, fossils pertaining to this earlier interval of anthropoid evolution have so far proven to be both scarce and fragmentary, being known mainly from two sites in Algeria. *Biretia piveteaui*, *Algeripithecus minutus*, and *Tabelia hammadae* appear to be early anthropoids of small body size, although their evolutionary significance remains uncertain because each species is known only from isolated teeth. Indeed, even the age of these poorly known Algerian species relative to the Egyptian anthropoids is disputed, although associated fossil mammals indicate that the Algerian sites are significantly older.

Early Asian anthropoids. Eocene fossil primates from Asia have enjoyed a long-standing, yet persistently controversial role in efforts to comprehend anthropoid origins. In 1927, *Pondaungia cotteri* was described from the Eocene of Myanmar (formerly, Burma) as a possible early anthropoid that was decidedly older than the anthropoids known from the Fayum region of Egypt. Additional specimens of *Pondaungia* and the closely related genus *Amphipithecus* were described sporadically over the rest of the twentieth century, and the significance of both of these Burmese primates remained equivocal as a result. Recently, more nearly complete specimens have been described for both *Pondaungia* and *Amphipithecus*, as well as a related form from southern Thailand called *Siamopithecus*. All three, which belong to the family Amphipithecidae, appear to be relatively large Eocene anthropoids characterized by robust jaws and bunodont, or relatively flat, cheek teeth.

Eosimias, recently described from middle Eocene sites (40–45 million years old) in eastern and central China, is probably the oldest and the most controversial discovery bearing on anthropoid origins in recent years. *Eosimias* is so primitive in certain anatomical details that it appears to represent the most basal branch of the anthropoid family tree discovered to date. For example, *Eosimias* retains an extra cusp on its lower molars, known as the paraconid, that is lost in most later anthropoids. At the same time, *Eosimias* possesses an advanced, anthropoidlike chin region, in which its vertically oriented front teeth, or incisors, are implanted. Ankle bones of *Eosimias* show features that are highly characteristic of anthropoids, although in some ways these bones are transitional between those of Eocene omomyid prosimians and later anthropoids. *Bahinia pondaungensis*, a larger and geologically younger relative of *Eosimias*, was recently discovered in the same Burmese fossil sites that yielded *Pondaungia* and *Amphipithecus*.

Taken together, this spate of recent discoveries of Eocene anthropoids in Asia bolsters the hypothesis that the anthropoid lineage is an ancient one that probably diverged from its nearest prosimian relatives by the early part of the Eocene. At the same time, the anatomy of these early Asian anthropoids makes it quite unlikely that anthropoids evolved from the lemurlike adapiforms. Instead, anthropoids appear to be more closely related to living tarsiers and their extinct relatives, the omomyids.

Paleobiogeography. By the second half of the Eocene, anthropoids are known to have occupied a vast area, ranging from coastal parts of China to the border region between Morocco and Algeria. Anthropoids clearly originated on either Asia or Africa, but their fossil record remains too poor to draw definitive conclusions about which continent harbored the first members of the anthropoid lineage. However, the oldest and most primitive anthropoids, consisting of *Eosimias* and its relatives, are so far known only from Asia. Intriguingly, the nearest living relatives of anthropoids are tarsiers, which occur in Southeast Asia today and are represented in the fossil record of Asia as far back as the middle Eocene. To date, undoubted fossil tarsiers have never been found in Africa. Therefore, a plausible hypothesis is that the initial divergence between the anthropoid and tarsier lineages occurred in Asia, and that anthropoids first dispersed to Africa only later in their evolutionary history.

Mosaic evolution. Paleontologists have long assumed that the numerous anatomical features distinguishing living anthropoids from prosimians must have evolved consecutively, rather than all at once. Not only is the fossil record of early anthropoids now sufficient to show that this was indeed the case, but it also suggests that the earliest anthropoids were more primitive in certain respects than anyone could have guessed a few years ago.

Although living anthropoid primates tend to be considerably larger than living prosimians, it now appears that the earliest members of the anthropoid lineage may actually have been smaller than the smallest living primate, the mouse lemur *Microcebus myoxinus* from Madagascar. Tiny primate ankle bones much smaller than those of *Microcebus* have been identified at one of the sites in China that has yielded fossils of *Eosimias* and related taxa. Some of these primate ankle bones are morphologically very similar to those of *Eosimias* and clearly pertain to an early anthropoid, yet they belong to animals much smaller than *Eosimias*. Using a variety of mathematical regression techniques, the body mass of these diminutive Chinese anthropoids has been estimated at about 15 grams (0.5 ounce), or roughly the size of a small mouse (see **illus.**). That early anthropoids could have been this small was entirely unexpected. Because all mammals are endothermic, or warm-blooded, such a tiny body mass poses

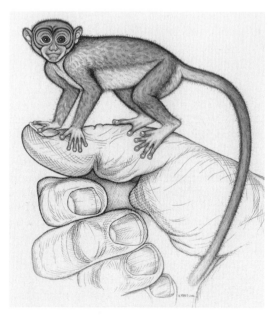

Small eosimiid anthropoid, depicted atop a human hand. This restoration is based on fossil jaws and postcranial bones from a middle Eocene site in eastern China. (*Drawing by Kim Reed-Deemer*)

significant physiological challenges, implying an active lifestyle with a shrewlike metabolic rate.

The anatomy of the earliest known anthropoid primates, the Chinese Eosimiidae, remains incompletely documented. However, the emerging picture is one of highly transitional animals that combine characteristics of living monkeys with features more typical of omomyid prosimians. As additional details of eosimiid anatomy are revealed by new paleontological discoveries, the fossil record is at last filling the void separating living anthropoids from their prosimian relatives.

For background information *see* ANIMAL EVOLUTION; FOSSIL; FOSSIL PRIMATES; METABOLISM; PALEONTOLOGY; PRIMATES in the McGraw-Hill Encyclopedia of Science & Technology.   Chris Beard

Bibliography. K. C. Beard et al., Earliest complete dentition of an anthropoid primate from the late middle Eocene of Shanxi Province, China, *Science*, 272:82–85, 1996; J. G. Fleagle and R. F. Kay, *Anthropoid Origins*, Plenum Press, New York, 1994; D. L. Gebo et al., The oldest known anthropoid postcranial fossils and the early evolution of higher primates, *Nature*, 404:276–278, 2000; J.-J. Jaeger et al., A new primate from the middle Eocene of Myanmar and the Asian early origin of anthropoids, *Science*, 286:528–530, 1999; R. F. Kay et al., Anthropoid origins, *Science*, 275:797–804, 1997.

Antifreeze protein

Freezing temperatures (below 0°C or 32°F) are potentially damaging to living things because growing ice crystals can puncture cell membranes or cause other detriment. Organisms that have developed mechanisms for surviving the cold either inhibit internal ice formation (freeze resistance) or withstand and perhaps even promote partial freezing (freeze tolerance). One adaptation developed by some cold-tolerant organisms is antifreeze proteins (AFPS). Unlike the antifreezes used in car radiators, they are effective at much lower concentrations and act by binding directly to the surface of an ice crystal, thereby disrupting its normal structure and growth pattern and inhibiting further ice growth (**Fig. 1**). This causes a decrease in the nonequilibrium freezing point without significantly affecting the melting point. The difference between these two temperatures is a measure of the strength of the antifreeze protein activity and is called thermal hysteresis.

Fish. The first antifreeze proteins to be characterized were those of marine fish. Since seawater contains high concentrations of dissolved salt, it freezes at around −1.9°C (28.6°F). However, the blood of most fish freezes at about −0.7°C (30.7°F). Therefore, some species synthesize antifreeze proteins with maximal thermal hysteresis activities approaching 1.5°C (2.7°F), which is sufficient to depress the freezing point by the required 1.2°C (2.2°F).

Five different types of fish antifreeze proteins have been characterized to date, and despite similar thermal hysteresis activities, their structures are very dif-

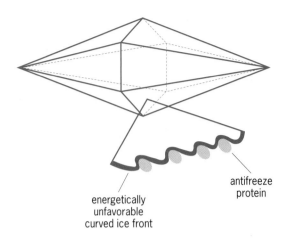

Fig. 1. Typical hexagonal bipyramidal ice crystal formed in the presence of fish antifreeze protein at temperatures below the melting point. The antifreeze protein is thought to bind to the ice surface, forcing the ice to grow in curved fronts, which lowers the freezing point. Further growth does not occur unless this new, nonequilibrium freezing point is exceeded. Under similar conditions in the absence of antifreeze protein, the ice would form a flat disc and grow continuously, since the melting and freezing points would be indistinguishable.

ferent (**Fig. 2a–c**). Type I antifreeze protein is a small, single α-helix (a very tight coil with about 3.6 amino acids per turn) rich in the amino acids alanine and threonine. Type II and type III antifreeze proteins are unrelated globular (structurally complex and somewhat spherical) proteins, and type II is similar to lectins (proteins that bind carbohydrate groups). Type IV is the most recently discovered fish antifreeze protein, and it may consist of four α-helical domains arranged in a bundle, since it is similar to certain apolipoproteins (proteins involved in lipid transport) containing this structure. Type V, the first to be discovered about 30 years ago, is an antifreeze glycoprotein. It is composed of variable numbers of three-amino-acid repeats (usually alanine-alanine-threonine) with two carbohydrate residues attached to each threonine residue.

An exciting finding in recent years was that very similar antifreeze glycoproteins have evolved from completely different genes (convergent evolution) in two species of fish (that is, both species developed virtually identical proteins independently). However, despite the large amount of information obtained on fish antifreeze proteins, the precise way in which they bind to the ice surface has eluded discovery. In addition, very high concentrations are required to produce thermal hysteresis activities in excess of 1°C (2°F), so their utility in biotechnological applications is limited.

Insects. Since insects are much smaller than fish, early attempts to characterize their antifreeze proteins were fraught with difficulty. However, modern microscale methods have recently enabled several groups to obtain pure samples of antifreeze protein from freeze-resistant insects. The gene sequences encoding these antifreeze proteins were isolated, and the proteins were produced in a bacterial expression system. These insect antifreeze proteins were 10 to 100 times more active (depending on concentration) than fish antifreeze proteins, with maximal thermal hysteresis activities in excess of 5°C (9°F). This was not surprising because insects are often exposed to temperatures below −30°C (−22°F).

The first insect antifreeze protein sequences reported (1997) were from the mealworm beetle, *Tenebrio molitor*, and the spruce budworm moth, *Choristoneura fumiferana*. Shortly afterward, proteins similar to those of the mealworm beetle were reported from the fire-colored bark beetle, *Dendroides canadensis*. Each species has multiple copies of its antifreeze protein genes, which produce many slightly different antifreeze protein variants, as was also found for fish. For example, mealworm beetles have more than 30 genes encoding at least 10 protein variants. Such increased gene dosages are thought to facilitate rapid production of the high levels of antifreeze protein required for adequate protection.

In 2000, the structures of both the mealworm beetle and spruce budworm moth antifreeze proteins were solved (Fig. 2*d–g*). Both form a helix containing β-strands (extended amino acid chains in which the orientations of the side chains alternate), but they coil in opposite directions. The spruce budworm protein has a triangular cross section, it contains a core of hydrophobic (water-repelling) residues, and in some places, adjacent loops are connected by disulfide bonds that covalently link the sulfur atoms of two cysteine residues. In contrast, the mealworm beetle antifreeze protein does not possess a hydrophobic core, has a smaller rectangular cross section, and is much richer in cysteine. Each repeat forms a virtually identical loop bisected by a disufide bond, but there are no disulfide bonds between adjacent loops.

Ice-binding mechanism. It is particularly exciting that similar ice-binding surfaces have arisen by convergent evolution in the unrelated mealworm beetle and budworm moth antifreeze proteins. Specifically, both proteins contain two rows of outward-facing threonine residues along one face, making both their sequences and structures somewhat repetitive. Although the precise orientation and spacing of these threonine residues has been determined only for the mealworm beetle protein, it is likely similar in the budworm moth antifreeze protein. Remarkably, the spacing between the two rows of threonine residues (**Fig. 3***a*) and between each threonine within a row (Fig. 3*b*) very closely matches the spacing of the water molecules on the surface of ice. This presumably allows the antifreeze protein to "freeze" tightly to the ice, thereby inhibiting further growth. Because fish antifreeze proteins do not contain such obvious icelike faces, this may help explain why these insect antifreeze proteins have such high thermal hysteresis activity. Thus, after a 30-year search, the mealworm beetle antifreeze protein provides the first clear view of how an antifreeze protein can bind to ice.

Plants. Several antifreeze protein sequences have also been obtained from plants. Carrot antifreeze protein, for example, is also repetitive, containing leucine-rich repeats of about 24 residues in length.

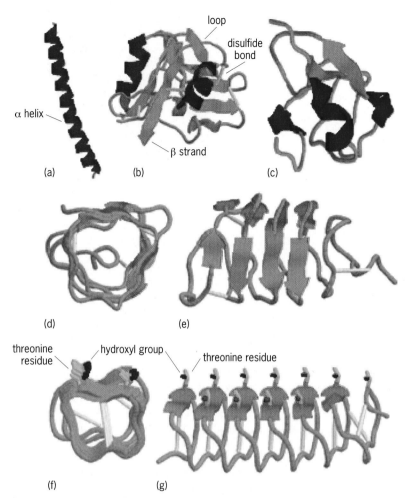

Fig. 2. Known antifreeze protein structures. (*a*) Type I (winter flounder). (*b*) Type II (sea raven). (*c*) Type III (eel pout). (*d–e*) Spruce budworm moth, end-on and side views. (*f–g*) Mealworm beetle, end-on and side views. [*Adapted from structures from the Protein Data Bank (PDB), http://www.rcsb.org/pdb/: 1WFA (type I), 2AFP (type II), 1MSI (type III), 1EWW (budworm moth), 1EZG (mealworm beetle)*]

It and several other plant antifreeze proteins have evolved from proteins involved in defense against pathogens such as fungi. Although the thermal hysteresis activity of known plant antifreeze proteins is considerably lower than that of fish antifreeze proteins, the *Lolium perenne* (ryegrass) antifreeze protein has been shown to be up to 200 times more active at inhibiting ice recrystallization. This process occurs in the frozen state, particularly when the temperature is within a few degrees of the melting point. Larger crystals grow slowly at the expense of smaller crystals because it is more energetically favorable for the same volume of ice to exist as one large crystal rather than numerous small ones. Therefore, some plants are thought to use antifreeze proteins to prevent damage due to the gradual growth of large crystals while frozen, rather than using them to prevent the initial formation of ice.

Applications. Many crops (for example, oranges) are sensitive to frost. Others suffer extensive loss of quality upon freezing (for example, strawberries) but may retain texture and last longer if stored unfrozen but at subfreezing temperatures. Other frozen goods, such as ice cream, vegetables, and meats,

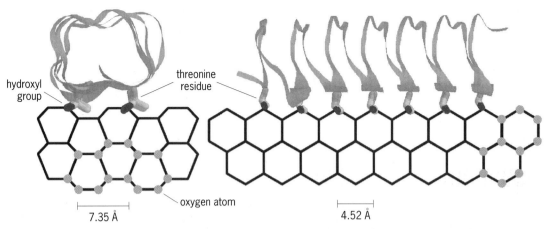

Fig. 3. Alignment of oxygen atoms on the ice-binding threonine residues of the mealworm beetle antifreeze protein with the oxygen atoms at the ice surface along two dimensions. Only a subset of the oxygen atoms of the ice lattice are shown. The hydrogen atoms lie along the black lines between oxygen atoms. The hydroxyl group of each threonine residue is thought to take the place of a water molecule at the ice surface. Å signifies Ångstrom units.

experience a loss of quality when ice recrystallizes. To protect crops from frost damage, antifreeze protein genes must be introduced into the organism through genetic engineering to be effective; for processed foods like ice cream, antifreeze protein could simply be added during manufacture. Experiments have shown that fish antifreeze proteins are either unstable or not expressed at high enough levels to confer sufficient thermal hysteresis activity to transgenic organisms. Therefore, if crops such as oranges are to be prevented from freezing during a frost, insect antifreeze proteins with their higher thermal hysteresis activity may be more effective. Conversely, if ice recrystallization inhibition is desired (as with ice cream), plant antifreeze proteins may be more effective. Antifreeze proteins would not be a new addition to the human diet because many foods consumed regularly, such as carrots and fish, already contain significant amounts of these proteins.

Organs and tissues used for transplant can be stored on ice only for limited periods of time. Their shelf-life might be prolonged by colder temperatures if antifreeze proteins could be introduced (for example, perfusion of hearts) to prevent either ice formation or recrystallization. Survival of frozen blood cells, pig eggs, rat sperm, and livers did increase in the presence of fish antifreeze proteins, presumably from the inhibition of ice recrystallization. Antifreeze proteins have also been found to inhibit the ion leakage that occurs when cell membranes are chilled. However, in other experiments, ion leakage or freezing damage increased. Further research is under way to optimize the efficacy of antifreeze proteins in these and other applications, as well as to develop methods to produce antifreeze proteins in sufficient quantity to permit their cost-effective use. There are still a large number of antifreeze proteins that have yet to be discovered and characterized from a wide range of organisms, including additional insects and plants as well as bacteria and fungi.

For background information *see* ANTIFREEZE (BIOLOGY); COLD HARDINESS (PLANTS); CRYOBIOLOGY; FOOD PRESERVATION; GENETIC ENGINEERING; PEP-TIDE; PROTEIN in the McGraw-Hill Encyclopedia of Science & Technology.   Laurie A. Graham

Bibliography. K. V. Ewart, Q. Lin, and C. L. Hew, Structure, function and evolution of antifreeze proteins, *Cell. Mol. Life Sci.*, 55:271–283, 1999; [Budworm moth (PDB ID:1EWW)] S. P. Graether et al., β-Helix structure and ice-binding properties of a hyperactive antifreeze protein from an insect, *Nature* 406:325–328, 2000; M. Griffith and K. V. Ewart, Antifreeze proteins and their potential use in frozen foods, *Biotechnol. Adv.*, 13:375–402, 1995; [Type II fish AFP (PDB ID:2AFP)] W. Gronwald et al., The solution structure of type II antifreeze protein reveals a new member of the lectin family, *Biochemistry*, 37(14):4712–4721, 1998; [Type III fish AFP (PDB ID: 1MSI)] Z. Jia et al., Structural basis for the binding of a globular antifreeze protein to ice, *Nature*, 384(6606):285–288, 1996; [Mealworm beetle (PDB ID:1EZG)] Y.-C. Liou et al., Mimicry of ice structure by surface hydroxyls and water of a β-helix antifreeze protein, *Nature*, 406:322–324, 2000; [Type I fish AFP (PDB ID:1WFA)] F. Sicheri and D. S. Yang, Ice-binding structure and mechanism of an antifreeze protein from winter flounder, *Nature*, 375(6530):427–431, 1995; K. E. Zachariassen and R. Lundheim, Applications of antifreeze proteins, in R. Margesin and F. Schinner (eds.), *Biotechnological Applications of Cold-Adapted Organisms*, pp. 319–332, Springer-Verlag, Berlin, 1999.

Asteroid

Launched on February 17, 1996, the *Near-Earth Asteroid Rendezvous* (*NEAR*) mission to asteroid 433 Eros has provided scientists with a detailed study of two very different asteroids.

NEAR mission sequence. Before settling into an orbit around asteroid Eros on February 14, 2000, the *NEAR* spacecraft had already flown past asteroid 253 Mathilde in June 1997 (**Fig. 1**) and Eros on December 23, 1999. The first flyby was planned, but not the

Fig. 1. Image mosaic of asteroid 253 Mathilde, constructed from four images acquired by the *NEAR* spacecraft on June 27, 1997. This was taken from a distance of 2400 km (1500 miles). Sunlight is coming from the upper right. The part of the asteroid shown is about 59 by 47 km (37 by 29 mi) across. Details as small as 380 m (1250 ft) can be discerned. The surface exhibits many large craters, including the deeply shadowed one at the center, which is estimated to be more than 10 km (6 mi) deep. The shadowed, wedge-shaped feature at the lower right is another large crater viewed obliquely. The angular shape of the upper left limb of the asteroid results from the rim of a third large crater viewed edge-on. The bright mountainous feature at the far left may be the rim of a fourth large crater emerging from the shadow. The angular shape is believed to result from a violent history of impacts. (*NASA/Johns Hopkins University Applied Physics Laboratory*)

second. Initial mission plans required the *NEAR* spacecraft to fire its main engine on December 20, 1998, to effect a rendezvous with Eros, but a malfunction of the engine caused the spacecraft to fly quickly past Eros on December 23, 1998, rather than going into orbit about it the following month. After quickly replanning the mission, flight controllers commanded a successful firing of the spacecraft's main engine on January 3, 1999, thus allowing the spacecraft to slowly catch up to Eros one year later. The final maneuver to place the spacecraft in orbit about the asteroid took place on February 14, 2000. Once in orbit about Eros, the spacecraft was renamed *NEAR-Shoemaker* to honor the recently deceased planetary astronomer, Eugene Shoemaker, who made significant contributions to the understanding of asteroids. After an initial 300-km (186-mi) circular orbit about Eros (**Fig. 2**), maneuvers on board the spacecraft were used to lower the orbits down in stages to 200, 100, 50, and 35 km (124, 62, 31, and 22 mi). After one year in orbit about Eros, the spacecraft underwent a series of controlled descents toward the surface on February 12, 2001, taking ever-higher-resolution images as it went closer to the surface (**Fig. 3**). After a successful landing, communications continued with the spacecraft until February 28, 2001, when the mission was terminated. This National Aeronautics and Space Administration (NASA) mission was carried out by the Applied Physics Laboratory of the Johns Hopkins University.

NEAR instrumentation. The *NEAR-Shoemaker* spacecraft carried a suite of scientific instruments to investigate the nature of asteroids Mathilde and

Eros. During the Mathilde flyby (minimum distance of 1212 km or 753 mi on June 27, 1997), the onboard television camera was used for imaging the asteroid's surface and determining its size, shape, and volume (Fig. 1). The infrared spectrometer was used to measure the reflectance properties of the surface materials in the near-infrared wavelength region. Various minerals have different reflectance properties (spectral signatures), so the near-infrared spectrometer provided information as to what types of minerals were resident on the asteroid's surface. The spacecraft radio tracking data that were used to precisely determine the spacecraft's position changes in space were also used to determine the asteroid's mass. This was achieved during the Mathilde flyby by noting the very slight amount of bending in the spacecraft's trajectory caused by the gravitational attraction of the nearby asteroid.

The camera, near-infrared spectrometer, and radio tracking of the spacecraft were also used to good advantage when the spacecraft went into orbit about Eros (Fig. 2) 32 months years after the Mathilde flyby. While the flyby distance at Mathilde precluded their use, a number of additional instruments were used to study Eros while the spacecraft was in close orbit about it. A magnetometer was used to investigate whether or not Eros had a magnetic field and hence a metallic core. Since no magnetic field was detected, Eros (unlike the Earth) is very unlikely to have a metallic core. A laser rangefinder was used to develop a detailed shape model for Eros, and the x-ray and gamma-ray spectrometers were used to determine the detailed chemical composition of the asteroid's surface materials. While the near-infrared spectrometer can infer what types of minerals are likely to be present on the asteroid's surface, the x-ray and

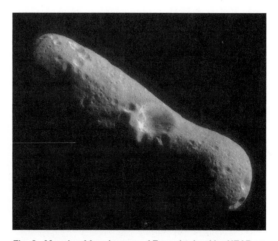

Fig. 2. Mosaic of four images of Eros obtained by *NEAR* on February 14, 2000, immediately after the spacecraft's insertion into orbit. The view is downward over the north pole of Eros at one of the largest craters on the surface, which measures 4 mi (6 km) across. Inside the crater walls are subtle variations in brightness that hint at some layering of the rock in which the crater formed. Narrow grooves that run parallel to the long axis of Eros cut through the southeastern part of the crater rim. A house-sized boulder near the floor of the crater appears to have rolled down the bowl-shaped crater wall. (*NASA/Johns Hopkins University Applied Physics Laboratory*)

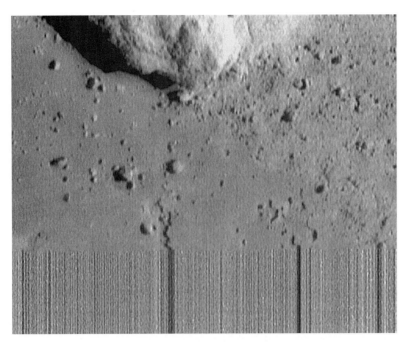

Fig. 3. Last image of Eros received from *NEAR-Shoemaker* spacecraft. Taken from a range of 120 m (394 ft), it measures 6 m (20 ft) across. What is visible of the rock at the top of the image measures 4 m (12 ft) across. The streaky lines at the bottom indicate loss of signal as the spacecraft descended upon the asteroid during transmission of this image. (*NASA/Johns Hopkins University Applied Physics Laboratory*)

ered with boulders of varying sizes. The surface of the asteroid is heavily cratered, indicating that Eros is relatively old. The largest crater is 5.5 km (3.4 mi) across (Fig. 2), but a 10-km (6-mi) saddlelike depression may well be a very old crater whose raised rim has been eroded by millions of years of impacts. Eros has several grooves and ridges that are probably due to the stresses resulting from frequent collisions. Results from the x-ray and gamma-ray spectrometer measurements suggest that the elements detected on the surface of Eros are very similar to those in ordinary chondrites, the most common meteorite found on the Earth's surface.

Mathilde. In contrast to Eros, Mathilde seems devoid of stress-related ridges and grooves, yet it has at least five craters between 19 and 33 km (12 and 20.5 mi) in diameter (Fig. 1). The first question that arises is how could this asteroid even survive the collisions that produced these enormous craters and, given that it did survive, why are there no stress fractures. Furthermore, why did not the debris from more recent collision events partially obliterate those craters formed by earlier such events? Another question is why Mathilde rotates so slowly when a series of major collision events might be expected to spin the object a bit faster with each impact. Although there is yet no good answer to the question of Mathilde's slow rotation, the extremely low bulk density of Mathilde may explain why it has been able to absorb such enormous impacts without breaking apart or even showing the stress features usually associated with these events. Because Mathilde has such a low bulk density, it must have interior voids where no rocky material resides. It may be a loose collection of material (rubble pile) so, when impacts do occur, the asteroid is better able to absorb the impact energy without shattering the object. For example, a solid brick hit with a hammer will shatter into many separate pieces, but a loose pile of sand or rocks would not suffer the same catastrophic breakup. Some researchers have even suggested that Mathilde is so effective in absorbing the energy of impacts that the colliding objects themselves are swallowed into the

gamma-ray spectrometers can go a step further and determine the detailed chemical makeup of these minerals.

Asteroid physical characteristics. The science instruments on board the *NEAR-Shoemaker* spacecraft have allowed scientists to study both asteroids Mathilde and Eros, and they could not be more different. Their physical characteristics are compared in the **table**.

Eros. The spacecraft measurements of Eros indicate that it is likely to be a uniformly dense collision fragment from a once larger body. It appears to be made of the same silicate minerals throughout its length and breadth, it is covered with a loose surface layer (regolith), and portions of the surface are cov-

Physical characteristics of Eros and Mathilde		
	Eros	Mathilde
Spectral class	S	C
Albedo (fraction of incident light reflected)	0.27	0.04
Color	Gray	Black
Rotation period, hours	5.27	418
Minimum diameter, km	8.7	46
Maximum diameter, km	31.6	66
Mass, kg	6.69×10^{15}	1.033×10^{17}
Volume, km^3	2503	78,000
Bulk density, g/cm^3	2.67	1.3
Escape velocity, m/s	3.1–17.2	23
Gravitational acceleration, mm/s^2	2.1–5.5	10
Suggested interior structure	Fractured rock	Rubble pile
Suggested chondrite meteorite analog	Ordinary	Carbonaceous
Composition	Silicate minerals	Silicate minerals and carbon-based mixtures

interior of Mathilde without forming the ejecta usually associated with impact events.

S- and C-type asteroids. The bulk of the objects in the asteroid belt between the orbits of Mars and Jupiter can be divided into two broad categories on the basis of their colors and their abilities to reflect light (albedo): S-type and C-type asteroids. The S-type asteroids are grayish and reflect about 20% of the sunlight that falls upon them. The C-type asteroids are black and reflect only about 4% of the incident sunlight. There are also subtle differences in the manner in which each spectral class of objects reflects light at particular wavelengths. That is, the spectra of each class differ. The S-class asteroids predominate in the inner asteroid belt between the orbits of Mars and Jupiter, while the C-class asteroids are far more numerous in the outer regions of the asteroid belt. Many asteroids are relatively primitive objects that have not changed significantly in terms of their chemical composition since the origin of the solar system some 4.6 billion years ago. Hence, knowledge of the chemical composition of the S-class asteroid Eros provided by the *NEAR-Shoemaker* mission provides evidence as to the chemical mix in the inner asteroid belt region when the solar system was first forming, and, by extension, to the conditions and composition of the primordial mixture from which the inner solar system (including Earth) formed.

Asteroid hazards. Eros is a member of a group of asteroids that can approach the Earth's orbit, called the near-Earth objects. While the orbit of Eros itself cannot approach that of the Earth too closely (currently, the closest approach of the two orbits is about 2.0×10^7 km or 1.2×10^7 mi), there are many near-Earth objects that can, and it is prudent to identify the objects that could, at some distant time, impact the Earth. Search efforts supported by NASA, together with some international efforts, are systematically cataloging those objects that can approach the Earth. Should one of these objects be found on an Earth-threatening trajectory, the knowledge gained on the *NEAR-Shoemaker* mission and future asteroid missions will be especially useful in mounting a successful mitigation effort. If a threatening asteroid is to be gently diverted from its Earth-threatening course, technologists must first know the type of object with which they are dealing, and in particular, whether the object is an S-class object (likely to be a fractured rock) or a C-type asteroid that has a very low density.

Asteroid resources. Ironically, those near-Earth objects that can most closely approach the Earth are also the most easily accessible in terms of exploiting their raw materials for the future colonization of the inner solar system. There are several near-Earth asteroids that are far richer than the Moon or Mars in terms of their wealth of minerals and metals, and it is also easier for spacecraft to land upon them. If major structures and habitats are to be built in space, these asteroid raw materials that are in near-Earth space will be needed. Future missions like the *NEAR-Shoemaker* effort should allow scientists to discern which among the millions of asteroids are the richest sources of raw materials. Apart from the Moon, the near-Earth asteroids are the Earth's nearest celestial neighbors and they will play an increasingly important role.

For background information *see* ASTEROID; METEORITE; SPACE PROBE in the McGraw-Hill Encyclopedia of Science & Technology.   Donald K. Yeomans

Bibliography. J. I. Trombka et al., The elemental composition of asteroid 433 Eros: Results of the *NEAR-Shoemaker* x-ray spectrometer, *Science*, 289: 2101–2105, 2000; J. Veverka et al., *NEAR* at Eros: Imaging and spectral results, *Science*, 289:2088–2097, 2000; J. Veverka et al., *NEAR's* flyby of Mathilde: The first look at a C-type asteroid, *Science*, 278:2109–2114, 1997; D. K. Yeomans et al., Estimating the mass of asteroid 253 Mathilde from tracking data during the *NEAR* flyby, *Science*, 278:2106–2109, 1997; D. K. Yeomans et al., Radio science results during the *NEAR-Shoemaker* spacecraft rendezvous with Eros, *Science*, 289:2085–2088, 2000.

Astrobiology

Astrobiology is a novel approach to the scientific study of the living universe. It seeks to understand the origin and evolution of life on Earth, to determine if life exists elsewhere in the universe, and to predict the future of life on Earth and in the rest of the universe.

Origin of life. In order to understand how life began on Earth, it is necessary to study the origin of the chemical compounds that make up living organisms as well as the physical factors needed to create an environment capable of supporting life.

Biogenic elements. One area of interest is the origin of biogenic elements (elements that are particularly important in living organisms such as carbon, hydrogen, oxygen, nitrogen, sulfur, phosphorus, iron, and magnesium) and organic compounds. From studies in cosmochemistry, it is known that all elements other than hydrogen, helium, and lithium are formed in stars (although helium and lithium can be formed in stars). Thus, all Earth organisms are literally made of stardust. Cosmochemists have demonstrated that a complex mixture of organic compounds can be produced in interstellar ice from water, methanol, ammonia, and carbon monoxide. These materials could have been transported to Earth and other bodies via meteorites, comets, or interplanetary dust particles. The European Space Agency's *Infrared Space Observatory* has detected benzene, a complex organic compound that forms the basis for many biological compounds, in a protoplanetary nebula (an old star that is about to become a white dwarf and is surrounded by a shell of heated, ionized gas known as a planetary nebula).

Physical factors. To a large extent, environmental factors determine whether life is possible and what evolutionary direction it can take. Thus, astrobiology also focuses on the environment that shaped the Earth and other bodies that could conceivably harbor life or at least some form of prebiotic (organic) chemistry. Since water is essential to life, it is critical to study the source of water and the environmental factors that permit liquid water to remain in a stable form.

Water on Earth coalesced as the planet was being formed, but it has also been delivered to Earth from comets. Early in 2001, an American team and an Australian team announced simultaneously that they had found evidence in ancient zircon crystals that liquid water and continental crusts existed on Earth as early as 4.3–4.4 billion years ago, a brief geological interval after the origin of the Earth 4.5 billion years ago. This discovery was exciting because the heavy bombardment of Earth by meteorites obliterated the rock record prior to 3.9 billion years ago.

It has been known for decades that water ice exists on Mars, but a recent interpretation of what appear to be flood plains and channels at the Martian surface suggests that near-surface liquid water may have existed there recently, and perhaps is still there even today. Current excitement is focused on magnetic data that suggest Europa and Ganymede, moons of Jupiter, harbor a liquid water ocean under a layer of water ice. The largest moon in the solar system, Ganymede, may also once have had volcanism and may even today have a subsurface ocean (although Ganymede's low heat production and less varied chemistry make it unlikely that this moon once harbored life).

Evolution of life. Life itself has yet to be created artificially, but certain reactions simulated in a laboratory suggest that, given adequate physical and chemical conditions, some steps in the origin of life are inevitable. For example, it is possible to demonstrate experimentally that lipids mixed with macromolecules, such as proteins or nucleic acids, can undergo drying and wetting cycles that simulate the tide pool environment. Under these conditions, the macromolecules are readily captured in membrane-bounded lipid vesicles, possibly as in the first cells. Ribonucleic acid (RNA) can perform both the information storage functions of deoxyribonucleic acid (DNA) and the self-processing functions of enzymes, suggesting that the first organisms were based on RNA (the so-called RNA world hypothesis). In addition, the evolution of biomolecules has been recreated in vitro, allowing observation of the evolution of new or improved functions.

Reconstructing early evolution has relied on three approaches: the fossil record, comparative biology, and actualistic paleontology.

Fossil record. The fossil record provides chemical fossils, such as the ratio of carbon-12 to carbon-13 (which indicates the level of photosynthesis) and morphological fossils. The earliest morphological record shows laminated microbial communities called stromatolites and cellularly preserved microfossils that are similar in structure to modern cyanobacteria.

Comparative biology. Comparative biology of modern organisms provides clues for reconstructing ancestral character traits. While in the past this approach was based on gross anatomy, comparative biology today relies primarily on DNA sequence comparisons and secondarily on biochemical and structural analyses. These studies currently suggest that the last common ancestor of all living organisms today may have lived in a high-temperature environment. This suggestion is consistent with the fact that the fossil record begins almost immediately after the ending of the late bombardment period, during which the oceans were heated.

Actualistic paleontology. Astrobiologists may use present analogs of ancient organisms under modern or simulated ancient conditions to learn how past communities functioned. Such studies are termed actualistic paleontology. Because of the prevalence of stromatolites in the fossil record from 3.5 to 0.6 billion years ago, microbial mats, which are modern analogs of such communities, are studied with particular intensity. These studies allow astrobiologists to utilize modern techniques such as molecular biology and geochemistry to reconst how life functioned in the past, from the molecular level of gene induction all the way to ecosystem function.

Search for extraterrestrial life. While astrobiology is not synonymous with the search for life in the universe, it is a vital part of the enterprise. This search includes determining the limits of life, the factors that make a planet habitable, and how to recognize signs of current and past life within our solar system (particularly on Mars and Europa).

Limits of life. Organic chemistry occurs in space. Meteorites, comets, and interplanetary dust rain down on Earth, providing a supply of organic material. During its first half billion years, Mars may have been more hospitable to life than Earth itself at that time, and it is known that rocks from Mars have struck Earth. Further, the European Space Agency in collaboration with European and American researchers has flown organisms into Earth's orbit, exposing them to the space environment, and found that some have survived (Biopan and Long Duration Exposure Facility experiments). All this has encouraged a reexamination of the possibility that life can travel from body to body within our solar system—a process called panspermia. Further experiments testing the survival of organisms in space are planned for the external Expose Facility on the International Space Station.

Current research on life in extreme environments on Earth has extended scientific knowledge regarding the physical and chemical limits of life. For example, until recently, it was thought that life could not survive at any temperature near the boiling point of water because high temperatures denature proteins and nucleic acids and melt lipids. However, it is now known that at least one organism, the archaea *Pyrolobus fumarii* (Crenarchaeota), can survive at 113°C. Similarly, low-pH environments (below a pH of 4–5) are considered extreme, and it has been discovered that a few unicellular organisms, such as the alga *Dunaliella acidophila* and several archaea and fungi, can survive below a pH of 1.

Habitable planets. Since 1995, the discovery of extrasolar planets (those orbiting other stars) has occurred at an increasing rate. The excitement lies in the fact that planets should provide more stable environments for life to take hold and evolve than more transitory bodies like comets. Although the extrasolar planets discovered to date have altered the conception of what constitutes a "normal" solar system, none are likely to be habitable. Recent theory suggests that habitable planets may orbit red dwarfs. While these stars are only 6–60% of the mass of the Sun, they may constitute 80% of the stars.

Signs of life on Mars and Europa. The main focus for searching for extraterrestrial life, both within and eventually outside our solar system, is on microscopic life-forms. This is both because larger organisms would have already been detected, and because microbes, in particular, are often adept at surviving in the extreme environments that the rest of the solar system has to offer. NASA's *Viking* missions were directed at detecting either visual or metabolic signals of life on Mars, but the results were negative or highly ambiguous at best. Current research focuses on the possibility of a former Martian biota that may be detected in meteorites (or other samples) from the planet or even an extant Martian biota, shielded from solar radiation and living deep under the ground, in the polar ice caps, or in salt deposits. Deep drilling on Mars is technically difficult but may be the most promising approach. Recent suggestions that liquid water may have been present on Mars in the recent past or even in the present are forcing a reevaluation of the deep-drilling strategy.

Europa has an advantage over Mars because a liquid water ocean is thought to exist there beneath a thick ice crust. Unfortunately, this thick crust and Europa's significant distance from Earth will also pose technical challenges.

Future of life. Just as organisms have evolved and diversified, the physical environment of Earth has changed. The Sun has become more luminous over time as it burns its supply of hydrogen. The atmosphere of Earth has changed from an anaerobic one dominated by carbon dioxide to an aerobic atmosphere with 21% oxygen. The continents have formed, drifted, and collided. Earth has been struck by comets and meteorites, some large enough to have caused the extinction of 90% of life at the end of the Permian Period 225 million years ago and of the dinosaurs 65 million years ago.

Currently, humans are altering the physical environment and the biodiversity of life on Earth to an unprecedented degree. Scientists are observing these changes from the ground and from the vantage point of space and are using this information to develop predictive models for the future.

Advances in astronomy and planetary science have revealed that in 1–2 billion years Earth's moon will be lost to space. Without the stabilizing effect of the moon, the obliquity of Earth will vary chaotically, resulting in large and frequent shifts in climate. Thus, an integrated research approach will be necessary to understand, predict, and adapt to the changing environment of Earth.

Roadmap for astrobiology. The field of astrobiology can be said to have been born in 1996, when David McKay of NASA's Johnson Space Center and his colleagues interpreted chemical and geological evidence from a meteorite that originated on Mars to argue that life was once present on that planet. Suddenly scientists from disparate fields coalesced under the new metadiscipline. Since that time, NASA has taken the initiative in establishing the field of astrobiology, with Ames Research Center in California designated as the lead center. In 1998, Ames hosted a workshop to develop a "roadmap" for astrobiology, with results that were not definitive or exclusive but suggested areas for research, including global climate forecasting, medicine, and biotechnology. Future roadmap workshops are likely; however, such planning is quickly being overcome by other events shaping the field.

NASA has founded a "virtual" Astrobiology Institute, which now includes 14 lead member institutions culled from universities, government labs, and research centers in the United States, and a growing number of affiliates from around the world. The first Astrobiology Science Conference was held at NASA Ames in April 2000, with over 600 participants from dozens of countries, and a second conference was scheduled for 2002.

Future destinations for scientific exploration will include Mars, Europa, and various comets and asteroids. Back on Earth, there will be further studies of environments that are at the physical and chemical limits for survival, which will give scientists a better sense of what are the "possible" conditions for life. The tools of molecular biology are also advancing at a tremendous pace, allowing rapid DNA sequencing, determinations of gene regulation, and analyses of protein function. To handle the flood of data, advances in computer systems have led to the creation of computational astrobiology. Access to space and the opportunity to perform biological experiments there have increased through unmanned missions as well as manned missions such as NASA's space shuttle and the Russian space station *Mir* and, in the future, the International Space Station. In this rapidly advancing metadiscipline, unanticipated breakthroughs will be increasingly common and undoubtedly startling in the way that they require reevaluation of our view of life.

For background information *see* ARCHAEBACTERIA; ASTRONOMY; COSMOCHEMISTRY; JUPITER; MARS; MOLECULAR BIOLOGY; PALEONTOLOGY; PREBIOTIC ORGANIC SYNTHESIS; SPACE BIOLOGY; THERMAL ECOLOGY in the McGraw-Hill Encyclopedia of Science & Technology.   Lynn J. Rothschild

Bibliography. C. F. Chyba and C. B. Phillips, Possible ecosystems and the search for life on Europa, *Proc. Nat. Acad. Sci. USA*, 98:801–804, 2001; K. Croswell, The right places to look for alien life, *New Scientist*, no. 27, January 2001; J. P. Dworkin et al., Self-assembling amphiphilic molecules: Synthesis in simulated interstellar/precometary ices, *Proc. Nat. Acad. Sci. USA*, 98:815–819, 2001; J. I. Lunine, The occurrence of Jovian planets and the habitability of planetary systems, *Proc. Nat. Acad. Sci. USA*, 2001; M. Malin and K. Edgett, Evidence for recent ground-water seepage and surface run-off on Mars, *Science*, 288:2330–2335, 2000; G. Marcy, Extrasolar planets: Back in focus, *Nature*, 391:127, 1998; G. W. Marcy and R. P. Butler, Detection of extrasolar giant planets, *Annu. Rev. Astron. Astrophys.*, 36:57–97, 1998; D. S. McKay et al., Search for past life on Mars: Possible relic biogenic activity in Martian meteorite ALH84001, *Science*, 273:924–930, 1996; L. J. Rothschild and R. L. Mancinelli, Life in extreme environments, *Nature (London)*, 409:1092–1101, 2001; P. M. Schenk et al., Flooding of Ganymede's bright terrains by low-viscosity water-ice lavas, *Nature*, 410: 57–60, 2001.

Bioceramics

Bone is formed by cells called osteoblasts, which arise from progenitor cells in a multistep process. At bone defect (injury) sites, osteoblast progenitors are recruited from thin cell layers surrounding the bone surface and tissue inside the bone (periosteum and bone marrow, respectively). That is, resected or injured bone sites induce the chemoattraction of multipotent progenitor cells that originally reside in the marrow or periosteum, and these cells become osteoblasts at the defect site through a series of controlled differentiation steps. These progenitor cells are referred to as stromal or mesenchymal stem cells (MSCs). When metals or their alloys are implanted at bone defect sites, corrosion may inhibit the differentiation of MSCs around the implanted metals, resulting in encapsulation of the implants by scar tissue or a fibrous membrane, with the thickness of the membrane proportional to the degree of metallic corrosion. In contrast, when alumina (Al_2O_3) ceramics or titanium devices are implanted into bone defects, the MSCs around the ceramics can undergo osteogenic differentiation. This results in functional regeneration of the bone tissue around the materials with minimal scar tissue or fibrous membrane formation. Still, the interface between the bone and the implant is not strong, and will detach easily upon shear and load stress.

Certain types of glass ceramics and calcium phosphate ceramics, when implanted, chemically bond with the regenerated bone tissue. The bone-ceramic

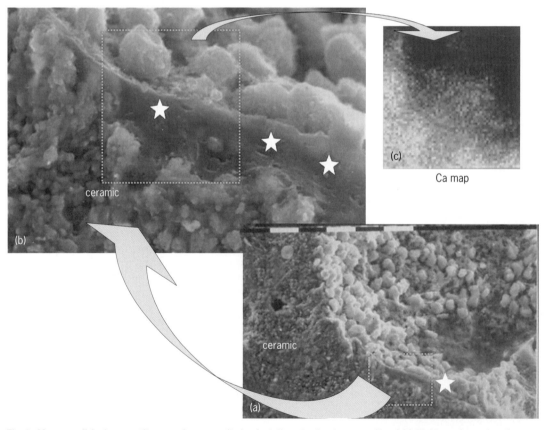

Fig. 1. Marrow cell–hydroxyapatite ceramic composite implantation at subcutaneous sites. (*a*) Initial bone formed on the ceramic surface as observed by scanning electron microscopy of the fractured surface of the composite. Preosteoblasts ∼ osteoblasts attach on the ceramic surface, and the osteoblasts fabricate initial bone (star) directly on the surface. (*b*) Higher magnification of the rectangular area of *a*. Osteoblasts (round cells) fabricate new bone (stars) on the ceramic surface. (*c*) Results of electron probe microanalysis for calcium of the rectangular area of *b*. The calcium content of the initial bone area is about one-third that of the ceramic area. The low content of calcium in the bone indicates that the bone is not fully mineralized. (*After J. Biomed. Mat. Res., 48:913–927, 1999, and 37:122–129, 1997*)

interface formed is very strong and stable. Upon loading, breakage usually occurs inside the bone or ceramic but not at the interface. Such interaction between bone and material (glass and calcium phosphate ceramics) is called bone bonding. Bone bonding was first demonstrated by L. L. Hench for glasses within a certain compositional range. These glasses contained silicon dioxide (SiO_2), sodium oxide (Na_2O), calcium oxide (CaO), and phosphorus pentoxide (P_2O_5) in specific proportions, and were later called bioactive glasses because they formed bonds with living tissues. (Although alumina ceramics may show bone deposition onto their surface without fibrous encapsulation, the bone cannot bond to the alumina surface. Alumina, therefore, is not bioactive.)

After the discovery of the bonding property of the glasses and glass ceramics, hydroxyapatite (HA) ceramic and other types of calcium phosphate ceramics were found to have this bioactive property as well. These bioactive ceramics show surface changes (dissolution and precipitation phenomena) that lead to carbonate-hydroxyapatite precipitation on the ceramic surface. Direct bone bonding has been reported via the apatite layer precipitated onto the bioactive ceramic. In contrast, nonbioactive ceramics show neither precipitation nor bone bonding. Since their discovery, many bioactive ceramics have been used in bone reconstruction surgery and have been reported to show excellent results.

Osteogenic differentiation in porous ceramics. The developmental cascade that results in functional bone bonding is initiated by the attachment of MSCs to the implant followed by osteoblastic differentiation. During the differentiation cascade, attached stem cells (fibroblastic cells) develop into small round cells (preosteoblasts). Later on, clusters of larger round cells (osteoblasts) form a thin layer of primary bone on the ceramic surface (**Fig. 1***a*). The primary bone is partially mineralized osteoid (uncalcified bone matrix) [Fig. 1*b, c*]. Later this tissue becomes fully mineralized bone, and bone bonding begins. As shown in Fig. 1*c*, the mineral content (calcium) in the thin primary bone is low (an indication of osteoid formation), but the calcium content and phosphorus content in the thick bone are comparable to those of the ceramic (**Fig. 2**). These high levels of mineral are found across the bone-hydroxyapatite ceramic interface, and there is no intervening fibrous tissue at the interface, again indicating bone bonding (Fig. 2). After the bone is fully mineralized, bone deposition follows. Invariably, bone formation always starts on the surface of the hydroxyapatite ceramic and proceeds toward the center of the pore. The cascade, called bonding osteogenesis, illustrates the importance of a bioactive ceramic surface to support the osteogenic differentiation of marrow MSCs. By analyzing the cell-ceramic composite graft, other bioactive materials such as glass ceramics (apatite wollastonite–containing glass ceramics) and tricalcium phosphate (TCP) and biphasic hydroxyapatite-tricalcium phosphate ceramics have been found to

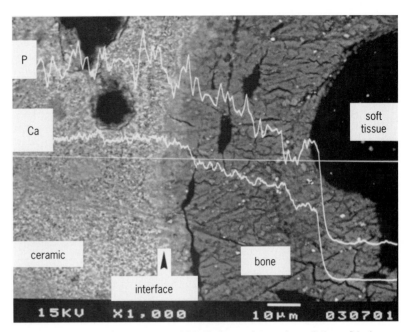

Fig. 2. Backscattered electron image which displays no intervening soft tissue (black area) between bone (gray area) and ceramic (white area). Simultaneous line analysis of phosphorus (P) and calcium (Ca) are derived from the electron beam pathway as indicated by the center white line. The calcium content of the bone is high (about 75% of that of the ceramic), which indicates fully mineralized bone compared to the low content of calcium in the initial bone on the ceramic surface (Fig. 1c). (*After Biomaterials, 12:411–416, 1991*)

exhibit the same cascade of bonding osteogenesis and to show the same performance relative to the cell differentiation sequence.

Manipulation of mesenchymal stem cells. Composites of fresh whole bone marrow cells and porous calcium phosphate ceramics exhibit osteogenic differentiation when implanted at subcutaneous or intramuscular sites. Although the number of marrow cells used in such implants is very large, only a small number of MSCs reside in the marrow preparation and eventually expand and differentiate into osteoblasts. Alternatively, marrow cells can be seeded into a culture in which MSCs can attach and mitotically expand before in vivo implantation. For example, the marrow obtained from one rat is sufficient to directly establish only two or three composite grafts that exhibit in vivo osteogenic potential. However, after in vitro expansion, there are sufficient cells (first passage cells) to establish 20 to 30 composites showing substantial osteogenic response. Thus, the number of stem cells can be increased about 10 times in a single culture. Due to the faster and more uniform bone formation that is observed with their use, in vitro expanded MSCs are now employed more often than fresh, unexpanded marrow cells in making osteogenic composites.

When fresh, whole marrow cells (**Fig. 3***a*) and culture-expanded MSCs (Fig. 3*b*) have been combined with porous ceramics, all the composites exhibited in vivo osteogenic potential, with those containing cultured MSCs observed to form bone much earlier. The fact that composites of culture-expanded MSCs and porous ceramics showed in vivo bone formation indicated that cell function was

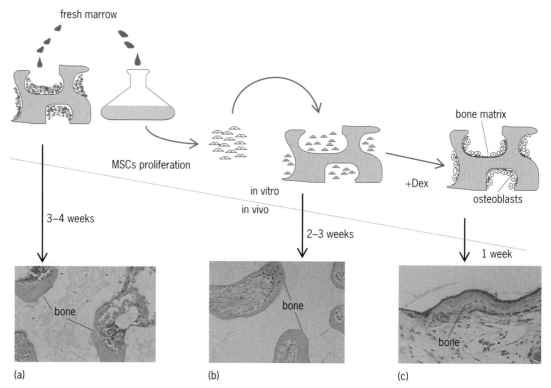

fresh marrow

MSCs proliferation

in vitro
in vivo

3–4 weeks

2–3 weeks

1 week

+Dex

bone matrix

osteoblasts

bone

bone

bone

(a)   (b)   (c)

Fig. 3. Schematic representation of the temporal events of bone formation of various composites of porous hydroxyapatite ceramic and fresh and cultured marrow cells (MSCs) after subcutaneous implantation. (*a*) Composite containing fresh dispersed marrow cells. It shows bone formation in pore areas of the ceramic 3–4 weeks after in vivo implantation. (*b*) Composite containing culture-expanded MSCs. (*c*) Composite containing culture-expanded MSCs allowed to differentiate in vitro into osteoblasts, which make bone matrix on the ceramic. The bone fabricated on the ceramic in vitro exhibits rapid (1-week) bone formation during the in vitro/in vivo transition. Dex indicates Dexamethasone.

maintained through in vitro culture expansion procedures (Fig. 3*b*). Taking this a step further, when MSCs were cultured in a porous framework in the presence of phosphate and the hormone dexamethasone, the stem cells not only expanded but differentiated into osteoblasts (which fabricate bone matrix), causing the porous areas of the hydroxyapatite to become covered with new bone and active osteoblasts (Fig. 3*c*). This in vitro culture-differentiated composite exhibited voluminous new bone growth one week after subcutaneous implantation, more than had ever been observed using composites of fresh marrow or cultured MSCs and porous ceramics (Fig. 3). These observations indicate that both culture-expanded MSCs and culture-differentiated osteoblasts survive the in vitro/in vivo transition. Such in vitro tissue engineered bone within the ceramic provides immediate new bone-forming capability after in vivo implantation and therefore has immense clinical significance.

Many osteoarthritic and rheumatoid arthritic patients need total joint replacements, and are compelled to use prosthetic devices that may have problems, including loosening of the implants. To prevent loosening, the prostheses have been fabricated with porous structures or coated with bioactive materials such as hydroxyapatite. A new concept to prevent such loosening involves coating joint prostheses with osteogenic cells or their precursors. Fresh

bone marrow would be collected from the patient and the MSCs isolated, expanded in number, and subsequently cultured on the surface of the prostheses under the osteogenic conditions described above. The surface of the prostheses would be covered with bone (osteoblasts and bone matrix) derived from the patient's own cells. This bone would possess the means for bone bonding as well as maximum new bone formation. As a result of this biologic surface reconstruction, loosening could be avoided and the postoperative rehabilitation program shortened due to early and secure bone formation around the implanted prosthesis. The concept, called osteogenic matrix coating, could also be used for other skeletal reconstruction surgery such as therapy for patients having bone tumors, fractures with massive bone defects, or delayed healing of fractured bone.

For background information *see* APATITE; BONE; CELL DIFFERENTIATION; CERAMICS; SKELETAL SYSTEM; TISSUE CULTURE in the McGraw-Hill Encyclopedia of Science & Technology.   Hajime Ohgushi

Bibliography. A. J. Friedenstein et al., Heterotopic transplants of bone marrow: Analysis of precursor cells for osteogenic and haemopoietic tissues, *Transplantation*, 6:230–247, 1968; L. L. Hench, Bioactive ceramics, *Ann. N.Y. Acad. Sci.*, 523:54–71, 1988; H. Ohgushi et al., Materials based cell differentiation: Osteoblastic phenotype expression on bioactive materials, in D. L. Wise (ed.), *Encyclopedia of*

Biomaterials and Bioengineering, 11:761–771, Marcel Dekker, 1995; H. Ohgushi and A. I. Caplan, Stem cell technology and bioceramics: From cell to gene engineering, *J. Biomed Mat. Res.*, 48:913–927, 1999; T. Yoshikawa, H. Ohgushi, and S. Tamai, Immediate bone forming capability of prefabricated osteogenic hydroxyapatite, *J. Biomed. Mat. Res.*, 32: 481–492, 1996.

Biocompatible semiconductor

Scientists are working toward the realization of implantable bio-interfaced electronics and bioneural computers. At the forefront of this research is the biocompatible semiconductor, nanostructured silicon. Bio-interfaced electronics rely on developing the ability to culture (grow) living cells on semiconductors—the semiconductor and cell must be mutually compatible—and being able to pass signals (information) between the cells and semiconductors. If the research is successful, a new generation of medical (optical, aural, and locomotory) and computing devices will emerge. But to be successful, the seemingly disparate technologies of computing, physics, and biotechnology must work together.

Biocompatibility of semiconductor materials. Biocompatibility is the ability of a material, such as a semiconductor, to interface with a natural substance without provoking a natural defensive response. The human body's typical response is to coat the surface of the material with proteins and cells from body fluids. This can result in infection and biological rejection of a device manufactured from noncompatible materials. The success of any implant is therefore dependent upon the behavior of cells in the vicinity of the biomaterial used to make the device. All biomaterials have shape, chemical, and electrical surface characteristics that influence the cellular response to the implant. However, the initial event that occurs is the adsorption of a layer of protein. The nature and quantity of the adsorbed protein depends on the pH, temperature, and chemical constituents of the biological fluid in question, and the nature of the surface. The in vitro deposition of hydroxyapatite (the mineral component of bone) onto the surface of a potential biomaterial from a simulated body fluid has become a standard indicator of potential bioactivity for materials for bone implantation.

Bulk silicon. Electronic devices that use bulk silicon semiconductors have been available for in vitro biosensor applications for some time. However, the bioincompatibility of this type of silicon has prevented its use in vivo. Bulk silicon-based integrated circuits will require packaging in biocompatible material if they are to be used in and interfaced with living cells.

Transistors and laser diodes. Two basic types of electronic devices constructed with semiconductor materials are the transistor and the laser diode. Transistors are silicon-based, and laser diodes are based on gallium arsenide. From an electronic point of view, it would be both practical and economic to construct transistors and laser diodes from the same material. For the purposes of developing biologically interfaced devices, the search for materials of this type is more urgent as gallium arsenide is toxic to biological systems. The search for an efficient, luminescent semiconductor that could fulfill the combined role of a laser diode and transistor moved closer to success when nanostructured silicon demonstrated these properties.

Nanostructured silicon. Nanostructured silicon is produced by etching bulk silicon with acid. The acid eats away the surface silicon to leave a complex and random pattern of pits and holes. The internal surface area of a nanostructured silicon layer varies from 200 to 600 m^2/cm^2. It is the movement of electrons (subatomic particles) within the fine structure of the nanostructured silicon that gives the material its luminescent properties. Nanostructured silicon also has properties that make it a very promising biomaterial, in particular for biologically interfaced sensing applications.

Nanostructured silicon has been shown to absorb the proteins human serum albumin (HSA) and fibrinogen. Hydration of the nanostructured surface significantly decreases the adsorption of HSA onto the surface but increases the amount adsorbed deeper in the nanostructured film. Hydration failed to affect the adsorption of fibrinogen, a protein essential in blood clotting processes. Nanostructured silicon has been shown to be reactive toward hydroxyapatite formation—the first indication of nanostructured silicon's potential as a biomaterial.

The etched surface of nanostructured silicon inevitably harbors impurities from both the surrounding air and the manufacturing process. For example, oxygen is normally adsorbed to levels of as much as 1% within minutes after air-drying, and over a period of a few days silicon oxides are formed. These oxides at the silicon surface are thought to play a crucial role in the biocompatibility of the material. Nanostructured silicon stored in atmospheric conditions can have an oxygen content up to 50%.

In addition to the characterization of the biocompatibility of the nanostructured silicon, the possibility of toxic effects must be considered. Silicon is essential in biological systems and affects development. However, little is known about the biological processes that handle silicon at the molecular level. When a system is exposed to excess silicon, the potential for silicon-induced toxicity exists. While there have been many studies regarding the possible toxicity of implanted silicone, there are few positive reports on the toxicity of silicon or its compounds. It has been recently demonstrated that cells grown on a plate containing silicon substrates grow to confluence (a complete continuous sheet) on the plastic surrounding the silicon material. Also, it is possible to culture cells on a nanostructured silicon substrate, and the cells are viable in terms of structure and metabolism. So, nanostructured silicon wafers are apparently not toxic to the cells and

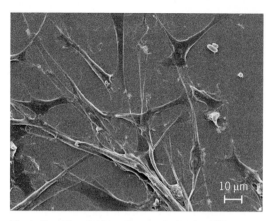

Fig. 1. Scanning electron micrograph showing B50 rat neurons growing on the surface of nanostructured silicon.

present an acceptable surface for growth (**Fig. 1**). Nanostructured silicon's lack of toxicity offers a distinct advantage over gallium arsenide and indium arsenide, as arsenic and indium are both toxic.

Cells are sensitive to the topological, chemical, and electrical properties of surface substrates on which they are grown. Cells cultivated on microstructures made by semiconductor technology grow normally on silicon surfaces covered with microelectrode arrays, as well as on microperforated silicon membranes with square pores (5, 10, or 20 micrometers edge length at the top and 1.2, 6.2, or 16.2 μm at the bottom). The cells spread over the 5- and 10-μm pores, but mostly failed to cover the 20-μm ones.

The ability to culture mammalian cells directly onto nanostructured silicon coupled with its apparent lack of toxicity offers exciting possibilities for the future of biologically interfaced sensing. This could involve the development of biologically interfaced neural networks or electronic sensing, with signal information being directly transduced from a living system to a nanostructured silicon device. None of the applications described above will be possible unless the technology is developed to allow the passage of information (in the form of a signal) across the interface between the silicon and a living cell. Developing this connectivity technology will be a great challenge.

Signaling across the silicon-cell interface. Biological signaling can take a number of forms. These include the chemical signals, such as hormones, that regulate growth and development, and the electrical signals that travel through the nervous system. The brain consists of a complex network of neurons that process information by transmission of electric potentials, making the brain the first, fastest, and best biological computer. Interfacing living neurons with silicon-based devices in a bioneural network poses a number of problems. Silicon devices operate between +5 and 0 V while neurons operate between −60 and +40 mV. Silicon devices are serial processors that have a hard-wired, fixed architecture, while neuronal networks are parallel processors that are plastic and self-organized. The biological requirements of neurons make them intolerant to large variations in temperature and pH, and in constant need of nutrition. Mammalian cells in culture require constant contact with a pH-controlled nutrient medium (solution) and must be maintained at 37°C (98.6°F). The nutrient medium contains electrolytes, sugars, amino acids, vitamins, growth factors, and dissolved gases, all of which may interfere with the passage of an electrical signal across the silicon-cell interface.

Signal propagation. Work in the area of bioneural networks in the past used leech neurons (approximately 500 μm in length). However, B50 rat neurons (which are much smaller, approximately 30 μm in length) have now been cultured on nanostructured silicon (Fig. 1). These cells were isolated from tumor tissue formed in a rat brain. They maintain some of the characteristics of neurons but, unlike neurons, they have the ability to divide. B50 cells seem to have greater contact affinity for nanostructured silicon than for amorphous or single-crystal silicon.

B50s have been used to investigate electrical signal propagation within a neuronal network. By mechanically stimulating a single neuron attached to a group, it is possible to record the propagation of a signal from its source to a second defined point. These time-based data provide information about the properties of signal transportation. A number of investigations have been carried out to target cell growth to specific points or to electrodes on the surface of a device by "plugging in" neurons where transistors used to be. However, in a biological environment the neurons are mobile and continually self-organize as part of the signaling process; consequently, problems have been experienced with cell mobility. Because of this, some groups are now moving toward the use of self-organized cultures and sampling signals at fixed points beneath the culture.

Direct two-way signaling. Direct two-way signaling between cell and device, using electrodes, has been problematic in that the present technology requires the cell membrane to be wholly or partially punctured. This shortens the lifetime of the cell to a couple of hours. A solution to this problem needs to be found if silicon devices are to be interfaced with animals (such as in human limb replacements). The possibility exists that the luminescent properties of

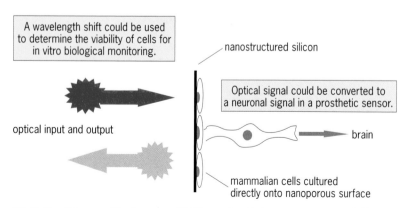

Fig. 2. Possible uses of luminescence (light) as a signal between nanostructured silicon and a living cell.

nanostructured silicon could allow it to be linked to a data logger by optical fibers, removing the need to perforate a cell to determine a response (**Fig. 2**). This could allow the development of new devices to replace damaged tissues in the eye.

Clearly, the success of bioneural network technology will depend on whether the electronic hardware can be adapted to the requirements of living cells. Over the next few years workers in the field may well move toward bridging the silicon-cell interface with signals of a different type, making use of current progress in the related technology of biosensors (chemical sensors using biological molecules).

For background information *see* BIOELECTRONICS; BIOSENSOR; INTEGRATED CIRCUITS; NEURAL NETWORK; NEUROBIOLOGY; NEURON; OPTICAL INFORMATION SYSTEMS; SEMICONDUCTOR; SEMICONDUCTOR DIODE; SILICON; TRANSISTOR in the McGraw-Hill Encyclopedia of Science & Technology.

Lorraine D. Buckberry; Sue C. Bayliss

Bibliography. S. C. Bayliss, L. D. Buckberry, and A. Mayne, Interaction of biomaterials with porous silicon, in L. Pavasi and E. Buzaneva (eds.), *Frontiers of Nano-Optoelectronic Systems*, pp. 199–207, Kluwer Academic, 2000; J. D. Bronzino (ed.), *The Biomedical Engineering Handbook*, 2d ed., CRC Press, Boca Raton, FL, 1999; A. E. G. Cass (ed.), *Biosensors: A Practical Approach*, IRL Press, Oxford, UK, 1990.

Cancer detection

Early detection is critical to the clinical outcome in the treatment of cancers. As an example, colon cancer accounts for 15% of all cancer-related deaths in the United States. Unfortunately, only 37% percent of colon cancers are found early enough for moderate treatment. Once this type of cancer reaches metastatic activity (spreads), there is only a 7% survival rate. Oral cancer is another example. Each year about 31,000 Americans develop oral cancer. Squamous-cell carcinoma (SCC) accounts for 95% of all malignant oral lesions, and it has a survival rate of only 50%. Yet when this type of cancer is detected in its earliest stages, the survival rate becomes approximately 80%. In cancers of the esophagus, the 5-year survival rate is only 5%. In contrast, if this type of cancer is detected when it is still contained in the mucosa, the 5-year survival rate becomes 90%.

Current detection methods. Current state-of-the-art means of detection often miss early-stage colon and oral SCC disease. These detection protocols use white light endoscopy with gross visualization. Yet the visual cues for determination of the disease state are small, especially the discrimination between nonmalignant, dysplastic (abnormal cells), and premalignant lesions. Visual assessment of early lesions within the colon and oral cavities depends on many factors, including the experience and ability of the clinician to identify the suspect lesions at an early stage of development and to select the suspect site that is to be biopsied. D. Rex and coworkers (1997) reported that an experienced colonoscopist performing back-to-back colonoscopies on patients will overlook as many as 15–24% of the neoplastic polyps smaller than 1 cm and up to 6% of the larger polyps present. Clearly there is a need to improve the detection of diseased tissue at its earliest stages.

Autofluorescence. Alternative techniques to aid in vivo diagnosis have recently been reported, including the use of techniques to detect changes in the native spectroscopic properties of tissue. Of particular interest has been the use of autofluorescence techniques to quantify changes in pathology. Some evidence suggests that tissue staging can be accomplished, allowing transformation from dysplasia to cancer to be detected. All tissue contains fluorophores, compounds that absorb light and subsequently emit light at a longer wavelength. Nicotinamide adenine dinucleotide (NAD[H]), flavins, collagen, and elastin are some of the tissue fluorophores. It is currently believed that autofluorescence primarily detects changes in concentration or distribution of these components. As normal tissue becomes dysplastic, the concentration or distribution of these endogenous fluorophores changes, leading to a detectable change in the resulting fluorescence spectrum. These changes are wavelength-dependent and correlate with changes in histology. While promising as a diagnostic tool, these signatures are generally not good candidates for the detection of early lesions. The inherent limitation of autofluorescence techniques for early detection is a low signal-to-noise ratio, stemming from a relatively low signal (small changes in concentration of the solutes detected in early disease) and a large background (scattering, reflected light), making the interpretation of the results of any quantitative measurement rather complicated.

Contrast enhancement. Because of the limitations associated with detection and diagnosis by native fluorescence, the use of contrast agents for diagnostic optical imaging has experienced increased attention.

Aminolevulinic acid. One way to increase the effectiveness of autofluorescence is to use 5-aminolevulinic acid, or ALA (**Fig. 1**), which is a precursor to the endogenous protoporphyrin IX. Protoporphyrin IX (PP IX) is a class of porphyrins that has been shown to overaccumulate in certain premalignant tissues and display somewhat attractive fluorescent properties: excitation in the blue region (about 450 nm) with emission in the red region (about 620 nm). ALA has been used primarily systemically, with the goal of increasing the endogenous concentration of PP IX in order to give contrast for disease detection through the production of and subsequent detection of the fluorophore PP IX. Some of the most successful uses of ALA have been the detection of dysplasia in Barrett's esophagus and colitis.

Toluidine blue. One of the most common nonendogenous compounds that has been used clinically as a contrast agent is toluidine blue (Fig. 1). This compound has been used as a contrast agent for the detection of occult malignancies of the cervix, has been found to provide some improvement for noninvasive

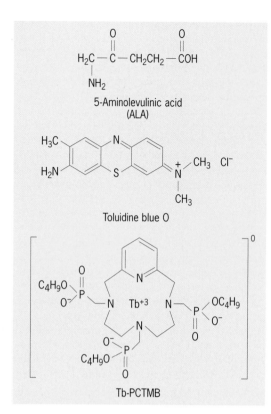

Fig. 1. Agents for contrast enhancement. ALA is a precursor to PP IX. Toluidine blue is used to mark lesions in the oral cavity. Tb-PCTMB is a lanthanide chelate for marking early lesions in the colon and oral cavity.

provement of sensitivity over conventional white light imaging, it suffers from a lack of specificity.

Photodynamic therapy. Another example of contrast enhancement agents or site-directed chemical agents that have seen recent success is the photodynamic therapy (PDT) class of markers. While these types of markers have shown promise in a diagnostic setting, they have limitations, including long delays for accumulation in tumors, prolonged photosensitization of skin, and phototoxicity of tissues being imaged.

Terbium chelate. Recently the use of a pyclen-based terbium chelate has shown promise in detecting chemically induced colon cancers in the Sprague Dawley rat. This particular molecule, Tb-[*N*-(2-pyridylmethyl)-*N′*, *N″*, *N‴*-tris(methylenephosphonic acid butyl ester)-1,4,7,10 tetraazacyclododecane] or Tb-PCTMB (Fig. 1), has excellent fluorescent properties, high specificity, and low toxicity. Some of the spectroscopic properties of Tb-PCTMB that are advantageous to its use as a fluorescent contrast enhancement marker include an extinction coefficient (molar absorptivity) of ~3000 L mole^{-1} cm^{-1}, a high quantum efficiency (ratio between the energy absorbed and the energy emitted) of 0.51, an extremely large Stokes' shift (shift between the absorption and emission spectrum to longer wavelength) of 280 nm, and a relatively long fluorescent lifetime (2.2 ms). **Figure 2** shows the absorbance and emission spectra of Tb-PCTMB. The figure illustrates that the emission signal is spectrally removed from the background, allowing inexpensive instrumentation to be used and low tissue dose levels to be administered. Sensitivity for this class of molecules is such that femtomole/pixel (picomolar) quantities have been quantified in intestinal tissue by endoscopy. This high sensitivity has allowed applications of millimolar solutions to be administered, very low light levels (TLC reader lamp) to be employed, and visual detection to be used for the detection of suspected sites in the colon of the Sprague Dawley rat. This preliminary work indicates that Tb-PCTMB can be used as an exogenous marker for dysplastic tissues with sensitivity as high as 94.7%. Using a bright green

detection of oral cancer, and has been used to detect SCC in the upper aerodigestive tract (oral cavity, pharynx, larynx, upper esophagus). While the exact mechanism of staining remains unclear, a study of the interaction of toluidine blue with tissue using electron microscopy suggests that the main factor governing selective uptake of this dye is the change in cellular membrane permeability. It has been noted that both injured and malignant lesions exhibit a greater permeability to the dye than that of normal mucosa. Even though toluidine blue does show im-

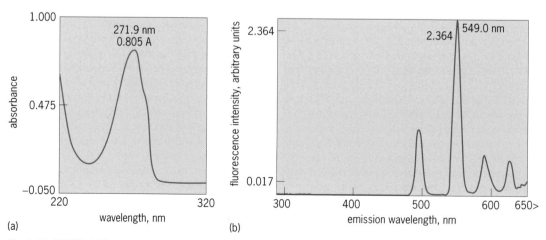

(a)   (b)

Fig. 2. Tb-PCTMB. (*a*) Absorbance spectrum. (*b*) Emission spectrum.

fluorescence from a lanthanide chelate in the same class as Tb-PCTMB, it has been possible to detect lesions in the Syrian Hamster Cheek Pouch Model not visible by white light imaging (**Fig. 3**). Histopathology confirms that these lesions are early-stage cancers.

Fluorophore conjugation. While autofluorescence and imaging agents transported by passive physiological mechanisms have shown promise in complementing traditional imaging modalities, active conjugation of site-directed molecules with fluorophores to produce a "flags on tags" continues to show promise. Many cancers overexpress certain receptors, allowing for the increased uptake of the corresponding ligand. The result is enhanced accumulation or association of this ligand with a certain type of cell, leading to exceedingly high detection specificity. Conjugation of a fluorophore to these ligands can potentially allow for increased specificity of the target cancerous tissue or cell, thus facilitating the use of fluorescence imaging.

For example, it has recently been demonstrated that implanted pancreatic acinar tumors, which overexpress the somatostatin (sst2) receptor, could be imaged using a DOTA (1,4,7,10-tetraazacyclododecane-N,N',N'',N-tetraacetic acid) and an indocyanine green (ICG) conjugate. In one case, a simple continuous-wave fluorescence imaging apparatus was used to image the lesions. Another example of site-directed contrast-enhanced imaging is the use of flovate to target several different kinds of cancer cells that are known to upregulate a receptor for this complex.

Bioluminescent markers. An alternative approach for contrast-enhanced biomedical imaging has recently emerged, involving the use of the bioluminescent marker luciferase which, in concept, can be imaged directly through the subject. This class of markers is unique because they produce the emission light without excitation radiation. (Luciferase is the compound that produces bioluminescence in fireflies and jellyfish.) One limitation, though, is that they require oxygen and the appropriate biological substrate to be present in order to emit light. The potential for using luminescent biological species remains to be seen, but is dependent in great part on the availability of high-sensitivity imaging cameras capable of detecting the attenuated signals emanating from the bioluminescent complex within the body (problematic for deep-tissue applications).

While a number of approaches have been devised for the detection of cancer, even some based on unique optical techniques such as light scattering, there will continue to be a need for contrast agents due to the inherently low signal-to-noise ratio in early lesions. There are a number of promising agents for detection of these early cancerous lesions, but it remains to be seen whether they will have the low toxicity, long-term stability, spectroscopic properties, and excellent specificity in humans needed for use as clinical tools.

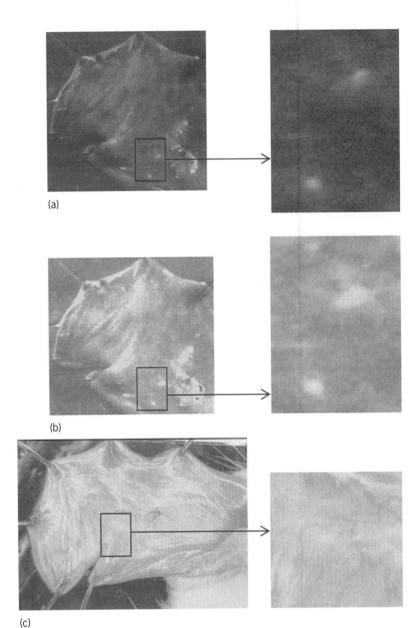

Fig. 3. Hamster cheek pouch (*a*) after treatment with lanthanide chelate and closeup of selected region, (*b*) with false colorization and closeup of selected region, (*c*) with white light, showing no observable lesions in selected regions, and closeup of selected region.

For background information *see* BIOLUMINESCENCE; CANCER; FLUORESCENCE; LUMINESCENCE; ONCOLOGY; PHOTOCHEMISTRY; SPECTROSCOPY in the McGraw-Hill Encyclopedia of Science & Technology.

Darryl J. Bornhop; John Griffin

Bibliography. D. J. Bornhop et al., Fluorescent tissue site-selective lanthanide chelate, Tb-PTCMB, for enhanced imaging of cancer, *Anal. Chem.*, 71:2607, 1999; *Cancer Facts and Figures*, no. 17, American Cancer Society, Atlanta, 1997; P. R. Contag et al., Bioluminescent indicators in living mammals, *Nature Med., New Technol. Sec.*, 4(2):245–247, 1998; P. B. Cotton, *Pratical Gastrointestinal Endoscopy*, 3d ed., Oxford Science, 1990; A. Gillenwater, R. Jacob, and R. Richards-Kortum, Fluorescence spectroscopy: A technique with potential to improve the early detection of aerodigestive tract neoplasia, *Head and Neck*, 20(6):556–562, 1998; M. P. Houlne et al., Imaging and quantitation of a tissue-selective lanthanide chelate using an endoscopic fluorometer, *J. Biomed. Opt.*, 3(2):145–153, 1998; J. Hung, S. Lam, and B. Placic, Autofluorescence of normal and malignant bronchial tissue, *Lasers Surg. Med.*, 11:99–105, 1991; A. Leunig et al., Detection of squamous cell carcinoma of the oral cavity by imaging 5-aminolevulinic acid-induced protoporphyrin IX fluorescence, *Laryngoscope*, 110(1):78–83, 2000; S. L. Marcus et al., Photodynamic therapy (PDT) and photodiagnosis (PD) using endogenous photosensitization induced by 5-aminolevulinic acid (ALA): Current clinical and developmental status, *J. Clin. Laser Med. Surg.*, 14:59–66, 1996; D. Rex et al., Colonoscopic miss rate of adenomas determined by back-to-back colonoscopies, *Gastroenterology*, 112:24–28, 1997; G. A. Wagniers, W. H. Star, and B. C. Wilson, *In-vivo* fluorescence spectroscopy and imaging for oncological applications, *Photochem. Photobiol.*, 68:603–632, 1998.

Candida

Candida species are fungi that cause important human diseases collectively referred to as candidiasis. The diseases are usually endogenous in origin, meaning that the source of the pathogen is within the host. *Candida albicans* is a common, usually harmless inhabitant of mucosal surfaces of the oral cavity, gut, and vaginal canal along with a variety of other microorganisms that are collectively called the normal flora. Most often, candidiasis occurs in the immunocompromised host; cancer, transplant, diabetic, AIDS, and surgical patients as well as premature infants are especially susceptible to infections. The reason for susceptibility varies. For the AIDS patient, a depletion in T-lymphocyte numbers triggers mucosal disease. The premature infant may develop candidiasis because of an underdeveloped immune system or because bacteria that are part of the normal flora of the human oral cavity or gut are not as protective. It is believed that competition among members of the normal flora of mucosal surfaces (oral cavity, vaginal system, gut) protects the host against the transgression of *C. albicans*. Consequently, if the competing bacteria at these locations are reduced in number by the prolonged use of antibiotics, an overgrowth of *C. albicans* can result in disease.

Other explanations are also likely. In the cancer or transplant patient, neutropenia (reduction in the absolute numbers of blood neutrophils) as a result of chemotherapy or the use of immunosuppressive drugs, respectively, can result in invasive, bloodborne infections caused by these organisms. Therapy is often delayed by the insensitivity of methods currently used to detect the presence of the organism in the patient's blood or tissues. Candidiasis in these patient populations represents the fourth most common hospital-acquired (nosocomial) disease in the United States. Complicating the management of these patients is the lack of alternative choices for therapy. The most common antifungal agent, amphotericin B, is usually toxic to the host, and the use of newer drugs (the triazoles) often leads to the development of drug resistance by either *C. albicans* or other species that are inherently resistant to the triazoles. Thus, the search continues to identify targets that might be exploited in the development of new antifungal agents.

Candida albicans has been extensively studied, especially during the past 20 years. In the early 1990s, methods were developed to construct strains of the organism that were missing a specific gene. Because gene-deleted strains display an altered phenotype (slower growth or an impaired ability to infect animals), the function of the gene in growth and virulence could be identified. Since *C. albicans* is a pathogen of the compromised patient, it was suspected that virulence factors (proteins or other types of biomolecules of an organism that promote its ability to cause disease) were not really necessary for invasion of the host. However, several factors have been found to be essential for *C. albicans*' virulence in animal models of candidiasis. They include (1) presence of adhesins (proteins that promote the colonization of host cells and tissues by the organism); (2) morphogenesis, or the conversion of the unicellular form (yeast) of the organism to a filamentous growth form; and (3) the presence of degradative enzymes, such as phospholipase and aspartyl proteinases, which are secreted by cells. Morphogenesis and the secreted enzymes are thought to promote the transgression of *Candida* during disease.

The identification and study of these virulence factors has several purposes. First, the studies have been done for "science's sake." Adhesins and degradative enzymes are biologically interesting proteins whose function and regulation of expression are still being studied. Second, the identification of these factors will be useful in explaining how disease begins and develops. For example, adherence of *C. albicans* cells to the gastrointestinal tract of humans should be a prerequisite for disease to occur. If the organism does not attach to host cells, then it will be likely cleared from the host and unable to cause disease. Third, it is likely that one or more of these factors may also provide a target for the development of new fungicidal drugs. Although the best targets are usually those that are essential for growth or reproduction of a pathogen, it is fairly certain that modern-day therapies will have to include new therapeutic approaches to control disease.

β_2 **integrin of vertebrate cells.** Vertebrate β_2 integrins are proteins located on the outer surface of cells that serve as a link between the extracellular space and the internal cytoplasm. The molecular structure of the β_2 integrins of vertebrates consists of two adjacent polypeptide subunits (α and β). The amino-terminus domain of each subunit is involved in the recognition of extracellular ligands that bind to the integrin proteins. The protein is oriented so that its amino terminus (ligand-binding domain) is on the external surface, while the carboxy terminus resides in the cytoplasm. Both subunits are anchored to the plasma membrane so that the overall arrangement of these proteins allows for extracellular and intracellular activities. The ligands that bind to the β_2 integrins are referred to as extracellular matrix (ECM) proteins, and include fibronectin, collagen, vitronectin, and laminin. Since the ECM proteins are found on numerous vertebrate cells, it is believed that the β_2 integrins are important in cell-to-cell interactions such as adherence. The signal transduction processes operate in a bidirectional manner. Thus, integrin proteins mediate communication between the extracellular and intracellular environments.

More recent data have shown that the integrins also act as conduits for intracellular signal transduction processes, and may be essential for cell shape and movements. The latter two processes are at least partially regulated through the direct role of the β_2 integrins in the assembly of the actin-containing cytoskeleton of cells.

β_2 integrin of C. albicans. Among the adhesins that are thought to promote the colonization of *C. albicans* on human cells is a protein that is homologous to the vertebrate family of β_2 integrin proteins. The *C. albicans* β_2 integrin-like protein is designated as Int1p (Int = integrin; 1 = the first integrin protein identified; p = protein), and its encoding gene is designated as *INT1*. In studies using reverse genetics approaches, the *INT1* gene of *C. albicans* was deleted from otherwise wild-type strains of the organism. Such altered strains were avirulent in a murine model of hematogenously disseminated (carried via blood) candidiasis. Further, the in vitro phenotype of the gene-deleted strain could not convert to a filamentous growth form and had reduced adherence to human epithelial cells. These studies imply that *INT1* plays an important role in virulence and growth of the organism.

iC3b-binding studies. The suggestion that a lower eukaryotic organism such as *C. albicans* expressed an integrin-like protein was first observed in binding studies of the organism with sheep erythrocytes coated with the ligand iC3b (i = inactive). This ligand is a degradative component of the alternative pathway of complement activation, a pathway that is critical to phagocytosis and removal of infectious microorganisms. (The fact that iC3b is a degradative product that is designated as inactive should in no way diminish its importance in immune clearance of pathogens.) The recognition of iC3b-coated microorganisms by phagocytic cells is essential to their clearance from the patient, and phagocytic cells utilize integrins to recognize iC3b-coated microorganisms. Thus, it was interesting to observe that *C. albicans* itself expressed a protein that could recognize iC3b much like mammalian cells could.

One of the important virulent properties of *C. albicans* is its ability to convert from a unicellular growth form (yeast) to a filamentous growth form (hyphae).

An early observation on the binding of iC3b was that hyphae forms of the organism were much more active in binding the ligand, leading researchers to believe that the invasive hyphal growth of the organism might be aided in some way by the presence of integrin(s). It was also postulated that Int1p might provide an adherence function for *C. albicans*. Further, its homology to mammalian cell proteins was thought to provide a molecular mimicry that allows *C. albicans* to escape immune detection since, conceivably, the host immune system should not recognize self-proteins. Of these possibilities, the last one has been the least developed experimentally.

ECM protein-binding studies. Since these initial studies of ligand binding, numerous other published reports have documented that whole cells or proteins extracted from whole cells of *C. albicans* can bind ligands such as fibronectin, laminin, collagen, and vitronectin in vitro. In addition to the observation that proteins of *C. albicans* bind ECM proteins, the apparent homology of the *Candida* integrin to that of the human protein was gleaned from studies that demonstrated the binding of antibodies to the human integrin to cells of *C. albicans*. Studies on fibronectin binding to *C. albicans* demonstrated that upregulation of the fibronectin-binding protein occurs in a growth medium supplemented with hemoglobin. In fact, hemoglobin also stimulates the binding of several other ECM proteins to the organism, suggesting a possible mechanism whereby β_2 integrin expression might be regulated during disease.

Assay studies. The integrin-like proteins of *C. albicans* have been identified using a variety of assays, including ligand-overlay blotting assays of electrophoresed proteins extracted from the organism, Western blotting using antibodies to the human β_2 integrins, and ligand (ECM proteins) affinity chromatography. To some extent, each of these assays for identifying integrin proteins is subject to several interpretations. For example, the observation that a protein from a crude extract can react with a specific ligand or antibody does not prove that this protein is a functional equivalent of the "correct" integrin. Extracts of proteins are notoriously complex, and the protein that binds to a ligand may in fact be a non–cell surface protein. More recently, a ligand-protection assay was used to identify a fibronectin-binding protein of *C. albicans*. Using this procedure, proteins with molecular masses of 55 and 30 kilodaltons were identified, but their role in pathogenesis has not been verified nor have encoding genes been cloned.

INT1 of C. albicans. The cloning and expression of a gene with low but significant homology to the human β_2 α-subunit (α_m) has been reported. A comparison of the structures of the human and *C. albicans* proteins showed that, among the conserved domains, a putative I domain (ligand-binding) was about 18% identical to the I domain of the human α_m protein (see **illus.**). The same region of the *C. albicans* protein was also 25% identical to that of the nonrepeat region of the fibrinogen-binding protein of the bacterium *Staphylococcus*

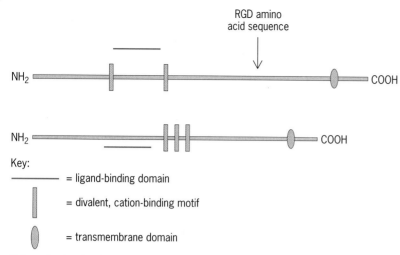

Schematic showing the polypeptides of the *Candida albicans* integrin-like protein Int1p (above) compared to the mammalian leukocyte integrin (below). NH$_2$ and COOH indicate the amino and carboxy termini of each protein. (*Modified from C. Gale et al., Closing and expression of a gene encoding an integrin-like protein in Candida albicans, Proc. Nat. Acad. Sci., 93:357–361, 1996*)

aureus. The arrangement of divalent cation-binding motifs (EF-hand motifs) was different for the two proteins: In the human protein, three EF-hand motifs are downstream of the I domain, while in *C. albicans*, one of two EF-hand motifs is located within the most amino-terminal end of the I domain, and the second EF-hand motif is downstream of the I domain, similar to the human integrin. Both proteins have a transmembrane location (anchored to the plasma membrane of cells). Antibodies to the second EF-hand motif reacted with whole cells, confirming that the *Candida* integrin is a surface protein.

The function of the *INT1* gene in *C. albicans* was investigated by constructing gene-deleted strains of the organism. This experimental approach requires that wild-type cells be compared phenotypically with single allele (heterozygote) and double allele-deleted (null) strains because the organism is diploid. Also, to ensure that the behavior of the gene-deleted strains is due to the deletion and is not a reflection of a transformation procedure that introduces nonspecific effects, the double allele-deleted strain must be transformed with a copy of the wild-type allele to construct a gene-reconstituted strain of the organism. The set of four strains (wild type, heterozygote, null, reconstituted) is then compared phenotypically to fulfill the molecular Koch's postulates, the criteria necessary for establishing gene function in this organism. In studies comparing these strains, it was observed that adherence of the null strain (*int1/int1*) to human epithelial cells was reduced by 39% in comparison with wild-type cells. Likewise, on an agar medium that normally induces morphogenesis, the null strain did not form filamentous colonies in comparison with wild-type cells. Most importantly, the null strain was comparatively avirulent in a hematogenously disseminated model of murine candidiasis.

Studies were also carried out with transformed *Saccharomyces cerevisiae* (bakers' yeast), which does not have a gene encoding an integrin protein. The *C. albicans INT1* gene was inducibly expressed in the presence of galactose and used to transform *S. cerevisiae*. Adherence of transformed *S. cerevisiae* increased when compared with nontransformed cells, and adherence of the transformed cells was reduced by preincubating cells with an antibody reactive with a domain of the Int1p protein. Transformed *S. cerevisiae* produced elongated filaments (not found in wild-type *S. cerevisiae*) that resembled germ tubes of *C. albicans* and expressed the Int1p on their cell surface. Thus, the *INT1* gene of *C. albicans* was hypothesized to provide a morphogenesis-like effect in *S. cerevisiae*.

These data indicate that *INT1* seems to be essential for more than one function in *C. albicans*. This observation may reflect a conservation of the genome, which is much smaller in *C. albicans* than in mammalian cells. Future studies will need to focus on the role of Int1p in signal transduction events in *C. albicans*.

For background information *see* CELLULAR IMMUNOLOGY; FUNGAL GENETICS; IMMUNOASSAY; IMMUNOLOGICAL DEFICIENCY; MEDICAL MYCOLOGY; MICROBIOTA (HUMAN); PHAGOCYTOSIS; YEAST in the McGraw-Hill Encyclopedia of Science & Technology.

Richard Calderone

Bibliography. W. L. Chaffin et al., Cell wall and secreted proteins of *Candida albicans*, *Microbiol. Mol. Biol. Rev.*, 62:130–180, 1998; C. A. Gale et al., Linkage of adhesion, filamentous growth, and virulence in *Candida albicans* to a single gene, *INT1*, *Science*, 279:1355–1357, 1998; C. Gale et al., Cloning and expression of a gene encoding an integrin-like protein in *Candida albicans*, *Proc. Nat. Acad. Sci.*, 93:357–361, 1996; M. K. Hostetter, An integrin-like protein in *Candida albicans*: Implications for pathogenesis, *Trends Microbiol.*, 4:242–246, 1996; S. A. Klotz et al., The fibronectin adhesin of *Candida albicans, Infect. Immun.*, 62:4679–4681, 1994; S. A. Klotz and R. L. Smith, A fibronectin receptor on *Candida albicans* mediates adherence of the fungus to extracellular matrix proteins, *J. Infect. Dis.*, 163:604–610, 1991; S. Yan et al., Hemoglobin induces binding of several extracellular matrix proteins to *Candida albicans, J. Biol. Chem.*, 273:5638–5644, 1998; S. Yan et al., Specific induction of fibronectin binding activity by hemoglobin in *Candida albicans* grown in defined medium, *Infect. Immun.*, 64:2930–2935, 1996.

Capillary electrochromatography

Capillary electrochromatography (CEC) is a separation technique that shares attributes of liquid chromatography and capillary zone electrophoresis. Currently developmental, it has promise for use in separations of multicomponent mixtures in complex matrices such as soils, ground and surface waters, food products, and biological samples. Like capillary zone electrophoresis (CZE), capillary electrochromatography is an efficient technique—that is, it produces a high number of theoretical plates per separation. Like high-performance liquid chromatography (HPLC), capillary electrochromatography is highly selective—it enables differential retention of one analyte over another.

Capillary electroseparation methods. Capillary electroseparation methods, such as capillary zone electrophoresis, micellar electrokinetic capillary chromatography, capillary gel electrophoresis, and capillary electrochromatography, have captured the interest of analytical chemists over the last few years. These techniques share several attributes: all use 25- to 200-micrometer inner-diameter capillaries, 20–100 cm long, across which high voltage are applied, and all generate very high efficiencies (>100,000 theoretical plates/m). The efficiency advantage is provided via the favorable flow characteristics of electroosmosis, a method of pumping a liquid by applying a high potential axially to a thin, fluid-filled tube. The speed at which analytes move through the separation conduit under the influence of electroosmosis is uniform throughout the cross section of the column, resulting in very little spreading of zones of analyte. Narrow, compact bands of analytes are therefore maintained, which results in high efficiency.

Capillary electrochromatography is the least developed of the capillary electroseparation techniques. In the most common form of capillary electrochromatography, a fused silica capillary is packed with silica-based reversed-phase particles for use as a separative column. Other possibilities include porous "monolithic" polymers as chromatographic media or surface-modified open tubular columns. In each of these formats, a surface charge on the capillary or sorbent surface is compensated by counterions in the eluent solution. By applying a potential across the capillary, a surface-originated electroosmotic flow (electrically induced fluid flow) is generated and used to transport analytes through the column.

Transport of analytes. Two phenomena act on analytes in capillary electrochromatography experiments to transport them through the column: electrophoresis and electroosmosis. These processes act in concert with adsorption or partitioning, which result in chromatographic separation.

Electrophoretic transport occurs along with electroosmotic transport when charged analytes are a part of the sample to be analyzed. Charged analytes are drawn toward the pole of opposite charge through the bulk solution at velocities which are dictated by their charge and size. Viscous drag on large solutes results in lower velocities, while increases in net charge result in higher velocities. This process of movement is electrophoresis. The electrophoretic velocity of a solute is given by $\mu_{ep} = 0.67[\varepsilon_o \varepsilon_r \zeta E \eta^{-1}]$, where ζ is the zeta potential in volts, E is the field strength in volts/cm, η is the solution viscosity in

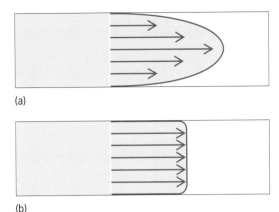

(a)

(b)

Fig. 1. Profiles of (a) laminar (parabolic) flow arising from a pressure-driven system and (b) bulk flow arising from an electroosmotically driven system. The flat flow profile of electroosmotic flow is responsible for making capillary electrochromatography a very efficient technique relative to high-performance liquid chromatography.

poise, ε_o is permittivity of a vacuum, and ε_r is the dielectric constant of the separation medium.

Critical to the operation of capillary electrochromatography is the fact that neutral species in solution are also moved through the capillary electrokinetically—through electroosmosis. The electrochemical double layer that exists at the interface between solid surfaces within the tube and the solution filling the tube is responsible for supporting electroosmosis. When a voltage gradient is applied axially to the tube, ions located near the wall of the tube (and near the surface of the particles filling it) will migrate toward their opposite pole. These moving ions drag along with them the surrounding bulk solution, owing to solvation and frictional forces in solution. This results in the transport of neutral species through the capillary. The flow profile of electroosmosis is flat (**Fig. 1**). The average velocity of electroosmosis is given by $v_{eo} = -\varepsilon \zeta E (4\pi \eta)^{-1}$. When a capillary is tightly packed with a typical high-performance liquid chromatography sorbent (about 5 μm in diameter), the interstitial spaces between the particles of packing material are sufficiently large that there will be no impedance of electroosmosis. The flat profile is the major source of the high efficiency typical of all capillary electroseparation techniques.

Resolution of mixtures. The attention given capillary electrochromatography has been driven by the need for a technique that can provide (1) higher sample loading capacity than capillary zone electrophoresis (for improved limits of detection), (2) the flexibility of working in organic operating solutions or mobile phases if desired, and (3) a greater capacity to separate complex mixtures containing many more analytes than is possible with high-performance liquid chromatography. High-performance liquid chromatography offers excellent selectivity through solvent optimization and careful stationary-phase (packing material) selection, high sample loading due to a high total stationary-phase surface area, and

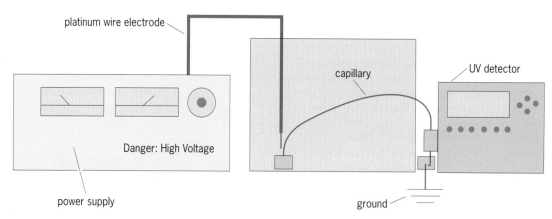

platinum wire electrode

capillary

UV detector

Danger: High Voltage

power supply

ground

Fig. 2. The instrumentation used in capillary electrochromatography is simple and inexpensive to home-build, though commercial instrumentation is also available and offers the advantages of automation and reproducibility. The instrument operator is protected from electrical shock by enclosing the high-voltage electrode in a Plexiglas box.

the ease of working in organic mobile phases. However, high-performance liquid chromatography offers only low efficiency. Capillary zone electrophoresis offers high efficiency, but cannot match the range of selectivity of high-performance liquid chromatography. Capillary electrochromatography is a marriage of capillary zone electrophoresis and high-performance liquid chromatography; it offers high efficiency, good selectivity, high sample loading, and good detectivity.

The resolving power of capillary electrochromatography arises from its HPLC-like selectivity and its CZE-like efficiency. According to general chromatographic theory, resolving power is described by the master resolution equation $R_s = 0.25[k(1 + k)^{-1}][(\alpha - 1)\alpha^{-1}]N^{0.5}$. This equation indicates that resolution arises as a function of three factors: k, the retention factor for an analyte; α, the selectivity of a column for one analyte relative to another; and

N, the efficiency of the separation system. The retention factor is a thermodynamic parameter which is directly related to the partition coefficient of the analyte between mobile (liquid solvent) and stationary (solid packing material) phases: $K = \beta k'$, where K is the thermodynamic partition coefficient and β is the ratio of the volume of mobile phase in the column to the volume of stationary phase. The partition coefficient, K, is the measure of the affinity of an analyte for the stationary phase relative to its affinity for the mobile phase. Since capillary electrochromatography is most often performed in capillaries packed with materials identical to those used in high-performance liquid chromatography, and since the mobile phases employed in the two techniques are essentially the same, no great change in capacity factor for a particular analyte is expected between the two techniques. Capillary electrochromatography will not offer improved resolving power (relative to high-performance liquid chromatography) in the capacity factor term of the master resolution equation.

Selectivity, embodied in the α term of the master resolution equation, is a measure of the affinity of one analyte for the stationary phase relative to the affinity of a second analyte for the stationary phase; it is the ratio of partition coefficients for the two analytes. It is in selectivity that high-performance liquid chromatography derives its power as a separations technique, since the selectivity of high-performance liquid chromatography systems is readily tunable by altering mobile-phase and stationary-phase composition. Once again, the similarities of capillary electrochromatography and high-performance liquid chromatography indicate that neither will have the edge in providing more resolving power than the other due to changes in selectivity.

The final term of the master resolution equation reflects the importance of efficiency in resolving mixtures. Efficiency (N) is measured as a total number of achievable theoretical plates; the greater the value of N, the better the system functions in terms of resolving power. Efficiency is calculated from

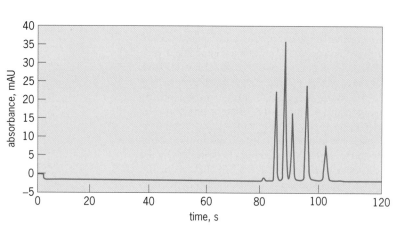

Fig. 3. Capillary electrochromatography separation of five polycyclic aromatic hydrocarbons. The peaks, in the order of elution, are naphthalene, acenaphthene, anthracene, pyrene, and chrysene. This separation demonstrates the rapidity and efficiency with which separations can be achieved using capillary electrochromatography—in excess of 300,000 plates per meter were generated with this column as measured for naphthalene. The eluent used was (20:80) 10 mM tris(tromethamine) pH 8.0: acetonitrile; separation voltage was +30 kV, generating a flow velocity of 3.1 mm s⁻¹. Injection: +3 kV, 3 s; detection: UV 230 nm; capillary dimensions: 75 μm inner diameter, 20 cm packed bed of a 5-μm reverse-phase sorbent, 28 cm total capillary length. Sample: 75 μg/mL of each compound prepared in the eluent.

chromatographic data as $N = (t_r/\sigma)^2$, where t_r is the retention time of an eluting analyte, and σ is the standard deviation in the analyte distribution. Efficiency is not the strong suit of high-performance liquid chromatography, where columns generate about 12,000 theoretical plates, in contrast to capillary zone electrophoresis and capillary electrochromatography which easily generate efficiencies of 100,000 theoretical plates. The efficiency term of the master resolution equation will be greater in capillary electrochromatography than it is in high-performance liquid chromatography; for this reason the total resolving power of capillary electrochromatography is superior to that of high-performance liquid chromatography.

Columns. There are a number of commercial sources of columns for capillary electrochromatography. Each vendor has its own method for producing columns, though there are some commonalities between them. The most popular columns are made using conventional high-performance liquid chromatography packing materials and fused silica capillaries with inner diameters of 25–200 μm. A section of capillary material is first cut to a length of 20–50 cm. A frit (fusible ceramic mixture) is installed in the outlet end of the column by sintering (softening and adhering together) particles of silica gel. This results in a porous ceramic filter that serves to retain the particulate packing material but readily permits the passage of liquid eluents. The capillary segment is then packed using procedures similar to those developed for preparing columns of larger inner diameter. A slurry of the packing material is often prepared in a ratio of $\sim$80:1 (mL:g) slurry liquid/packing material. The slurry liquid may be an aqueous buffer or an organic solvent such as acetonitrile. Silica-based reverse-phase packing material is most often used in producing the packed capillaries. The slurry is treated ultrasonically for 10–15 min and transferred to a reservoir. The reservoir is then connected to a high-pressure pump which is used to force the slurry into the capillary at 3000–10,000 psi (21,000–69,000 kPa). When the desired length of the column has been filled, an outlet frit is made using a source of highly focused heat to sinter a portion of the packing material. Following this, the column is removed from the packing apparatus and flushed with mobile phase (solvent), and a narrow (0.5-mm) detection window is made immediately adjacent to the outlet frit, again using a focused heat source.

Instrumentation. Capillary electrochromatography instrumentation is little different from that used in capillary zone electrophoresis. A high-voltage power supply (30 kV, 10 mA), an on-column capillary format UV detector, and a safety interlock box are occasionally supplemented with a source for application of external pressure to the inlet and outlet buffer vials to prevent formation of gas bubbles. The ends of the packed capillary are immersed in separate reservoirs containing an eluent solution that is also flushed through the capillary. A high potential (5–30 kV) is applied at one end of the capillary, while a ground connection is made at the other. This spurs electroosmotic flow through the a capillary, and also causes charged analytes to move as a result of electrophoresis. In accord with the equation given earlier for determination of electroosmotic flow velocity, a linear relationship exists between field strength and linear velocity. By controlling the applied potential, a range of flow velocities can be achieved. The moving analytes are transported past the detection window near one end of the capillary, where a UV detector is typically placed. Peaks are recorded indicating the presence and relative quantity of the given analyte. A schematic of the instrumentation is shown in **Fig. 2**. This represents a home-built instrument; a number of commercial instruments are also available. **Figure 3** is an example of a separation of a mixture by capillary electrochromatography. The very narrow peaks representing each analyte are a manifestation of the very high efficiency of capillary electrochromatography.

For background information *see* CHROMATOGRAPHY; ELECTROKINETIC PHENOMENA; ELECTROPHORESIS; LIQUID CHROMATOGRAPHY in the McGraw-Hill Encyclopedia of Science & Technology.

Vincent T. Remcho

Bibliography. L. A. Colon, Y. Guo, and A. Fermier, Capillary electrochromatography, *Anal. Chem.*, 69:461A, 1997; I. S. Krull (ed.), *Capillary Electrochromatography and Pressurized Flow Capillary Electrochromatography: An Introduction*, HNB Publishing, New York, 2000; P. Myers and K. D. Bartle (eds.), *Capillary Electrochromatography*, Royal Society of Chemistry, Cambridge, 2000; P. T. Vallano and V. T. Remcho, Capillary electrochromatography: A powerful tool for the resolution of complex mixtures, *J. AOAC Int.*, 82:1604, 1999.

Chemical imaging

Imaging mass spectrometry (MS) has a unique ability to acquire molecule- and element-specific pictures on a micrometer to submicrometer spatial regime. It has created fascinating opportunities in many research fields, including biochemistry, biotechnology, microelectronics, materials science, and geochemistry. Indeed, the trend toward miniaturization in many areas has driven the mass spectrometry community to improve the sensitivity and the spatial (position) resolution of imaging techniques.

More than 35 years ago, R. Castaing and G. Slodzian first used mass spectrometry to provide spatially resolved information. In their ground-breaking experiments, ions were created by bombarding solids with energetic particles. They designed an ion-optical collection system, analogous to a lens used in a light microscope, to preserve the spatial relationship of the desorbed ions from sample to detector. Another complementary method is the ion microprobe, with which an image is acquired by rastering (scanning) a focused ion beam across a target or vice versa. The latter concept has gained popularity because it has fewer restrictions on the nature of mass analyzers and higher lateral resolution.

Among the imaging techniques are secondary ion mass spectrometry (SIMS) and matrix-assisted laser desorption/ionization mass spectrometry (MALDI). Although their detection and imaging schemes are similar, their desorption and ionization concepts are different.

SIMS was first developed by R. F. K. Herzog and F. Viehboeck in the 1940s. Similar to Castaing and Slodzian's experiments, a beam of energetic ions [called primary ions, such as oxygen (O_2^+), cesium (Cs^+), gallium (Ga^+), and indium (In^+)], typically with a few kiloelectronvolts of kinetic energy, is directed at a target. This projectile transfers some of its momentum to the sample surface, causing desorption of ions (called secondary ions), neutral atoms, and molecules from the sample. A beam emitting from a liquid metal ion gun source can be focused onto the target with a diameter of less that 20 nanometers, providing very high lateral resolution.

Compared to SIMS, MALDI is a newer technique that was first introduced by M. Karas and F. Hillenkamp in 1988. A laser beam is directed onto a sample, which is a mixture of analytes (components being analyzed) and an organic compound that acts as a matrix to facilitate ablation and ionization of the analytes. Because MALDI normally produces molecular ions without extensive fragmentation, it is one of the most versatile techniques for qualitative characterization of large biological and synthetic polymers. However, the best reported lateral resolution of MALDI is about 25 micrometers, approximately 1000 times larger than SIMS.

SIMS. Two regimes, termed dynamic and static SIMS, are generally considered. In dynamic SIMS, a high-flux primary ion beam ($>10^{13}$ primary ions/cm^2) causes chemical damage to the surface and erosion of the sample, yielding primarily elemental information with very high sensitivity. In static SIMS, a low dose of primary ions ($<10^{13}$ primary ions/cm^2) probes less than 1% of the top layer, ensuring that molecules on the surface are relatively undisturbed. As a result, both elemental and molecular species are observed.

There are many available choices of ion sources and mass analyzers. Typically, the liquid metal Ga^+ ion gun provides the best possible lateral resolution of any ion source, but the sputtering (ejection) yields of molecular species are lower than other sources such as surface ionization Cs^+ ion sources and duoplasmatron sources. For mass analyzers, the time-of-flight (TOF) analyzer is best suited to static SIMS imaging because of its high transmission. The magnetic sector mass analyzer is found mostly in dynamic SIMS where the primary ion beam is not pulsed and one mass at a time can be imaged.

SIMS imaging is applied to metals, ceramics, oxides, composites, semiconductors, and minerals. In this article, some of the latest contributions to the fields of biology, biochemistry, and pharmaceutical chemistry are presented.

Chemical imaging of plant and animal cells and tissues using SIMS has provided the spatial distribution of both endogenous and exogenous species within these structures. Nutrients such as calcium, potassium, magnesium, phosphorus, sulfur, and nitrogen, and toxins such as aluminum, copper, iron, chromium, and lithium have been monitored. In addition, the distribution of pharmaceuticals in the cellular to subcellular scale has been studied with drugs that contain an atom, such as boron or fluorine, which is not native to cells or tissues. Using isotopic labeling, ion transport in organisms and defects of genes have also been identified. These experiments offer unique information about the structure and function of some organisms, the metabolites of pharmaceuticals, and the mechanisms of certain diseases.

Only recently has imaging SIMS been applied to acquire molecular images of biological materials. Freeze-fractured frozen-hydrated membranes, whose components, chemical domain, and dynamics are of fundamental importance to cell biology, are of special interest. In preliminary experiments, molecule-specific images of lipids from liposomes and red blood cells have been reported. In addition, the liposome model system has been used to capture the stages of membrane fusion between two merging bilayer systems. The positive ion TOF-SIMS images are obtained for freeze-fractured frozen-hydrated dipalmitoyl phosphatidylcholine (DPPC)/cholesterol liposomes mixed with dipalmitoyl phosphatidyl-*N*-monomethylethanolamine (DPPNME)/cholesterol liposomes (**Fig. 1**). The spatial relationship between the distributions of the two liposomes establishes the sequence of membrane fusion.

With the use of high-density arrays, SIMS imaging is capable of chemically characterizing a large number of diverse sample sets within a single chemical image. Fast and simple, this approach meets the need of high-throughput analytical methods in the pharmaceutical industry. For demonstration, a binary array of melatonin and uridine was built by loading 50 μm $\times$ 50 μm vials with 2.4 picomoles (pmol) of the respective molecules (**Fig. 2**). A Ga^+ ion beam probe was rastered across the sample, and a single image was taken for the whole field of view. The mass spectra from each individual vial were recovered later during data acquisition. Figure 2*b* is a representative mass spectrum abstracted from a single 50-μm nanovial containing 2.4 pmol of melatonin. The researchers claimed that the acquisition times are approximately 100 milliseconds per nanovial and the sensitivities for the molecules investigated are in the femtomole regime.

MALDI. Commercially available MALDI instruments are normally equipped with a nitrogen laser (337 nm) and a linear TOF mass analyzer. Since the signal of matrices (mostly acids) complicates the mass spectra below 300 atomic mass units (amu; dalton), MALDI is not preferred for low-mass analysis. However, MALDI could detect molecules with molecular weights up to several hundred kilodaltons,

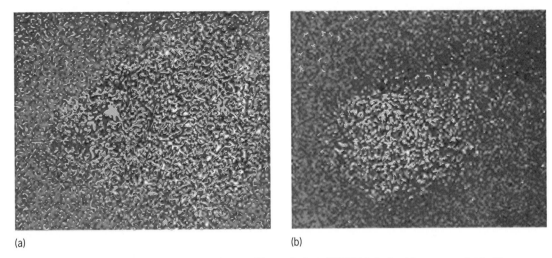

Fig. 1. Positive ion TOF-SIMS images of freeze-fractured frozen-hydrated DPPC/cholesterol liposomes mixed with DPPNME/cholesterol liposomes. All images are overlaid with water in gray. DPPC headgroup is pictured in dark color and DPPNME in light color. (*a*) The distribution of membrane components is heterogeneous before membrane fusion is completed. (*b*) Membrane fusion results in the homogeneous distribution of membrane components. (*From D. M. Cannon et al., Molecule specific imaging of freeze-fractured, frozen-hydrated model membrane systems using mass spectrometry, J. Amer. Chem. Soc., 122:603–610, 2000*)

resulting in wide applications to oligonucleotides, lipids, oligosaccharides, proteins, and peptides.

In MALDI experiments, sample preparation can be as simple as diluting the analyte in a suitable matrix solution and then depositing the solution onto the sample probe. Recently, protocols that modify the surface of MALDI probes with membranes and films, such as polyethylene, nonporous polyurethane, and polyvinylidine fluoride (PVDF), as well as C_{18}-coated microbeads have been developed. Carrying out the sample purification and other treatment processes, including proteolytic digestion, on the surface of the MALDI probe avoids many sources of sample loss and thus enhances MALDI signals.

First described in 1995, MALDI-based imaging was developed as a detection method for thin-layer chromatographic separation. Later, the method was used to study the localization of peptides and proteins in a tissue section. The direct analysis of tissue itself is not recommended because of the interference from signals from other abundant molecules such as lipids. Thus, a blotting procedure was developed whereby freshly cut tissue is placed onto C_{18}-coated microbeads for 10–30 s and then carefully lifted off.

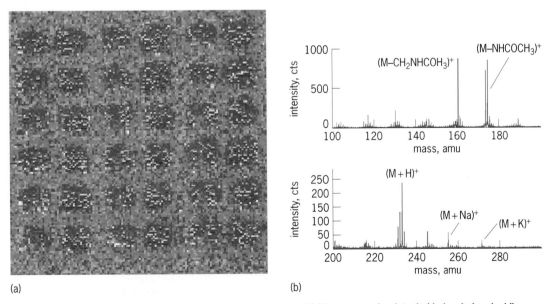

Fig. 2. TOF-SIMS image and mass spectrum of a high-density array. (*a*) Binary array of melatonin (dark color) and uridine (light color) Each 50 μm × 50 μm vial was loaded with 2.4 pmol of the respective molecules. (*b*) Representative mass spectrum obtained from 2.4 pmol of melatonin localized within a single nanovial. (*From R. M. Braun et al., Spatially resolved detection of attomole quantities of organic molecules localized in picoliter vials using time-of-flight secondary ion mass spectrometry, Anal. Chem, 71:3318–3324, 1999*)

The blotted surface is rinsed with MilliQ (ultrapure) water followed by electrospraying matrix onto the surface. Since proteins and peptides are bound fairly strongly to the beads by hydrophobic interactions, the blotted surface reflects the composition and localization of proteins and peptides in the original tissue. This procedure allows localization with a special resolution of about 30 μm.

Quantitation with SIMS and MALDI imaging. Quantitation is a difficult task for both imaging techniques because signal intensity is a function of many experimental variables. Such variables include the chemical environment of the analytes, sample morphology, matrix and primary ion beam interactions, instrumental transmission, and detector response. To further complicate matters, some of the above factors are dependent upon each other. A common approach to quantify the data is to use internal standards, which have to be chemically similar to the analytes. The relative sensitivity factors (RSFs) of analytes are also used in SIMS imaging to quantify the data.

Conclusions. Due to their simplicity and relatively high sensitivity (~femtomole) for diverse samples, SIMS and MALDI are the most popular tools for chemical imaging. Currently, SIMS imaging is faster and has far better lateral resolution (best reported value, 100 nm) than MALDI (best reported value, 25 μm). However, the mass range with best analysis results is around 1000 daltons for SIMS, compared to greater than 500,000 daltons for MALDI. For these reasons, SIMS imaging is widely applied to inorganic samples and small organic molecules, and MALDI imaging is almost exclusively performed on large biomaterials.

For background information *see* ANALYTICAL CHEMISTRY; ION SOURCES; LASER; MASS SPECTROMETRY; MASS SPECTROSCOPE; SECONDARY ION MASS SPECTROMETRY (SIMS); SPUTTERING; TIME-OF-FLIGHT SPECTROMETERS in the McGraw-Hill Encyclopedia of Science & Technology.   Nicholas Winograd

Bibliography. R. M. Braun et al., Spatially resolved detection of attomole quantities of organic molecules localized in picoliter vials using time-of-flight secondary ion mass spectrometry, *Anal. Chem.*, 71: 3318–3324, 1999; D. M. Cannon et al., Molecule specific imaging of freeze-fractured, frozen-hydrated model membrane systems using mass spectrometry, *J. Amer. Chem. Soc.*, 122:603–610, 2000; R. M. Caprioli et al., Molecular imaging of biological samples: Localization of peptides and proteins using MALDI-TOF MS, *Anal. Chem.*, 69:4751–4760, 1997; A. I. Gusev et al., Thin-layer chromatography combined with matrix-assisted laser desorption/ionization mass spectrometry, *Anal. Chem.*, 67: 1805–1814, 1995.

Climate change impacts

Climate is an important influence on both the environment and society. Year-to-year climate variations are reflected in such things as the number and intensity of storms, the amount of water flowing in rivers, the extent and duration of snow cover, ocean-current-induced changes in sea level, and the intensity of waves that strike coastal regions and erode the shoreline. These factors in turn determine agricultural productivity, the occurrence of floods and droughts, the safety of communities, and the general productivity of society. Science now suggests that human activities are causing the natural climate to change, mainly by inducing global warming and an associated intensification of the global hydrologic cycle. Although details are still emerging about the magnitude, regional pattern, and timing of the changes projected for this century, it is widely recognized that climate will be changing. Indeed, temperatures have already increased in many areas, Arctic sea ice is much thinner, continental snow cover is not lasting as long in the spring, and total precipitation is increasing, with more rainfall occurring in intense downpours. These changes also appear to be affecting the distribution of plants and wildlife. There is evidence of a longer growing season in northern areas, and of changing ranges for butterflies and other species.

The Global Change Research Act of 1990 (Public Law 101–606) gave voice to early scientific findings that human activities were starting to change the global climate: "(1) Industrial, agricultural, and other human activities, coupled with an expanding world population, are contributing to processes of global change that may significantly alter the Earth habitat within a few generations; (2) Such human-induced changes, in conjunction with natural fluctuations, may lead to significant global warming and thus alter world climate patterns and increase global sea levels. Over the next century, these consequences could adversely affect world agricultural and marine production, coastal habitability, biological diversity, human health, and global economic and social well-being."

Assessment. To address these issues, Congress established the U.S. Global Change Research Program (USGCRP) and instructed the federal research agencies to cooperate in developing and coordinating "a comprehensive and integrated United States research program which will assist the Nation and the world to understand, assess, predict, and respond to human-induced and natural process of global change." Further, Congress mandated that the USGCRP "shall prepare and submit to the President and the Congress an assessment which (1) integrates, evaluates, and interprets the findings of the Program and discusses the scientific uncertainties associated with such findings; (2) analyzes the effects of global change on the natural environment, agriculture, energy production and use, land and water resources, transportation, human health and welfare, human social systems, and biological diversity; and (3) analyzes current trends in global change, both human-induced and natural, and projects major trends for the subsequent 25 to 100 years."

The cycle of climate change assessments began in 1990 with USGCRP support for international assessments by the Intergovernmental Panel on

Examples of important consequences of climate change affecting particular areas of the United States

Regions and subregions	Examples of key consequences affecting:		
	Environment	Society/economy	Daily life
Northeast New England and upstate New York Metropolitan New York Mid-Atlantic	Northward shifts in the ranges of plant and animal species (such as colorful maples). Coastal wetlands inundated by sea-level rise.	Reduced opportunities for winter recreation such as skiing; increased opportunities for warm-season recreation such as hiking and camping. Coastal infrastructure will need to be buttressed.	Rising summertime heat index will make cities less comfortable and require more use of air-conditioning. Reduced snow cover.
Southeast Central and Southern Appalachians Gulf Coast Southeast	Increased loss of barrier islands and wetlands, affecting coastal ecosystems. Changing forest character, with possibly greater fire and pest threat.	Increased productivity of hardwood forests, with northward shift of timber harvesting. Increased intensity of coastal storms threaten coastal communities.	Increased flooding along coastlines, with increased threat from storms. Longer period of high heat index, forcing more indoor living.
Midwest Eastern Midwest Great Lakes	Higher lake and river temperatures may cause a trend in fish populations away from trout toward bass and catfish.	Increasing agricultural productivity in many regions, ensuring overall food supplies, but possibly lowering commodity prices.	Lowered lake and river levels, impacting recreation opportunities. Higher summertime heat index reduces urban quality of life.
Great Plains Northern Central Southern Southwest/Rio Grande Basin	Rising wintertime temperatures allow increasing presence of invasive plant species, affecting wetlands and other natural areas. Disruption of migration routes and resources.	Increasing agricultural productivity in north; more stressed in the south. Summertime water shortages become more frequent.	Altered and intensified patterns of climatic extremes, especially in summer. Intensified springtime flood and summertime drought cycles.
West California Rocky Mountains/Great Basin Southwest/Colorado River Basin	Changes in natural ecosystems as a result of higher temperatures and possibly intensified winter rains.	Rising wintertime snowline leads to earlier runoff, stressing some reservoir systems. Increased crop yields, but with need for greater controls of weeds and pests.	Shifts toward more warm season recreation activities (such as hiking instead of skiing). Greater fire potential created by more winter rains and dry summers. Enhanced coastal erosion.
Pacific Northwest	Added stress to salmon populations due to warmer waters and changing runoff patterns. Enhanced coastal erosion.	Earlier winter runoff will limit water availability during warm season. Rising forest productivity.	Reduced wintertime snow pack will reduce opportunities for skiing, increase opportunities for hiking.
Alaska	Forest disruption due to warming and increased pest outbreaks. Reduced sea ice and general warming disrupts polar bears, marine mammals, and other wildlife.	Damage to infrastructure due to permafrost melting. Disruption of plant and animal resources supporting subsistence livelihoods.	Retreating sea ice and earlier snowmelt alter traditional life patterns. Opportunities for warm-season activities increase.
Coastal and Islands Pacific Islands South Atlantic Coast and Caribbean	Increased stress on natural biodiversity as pressures from invasive species increase. Deterioration of corals reefs.	Increased pressure on water resources needed for industry, tourism, and communities due to climatic fluctuations, storms, and saltwater intrusion into aquifers.	Intensification of flood- and landslide-inducing precipitation during tropical storms. More extreme year-to-year fluctuations in the climate.
Native People and Homelands	Shifts in ecosystems will disrupt access to medicinal plants and cultural resources.	The shifting climate will affect tourism, water rights, and income from use of natural resources.	Disruption of the religious and cultural interconnections of indigenous people and the environment.

Climate Change (IPCC). The IPCC assessments from 1990, 1996, and 2001 document existing global-scale changes and project that these changes will increase in magnitude over the next 100 years. As a consequence of the changes in climate, the IPCC also projects significant environmental change, generally at the continental scale. To provide a more focused picture of what climate change might mean for the United States, the USGCRP initiated the *National Assessment of the Potential Consequences of Climate Variability and Change* in 1997. This Assessment is focused on how people in the United States are likely to be affected by climate change and how actions might be taken to effectively prepare for an average national warming of 5-9°F (3-5°C) and significantly altered patterns of precipitation and soil moisture. It is just such changes that are simulated by global climate models assuming that global emissions of carbon dioxide and other greenhouse gases continue to climb as projected in response to the growing world population, rising standard of living, and changing technologies and sources of energy.

The overall goal of the Assessment has been to analyze and evaluate what is known about the potential consequences of such changes in the context of other pressures on the public, the environment, and the nation's resources. By building a broader understanding of the prospects for climate change and of the importance of these changes for the nation, the USGCRP is aiming to promote an intensifying exploration of options that can help to reduce the vulnerability of individuals, public and private-sector organizations, and the economy and resource base on which society depends. These responses should be able to help build resilience to climate variations and, to at least some extent, avoid or reduce the deleterious consequences of climate change while taking advantage of conditions that may be more favorable.

Assessment process. The Assessment process has been broadly inclusive in its approach, drawing on inputs from many sources. Support has been provided in a shared manner by the set of USGCRP agencies, including the departments of Agriculture, Commerce (National Oceanic and Atmospheric Administration), Energy, Health and Human Services, and Interior, plus the Environmental Protection Agency, National Aeronautics and Space Administration, and National Science Foundation. Although support for various activities has come mostly from the federal agencies, the conduct of the Assessment has been carried out in a highly distributed manner. Each of a diverse set of activities has been led by a team of experts drawn from universities and government, the public and private sectors, and the spectrum of stakeholder communities. Through workshops and assessments, a dialogue has been started about the significance of the scientific findings concerning climate change and the degree to which existing and future changes in climate are likely to affect issues that people care about, both at present and in the future. The reports that have been prepared have gone through an extensive review process involving scientific experts and other interested stakeholders, ensuring both their technical accuracy and balance.

Regional analyses and assessments. An initial series of workshops provided the basis for characterizing the potential consequences of climate variability and change in regions spanning the United States. A total of 20 workshops were held around the country in 1997 and 1998; 16 of these groups then went on to prepare assessment reports focusing on the most critical issues identified. These activities focused on the implications of the patterns and texture of changes where people live. Although issues considered often seemed to have a common thread, the implications often played out in different ways in different places. For example, various manifestations of the issue of water arose in virtually all regions. In some regions, it was changes in winter snowpack that project the need to adjust water allocations and the operational procedures for managing reservoir systems to ensure safety and supplies for electric generation, irrigation, industry, and communities. In other regions, issues related to the potential influences of changes in precipitation amount on water quality, summertime drought, or river and lake levels were paramount. The **table** highlights examples of issues as they arose across the country in nine consolidated regions.

Sectoral analyses. To explore the potential consequences for sectors of national interest that cut across environmental, economic, and societal interests, the Assessment examined implications for Agriculture, Forests, Human Health, Water, and Coastal Areas and Marine Resources. These sectoral studies analyzed how the consequences in each region would affect the nation, and how national level changes would affect particular areas. Key findings from each of the sectoral studies contributed to the findings at both the regional level and the national level.

National overview. A 14-member National Assessment Synthesis Team (NAST) drawn from academe, industry, government, and nongovernmental organizations was responsibile for summarizing and integrating the findings of the regional and sectoral studies and then drawing conclusions about the importance of climate change and variability for the United States. To document their findings, an extensive Foundation report was prepared that ties the findings to the scientific literature. To convey their message to the broader public and leading decision makers, an Overview report was prepared that describes the key issues facing nine regions and five sectors across the United States. The key findings from their report that apply to the nation as a whole are summarized as follows:

1. Increased warming and more intense precipitation will characterize this century.
2. Differing regional impacts will occur, with greater warming in the western United States, but a greater rise in heat index in the eastern and southern United States.

3. Vulnerable ecosystems, particularly alpine areas, barrier islands, and forests in the southeast are very likely to be significantly impacted.

4. Water will be a concern across the country, with increased competition for available resources, and the potential for more droughts and floods and reduced winter snowpack in some areas.

5. Food availability will increase because of increased crop productivity, although lowered commodity prices will stress farmers, especially in marginal areas.

6. Forest growth will increase in the near term, but some forests will be threatened over the long term by increased susceptibility to fire, pests, and other disturbances.

7. Increased damage is very likely in coastal regions due to sea-level rise and more intense storms, while damage in other areas will result from increased melting of permafrost.

8. Adaptation will determine the importance of health outcomes, so that strengthening of the nation's community and health infrastructure will be increasingly important.

9. The impact of other stresses will be magnified by climate change, with multiple factors causing adverse impacts on coral reefs, wildlife habitats, and air and water quality.

10. Uncertainties remain in current understanding, and there is a significant potential for unanticipated changes.

An important advance was consistent use of a set of well-defined terms to indicate the relative likelihood of various outcomes based on the considered judgment of the experts serving on the NAST. While some types of changes were found to be highly likely or unlikely, many were judged to be only possible based on current understanding. To gain better information, the NAST summarized key directions for research, urging particularly the strengthening of efforts to take an integrated look at the changing set of stresses facing regions and resource managers.

Outcome. With the increasing level of understanding of global-scale environmental challenges, the conduct of assessments provides an important means for linking the emerging findings of the scientific community with the information needs of stakeholders. While also creating an urgency for new and elaborated scientific findings, these couplings provide insights about potential vulnerabilities and response options for those handling economic development and societal welfare. The progress made in the initiation of the National Assessment and related activities in other nations is stimulating the beginning of a greater number of such activities around the world, building a society that is better informed and better prepared not only for climate change but also for other long-term issues of sustainability.

(This paper is based in part on material extracted from the Overview report of the National Assessment Synthesis Team; for a more complete, and therefore more balanced, discussion, the reader is referred to that report. This paper was prepared by the author under the auspices of the Department of Energy, Environmental Sciences Division by the Lawrence Livermore National Laboratory under contract W-7405-ENG-48.)

For background information *see* CLIMATE HISTORY; CLIMATE MODELING; CLIMATE PREDICTION; CLIMATOLOGY; CROP MICROMETEOROLOGY; WEATHER MODIFICATION in the McGraw-Hill Encyclopedia of Science & Technology.   Michael C. MacCracken

Bibliography. Environment Canada, *The Canada Country Study: Climate Impacts and Adaptation*, Ottawa; Intergovernmental Panel on Climate Change, *Third Assessment Report* (Synthesis report and three Working Group reports), prepared for the United Nations Environment Programme and World Meteorological Organization, Cambridge University Press, 2001b; National Assessment Synthesis Team, *Climate Change Impacts on the United States: The Potential Consequences of Climate Variability and Change*, US Global Change Research Program, Washington, DC, 2000; United Kingdom Climate Impacts Programme, *Climate Change: Assessing the Impacts—Identifying the Responses*, Department of the Environment, Transport and Regions, West Yorkshire, UK, 2000.

Cold plasma

Plasma is commonly referred to as the fourth state of matter, the other three being the solid, liquid, and gaseous states. The transition from one state to another is obtained by applying external energy to a medium. If a gas is heated, there comes a point when some of the neutral atoms and molecules constituting the gas become ionized (that is, some of their loosely bound electrons get stripped away). The mixture of neutral atoms and molecules, ions, and electrons is called plasma. Plasma generates radiation spanning the spectrum from the infrared to the ultraviolet. Plasmas also may contain chemically reactive free radicals, which are atoms or molecular fragments that possess one unpaired electron. Examples are atomic hydrogen (H) and hydroxyl (OH). Another important property that makes plasmas unique as compared to an ordinary gas is their ability to interact with electric and magnetic fields. Using this characteristic of plasma, scientists are able to confine it within an enclosed space and heat it by driving an electric current through it or by injecting electromagnetic energy.

Plasma occurs naturally, and can also be manufactured. In nature, plasma can readily be seen when lightning strikes and ionizes the air. Solar winds induce large currents in the Earth's upper atmosphere and generate beautiful glowing plasmas known as the aurora borealis (northern lights), which can be seen from the northern latitudes. In fact, more than 99% of the visible matter in the universe is in the plasma state, including the Sun, the stars, and much of the

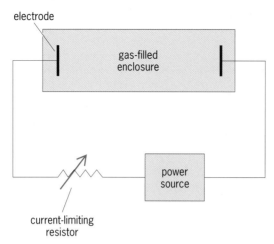

Fig. 1. Gaseous discharge apparatus.

intergalactic matter. Manufactured plasmas can be found in lighting devices, flat-screen TVs, nuclear fusion reactors, and so on. Plasma also plays an important role in the manufacture of semiconductor devices.

Thermal versus nonthermal plasmas. Energetically, plasmas can be classified as thermal (or hot) plasmas and nonthermal (or cold) plasmas. Thermal plasmas are found, for example, in fusion reactors where the particles constituting the plasma are heated to several million degrees Kelvin. Nonthermal plasmas can be generated via electrical discharges in gases (**Fig. 1**). These discharges can be operated in a regime where the electric current is not allowed to rapidly increase and cause excessive heating of the gas. A common example is the glow regime, where the discharge emits visible light and appears to be glowing to the naked eye. Cold plasmas are also known as nonequilibrium plasmas, meaning that the various particles in the discharge are not equilibrated to the same temperature. Generally, the electrons are the energetic particles, while the energy of the ions is close to the energy (or temperature) of the background neutral gas.

Recently, scientists developed several methods for producing large volumes of cold plasmas at atmospheric pressure. These developments eliminated the need for vacuum systems in the generation process. Cumbersome, bulky, and tedious to work with, vacuum systems constitute the major expense in plasma devices. A uniformly diffuse cold plasma (as opposed to a filamentary plasma, which is a collection of discrete thin filaments) can be generated in a mixture of air and helium at atmospheric pressure. The device for generating such a plasma is called a dielectric barrier discharge (DBD). It consists of a gas-filled enclosure containing two electrodes made of two metallic plates covered by a dielectric (nonconducting) material (**Fig. 2**). The electrodes are energized by a high-voltage radio-frequency (RF) power supply (typical voltages are 1–10 kilovolts, at frequencies from kilohertz to megahertz). To optimize the amount of power deposited in the plasma,

an impedance-matching network is generally introduced between the RF power supply and the electrodes. The dielectric material covering the electrodes plays the crucial role in keeping the plasma "cold." This is achieved as follows. When the RF voltage is applied between the electrodes, the gas breaks down (ionization occurs) and an electric current starts flowing in the gas. Soon after, electrical charges start accumulating on the surface of the dielectric. This creates an electrical potential, which counteracts the externally applied voltage and therefore limits the flow of current. This process prohibits the transition of the discharge to an electrical arc. If the electric current were allowed to increase to high levels, an arc would occur and the discharge would transform into a thermal plasma.

Cold plasmas have traditionally been used under low-pressure conditions for various materials-processing applications such as surface modification, deposition of thin films on substrates, and materials etching. Recently, novel applications of cold plasmas under high-pressure conditions have emerged. These include the use of cold plasmas in biological and chemical decontamination of matter, as spectrally selective radiation sources, and as absorbers or reflectors of electromagnetic waves.

Biological decontamination. Plasmas contain a variety of neutral and charged particles as well as electromagnetic radiation. Therefore, plasmas exhibit an environment that in principle can be quite rich in reactive agents for interacting with living organisms. For the past few years, scientists have been investigating the effects of high-pressure cold plasmas (helium/oxygen and helium/air mixtures as well as other gas mixtures) on the cells of microorganisms. It was found that indeed these types of plasmas have strong germicidal effects. Gram-negative bacteria such as *Escherichia coli* and gram-positive bacteria such as *Bacillus subtilis* were killed after a few tens of seconds' exposure to the plasma. Other microorganisms, such as surrogates (less aggressive strains that do not require high levels of biosafety) to anthrax (*Bacillus anthracis*), were also killed by the plasma. These results suggest that cold plasmas could be used to sterilize medical tools (especially those sensitive to heat), to decontaminate food packaging (for safety and longer shelf life), and to decontaminate surfaces

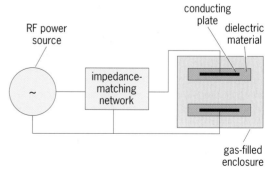

Fig. 2. Dielectric barrier discharge.

that have been exposed to biological warfare agents. The advantages of using cold plasmas for these applications are the relatively short treatment times (as compared to conventional methods, such as autoclaving or treatment by ethylene oxide), the lack of toxic residues after treatment, the ability to treat heat-sensitive materials, and the absence of a vacuum system. Presently, more comprehensive research is needed to further characterize the potential of high-pressure cold plasmas as decontamination devices. What causes the demise of the microorganisms is still a matter of scientific debate, as are the effects of cold plasmas on cell morphology and cellular biochemical pathways.

Radiation sources. High-pressure, nonequilibrium gas discharges such as the dielectric barrier discharge are excellent sources of excimers. Excimers are energetically unstable molecules that tend to decay rapidly and, in the process, yield incoherent radiation in the ultraviolet (UV) and vacuum ultraviolet (VUV, wavelengths shorter than 200 nm) range. Lamps generating intense UV radiations (called excimer lamps) are commercially fabricated using the dielectric barrier discharge configuration. Excimer lamps are filled with a rare gas, such as argon (Ar), krypton (Kr), and xenon (Xe), or a rare gas and halogen gas mixture. Nearly monochromatic radiation emission spanning the 127–308-nm-wavelength range has been routinely produced. Examples include emission at 127 nm for Ar_2, 172 nm for Xe_2, 222 nm for KrCl, and 308 nm for XeCl.

Applications of excimer UV lamps include the surface modification of materials, decontamination of surfaces and transparent liquids, pollution control, UV curing of paints, material deposition, and short-wavelength radiation sources for microlithography.

Interaction of electromagnetic waves with cold plasmas. A unique property of plasmas is their ability to interact with electric and magnetic fields. Scientists take advantage of this characteristic to trap plasmas and control them. In addition, electromagnetic energy can be coupled to the charged particles present in the plasma to increase their energy. Hence, as an electromagnetic wave gives off its energy to a plasma, it is attenuated. Based on this principle, plasmas can be used as absorbers of electromagnetic radiation. However, for the absorption to occur, the electromagnetic wave has to able to penetrate the plasma first, and then there has to be a physical mechanism through which the energy of the wave can be coupled to the charged particles (through frequent collisions, for example). Wave penetration occurs only if the number density (number per unit volume) of the free electrons in the plasma is below a certain critical number density, n_c. The value of n_c is directly dependent on the frequency of the incident electromagnetic wave. If the above condition is not met, the electromagnetic wave is completely reflected at the surface of the plasma (that is, the plasma becomes a mirror). Both situations (absorption and reflection) can be useful depending on the

particular application. For example, if the intention is to use the plasma as a mirror to deflect an incoming wave, an adequately dense plasma is needed. Because of their relatively low temperatures, cold plasmas are well suited for such applications. They can be used as agile mirrors to guide microwaves in a radar communication system. Among the advantages of cold plasma mirrors, they can be controlled within very short times (as compared to solid material mirrors), they can be shut off when not in use, and they do not destroy the materials or equipment with which they may come in contact.

For background information *see* ELECTRIC FIELD; FREE RADICAL; IONIZATION; MAGNETIC FIELD; PLASMA (PHYSICS) in the McGraw-Hill Encyclopedia of Science & Technology.   Mounir Laroussi

Bibliography. B. Eliasson and U. Kogelschatz, Nonequilibrium volume plasma chemical processing, *IEEE Trans. Plasma Sci.*, 19(6):1063–1077, 1991; H. W. Herman et al., Decontamination of chemical and biological warfare agents using an atmospheric pressure plasma jet, *Phys. Plasmas*, 6(5):2284–2289, 1999; M. Laroussi, Interaction of microwaves with atmospheric pressure plasmas, *Int. J. Infrared Millimeter Waves*, 16(12):2069–2083, 1995; M. Laroussi, Sterilization of contaminated matter with an atmospheric pressure plasma, *IEEE Trans. Plasma Sci.*, 24(3):1188–1191, 1996; M. Laroussi, I. Alexeff, and W. Kang, Biological decontamination by non-thermal plasmas, *IEEE Trans. Plasma Sci.*, 28(1):184–188, 2000; M. A. Lieberman and A. J. Lichtenberg, *Principle of Plasma Discharges and Materials Processing*, John Wiley, New York, 1994.

Colloids

Chemists long assumed that all matter existed in one of three states (gaseous, liquid, or solid) depending on the temperature and pressure. However, during the nineteenth century it was recognized that many chemical systems did not fall into these categories. For example, fine clay suspensions can be considered neither true solids nor true solutions. Gels and smoke are other examples of states of matter that are difficult to define. Since many of these apparently anomalous systems are amorphous, they were called colloid, which means glue. As knowledge of these systems progressed, they were divided into two categories: hydrophilic (water-loving) colloids, which are stabilized by strong solute-solvent interactions; and hydrophobic (water-hating) colloids, which are stabilized kinetically. Polymer solutions and gels are typical examples of the first category, and hydrosols of the second. While these terms are still used, they are not representative of the large number of systems now considered colloids. This is particularly true for self-assembly systems, recognized as part of the colloid domain in the midtwentieth century only. Furthermore, these terms can in some cases be misleading since, for example, the so-called hydrophobic colloids are not really hydrophobic.

The systematic study of colloids began around 1850, but since they were considered to be largely an oddity, most chemists had a tendency to ignore this family of substances. Since about 1970, with the development of modern instrumentation and new industrial and medical applications, a major effort has been made to understand and make full use of colloid systems. The field of colloid science has become one of the most important areas of physical chemistry and overlaps with many other disciplines and fields of application such as biology, medicine, pharmacology, food science, cosmetics, coating industry, and oil recovery. This is partly due to the realization that the forces and structure of living matter are quite close to those of colloid systems.

While the field of colloids is broad and covers all systems in between pure liquid, gaseous, or solid states, there are two main characteristics that stand out in all these systems: the size of the particles, typically between 1 nanometer and 1 micrometer; and the importance of interfacial phenomena or interactions in their stability and structure.

Hydrotropes. Polar organic substances having both hydrogen-bonding capabilities and hydrophobic character, often called hydrotropes, are the first group of systems which begin to show some anomalous solution behavior. Medium-chain alcohols are typical of such systems. Alcohols such as 1-propanol, *tert*-butanol, or 2-butoxyethanol are completely miscible with water and also with alkanes. However, in both cases the physicochemical properties of these mixtures are anomalous when compared with real solutions. For example, the enthalpies of solution of *tert*-butanol at very low concentrations vary considerably in water and in cyclohexane (**Fig. 1**), reflecting the large difference in interactions between the alcohol solute and the two solvents. More spectacular is the rapid observed change with concentration in both solvents. This behavior is approaching a phase change, suggesting that, at finite concentra-

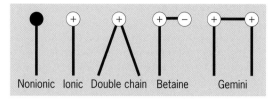

Fig. 2. Schematic representation of the structure of surfactants.

tion, alcohols in solution are highly organized systems which depend on the dual aspects of their structure: a polar group which has a high tendency to form hydrogen bonds with water or with itself; and a hydrophobic chain which prefers to stay away from water but close to hydrocarbons. Such systems are said to display microheterogeneities in solution.

These remarkable properties become very important when a hydrotrope is combined with other solutes such as surfactants (soaps or detergents). For example, *n*-butanol, which is partially soluble in water, becomes completely miscible when a small quantity of medium-chain surfactant, such as sodium dodecylsulfate, is added. When combined with surfactants in the proper ratio, and adjusting the salinity if necessary, such systems can solubilize or disperse large quantities of oil in water. These mixtures are known as microemulsions; they have some of the features of emulsions but are transparent and thermodynamically stable.

Micellar systems. If the hydrophilic character of a hydrotrope is increased, either by increasing the number of polar groups or by introducing ionic groups, the hydrophobic chain can also be increased significantly while remaining soluble in water or in an oil. The changes in the solution properties with concentration become much sharper and occur at lower concentrations than with alcohols. This family of amphiphilic solutes (water- and oil-loving) typically contain 6–20 CH_2 groups with various chemical structures. Some typical examples are shown in **Fig. 2**.

One characteristic property of this class of solutions is the surface tension (γ). As shown in **Fig. 3**, this parameter, which is directly related to the surface free energy, drops rapidly as the concentration increases from its value in pure water to a value which is close to that of alkanes. This change occurs because these solutes have a tendency to concentrate near the surface of the solution to minimize the unfavorable chain-water interactions. For this reason, these solutes are called surface-active substances or surfactants. Once the surface is saturated, the system will reduce the hydrophobic chain-water interactions through self-association or aggregation. These self-assembly systems, or micelles, can vary largely in size and shape (spheres, cylinders, or plates), depending on the chemical structure of the surfactant, chain length, concentration, and temperature. At very high concentrations, such self-assembly systems can form liquid crystals.

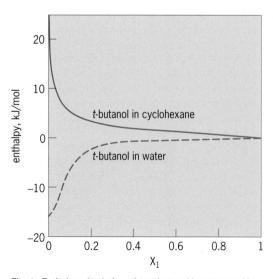

Fig. 1. Enthalpy of solution of *tert*-butanol in water and in cyclohexane at 25°C (77°F) as a function of mole fraction.

Gemini surfactants, which have been studied extensively in the last 10–15 years, are made up of two or more surfactants joined together by a spacer (two or more CH$_2$ groups). With these surfactants many structures are possible, including threadlike micelles, which in some cases can be seen with atomic force microscopy. Some are used in the formulation of powerful detergents.

The transition zone or critical micellar concentration (CMC) will occur anywhere from 1 to below 10^{-6} mol L^{-1}, and the micelle will contain from 8 to over 100 monomers. The transition can be gradual (short-chain surfactants) or quite sharp. In the first instance, the system can be treated as an association process, while in the second case it is closer to a pseudo-phase transition.

Surfactants can also aggregate in hydrocarbons. In this case, the polar head groups tend to concentrate in the core of the micelle, and the hydrocarbon chains remain in contact with the solvent. Often, a small quantity of water helps in forming these structures, called reversed micelles. The kind of surfactants that favor their formation are molecules with a predominant hydrophobic character, or even better, surfactants with a double chain (Fig. 2).

A large number of techniques are now available to determine the CMC, the structure, the aggregation number, and the kinetics of micellar systems.

Solubilization and detergency. One principal use of a surfactant system is as a detergent. Soaps have been used for thousands of years, but the formulation of modern detergents and cleaners is a fine art. The basic principle is simple: stains are soluble in alkanes, alkanes are soluble in alcohols, and alcohols are soluble in water in the presence of surfactants. The composition of this multicomponent system can be optimized through phase diagrams. In addition, detergents require a proper wetting agent, and must avoid foaming agents for automatic washing machines. The detergent should not cause skin irritation, and should be able to soften fabric fibers. Finally, the components must be environmentally safe. Similar formulations of cleaners are also used in industrial processes such as tertiary oil recovery, extraction of tar sands, restoration of contaminated sites, and elimination of oil spills.

Foams, emulsions, liposomes, and vesicles. It is not always necessary or even desirable that dispersed systems be macroscopically homogeneous, as they are in micellar systems or microemulsions. Surface-active agents can also be used to stabilize systems such as gases in liquids (foams) or liquids in liquids (emulsions). The role of the agent is to render both phases as compatible as possible and to slow down the rate of coalescence. With emulsions, the continuous phase can be either water (as in cream) or oil (as in butter), and the structures of the surfactants are responsible for the stabilization of oil in water (O/W) or water in oil (W/O) emulsions. This depends on the balance between the hydrophobic and hydrophilic character; semiempirical parameters have been developed to characterize the surfactants for this pur-

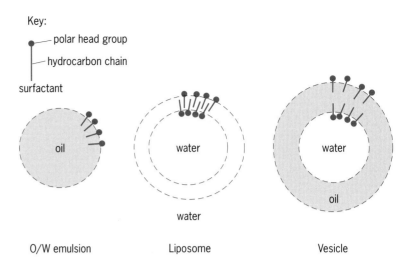

Fig. 3. Variation of the surface tension with concentration of a surfactant solution.

pose. Nearly always, mixtures of surfactants are used for this purpose. In certain cases, emulsions can be inverted (for example, cream to butter). It is also possible to form double emulsions, known as liposomes or vesicles. With these systems, the inner aqueous phase is separated from the outer aqueous phase by a semipermeable film. Such structures have enormous potential in agriculture (pesticides and herbicides), pharmacology (drugs), and cosmetics since the passage of chemicals from the inner to the outer phase can be controlled by diffusion through the membrane.

Sols. Finely divided inorganic solid particles can also be dispersed in liquids. The most common systems are hydrosols. The stability of such systems comes from the presence of charges on the surface of the particles. The repulsion between these charges prevents the particles from coagulating, and the system is said to be metastable. Sols are sometimes used as precursors for the preparation of certain materials (such as in deposition of fine magnetic particles, and preparation of fine porcelain), while in other cases they are a nuisance (for example, in wastewater).

Polymer colloids, gels, and membranes. Originally, polymer solutions were included with colloids, but as the science of polymers progressed in the twentieth century, polymer science and colloid science were considered different scientific fields. However, there are many examples where polymer systems have colloidal properties. For example, copolymers are often used to stabilize hydrosols. They are also combined with surfactants in hair conditioners. Gels are also examples of colloidal polymers; in this case, a small quantity of crosslinked polymers (as low as 1%) can immobilize a large quantity of solvent. There are many ingenious applications of gels. For example, some are prepared with a gel temperature close to 37°C (98.6°F); this property can be used for automatically controlled drug delivery when a patient develops a fever.

Monomers can also be dispersed as a microemulsion or emulsion before polymerization. For example, this is how latex paints are prepared. As in the case of hydrosols, the stability of such dispersions depends largely on the presence of charges on the latex particles. The coating industry makes full use of such systems. For example, the anticorrosion treatment of automobiles comes from the dipping of the car body in such a dispersion; the particles are deposited on the metal surface by applying a strong electric field between the metal and the solution (electrophoretic deposition). Polymerization of monomers, dispersed as microemulsions, are used to prepare membranes having very specific pore sizes. These membranes are then used to filter supramolecules or very fine particles.

Future developments. The development of new colloid systems and their potential uses in biomedical and industrial systems are limited only by the imagination of the scientists. For example, in recent years a new field called nanotechnology, by which ultrasmall devices are created, has been developed. Many of these materials, prepared as supramolecules or fine self-assembly colloidal systems, have important potential applications, for instance as extremely small switching devices in microelectronics. The computers of the future will probably be based on nanomaterials containing colloids.

For background information *see* COLLOID; EMULSION; FOAM; GEL; MICELLE; NANOTECHNOLOGY; POLYMER; SOL-GEL PROCESS; SURFACTANT in the McGraw-Hill Encyclopedia of Science & Technology.

Jacques Desnoyers

Bibliography. K. S. Birdi (ed.), *Handbook of Surface and Colloid Chemistry*, CRC Press, New York, 1993; D. F. Evans and H. W. Wenneström, *The Colloidal Domain: Where Physics, Chemistry, Biology and Technology Meet*, VCH Publications, New York, 1999; R. M. Fitch (ed.), *Polymer Colloids*, Academic Press, New York, 1997; D. Myer, *Surfactant Science and Technology*, 2d ed., VCR Publications, New York, 1992; *Surfactant Science Series* (a collection of some 90 manuscripts covering various aspects of colloid science), M. Dekker, New York.

Communications standards

Technical standards provide the syntax for all technical communications. The level of technical standards (number system, measurement system, monetary system, navigational references, communication systems, and process systems) that emerge in each tribe or society is an indicator of its technical sophistication and technological attainment. Technical standards appear to be inherent in all complex systems, to be fundamental to almost all forms of commerce, and to be required for any communications. The process of creating a technical standard is termed standardization.

Standardization. Standardization organizations may be classified in a hierarchy of levels (**Table 1**).

Understanding the level of the standards development organization (SDO) is as important as knowing the level of any source of information. While any standards organization can produce standards of economic import, more formal standards organizations (higher up in Table 1) are more likely to produce standards that have more significant economic or technological impact. The three major international organizations are the ITU (International Telecommunications Union), the ISO (International Organization for Standardization), and the IEC (International Electrotechnical Commission). These organizations develop and approve worldwide technical standards for almost everything.

Accredited national and regional standards organizations are in the second level. ETSI (European Telecommunications Standards Institute) is a regional telecommunications organization. Other examples include the ASTM (American Society of Testing and Materials), EIA (Electronic Industry Alliance), and ATIS (Alliance for Telecommunications Industry Solutions) Committee T1 in the United States, and TTC (Telecommunication Technology Committee) in Japan. These accredited organizations produce standards which have national impact. The IEEE (Institute of Electrical and Electronic Engineers) and the IETF (Internet Engineering Task Force) are international in scope. Sometimes the standards produced in any of these organizations are also submitted to the top-level international organizations or to each other.

Some independent consortia such as ATM (Asynchronous Transfer Mode) Forum, Frame Relay Forum, and DSL (Digital Subscriber Line) Forum have been formally recognized by accredited standards organizations. Thus, work from these consortia can be directly submitted to the accredited standards development organizations, and the standards development organizations are willing to have liaison relationships with these consortia.

Consortia, which are controlled by one or more companies, are often vehicles for aggrandizement and increasing market share. They may create confusion if they try to give the appearance of belonging to a higher level in Table 1 in order to further their commercial interests. However consortia can develop useful de facto standards, such as the Microsoft Telephony Applications Programming Interface (TAPI).

Standards and economics. A technical standard is, in essence, an independent description used by a society for the purpose of comparison. For example, a compass is used to compare current direction to destination. A telephone modem compares received signals to signals used to decode them. A scale is used to compare the weight of our food to our nutritional needs. Sometimes the comparison has major economic consequences for the society. For example, the use of weights, measures, and currency enables trade to become commerce and increases government revenue with new forms of taxation. Adam Smith in *Wealth of Nations* (1776) explains that the

TABLE 1. Examples of standardization organizations

Organizations	Type	Controlled by
ITU, ISO, IEC	Governmental standards authority	Governments
ATIS, EIA, TTC, ETSI, IEEE, ASTM, IETF	Membership standards authority, government-accredited	Trade associations
ATM Forum, DSL Forum, Frame Relay Forum	Independent consortia with ties to standards authorities	Large number of companies
Instant Messaging Unified, Microsoft TAPI, Sun Java	Consortia	Single or small groups of companies

division of labor provides the greatest improvement in productivity, and that the power of exchange gives occasion to the division of labor. Technical standards make possible more complex transactions (increasing the power of exchange), which expands the division of labor, resulting in increased productivity and economic growth.

J. A. Schumpeter in *Business Cycles* (1939) provides additional understanding of the effect of technology on economic growth. He develops the concept that economic value is generated by invention and innovation. Each new technology is created by invention and deployed using innovation. The more significant changes in technology provide the basis for new forms of communications which support the creation of new forms of value. These major technology changes represent a paradigm shift that may be identified by the emergence of a new class of technical standards. In **Table 2** the relationship between standards and value systems is shown as a taxonomy relating standards to the technology, communications, and value systems in five historic eras.

Each historic era (such as pastoral, agrarian, or industrial) has a related succession of standards (such as reference, unit, or similarity). Each succession of standards remains in use as the foundation for the next succession. Then the next succession of standards codifies the communications technologies basic to that new era. A specific technical standard may include characteristics from multiple succes-

sions. The new forms of ownership that emerge with each paradigm shift create a significant increase in the economic wealth of that society.

Private good versus public gain. Successful standards represent a balance between two conflicting societal objectives: incentives for innovation (enabling private gain); and the diffusion of new products, services, and processes (enhancing the public good by enabling lower prices, greater usage, and economic growth). The first succession of standards, reference, evolved very slowly as agreement is the only incentive. In the next succession, the creation of unit standards has been the providence of governments as unit standards benefit the buyer (user) but not the seller (provider). Subsequent successions of standards offer increasing incentives for innovation. As the incentives for innovation increase, the need for government promotion of standards decreases and the marketplace emerges as the determinant of the success of standards. However, a balance between private gain and public good still should be maintained by any standards development organization.

Reference, unit, and similarity standards. Standards that define references and units define virtual and physical properties, respectively. The first technical standard may have been a reference standard: marks indicating a count. Possibly the first unit standards were for length, using an arm, hand, or foot as the reference. While unit standards (for example, liter) define the units to measure the carrying capacity of

TABLE 2. Taxonomy of standards through historic eras

	Pastoral	Agrarian	Industrial	Information	
				Sequential	Adaptive
Value system (form of ownership)	Private property ownership	Private land ownership	Invention ownership (such as patents)	System ownership (such as utility company)	Concept ownership (such as trademarks and brands)
Communications	Written language	Trade routes	Mechanized transport	Telegraph and telephony	Internet
Technology	Counting	Measuring and currency	Manufacturing (such as powered machines)	Linear processes (such as assembly line, railroad)	Adaptive processes (such as computers)
Succession of standards	Reference	Unit	Similarity	Compatibility	Adaptability

a barrel, similarity standards define how similar in construction one barrel is to the next. Making each barrel similar can offer economic advantages in manufacturing as well as distribution, selling, and use. Such economic advantages are marketplace incentives (self-reinforcing effects) that increase the use of existing similarity standards.

Compatibility standards. Standards for compatibility define a local or remote physical relationship between two or more independent entities for the purpose of interworking (physical connection) or compatibility. Compatibility standards or specifications (the term for proprietary "standards") provide the basis of all technical communications by defining an interface between two dissimilar devices. Once accepted public interfaces are available, public compatibility standards become essential parts of day-to-day life, for example, alternating-current plugs and outlets, Edison light bulb sockets, telephone jacks, electronic mail, the air interface between a cellular base station and a cellular mobile station, and the World Wide Web (WWW). The amazing and ever-changing array of devices and software that attach to these compatible interfaces give evidence of the powerful self-reinforcing effects that may be associated with public compatibility standards.

Adaptability standards. The widespread use of programmable computers complicates communications by increasing the possible means of communications. Adaptability standards are emerging to negotiate which compatibility standard is utilized in a multimode communications device. The emerging era of adaptive information is exemplified by the Internet, which is built from a compact series of protocol standards [such as Transmission Control Protocol (TCP), Internet Protocol (IP), and User Datagram Protocol (UDP)] which enable end-to-end communications between multiple programmable computers. The Internet allows for a level of adaptability that was not possible in earlier public communications systems such as the telephone system. Using the Internet, computers may use a specific type of protocol termed an etiquette or discovery protocol to negotiate the desired communications protocols (the syntax of electronic communications) for a specific application.

Early examples of etiquettes used to negotiate with remote systems include the International Telecommunications Union (ITU) V.8 used by telephone modems to negotiate remote compatible operation with the far-end modem. This is how older and newer telephone modems (for example, V.34 and V.90) find a common way to communicate. In Group 3 facsimile, the negotiating protocol ITU T.30 is an etiquette that has also been very successfully extended (that is, from 4800 to 9600, 14,400, and 28,800 bit/s) for over 30 years.

Creating value using etiquettes. Etiquettes can offer new ways to control proprietary value added to communications systems. The negotiation defined by an etiquette can support all types of compatibility, and can also support proprietary enhancements using a standardized way of passing proprietary information. In the etiquette, a character string (or similar) may be used to identify the legal owner of proprietary features or information. For example, the character string representing IBM is the property of that company. Transferring the character codes for IBM electronically can indicate that other, following codes are also proprietary to IBM and not for use by other companies. In Group 3 facsimile, the standardized way to support proprietary enhancements is called ITU T.30 Non-Standard Facilities (NSF). In the IETF SIP (Session Initiation Protocol), a reverse domain name is used to provide the unique identity. An earlier example is the ASCAP (American Society of Composers, Authors, and Publishers) requirements that identify copyright via fixed-length expressions (for example, five bars of music). Each of these different, unique identifiers is legally owned by the organization that implements them.

Using such proprietary identification, etiquettes can increase profits by offering specific and proprietary capabilities to particular market segments. For example, the banking industry may negotiate better encryption, the radiologist market may negotiate higher resolution, and the wireless market may negotiate better error control. Market segmentation via the etiquette can also be applied to the distribution channel, allowing individual equipment dealers and distributors to automatically poll their specific customers' equipment for usage billing (for example, copier market), problem analysis, and maintenance support (automatic ordering of replacement parts). In some automatic polling cases, the equipment dealer may include a unique identification of itself or even the end customer to facilitate and control transactions.

Commercial organizations that create communications products and services and wish to maximize their profits may find that etiquettes are a new way to control their markets. Etiquettes transport proprietary information in addition to the information necessary to select compatible modes of operation. In this manner, etiquettes support a new means to achieve monetary gain based on invention or innovation, while supporting the compatibility vital for public communications.

For background information *see* DATA COMMUNICATIONS; ELECTRICAL COMMUNICATIONS; ELECTRONIC MAIL; FACSIMILE; INTEGRATED SERVICES DIGITAL NETWORK (ISDN); LOCAL-AREA NETWORKS; MODEM; PHYSICAL MEASUREMENT; TELEGRAPHY; WIDE-AREA NETWORKS in the McGraw-Hill Encyclopedia of Science & Technology.   Ken Krechmer

Bibliography. C. Cargill, *Open Systems Standardization: A Business Approach*, Prentice Hall, Upper Saddle River, NJ, 1997; K. Krechmer, The Fundamental Nature of Standards: Economic Perspective, *International J. A. Schumpeter Society Economics Conference*, Manchester, June 28–July 1, 2000; K. Krechmer, The Fundamental Nature of Standards: Technical Perspective, *IEEE Commun. Mag.*, 38(6): 70–80, June 2000, IEEE Communications Society; K.

Krechmer, papers on the theory of standards and standardization; A. Updegrove, Consortia and the role of government in standards settings, in B. Kahin and J. Abbate (eds.), *Standards Policy for Information Infrastructure*, MIT Press, 1995; G. Wallenstein, *Setting Global Telecommunications Standards*, Artech House, Norwood, MA, 1990.

Complex adaptive systems

Complex adaptive systems (CAS) constitute an emerging field of study in science, technology, the behavioral sciences, and management. This field has evolved from major knowledge areas, including mathematics, chemistry, physics, biology, organizational science, computer science, and engineering. Fundamentally, a system is complex when it is not amenable to linear understanding, simple cause-and-effect relationships, or other standard methods of systems analysis. Complex systems are ones in which patterns can be seen and understood, but the interplay of individual elements cannot be reduced to the study of such elements considered in isolation from one another.

Complex adaptive systems involve phenomena that may be characterized by the interactions of numerous individual agents or elements, which self-organize at a higher systems level, and this organization results in evolutionary, emergent, and adaptive properties that are not exhibited by the individual agents themselves. These systems obtain data and information from their internal and external environment. They find patterns in this environment, and ultimately process and represent these patterns as internal models that are then used to describe and predict the potential outcome of future actions as a basis for decision making. Complex systems will often exhibit evolutionary, emergent, and adaptive behavior when these internal models are subjected to revision as the impacts of decisions taken are observed. There are complex nonadaptive systems, such as atoms and galaxies, which do not form internal models that evolve over time, and complex adaptive systems, such as biological organisms and human organizations, which do form internal models that selectively change.

The emergent and evolutionary characteristics of complex adaptive system often result in self-organization into a higher-level complex system. For example, an animal may be an agent in a formation of a herd of animals, herds of animals may become a species, and the species may be part of a particular ecosystem.

There are a number of "sciences of complexity," and they generally deal with approaches to understanding of the dynamic behavior of units that range from organisms to the largest economic, social, and political organizations. Complexity studies are generally multidisciplinary and transdisciplinary in nature. They attempt to pursue knowledge and discovery of the features that are shared by all the different systems that are described as somehow "complex." These studies are known by such names as complex adaptive systems, complex systems theory, complexity theory, dynamical systems theory, complex nonlinear systems, and computational intelligence.

Central ideas and concepts. There are no uniformly accepted definitions for the central ideas constituting complex adaptive systems. Nevertheless, some widely used terms have a relative common meaning. These terms collectively form an interconnected system of ideas.

Agents. Agents are individual elements of a complex adaptive system. Each agent will have its own set of internal states, skills, rules, and strategies that determine its behavior. Agents generally exist in hierarchies. For example, the cells in a human body may be regarded as agents while the human interacts at a higher hierarchical level with other agents in the environment. Agents receive data and information from their environments. They find regularities in the data and compress these perceived regularities into internal models that are used as a basis for action. These internal models sometimes lead to changes in the internal structure of an agent. Sometimes the term "schema" is used as a synonym for this internal model.

Adaptation. Adaptation occurs when the function or structure of an agent changes in a way that improves its chances for survival and reproduction in its environment. In some areas, such as evolutionary biology, these functional and structural changes may not occur during the lifetime of a particular agent. Here, the ancestor of an agent was born with a random mutation gene that permitted it to perform some function better than most agents, and this change improved its chances to survive sufficiently long to reproduce and pass on this gene to its offspring (natural selection). In other agents, such as organizational agents, the structural and functional changes may occur during the lifetime of a given organization, and this may lead to changes in the organization through reengineering.

Artificial life. Artificial life, sometimes called ALife, is a related field of study that investigates important aspects of living systems by using computer simulations. ALife investigates natural life by trying to capture the behavioral essence of the components of a living system, and then endowing a collection of artificial components with similar behavioral characteristics such that the ALife representation or model exhibits the same behavior as the natural system. ALife can be applied at any hierarchical level, from modeling molecular dynamics on millisecond time scales to modeling population evolution over centuries.

Chaos. The key idea behind chaos and chaos theory is that small changes in initial conditions or parameters of a system will often be magnified considerably over time and may have an enormous influence on the long-term behavior of a system. The butterfly effect, in which a butterfly that flaps its wings in South America causes atmospheric perturbations such as to cause a snowfall in Alaska, is often cited as illustrative of this notion. The metaphors "chaos" and "edge of chaos" have been adapted from physics and chemistry. Water, for example, exists in three phases: solid ice, liquid water, and gaseous steam. The point at which a change occurs from one phase to another is called a phase transition. Complex adaptive systems may exist and evolve in phaselike transitions. They generally function best "at the edge of chaos" where behavior is intermediate between that characteristic of a rigid ordered phase and a chaotic phase.

Emergence. Emergent phenomena are probably the most inscrutable aspect of complex adaptive systems. The fact that individual agents behaving according to relatively simple rules may interact and self-organize at higher system levels, and that the resulting shared properties are different than those of an individual agents, is of major interest in science, engineering, and social and behavioral research concerning organizations and their management.

Selection. Selection is concerned with the suitability of particular traits of an agent for surviving long enough to reproduce in a particular environment. Biologists often use the expression "fitness function" to describe this suitability. An agent may have a high fitness function in one environment and a low fitness function in another.

Self-organization. Concepts of self-organization are fundamental in complex adaptive systems. Self-organization occurs when individual independent agents in a complex adaptive system interact in a jointly cooperative manner that is also individually appropriate, so as to generate a higher-level organization. Self-organization is intrinsically decentralized as it is due to the local interactions of many individual agents, with order emerging or evolving without any centralized command-and-control-like influences.

Models. A complex adaptive system may be described by an analogous system that, in some ways, replicates the behavior of the original system for a given class of inputs. The complexity of the system may be determined by a measure of the complexity of the model necessary in order to effectively predict its behavior. The more the simulation model must look like the actual system in order to yield the same behavior, the more complex the system is said to be.

With a complex nonlinear system, very small changes in the initial conditions of the system states, or inputs to the system, will often result in very different outputs for that system. A general rule for complex systems is that we cannot create a model that will accurately predict the outcomes of the actual system. However, we can create a model that will accurately simulate the processes that the system will use in order to create a given output. This awareness has profound impacts for organizational efforts. It also calls forth many concerns relating to the real value of creating organizational plans and mission statements and suggests instead the creation of a model of the planning process itself and subjecting this to various inputs in order to generate output scenarios.

One of the major behavior patterns in a complex adaptive system is that agents recognize patterns, as well as features similar to previously recognized situations. They simplify the problem by using these patterns and features to construct internal models, hypotheses, or schemata to use on a temporary basis. They attempt simplified deductions based on these hypotheses and act accordingly. The resulting feedback of results from these interactions enables the agents to learn more about the environment and the nature of the task at hand. Based on this, hypotheses are revised, and this leads to reinforcing appropriate hypotheses and discarding poor ones. This use of simplified models is inductive behavior.

Characteristics. John Holland described seven basic concepts in the field of complex adaptive systems: four properties (aggregation, nonlinearity, flows, and diversity) and three mechanisms (tags, internal models, and buildings blocks).

1. *Aggregation.* Aggregation is a basic mechanism in object modeling and is the basis for identity, a fundamental object concept. Forming components out of objects and enterprise systems from components is higher-level aggregation.

2. *Tagging.* Tagging is a mechanism that consistently facilitates the formation of aggregates. Complex adaptive systems use tags to manipulate symmetries and to facilitate selective interaction. Tagging is a mechanism that agents utilize for aggregation and flows of information. Tags facilitate selective mating and, as a consequence, preserve boundaries between aggregates. They allow us to specify a variety of object models and enable such efforts as information filtering and agent cooperation.

3. *Nonlinearity.* Nonlinear systems may exhibit catastrophic and chaotic behaviors. The underlying assumption in a complex adaptive system is that there is not a direct and easily predictable linear relationship between an agent's actions and the consequence of that action. Nonlinear interactions almost always make the behavior of the aggregate more complicated than would be predicted by the traditional linear actions of summing or averaging.

4. *Flows.* Flows, such as work flows and message and transaction routing, typically have a multiplier effect. A network encourages flow of information and knowledge or other entities of importance.

5. *Diversity.* Complexity implies diversity. Diversity supports the notion of agents seeking differing goals within a larger environment. In complex adaptive systems, diversity is found to increase stability at the edge of chaos.

6. *Internal models.* The utility of complex systems is enhanced if the system can learn from experience and adapt its behavior. The ability of the system to develop and act on internal models, or schema, that simplify the external world is basic to this mechanism.

7. *Building blocks.* Building blocks are the basis for generation of internal models and are essential to the construction of adaptive systems.

Studies of complex systems often run counter to the trend toward increasing fragmentation and specialization in most disciplines. It is not at all a large number of parts in a system that makes the system complex; it is the way that the parts interact. A product may consist of abundant parts, but if these parts interact only in a known, designed, and structured fashion, the system is not complex although it may be big. Complexity exists when the interconnected parts of a system interact in unanticipated ways. One defining characteristic of complex systems is emergence, whereby the behavior of the overall system is different from the aggregate behavior of the parts and knowledge of the behavior of the parts will not allow us to predict the behavior of the whole system. The emergence property is a form of control. It allows distributed agents to organize together to determine higher-order system behavior, so that structure and control emanate or grow from the bottom up. Thus, the reductionist scientific approach generally does not work with complex systems. Virtually all organizational behavior in such systems comprises agents adapting to their environments and, in the process of so doing, affecting the environments of all other agents. In some situations, when systems are driven sufficiently far from equilibrium, "bifurcations" occur and chaotic behavior may result.

Steps in building systems. Clearly, there are many considerations involved in such efforts. The prototypical steps in building an experimental and exploratory model of a complex adaptive system might be described as follows:

1. Simplify the problem as much as possible, being sure to retain the essential features of the situation.

2. Identify a potentially appropriate model of the situation that represents agents that follow simple rules with specified interactions and randomizing elements.

3. Construct a simulation based on this model.

4. Run the simulation many times with appropriately different random variables and collect data and compute statistics from the different runs.

5. Identify how simple behavioral rules result in observed behavior.

6. Study the responses obtained by sensitivity studies and appropriate parameter changes to determine critical parameters, sources of behavior, and effects of different parameters on systems responses.

These prototypical steps have been used to construct a number of complex adaptive system models. Such studies reintegrate the fragmented interests of most disciplines into a common pathway, enhanced by the capabilities associated with modern modeling and simulation.

For background information *see* ADAPTIVE CONTROL; ARTIFICIAL INTELLIGENCE; AUTOMATA THEORY; CATASTROPHE THEORY; CHAOS; GENETIC ENGINEERING; INTELLIGENT MACHINE; MODEL THEORY; NEURAL NETWORK; SIMULATION; SYSTEMS ENGINEERINGS in the McGraw-Hill Encyclopedia of Science & Technology.   Andrew P. Sage

Bibliography. R. Axelrod and M. D. Cohen, *Organizational Implications of a Scientific Frontier*, Free Press, New York, 1999; J. L. Casti, *Would-Be Worlds: How Simulation Is Changing the Face of Science*, John Wiley, New York, 1997; J. H. Holland, *Emergence: From Chaos to Order*, Addison Wesley, Reading, MA, 1998; S. Kauffman, *At Home in the Universe: The Search for the Laws of Self-Organization and Complexity*, Oxford University Press, New York, 1995; S. Kauffman, *Investigations*, Oxford University Press, 2000; S. Kelly and M. A. Allison, *The Complexity Advantage: How the Science of Complexity Can Help Your Business Achieve Peak Performance*, McGraw-Hill, 1999; A. P. Sage and W. B. Rouse (eds.), *Handbook of Systems Engineering and Management*, John Wiley, New York, 1999.

Computer design

Modern computer design is a recursive procedure in which teams of specialists employ a myriad of computer-aided design (CAD) tools in the conception, layout, and fabrication of a new machine. The top-level architecture or logical design is still based on the stored-program concept proposed by John von Neumann in 1945, but the physical implementation has evolved rapidly to produce faster and smaller computers with greater capacity and reliability.

Architecture. The von Neumann computer architecture consists of five parts: input, output, memory, arithmetic logic unit (ALU), and control. Typical input devices are the keyboard and mouse. Examples of output devices include printers and visual display units. Hard disks, floppy disks, and modems are examples of input/output (I/O) devices. The memory, or main store, contains both machine instructions and data. The ALU performs arithmetic and logical operations upon data as directed by the control unit, which fetches instructions from the memory and activates appropriate sequences of data transfers and ALU operations.

The computer's central processing unit (CPU), or processor, contains the control unit and the datapath. The datapath consists of the ALU and all of the support circuitry required for data interconnections, routing, and temporary storage. All of the machine instructions, numeric and character data, and control signals are encoded as binary digits (bits) or arrays of bits. The physical representation of the binary digits depends upon the implementation technology, but is usually 0 volts for binary 0 and a positive voltage between 2 and 5 volts for binary 1. A system bus interconnects the CPU, memory, and I/O units, allowing parallel data transfer of N bits, where N is the bus width, which has followed the progression 4, 8, 16, 32, and 64 bits in successive design generations.

Temporary data storage in the datapath is provided by registers, which are arrays of one-bit storage elements called flip-flops. Commonly used registers are the instruction register, which holds the current machine instruction; the program counter, which contains the address of the next instruction to be fetched from memory; the memory address register, which contains the address of data to be accessed in memory; the memory data register, which contains the data fetched or to be stored in memory; and a set of general-purpose registers, which store inputs and outputs for ALU operations.

CPU operations and data transfers are synchronized by a system clock whose speed is limited by propagation delays in the electronic circuits. Advances in integrated circuit technology have allowed the clock rate of personal computers to increase more than two orders of magnitude, from 8 MHz in 1980 to over 1 GHz in 2000. Computer performance has been improved not only by increasing clock rates but also by increasing bus widths, and by architectural enhancements such as instruction pipelining, cache memories, and multiprocessing.

The most distinguishing feature of a processor is its instruction set architecture (ISA). Processors such as the Digital Equipment Corporation VAX or the Intel Pentium employ complex instruction set computing (CISC), which provides hundreds of different machine instructions with a great variety of data addressing modes. Reduced instruction set computing (RISC) processors, such as the Sun Microsystems SPARC, the MIPS, and the Power PC, have fewer instructions, and require that all input and output data for ALU operations be associated with general registers. The RISC architecture provides a simpler hardware implementation at the expense of longer machine instruction sequences.

System design. The design of a computer and its constituent elements proceeds through the phases typical of any electronic system design: specification of requirements; design specification; behavioral description of a functional design; structural representation of a logic design; circuit design in a specific technology; physical layout; and finally fabrication (**Fig. 1**). CAD tools are available at each phase to provide simulation, verification, and feedback for revision of the current and previous phases.

In the first phase, design specifications are developed in view of the system requirements, past experience, cost, and feasibility with respect to current technology. This would include selection of an existing processor or specification of the instruction set architecture and characteristics of a new processor, memory size and access time, system bus size, I/O device support, and interface properties for all modules. The module implementations can be performed concurrently, carrying each one through the remaining phases of the design process.

The functional design phase begins with the development of an algorithm to implement the specified behavior of the module. This could be simply a truth table for a combinational logic module such as a decoder or multiplexer. An N-bit binary multiplier module, on the other hand, could be implemented with combinational logic for small N, but may require a sequential algorithm for large N. The behavior of a sequential module may be described by an algorith-

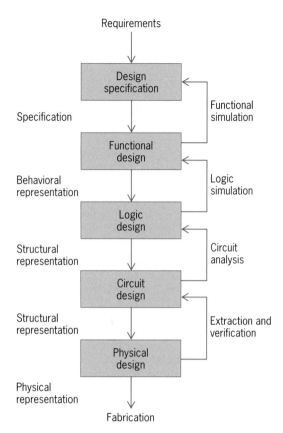

Fig. 1. Phases of electronic system design.

mic state machine (ASM) chart, which depicts graphically the state transitions as determined by module inputs, as well as the actions (outputs) associated with each state. For example, the ASM chart for a counting multiplier contains only two states, and indicates the appropriate actions to initially reset the up-counter (product) and load the multiplicand M1 and multiplier M2 into down-counters 1 and 2, and then decrement counter 1 from N1 to 0 N2 times while incrementing the product (**Fig. 2**).

A hardware description language (HDL) is a powerful CAD tool which will both describe and simulate the behavior of digital systems. The most commonly used HDLs are Verilog and VHDL [VHSIC (Very High Speed Integrated Circuit) HDL]. These are high-level programming languages containing constructs to describe and simulate the concurrent activity and signal propagation delays associated with complex digital circuits. For example, the VHDL behavioral description of the counting multiplier contains declarations for a library, an entity, and an architecture (**Fig. 3**). The library contains type declarations and function definitions. The entity declaration defines the name of a module as well as the names and types of all the input and output signals. The architecture code describes the behavior of the multiplier in accordance with its ASM chart. After compiling the VHDL program, the designer can invoke the simulator through a graphical user interface (GUI), which allows interactive specification of time

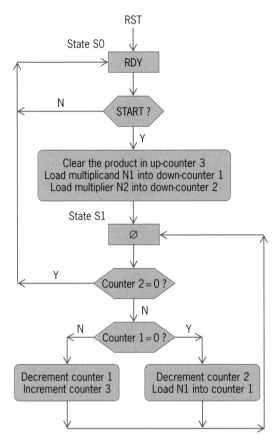

RST

State S0

RDY

N

START ?

Y

Clear the product in up-counter 3
Load multiplicand N1 into down-counter 1
Load multiplier N2 into down-counter 2

State S1

∅

Y

Counter 2 = 0 ?

N

N   Counter 1 = 0 ?   Y

Decrement counter 1
Increment counter 3

Decrement counter 2
Load N1 into counter 1

Fig. 2. ASM chart for a counting multiplier.

on propagation delay and area. The output of the synthesis and optimization is a file in the electronic data interchange format (EDIF) describing the structural interconnection of components in the target technology in a standard format acceptable as input to all vendor CAD tools.

The circuit design and analysis phase is initiated by invoking the target technology vendor's CAD tools with the EDIF file as input. This software will configure PLDs or allow interactive layout of ASIC standard cells, provide data for simulation with accurate propagation delays, and produce files to program PLDs or fabricate ASICs.

Fabrication. The final phase of physical design for fabrication may involve the custom layout of an ASIC or the printed circuit board (PCB) layout for a circuit containing a large number of components. Interactive CAD tools are used in both cases. IC design programs allow the designer to graphically lay out the various geometric regions of polysilicon, diffusion, and metal which define the transistors and their interconnections, to verify the correspondence with a logic schematic diagram, and to produce a standard data file for IC fabrication at the foundry. Similarly, PCB software operates on an appropriately annotated schematic circuit diagram to package gates into ICs, lay out the components on the board, route the connections in accordance with the schematic, and produce a standard output file for fabrication of the board.

histories for the input signals and observation of the corresponding output waveforms or lists.

A single VHDL entity may have more than one architecture to describe different implementations of the same module. For example, a structural architecture would describe the entity in terms of the interconnection of other entities. Thus, with VHDL descriptions of the counters for the multiplier, a structural architecture for the multiplier can be written using port mapping statements to show their connections. Accurate behavioral models with realistic signal propagation delays for commonly used entities are stored in libraries to be accessed as components in structural architectures of higher-level entities.

Synthesis. After verification of functional validity, the next phase in the design process is logic design which maps the module into a structural interconnection of logic elements including basic gates (AND, OR, NOT, NAND, NOR, XOR), multiplexers, decoders, adder/subtractors, comparators, flip-flops, registers, counters, and memories. The target technology for implementation of this module may be commercial integrated circuits or programmable logic devices (PLDs) or a new application-specific integrated circuit (ASIC). CAD synthesis tools will accept an HDL description, map it into a generic logic design, and then optimize it into a specified target technology. The designer can specify constraints

```
-- mult.vhd 4-bit binary counting multiplier

library IEEE;
use IEEE.std_logic_1164.all;
use IEEE.numeric_std.all;

entity mult4 is
  port (start, reset, clk : in std_logic;
        n1, n2 : in unsigned (3 downto 0);
        ready : out std_logic;
        ct1, ct2 : inout unsigned (3 downto 0);
        p : inout unsigned (7 downto 0) );
end entity mult4;

architecture behav of mult4 is
  signal state : std_logic;
begin
  process (reset, clk)
  begin
    if reset = '1' then state <= '0';
    elsif rising_edge(clk) then
    case state is
      when '0' =>
        if start = '1' then
          p <= "00000000"; ct1 <= n1; ct2 <= n2;
          state <= '1';
        else state <= '0';
        end if;
      when '1' =>
        if ct2 = "0000" then state <= '0';
        elsif ct1 = "0000" then
          ct2 <= ct2 - 1; ct1 <= n1; state <= '1';
        else
          ct1 <= ct1 - 1; p <= p + 1; state <= '1';
        end if;
      when others => null;
      end case;
    end if;
  end process;
  ready <= not state;
end architecture behav;
```

Fig. 3. VHDL behavioral description of a counting multiplier.

CAD software is available to support hierarchical models in which some elements are described by HDL and others by schematic diagrams. Other programs will automatically create HDL models from state transition diagrams or from schematic diagrams. The development and utilization of advanced CAD tools, together with improvements in fabrication technology, allowed Sun Microsystems, for example, to produce the Ultra SPARC III processor with a 64-bit data bus, 750-MHz clock, and 29 million transistors on a single chip.   Euqene W. Henry

Bibliography. M. Campbell-Kelly and W. Aspray, *Computer: A History of the Information Machine*, BasicBooks, New York, 1996; K. C. Chang, *Digital Systems Design with VHDL and Synthesis: An Integrated Approach*, IEEE Computer Society Press, 1999; G. De Micheli, *Synthesis and Optimization of Digital Circuits*, McGraw-Hill, 1994; J. P. Hayes, *Computer Architecture and Organization*, 3d ed., McGraw-Hill, 1998; D. A. Patterson and J. L. Hennessy, *Computer Organization & Design*, Morgan Kaufmann, 2d ed., 1997; C. H. Roth, Jr., *Digital Systems Design Using VHDL*, PWS Publishing, 1998; J. P. Uyemura, *A First Course in Digital Systems Design*, Brooks/Cole Publishing, 2000; M. Zwolinski, *Digital System Design with VHDL*, Prentice Hall, 2000.

Continental roots

Improvements in modern diamond exploration practice and our knowledge of Earth's early evolution are closely tied to a better understanding of the structures known as continental roots. These structures underlie the ancient crust of a majority of Earth's continental shields, including the Slave craton (**Fig. 1**) in Canada. Geophysical studies of continental roots in combination with petrological and geochemical studies of diamondiferous kimberlites, both in Canada and abroad, are leading to new insights into the forces that shaped the origin of the very first landmasses.

Root Formation

Seismologists have recognized since the 1970s that the mantle lithosphere underlying continental shields is characterized by higher (to several percent) seismic velocities than its surroundings to depths of 200–250 km (125–155 mi). These velocity anomalies define the continental roots. Seismic velocity is a proxy for thermal state, implying that the roots are colder than ambient mantle—an inference supported by heat flow studies and thermobarometric analyses of xenolith suites. Early explanations for the origin of continental roots drew upon a direct analogy with the evolution of oceanic lithosphere. It was postulated that the roots grew as thermal boundary layers which thickened as a result of gradual cooling over time. However, several objections were raised to this hypothesis, perhaps the most important being that cooler temperatures lead to

increased density and, eventually, gravitational instability. In the oceans, this process is manifest through the development of subduction zones and convective recycling of lithosphere into the deeper mantle. In contrast, continental roots, through their association with ancient crust, appear to be characterized by stability and a resistance to mantle recycling. Consequently, a purely thermal origin cannot account for the origin of continental roots. Petrological and physical property studies of xenoliths have further established that this stability is developed through compositional contrasts between the continental roots and ambient mantle. In particular, the roots are more harzburgitic in bulk composition (that is, garnet- and clinopyroxene-poor) and depleted in iron. These factors render the roots less dense than normal mantle, thereby effectively canceling the negative buoyancy induced by temperature.

Geochronology. Diamonds can be used to place constraints on the timing and nature of root formation. In particular, garnet inclusions within diamonds can be dated using neodymium isotope geochronology to infer an age of diamond crystallization, and of root formation. In South Africa, 3-billion-year neodymium-model ages derived using this approach suggested that diamonds and continental roots formed synchronously with or soon after the stabilization of the overlying crust. Studies using rhenium-osmium model ages also support a close temporal association between the timing of crustal stabilization and melt depletion of the underlying mantle lithosphere. In addition, rhenium-osmium dating of xenoliths at different levels within the lithospheric column has failed to reveal any clear depth dependence with age, possibly indicating rapid root formation. Both of these results are, again, inconsistent with an origin for the roots in terms of gradual thickening of a thermal boundary layer.

Seismological analysis. Recent geophysical and petrological work on the Slave craton has shed new light on the conundrum of root formation. Using elastic waves generated by remote earthquake sources, it is possible to characterize the internal structure of continental roots using seismological techniques. This approach has been applied to data recorded on the Yellowknife seismic array operated by the Geological Survey of Canada and located toward the southern end of the Slave craton. The images reveal a well-defined stratigraphy within the mantle-lithosphere below the Slave craton which extends to depths of 220 km (137 mi). This stratigraphy is characterized, in part, by layers of order 10 km (6 mi) in thickness which exhibit elastic anisotropy; that is, the velocity at which waves propagate through the layering is dependent upon the propagation direction.

The nature of the stratigraphy is further constrained by a deep seismic reflection profile undertaken as part of the Lithoprobe Slave–Northern Cordillera Lithosphere Evolution (SNORCLE) transect. The SNORCLE transect is remarkable for the clarity and continuity with which two sets of mantle reflections are evident in the upper 100 km (60 mi) of lithosphere. One set near 60 km (36 mi) depth below Yellowknife coincides closely with the shallow layering documented in the teleseismic study. The second set merges with lower crustal reflectivity 500 km (310 mi) to the west and would project 150–200 km (93–124 mi) beneath Yellowknife, a depth interval where several distinct layers are evident in the image derived from earthquake seismology. Based on surface geology and aeromagnetic signatures, this second set of reflectors can be confidently ascribed to a Proterozoic (1.84 billion years ago) subduction episode.

Subduction of oceanic plateaus. Several additional factors imply that subduction may have played an important role in continental root development beneath the Slave craton. The character of the shallow subhorizontal mantle layering observed below Yellowknife suggests that it may represent former oceanic crust. Xenolith suites assembled from diamondiferous kimberlite pipes at a range of locations across the Slave craton provide supporting evidence for this conjecture. They are characterized by a high proportion (up to 30%) of eclogite, which is the high-temperature/pressure form of basalt, the dominant rock type in oceanic crust. The well-developed anisotropic character of the seismic layering can be explained by the observation that the eclogite samples frequently exhibit mineral banding of garnet and clinopyroxene, which would be expected to produce a macroscopic dependence of seismic wave velocity on direction.

These observations are consistent with a model for continental root formation that involves shallow subduction. This model, a variant of one originally proposed by H. Helmstaedt and D. J. Schulze (1989), posits that the early continents developed through successive underplating of anomalously buoyant oceanic lithosphere, capable of resisting mantle recycling. The Ontong-Java plateau in the southwestern Pacific may serve as a modern example of such lithosphere. Oceanic plateaus are thought to form where mantle plumes interact with mid-ocean ridges to produce abnormally thick oceanic crust, underlain by mantle lithosphere that is iron-depleted and harzburgitic in composition. These latter characteristics are shared by peridotite xenoliths erupted from Archean cratons, and result in mantle lithosphere that is less dense than normal. Consequently, an oceanic plateau will resist normal subduction at convergent margins and may subduct horizontally, as is presently inferred to occur along portions of the South American and western Pacific subduction zones. Higher temperatures earlier in Earth's history may have resulted in more widespread generation of oceanic plateaus, but they also serve to refocus attention on how thick, cool continental roots could have formed so soon after stabilization of overlying crust, as required by geochronological studies. In fact, shallow subduction of oceanic plateaus, perhaps in successive episodes, affords an efficient means of continental root assembly through rapid advection of

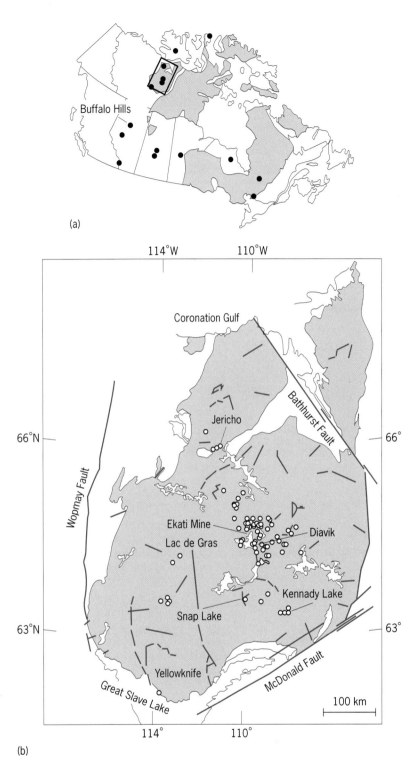

(a)

(b)

Fig. 1. Locations of Canadian kimberlite occurrences and advanced projects. (*a*) Outline map of Canada showing cratons (tinted areas) and kimberlite fields (dots). The rectangle marks the (*b*) Archean Slave craton, including major geological faults (bold lines) and individual kimberlite pipes (open circles).

cool, near-surface material to greater and hotter depths.
                                                    Michael Bostock

Canadian Diamond Deposits

Although Canada began to produce diamonds only in 1998, by 2003 its diamond production will

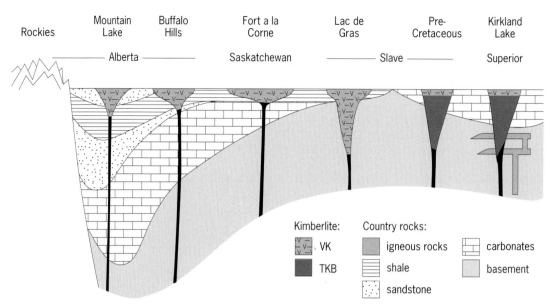

Fig. 2. Schematic section across Canada summarizing pipe shapes, pipe infill, and geological setting for many kimberlites reconstructed for the time of emplacement. **VK** = volcaniclastic kimberlite formed in an upper part of a kimberlite pipe (in a crater zone); **TKB** = tuffisitic kimberlite breccia formed in the middle part of a kimberlite pipe (in a diatreme zone).

account for about 10% of the world's diamonds by value and 8% by weight. Diamonds in Canada are found in primary deposits associated with magmatic bodies of kimberlite. Kimberlites are carbon dioxide- and water-rich mafic alkaline rocks that originate deep (200 km or 125 mi) in the mantle and are emplaced as volcanic pipes, dikes, and sills (types of magmatic formations).

Diamondiferous kimberlites. Diamonds, although found in kimberlites, do not crystallize from the kimberlite magma. They originate in other deep mantle rocks picked up by kimberlite magma on its ascent to the surface. Diamonds form in eclogites, peridotites, and ultradeep lower mantle rocks. Solid fragments of these rocks disintegrate in the host magma and liberate diamonds into the kimberlite, which transports them to the surface. Only a few kimberlite intrusions are economically diamondiferous. These minable kimberlites are restricted to stable geological terranes that have not been deformed since the Archean (that is, Archean cratons). Economic kimberlites in Canada are found on the Slave craton (Fig. 1). Because Canadian kimberlite pipes are much smaller than South African pipes, several adjacent pipes need to be mined to make a deposite economical. All economically diamondiferous kimberlite in Canada is type I kimberlite (nonmicaceous), according to chemical and mineralogical classification. Type I kimberlite magmas have an asthenospheric origin.

Most of the kimberlite occurrences on the Canadian craton are located in the Lac de Gras area, where kimberlites were emplaced during the Tertiary, 47–84 million years ago. The Lac de Gras kimberlite is found in relatively small steep-sided pipes that were probably originally less than 600–700 m (1980–2220 ft) deep (**Fig. 2**). Most of the pipes are infilled with pyroclastic and epiclastic kimberlite. The most notable feature of these pipes is the lack of tuffisitic kimberlite and tuffisitic kimberlite breccia—

two varieties of kimberlite formed within the diatreme zone by a fluidization process. The Lac de Gras pipes are different from the majority of pipes in southern Africa.

Outside the Lac de Gras area, kimberlites are older (540 million years for the Kennady Lake cluster, 172 Ma for the Jericho pipes, 450 Ma for the Drybones kimberlite). These pre-Cretaceous kimberlites have the morphology of classical southern African kimberlites, with crater, diatreme, and root zones from top to bottom of volcanic pipes, respectively. The kimberlite pipes are eroded to diatreme, lower diatreme, and root zones.

Economic kimberlite pipes in Canada were discovered in 1991 by Charles E. Fipke. For 10 years, he looked for the primary source of anomalous till samples rich in diamond indicator minerals found in the McConnel Ranges of the MacKenzie Mountains. Systematic heavy mineral sampling advanced him 1200 km (750 mi) west–east across the Northwest Territories. Fipke backtracked the glacier movement that sheared off the tops of the pipes, spreading the kimberlite material for miles and ultimately leaving trails of indicator minerals. Drilling at a site where Fipke found a single particle of chrome diopside (an indicator mineral) intersected a diamondiferous kimberlite in November 1991. Subsequent magnetic and electromagnetic airborne surveys helped to identify many more kimberlite pipes. The most recent discovery, the Buffalo Hills kimberlite province, was made as a result of fixed-wing airborne magnetic surveys conducted during oil and gas exploration.

Canadian diamond characteristics. Diamonds mined in Canada are extraordinarily high-quality; the bulk of their value is in white clear stones. They exhibit all the common crystal shapes; broken stones and fragments are very common. High-quality white tetrahexahedra and octahedra are present up to the largest diamond sizes. Cubes, which are exclusively fibrous (contain inclusions), are confined mainly to the smaller size fractions. Brown and gray types of diamond due to crystal irregularities are very common. These diamonds represent more than one diamond-forming episode and have experienced a range of deformation and resorption events as is common for kimberlite diamond production elsewhere. Fluids trapped by fibrous diamond coats on octahedral cores and involved in diamond formation differ significantly from fluids in African diamonds.

Diamonds from the DO-27 pipe (Lac de Gras area) display a wide range of morphology with a high (14%) proportion of cubo-octahedral stones, half of which are fibrous. Resorption removed 25–50% of the original mass in more than 50% of the diamonds. Plastic deformation is evident in about half the stones. Three-quarters of the stones are colored [shades of brown (55%) to yellow and gray] because of impurities (such as nitrogen and boron) or crystal irregularities. Syngenetic inclusions in diamonds suggest that they came from three types of parent rocks: eclogite (50% of the sample), peridotite (25%), and superdeep mantle at depths >670 km [420 mi] (25%). Eclogitic diamonds have lower nitrogen aggregation states than those of peridotitic diamonds. This may indicate younger ages of eclogitic diamonds, lower average mantle storage temperatures, or less plastic deformation. Eclogitic diamonds have the lightest carbon δ^{13}C values of $-(22-3)‰$, which suggest a biogenic source of carbon and the origin of eclogites in a subducted oceanic crust. In contrast, peridotitic diamonds [δ^{13}C = $-(6-1)‰$] may have formed from juvenile carbon degassed from the mantle. Diamonds from the DO-27 pipe have similar characteristics to diamonds from the Ekati mine 50 km (30 mi) to the north and may be derived from a similar mantle source.

Diamond macrocrysts (crystals, typically 0.5–10 mm in diameter) from the Buffalo Hills area are dominated by tetrahexahedra and resorbed fragments, vary in color from white through various shades of yellow, brown, and pale pink and, according to carbon isotope analysis, may have formed in peridotite.

The occurrence of abundant eclogitic diamonds suggests the presence of former oceanic basaltic crust (the precursor of eclogite) in the subcratonic mantle. As argued by M. G. Bostock (1998), these diamonds support a model of continental root formation that involves shallow subduction and slab stagnation.

For background information *see* BRECCIA; CHEMOSTRATIGRAPHY; CRATON; CRYSTAL STRUCTURE; DIAMOND; ECLOGITE; GARNET; LITHOSPHERE; MAGMA; PERIDOTITE; PYROXENE; SEISMOLOGY; STRATIGRAPHY; SUBDUCTION ZONES; XENOLITH; VOLCANOLOGY in the McGraw-Hill Encyclopedia of Science & Technology.   Maya G. Kopylova

Bibliography. M. G. Bostock, Mantle stratigraphy and evolution of the Slave province, *J. Geophys. Res.*, 103:12,183–12,200, 1998; F. A. Cook et al., Frozen subduction in Canada's Northwest Territories: Lithoprobe deep lithospheric reflection profiling of the western Canadian Shield, *Tectonics*, 18:1–24, 1999; J. J. Gurney et al. (eds.), *Proceedings of the 7th International Kimberlite Conference*, Red Roof Design, Cape Town, 1999; H. Helmstaedt and D. J. Schulze, Southern African kimberlites and their mantle sample: Implications for Archean tectonics and lithosphere evolution, in J. Ross (ed.), *Kimberlites and Related Rocks*, pp. 358–368, Blackwell, Cambridge, MA, 1989; T. H. Jordan, Structure and formation of the continental tectosphere, *J. Petrology*, Special Lithosphere Issue, pp. 11–37, 1988; M. Kirkley, J. J. Gurney, and A. A. Levinson, Age, origin, and emplacement of diamonds: Scientific advances in the last decade, *Gems and Gemology*, 27(1):2–25, 1991; D. G. Pearson, The age of continental roots, in R. D. Van der Hilst and W. F. McDonough (eds.), *Composition, Deep Structure and Evolution of Continents*, pp. 171–194, Elsevier, Amsterdam, 1999; J. Pell, Kimberlites in the Slave craton, Northwest Territories, Canada, *Geosci. Can.*, 24:77–90, 1997; J. Picton et al., Diamond production in the 21st century, *Gems and Gemology*, 35(3):114–115, 1999; S. H. Richardson et al., Origin of diamonds in old enriched mantle, *Nature*, 310:198–202, 1984.

Degenerative neural diseases

The number of people with neurodegenerative disorders associated with aging, such as Alzheimer's disease and Parkinson's disease, is increasing rapidly as life expectancy in developed countries continues to rise. Without a medical breakthrough, the number of people afflicted will reach 14 million for Alzheimer's disease and 4 million for Parkinson's disease by 2050, threatening to make these conditions an unmanageable burden not only for the patients but also for society. Neurodegenerative disorders comprise a wide range of neurological diseases characterized by loss of synaptic connections among neurons, neuronal dropout, astrogliosis (proliferation of astrocytes within the brain), and abnormal accumulation of neuronal proteins inside neurons or in the extracellular space. Alzheimer's disease is among the most common neurodegenerative disorders and the leading cause of dementia in the elderly. Other common neurodegenerative disorders, such as Parkinson's disease, amyotrophic lateral sclerosis, and Huntington's disease, are characterized by motor function abnormalities. *See also* DEMENTIA.

Pathogenesis. Although the cause of many of these disorders is not completely clear, recent studies have shown that mutations in specific neuronal proteins are responsible for familial (inherited) forms of the diseases (**Table 1**). The causes of sporadic forms are still under intense investigation; however, a common feature among genetic and sporadic forms is the abnormal accumulation of proteins in the synaptic connections of specific neuronal populations, leading to impaired communication among cells in specific regions of the brain. For example, in Alzheimer's disease, amyloid beta protein, a 42-amino-acid proteolytic product of the amyloid precursor protein (APP), accumulates in the frontal cortex and the

TABLE 1. Familial forms of neural degenerative diseases

Disease	Genes involved
Alzheimer's disease	Amyloid precursor protein, presenilin 1 and 2, apolipo-protein E4
Parkinson's disease	α-synuclein, Parkin, ubiquitin c-terminal hydroxylase L1
Amyotrophic lateral sclerosis	Superoxide dismutase
Huntington's disease	Huntingtin
Frontotemporal dementia	Tau
Creutzfeldt-Jakob disease	Prion protein

hippocampus, areas of the brain involved in memory formation and learning. In Parkinson's disease, an abundant synaptic protein known as α-synuclein accumulates in the substantia nigra and the temporal cortex areas involved in regulation of movement.

Experimental animal models. Experimental animal models have been developed recently to better understand the pathogenesis of neural degenerative diseases and to develop new treatments. The models are developed either to study sporadic forms by reproducing several aspects of the disease or to study inherited forms by reproducing some central features of the disease based on the known genetic basis. Since human neurodegenerative disorders are very complex in nature, the closer the model species is to humans, the closer the alterations that will be reproduced. For example, among the animal models that more closely mimic Alzheimer's disease and Parkinson's disease is the aged monkey. However, experimental studies in higher species, such as the rhesus monkey, possess several logistical and technical problems, especially since the neural disorders take years to develop. Thus, new studies have been focused toward modeling neurodegenerative disorders in mice by overexpressing specific mutant proteins under the regulation of brain-specific promoters. The development of such models has been significantly bolstered by the dramatic progress in research identifying mutations associated with familial forms of the diseases. Among the first disorders to be successfully modeled were amyotrophic lateral sclerosis and prion diseases. The main postulate of several of these models is that a single molecular alteration might trigger a cascade of events that eventually results in the full spectrum of neurodegenerative and cognitive alterations observed in neural degenerative diseases. Therefore, overexpression of mutant proteins in transgenic mice (mice with experimentally

transferred foreign genes) might mimic neurotoxicity, while targeted deletion of selected genes might mimic the loss of a nutritional or protective function. However, overexpression of a single molecule associated with neural degenerative diseases might not necessarily result in the development of all the aspects of the disease, but in an increased susceptibility to developing the disease process if the appropriate pathogenic factors are present. In this context, the main objectives of developing transgenic animal models of neural degenerative diseases are (1) to define the time course of molecular events associated with the development of neural degenerative diseases, (2) to study the progression of the neurodegenerative process, (3) to determine the individual pathogenic role of specific molecules or specific mutations in vivo, (4) to study novel disease mechanisms, and (5) to test novel therapies.

Amyotrophic lateral sclerosis models. Amyotrophic lateral sclerosis is a progressive neurological disorder characterized by loss of connection and death of motor neurons in the cortex and spinal cord. Familial forms of amyotrophic lateral sclerosis are associated with mutations in the superoxide dismutase gene. Overexpression of the mutant superoxide dismutase gene in transgenic mice results in paralysis and motor deficits strikingly similar to those associated with amyotrophic lateral sclerosis. Similarly, overexpression of the mutant prion protein in transgenic mice results in spongiform encephalopathy and death, mimicking familial forms of Creutzfeldt-Jakob disease, a disorder analogous to bovine spongiform encephalopathy (mad cow disease). Studies of these genetically manipulated mice have led to a better understanding not only of the contribution of the prion protein to the development of spongiform encephalopathy but also of how abnormal conformation of proteins in the brain might lead to neurodegeneration.

Alzheimer's disease models. Dramatic progress has also been made toward modeling Alzheimer's disease (**Table 2**). Overexpression of mutant hAPP (human APP) under the platelet-derived growth factor (PDGFβ) promoter, which specifically targets APP expression to neurons, resulted in extensive amyloid deposition in the hippocampus and neocortex, accompanied by synaptic loss and memory deficits. Since then, several new transgenic models have been developed, including the expression of APP under the control of other neuron-specific promoters, such as protease-resistant prion protein

TABLE 2. Transgenic animal models of Alzheimer's disease

Promoter	Transgene	Mutation (amino-acid substitution)	Phenotype
PDGFβ	hAPP (770/751/695) minigene	V717F	Diffuse and mature amyloid plaques
Human or mouse Thy1	hAPP751	K670M/N671L and V717I	Dense amyloid plaques
Hamster PrP	hAPP695	K670M/N671L	Amyloid deposits
Human Thy1	hAPP695	V717F	Amyloid deposits

(PrP) and thymus cell antigen 1, theta protein (Thy1) [Table 2]. These studies have shown that neuropathologic features characteristic of Alzheimer's disease (namely plaques and synapse damage) begin to develop in 6- to 8-month-old mice that express APP under Thy1, PDGFβ, or PrP promoters if levels of APP are more than five to seven times greater than endogenous levels. After these mice reach 12 months of age, tau-immunoreactive neurites (abnormal neurites with Alzheimer's disease-like characteristics) in plaques and astroglial/microglial activation are observed. However, none of these models have shown the presence of neurofibrillary tangles. Neuronal loss and alterations of synaptic function and connectivity are observed in specific regions of the hippocampus where Alzheimer's disease pathology is often found.

Since mutations and polymorphisms (alterations) in other genes have also been associated with early-onset familial Alzheimer's disease and sporadic Alzheimer's disease, more recent studies have been focused on investigating the interactions among these proteins by crossing APP transgenic mice with other genetically modified mice or with knockout mice (in which the target gene is deleted). Coexpression of APP with other genes associated with Alzheimer's disease modifies the typical APP phenotype. For example, addition of the mutant presenilin 1 gene accelerates the onset of plaque formation, transforming growth factor β enhances vascular amyloidosis, and apolipoprotein E decreases the deposition of amyloid (**Table 3**). These transgenic animal models are currently being used to develop new treatments for Alzheimer's disease such as an anti–Alzheimer's disease vaccine.

Other models. Another mysterious set of disorders that are mostly inherited and are characterized by complex movement alterations and mental deficits are the trinucleotide repeat disorders such as Huntington's disease and spinocerebellar ataxia. Several lines of transgenic mice models have been developed, providing new insights into the role of polyglutamine repeats and nuclear transport in the pathogenesis of neurodegeneration. More recently, two other important neurodegenerative disorders have been modeled in transgenic mice, namely Parkinson's disease and frontotemporal dementia.

Parkinson's disease is associated with accumulation of the synaptic protein α-synuclein, and frontotemporal dementia with the accumulation of cytoskeletal protein tau (Table 1). These models are being used in the development of novel therapies for neural degenerative diseases.

Future research. Genetically manipulated mouse models of neural degenerative diseases have enriched the understanding of the disease process and its mechanisms. The new generations of these transgenic models will attempt to test the hypothesis of the pathogenic role of various molecules when targeted to specific cell groups or expressed under the control of regulatable promoters. Furthermore, the technology is now available to use retroviral vectors to deliver transgenes to specific areas of the brain at predetermined time points. Transgenic rats, flies, and worms are also beginning to be widely used. The types of promoters, genetic background, and environmental factors are being extensively investigated. Emphasis continues to be placed on the role of amyloid production and amyloid toxicity, with less attention paid to alternative mechanisms that induce cell death and neuronal vulnerability. However, more research into the latter is critical, since the clinical alterations seen in neural degenerative disease are associated with cell death and synapse loss. Better understanding of the mechanisms of neurodegeneration and developing models for this process will help in the future development of novel therapeutic targets.

For background information *see* ALZHEIMER'S DISEASE; HUMAN GENETICS; HUNTINGTON'S DISEASE; MUSCULAR SYSTEM DISORDERS; NERVOUS SYSTEM DISORDERS in the McGraw-Hill Encyclopedia of Science & Technology.   Eliezer Masliah

Bibliography. D. Games et al., Alzheimer's type neuropathology in transgenic mice overexpressing V717 beta amyloid precursor protein, *Nature*, 373:523-527, 1995; M. E. Gurney et al., Motor neuron degeneration in mice that express a human Cu,Zn superoxide dismutase mutation, *Science*, 264:1772-1775, 1994; T. Ishihara et al., Age-dependent induction of congophilic neurofibrillary tau inclusions in tau transgenic mice, *Amer. J. Pathol.*, 158:555-562, 2001; L. Mangiarini et al., Exon 1 of the HD gene with an expanded CAG repeat is sufficient to cause a progressive neurological phenotype in transgenic mice, *Cell*, 87:493-506, 1996; E. Masliah et al., Dopaminergic loss and inclusion body formation in alpha-synuclein mice: Implications for neurodegenerative disorders, *Science*, 287:1265-1269, 2000; E. Masliah and E. Rockenstein, Genetically altered transgenic models of Alzheimer's disease, *J. Neural Transm.*, 59:175-183, 2000; S. B. Prusiner, Transgenetics of prion diseases, *Curr. Top. Microbiol. Immunol.*, 206:275-304, 1996; F. Theuring et al., Transgenic animals as models of neurodegenerative diseases in humans, *TIBTECH*, 15:320-325, 1997; F. van Leuven, Single and multiple transgenic mice as models for Alzheimer's disease, *Prog. Neurobiol.*, 61:305-312, 2000.

TABLE 3. Pathogenesis of Alzheimer's disease in crosses of transgenic mice	
Crosses	Phenotype
APP $\times$ PS1 transgene	Accelerated plaque formation
APP $\times$ TGFβ transgene	Amyloid angiopathy
APP $\times$ apoE 3, apoE4 transgene	Decreased amyloid deposition
APP $\times$ APOE KO (knockout)	Decreased amyloid deposition
APP $\times$ ACT transgene	Increased amyloid deposition
APP $\times$ SOD transgene	Decreased mortality
APP $\times$ FGF2 transgene	Increased mortality
APP $\times$ α-synuclein	Increased inclusion formation (Lewy bodies)

Dementia

Dementia is a syndrome characterized by a generalized decline in cognition that is severe enough to cause functional impairment in daily activities. Dementia can be caused by a variety of conditions common in late life, including vascular disease or strokes, Parkinson's disease, Lewy body disease, alcoholism, and Alzheimer's disease (by far the most common cause of dementia).

Alzheimer's disease. Alzheimer's disease was first described by Alois Alzheimer in a 1907 case report of presenile dementia in a 51-year-old woman. Until the 1970s, it was generally believed that the condition was a relatively rare phenomenon that occurred in middle age. Today it is known that Alzheimer's disease and related dementias are very common, especially in advanced industrial countries that have rapidly growing geriatric populations. The rising number of older persons with dementia who are being cared for by family members and in institutional settings has captured the attention of the public and policymakers around the globe. The high cost of long-term care alone is a compelling reason for medicine and society to be interested in the "epidemic" of Alzheimer's disease.

Pathophysiology and diagnosis. Alzheimer's disease affects the structure of the brain. Two main types of neuropathologic changes, or lesions, are seen in the brains of affected persons, particularly areas of the brain that are used for memory and other cognitive functions. The disease is characterized by the formation of amyloid plaques (dense, largely insoluble deposits of protein and cellular material) outside and around the brain's neurons, and neurofibrillary tangles (insoluble twisted fibers) that build up inside neurons. Definitive diagnosis of Alzheimer's disease by autopsy involves counting the number of plaques and tangles found in specific brain structures. Increased brain volume atrophy that occurs secondary to neuronal death is another diagnostic characteristic of the disease. Although the postmortem analysis of brain tissue is currently the "gold standard" for diagnosing Alzheimer's disease, it has two significant limitations. First, Alzheimer's disease plaques and tangles are sometimes found in persons who did not exhibit symptoms of the disease before death. Second, both brain atrophy and the development of plaques and tangles are associated with general aging, making it difficult to discriminate the pattern of neuropathologic lesions that occurs with Alzheimer's disease from the pattern of atrophy and lesions that may occur in normal older adults.

The most common early clinical symptoms of Alzheimer's disease include a loss in recent (short-term) memory that is accompanied by problems with language and judgment. The affected individual may exhibit personality changes and begin to have difficulty with familiar activities, such as operating household appliances, meal preparation, or managing finances. As the disease progresses, disturbances in behavior, such as emotional outbursts, wandering, or agitation, may occur. Eventually, as more areas of the brain become affected, the patient may become bedridden, incontinent, and totally dependent on others for all aspects of daily care.

Clinicians rely on a variety of tools to diagnose "probable" or "possible" Alzheimer's disease in persons experiencing progressive changes in their memory and cognition. Once dementia is suspected, physicians obtain a comprehensive medical history and conduct a thorough physical examination, including laboratory tests to rule out other causes of cognitive decline. Today there is still no accepted single test for diagnosing Alzheimer's disease. In many cases, neuropsychological testing is conducted to measure memory, language skills, and other abilities related to brain functioning. Brain imaging tests such as computed tomography scan and magnetic resonance imaging are often conducted to look for evidence of stroke or other structural brain disease.

Causes. The most important risk factor for Alzheimer's disease is age. However, a growing body of research into the etiology and risk factors for development indicates that Alzheimer's disease is most likely caused by the interaction of a number of genetic and environmental factors. Gene mutations on three different chromosomes (1, 14, and 21) have been identified as the cause of many familial (inherited), early-onset cases. This form of Alzheimer's disease, which generally affects persons between the ages of 30 and 60, is rare, occurring in fewer than 1% of all cases. The condition often runs in families, since inheritance of any one of these mutations means that an individual will definitely develop the associated form of early-onset Alzheimer's disease. Cases of familial, early-onset Alzheimer's disease tend to progress more rapidly than the more common, late-onset form.

There is currently no evidence that the gene mutations associated with early-onset Alzheimer's disease play a major role in the more common, nonfamilial (or sporadic) form of late-onset Alzheimer's disease. However, researchers have found evidence that genetics do contribute to the risk of development of sporadic Alzheimer's disease. The apolipoprotein E (ApoE) gene, which is found on chromosome 19, codes for a protein that helps carry blood cholesterol through the body. It is also found in excess amounts in the plaques that accumulate in the brains of persons with sporadic Alzheimer's disease. There are three common forms (alleles) of the ApoE gene: $\varepsilon 2$, $\varepsilon 3$, and $\varepsilon 4$. Scientists have found that persons who have one or two copies of the ApoE $\varepsilon 4$ allele have an increased risk of developing sporadic Alzheimer's disease. In contrast, the relatively rare ApoE $\varepsilon 2$ allele may be associated with a decreased risk of developing sporadic Alzheimer's disease and with a later onset when the disease does develop. Unlike the inheritance of gene mutations associated with early-onset, familial Alzheimer's disease, the mere inheritance of one or two ApoE $\varepsilon 4$ alleles does not guarantee that an individual will develop late-onset, sporadic Alzheimer's disease, nor does the presence of ApoE $\varepsilon 2$ alleles ensure that sporadic Alzheimer's disease will not occur. In addition to the ApoE gene on chromosome 19, another susceptibility locus for sporadic Alzheimer's disease was recently found on chromosome 10, providing further evidence that genetic susceptibility factors play a role in Alzheimer's disease development. The finding that risk of disease development is to some degree linked with inheritance of various alleles may help explain some of the variations that occur in age of disease onset and on disease prevalence among various racial and ethnic groups. It also provides new clues as to the possible mechanisms that cause Alzheimer's disease and to new avenues for research into its diagnosis, treatment, and prevention.

In addition to research into the genetics of Alzheimer's disease, there is considerable interest in possible nongenetic or environmental factors that may alter disease risk. Risk factors that have been identified in the scientific literature include a history of serious head trauma, depression, and occupational exposure to neurotoxins. Lower levels of education are associated with earlier onset, and advanced levels of education seem to delay onset. Recent studies have suggested that lower levels of brain maturation related to social and economic deprivation during early childhood development—when brain development, particularly the regions responsible for memory and higher-order function, occurs—is associated with increased risk of Alzheimer's disease. There is also growing evidence that the presence of other neurological, cerebrovascular, or inflammatory diseases may increase the clinical severity of Alzheimer's disease or perhaps even its development. In combination, these findings suggest that the onset and progression of Alzheimer's disease is due to a combination of genetic and nongenetic factors that affect total brain reserve and age-related decline (for example, cell loss and interneuronal activity). Thus, factors that reduce brain reserve and speed cognitive decline, such as exposure to neurotoxins or development of multiple small brain strokes, may increase the risk of development of Alzheimer's disease, particularly in genetically predisposed individuals. Conversely, other factors that promote general health and reduce risk for age-related disease such as stroke may delay Alzheimer's disease onset or slow the rate of decline in affected individuals.

Other dementia subtypes. Vascular dementia, related to cerebrovascular disease, is the second most common form of age-related dementia. This type of dementia is not caused by a single condition; it may result from one or more large-vessel infarcts or strokes, or from the presence of widespread small-vessel disease or infarcts, particularly in the frontal and subcortical portions of the brain. Vascular dementia also frequently coexists with other causes of dementia, especially Alzheimer's disease. The type, extent, and severity of cognitive impairment found in vascular dementia cases are associated with the location and size of the strokes and vascular changes in the brain. For example, patients with left-brain lesions are more likely to show problems with verbal naming and fluency tasks, whereas patients with right-brain lesions exhibit more problems with visuospatial skills. Frontal or subcortical lesions may produce changes in personality and judgment. Diagnosis is generally made in cases in which there are external changes detected during a neurologic examination, there is evidence of vascular disease or infarcts on neuroimaging tests, and there is a history of vascular risk factors accompanied by clinical symptoms of progressive dementia. However, diagnostic differentiation of vascular dementia from Alzheimer's disease can be difficult, particularly in cases in which the individual experiences gradual cognitive decline and has a combination of features of vascular disease, Alzheimer's disease, and other causes of dementia.

Frontotemporal dementias (including Pick's disease) are associated with significant atrophy of the frontal and temporal lobes that is usually evident on neuroimaging. The frontotemporal dementias are the second most common degenerative dementias (after Alzheimer's disease) in persons who develop dementia before the age of 65. Subcortical dementia refers to the progressive cognitive changes associated with a variety of conditions such as Parkinson's disease, Lewy body disease (dementia associated with abnormal brain cell structures called Lewy bodies), supranuclear palsy, combined systems disease, and Huntington's disease. These diseases are characterized by changes in subcortical or deep-brain structures, including the basal ganglia, caudate, substantia nigra, and putamen. Diagnosis is made primarily based on the neurologic examination and clinical course. Huntington's disease, which is caused by an abnormal gene on chromosome 4, can also now be confirmed by genetic testing. Compared with Alzheimer's disease patients, individuals with frontotemporal or subcortical dementias tend to have more prominent impairments in executive functioning (for example problem solving or planning), visual perceptions, mental and motor speed, and language fluency. Mood and personality changes, particularly depression, are common. Memory difficulties, although present, may be less severe than for the Alzheimer's disease patient and can often be compensated for with recognition cues and repetition. Patterns of cognitive decline can vary widely among individuals with frontotemporal or subcortical diseases. In older persons who are at risk for developing multiple dementia-related illnesses, differential diagnosis can be particularly difficult. In fact, "mixed" dementia (usually referring to a common form of dementia caused by a combination of Alzheimer's disease and cerebrovascular disease) frequently coexists with Parkinson's disease and Lewy body disease in varying combinations. There is a general need for research to improve diagnostic methods for differentiating vascular, frontotemporal, and subcortical dementias. More accurate diagnoses are particularly important if more effective treatments become available.

Treatment. It is important to provide ongoing care for individuals with dementia. Virtually every dementia patient needs safety and security issues addressed. Treatment generally involves a combination of medication and behavior-management approaches. Medication treatments are geared toward treating symptoms of cognitive decline, slowing disease progression, and delaying or preventing disease onset. The Food and Drug Administration has approved three medications (Tacrine, Donepezil, Rivastigmine) for treating the cognitive symptoms of Alzheimer's disease. These drugs act to prevent the breakdown of acetylcholine (a neurotransmitter) in the central nervous system and have been shown in double-blind, placebo-controlled trials to produce modest improvements in memory and cognition in some patients. Epidemiologic studies of aging and dementia have also led to interest in the use of drugs with antioxidant activity or anti-inflammatory agents in the prevention and treatment of dementia. Recent studies by the National Institutes of Health suggest that the use of estrogen replacement therapy and regular use of some nonsteroidal antiinflammatory drugs may be associated with lower risks of developing Alzheimer's disease. Results from a recent blinded, controlled trial conducted by the Alzheimer's Disease Cooperative Study showed that both selegiline and vitamin E delayed progression of Alzheimer's disease (as measured by the end points of severe functional decline, nursing home placement, or death). A variety of medications are also used to treat some of the behavioral symptoms that are common in Alzheimer's disease patients, such as agitation, depression, and sleep disorders.

Nonmedication treatments for dementia are typically focused on helping the patient and care-giver maintain an optimal quality of life by reducing patient behavior problems and care-giver burden. Behavioral approaches to understanding, preventing, and intervening with psychiatric symptoms and behavior problems have been successfully used in both community and institutional settings. Good medical care to treat or prevent other common age-related illnesses can also help patients with dementia remain functional for as long as possible. For example, patients who maintain an appropriate daily level of physical activity may reduce risk for fall-related injury and have fewer problems with sleep disturbance, depression, and motor restlessness. Other age-related conditions that interfere with patient quality of life and are exacerbated by dementia, such as daytime urinary incontinence, vision and hearing loss, and muscular or arthritic pain, may be improved with proper medical evaluation and treatment.

For background information *see* ACETYLCHOLINE; AGING; ALZHEIMER'S DISEASE; HUMAN GENETICS; HUNTINGTON'S DISEASE; MUTATION; NERVOUS SYSTEM DISORDERS; PARKINSON'S DISEASE in the McGraw-Hill Encyclopedia of Science & Technology.

Susan M. McCurry; Eric B. Larson

Bibliography. J. L. Cummings et al., Dementia, in C. K. Cassel et al. (eds.), *Geriatric Medicine*, 3d ed., pp. 897–916, Springer, New York, 1996; C. Kawas et al., Age-specific incidence rates of Alzheimer's disease: The Baltimore Longitudinal Study of Aging, *Neurology*, 13:2072–2077, 2000; V. M. Moceri et al., Early-life risk factors and the development of Alzheimer's disease, *Neurology*, 54:415–420, 2000; G. W. Small et al., Diagnosis and treatment of Alzheimer disease and related disorders, *JAMA*, 278: 1363–1371, 1997; D. A. Snowdon et al., Brain infarction and the clinical expression of Alzheimer disease: The Nun Study, *JAMA*, 277:813–817, 1997; M. X. Tang et al., The APOE-Epsilon4-allele and the risk of Alzheimer disease among African Americans, White, and Hispanics, *JAMA*, 279:751–755, 1998.

Depression

Depression is not a normal consequence of growing old. It is, instead, a treatable illness, a source of excess disability, and the major cause of suicide in the elderly. Depression occurs in 2–3% of persons aged 60 and above residing in communities, 8–10% of elderly patients attending general medical clinics, and 20% or more of elderly patients in acute medical hospitals or long-term care facilities. The World Health Organization has estimated that depression and suicide constitute the fourth leading contributor to the global burden of illness-related disability and death. It expects that the importance of depression as a source of disability and impaired quality of life will increase in the next 10 years, especially in the developed economies, because of the high proportion of elderly persons in the general population.

Depression in old age usually coexists with other chronic medical disorders, such as heart disease, cerebrovascular disease, arthritis, diabetes, cancer, and neurodegenerative disorders such as Alzheimer's disease and Parkinson's disease. Depression amplifies the disability associated with these illnesses and leads to premature mortality. Furthermore, suicide rates are highest in the elderly, especially among white males aged 85 and above. Many elderly victims of suicides have seen a primary care physician in the month before death, suggesting that their depression went unrecognized or was inadequately treated. The epidemiology of late-life depression and suicide points strongly to a need to improve its recognition and treatment in general medical practice, as the U.S. Surgeon General emphasized in his 1999 report.

Causes. Scientists think of depression in later life as being caused by many different factors, rather than a single factor. Risk factors include one or more occurrences of clinical depression at an earlier age, cardiovascular and cerebrovascular disease, other chronic debilitating illnesses, severe life events such as bereavement, abuse of alcohol, neuroticism, and chronic insomnia. Scientists speculate that imbalance of activity or "tone" in certain neurotransmitter systems is associated with depression. For example, there is evidence that an excess of cholinergic activity over serotonin or norepinephrine activity occurs in depression. It has been difficult to test these hypotheses directly until recently. The availability of modern brain imaging tools, such as positron emission tomography, together with ligands that bind to specific neurotransmitter receptors in the brain, promises new insights into the biology of depression and into the mechanisms of treatment response. Genetic predisposition to depression does not seem to be a major causal factor in late life but is thought to be important in early-onset cases.

Clinical signs and course of illness. Depression in old age is heralded by low mood, lack of pleasure in everyday activities, anxiety, social withdrawal, pessimism, hopelessness, suicidal ideation, difficulties with sleep, decreased appetite and weight loss, preoccupation with bodily aches and pains, and difficulties with memory and concentration. In some cases depression is chronic, while in others it is episodic, typically following a relapsing, recurrent course. While depression in old age can be a severe, life-threatening illness, it can also occur in milder forms. The interval between episodes of clinical depression becomes shorter as people age. Depression often coexists with anxiety in old age.

Diagnosis. The diagnosis of late-life depression is made by examining patient history and mental status; there are no specific laboratory features. An accurate diagnosis often depends upon the availability of information from caregivers or other significant persons in the patient's life. According to the American Psychiatric Association's *Diagnostic and Statistical Manual of Mental Disorders* (4th ed.), for the diagnosis of a major depressive episode, the patient must demonstrate core symptoms, such as low mood and lack of pleasure, on most days for a minimum of 2 weeks, together with accompanying symptoms, such as anxiety, sleep disturbance, diminished appetite and weight loss, inappropriate guilt, death wishes or suicidal ideation, and either agitation or slowing of activity. The presence of these symptoms must be associated with impairment of functioning. The core symptoms of depression in old age do not differ substantially from depression earlier in life. Severe or so-called melancholic or endogenous clinical presentations, in which a patient's mood shows little reactivity to good news and is typically worse in the morning than in the evening, are common. The diagnosis of depression in old age can be overlooked because of the tendency of patients to complain about bodily discomfort or the proclivity of patients and clinicians alike to misattribute symptoms to physical illness.

Many older people are reluctant to acknowledge that they are depressed because they will feel stigmatized. Clinicians may be reluctant to diagnose depression for the same reason—stigmatizing the patient—or because they feel uncertain about what treatment to offer. Clinicians often make the error of assuming that depression is a normal part of growing old and therefore not worthy of treatment. The consequences of failure to diagnose and treat depression in late life are extremely serious. They include increased health-care utilization, more outpatient clinic visits, prescription of a greater number of medications, prolonged hospital stays, failure to engage effectively in rehabilitation programs (for example, following surgical repair of a fractured hip), premature mortality following medical events such as heart attack, increased mortality rates in the nursing home, and suicide.

Treatment. Depression in old age is no less treatable than in earlier life, despite widely held beliefs to the contrary. As emphasized by the 1991 National Institutes of Health Consensus Development Conference on Depression in Late Life, depression in old age is very treatable by both medication and psychotherapy. As first-line treatment, physicians typically offer a selective serotonin reuptake inhibitor (SSRI) antidepressant medication. These newer medications are just as effective as the older tricyclic medications, but they are tolerated much better, safer in overdose, and easier to prescribe because the starting dose is often the appropriate dose, or close to it. Psychotherapy is also effective for depression in old age, especially interpersonal psychotherapy, cognitive behavioral therapy, and problem-solving therapy.

Because late-life depression tends to be a relapsing illness, treatment goals need to encompass not only symptom reduction and restoration of function in the acute phase but also prevention of relapse in the long run. The combination of antidepressant medication and psychotherapy is probably the optimal approach for preventing recurrent episodes of depression. The dosage of antidepressant medication that is needed to help a patient recover should be maintained to keep depression from recurring. If the dose is lowered after the patient feels better, the risk for return of symptoms increases.

Much of the evidence for the effectiveness of depression treatment in old age has been collected in university mental health settings. More studies are needed to show whether these same treatments work and have practical value in general medical settings, where the elderly usually go if they decide to seek help for depression.

For background information *see* AFFECTIVE DISORDERS; MEDICAL IMAGING; NORADRENERGIC SYSTEM; PSYCHOPHARMACOLOGY; PSYCHOTHERAPY; SEROTONIN in the McGraw-Hill Encyclopedia of Science & Technology.   Charles Reynolds

Bibliography. American Psychiatric Association, *Diagnostic and Statistical Manual of Mental Disorders (DSM-IV)*, 4th ed., American Psychiatric Association, Washington, DC, 1994; B. D. Lebowitz et al., Diagnosis and treatment of depression in late-life: Consensus statement update, *JAMA*, 278:1186–1190, 1997; National Institutes of Health, NIH Consensus Conference: Diagnosis and Treatment of Depression in Late Life, *JAMA*, 268:1018–1024, 1992; C. F. Reynolds et al., Nortriptyline and interpersonal psychotherapy as maintenance therapies for recurrent major depression: A randomized controlled trial in patients older than 59 years, *JAMA*, 281:39–45, 1999; U.S. Department of Health and

Human Services, *Mental Health: A Report of the Surgeon General*, 1999.

Design and build

Various cultures have left their imprint on history through their great monuments and buildings, such as the Egyptian pyramids, the Parthenon in Athens, the Roman Pantheon, and Hagia Sophia in the "New Rome," Constantinople. These great structures were by "master builders" who developed the project concept, designed the appearance and technical details of the finished building or monument, and mobilized the resources needed to realize the final structure. This classical approach was used to build the great castles and cathedrals of the Middle Ages, the palaces of the Renaissance, and the civil engineering infrastructure of the industrial revolution. Great architects and engineers followed an integrated process of conception, design, and construction. This integrated construction process persisted through the end of the nineteenth century, when the Roebling family (John, Washington, and Emily) designed and built the Brooklyn Bridge.

During the twentieth century, the processes of designing and building (construction of a project) were gradually separated. Design and construction were viewed as separate endeavors. A design professional prepared the project plans, and then a separate company was contracted for the actual construction of the structure. This separation of activities also led to a chronological sequencing of the business of building. A comprehensive set of plans and specifications were first developed, providing a graphical (schematic) and verbal definition of the project to be constructed. Then, after fully defining the project, contractors were solicited to provide quotations or bids for the construction of the facility. Finally, the construction phase of the project commenced, and was carried through to completion. This led to the development of a design, bid, build (DBB) approach that is still widely used on the majority of projects in the United States today, particularly in the public sector.

Over the last 30 years, a number of new concepts for project delivery have been developed to compress the time required to realize a constructed facility. It has been recognized that the design, bid, build method, with its sequential emphasis, leads to longer-than-necessary project delivery time frames. It is advantageous from a time perspective to have design and construction proceed simultaneously. This has led to a reconsideration of the classical master builder concept and the expedited project realization available by proceeding with design and construction in parallel.

Single-group approach. Within the past 15 years, this return to the classical approach in which design and construction teams simultaneously work together to maximize functional performance and minimize time frame has come to be called design-build (DB).

In the design-build format, design and construction professionals work together from the time that the contract is signed. Therefore, design and construction can proceed at the same time. This is sometimes called phased construction. If a high-rise building is being constructed, excavation of the foundation can be in progress while design of the rooftop restaurant and tower is still being developed. This compresses the overall schedule by allowing design of later portions of the work to be done at the same time that construction of the earlier portions of the project is in progress (see **illus.**).

Design-build also has the advantage that the client works with a single service provider. From the client's point of view it is advantageous to have a single point of contact rather than contracting with an architect, an engineer or multiple engineering firms, and separate construction contractors. Design-build simplifies the purchasing of construction projects and implements the idea of "one-stop shopping." In the 1970s, design-build contracts became popular in the manufacturing and industrial construction sectors. At that time, normally only large construction companies, which were able to support large design divisions within their firm or to supervise design professionals under subcontract, were active in providing clients a single source for concept development, detailed design, and construction, as well as start-up of the facility.

In addition, when a single entity develops the design and does the construction, the design can be optimized in terms of long-term functionality and of the production or realization phase of the project. That is, design for function and production are coordinated. From a manufacturing or production perspective, the firm designs the item so that it can be assembled in the most cost-efficient and timely manner so as to reduce production costs. Reduced construction cost based on improved "constructability" of the facility translates into a lower total cost to the client.

In the past 10 years, design-build contracts have become more common in the building construction sector. A number of firms are marketing this approach to private entrepreneurs in the building sector as a way to receive the best product in the most timely way at the best price. Since most building construction contractors do not have an in-house design capability, lead contractors typically form a team or consortium of designers and specialty contractors who meet the needs of the client as a single group providing the total project package.

Each member of the consortium is at risk and is motivated to work with other members to minimize delays and disputes. In effect, a group of designers and constructors form a consortium to build a project based on conceptual documentation provided by the owner. They agree to work together to

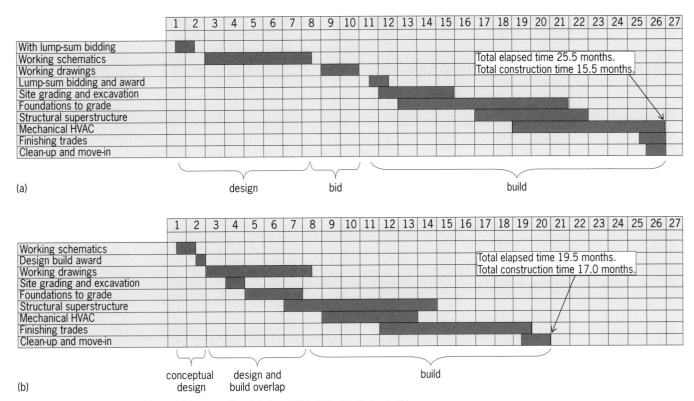

Schedule compression in moving from (a) conventional design, bid, build to (b) design-build.

achieve the project and, therefore, implicitly agree to avoid developing an adversarial relationship with one another.

Project cost. The owner/client is given a stipulated-sum price for the project early in the development process (typically with only conceptual design complete). The design-build team locks in on a specific price at this early time in the life cycle of the project and guarantees that, barring major changes to the scope of the project, the owner can rely on the original price quoted. This fixed purchase price at an early stage in the project cycle allows the client to make financial plans without concerns that a contingency must be maintained to deal with cost escalation or changes at a later time.

With a fixed-price commitment, the members of the design build team must ensure that the detailed design conforms to the price quoted. All members are "in the same boat," so conflicts must be reconciled quickly to avoid delay and additional cost to the team. Bickering among the members may lead to cost escalation, which impacts overall profits to the consortium. Therefore, the potential for an adversarial relationship to develop among the members of the consortium is eliminated.

There is also an incentive to be innovative in finding cost-efficient solutions to constructing the project in the most timely way. Value engineering becomes a way of increasing the profits to all members of the design-build team. Value engineering aims to seek solutions to design problems that maintain the function of a facility while reducing the cost of the labor and materials needed to construct the facility. If improved methods can be found to implement the design at a lower cost during the construction, this is money in the pocket of the design-build team. This incentive stimulates innovative thinking by both design and construction professionals since they are working together to gain performance at a lower production cost.

The fixed sequential order of design preceding construction, as in the design, bid, build approach, defeats the idea of cooperation between the design professionals and the construction contractor. There is no benefit to either party in collaborating unless the client rewards such collaboration. In the design-build format, collaboration is mandated and automatically rewarded.

Results and benefits. Estimates vary, but trade organizations such as the Design-Build Institute of America estimate that as much as 15% of the construction in the United States today is done using the design-build contract format. Many state highway departments have adopted design-build as a format for the construction of major transportation systems. A number of agencies within the federal government are using design-build contracting as a framework for getting projects constructed more quickly and in a more cost-effective way. A large number of federal buildings, such as Internal Revenue Service processing centers, have been successfully constructed using design-build. Legislation is pending in a number of states throughout the United States to allow use of design-build contracting for all capital facilities. Momentum appears to be building for wide acceptance of design-build as a method of contracting in the public sector.

Private entrepreneurs have been using the design-build approach for many years, preferring to work with proven design-build consortia to build office and apartment complexes. In many cases, low cost is not the motivator. Since "time is money" in the private sector, using design-build to bring a sophisticated facility (such as a computer chip manufacturing installation) on line more quickly is critical and much more important than lowest cost. If a computer manufacturer can start producing to meet market demand 6 months earlier than its competition, this alone may translate to $50–100 million profit for the client company. Therefore, the potential for time compression and early availability of a manufacturing facility overwhelms all other cost considerations.

Outlook. Major trade and professional organizations such as the Associated General Contractors (AGC) and the American Institute of Architects (AIA) have recognized the trend toward more use of the design-build delivery method by developing and issuing standard guidelines and contract forms for design-build projects.

Design and build, according to Jack Rizzo, President, Corporate Development, Perini Corporation, "has become a popular alternative to the more traditional method of awarding separate contracts for design and construction. Its many benefits, including single-source accountability, fast-track scheduling, early guarantee of project costs and schedule, ongoing value engineering and construction review, to name a few, indicate that this alternative delivery system is bound to become even more mainstream in the future."

Design-build is a method that invokes valuable lessons of the past (and the master builders of old) and provides a time- and cost-effective way of meeting demanding schedules for new construction. It has been estimated that 80% of the buildings and infrastructure needed worldwide to meet societal needs for the year 2020 have not yet been constructed. This is a tremendous backlog of work and a great challenge to the architecture, engineering, and construction community. It is clear that the design-build project delivery approach will play a major role in meeting this challenge.

For background information *see* ARCHITECTURAL ENGINEERING; BUILDINGS; CONSTRUCTION ENGINEERING; ENGINEERING DRAWING; STRUCTURAL DESIGN in the McGraw-Hill Encyclopedia of Science & Technology.   Daniel W. Halpin

Bibliography. R. Dorsey, *Project Delivery Systems for Building Construction*, Associated General Contractors, Washington, DC, 1997; D. Halpin and R. Wood, *Construction Management*, 2d ed., John Wiley, New York, 1997.

Differential GPS

Differential global positioning systems (DGPS) improve the positioning accuracy of GPS users who are equipped to receive correction messages broadcast over radio links from a GPS reference receiver or a network of receivers at a fixed, known location or locations. The method exploits the fact that the leading error sources for civil GPS receivers will be highly correlated at the reference receiver and at nearby user receivers. These errors include the propagation delay of radio signals through the ionosphere and the troposphere as well as errors that are characteristic of the satellite broadcast signals. The latter include satellite timing (clock) errors and errors in the broadcast satellite ephemerides (which provide precise satellite positions at any given time). A reference station estimates the combined effect of these errors as the difference between the measured value of the line-of-sight range to each satellite and the value of that range predicted from the known locations of a satellite and the reference station. Public-use DGPS transmit the current value of these ranging errors for all satellites that are in view. Some systems separate out the individual error contributions attributable to satellite clock offsets, ephemeris errors, and propagation errors, whereas others transmit only the composite line-of-sight errors. These two approaches are generally characteristic of wide- and local-area DGPS respectively. An intermediate approach whereby each station transmits local, scalar corrections but a weighted average of corrections from multiple transmitters is used in the user receiver, is called regional DGPS. This concept is used in the Northwest European LORAN (NELS) Eurofix system, where DGPS corrections are transmitted by pulse-position modulation of the LORAN signal.

For most of the life of GPS the civil accuracy (2DRMS or 95% confidence level) was maintained at approximately 100 m (330 ft) horizontal and 150 m (500 ft) vertical through the use of selective availability (SA), which mechanized the intentional degradation of accuracy of the signal available to civilian users. This was accomplished by dithering the satellite clocks and truncating the broadcast ephemeris data bits. DGPS improved this accuracy to approximately 1 m (3.3 ft) over distances up to 50 km (31 mi). In his Presidential Decision Directive in 1996, President Clinton committed to terminating SA by 2006. On May 1, 2000, in a White House press release, the termination of SA was announced and a few hours later it was turned off. After this termination some experiments have shown that absolute positioning accuracy to 7 m (23 ft) 2DRMS (horizontal) could be obtained by civil GPS users, thereby putting the need for DGPS in question. A new specification for the accuracy of the civil GPS signal is in development, but it is not expected that it will guarantee performance that meets requirements for aviation or maritime operations. The U.S. Department of Transportation position is that DGPS will continue to be needed to meet both accuracy and integrity requirements for both maritime and aviation users.

Maritime DGPS. The simplest DGPS concept, and first to become operational, was maritime DGPS. Existing radiobeacons in the 285–325-kHz band were converted to transmit differential corrections via a minimum-shift-keyed (MSK) frequency modulation scheme. A base station receiver sited at a fixed, known location measures pseudorange errors relative to its own clock and known position. Messages containing the time of observation, these pseudorange errors, and their rate of change are then transmitted at either 100 or 200 baud. Parity bits are added to these messages to check for errors, but forward error correction is not used. If the parity checks fail, the receiver ignores the current message and waits for the next message. The user receiver extrapolates these corrections from their time of validity to the present based on the age of the correction and its rate of change.

With the termination of selective availability, the excess radio propagation delay through the ionosphere and the troposphere is the largest error source for single-frequency GPS users. However, these delays should be highly correlated over the limited distance that marine DGPS signals will propagate from a reference station to users. Their effect should be considerably reduced by DGPS if both the reference receiver and the user receiver process raw delay measurements and disable internal algorithms that are built in to predict the propagation delays for use

when DGPS is not available. In the long term, civil GPS may become a dual-frequency system. In that event each user will be able to measure line-of-sight propagation delay from the received signals. At that time it will be necessary again to reconsider the need for DGPS.

Because the line-of-sight corrections are measured relative to the base-station clock, they are relative and not absolute corrections. This means a user receiver must have corrections for all the satellites that it uses to determine its position. It also means that the user receiver will compute corrections to its clock time such that it will track offsets of the base station clock. Therefore maritime DGPS provides no improvement in timing accuracy for fixed-position users for whom the system functions as a time standard. For moving users, their solution for time will also track the base station clock, and since this base station receiver is in a fixed known location, the time solution in the moving receiver will see improved accuracy as well.

The U.S. Coast Guard operates maritime DGPS along the coasts and rivers. In addition, the system is being expanded inland to support land users. The locations of existing and proposed United States stations are shown in **Figs. 1** and **2**. The Coast Guard is utilizing a number of former U.S. Air Force Ground Wave Emergency Network (GWEN) sites. In some cases, the DGPS transmitter is being located at the former GWEN site and will use the existing tower and some of the transmitting equipment. Other new sites are being established with a mixture of newly

Fig. 1. Existing and planned DGPS sites in the contiguous United States (CONUS) as of September 2000. (*U.S. Coast Guard Navigation Center*)

Key:

	Operating	Planned
New site		△
Coast Guard site	●	○

Fig. 2. Existing and planned DGPS sites in Alaska and Hawaii as of September 2000. (*U.S. Coast Guard Navigation Center*)

procured equipment and equipment taken from disestablished GWEN sites. In addition to the United States stations shown in Figs. 1 and 2, many foreign governments worldwide operate compatible systems primarily for maritime applications.

Aviation wide-area DGPS. Three satellite-based augmentation systems (SBASs) for aviation users are in various stages of development: the Wide-Area Augmentation System (WAAS), developed by the U.S. Federal Aviation Administration (FAA); the European Geostationary Navigation Overlay System (EGNOS), developed jointly by the European Union, the European Space Agency (ESA), and EUROCONTROL; and the MTSAT Satellite Based Augmentation System (MSAS), developed by the Japan Civil Aviation Bureau (JCAB). These systems are intended for the enroute, terminal, nonprecision approach, and category I (or near-category I) precision approach phases of flight. All of these systems are designed to be compatible.

The SBAS corrections are transmitted from geostationary satellites at the frequencies designated L1 in GPS (centered at 1575.42 MHz), with signal characteristics similar to GPS. The SBASs provide additional ranging signals and transmit their own ephemeris information to improve fix availability.

The SBASs corrections need to be useful throughout the very large geographic area covered by the satellite footprint. Earth coverage by geostationary satellites spans thousands of kilometers. At these large distances the errors due to satellite ephemeris and ionospheric delay vary with user location and cannot be combined into one overall pseudorange correction. Separate messages must be provided for satellite clock corrections, vector corrections to satellite position and velocity, and integrity information. These messages are valid worldwide. However, the appropriate propagation corrections are different at different user locations. The SBAS accommodate this variation by broadcasting estimates of the propagation delay of the signal from a satellite at zenith (vertical elevation) at selected grid points. These grid points are spaced at 5° increments in latitude and longitude except that larger increments are used at extreme northern or southern latitudes. The user receiver estimates the ionospheric delay for each satellite by first determining the point at which the propagation path from that satellite to the receiver enters the ionosphere (called the ionospheric pierce point), and then interpolating the zenith delay corrections that are broadcast for the grid points that surround the pierce point; then the user corrects for the slant angle to the satellite to convert from zenith delay to line-of-sight delay.

In the WAAS, message symbols at 500 symbols per second are modulo-2 added to a 1023-bit pseudorandom noise code. The baseline data rate is 250 bits per second. These data are rate-1/2 convolutional-encoded with a forward error correction code, resulting in 500 symbols per second. The data are sent in 250-bit blocks, or one block per second. The data block contains an 8-bit preamble, a 6-bit message type, a 212-bit message, and 24 bits of cyclical redundancy check (CRC) parity.

The fact that the WAAS signals can be used by conventional GPS hardware with only minor software changes, as opposed to a separate medium-frequency antenna and receiver as for the maritime DGPS, probably means that high-volume, low-cost applications, such as recreational boating, automobiles, hikers, and so forth, will migrate to WAAS receivers if DGPS is used at all. The major advantage of maritime DGPS for higher-end or safety-critical nonaviation users is that the low-frequency signal will penetrate into real and urban canyons where the signal from the WAAS geostationary satellite may be obscured even though other GPS satellites are visible. This will be particularly true at higher latitudes.

Since each WAAS reference station has three cesium time standards, the transmitted corrections are relative to a precise WAAS master clock. Unlike maritime DGPS, timing receivers in fixed known locations using WAAS should see improved performance.

Aviation local-area DGPS. For category II and III precision approaches, the FAA will implement the Local Area Augmentation System (LAAS). The transmitted data will include pseudorange correction data, integrity parameters, approach data, and the ground-station performance category. The signal will also be used for other terminal-area applications such as instrument departure, airport surface situational awareness, and as the position sensor for automatic dependent surveillance (ADS). The broadcast will be in the 108–117.95-MHz band presently used for very high frequency omnidirectional range (VOR) systems and instrument landing systems (ILS). The high data rate of the system, 31,500 bits per second, is necessary mainly to meet the stringent 2-second time-to-alarm requirement. The most difficult aspect of implementing this requirement is to ensure that the probability is less than one in 10^{-9} that the height of the aircraft, as indicated by flight instrumentation, is in error by more than 5.4 m (18 ft). This requires that the error in estimating aircraft altitude must be less than 1 m (3.3 ft) at least 95% of the time. Provision will be made for local pseudolites to provide additional ranging signals which may be necessary to meet accuracy and availability requirements at certain major airports. These pseudolites will be pulsed at a low duty cycle so as not to interfere with a user receiver's ability to track conventional GPS signals.

For background information *see* AIR NAVIGATION; AIR TRAFFIC CONTROL; ATOMIC CLOCK; ELECTRONIC NAVIGATION SYSTEMS; INFORMATION THEORY; MODULATION; SATELLITE NAVIGATION SYSTEMS in the McGraw-Hill Encyclopedia of Science & Technology.

Benjamin B. Peterson

Bibliography. R. Braff, Description of the FAA's Local Area Augmentation System (LASS), *Navigation*, 44(4):411–424, Winter 1997; *1999 Federal Radionavigation Plan*, U.S. Departments of Defense and Transportation, NTIS Rep. DOT-VNTSC-RSPA-98-1/DoD-4650.5, 1999, available in electronic form (Adobe Acrobat) from USCG NAVCEN; T. Walter and M. Bakery El-Arini (eds.), *Selected Papers on Satellite Based Augmentation Systems (SBASs)*, vol. VI in the GPS Series, Institute of Navigation, Alexandria, VA, 1999.

Direct-current transmission

Since the beginning of the twentieth century, alternating current (ac) of frequency 50 or 60 Hz has been the dominant method for transmission and distribution of electric power. However, high-voltage direct-current (HVDC) transmission can provide economic solutions for very long distances and enables interconnections between unsynchronized ac networks to be realized. Worldwide, HVDC schemes totaling 59 GW with individual scheme ratings between 50 and 6300 MW have been installed, and many more applications for this technology are being considered.

A new technology called voltage-sourced converter (VSC) transmission has been introduced and offers the same advantages as conventional dc transmission, plus a number of new advantageous features. These include installations that occupy a smaller surface area, and the ability to feed dc power into a passive ac network. At present, the total of capital costs and capitalized losses for this solution is higher than for conventional dc transmission, and the maximum rating per converter is presently less than 100 MW.

Development work is in progress to achieve higher power capability for the VSC transmission solution and to reduce losses. The next few years are likely to reveal developments in this technology, enabling it to compete with classical HVDC transmission.

Power transmission. In an ac network the power flowing in each transmission line is determined by the impedance of the line, as well as the voltage magnitude and phase angle at the two ends. The system operator can influence these quantities by controlling generation, system transformer tap changers, and shunt or series reactive power elements.

In an HVDC system, power electronics (a converter) is used to convert ac power to dc power at one terminal (the rectifier) and vice versa at the other terminal (the inverter) [**Fig. 1**]. Normally, each converter is able to work either as a rectifier or as an inverter, so power can be transmitted in either direction. Since dc power is independent of the ac supply frequency and phase, two ac networks connected by a dc link need not be synchronized and can be operated independently of each other.

HVDC equipment. Virtually all converters for HVDC use thyristors. When forward-biased, a thyristor can be turned on by applying a pulse to its gate. Current then continues to flow until the external circuit causes the current to become zero. This is called a line-commutated converter.

Direct voltages up to 500 kV are often used. Converters with such high voltages require that many thyristors with their grading-damping networks and other support components be connected directly in series. The resulting assembly is called a thyristor

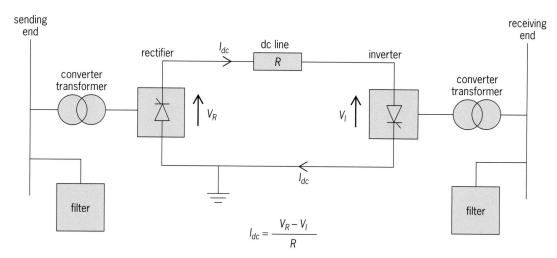

$$I_{dc} = \frac{V_R - V_I}{R}$$

Fig. 1. Basic high-voltage direct-current (HVDC) system. (*ALSTOM*)

valve. The converter transformers in Fig. 1 provide galvanic isolation and suitable adjustment of the voltage magnitude by means of their tap changers.

The converters naturally absorb reactive power. Also, as the current waveform on the ac side is not a pure sinusoid, the converter is a source of harmonics, which could pollute the ac networks. Therefore, harmonic filters, which are capacitive at the fundamental frequency, are installed on the ac-line side of the converter transformer. They are energized and deenergized to match the reactive power demand of the converter. The converters are controlled to regulate the direct current during each power-frequency cycle to give the desired power transfer.

Figure 2 shows a 300-MW HVDC station in Korea, with gas-insulated metal-clad switchgear for filter and station switching on the left, ac filters in the foreground, and the converter transformer with its valve winding bushings protruding into the building, which houses the converter and the control and auxiliary plant.

Applications. HVDC lines or cables are cheaper than equivalently rated ac versions. However, HVDC stations are more expensive than ac substations. The break-even distance is the distance beyond which it is more economic to transport bulk power as dc instead of ac. This distance depends on local factors, but typically ranges 800–1000 km (500–600 mi) for overhead lines and 50–80 km (30–50 mi) for cables.

When two ac systems operate at different frequencies, the only practical way to achieve power flow between them is by an HVDC connection.

Because the dc power flow is fully controllable, an HVDC link does not suffer from power swings giving rise to overload tripping, which may afflict a corresponding ac connection in the event of a fault in one ac system. In fact, an HVDC link can enhance the transient stability of the ac systems.

The ability directly and instantaneously to control the power flow on an HVDC link provides a mechanism for damping power oscillations between rotating machines. HVDC transmission embedded within an ac network may allow additional power to flow through parallel ac interconnections without jeopardizing the steady-state stability of the ac system.

In the event of a fault in the receiving ac network, the current delivered by HVDC transmission is limited very swiftly to no more than the rated current of the HVDC link. When the short-circuit capacity of a network approaches the capability of the existing ac switchgear, importing power with dc may avoid upgrading such switchgear.

In many countries, the demand for power is rising, but there are increasing environmental objections against constructing new overhead lines. Converting an overhead line from ac to dc transmission permits more power to be delivered through a given corridor without affecting its visual impact, offering a possible solution.

Fig. 2. A 300-MW converter station in Korea. (*ALSTOM*)

Advances in line-commutated dc transmission. Since their introduction to HVDC transmission in the 1970s, the rating of thyristors has increased steadily, reducing the number of series-connected devices needed in converters. Control technology has improved, making it possible to connect HVDC systems to ac networks with lower short-circuit capacity. Costs have fallen and scheme capability has increased, notably by enabling the converters to exercise some control of reactive power, minimizing the necessary subdivision of harmonic filters. HVDC using thyristors is now a mature technology, is very reliable, and requires relatively little maintenance. Several converter stations now operate without human supervision.

However, the line-commutated technology has limitations. For example, to feed power to a passive network (one that has no generation), it is necessary to provide a commutating emf from a synchronous compensator.

VSC technology. This limitation can be overcome by using voltage-sourced converters (VSCs), which are equipped with devices capable of turning on as well as off in response to a gate signal. This technology is called VSC transmission. The turn-off capability means that the converter is self-commutating; that is, it can be turned off before the current naturally reaches zero. Additionally, the phase-angle difference between the ac current and the ac voltage can be dictated by the control system; that is, the converter can operate at any desired power factor.

In the 1980s, such devices began to find applications in motor drives, and efforts to develop high-power devices were increased. Today the most significant device is the insulated-gate bipolar transistor (IGBT).

The modern IGBT requires relatively little gate power. It is also capable of being switched on and off many times each power frequency cycle. This capability can be used to control the harmonic output from the converter, by a technique called pulse-width modulation (PWM). Using this technique, the converter output switches between two voltage levels, with the duration at each level being chosen such that the average converter terminal voltage approximates a sinusoidal wave shape. The harmonics are shifted to high orders and filtered easily at the ac terminal.

Advantages and drawbacks. The ability of the converter to control reactive power and to reduce selected harmonics means that the filters at the ac terminals can be small. Accordingly, the surface area occupied by the installation is considerably smaller than that of a line-commutated installation.

The major disadvantage of VSC transmission constructed using present-day IGBTs is that the conduction and switching losses are much larger than those of a thyristor of comparable rating. In addition, several present-day IGBTs are required to equal the rating of one large thyristor. Notwithstanding these limitations of present VSC transmission technology, its introduction has created a new niche market, where the benefits of its technical characteristics outweigh any additional capital cost and the cost of its power losses.

Initially, this new technology is being applied on schemes having lower ratings than the traditional HVDC schemes. By the end of 2000, five schemes had been put into service, using converters with a maximum rating of 60 MW, one scheme having three such converters in parallel.

Prospects. For VSC transmission to compete with established and mature technology in the traditional

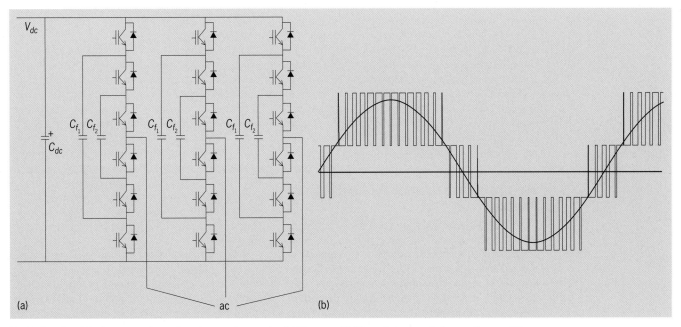

Fig. 3. Four-level thyristor capacitor voltage-sourced converter. (*a*) Diagram. (*b*) Alternating-current phase voltage and its fundamental component.

HVDC market, converters for much higher voltage must be developed and ways found to reduce the power losses. The technical challenge of building an efficient and reliable series-connected string of IGBTs increases dramatically with voltage. While advances in IGBT technology may lead to some reduction, other methods to reduce power losses are also being examined.

Multilevel topology. One possible solution to both of these limitations may come from using a new topology for the converters. By using a multilevel topology, the switching voltage for each valve can be reduced (**Fig. 3**). The minimally switched waveform also resembles a sinusoidal wave shape more closely, minimizing the extent of pulse-width modulation switching necessary to achieve an acceptable waveform. Since the switching losses of the IGBT account for at least 50% of the converter losses in a 2-level converter, a reduction in switching frequency is obviously most beneficial. A 3-phase, 4-level converter (giving 7 levels phase-to-phase) is shown in Fig. 3.

One drawback of the multilevel topology shown in Fig. 3 is the considerable space occupied by the dc capacitors C_{f1} and C_{f2}. This limits the practical number of levels in such a converter.

For background information *see* CONVERTER; DIRECT-CURRENT TRANSMISSION; ELECTRONIC POWER SUPPLY; PULSE MODULATION; SEMICONDUCTOR RECTIFIER in the McGraw-Hill Encyclopedia of Science & Technology.   Bjarne R. Andersen; K. T. G. Wong

Bibliography. B. R. Andersen et al., Korean mainland to Cheju island HVDC link, HVDC transmission, *Mod. Power Sys.*, pp. 29–36, January 1994; N. Aouda et al., A multilevel rectifier with unity power factor and sinusoidal input current for high voltage applications, *EPE J.*, 6(3–4):27–35, December 1996; J. Arrillaga, *High Voltage Direct Current Transmission*, 2d ed., Institution of Electrical Engineers, London, 1998; R. L. Cresap et al., Operational experience with modulation of the Pacific HVDC Intertie, *IEEE PES Summer Meeting*, Mexico City, July 17–22, 1977; A. Lindberg and T. Larsson, PWM and control of a three level voltage source converters in an HVDC back-to-back station, *AC and DC Power Transmission Conference*, April 29–May 3, 1996, Publ. no. 423, pp. 297–302, Institution of Electrical Engineers, London, 1996; Y. Shakweh, Power devices for medium voltage PWM converters, *IEE Power Eng. J.*, pp. 297–307, December 1999.

Disease

In recent years, the growing problem of antibiotic resistance has enabled diseases once on the verge of eradication, such as tuberculosis, to reemerge as major threats to human health. The resurgence of old diseases, as well as the appearance of new ones, has prompted scientists to examine closely the factors that enable pathogens (disease-causing organisms) to invade and persist in human populations. Together, the problems of emerging infectious diseases and an-

tibiotic resistance have fueled the development of the field known as evolutionary epidemiology. Fundamental to this new discipline is the recognition that pathogens are themselves biological entities that evolve in response to the challenges of their environment.

Evolution of virulence. Pathogens can quickly evolve to take advantage of ecological or environmental changes that promote their transmission. An important feature of pathogens is that their life cycle is so short relative to that of their hosts. Most viruses and bacteria can replicate many times a day, enabling significant evolutionary change to accumulate over relatively short periods of time. Virulence, defined here as increased mortality due to infection, also evolves, causing previously mild diseases to become life-threatening ailments, or vice versa. In the midnineteenth century, for example, scarlet fever was one of the leading causes of childhood mortality. Today, it is rarely diagnosed—in part because it can be controlled with antibiotics, but also because its causative organism, *Streptococcus pyogenes*, has grown inexplicably milder over the last century.

Clearly, virulence is a trait that varies from pathogen to pathogen. However, virulence can also vary greatly from strain to strain. For example, the gram-negative bacterium *Escherichia coli* has been the cause of death in some food-borne illnesses, and yet harmless strains of the very same bacterium inhabit nearly every human gut. Likewise, different types of the human immunodeficiency virus (HIV) have different disease characteristics. HIV-1 is predominant in the United States, Europe, and central and eastern Africa, whereas HIV-2 is mostly restricted to west Africa. Compared to HIV-1, HIV-2 is a slower-progressing disease, associated with lower viral loads, less efficient transmission, and a longer latent period. Given such broad differences among closely related diseases, one goal of evolutionary epidemiologists is to explain the factors contributing to the evolution of virulence.

Originally, it was believed that all pathogens would evolve to be less virulent over time. The assumption was that natural selection would favor pathogens that cause the least amount of harm to their host, because in driving their host to extinction, they drive themselves to extinction as well. It was believed that exceptionally virulent diseases were caused by novel pathogens that had not had sufficient time to adapt to human populations. This theory, however, has not held up to theoretical or empirical analyses. In fact, virulence may evolve to greater or lesser levels, depending on the pathogen's mode of transmission and the degree to which virulence impedes or enhances transmission.

Modes of transmission. Pathogens are transmitted by two modes. Horizontal transmission occurs within a host generation, among unrelated individuals. Vertical transmission occurs from parent to offspring, either during reproduction (so that the offspring is born with the disease) or afterward through parental care or nursing. Virulence evolves differently, depending on whether the pathogen relies on vertical or horizontal transmission.

With vertically transmitted pathogens, the reproduction of the pathogen and that of its host are interdependent: Unless the host lives to reproduce (as do its offspring), the pathogen will go extinct. In this case, natural selection is believed to favor virulence, because low/high virulence impedes vertical transmission. With horizontally transmitted pathogens, however, the degree to which the pathogen is dependent on its host is less straightforward. As with vertical transmission, natural selection favors pathogens that are readily transmitted—but high virulence does not necessarily impede horizontal transmission. In fact the opposite may be true: Virulence may be a side effect of processes designed to ensure transmission. For instance, a high serum concentration of virus may be associated with illness, but it ensures that the transfer of even small amounts of bodily fluid (as from a mosquito bite) will be sufficient to establish a new infection. Likewise, the symptoms associated with respiratory illnesses (such as runny nose and coughing) are directly responsible for the transmission of that pathogen to others. Some pathogens are even thought to alter the behavior of their host so as to make transmission more likely. For instance, infection with rabies causes animals to exhibit aggressive behaviors (such as biting) that result in transmission. In these cases, the symptoms associated with virulence enhance transmission and, thus, are likely to be favored by natural selection.

Despite evolution toward increased virulence, there is an upper limit at which further increases in virulence will no longer enhance transmission. For instance, a pathogen will go extinct if it is so virulent that its hosts tend to die before they can infect other people. This reasoning is often used to explain why HIV can cause a global pandemic but the Ebola virus causes self-limited, local epidemics. The asymptomatic nature of HIV means that transmission can occur for 10 years or more. The Ebola virus kills its host in a matter of days, allowing little time for transmission to occur. Thus, the trade-off between increased transmission and increased host mortality predicts that horizontally transmitted pathogens will evolve intermediate virulence.

Unfortunately, the relationship between transmission mode and virulence is considerably more complicated. It may depend on many other factors, such as the population density of uninfected hosts. When population density is high, a pathogen can sustain itself at a higher level of virulence, since the probability that transmission will occur before host death is also higher. Presumably the pathogen itself can also affect host density; therefore, the effect of natural selection on virulence might fluctuate over time. Because of this relationship, increasing human population sizes, urban crowding, and long-distance travel (effectively the same as higher population densities) are of great concern to epidemiologists regarding the spread of diseases and the evolution of virulence.

Origin of new diseases. Pathogens evolve not only to be better adapted to their current hosts but also to infect new hosts. The movement of wildlife diseases to humans in recent years has resulted in an alarming number of new diseases, including hantavirus, Lyme disease, and Ebola virus disease. This broadening of a pathogen's host range to include other species can be attributed to both evolutionary and ecological causes. Evolutionary factors involve an actual change in the pathogen's ability to cause an infection. In a virus, this could be as simple as a subtle change in the viral receptor that enables it to bind to the cells of its new host. Other challenges for the pathogen relate to its ability to be transmitted before its host either dies or recovers from the infection. Thus for every disease such as Ebola, with its extremely rapid host mortality, there may have been many other cross-species transmissions that were unsuccessful because they were easily cleared by the immune system. Ecological factors may be as important as evolutionary factors in determining how often expansion of the host range to include humans occurs. Large-scale ecological changes, particularly land-use changes such as deforestation, have been determined to play an important role in the emergence of these new diseases.

Evolution of antibiotic resistance. Nowhere has evolution been more apparent than in the emergence of resistant strains of bacteria. The evolution of drug resistance continually outpaces the development of new antimicrobial drugs, and for the first time in over half a century there are strains of bacteria that are virtually untreatable. This escalating cycle between the creation of new drugs and the evolution of resistance to counter these drugs is often referred to as an "arms race," in which the only way to survive is to stay one step ahead.

The speed with which resistance evolves reflects the fact that these bacteria are being subjected to unusually strong selective pressures—when faced with an antibiotic-rich environment, only those naturally occurring variants with resistance can persist. And in the absence of all of their susceptible competitors that have been killed off by the antibiotic, they can quickly predominate. The problem of resistance is worsened by the fact that antibiotic resistance genes can easily be exchanged between unrelated bacteria.

Evolutionary epidemiologists would like to find ways to slow the evolution of resistance. Often this includes considering the effectiveness of cycling or combining drugs. In drug cycling, an antibiotic is used for only a short period of time before it is replaced with another drug, giving the bacteria as little time as possible to evolve resistance: just as they stumble on resistance, the environment changes and they must try to adapt anew. Another solution has been the use of "drug cocktails," in which multiple drugs are given at once so that a pathogen must simultaneously evolve resistance to all of the drugs in order to survive. This strategy has been used to control HIV; however, its success has been mixed. Although it may extend the life of any given drug, eventually the virus does seem to be able to make those improbable leaps and evolve resistance to multiple drugs.

Recommendations for reducing virulence evolution. Ultimately, the goal of evolutionary epidemiology is to provide an understanding of the evolution of diseases that can guide the development of innovative methods for controlling them. Paul Ewald, in particular, has advocated low-technology behavioral changes that may encourage the evolution of reduced virulence. Thus, while not eradicating the disease, these actions may help to ease the burden of infected persons. Simple measures such as screening houses in high malarial areas and rigorous enforcement of hand washing in hospital settings can reduce transmission and may encourage the evolution of reduced virulence. Likewise, ensuring that patients finish their complete course of antibiotics so as to limit the perpetuation of resistant pathogens, and restricting the unnecessary use of antibiotics may also slow the evolution of resistance in bacteria so that medicine will be available when it is really needed.

For background information *see* ANTIBIOTIC; BACTERIAL GENETICS; DISEASE; DRUG RESISTANCE; INFECTIOUS DISEASE; MEDICAL BACTERIOLOGY; VIRULENCE; VIRUS in the McGraw-Hill Encyclopedia of Science & Technology.   Elizabeth Ostrowski

Bibliography. R. M. Anderson and R. M. May, *Infectious Diseases of Humans: Dynamics and Control*, Oxford University Press, New York, 1991; P. W. Ewald, *Evolution of Infectious Disease*, Oxford University Press, New York, 1994; L. Garrett, *The Coming Plague: Newly Emerging Diseases in a World out of Balance*, Farrar, Strauss, and Giroux, New York, 1994; R. M. Nesse and G. C. Williams, *Why We Get Sick: The New Science of Darwinian Medicine*, Times Books, New York, 1994; C. Willis, *Yellow Fever, Black Goddess: The Coevolution of People and Plagues*, Addison-Wesley, Reading, MA, 1996.

Dry machining

A cutting fluid, usually an oil or emulsion of a lubricating phase (oil, graphite, and so on) in water, is traditionally used in machining operations to reduce heat generated by friction, to lubricate, and to flush away chips (residual cut metal). The costs of maintenance, record keeping, and compliance with current and proposed regulations are rapidly boosting the expense of using cutting fluids. In 1980, cutting fluids and their disposal accounted for less than 3% of the cost of most machining processes. Today, cutting fluids account for as much as 16% of the cost of a machined part because of increased costs and stringent environmental regulations. In contrast, tooling accounts for only about 4% of that same part.

The pressures of cost reduction and environmentally conscious manufacturing have forced many manufacturers to consider changing from wet machining to dry machining, in which the cutting fluid is eliminated. Additional reasons for eliminating cutting fluids are to reduce waste dumped in landfills, mist in the factory atmosphere, dirty shop floors, and dermatological problems for machine operators.

The ease of the transition from wet to dry machining usually depends on the work material. Eliminating coolant from steel and cast iron machining operations typically can be accomplished with little problem. Stainless steel machining can be done dry, but special attention must be given to surface finish. Dry machining of superalloys (cobalt-, nickel-, chromium-, molybdenum-, or tungsten-based alloys that are able to withstand high stress at high temperatures) may not be possible due to the extremely high amount of heat generated during metal removal operations.

The type of operation also has some influence on the ease of transition from wet to dry machining. It is usually possible, even preferable, to turn and mill without the use of coolant. Most of today's tools, such as coated carbide grades, ceramics, cermets, polycrystalline cubic boron nitride, and polycrystalline diamond, are brittle. They chip, fracture, and crack, especially in facing and milling operations where rapid repetitive thermal fluctuations lead to expansion and contraction of the cutting tool edge. Keeping the tool hot is beneficial, because a hot tool is often a tougher, more reliable tool. In this case, the presence of coolant may actually exacerbate the tool cracking problem. (**Figs. 1** and **2**).

Tapping, reaming, and especially drilling present greater challenges. Cutting fluids are often necessary while drilling because they provide lubrication at the drill tip and flush chips from the hole. Without fluids, chips can bind in the hole, and the inner surface roughness can average twice as high as when machining wet. Cutting fluids also can reduce the required machine torque by lubricating the point at which the drill touches the hole's wall.

There are some techniques that can make dry machining easier. For example, some advantage is realized by using sharper drill edges. Studies of drilling steel by Guhring, Inc., have shown that the cutting-edge temperature of a drill can be decreased from $200°C$ to $120°C$ ($395°F$ to $250°F$) when the edge hone is reduced from 0.25 mm to 0.02 mm (0.01 in. to 0.001 in.). A sharp edge minimizes runout (concentricity error) and improves surface finish as well.

Another effective technique is to apply coatings. In particular, physical vapor deposition (PVD) coatings are beneficial in decreasing cutting temperature and increasing tool life.

Challenges. Most of the problems associated with dry machining are related to heat. Deformation occurs earlier, degrading tool life. Thermal expansion of the workpiece makes it difficult to maintain tight dimensional tolerances; and increased cutting temperatures produce softer, more ductile chips, particularly in ferrous materials. As a result, chip control becomes more difficult. Additionally, the absence of lubrication can deteriorate the surface finish. Thus, dry machining requires learning new techniques and adjusting to new sets of problems.

Chip control and evacuation. Except for drilling operations, this problem may be the easiest one to solve. Air blast equipment can blow chips into chip conveyors. Inverting a turning tool holder allows gravity to pull the chips down and away from the tool and workpiece.

Breaking chips into manageable shapes and sizes can also be accomplished with modern and versatile chip grooves built into the surface of tools. Chip grooves are depressions formed in the top of a tool for the purpose of bending and breaking the chip that is sheared off the workpiece. Hot chips are more difficult to break because of their increased ductility, which can produce dangerous chip tangles and a poor surface finish. Changing to a groove designed to control stringy materials will usually solve this problem. These grooves usually have a more positive rake (the angle at which chips enter the groove) and are more fragile. However, when the tool temperature is high, toughness increases and chippage is not likely to be a problem.

If a change in the chip groove is not adequate to control chips, adjusting the feed rate should follow. Increasing the feed rate usually provides the greatest benefit, but the feed rate should be both increased and decreased to see which gives the better result.

Tool degradation. When coolant is removed from a cutting operation, the temperature of that operation will increase. Consequently, tool deformation and cratering (chemical dissolution of the tool edge) become more prevalent. This can be effectively counteracted by using harder, more stable coated carbides or advanced tool materials. Hard materials resist deformation; and coatings, especially those containing aluminum oxide or titanium aluminum nitride, are chemically stable and function effectively at very high temperatures. Coated tools are usually more brittle. However, since increasing operating temperatures results in increased tool toughness, moving to a slightly harder tool concurrent with a change from wet to dry machining almost never reduces tool life or dimensional consistency. In fact, usually the opposite occurs, especially in turning and milling operations.

In the rare instance that a move to a harder, more stable tool grade is not possible, a reduction in the cutting conditions can be effective in reducing heat and pressure. As shown in the **table**, lowering the speed is the most effective change. Typically the speed is decreased by some amount (about 15%) and the feed rate is increased by the same amount. No loss in productivity results, but the overall temperature of the operation goes down and deformation is minimized. Increasing the feed rate can degrade

Fig. 1. Dry machining significantly reduces the cost of metal removal.

Fig. 2. Due to the increased temperatures associated with dry machining, special techniques must be used to ensure tight dimensional control.

surface finish, but increasing the tool's nose radius (that is, the radius of the portion of the tool engaged in the cut) will correct for this.

Free cutting (positive rake) edges can also reduce the temperature of the cut because the metal chip is deformed less. When machining dry, a change to a freer cutting (and more brittle) edge can be made

Effect of operating conditions on tool life		
Operating condition	Change	Effect on tool life
Speed	Increase by 50%	−90%
Feed	Increase by 50%	−50%
Depth of cut	Increase by 50%	−5%

for the same reasons described above; that is, a hot tool is a tougher tool.

Lubricous coatings can also reduce temperature by decreasing friction. Coatings such as molybdenum disulfide and tungsten carbide/carbon have low coefficients of friction. Unfortunately these coatings are usually soft and have relatively poor tool life. To compensate, the coatings are often used in conjunction with hard underlayers such as titanium carbide, titanium aluminum nitride, or aluminum oxide.

Part tolerance. Another problem associated with dry machining is the drift in part size caused by increasing temperature. In the absence of coolant, thermal expansion often may cause the cooled parts or cuts to be undersized. Confounding the problem is the fact that the temperature does not always increase at the same rate day to day and is not maintained consistently throughout a workday.

There are three types of corrective actions to minimize dimensional tolerance problems. First, decrease the contact time between the tool and the component by increasing the speed or feed. Second, switch to a freer cutting edge. If the chip can be removed with a minimum of chip deflection or deformation, less heat is generated. Today's chip grooves are freer-cutting and generate less heat than those of a generation ago. Third, make as light a depth of cut as possible for the finish cut. The less metal bent during chip formation, the lower the temperature of the cut.

It may also be possible to study the way in which the metal removal system heats up and to compensate accordingly. At the beginning of a workday or shift, the machine, tool, and tool holder as well as the parts will be cool. As metal cutting continues, friction will heat the system. If the expansion of the various components can be accurately defined, it will be possible to adjust the depth of cut (especially in milling and drilling operations) to bring the parts to within dimensional specification.

Lubrication. Heavy lubricants such as oil work best in difficult operations (drilling, grooving, parting-off) and on difficult materials such as stainless steels and high-temperature alloys. Cutting speeds and feeds are usually low and do not generate high cutting temperatures. The oil is used to lubricate (not cool) and resist metal buildup. Under these conditions, it may be possible to remove the lubricant. Increasing the sharpness of the cutting edge, increasing rake angle, and adjusting the lead angles may result in decreased friction, eliminating the need for an oil. Under conditions where lubrication is needed, minimum quantity lubrication, which consists of drops or droplets of oil suspended in compressed air, may be all that is needed.

Implementation. Machinists and manufacturing engineers who switch to dry machining can reap tremendous economic and environmental benefits but must solve significant problems first. In general, tool life issues are easily handled. Problems with deformation at higher speeds can be dealt with by a simple tool material change. At lower machining speeds, tool life will be more consistent and can increase due to increased toughness.

Controlling part size is a more complicated issue. Thermal expansion can result in undersized parts, but can be compensated for by making the appropriate adjustments in cutting conditions, tool geometry, or depth of cut.

If chip handling is a problem, changing the chip groove will often suffice; making changes in the cutting conditions, especially feed rate, also may help. And minimum-quantity lubrication in difficult operations such as drilling and grooving may provide as much benefit as using copious amounts of oil.

For background information *see* CAST IRON; HIGH-TEMPERATURE MATERIALS; LUBRICANT; MACHINABILITY OF METALS; MACHINING; METAL, MECHANICAL PROPERTIES OF; STEEL in the McGraw-Hill Encyclopedia of Science & Technology.   Don Graham

Bibliography. *ASM Metals Handbook*, vol. 16: *Machining*, p. 83, 1989; D. Graham, Dry out, *Cutting Tool Eng.*, pp. 56–65, March 2000; P. Zelinski, The fast track to high speed drilling, *Modern Machine Shop*.

Earthquake engineering

Earthquake engineering is the branch of engineering concerned with reducing earthquake or seismic risk. Because strong earthquakes are rare events, building codes have traditionally allowed a significant degree of damage. Even high-seismic regions, such as San Francisco or Tokyo, typically experience a strong earthquake only once in many decades. If an earthquake occurs, buildings and other structures are designed such that most will be damaged (but should not collapse), and will have costs for repairs, business interruption, and potentially casualties. While building collapse is the primary cause of loss of life in most earthquakes, other contributors to earthquake loss include equipment and contents damage, business interruption, and damage to lifelines, such as water, power, gas, communications, and transportation. To limit these losses to acceptable levels, earthquake engineering involves a process of (1) seismic hazard identification, (2) structural analysis, design, and/or retrofitting to prevent structural collapse and reduce property damage, and (3) review of equipment and operations to prevent disruption due to earthquakes—that is, an integrated, comprehensive program of facility seismic review, analysis, retrofit, emergency planning, and risk transfer, involving the expertise of mechanical engineers, operations specialists, emergency planners, and insurers, in addition to geoscientists and structural engineers.

Hazards. Most earthquakes are caused by the fracture and sliding of portions of the Earth's crust along faults, which may be hundreds of miles long, from 1 mi (1.6 km) to over 100 mi (160 km) deep, and sometimes not readily apparent on the ground surface. Earthquakes can occur anywhere on Earth, but most occur along major tectonic plate boundaries ("Ring of Fire"), especially on the circum-Pacific plate boundary, the Caribbean, and the Trans-Alpide belt (stretching from southern France through the Mediterranean and the Middle East, along the Himalayan foothills and the Indonesian archipelago). In the United States, 39 of the 50 states are considered at moderate to high seismic risk, with major earthquake potential in the western states, Alaska, the central United States (St. Louis–Memphis region), and portions of the east coast (South Carolina, Massachusetts).

Earthquakes can cause significant damage to the built environment due to fault rupture, vibratory ground motion (shaking), inundation (tsunami, seiche, dam failure), various kinds of permanent ground failure (liquefaction, landslide), and fire or hazardous materials release. In a particular event, any of these hazards can dominate, and historically each has caused major damage and great loss of life. The expected damage given a specified value of a hazard parameter is called vulnerability, and the product of the hazard and the vulnerability is the expected loss or seismic risk.

For most earthquakes, shaking is the dominant and most widespread cause of damage. Shaking near the actual earthquake rupture lasts only during the time when the fault ruptures, a process which takes seconds or at most a few minutes. The seismic waves generated by the rupture propagate long after the movement on the fault has stopped, spanning the globe in about 20 min. Typically earthquake ground motions are powerful enough to cause damage only in the near field (that is, within a few tens of kilo-

meters from the causative fault). In a few instances, long period motions have caused significant damage at great distances to selected structures. A prime example was the 1985 Mexico City earthquake, where a magnitude 8.1 earthquake occurring at a distance of approximately 400 km from Mexico City caused numerous collapses of mid- and high-rise buildings.

The most important scales for measuring earthquake magnitude are surface-wave magnitude M_s, body-wave magnitude m_b, and moment magnitude M_W (the last is the more commonly preferred scale today). Magnitude can be related to the total energy release by $\log_{10} E = 11.8 + 1.5M_S$, where E is the total energy in ergs. Since $10^{1.5} = 31.6$, an increase of one magnitude unit is equivalent to 31.6 times more total energy release, an increase of two magnitude units $\sim$1000 times more total energy, and so forth. Whereas magnitude is a measure of the overall size of a single earthquake, intensity is a measure of the effect, or the strength, of an earthquake hazard at a specific location. Intensity scales in use include the Modified Mercalli Intensity (MMI) in the United States, Medvedev-Sponheur-Karnik (MSK-81) in Europe, and the Japan Meteorological Agency (JMA) scale. Roman numerals are traditionally used for intensity scales to indicate the qualitative nature of the scales, which are based on subjective observations rather than instrumental records. For the MMI and MSK scales, 0 is no earthquake, VI is the initiation of damage to poor-to-average structures, and XII is total destruction. Earthquake shaking at a site is recorded on a seismometer, which digitally or optically records a time history of ground accelerations at the site. Statistical analysis of hundreds of such

Fig. 1. Relatively new reinforced concrete buildings that collapsed in the Marmara, Turkey, earthquake, August 17, 1999.

accelerograms provide attenuation regressions, which permit estimation of the maximum acceleration, velocity, or displacement that a structure will experience at a site as a function of the structure's natural period, the site-specific soil properties, and the magnitude, distance, and depth of the hypothesized earthquake. In the United States, ground motions due to all potential earthquakes have been synthesized by the U.S. Geological Service, so that any site's seismic hazard (based on an average soil type) can be quickly determined. Based on this hazard, or on building code requirements mapped at a regional scale, the seismic lateral force requirements against which a structure must be designed are determined.

Structural analysis and design. In the developed world, existing buildings and infrastructure constitute by far the preponderance of earthquake risk, so that a major focus of earthquake engineering is on identification, analysis, and mitigation (reduction) of this risk. New construction is generally safer, but is still a focus for earthquake engineers, especially larger or unusual structures. In the developing world, even recent and new construction is a significant contributor to seismic risk, due to lack of building code enforcement (**Fig. 1**). The most seismically hazardous existing building structures are low-strength masonry (such as adobe in developing countries), unreinforced brick masonry (such as exists in older portions of United States cities), nonductile reinforced concrete (typically, concrete frames constructed prior to the 1980s), and certain kinds of precast concrete buildings. Steel and wood structures are not immune to earthquake damage, but typically are less collapse-prone. However, the tallest high-rise building to collapse, for any reason, was the 23-story Piño Suarez steel-framed building in the 1985 Mexico City earthquake. Building vulnerability is typically mitigated via structural retrofits involving strengthening wall-diaphragm connections, adding shear walls or bracing, and improving ductility of columns by steel jacketing or other methods. A diaphragm is a horizontal force-distributing element, such as a roof or floor.

Earthquake analysis and design of a structure is commonly performed assuming the structure remains elastic (that is, the design neglects the effects of damage), though inelastic analyses (which takes into account material inelasticity) are increasingly used. For ordinary design, the actual earthquake forces are reduced by a response modification factor that varies depending on the assumed inherent ductility of the structure, based on its lateral force-resisting system (moment frame, braced, shear wall) and material (wood, steel, concrete). This force reduction, along with appropriate safety factors, is combined with gravity and other loads in the design of the overall structure. Pseudo-static or linear dynamic analytical methods are used to determine structural member forces. Pseudo-static methods are based on the structure's natural period (first mode of vibration) and are appropriate only for simple structures. Linear dynamic methods account for the first and higher modes, which analyses are usually more quickly performed in the frequency domain, using techniques based on the fast Fourier transform. For larger and more important structures, nonlinear dynamic analyses are used, typically in the time domain. In a nonlinear dynamic analysis, the structure is subjected to earthquake acceleration time histories (actual or synthesized records, scaled to match the site hazard), and member response into the inelastic range is taken into account, including P-Δ effects (that is, the increase in overturning moment due to the structure's weight P times its lateral deflection Δ). Dynamic analysis has become common due to the advent of more powerful computers and specialized software—ETABS, SAP2000, ANSYS, STAAD, and LARSA are some of the structural analysis packages commonly used today.

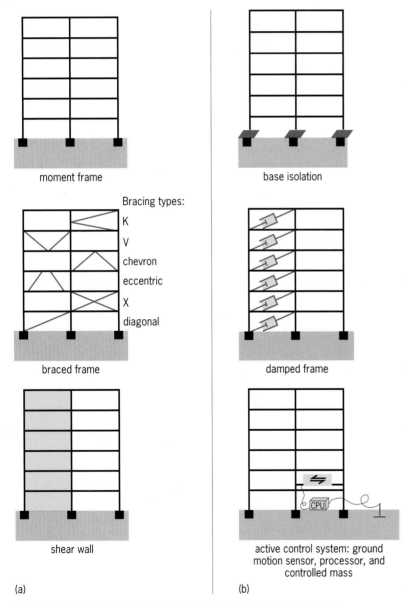

Fig. 2. Structural designs: (*a*) Traditional earthquake force-resisting systems.
(*b*) Emerging technologies for earthquake force-reducing systems.

New design approaches. Traditionally, engineers have designed structures to resist earthquake forces, intending the strength of structural members to be equal to or greater than the seismic demand placed on them (**Fig. 2**). During the 1980s and 1990s, new approaches to seismic design emerged, which involved modifying the structural response to reduce earthquake loads to more tolerable levels. These included base isolation, supplemental damping, and active control. Base isolation involves placing special components, termed isolators (not always at the base, so that the technique is more properly termed structural isolation), within the structure that are relatively flexible in the lateral direction, yet can sustain the vertical load. When the earthquake causes ground motions beneath the structure, the isolators allow the structure to respond more slowly than it would without them, resulting in lower seismic demand on the structure. Isolators may be laminated steel with high-quality rubber pads, sometimes incorporating lead or other energy-absorbing materials, or parabolic dish-shaped base plates, which rely on the structure's own weight trying to "climb" the sloping sides of the "dish" to counteract the lateral force of the earthquake. Supplemental damping involves placing dampers within the structure, which retard the structural response of the normal or lightly damped structure, again resulting in lower seismic demand on the structure. Active control involves placing hydraulic rams or other actuating devices within the structure, which introduce forces counter to those caused by the earthquake, thus negating some or all of the seismic demand. To determine how much force to apply to an actuator involves a real-time solution of the dynamics of the structure. Sensors must measure the ground motion at the base of the structure, a structural analysis for these accelerations must be performed on a dedicated processor, and the optimum pattern of forces must be determined and applied via quick-response actuators during a time interval equal to or less than the dynamic response of the structure. Base isolation and supplemental damping are currently entering the mainstream of structural engineering, while active control is still largely the subject of research.

The most recent development in earthquake engineering is performance-based design, in which the expected structural damage due to the maximum expected earthquake is quantified. If the expected damage is unacceptably high, the owner and the engineer may agree on a design in excess of building code requirements.

For background information *see* ARCHITECTURAL ENGINEERING; BUILDINGS; CIVIL ENGINEERING; EARTHQUAKE; LOADS, DYNAMIC; SEISMIC RISK; SEISMOLOGY; STRUCTURAL ANALYSIS in the McGraw-Hill Encyclopedia of Science & Technology.

Charles Scawthorn

Bibliography. B. A. Bolt, *Earthquakes*, W. H. Freeman, San Francisco, 1993; W. F. Chen (ed.), *Handbook of Structural Engineering*, CRC Press, 1997; A. K. Chopra, *Dynamics of Structures*, Prentice Hall, Englewood Cliffs, NJ, 1995; S. L. Kramer, *Geotechnical Earthquake Engineering*, Prentice Hall, Englewood Cliffs, NJ, 1995; C. H. Scholz, *The Mechanics of Earthquakes and Faulting*, Cambridge University Press, New York, 1990; P. I. Yanev, *Peace of Mind in Earthquake Country*, Chronicle Books, San Francisco, 1991.

Earth's surface

The complex array of processes that govern the evolution of the Earth's surface affects nearly all aspects of today's society. Topical examples include weather-related issues such as forecasting, severe storms, floods, and extreme events; global change issues such as climate change, loss of biodiversity, and ozone depletion; land-use issues for both managed and unmanaged ecosystems; and other natural hazards. Addressing such problems requires comprehensive programs of systematic observation, advanced data analysis, and numerical modeling, often using state-of-the-art computational facilities.

Measurements of the land and ocean surfaces have historically been made using direct sampling (in-situ) mechanisms. Through such observations, scientists have developed an understanding of the many processes that control the Earth's land surface, and have illuminated regional processes such as tectonics, hydrology, and land-use change. Similarly, for many decades measurements of the ocean have been made from ships, drifting buoys, or other platforms using a variety of instruments to provide useful information about ocean physics, chemistry, and biology.

For the past 40 years, Earth-orbiting satellites have provided the capability to make remote observations that complement in-situ measurements. Many passive instruments (such as radiometers and spectrometers) have been devised and flown to gather observations of the Earth's atmosphere and surface at various wavelengths of the electromagnetic spectrum (visible, infrared, microwave). More recently, active instruments, those that both transmit and receive electromagnetic signals (for example, radar), have been employed to great effect.

A judicious combination of satellite and in-situ measurements currently provides scientists with the capability to observe the atmosphere, the ocean surface, and the land. These observations are used to develop scientific theories and to provide input into numerical and analytical models that are subsequently used to explain and predict Earth phenomena. The capability to understand and to make useful predictions about the Earth depends on obtaining accurate and complete information from a variety of observing systems, as well as using this information in appropriate models.

International projects and programs. Collaborative international projects and programs collect and use relevant observations (principally related to meteorology, oceanography, and climatology) to meet a variety of scientific or societal purposes. International projects and programs have arisen through governmental agencies seeking to meet a societal need such as weather forecasting, and in the international scientific community to understand large-scale processes. Other projects have arisen as a result of international conventions and protocols.

In meteorology, for example, the need for international cooperation in data collection and dissemination of meteorological information led to the establishment of a World Meteorological Organization (WMO) in 1950. The WMO has assisted the national weather services and the broader atmospheric community in gaining access to an internationally coordinated system for making and sharing observations as well as forecasts and warnings. The WMO established the World Weather Watch (WWW), which currently facilitates and coordinates surface and space-based observations from member states. The Global Observing System of the WWW includes several thousand surface observation sites, including oceanographic ships and buoys, several hundred sites for daily balloon releases to measure upper-air variables, and a fleet of geostationary and polar-orbiting satellites for clouds, temperatures, and other derived variables (see **illus.**). In addition, the WMO sponsors the Global Atmosphere Watch, which includes about 20 major sites to collect comprehensive information on the global distribution of important atmospheric constituents and a large number of sites to collect data on variables of regional or local importance.

In the scientific community, early examples of such international collaboration include the International Geophysical Year and the Global Atmospheric Research Program. These pioneering efforts indicate the willingness of scientists around the world to cooperate in obtaining observations at the appropriate space and time scales to address pressing global science questions. This legacy continues in the current group of international research programs such as the World Climate Research Program (WCRP) and the International Geosphere-Biosphere Program (IGBP). It must be noted that, while collaborative, these international programs are conducted by the scientists in their various countries, using national resources and capabilities.

The Montreal Protocol provides an example of international observational cooperation to meet a global problem. The Protocol requires nations to conduct systematic observations that monitor key processes related to atmospheric ozone and its destruction. Similarly, the United Nations Framework Convention on Climate Change (UN/FCCC) is developing a plan of action to undertake observations related to long-term climate change. Other conventions (such as biodiversity and deforestation) are in various stages of development, but will likely become established observational requirements, and may lead to a program of systematic measurements.

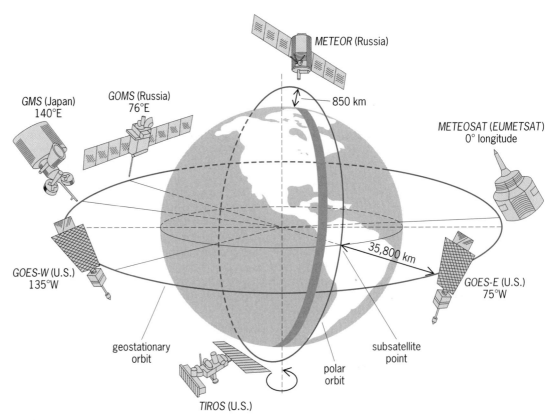

World Weather Watch meteorological satellite network. (*World Meteorological Organization*)

Operational and research observing systems. While observing systems serve a number of diverse needs, it is useful to distinguish them according to their purposes or to the type of organization responsible for conducting the observations. Operational observing systems, for example, are designed to meet a specific purpose such as weather forecasting or flood forecasting. Although often focused on local areas, these observing systems are usually global in scope, and must function continuously and reliably. The instruments in use are often well proven and well understood. The data are distributed widely and systematically. Such systems are usually operated by government agencies with mandates to provide specific services to citizens.

Research observing systems are designed to meet a specific scientific purpose, such as understanding El Niño or the carbon cycle. These systems are often deployed regionally for limited times. They often consist of experimental instruments deployed to discover or understand phenomena not yet well characterized. The nature of the scientific issue governs the type of instrument, its complexity, and the duration of its deployment.

Obviously, these two systems do not function independently. The operational systems are continually being improved through scientific investigation and research, and the scientific community depends on the consistent observations obtained through operational systems. The two are synergistic.

Current observational systems. Two major international research programs are undertaking extensive observing activities as part of their implementation plans.

The World Climate Research Program is jointly supported by the World Meteorological Organization, the International Council for Science (ICSU), and the Intergovernmental Oceanographic Commission (IOC) of UNESCO. The WCRP is developing the fundamental scientific understanding of the physical climate system and climate processes needed to predict climate variations on scales from seasons to centuries, and to assess the extent of human influence on climate. The program encompasses five programs to study the global atmosphere, oceans, sea ice and land ice, and the land surface, which together constitute the Earth's physical climate system. Its programs have developed a number of global data sets to address climate problems.

The International Geosphere-Biosphere Program, sponsored by the International Council for Science, describes and studies the interactive physical and biological processes that regulate the Earth system, the unique environment that it provides for life, the changes that are occurring, and how they are influenced by human actions. It has established an integrated multidisciplinary research program of core projects: past global changes, global atmospheric chemistry, biospheric aspects of the hydrological cycle, global change and terrestrial ecosystems, land-use and land-cover change, land-ocean interactions in the coastal zone, joint global ocean flux study, and global ocean ecosystem dynamics. These projects are providing a growing base of observational information on the Earth's surface and related processes.

Global observing system. Recently, international organizations established observing systems to address needs for climate, the oceans, and the land surface. These are the Global Climate Observing System (GCOS), the Global Ocean Observing System (GOOS), and the Global Terrestrial Observing System (GTOS). These systems are intended to provide long-term sustained observations that will serve a variety of needs. They are being implemented through the expansion of existing observing elements, while developing new components where needed.

With other sponsors [such as the IOC, United Nations Environment Program (UNEP), and ICSU], the WMO established a Global Climate Observing System shortly after the Second World Climate Conference (1990). GCOS was intended to address the comprehensive observations needed for climate variability, climate change, and climate impacts. GCOS is assisting the UN/FCCC to meet its needs for climate information, and is also supplying information to national organizations providing climate predictions. GCOS consists of a number of existing international operational programs that are themselves composed of national contributions (such as the Global Upper-Air Network and Global Surface Network of WWW, and GAW). It also includes other specific observing elements (such as the World Glacier Monitoring Network and International Permafrost Association).

About a decade ago, ocean scientists proposed a Global Ocean Observing System. The GOOS is being developed by the IOC with participation from WMO, UNEP, and ICSU. While GCOS is focused on the issue of climate change and includes elements of the atmosphere, ocean, and land sufaces, GOOS focuses on the comprehensive oceanographic observational needs. These include not only climate but also coastal processes, marine living resources, and ocean pollution. GOOS comprises a number of existing programs (such as the Global Sea Level Observing System, and Data Buoy Coordination Program).

Some years later, these same international organizations and the Food and Agriculture Organization (FAO) established a Global Terrestrial Observing System (GTOS) to obtain terrestrial observations to address water availability, loss of biodiversity, toxic pollutants in the environment, and climate change. GTOS has also incorporated existing observations, particularly related to ecosystems and hydrology, into its program.

Integrated global observing strategy. The development of these global observing systems poses challenges for the nations of the world, since the systems call for sustained in-situ and space-based observations on global scales. The research programs and the global observing systems, along with their sponsoring organizations, have joined with the Committee on Earth Observing Satellites (CEOS) in a partnership to develop a comprehensive strategy to meet these integrated observing needs. To date, the Integrated Global Observing Strategy (IGOS) partnership has identified several themes for observational focus. These include the ocean, terrestrial and global carbon, water resources, and atmospheric constituents. The theme approach permits the observing programs and the space agencies to plan future deployments in close consultation.

Data and information management. Data management and information management are important elements in all research, operational, and observing programs. In global programs, the data volume is obviously enormous, so resources have to be made available to collect, calibrate, assimilate, disseminate, and archive the resulting information. Many United States agencies have established data distribution and archival sites for national and international information.

In the earth sciences, georeferencing is particularly important. Efforts have been under way for some time to develop geospatial information methods, such as by the National Spatial Data Infrastructure in the United States. In the future, diverse data sets may be used more effectively to address critical scientific and societal issues.

For background information *see* APPLICATIONS SATELLITES; CLIMATE PREDICTION; CLIMATOLOGY; EARTH; HYDROLOGY; METEOROLOGICAL INSTRUMENTS; METEOROLOGICAL SATELLITES; METEOROLOGY; OCEANOGRAPHY; REMOTE SENSING; SATELLITE METEOROLOGY; WEATHER FORECASTING AND PREDICTION; WEATHER OBSERVATIONS in the McGraw-Hill Encyclopedia of Science & Technology.

Thomas W. Spence

Ecosystem development

Landscapes change dramatically with climate on short time scales (for example, during the Dust Bowl of the 1930s) and long time scales (for example, swamps existed on Antarctica 50 million years ago). But we know relatively little about how landscapes change during climatic transitions and, particularly, what processes control ecosystem dynamics during these transitions. Recent geochemical studies have begun to shed light on this process. These studies have focused on how soil development and landscape evolution control the availability of essential plant nutrients. They also reveal how landscapes and ecosystem nutrient supplies respond to climate change on rapid time scales.

Among the most dramatic climate shifts the Earth has experienced are ice ages, which have arrived like clockwork (about every 100,000 years recently) to drastically modify landscapes with the buildup of ice equatorward, covering large areas of North America, Europe, and Asia. Ice ages are caused by irregularities in the Earth's orbit around the Sun; the glaciers and ice sheets become the dominant mechanisms of erosion at high and mid latitudes, modifying the landscape by direct and massive physical weathering. Even where no ice sheets exist, climatic change is seen during glacial intervals, even at the Equator. Along with landscape changes come substantial shifts in plant habitats, with entire ecosystems creeping north and south in response to ice and climate.

Phosphorus as a plant nutrient. Plants need many nutrients to live and grow, as do animals. Plant photosystems, however, require one particular element in high quantities—phosphorus. Phosphorus is the primary component of adenosine triphosphate (ATP), the molecule responsible for trapping and shuttling the energy gained by the photosynthetic capture of sunlight, and ultimately contributing to the fixation of carbon and the modulation of the global climate. The availability of this nutrient is central to life on Earth, but nature made a puzzling choice in its dependence on phosphorus. Phosphorus is only a minor element in rocks (averaging about 0.1% by weight), and it is relatively immobile in the environment. As a result, plants do not have much to start with, and what exists is hard to release from its source. However, the phosphate ion (PO_4^{3-}) has ideal bonding characteristics to serve as the critical energy shuttle, and has likely been the cornerstone of photosynthesis for billions of years. How the availability of phosphorus changes when climate changes, and how this might influence the development and stability of ecosystems on land, is an important question.

Lake records of nutrient dynamics. To explicitly examine ecosystem and soil responses during climatic transitions, a new approach based on examining the geochemistry of phosphorus in lake sediments was used to determine the nutrient status of the soil surrounding watersheds, taking advantage of the geochemical transformations of phosphorus during soil development. The assumption implicit to this approach is that lakes reflect the surface soil characteristics of their watersheds through the process of erosion. This assumption seems reasonable, given that records of erosion found in lake sediments match closely with stream sediment influx records. Determining the timing of changes in nutrient status and ecosystem structure in this way and comparing them to known climatic changes provides insight into temporal responses to climate transitions.

All ecosystems lose phosphorus during soil development and over time, but there are several geochemical transformations that act to reduce this net loss. The initial source of nearly all phosphorus in ecosystems is held in rocks, particularly in the form of the mineral apatite (termed mineral phosphorus). This mineral phosphorus is soluble in acidic soil. In addition, plant roots and microbes secrete an enzyme specific for phosphorus (phosphatase) that speeds the dissolution of mineral phosphorus in the soil. This pool of mineral phosphorus is considered very bioavailable, as is an additional geochemical pool, called nonoccluded phosphorus, associated with easily dissolved or removed surface coatings. If incorporated into plants, this nutrient is transformed into an organic phosphorus form which is released only slowly when the plant matter decays in soils and, short term, is thus less bioavailable. Some released phosphorus precipitates along with soil oxide minerals; this is called the occluded phosphorus pool, and is also less bioavailable. Although organic phosphorus and occluded phosphorus are available to the ecosystem, their release and availability is on the decade time scale. Hence, geologists would consider them bioavailable to the ecosystem as a whole, whereas plant biologists would consider them unavailable to an individual plant during the growing season.

As soil matures, a process that takes time, the geochemistry of phosphorus in soils changes in some predictable ways. For example, the proportion of mineral phosphorus decreases, while that of organic and occluded phosphorus increases. Soil maturity is a relative term. Maturation occurs more rapidly in warm and wet environments than in cold and dry ones. On stable landscapes with mature soils, nearly all of the available phosphorus is bound up in organic and occluded forms, a situation that can limit the amount of plant production inherent to that ecosystem. In contrast, landscapes with a high amount of erosion constantly lose soil material, and the remaining soil and ecosystem are stuck in an immature state.

Lake sediments, by reflecting soil material from surrounding landscapes, record the history and speed of soil and ecosystem development. Contrasting environments reflect this history well. In the western Appalachian Plateau, glacial-type conditions existed at the Last Glacial Maximum about 20,000 years ago. Although no glaciers were present, the landscape was marked by arid and cold conditions, leading to ecosystems dominated by sparse pine forest cover and thin, poorly developed soils (**Fig. 1**). These conditions are reflected in the lake sediment records of phosphorus chemistry, which reveal a landscape with a significant proportion of mineral phosphorus in the soils and smaller proportions of organic and occluded phosphorus. As temperatures warmed and precipitation increased, soils started to deepen and mature, and deciduous forest cover began to replace the pine forests, leading to the established and stable southern hardwood forests seen today. This transition is displayed clearly in the phosphorus geochemical records from this region, which reveal a transition to more organic and occluded phosphorus and relatively stable soil nutrient status from 8000 years ago to today.

The steep slopes of the Coast Mountains of western British Columbia were completely glaciated until about 12,500 years ago, when the ice finally retreated to the point where the glaciers remained only on the higher peaks of these mountains. Although this warming is reflected in a number of biological records, the main result was the establishment of pine forest ecosystems in this region, an ecosystem relatively unchanged to the present (**Fig. 2**). Continual erosion from these steep slopes is seen in lake sediment records of phosphorus geochemistry, which reveal a dominance of mineral

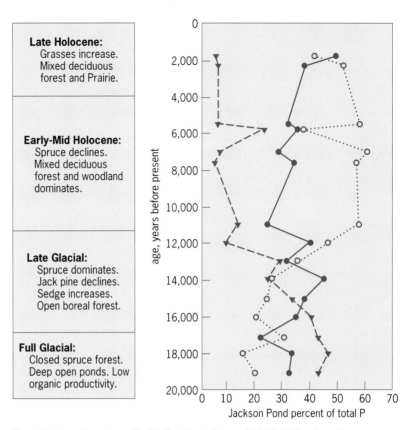

Late Holocene:
  Grasses increase. Mixed deciduous forest and Prairie.

Early-Mid Holocene:
  Spruce declines. Mixed deciduous forest and woodland dominates.

Late Glacial:
  Spruce dominates. Jack pine declines. Sedge increases. Open boreal forest.

Full Glacial:
  Closed spruce forest. Deep open ponds. Low organic productivity.

Fig. 1. Lake sediment records of soil nutrient changes in Jackson Pond, western Appalachian Plateau, United States. As climate changes from the Last Glacial Maximum 20,000 years ago to present, the geochemistry of phosphorus on the surrounding landscapes also changes, with the conversion of mineral phosphorus (triangles) to organic (open circles) and occluded (filled circles) phosphorus forms as soil maturity progresses and the ecosystem stabilizes by about 12,000 years before present.

phosphorus forms over the last 12,500 years indicative of poorly developed soils in this region. Superimposed on this general pattern, however, is a critical detail. For the first 2000 years of ecosystem development, the soil nutrient status did evolve from >90% mineral phosphorus forms to a small but significant proportion of organic and occluded phosphorus forms, hinting at weak soil development in the region correlating with observed warming. This region, however, has experienced several glacial readvances. The most recent, called the Little Ice Age, began about 600 years ago and can be seen across much of the high-latitude Northern Hemisphere. The lake sediments reflect a startling effect of this climatic shift—soil nutrients shifted backward to a near-glacial state, with mineral phosphorus forms once again dominating the landscape.

Several important lessons arise from this new approach to looking at ecosystem dynamics. First, although phosphorus is relatively immobile and only a trace component of soils, ecosystems are incredibly dependent on this element and a lot can be learned by examining its geochemistry. Second, lake sediment records may provide an ideal window into the past for quantifying the effects of climate on soil, landscape development, and biogeochemical cycles. Finally, the rapid changes observed in the terrestrial

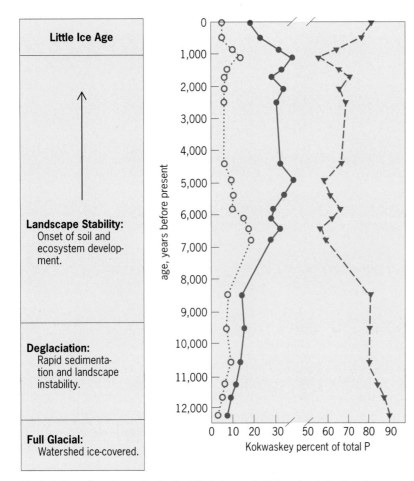

Little Ice Age

Landscape Stability:
Onset of soil and ecosystem development.

Deglaciation:
Rapid sedimentation and landscape instability.

Full Glacial:
Watershed ice-covered.

Fig. 2. Lake sediment records of soil nutrient changes in Kokwaskey Lake, Coast Mountains, British Columbia, Canada. As the landscape is deglaciated and stability is achieved by 7000 years before present, phosphorus geochemistry reflects a transition from bare rocky slopes to a poorly developed soil. This conversion is reversed during the Little Ice Age (~last 600 years) to near-glacial conditions on the landscape.

cycling of phosphorus indicate that the global cycle of this element is highly variable, even on glacial time scales. This is an important feature that impacts models of past and future global change.

For background information *see* ADENOSINE TRIPHOSPHATE (ATP); APATITE; BIOGEOCHEMISTRY; CLIMATE HISTORY; ECOSYSTEM; GLACIAL EPOCH; LANDSCAPE ECOLOGY; PHOSPHATE MINERALS; PHOSPHORUS; POSTGLACIAL VEGETATION AND CLIMATE; SOIL CHEMISTRY in the McGraw-Hill Encyclopedia of Science & Technology.   Gabriel Filippelli

Bibliography. O. A. Chadwick et al., Changing sources of nutrients during four million years of ecosystem development, *Nature*, 397:491–497, 1999; G. Filippelli et al., Terrestrial records of Ge/Si cycling from lake diatoms, *Chem. Geol.*, 168:9–26, 2000; G. M. Filippelli and C. J. Souch, Effects of climate and landscape development on the terrestrial phosphorus cycle, *Geology*, 27:171–174, 1999; T. W. Walker and J. K. Syers, The fate of phosphorus during pedogenesis, *Geoderma*, 15:1–19, 1976; F. H. Westheimer, Why nature chose phosphates, *Science*, 235:1173–1178, 1987.

Electric propulsion

Electric propulsion systems are widely used to maintain and adjust the orbits of communications satellites. Systems with higher exhaust velocities are being developed for extended missions by space probes, and such a system was used on the *Deep Space 1* mission.

A key measure of space-based thruster performance is the thrust produced per unit of propellant consumed, a parameter that is directly proportional to the propellant exhaust velocity. Electric thrusters are more attractive than conventional chemical ones in many applications because they make it possible to add more energy to each kilogram of propellant, thereby inducing higher exhaust velocities and better performance. Because electrical power is limited on typical spacecraft, the thrust of electric rockets is also limited to values that are typically in the range of a few newtons to a few hundredths of a newton. Hence, they are not used for launch applications. Rather, they operate for long times (typically years) and are used for in-space applications related to communication satellite orbit changes and stationkeeping and for acceleration of spacecraft in interstellar environments, where gravitational forces are truly negligible.

Advantages. The capacity of an electric rocket to produce a greater exhaust velocity than a chemical one translates into a lower propellant mass requirement to achieve each increment of mission objective. For typical spacecraft maneuvering and stationkeeping missions, the resultant total savings in propellant mass per kilogram of payload is advantageous for commercial spacecraft services. This is because a smaller, less costly launch vehicle can be used to place the spacecraft in orbit; more revenue-producing equipment can be installed as payload; or the satellite service lifetime can be increased by retaining the additional propellant.

For many deep-space missions, electric propulsion is enabling. For others, it results in lower mission costs; a greater payload fraction yielding greater scientific or other capabilities; or in some cases, shorter payload delivery times than those associated with chemical propulsion.

Thruster types. Power generation, power conditioning, and propellant storage and delivery subsystems, which are essential components of complete electric propulsion systems, affect the selection of the electric propulsion option that should be used for a particular mission. These thruster options fall into three classes defined by the mechanism of propellant acceleration: (1) electrothermal thrusters that expand an electrically heated plasma in a nozzle; (2) electromagnetic thrusters that accelerate a plasma through the application of body forces induced by combined electric and magnetic fields; and (3) electrostatic thrusters in which ions are produced and then accelerated in a steady electric field. Many alternative designs are being researched and developed in each of these classes. This article focuses on

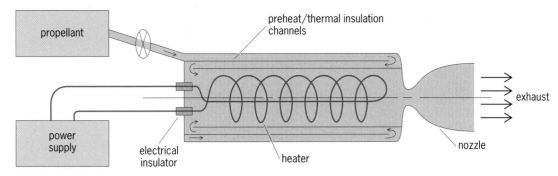

Fig. 1. Resistojet thruster system.

those that have been developed in the United States and are being used commercially in Earth orbit and on the *Deep Space 1* mission. About 150 Earth satellites employ electric thrusters, mostly of the electrothermal class. New ones are being launched at a rate of the order of one per month.

Resistojet. The simplest of electric thrusters is the resistojet, in which propellant is heated by passing it over resistively heated surfaces and then through a converging-diverging nozzle (**Fig. 1**). Hydrazine, which decomposes exothermally and has a low molecular mass, is a typical propellant. A primary design goal is to husband the electrical power so as little as possible is lost and the propellant temperature is as high as possible. The temperature is limited to prevent melting, phase changes, and chemical reactions in the refractory metal heater. Typical resistojets have a nozzle orifice diameter near 1 mm (0.04 in.) and operate at powers of several hundred watts and exhaust velocities around 3000 m/s (10,000 ft/s). A resistojet was the first commercial electric thruster used in space (1983), probably because the transition to it from an existing conventional hydrazine thruster involved only a relatively straightforward increase in exhaust velocity achieved by heating the propellant.

Arcjet. Further increases in propellant velocity can be realized in electrothermal thrusters by passing electrical current directly through the propellant rather than through a heater. The resulting thruster is called an arcjet (**Fig. 2**). Electrical heating involves electron emission from the refractory metal cathode, passage of electric current through and joule heating of propellant flowing between the anode and cathode, and electron collection on the refractory metal anode, which also serves as the diverging section of the nozzle. Typically, propellant swirl is introduced to stabilize the discharge, control of heat losses is a primary goal, and plasma-gas temperatures near 1 eV (around 12,000 K or 22,000°F) are sustained. Hydrazine propellant, a nozzle with a throat diameter less than 1 mm (0.04 in.), a 2-kW power source and a 6000-m/s (20,000-ft/s) exhaust velocity typically characterize resistojets used on geosynchronous communication satellites. They have electrical efficiencies (the ratio of directed kinetic exhaust-jet power to input electrical power) in the range of 30–40%.

Hall plasma thrusters. Exhaust velocities are increased further by applying electromagnetic body forces to accelerate a plasma. Among many possible design concepts in this class, the Hall or stationary plasma thruster is closest to commercial application. This thruster concept (**Fig. 3**) evolved initially in the former Soviet Union. The electron source that it employs (a hollow cathode) produces electrons in an orificed, refractory metal tube through which a small fraction of the propellant flows. Electrons are drawn through the orifice from a metal surface with a low work function (so that little energy is required

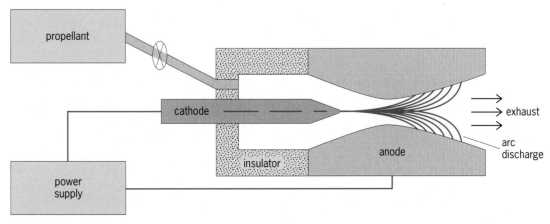

Fig. 2. Arcjet propulsion system.

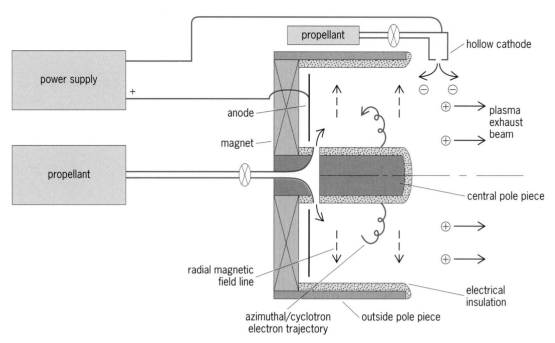

Fig. 3. Hall thruster system.

to liberate electrons from it) and from a propellant plasma, both of which are within the tube. As these electrons are drawn through the annular discharge chamber toward the anode at the upstream end, they feel the effect of a radial magnetic field that exists between the central and outside pole pieces that bound the discharge region. This causes the electrons to follow azimuthal cyclotron (Hall-effect) trajectories that increase the probability that they will have ionizing collisions with propellant atoms being injected at the upstream end of the chamber. Positive ions produced via these collisions feel an electrostatic force due to the axial electric field between the cathode and anode. Because their mass is much greater than that of an electron, they are not deflected substantially in the magnetic field and are, as a consequence, accelerated axially downstream, thereby producing thrust. Alternatively, the acceleration process may be viewed in terms of an azimuthal electron current $\vec{j}$ through the plasma interacting with a radial magnetic field $\vec{B}$ to induce an electromagnetic ($\vec{j} \times \vec{B}$) body force that accelerates the plasma downstream.

The first commercial Hall thrusters, which are scheduled for stationkeeping use in 2001, will use xenon propellant; will have a 10-cm (4-in.) diameter; and will operate at an exhaust velocity, power, and electrical efficiency of 16,000 m/s (52,500 ft/s), 1350 W, and 50%, respectively. Since some propellant in this class of thruster escapes un-ionized and is, therefore, not accelerated, the fraction of propellant atoms that are accelerated (the propellant utilization efficiency) is also of interest. For this thruster it is estimated to be approximately 95%.

Ion thrusters. The greatest exhaust velocities are achieved in gridded electrostatic ion thrusters (**Fig. 4**). A key component of the device is the main hollow cathode, which operates like the one described for the Hall thruster. It supplies electrons into the discharge chamber through the cathode orifice with the kinetic energy needed to bombard and ionize propellant atoms that are also being supplied into the chamber. Both these electrons and the positive-ion/electron pairs they produce via ionization are confined and directed by judiciously placed magnetic fields. In particular, ion diffusion toward the screen and accelerator grids is enhanced so that the current of ions accelerated through the matched hole pairs in these grids can be maximized. The voltage difference applied between the grids can be set to yield a desired ion kinetic energy or exhaust velocity. Proper ion focusing is assured by selecting hole sizes and the grid spacing for the applied voltage difference. The second hollow cathode shown downstream of the grids serves to supply electrons at the rate needed to neutralize the charge and current associated with the extracted ion beam.

Ion thrusters being used for Earth orbit applications operate on xenon propellant; have diameters of 13 and 25 cm (5 and 10 in.); have exhaust velocities over the range 25,000–40,000 m/s (80,000–130,000 ft/s); and have powers of 1–2 kW and efficiencies for utilization of power and propellant of approximately 80% and greater than 90%, respectively.

The ion thruster used to propel the *Deep Space 1* payload past the asteroid Braille and to a rendezvous with Comet Borrelly exhausts xenon in a 30-cm-diameter (12-in.) beam at a velocity that is generally over 30,000 m/s (100,000 ft/s). Because the solar flux decreases as distance from the Sun increases during the mission, the available electrical power for thrusting drops from 2.5 kW at the start of the mission to 500 W as the mission proceeds. As a consequence,

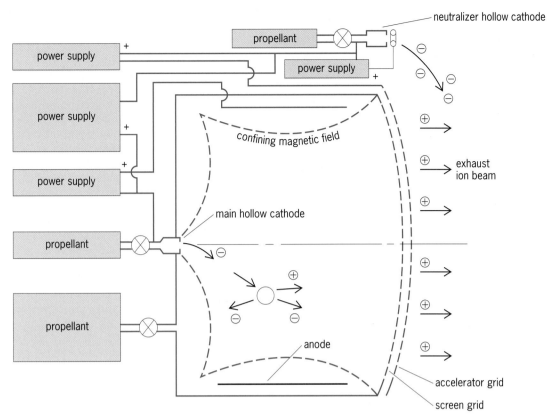

Fig. 4. Ion thruster system.

the electrical and propellant utilization efficiencies, which were both about 80% at mission start, also drop until they are both about 60%.

For background information *see* ELECTROTHERMAL PROPULSION; ION PROPULSION; PLASMA PROPULSION; SPACECRAFT PROPULSION; SPECIFIC IMPULSE; WORK FUNCTION (ELECTRONICS) in the McGraw-Hill Encyclopedia of Science & Technology.   Paul J. Wilbur

Bibliography. G. R. Brewer, *Ion Propulsion, Technology and Applications*, Gordon and Breach, 1970; L. H. Caveny (ed.), *Orbit-Raising and Manuvering Propulsion: Research Status and Needs*, vol. 89, Progress in Astronautics and Aeronautics Series, American Institute of Aeronautics and Astronautics, 1984; R. C. Finke (ed.), *Electric Propulsion and Its Applications to Space Missions*, vol. 79, Progress in Astronautics and Aeronautics Series, American Institute of Aeronautics and Astronautics, 1981; R. G. Jahn, *Physics of Electric Propulsion*, McGraw-Hill, 1968.

Electronic chart

A number of modern technologies, including computer, satellite navigation, and in some cases radar, are combined to create electronic charts. These charts display where a conveyance (marine, aerospace, or land vehicle) is located, as well as where it has been and where it is headed. Information is also provided to guide the conveyance to its destination while avoiding hazards that lie along the route. In these systems, a nautical chart or a land map is displayed on a computer's screen or on a hand-held device, showing appropriate detail. A symbol that marks the vehicle's present location and a track of its previous positions is drawn on the chart or map. In the case of marine vessels, radar images of objects in the water show the location of shore, lights, and buoys, as well as the presence and tracks of other ships in the area. Thus, the electronic chart shows a complete picture of the situation facing a pilot or driver.

Development. The earliest examples of electronic charting systems arose in the marine field. The first appeared in arctic and harbor navigation during 1979 and 1980. The first fully integrated system involving computer, electronic positioning, and radar dates to 1983 during demonstrations in Baltimore harbor using full-color displays combined with a digitized radar image on the same display. Soon afterward, electronic charts were used in automobiles and other land vehicles, followed by their use in the air.

All current systems use the Global Positioning System (GPS) for vehicle location. It provides accuracies of about 5 m (16 ft) with a degree of precision that is sufficient for virtually all modes of transportation. This system is composed of computerized ground receivers and 24 satellites in orbit around the Earth. The satellites emit signals that allow a receiver on Earth to accurately determine real-time position information instantaneously. The location of an object

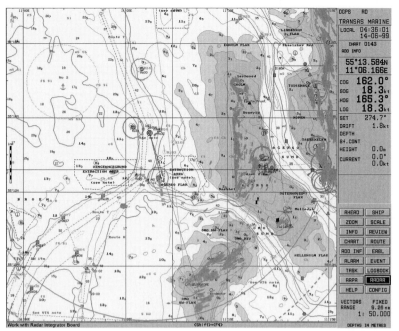

Fig. 1. Typical vector chart used in marine applications. This is a combination of nautical chart with radar shown in a single display. (*Transas Marine*)

navigational services are provided by GPS: precise positioning service (PPS) for military use, and standard positioning service (SPS) for civilian applications.

System. Electronic chart systems, in use throughout the world, have proliferated as advanced digital versions of maps and charts have become available. Both vector and raster types of charts are in use, produced by government hydrographic and cartographic offices as well as private companies.

Vector charts. Digitizing existing paper maps and charts has become a relatively simple task by employing computer programs that trace features on paper maps and charts and automatically transform them into digital files (that is, vector data files). Vector data files store information about individual features, such as shorelines and buoy, or light locations, separately. The computer in the vehicle translates these files and combines them to produce a unified graphic display. These displays are called vector charts (**Fig. 1**).

Raster charts. Another scans an entire prototype map or chart and produces a single digital file (that is, a raster data file). The computer's program creates a replica of the paper version from this file and displays it on the screen. These displays are called raster charts (**Fig. 2**).

Updating chart data. In critical applications, such as in regulated marine or aviation use, where accuracy, adequate content, and timely updates are required, the equipment and the chart data must meet

(that is, latitude, longitude, and elevation) is determined through a mathematical method known as triangulation. This same information can be used to guide a vehicle from one location to another. Two

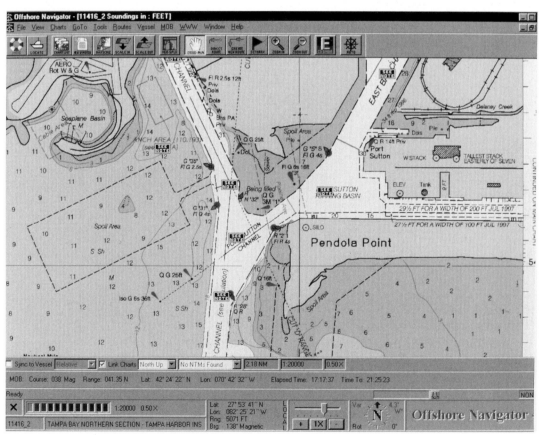

Fig. 2. Raster chart produced as a digital facsimile of a paper nautical chart. (*Maptech*)

stringent international standards. This tends to make the equipment larger, rugged, and more expensive. Many other applications, which are unregulated, are typically those used in automobiles, small boats and yachts, and farm tractors, or in pocket-sized systems used by hikers and cyclists. Mass production of these latter types of units have resulted in low prices and widespread application.

Chart or map data provide details down to the level of streets in many metropolitan areas. Chart and map coverage is expected to grow rapidly, accompanied by improvements in the means for installing updates to databases. For example, a system connected temporarily to the Internet can be given corrections to its database that update or correct various details of roadway, airway, or harbor situations. Updating these details is essential to preserve the safety-enhancing quality of the electronic chart. Accidents can result if the displays contain errors; they can be prevented if timely corrections are incorporated.

Uniqueness. The combination of electronic charting and position finding in these systems gives rise to some unique qualities. For example, in a conventional automatic pilot, a planned heading is inserted, and the navigator checks and corrects the vessel's direction along the planned route. When the autopilot is based on the electronic chart/GPS combination, the check of geographic location is measured by the GPS and position corrections are automatic. The electronic chart provides the desired location and the GPS senses the actual location; the difference between them is the needed correction and is applied automatically. In the old-fashioned autopilot that accepts only the desired heading, deviations are measured and corrections are made by human intervention.

Radar may be added to the combination. A radar image of a shoreline, buoy, or lighthouse is added to a chart display that also shows the drawn image of these objects. If the objects are properly superimposed—the radar image on top of the drawing of the object—it is evident that the whole system is functioning properly. This is a unique test resulting from the combination.

Future development. Continued miniaturization of all electronic components results in the development of smaller units (usually portable) that require less power for their operation. These smaller units require increased integration with systems that provide added function. One example is the combination of electronic chart, GPS, and cellular telephone contained in one hand-held, telephone-sized unit (**Fig. 3**). This triple combination can function in many applications where it is desirable for one vehicle to talk to another vehicle or fixed station while transmitting its location via the telephone connection. Other combinations result in more applications and continued integration of function.

Other future developments involve the combination of computer processing and map or chart display. Computers produce a variety of displays that give the user a more graphic portrayal of the sur-

rounding area. For example, there are three-dimensional displays of terrain for mobile land users and aircraft that need to avoid terrestrial hazards; and shipboard processors that display a three-dimensional view of the sea bottom in the vicinity of a vessel (replacing the two-dimensional contours now provided on paper nautical charts).

Dynamic displays are being developed to assist in the identification of various objects. In a harbor, especially at night, buoys and lighthouses provide navigational information via various intervals and colors of flashing lights visible in different sectors. All of these colors and flashes can be duplicated, and even synchronized with the real counterparts to provide an accurate replica of the surroundings.

The future utility of electronic chart systems depends upon the existence of a high-quality digital

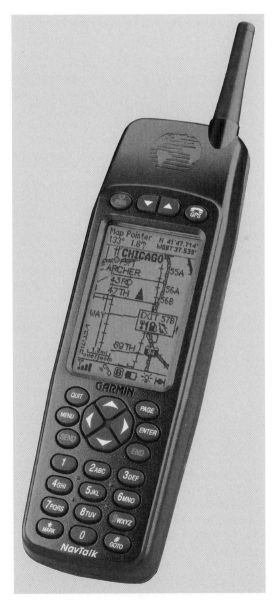

Fig. 3. Portable unit that contains the combination of GPS, cellular telephone, and electronic chart. It is hand-held and battery-operated. Both the GPS and cellular antennas are contained within this small package. (*Garmin International*)

chart database. If a road map used for highways is inaccurate, the display may show a vehicle off the side of the road. If a nautical chart contains an error, it can show a ship outside a dredged channel or on shore when it is actually perfectly positioned in the water. If the chart is out of date, it may show a lighted buoy in the wrong position. As a result, much of the future promise of electronic chart systems is dependent on attaining an infrastructure of accurate, timely data. This is a case where the value of the technology can be compromised by inadequate support from sources of data. The proliferation of these systems around the world, and the increasing demand for them in many types of application, will undoubtedly spur the investment needed to maintain an increasing supply of high-quality data.

For background information *see* AUTOPILOT; CARTOGRAPHY; COMPUTER GRAPHICS; COORDINATE SYSTEMS; DIGITAL COMPUTER; ELECTRONIC DISPLAY; ELECTRONIC NAVIGATION SYSTEMS; GEOGRAPHIC INFORMATION SYSTEMS; MAP DESIGN; NAVIGATION RADAR; SATELLITE NAVIGATION SYSTEM; TERRESTRIAL COORDINATE SYSTEM in the McGraw-Hill Encyclopedia of Science & Technology.   Mortimer Rogoff

Bibliography. C. R. Drane and C. Rizos, *Positioning Systems in Intelligent Transportation Systems*, Artech House ITS Series, 1998; M. Ferguson, R. Kalisek, and L. Tucker, *GPS Land Navigation: A Complete Guidebook for Backcountry Users of the NAVSTAR Satellite System*, 1997; Y. Zhao, *Vehicle Location and Navigation Systems*, Artech House ITS Series, 1997.

gin of gems, especially emeralds of excellent quality that are poor in inclusions. Thus, doubt or ambiguity always exists when determining the geographical origin of cut or carved emeralds removed from the environment in which they formed.

Formation. Emerald is the green variety of beryl (see **illus.**), beryllium aluminum silicate ($Be_3Al_2Si_6O_{18}$) with trace amounts of chromium and vanadium, which produce its characteristic color. Emerald is rare because chromium (and vanadium) and beryllium have antagonistic geochemical behavior in the Earth. Chromium is concentrated in mafic rocks from the mantle (the zone of the Earth located under the continental or oceanic crust), while beryllium has an affinity for the continental crust and concentrates in granitic magmas. The formation of emerald necessitates the bringing together of chromium (and vanadium) and beryllium in the same geological site.

Two types of emerald deposits have been recognized worldwide: One type is associated with granites and pegmatites (rocks composed of quartz, feldspar, and mica, with crystals sometimes having a size of several meters). These emeralds result from the circulation of hot alkaline fluids that developed an alteration halo around pegmatite veins. The fluids, at temperatures between 400 and 600°C, dissolve the main constituents of pegmatite (beryllium) and mafic rock (chromium and vanadium), transforming the two rocks into plagioclasite (rock composed of white feldspar rich in sodium) and phlogopite (biotite schist). Thus chromium dissolved from

Emerald

Emerald with lapis-lazuli is the oldest known gemstone, having been mined since ancient times in Egypt and probably in Bactria and Scythia. Emerald is rarer than diamond and may have a value thousands of times that of gold. Colombia produces 60% of the world's emeralds (total world production in 1986 was estimated to be 15 million carats), followed by Zambia (15%), Brazil (10%), Russia (5%), Madagascar (3%), Zimbabwe (2%), and Pakistan and Afghanistan (together 5%). Colombian emeralds are prized for their exceptional color, clarity, and carats, as are those from Afghanistan, and to a lesser extent Zambia and Russia. The mineralogical and gemological properties that are normally used to determine the origin of emeralds are their optical features (refractive indices and birefringence), density, adsorption spectra (ultraviolet and near-infrared), internal characteristics (growth phenomena and solid and fluid inclusions), and chemical composition. The diagnostic value of these properties is often restricted because there may be an overlap for emeralds originating from different deposits. However, a combination of mineralogical and gemological properties can be used, in many cases, to accurately identify emeralds from specific localities. Nevertheless, gemological features are often insufficient to certify the ori-

(a)

(b)

Colombian trapiche emeralds, from the Coscuez mine. (*a*) Crystallized within black shale. The width of the trapiche crystal on the left is 4 cm. (*b*) 48-carat gemstone showing solid and fluid inclusions. (*Photograph by Omar Bustos*)

the mafic rock and beryllium dissolved from the pegmatite are concentrated in phlogopitite and plagioclasite where emeralds have crystallized. Emeralds of this type occur in most of the deposits of the world (Zambia, Brazil, Madagascar, Russia, Zimbabwe, Australia).

The other emerald type is linked to plate tectonics (thrusts, faults, and shear zones) affecting chromium- and vanadium-bearing rocks. These emeralds correspond to deposits hosted by (1) mafic rocks from the mantle, such as the Brazilian deposits of Santa Terezinha de Goiás and Itaberaí, the emerald mines of Djebel Sikaït, Zabara, and Umm Kabo in Egypt, the Pakistani (Swat-Mingora), and the Austrian (Habachtal) emeralds; (2) metamorphic rocks of sedimentary and magmatic origin, as in the Pansjhir valley (Afghanistan); and (3) sedimentary rocks, such as the black shales of the Colombian deposits. The chromium and vanadium in these emeralds were extracted from source rocks by fluids at temperatures between 300 and 600°C. The source of the beryllium in these deposits is still in debate, with the exception of the Colombian emeralds, the beryllium for which is from the black shales in which they are found.

Oxygen isotopes. Oxygen is the most abundant chemical element in the Earth's crust and is the main constituent of emerald (up to 45% by weight). The variation among rocks of the isotopic ratio of the most abundant oxygen isotopes ($^{16}O = 99.756\%$ and $^{18}O = 0.205\%$) is commonly expressed as $\delta^{18}O$ (per thousand, ‰), which is the relative difference between the $^{18}O/^{16}O$ ratio of the sample and that of the Standard Mean Ocean Water (SMOW), whose $^{18}O/^{16}O$ ratio is 2.0052×10^{-2}. Positive values of $\delta^{18}O$ indicate higher ^{18}O compared to SMOW, and vice versa. Extraction of oxygen from emerald can be performed under vacuum, using the oxidizer BrF_5 (a destructive technique). In this technique, the oxygen released from emerald is converted to carbon dioxide (CO_2) in a graphite furnace. The $^{18}O/^{16}O$ ratio of the CO_2 is then analyzed in a mass spectrometer. The overall reproducibility is 0.2‰. Natural emeralds collected from 62 deposits from 19 countries have been studied by this technique. Their $\delta^{18}O$ values range from $+6.2$ to $+24.7‰$. Emeralds can be classified in three groups according to their $\delta^{18}O$ values. The first group, with $\delta^{18}O$ greater than $+6.2‰$ but less than $+7.9‰$, corresponds to emeralds from Brazil (Quadrilatero Ferrífero and Anagé districts), Austria (Habachtal, $\delta^{18}O = +7.1 \pm 0.1‰$), Australia (Poona), and Zimbabwe (Sandawana). The second group, with $\delta^{18}O$ greater than $+8.0$ but less than $+12‰$, encompasses most of the deposits in the world, such as those from Zambia, Tanzania, Russia, Madagascar, Egypt ($\delta^{18}O = +10.3 \pm 0.1‰$), Pakistan (Kaltharo), and Brazil (Carnaíba and Socotó). The third group, with $\delta^{18}O$ greater than $+12‰$, includes the emerald deposits from Brazil (Santa Terezinha de Goiás), Afghanistan, Pakistan (Swat-Mingora district), and Colombia (eastern zone, $\delta^{18}O = +16.8 \pm 0.1‰$, and western zone, $\delta^{18}O = +21.2 \pm 0.5‰$).

The isotopic oxygen composition of emeralds also can be determined by a near-nondestructive technique using an ion microprobe. In this approach, the $^{16}O^-$ and $^{18}O^-$ ions are sputtered off an emerald crystal by bombardment with Cs^+ ions and negative ions, and analyzed in an ion probe mass spectrometer. Cut or crystal emeralds having flat surfaces can be analyzed after gold metallization. This technique allows the determination of the $^{18}O/^{16}O$ ratio with a precision of $\pm 0.4‰$ (1 sigma). The size of the craters produced on the emerald is about 10–20 micrometers in diameter and a few angstroms deep (that is, about 2×10^{-11} g sputtered for one analysis). These spots are invisible to the naked eye, so this method is considered nearly nondestructive and can be used on gems of high value.

Trade routes. The recent examination by ion microprobe of a set of historical emeralds has revealed unexpected origins, hinting at previously unknown trade routes for the gems. Four emeralds from the Nizam of Hyderabad (India), cut in the eighteenth century in India, were analyzed; three of them came from Colombia and one originated from Afghanistan. This result contradicted their proposed origin as old-mine emeralds from long-lost Indian mines, and showed that Afghani mines had been exploited as early as the eighteenth century, although their recorded discovery was in 1976. In another study, four emeralds belonging to the National Museum of Natural History in Paris were examined. The oldest is an emerald set in a Gallo-Roman earring which originated from Swat in Pakistan, a region along the ancient silk road route. The 51.5-carat emerald, mounted in the Holy Crown of France by Louis IX in the thirteenth century, was shown to have come from Habachtal in Austria. The other two emeralds, belonging the eighteenth-century French mineralogist René Just Haüy, originated in Austria and Egypt. These results showed that during ancient times emeralds of Pakistan and Egypt were traded by way of the silk route. These two places and Austria were the only sources of emeralds. After the discovery of the Colombian mines by the Spaniards in the sixteenth century, a new trade route was established via Spain to Europe and India. Emeralds of exceptional quality have been cut in India since the Bobur Moghul dynasty and sold as old-mine emeralds worldwide. One emerald, property of the Mel Fisher Maritime Heritage Society, was recovered from the Spanish galleon *Nuestra Señora de Atocha*, which sank off the Florida coast in 1622. The galleon was coming from Colombia, but isotopic analysis of the emerald demonstrated that the crystal came from the old Tequendama mine, exploited during Pre-Colombian times.

Tracing the sources of gemstones. Other techniques can be used to determine the geographic origin of precious gemstones. An ion-beam technique, external beam proton-induced x-ray emission (PIXE), is suitable for chemically analyzing gems. The ion beam is external (diameter of the proton beam on target is less than 0.5 mm), and the sensitivity is excellent for the analyses (even in trace amounts) of the transition elements (vanadium, chromium, titanium, iron, magnesium) that are found in emerald. The chemical composition obtained from artifacts compared to a collection of standards from different well-known sources allows in many cases the probable identification of their geographical origin. This technique has been applied successfully to antique jewelry by ion-beam analysis at the Laboratoire de Recherche des Musées de France at the Palais du Louvre in Paris.

Another technique used to identify emerald origin is Fourier-transform infrared spectroscopy. An infrared beam is transmitted through the crystal, and an absorption spectrum representing the nature of impurities and the water, carbon dioxide, and nitrogen content is obtained. Comparison of the different absorption bands found for the gem is compared to emerald standards from known world sources. This technique is nondestructive. The geographic determination of emeralds by infrared spectroscopy is used at the Laboratoire Environnement et Minéralurgie in Nancy, France.

For background information *see* BERYL; EMERALD; GEM; INFRARED SPECTROSCOPY; MASS SPECTROMETRY; MINERAL; MINERALOGY; OXYGEN; PRECIOUS STONE in the McGraw-Hill Encyclopedia of Science & Technology.   Gaston Giuliani

Bibliography. T. Calligaro et al., PIXE/PIGE characterisation of emeralds using an external micro-beam, *Nucl. Instrum. Meth. Phys. Res. B*, 161–163, 769–774, 2000; D. Giard et al., *L'émeraude*, Association Française de Gemmologie, CNRS et ORSTOM Editions, AFG Press, Paris, 1998; G. Giuliani et al., Les gisements d'émeraude du Brésil: Genèse et typologie, *Chronique de la Recherche Minière*, 526: 17–61, 1997; G. Giuliani et al., Oxygen isotope systematics of emerald: Relevance for its origin and geological significance, *Mineralium Deposita*, 33:513–519, 1998; G. Giuliani et al., Oxygen isotopes and emerald trade routes since antiquity, *Science*, 287: 631–633, 2000; E. Gubelin and J. Koivula, *Photoatlas of Inclusions in Gemstones*, 2d ed., ABC Editions, Zurich, 1986; La gemmologie, *Analusis Mag.*, 23(1):10–54, 1995; J. Sinkankas, *Emeralds and Other Beryls*, Chilton Book, Radnor, PA, 1981, reprinted by Geoscience Press, Tucson, 1989.

Endophyte grasses

As an endophyte, a fungus may grow within a plant in a mutualistic relationship. Two notable examples are tall fescue (*Festuca arundinacea*) infected with the fungus *Neotyphodium coenophialum* (formerly *Acremonium coenophialum* and *Epichloe typhina*) and perennial ryegrass (*Lolium perenne*) infected with the fungus *Neotyphodium lolii*. This mutualistic relationship benefits the fungus through provision of energy, nutrients, shelter, and a means of propagation, while it benefits the plant with mechanisms for improving persistence through biochemical deterrents to overgrazing and insect damage.

Identification of the fungus. Tall fescue was widely planted in the United States during the 1940s and 1950s following the release of the variety Kentucky-31 in 1943 by E. N. Fergus of the University of Kentucky. This grass filled a void in the southern states, where few other cool-season forages were adapted. However, cattle grazing tall fescue often developed a chronic health condition, which was especially apparent in the summer (called summer syndrome or summer slump). Beginning in 1973, a farm near Mansfield, Georgia, provided the background for J. D. Robbins and C. W. Bacon to discover the reason why one cattle herd on the farm exhibited summer syndrome while another herd did not. Animals suffering from fescue toxicosis were consuming tall fescue with all plants infected with the endophytic fungus, while unaffected animals were consuming tall fescue in a pasture with only 10% of the plants infected with the endophyte.

Spread of the fungus. The endophytic fungus spreads to other plants only through seed dispersal of infected plants and is not known to be culturable outside the host plant. A common method of removing the fungus from host plants is to store seeds at room temperature for a year following harvest, so that the fungus dies, and then to plant the remaining viable seeds. The fungus does not change the growth or appearance of the host grass, and therefore detection of the fungus requires laboratory analysis, often by staining and observation of leaf sheath tissue under a microscope (**Fig. 1**).

Tall fescue is grown on about 14 million hectares (35 million acres) of land in the United States and at least the same area in other parts of the world, including Europe, Asia, and northern Africa. Surveys in the United States indicate a high frequency of endophyte infection on most tall fescue pastures. Seeds from a number of varieties of tall fescue are currently being

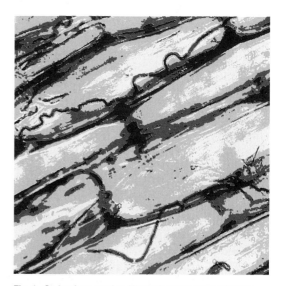

Fig. 1. Stained serpentine *Neotyphodium coenophialum* mycelia occupying the spaces between cells of a tall fescue leaf sheath. Magnification 400×. (*Courtesy of Nicholas Hill*)

sold with low infection frequency of the endophyte for those producers that want to avoid the negative impacts of endophyte infection.

Animal response. Farm animals grazing endophyte-infected tall fescue can suffer from a number of health disorders, including fescue foot, fat necrosis, and fescue toxicosis. Documented animal responses to consumption of endophyte-infected tall fescue include (1) lower feed intake, (2) lower weight gain, (3) lower milk production, (4) higher respiration rate, (5) higher core body temperature, (6) rough hair coat, (7) more time spent in ponds or puddles, (8) more time spent in shade, (9) less time spent grazing, (10) excessive salivation, (11) reduced blood serum prolactin level, and (12) reduced reproductive performance. Reduced animal performance and productivity are linked to toxic ergopeptine alkaloids that accumulate in endophyte-infected leaf tissue. Symptoms of toxicity are most prevalent during the hot summer months and less prevalent during the cool autumn and spring months.

Ecological benefits of association. Both tall fescue and perennial ryegrass are cool-season grasses, which begin to grow during the cool autumn months, survive the winter in a green state depending upon the severity of cold, resume growth in early spring, and flower in late spring prior to the heat of summer. The long hot summers in the southeastern United States limit the ability of most cool-season grasses to persist for more than a few years. Tall fescue, however, persists with cattle grazing in the southeastern United States better than most other cool-season forages, probably because of association with the endophyte. Research suggests that endophyte-infected tall fescue is more productive and able to survive drought better than uninfected tall fescue.

There have been a number of documented cases of resistance to pest infestation and damage in endophyte-infected perennial ryegrass and tall fescue stands compared with endophyte-free stands. These cases include responses to Argentine stem weevil, black beetle, sod webworm, aphids, fall armyworm, pasture mealybug, cutworm, and spiral nematodes.

The effect of the endophyte may go far beyond that solely in the immediate vicinity of the plant. Recent research from long-term tall fescue pastures in Georgia indicates that endophyte-infected pastures could be putting more of the carbon captured by photosynthesis of the plant into the soil as organic matter, rather than respiring the carbon back into the atmosphere as carbon dioxide.

Carbon dioxide is fixed by plants during photosynthesis into simple sugars, which are then metabolized into higher-level organic compounds needed by the plant. Plants respire part of the carbon back into the atmosphere as carbon dioxide to generate metabolic energy, and the remainder leads to growth of the plant. Following death, plant tissues are attacked by soil bacteria, fungi, and various other organisms that use this organic material as energy and carbon sources for their own growth. These organ-

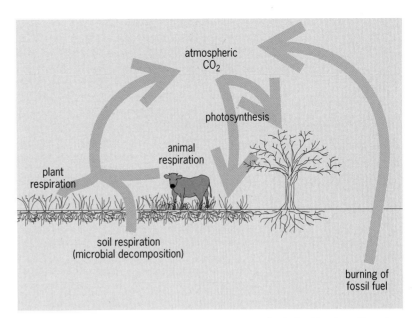

Fig. 2. Simplified diagram of the carbon cycle. Carbon dioxide is fixed by plants through photosynthesis and returns to the atmosphere when organisms (plants, animals, and microbes) respire.

isms decompose organic material, returning a portion of the organically bound carbon to the atmosphere as carbon dioxide (microbial respiration) and transforming the remainder into microbial biomass and more inert forms of soil organic matter or humus. Photosynthesis and respiration are major components of the global carbon cycle that have kept the level of carbon dioxide in the atmosphere at a relatively constant level (**Fig. 2**). Rising atmospheric carbon dioxide levels during the past century indicate an imbalance in the carbon cycle that could lead to global warming and widespread ecological changes. Sequestration of carbon in the soil as organic matter is a management strategy that could be employed by individual land owners on a large enough scale to help mitigate this rise in atmospheric carbon dioxide.

In pastures near Watkinsville, Georgia, cattle had been grazing Kentucky-31 tall fescue with low and high endophyte infection in one set of replicated paddocks for 8 years and another set of paddocks for 15 years. During these periods, total organic carbon had accumulated near the soil surface under both low- and high-endophyte-infected tall fescue, but more so with high endophyte infection (**Fig. 3**). Greater total soil organic carbon under high endophyte infection was also associated with greater total nitrogen in soil and lower soil bulk density, but lower potential soil microbial activity (see **table**). Commonly however, potential soil microbial activity changes proportionally with changes in total soil organic carbon. The observation that total soil organic carbon was greater with high endophyte infection, but potential soil microbial activity was lower than with low endophyte infection, suggests that various ecological changes occurred in the soil in response to the presence of the endophyte. Possible responses

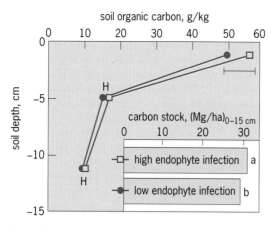

Fig. 3. Concentration of soil organic carbon with depth under low and high endophyte infection of tall fescue and carbon stock at a depth of 0–15 cm. Error bars within a soil depth indicate significance at probability (p) $\leq$ 0.05. Letters a and b beside bars in inset indicate significant difference at $p = 0.01$. (*Data from A. J. Franzluebbers et al., Soil carbon and nitrogen pools under low- and high-endophyte-infected tall fescue, Soil Sci. Soc. Amer. J., 63:1687–1694, 1999*)

include (1) direct inhibition of the ability of soil microorganisms to decompose the additional alkaloid compounds present in endophyte-infected tall fescue residues; (2) direct selection of specific microbial groups that could have formed specific intermediate degradation products, which may have resisted further breakdown; (3) indirect changes in the community of soil microorganisms, of which the multitude of bacteria, fungi, and protozoa have different abilities to coexist and process soil organic matter; and (4) indirect changes in the quantity, quality, and timing of carbon translocated to roots and root exudates, which provide a portion of the carbon source for soil microbial and faunal activity. All of these possible responses could have led to accumulation of total soil organic carbon and nitrogen in soil with endophyte infection.

One response of animals to the consumption of endophyte-infected tall fescue is to spend more time in the shade and less time foraging. This behavior can lead to an accumulation of nutrients (such as phosphorus and magnesium) in soil near the shade and water sources of pastures. Redistribution of nutrients within pastures is common as animals consume forage throughout the pasture, but spend more time drinking, lounging, and defecating near shade and water sources. Concentration of nutrients near shade and water sources in endophyte-infected pastures is, therefore, exacerbated due to animal behavior.

Endophyte-enhanced varieties of grasses have recently become of great interest to researchers trying to retain the beneficial aspects of the association while removing the negative aspects. Plant-fungi genetic combinations are being sought that keep the biochemical codes necessary for maintaining plant persistence but reduce or delete the biochemical compounds responsible for causing animal health disorders. A great deal has been learned about plant and animal responses to endophytes during the past 30 years; however, short- and long-term ecological impacts of endophytes are only beginning to be discovered.

For background information *see* ALKALOID; BIO-DEGRADATION; FESCUE; FUNGI; GRASS CROPS; SOIL CHEMISTRY in the McGraw-Hill Encyclopedia of Science & Technology.   Alan J. Franzluebbers

Bibliography. C. W. Bacon et al., *Epichloe typhina* from toxic tall fescue grasses, *Appl. Environ. Microbiol.*, 34:521–581, 1977; A. J. Franzluebbers et al., Soil carbon and nitrogen pools under low- and high-endophyte-infected tall fescue, *Soil Sci. Soc. Amer. J.*, 63:1687–1694, 1999; H. H. Schomberg et al., Spatial distribution of extractable P, K, and Mg as influenced by fertilizer and tall fescue endophyte status, *Agron. J.*, 92:981–986, 2000; J. A. Stuedemann and C. S. Hoveland, Fescue endophyte: History and impact on animal agriculture, *J. Prod. Agri.*, 1:39–44, 1988.

Soil properties (0–15-cm depth) under tall fescue as affected by endophyte infection level*

Soil property	Endophyte level		
	Low	High	Significance[†]
Soil bulk density, Mg · m^{-3} soil	1.27	1.25	0.02
Total organic carbon (TOC), g · kg^{-1} soil	15.3	16.8	<0.01
Total nitrogen, g · kg^{-1} soil	1.12	1.23	<0.01
Potential microbial activity, mg CO$_2$-C · kg^{-1} soil	28.1	25.7	0.16
Specific microbial activity, g CO$_2$-C · kg^{-1} TOC	1.87	1.57	0.01

*Data from A. J. Franzluebbers et al., Soil carbon and nitrogen pools under low- and high-endophyte-infected tall fescue, *Soil Sci. Soc. Amer. J.*, 63:1687–1694, 1999.

[†]"Significance" is probability that the magnitude of difference between means would occur again if values were random.

Engineering education

The primary role of most universities has changed over the last 40 years as they have become research institutions with little faculty time to teach undergraduate students. At the same time, the role of engineers in society and the skills required to fill this role have also changed. Dramatic advances in computers and communications are responsible in part for these changes, while providing new teaching tools and an improved knowledge of the learning process. As a result, there has been an extensive and intensive reevaluation of engineering curricula and teaching methodologies. Computers and multimedia technology will profoundly affect teaching and learning in the future. They are, however, only tools, and there will always be a need for faculty involvement in mentoring, lecturing, and the more traditional methods of instruction.

Role of engineers. In the past, engineers were trained to be efficient solvers of routine problems, performing analyses with various methods and analogies, dimensioning elements or subsystems following appropriate codes, supervising construction or fabrication processes, managing the operation of facilities, or dealing with equipment and labor issues. Initially, engineers had to perform all these duties because they were in charge of complete projects. As the complexity of these tasks increased, engineers increased their specialization, with some devoting their lives to such work as computation, fieldwork, or project management. In most schools, the desire was to produce engineers at the bachelor's degree level who could be of immediate use to industry. Only a few universities perceived their role as that of educating engineers, rather than training them. It was at the graduate level where the reasons for different methods and approaches were explained—teaching the "why" instead of the "how." The implication was that only a small number of engineers had to know this, and that the demand was mostly for highly educated technicians. With the emphasis on science and scientific research in the 1950s, major universities concentrated primarily on graduate education, and the production of Ph.D.'s in particular, preparing engineers for academic careers rather than actual practice.

The engineer of the future cannot be just an academician or an efficient, specialized performer of routine computations. An engineer cannot be only a simple user of existing computer software, a proficient chemical analyst, a traffic counter, a construction supervisor, a fabrication inspector, a systems operator, or an administrator. While all these jobs are necessary, they do not require four years of undergraduate school or most of the other skills that engineers receive. An engineer is not a scientist, but needs to understand science and be aware of scientific developments. An engineer is not a technician, but needs to understand the practical aspects of technology. An engineer is not an economist or a sociologist, but needs to understand the basic concepts of economics and finances, as well as social issues. J. Bordogna (1998) saw future engineers as "master integrators."

Impact of computers and new technologies. With the advent of computers, the need for approximate methods of analysis disappeared, and even the most conservative institutions have phased out the teaching of clearly obsolete methodologies. For a while, the number of analysis packages available in practice was small, and the programs required extensive training to be used correctly. Several institutions replaced the training in hand computations with training in the use of specific software (particularly their own). The traditional routines of analysts and designers are rapidly becoming obsolete as powerful, user-friendly, general-purpose analysis and design software evolves and visual computer-aided design, three-dimensional walk-through programs, and virtual construction or operation labs are developed. The problem is compounded by the fact that projects can now be performed by international teams, communicating almost instantaneously by electronic means, and working often in other countries at a lower cost. Highly trained and experienced users of available software do not even have to be engineers as long as there are supervisors who check the reasonableness of the input models and the results.

The same advances in computer technology and engineering software causing the loss of many jobs have also allowed properly conceived and nontraditional curricula (particularly design courses) to include much larger and more realistic projects, eliminating the need for cumbersome hand computations. The development of user-friendly software and computer-aided design (CAD) packages has allowed students to conduct extensive parametric studies that provide, in a short time, experience that before could have taken several years of office work. The Web and other recent multimedia developments have already facilitated and increased communication between students and faculty. It is common now for faculty to put the course syllabi, homework assignments and, later, their solution, and a number of messages and class notes on interactive Web pages.

The development of electronic textbooks on the Web or CDs, with visual and simulation capabilities, can revolutionize teaching. Students can now conduct an actual physical experiment in a laboratory, then recreate the same experiment or change geometry and conditions in their computers, doing it on their own time as often as they want. These are all extremely valuable tools that can enhance engineering education. Their application in distance learning programs is very promising. Still, they are only tools. Changes in teaching methods must be accompanied by appropriate changes in the substance of what we teach.

Curriculum changes. J. Bordogna et al. (1993) stated that a strong foundation in science is needed continuously in a broad and general engineering education. The Engineering Deans Council and Corporate Roundtable (Engineering, 1994) recommended that universities continue to teach fundamentals and prepare students for the broadened world of engineering by incorporating teamwork, communication skills, leadership and system perspective, integrating knowledge throughout the curriculum, with a commitment to quality and ethics. Most panels and committees seem to agree that the future engineer must have not only a solid knowledge of basic science and engineering but also a good understanding of economics, risk and decision analysis in the face of uncertainty and the sociopolitical implications of large engineering projects, skill in technical communications involving oral and written communications as well as the latest multimedia tools, and an early exposure to practice particularly in engineering design. The problem is how to accommodate all this material within shrinking curricula. Course requirements are being continuously reduced either by the desire to compete with other universities that market aggressively new degrees with reduced requirements, or by government mandates at the state level, following the new concept that all citizens are entitled to a college degree. The only viable solution is a complete revision of the curricula starting from scratch, rather than the usual piecemeal adjustment of specific courses. It is necessary in particular to ensure that courses are properly coordinated and the material integrated throughout the curriculum. Courses should include realistic case studies that are fully documented to take advantage of the new available tools and to increase the participation of practicing engineers in the education process. At one time, engineering professors had extensive practical experience and continued to keep in touch with the real world. At present in the research university, faculty members are hired upon completion of their Ph.D. and are required to generate a substantial amount of research funding and publish a large number of papers within 5 years. This leaves very little time for exposure to practice. As a result, it is increasingly difficult to teach meaningful design courses with only the available faculty. A new paradigm of collaboration between industry and academia must be developed if universities are going to continue to produce engineers with the desired qualifications. Otherwise, industry will have to take over the education and training of their personnel.

(The authors wish to thank the Wofford Cain '13 Senior Chair of Engineering in Offshore Technology and the Lohman Professorship in Engineering Education at Texas A&M University for their support in preparing this paper.)

For background information *See* COMPUTER; COMPUTER-AIDED DESIGN AND MANUFACTURING; ENGINEERING; MULTIMEDIA TECHNOLOGY in the McGraw-Hill Encyclopedia of Science & Technology.

James T. P. Yao; Jose M. Roesset

Bibliography. J. Bordogna, Tomorrow's civil systems engineer—the master integrator, *J. Prof. Issues Eng. Educ. Prac.*, 24(2):48–50, April 1998; J. Bordogna, E. Fromm, and E. W. Ernst, Engineering education: Innovation through integration, *J. Eng. Educ.*, pp. 3–8, ASEE, January 1993; *Engineering Education for a Changing World*, Joint Report of the Engineering Deans Council and Corporate Roundtable, February 24–25, 1994.

Evolutionary relationships

Understanding evolutionary relationships is an important part of the discipline of systematics (the scientific classification of organisms). Hypotheses of relationship include ideas about the common ancestry of species, often referred to as phylogenetic trees. Historically much of the understanding of species relationships has been derived from comparative anatomy (morphology) of both living and extinct organisms. With the advent of molecular sequencing, however, there are also vast amounts of gene data from living organisms (and the very small viable amount recovered to date from extinct organisms) to consider. A hypothesis of relationship that has been well tested by comparing character data (heritable features of an organism, be they morphological, molecular, or behavioral—such as skeletal structure, nucleotide sequence, or child-rearing behavior) makes the most solid foundation for naming and classifying organisms. Such a classification reflects natural groupings or clades, that is, groups of organisms that contain an ancestor and all of its descendants.

Systematists use two different types of branching diagrams, cladograms and phylogenetic trees, to illustrate hypotheses of evolutionary relationship. Cladograms describe relationships among species—that is, who is more closely related to whom; phylogenetic trees are cladograms that incorporate hypotheses of ancestry—that is, who is descended from whom.

Use of character data. To construct cladograms and phylogenetic trees, character data for the species under consideration are collected and codified for assembly into a character taxon matrix. Such character data can be gleaned from extinct organisms, for example via fossil bones or footprints, as well as from living ones. The matrix is subjected to some sort of optimality criterion (for example, parsimony analysis, neighbor joining, and maximum likelihood), which is essentially an algorithm that compares possible trees and selects one that best explains the data according to the rules of the algorithm. However, all characters cannot be sampled for all taxa. (For example, certain molecular or non-fossilizable morphological data may not be available.) This common problem means that the summary of data assembled often contains many missing observations. Systematists are still pursuing the implications of missing data for phylogenetic analysis.

Character data and the evolution of whales. It is difficult to compare the morphological and functional attributes evolved by organisms without relying to some degree on an underlying hypothesis of their evolutionary relationship. An example of a hypothesis of relationship that is based on the observation of character data is the hypothesis that whales and sharks developed their torpedo shape independently (**Fig. 1**). While a whale shares some characters with a shark (for example, lack of hair, general body shape), it shares many more characters with mammals (including mammary glands, warm-bloodedness, complex cerebral cortex, and numerous genetic similarities). The simplest explanation of these observations is that whales are mammals and developed their torpedo-shaped body independently of sharks. It was not known in advance of this hypothesis that body shape and hairlessness are more prone to evolve convergently (develop independently multiple times) than other characters. Instead, this was discovered from the weight of the character evidence collected. New evidence can be introduced at any time to test a hypothesis of relationship such as that shown in Fig. 1.

Fossil discoveries such as the "walking whale" *Ambulocetus* have reinforced the hypothesis that whales are mammals and evolved independently of sharks (Fig. 1). The position of *Ambulocetus* was

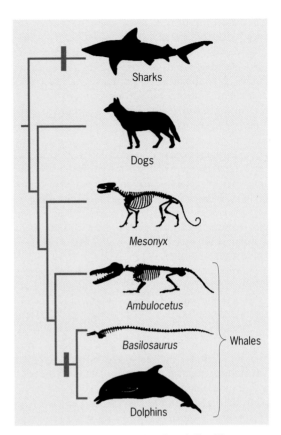

Fig. 1. Phylogenetic tree showing the relationships among six vertebrate organisms. Depicted are a shark and the mammals dog, dolphin, and three fossil organisms from the Eocene Epoch (34–55 million years ago) that are thought to be closely related to living whales. Note that *Ambulocetus* had fully developed legs. Two groups of vertebrates, sharks and whales, have torpedo-shaped bodies. Comparative data, however, suggest that this body shape evolved convergently (independently) in each of these groups.

derived by comparing the morphology of its skeleton with the skeletons of other vertebrates. Although it had legs equal in size to those of living mammals that walk on land, suggesting that it was capable of the same, *Ambulocetus* falls at the base of the whale tree because it shares numerous features of the bony ear region and dentition with other whales.

Conclusions drawn from comparing character data among one group of species should not be used to analyze the relationship between others. For example, if a character is found to show convergent evolution (independent origin) in two groups of organisms, such as a torpedo-shaped body in whales and sharks, it does not follow that the character exhibited convergent evolution in all species. The character "torpedo-shaped body" would not be eliminated from an analysis of other groups simply because it developed independently in sharks and whales. In fact, a preponderance of other character data show that two different groups—a fossil whale, *Basilosaurus*, and a living whale, a dolphin—share this character because they share a common ancestor. Within mammals, whales (except the transitional forms at the base of the tree like *Ambulocetus*) are distinctive in this respect.

In constructing a phylogenetic tree, no particular character or characters, and no potential hypothesis of relationship, should be discounted. Nor should any particular category of characters be considered more important than others (for example, morphological characters are not necessarily preferable to molecular characters, or vice versa). In advance of knowing a phylogenetic tree, we cannot know that a particular set of characters will find the best hypothesis of relationship whereas another set will mislead us.

Amniote relationships and the importance of fossils. Amniotes are a group of tetrapods that produce watertight eggs. This group includes turtles, crocodiles, birds, and mammals, as well as a large number of extinct animals such as dinosaurs (**Fig. 2**). Two questions regarding the amniote family tree have

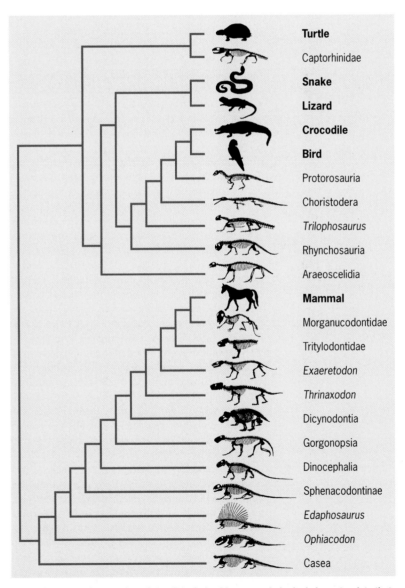

Fig. 2. Phylogenetic tree of amniotes. It is derived from morphological character data that come from both fossil and living organisms. The same tree was recovered after molecular sequences were added to the morphological data. Skeletal silhouettes represent fossil species; body silhouettes represent living species. Although birds and mammals are both warm-blooded, this phylogeny suggests that each developed that feature convergently (independently). (*Adapted from J. Gauthier et al., 1988*)

drawn particular attention and highlight the interaction of fossil and molecular data in the reconstruction of evolutionary relationships: (1) are birds more closely related to crocodiles or mammals? and (2) are turtles a type of basal amniote, embodying the primitive biology of this group, or are they a group of derived species that share a closer relationship to squamates (lizards, snakes, and their close relatives) or crocodiles? The broader importance of these questions touches on such issues as whether birds and mammals developed their warm-bloodedness convergently.

Placement of birds. The classic understanding of the evolutionary relationships of amniotes is that birds are more closely related to crocodiles (Fig. 2), indicating that birds developed their warm-bloodedness independently from mammals. Some analyses that were based on comparative anatomy of living amniotes alone have suggested that birds are more closely related to mammals than they are to crocodiles and dinosaurs. The characters used to support this hypothesis came from soft tissues (including aspects of the circulatory system and brain) and hard tissues such as the bony structure of the nasal passageways. Comparison of nucleotide data, such as the 18S ribosomal ribonucleic acid (rRNA), also supported this novel hypothesis. However, once fossil species, especially many that fall along the branch of the tree leading to living mammals (Fig. 2), and more characters were included in the analysis, the hypothesized close relationship of birds and mammals was rejected. The hypothesis was rejected simply because more of the character data supported the close relationship between crocodiles and birds. Because there is no justification for exclusion of characters or species, in the wake of this new data the classic hypothesis of relationship must be accepted as the hypothesis that best explains the data. The fossils on the tree leading up to living mammals have been recognized as being critical to supporting the hypothesis that birds and crocodiles are more closely related to each other than either is to mammals. This hypothesis is supported even though there were more nucleotide sequence characters analyzed, which collectively supported a different result, than morphological characters. In other words, a particular category of data that encompasses the highest number of characters will not necessarily dominate when all data are combined and analyzed simultaneously.

Placement of turtles. The traditional hypothesis of amniote origins argues that turtles are at the base of the amniote tree, but this has been challenged with new fossil and molecular data. Analyses of new morphological data that include fossils absent from the first morphological analysis—such as those from a group of extinct marine reptiles called Sauropterygia—place turtles within the clade that includes squamates, crocodiles, and birds, putting them closest to squamates. Certain molecular data also place turtles in this general position but possibly closer to birds and crocodiles than to snakes and lizards. This research brings us closer to answering the question of

whether or not turtles can be considered representative of the basal reptile.

Conclusion. To create a complete tree of life, biologists would have to discover relationships among at least 1.5 million named living species and hundreds of thousands of named fossil species. The numbers of species in both categories continue to grow with new discoveries, and are generally considered to be extreme underestimates of the total diversity of life that has ever occupied the planet. In fact, it has been estimated that 99% of the organisms that have ever lived are extinct, and if that is true only a small sampling of these have been found as fossils. Given the demonstrated importance of fossils in the reconstruction of phylogenetic relationships, their scarcity would seem to make completion of the tree of life difficult indeed.

For background information *see* AMNIOTA; ANIMAL SYSTEMATICS; FOSSIL; MACROEVOLUTION; PHYLOGENY in the McGraw-Hill Encyclopedia of Science & Technology.   Maureen A. O'Leary

Bibliography. R. L. Carroll, *Vertebrate Paleontology and Evolution*, W. H. Freeman, New York, 1988; J. Gatesy and M. A. O'Leary, The origin of whales: Deciphering dramatic evolutionary transformations with molecules and fossils, *Trends Ecol. Evol.*, 2001; I. J. Kitching et al., *Cladistics, Second Edition: The Theory and Practice of Parsimony Analysis*, Oxford University Press, 1998; A. B. Smith, *Systematics and the Fossil Record: Documenting Evolutionary Patterns*, Blackwell Science, London, 1994; C. Zimmer, *At the Water's Edge*, Free Press, New York, 1998.

Eye development

The vertebrate eye is an extraordinary organ in terms of structure, function, and development. Vision is acquired during embryonic development as a result of coordinated patterning and growth of many different eye tissues. The mature eye consists of anterior and posterior sectors (**Fig. 1**). The major tissues of the anterior sector are the cornea, anterior chamber, iris, and crystalline lens. The posterior sector contains the posterior chamber, retinal pigment epithelium, and neural retina, which projects neurons through the optic nerves to visual centers in the brain. The eye tissues are derived from different embryonic sources. The lens and external part of the cornea originate from ectoderm on the surface of the embryo. The interior part of the cornea and a portion of the iris are derived from neural crest cells, which migrate into the developing eye from the neural tube. The retina, retinal pigment epithelium, and part of the iris are formed from the optic vesicle, an outpocketing of the brain.

The first morphological indication of the posterior sector of the eye is formation of the optic vesicles as lateral protrusions from the forebrain (Fig. 1). The optic vesicles invaginate to form the optic cups; the neural retina develops on the outer side and the retinal pigment epithelium on the inner side of the optic cup. Three distinct cell layers will differentiate in the

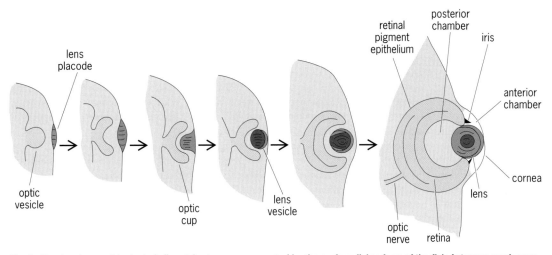

Fig. 1. Eye development typical of all vertebrates, as represented by the surface-living form of the fish *Astyanax mexicanus*.

retina: an outer layer of neural ganglion cells, which connect to visual centers in the brain; a middle layer of glial, bipolar, and horizontal cells; and an inner layer of rods and cones, the photoreceptor cells.

Lens formation is the first step in development of the anterior sector (Fig. 1). The lens develops from the lens placodean, ectodermal thickening that moves into the space within the optic cup to form the lens vesicle. The lens vesicle consists of lens epithelial cells, stem cells that divide and differentiate during lens growth, and precursors of the lens fiber cells, which produce the lens crystallin proteins. Later, the cornea forms as a thickening in the ectoderm above the lens, and the iris develops at the margin of the optic cup. Migrating neural crest cells make a large contribution to both of these tissues. In some vertebrates, such as fishes, the eye grows continuously through life, whereas in other vertebrates, such as mammals, the eye reaches its maximum size shortly after birth.

Intimate contact with the optic cup was once thought to be both sufficient and necessary for lens induction. However, it is now understood that the lens is induced by a more complex series of tissue interactions beginning during gastrulation (formation of the ectoderm, mesoderm, and endoderm by a complex series of cell movements) and continuing until the time of first contact between the lens placode and optic vesicle. Once contact is established between the lens and optic cup, further development of the anterior and posterior sectors is mediated by reciprocal signaling between the lens, retina, and retinal pigment epithelium. These tissues transmit and accept signals from each other in a concerted effort to orchestrate eye growth and differentiation.

Eye degeneration. Many optic diseases involve degeneration of the eye or one of its components. Eye degeneration also occurs naturally in some burrowing mammals and a variety of cave animals which have lost their eyesight while adapting to life in perpetual darkness. The mechanisms involved in eye degeneration in these animals may be similar to those responsible for abnormal eye development in sighted animals, and thus may shed light on this process.

Astyanax mexicanus. The blind cavefish *Astyanax mexicanus* is being developed as a model system to study the molecular and cellular basis of eye degeneration. This species, a native of northeastern

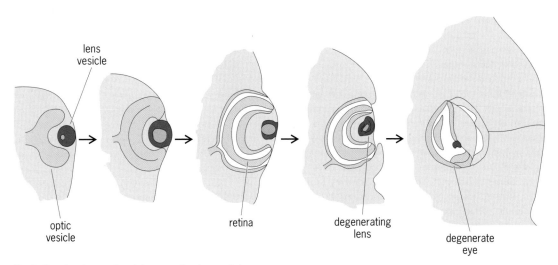

Fig. 2. Eye development and degeneration in cavefish.

Mexico, exists in two forms: a surface-dwelling form (surface fish) with eyes and pigmentation, and a cave-dwelling form (cavefish) that has lost its eyes and pigmentation. It is estimated that cavefish evolved from surface-fish ancestors about 10,000–100,000 years ago.

Eye degeneration is a normal process during cavefish development (**Fig. 2**). Cavefish embryos develop eye primordia at the same time as surface-fish embryos, although these primordia are somewhat smaller than their surface-fish counterparts. The lens and optic cup are formed, and both tissues begin to differentiate. Up to this point in development, eye formation appears to be normal. Eventually, the cavefish eye stops growing, and at maturity the crystalline lens, cornea, iris, and retinal photoreceptor cells are rudimentary or absent. With the exception of photoreceptor cells, the other cell types of the retina develop normally. The degenerate eye eventually sinks into the orbit and is covered by a thick flap of skin.

Role of apoptosis. The first defect to appear in the cavefish eye is the absence of lens cell differentiation. Instead of dividing and differentiating into lens fiber cells, the lens epithelial cells undergo apoptosis, or programmed cell death. At the time of lens apoptosis, there is little or no cell death in other parts of the developing eye, including the optic cup. Later in cavefish eye development, after the lens begins to disintegrate, apoptosis also occurs in the retina, although this structure does not completely degenerate. The lens is thought to regulate later eye degeneration, because it is the first structure to undergo developmental arrest.

Lens apoptosis might be controlled by signals emanating from the optic cup, which is in intimate contact with the lens during the critical developmental stages, or it might be controlled within the developing lens itself. Lens apoptosis has been studied by transplanting a cavefish lens vesicle into a surface-fish optic cup (after removal of the surface-fish lens vesicle; **Fig. 3**). The transplanted cavefish lens shows apoptosis on schedule in the surface-fish host. In the reciprocal experiment, a surface-fish lens vesicle was transplanted into a cavefish optic cup. In this experiment, the surface-fish lens vesicle did not undergo apoptosis, but developed into a mature crystalline lens in the cavefish host, indicating that signals from the optic cup do not control apoptosis.

Lens signaling. The success of lens transplantation opened the possibility of studying reciprocal signaling in cavefish (Fig. 3). When a surface-fish lens is transplanted into a cavefish optic cup, eye degeneration is prevented and an eye is restored in the host. The restored eye exhibits a complete anterior sector, with a cornea, anterior chamber, iris, and an enlarged retina containing photoreceptor cells. An optic nerve also forms, connecting the retinal ganglion cells to visual centers in the brain, although it remains to be determined whether the cavefish host regains its sight. Thus, the surface-fish lens can stimulate growth and differentiation of eye parts in cavefish, indicating that the latter have retained the ability to respond to lens signaling.

Conversely, when a cavefish lens is transplanted into a surface-fish optic cup, or a surface-fish lens is extirpated, the cornea, anterior chamber, and iris fail to form, retinal growth is retarded, and the eye sinks into the orbit, just as it does in cavefish. Some retinal growth and differentiation does persist in the surface-fish host with an extirpated lens, however, showing that the lens does not control all aspects of eye formation.

The lens is thus a major signaling center for anterior sector development, and also has a significant influence on retinal growth and photoreceptor differentiation. The transplantation studies also demonstrate that evolutionary changes in the lens are the primary (but not the only) cause of cavefish eye degeneration.

Genetic control. Cavefish are also being used as a model system to identify the genes involved in eye degeneration. Matings between surface-fish and cavefish show that multiple genes control cavefish eye development. To identify these genes, the expression patterns of eye regulatory genes have been compared during surface-fish and cavefish development.

The *Prox 1* gene, encoding a DNA transcription factor, controls the differentiation of lens fiber cells and several retinal cell types during vertebrate eye development. No differences in *Prox 1* expression

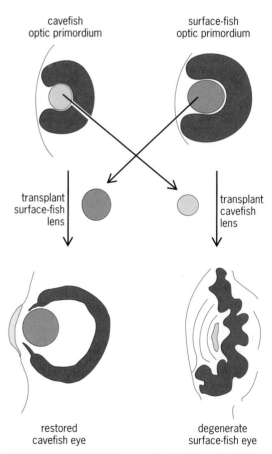

Fig. 3. Lens transplantation experiments. Transplantation of a surface-fish lens vesicle into a cavefish optic cup results in a restored eye in the cavefish host. Transplantation of a cavefish lens vesicle into a surface-fish optic cup results in eye degeneration in the surface-fish host.

are found between surface fish and cavefish during eye development, suggesting that the important genetic changes must occur before the function of the *Prox 1* gene. Alternatively, the genetic changes in eye development could occur in genes that function independently of the *Prox 1* gene. *Prox 1* gene expression is initiated relatively late in eye development, after size differences are already apparent between the surface-fish and cavefish optic primordia. In contrast, the *Pax 6* gene, another transcription factor which is expressed in all eye tissues and controls eye development throughout the animal kingdom, is expressed earlier in the optic primordia.

Two striking differences in *Pax 6* expression have been discovered in early cavefish embryos as compared with the surface fish. First, the regions of *Pax 6* expression corresponding to the eye primordia are reduced in cavefish, consistent with the smaller eyes formed later in development. Second, the regions of *Pax 6* expression are widely separated at the anterior midline of early cavefish embryos. Signaling molecules emanating from the anterior midline of the embryo control the size and bilateral positioning of the eye primordia by suppressing the *Pax 6* gene. The wide gap in *Pax 6* expression in cavefish suggests that the genes that control midline signaling may be more active in cavefish. More extensive midline signaling results in *Pax 6* suppression and a smaller eye and lens, which later degenerates. Thus, enhancement rather than suppression of gene activity may ultimately explain lens and eye degeneration not just in *Astyanax mexicanus* but in all vertebrates.

For background information *see* CELL DIFFERENTIATION; CELLULAR ADHESION; EYE; GASTRULATION; GENE; TRANSPLANTATION BIOLOGY in the McGraw-Hill Encyclopedia of Science & Technology.

William Jeffery

Bibliography. L. Browder et al., *Developmental Biology*, 3d ed., pp. 685–721, Saunders College Publishing, 1991; W. J. Gehring and K. Ikeo, *Pax 6*: Mastering eye morphogenesis and evolution, *Trends Genet.*, 15:371–377, 1999; W. R. Jeffery, Cavefish as a model system in evolutionary developmental biology, *Dev. Biol.*, 231:1–12, 2001; W. R. Jeffery et al., *Prox 1* in eye degeneration and sensory organ compensation during development and evolution of the cavefish *Astynanx*, *Dev. Genet. Evol.*, 210:223–230, 2000; W. R. Jeffery and D. P. Martasian, Evolution of eye regression in the cavefish *Astyanax*: Apoptosis and the *Pax-6* gene, *Amer. Zool.*, 38:685–696, 1998; M. S. Saha et al., Vertebrate eye development, *Curr. Opin. Genet. Dev.*, 2:582–588, 1992; Y. Yamamoto and W. R. Jeffery, Central role for the lens in cavefish eye degeneration, *Science*, 289:631–633, 2000.

Fish hydrodynamics

Interest in fish swimming stems not only from the agility of certain species of fish and mammals in water but also from the fact that fish propulsion is so different from that of ships and submarines. Whereas these craft are designed to avoid unsteady motions as much as possible, fish employ rhythmic motions as their principal mode for swimming and turning. Recent studies have demonstrated that fish exploit unsteadiness to generate large, short-duration forces; coordinate the rhythmic unsteady body and tail motion to minimize the energy required for steady propulsion; and coordinate the motion of the body and tail to minimize the expended energy during maneuvering.

Fish propulsion. The airflow around a car moving in the rain can be observed thanks to the droplets that make the flow patterns visible. The flow around and behind the car appears to be a confused mess of irregular patterns. It comes as a surprise, then, to observe the flow around a swimming fish, made possible by the development of new visualization techniques. It is a very orderly flow, consisting of large vortices, that is, large masses of swirling flow, positioned at predictable distances from each other. The large, regular—but invisible to the unaided eye—flow patterns surrounding a fish, produced through precise body motion control, are intriguing and have shed light on a number of flow-control mechanisms that were previously unexplored.

Propulsion physics. In order to move forward at steady speed, a fish must push a certain mass of water in the opposite direction so as to generate, by reaction, the thrust force that will overcome the drag on its body. Hence, a jet forms behind a moving fish in the same way as a propeller in the stern of a ship generates a jet of water. The propeller jet consists of helical vortices, shed by the propeller blades, especially near the tips. These vortices have the same helical shape as the grooves around a wood screw, and they are stable and persistent, decaying slowly downstream from the propeller. By contrast, the jet behind a fish contains large, regularly shaped vortices, which resemble interconnecting vortex rings of fluid (vortices in which the vortex lines, about which fluid particles circulate, are themselves closed curves). The wake can be observed by shining, for example, a sheet of laser light in a flow seeded with small particles, in a plane that contains the direction of forward motion and the direction of transverse motion of the fish at about the midheight level. A cut is then seen through the vortex rings, consisting of large alternating-sign vortices arranged in such a way as to induce a jet flow, forming a reverse Kármán street (**Fig. 1**). Such flows are orderly, but the vortices interact strongly with each other and hence are dynamically evolving, causing continuous changes in the shape of the jet flow.

Vortex description. Regular Kármán streets form spontaneously behind bluff bodies—such as behind a long cylinder placed with its axis transverse to an oncoming stream—and also consist of two rows of alternating-sign vortices (but rotating in the direction opposite to that behind fish). They are characterized by a universal parameter, the Strouhal number, St, given by Eq. (1), where f is the frequency of vortex

$$\text{St} = \frac{fd}{U} \qquad (1)$$

formation, d the diameter of the body, and U the stream velocity. The value of the Strouhal number is surprisingly constant, around St $= 0.20$, over a very wide parametric range.

The similarity between the Kármán street behind bluff bodies and the vortical street behind a swimming fish is not fortuitous. Vortices and their interactions are the key mechanism in both flows, so a similar Strouhal number can be formed, given by Eq. (2), where f is the frequency of vortex forma-

$$\text{St} = \frac{fA}{U} \qquad (2)$$

tion, A the excursion of the fish tail, and U the speed of the fish. Fish, from small goldfish to large sharks and dolphins, swim with Strouhal numbers between 0.25 and 0.35, suggesting a new universal quantity. The explanation is that a reverse Kármán street, consisting of a pair of oppositely rotating vortices per cycle, requires the minimum energy for a given thrust, compared to streets with multiple vortices per cycle.

Fish motion. While this global picture is satisfactory in explaining the appearance of the vortical structures in the wake behind a fish, it does not provide the details of how vortices form near the body of the fish. Most fish flex their body from side to side, in addition to the tail, despite the fact that they are perfectly capable of propulsion with the fins alone—like the triggerfish.

Depending on the portion of the body that participates in the unsteady motion, fish motion can be classified into three broad categories. An eel (Latin

name, *anguilla*) moves its entire body, from head to tail, with considerable amplitude, although some increase in amplitude is noted going from head to tail; this motion is called anguilliform. Fish which move their head slightly but build considerable amplitude of motion toward the tail belong to the carangiform category. A tuna (*thunnus*) which moves only the latter third of its body is in the thunniform mode. These are broad characterizations, and sometimes it is not clear how to characterize a fish species.

The motion of the body can produce a sustained thrust only if vortices are shed from it. The shedding of vortices from the fish can occur either at the edges of fins and finlets attached to the body, or directly from the body, especially in the region near the tail, where the transverse dimensions of the fish decrease. The lateral motion of the body is in the form of a traveling wave of increasing amplitude from head to tail. The ratio, c/U, of the speed at which the wave travels to the speed of travel is a significant parameter, and takes a value around 1.1–1.2 for the faster fish and somewhat higher for other fishes.

Vorticity control. To explain the body motion requires complex flow visualizations from simulation or experiment. Although the length of the fish is several times larger than the transverse dimensions, the body does not behave like a low-aspect-ratio airfoil as might be expected, which would entail substantial velocity in the transverse direction, especially around the edges of the fish. Instead, there is significant velocity in the direction along the length of the fish, a characteristic of a higher-aspect-ratio foil. A qualitative picture emerges (Fig. 1) which shows the flow around the body to have a form as if there were vortices whose core lies inside the body (called descriptively bound vorticity), traveling down the body of the fish. Once these vortices reach near the tail, prior to the peduncle, they are shed into the flow and then interact with the tail. The tail sheds additional vorticity and repositions the body-shed vortices through direct action, or indirectly through pairing them with additional vortices. Thrust develops optimally when a jet is formed with two vortices of opposite sign being generated per cycle and positioned such as to form a reverse Kármán street when viewed in a planar cut. This is ensured by the action of the tail, which repositions the oncoming vorticity at the proper side of the wake, depending on its sign, while vortex pairing and vortex destructive interference make sure that the dominant structure is that of a reverse Kármán street.

The principal technique, then, for understanding the gross features of fish swimming is to study the interaction of vortices with each other and with other flow features so as to derive rules for manipulating a flow (vorticity control). The principal modes of vortex-to-vortex interactions are (1) constructive interference when two vortices of the same sign coalesce into a larger eddy, and (2) destructive interference when two vortices of opposite sign pair up to form an effectively weaker vortex. Fish

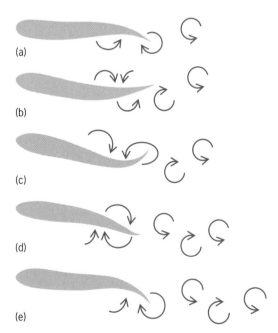

Fig. 1. Flow patterns around a swimming fish along a laser-illuminated plane. (a–e) Sequential views. Vortices are initially body-bound and are ultimately released in the wake to form a reverse Kármán street. (*After M. J. Wolfgang et al., Near-body flow dynamics in swimming fish, J. Exp. Biol., 202:2303–2327, 1999, The Company of Biologists Limited*)

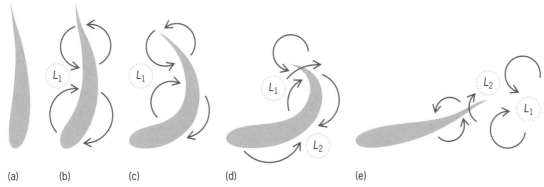

Fig. 2. Formation of a pair of vortices, L_1 and L_2, in the wake of a maneuvering fish. (*a–c*) Generation of vortices, primarily by the body. (*d–e*) Shedding of vortex ring structure (including L_1 and L_2) in the fluid. (*After M. J. Wolfgang et al., Near-body flow dynamics in swimming fish, J. Exp. Biol., 202:2303–2327, 1999, The Company of Biologists Limited*)

employ the latter for optimal propulsion and the former for strong thrust production, when, for example, they accelerate. Indeed, the generation of body-bound, traveling vorticity, its shedding into the fluid, and the subsequent tail manipulation of the oncoming vorticity to generate a thrust-producing jet provide a good overall description of fish swimming.

Boundary-layer turbulence. Still, there are further details that shed light on a flow mechanism that is at least as important: The active motion of the body alters the small region close to the skin, the boundary layer. This region, despite its relatively small size compared to the body volume, has critical importance because it is the origin of the force that opposes the motion of the fish. The boundary layer, at very small body speeds, consists of orderly flow patterns parallel to the body contour. As the speed increases, the patterns spontaneously become disorderly, and the flow is called turbulent. The energy sustaining this disorderly motion comes from the body itself, causing a large increase in the retarding force. Despite efforts for several decades, the exact mechanisms of turbulence production, and especially turbulence control, remain only partially understood. It is known, though, that turbulence can be controlled in a number of ways, such as blowing or sucking fluid from the surface of the body, or moving appropriately the surface of the body.

It turns out that the transverse body flexing of the fish is another way to reduce turbulence. When the body surface moves sideways in the form of a traveling wave, and when the speed of travel of the wave is comparable to the flow speed, it is observed that the intensity of turbulence is markedly reduced. This, in turn, causes the form drag to be reduced.

Fish maneuvering. Fish have outstanding agility, being capable of turning rapidly within a fraction of their length, and are also capable of fast starting with accelerations several times the acceleration of gravity. Simple momentum arguments show that a large transverse force must develop to produce this performance, and also that the drag force from the fluid must be small.

Again, to produce the required large force, the fish must push back a certain mass of fluid at a corresponding speed. This requires the generation of vortices, which interact with each other to move away from the body. A vortex ring is one such arrangement of distributed vortices which causes them to move in a direction perpendicular to the plane containing the ring. Fish generate a vortex ring by flexing their body—the details of this process are quite complex. By creating a laser sheet of light, it is possible to visualize the most important features of the generation of this ring (**Fig. 2**). The intersection of a vortex ring with the plane of a sheet of light shows two vortices of opposite rotation. Initially, the flow appears as if these two vortices are "inside" the body of the fish. To complete the three-dimensional picture of a ring, it is necessary to imagine these two "body-bound" vortices connected with vortices at the edges of the body. The body-generated vorticity is repositioned by the tail and interacts with tail-shed vorticity to form a precisely controlled ring.

The timing of formation, shedding, and positioning of the vortices is crucial to an effective maneuver. In fact, flow visualization shows the absence of any other (parasitic) shed vorticity, hence the absence of separation drag. A rigid streamlined body undergoing a similar maneuver would be subject to considerable flow separation and resulting drag force. The rapid generation of a vortex ring through body flexing and tail manipulation, and the absence of separation drag, explain the agility of fish.

As in the case of a straight-swimming fish, when a planar cut is visualized the flow has a two-dimensional appearance except close to the edges of the fish. Hence, vorticity control by the tail may be described qualitatively in terms of two-dimensional vortex manipulation concepts, although the detailed flow picture and force estimation require a three-dimensional procedure.

Foils and fins. Fish employ their fins and tails to produce forces for propulsion and maneuvering. The process of generation of these forces is very similar qualitatively to that described for the overall fish. The primary hydrodynamic feature behind propulsive foils is the generation of a jet coexisting with large vortical structures which, under optimal conditions, have been visualized to have the form of a reverse Kármán street, consisting of two vortices per cycle, when viewed in a planar cut. The generation of unsteady forces to effect sharp turning or maneuvering involves the optimal generation of vortex rings, first through the formation of leading-edge vortices, and then through manipulation of these vortices by additional vortices shed by the trailing edge. A nondimensional frequency, the Strouhal number, is again a basic parameter controlling unsteadily operating foils.

Biomimetics. The agility of marine animals moving underwater has always generated great interest in identifying the basic principles and the details of their hydrodynamics, sensing, and control, to devise improved technological apparatus. Biomimetics, the field of extracting basic principles from biological systems in order to improve technological systems, has been employed to design a number of devices and vehicles for underwater use. *See* NEUROMORPHIC ENGINEERING.

Fish employ rhythmic unsteady motions for propulsion and maneuvering, a totally different paradigm from human-made vehicles. For flow control they exploit the basic flow principles outlined above. Their body control functions constitute a different paradigm as well, since the basic rhythm of oscillation must be continuously and smoothly adapted to a changing environment as sensed through an elaborate system, including the lateral line.

Fish employ muscles and tendons capable of storing and releasing energy: Under sustained swimming conditions, they tune their frequency of operation to the natural frequency of their body (the mass of their body and the elasticity of their muscular system can be thought of as a mass-spring system with an associated natural frequency of oscillation) to minimize energy losses. This same principle should be applied to any technological device employing a fishlike swimming mode for propulsion and maneuvering.

For background information *see* ANGUILLIFORMES; BIOMECHANICS; BOUNDARY-LAYER FLOW; FLIGHT; FLUID FLOW; HYDRODYNAMICS; KÁRMÁN VORTEX STREET; LATERAL LINE SYSTEM; MUSCULAR SYSTEM; TUNA; TURBULENT FLOW; VORTEX; WAKE FLOW in the McGraw-Hill Encyclopedia of Science & Technology.

Michael S. Triantafyllou

Bibliography. U. Mueller et al., Fish foot prints: Morphology and energetics of the wake behind a continuously swimming mullet (*Chelon labrosus Risso*), *J. Exp. Biol.*, 200:2893–2906, 1997; M. S. Triantafyllou and G. S. Triantafyllou, An efficient swimming machine, *Sci. Amer.*, 272(3):64–70, March 1995; M. S. Triantafyllou, G. S. Triantafyllou, and D. K. P. Yue, Hydrodynamics of fishlike swimming, *Annu. Rev. Fluid Mech.*, 32:33–53, 2000; M. J. Wolfgang et al., Near-body flow dynamics in swimming fish, *J. Exp. Biol.*, 202:2303–2327, 1999.

Fisheries ecology

Climate has shaped the course of history. Subtle variability in climate often dictates when there is famine and when there is plentitude. This is conveyed most strongly when normal patterns suddenly change. In Peru, for example, farmers have historically recognized *años de abundancia* (years of abundance) every 2 to 7 years. Unusually warm temperatures are accompanied by heavy rains, turning the coastal desert to abundant pasture. Crops and livestock thrive in regions where they typically languish. However, the sea, usually teeming with fish, becomes anomalously warm and dormant. This phenomenon, called El Niño (Little Boy or Christ Child) by Peruvians for its tendency to begin around Christmas, is now recognized as a disruption of the ocean-atmosphere system in the tropical waters of the Pacific with consequences felt around the globe. El Niño has received widespread attention because the interplay between winds and ocean currents affects not only the climate but also fisheries, marine biology, and, ultimately, human welfare.

In normal, non–El Niño conditions, winds moving from the Poles toward the Equator act in combination with the spinning of the Earth to move surface waters offshore and draw cold, deep water to the surface. This upwelling of deep water occurs along the eastern margins of all ocean basins. Upwelled water infuses surface waters with essential plant nutrients such as nitrate, phosphate, and silicic acid; and this often leads to blooms of phytoplankton, which form the foundation of the food chains that support coastal fisheries, seabirds, and marine mammals. Along the North American Pacific coastline, upwelling occurs during periods of strong northwesterly winds in spring and early summer, producing a band of cold water along the coast. This band is typically tens of kilometers wide and is separated from offshore warmer water by a series of highly variable jets, plumes, and eddies. During El Niño events, trade winds in the central and western Pacific relax, reducing the amount of cool, nutrient-rich upwelled water. As a consequence, temperatures on the sea surface rise, and the lack of nutrients leads to drastic declines in phytoplankton and, subsequently, all animals in the food web. Eventually, usually after a period of 1 or 2 years, normal trade winds resume, and cool, nutrient-rich water again dominates the system. While the Atlantic and Indian oceans experience events similar to the Pacific's El Niño, these events are not nearly as strong, nor do they influence marine life to the extent El Niño does.

El Niño impact. The impacts of El Niño on the world's fisheries are profound and mostly detrimental. By decreasing the food resources of the fish, El Niño episodes may dramatically reduce the size of fish stocks. Additionally, changes in sea surface temperature and ocean current patterns may force fish to move to new areas or to shift their patterns of migration.

The most notorious impact of El Niño has been its devastating effect on the Peruvian anchoveta (anchovy) fishery. The anchoveta is a small filter feeder that lives in tremendous schools in the upwelled waters of the Peruvian Current, along the west coast of South America. In the past, enormous stocks of anchoveta fed upon the abundant plankton there and served as prey for large colonies of seabirds.

Direct human exploitation of the anchoveta began in the 1950s, and the fishery increased at a phenomenal rate through the 1960s, when it ranked as the largest fishery in the world. Fisheries biologists projected that the maximum sustainable yield of anchoveta was 10–11 million metric tons. After subtracting 2 million metric tons to support seabird populations, biologists estimated that a harvest of 9–10 million metric tons could be sustained. From 1964 to 1971, the catch of anchoveta was close to this theoretical maximum, even exceeding it in 1970, when the catch was 12.3 million metric tons, more than 18% of the entire world catch.

An El Niño event began early in 1972, and the upwelling system off the Peruvian coast weakened. The resulting reduction in food resources forced the anchoveta to move to patches of cooler water to the south. Although the population had declined by 80% of normal levels, the remaining fish were concentrated in small areas and heavy fishing of these aggregations produced large catches. By June 1972, catches declined drastically and young fish were no longer entering the fishery. Protective measures that greatly reduced fishing pressure resulted in the recovery of the anchoveta stock by the early 1990s. By the late 1990s, however, fishing pressure once again exceeded the maximum sustainable yield. With the unusually severe 1997–1998 El Niño, there was another crash in the anchoveta fishery with reverberations to other industries. For example, anchoveta are typically exported as fish meal for chicken feed. The crash of the anchoveta fishery caused United States farmers to use higher-priced chicken feed, and ultimately caused a rise in egg prices.

While the impact of El Niño on the Peruvian anchoveta fishery is the most well known, El Niño events clearly affect other fisheries worldwide. Like the anchoveta, Pacific salmon respond strongly to changes in their food supply. El Niño episodes off the Oregon coast not only reduce the food supply but also alter the species composition of the fish prey. For example, during warm El Niño years the zooplankton are dominated by krill and copepod species typically found in warmer waters to the south. Studies of coho salmon indicate that such changes in zooplankton are associated with a nearly 60% decrease in survival. Moreover, those salmon that manage to survive tend to be smaller.

El Niño events also impact migrations of fish by changing current patterns and sea surface temperature. Bluefin tuna, for example, undergo fantastic migrations throughout the Pacific. During normal years, juvenile bluefin tuna migrate at roughly 40°N latitude from Asia eastward across the North Pacific as far as California. After traveling south, they make a return westward migration along 20°S and spawn near Japan. During El Niño years, however, bluefin tuna shift their east-west migratory loop southward by 10–15° in response to unusually cold water that develops in the central North Pacific. Similar changes in distribution occur in albacore and skipjack tunas. Additionally, it appears that the locations where the preferred temperature conditions of tuna occur become limited during El Niño episodes; this forces fish into smaller areas and increases the ease with which they are captured by fishers.

Since nearly all fish produce larvae that are pelagic (live in the surface waters of the ocean) and may be dispersed by currents, changes in currents also impact where and when young fish develop. Such changes in currents are particularly a problem for fish that live in habitats close to shore (for example, kelp forests, coral reefs, seagrass meadows) because their larvae eventually must manage to get back to shore. Comparisons among rockfishes (which include important food fish marketed as Pacific snapper) that live in kelp forests along the California coast have been particularly instructive in highlighting how El Niño events affect the transport of fish larvae. Some species of rockfishes produce larvae that reside in deep, offshore waters. In normal years, upwelling causes deep offshore water to flow toward shore. As a result, the rockfish larvae can flow with currents to kelp forests along the shore. During El Niño episodes, the disruption of upwelling stops the deep onshore current. Rockfish larvae that depend on this current are stranded offshore, away from their adult habitat, and eventually die. In contrast, other species of rockfish produce larvae that reside near the surface of the water. Because upwelling normally forces surface waters offshore, these larvae tend to be transported away from adult habitats along the coast. During non-El Niño years, larvae of these species only rarely make it to the coast, and most die in offshore waters. However, during El Niño events the offshore surface currents are disrupted and the surface-dwelling rockfish larvae are retained in coastal waters near the kelp forests they will use as adults. Interestingly, rockfish are very long lived fish (some live more than 90 years). Since El Niño events occur every 2–7 years, both groups of these rockfish have many opportunities to successfully reproduce.

Fisheries management during El Niño. The twists and turns in the ongoing dialogue between the ocean and fish populations have important implications for how fishery resources are managed around the globe. Scientists frequently determine the maximum sustainable yield of fish stocks based on average conditions over a number of years. Clearly, conditions during El Niño episodes are not average, making predictions based on average environments prone to major mistakes. What is needed to better manage the world's fishery resources in the face of irregular climatic rhythms such as El Niño? First, accurate predictions of El Niño events must be achieved. Then, an understanding of how different fish species respond to changing ocean conditions must be developed. Finally, management schemes must be altered accordingly.

Indeed, scientists have made great strides in recent years. Climatologists now use computer models to predict what ocean conditions will prevail several years into the future. The results of such models are by no means perfect, but they give a better indication of the conditions that fish are likely to face, rather than simply assuming conditions will be average. The

Peruvian anchoveta provides a good example of how such El Niño forecasts might be used in fisheries management. Since 1983, El Niño forecasts have been issued in Peru. Typically, these forecasts would predict one of four possibilities: (1) average conditions; (2) a weak El Niño with some disruption of the upwelling system; (3) a major El Niño that strongly interferes with upwelling; or (4) cooler-than-normal waters offshore (also known as La Niña). Once the forecast is issued, fisheries scientists and government officials can decide on the appropriate levels of harvest.

The ability to use forecasts to alter harvest levels requires some understanding of the mechanisms by which El Niño affects fish stocks being managed. The response of fishery managers to an impending El Niño would certainly vary if the effects of El Niño were based on changes in food resources, migratory patterns, or juvenile survival. While there is a copious literature describing the impacts of El Niño on the productivity of fishes, much of this work has simply described patterns, rather than the underlying processes. Clearly, the management of fisheries affected by El Niño needs to be tuned to the biology of the fish. Attention to such biological processes, in concert with accurate climate models, will begin to give fishery managers the tools needed to protect the world's fisheries.

For background information *see* CLIMATE MODELING; EL NIÑO; FISHERIES ECOLOGY; MARINE CONSERVATION; MARINE FISHERIES; WEATHER FORECASTING AND PREDICTION in the McGraw-Hill Encyclopedia of Science & Technology.   Phillip S. Levin

Bibliography. M. H. Glantz, Currents of change: El Niño's impact on climate and society, Cambridge University Press, New York, 1996; P. J. Harrison and T. R. Parsons, *Fisheries Oceanography: An Integrative Approach to Fisheries Ecology and Management*, Blackwell Science, 2000; National Research Council, *Learning To Predict Climate Variations Associated with El Niño and the Southern Oscillation: Accomplishments and Legacies of the TOGA Program*, National Academy Press, Washington, DC, 1996; W. G. Pearcy, *Ocean Ecology of North Pacific Salmonids*, University of Washington Press, Seattle, 1992; G. S. Philander, *El Niño, La Niña, and the Southern Oscillation*, Academic Press, San Diego, 1990.

Fission yeast

Biologists typically reduce a process to its simplest form in order to identify the fundamental mechanisms driving it. Cell division is one such process, and understanding the mechanisms controlling the cell division cycle is essential for understanding human growth and development. Diseases such as cancer are characterized by cell division in which these controls have gone awry. The simplest form of cell division likely to give insight to the process in humans is that occurring in unicellular eukaryotes, such as yeast. Yeast is a general term that covers a broad range of unicellular fungi. One of the most useful species is the fission yeast, *Schizosaccharomyces pombe*, which has become a model system for studies of growth control and cancer genetics.

Cell cycle. *Schizosaccharomyces pombe* is considered to be as distant in evolution from animals as from baker's yeast, *Saccharomyces cerevisiae*. Originally isolated as an African brewing yeast, *S. pombe* is harmless and fast-growing, as well as genetically and molecularly tractable. The cell cycle of the fission yeast is similar to that of animal cells (**Fig. 1**). It consists of an S phase (DNA synthesis phase), during which the chromosomes are duplicated, and an M phase (mitosis), during which the chromosomes are segregated. These are separated by G1 and G2 phases, during which cell growth occurs. *Schizosaccharomyces pombe* normally grows as a haploid, and when growing rapidly in the laboratory, it spends about 70% of its cycle time in the G2 phase.

When cells are starved for nutrients, they exit the cell cycle and enter a stationary phase: a period of dormancy that is more severe than the G0 phase in mammalian cells. However, if a cell of the opposite mating type is available, two cells will fuse, forming a transient diploid that will immediately progress into meiosis (not shown in Fig. 1). The four spores packaged in the yeast ascus (spore sac) are the fungal equivalent of human gametes. Thus, the entire life cycle of *S. pombe* provides a simple model for events occurring in human cells.

Cell cycle regulation. Classical genetics is the greatest tool in the study of cell cycle regulation in *S. pombe*. Investigators can isolate conditional mutations affecting cell division genes, such as mutations in which the affected protein is functional at a low temperature but defective at a high temperature. By examining the appearance of cells with these mutations (their phenotype), predictions can be made about the function of the mutated gene and then can be tested biochemically. Such analysis has identified important control mechanisms that are remarkably conserved in human cells. These mechanisms are particularly relevant to cancer, in which cells have lost control of normal division. Aberrant activity of the human equivalents to these yeast genes may cause or contribute to cancer, and thus could be targets for therapy or prevention.

The first step toward identifying such control genes was to isolate yeast cells defective in normal cell cycle regulation. Mutations that block the cell division cycle but do not affect overall cell growth lead to elongated morphology and eventual cell death. These *cdc* mutants were among the first identified (**Fig. 2**). Most crucial, however, was identification of cells that divided at a smaller than normal size: the *wee* phenotype. These mutants changed the rate of progression through the cell cycle without otherwise affecting cell viability. This led to a model proposing that rate-limiting factors control cell division, because some of the *cdc* mutants identified positive regulators (the cell cycle was blocked in their absence), while the *wee* mutants identified negative regulators (the cell cycle speeded up when they were mutated).

The next step in this early analysis was the cloning and characterization of the genes involved. Fortunately, fission yeast lends itself readily to molecular analysis, which became even easier when the complete genome was sequenced in 2000. Once the sequence of a gene identified by mutation is determined, it is possible to predict the sequence and sometimes the structure of the protein it encodes. Significantly, the gene can be replaced in the cell with selectively mutated versions, or even deleted entirely, to assess the phenotypes associated with a variety of changes. Using this sort of study, some of the major regulators of the cell cycle were identified as Cdc2 kinase, its partner the cyclin Cdc13, Wee1 kinase, and Cdc25 phosphatase.

Wee1 kinase and Cdc25 phosphatase antagonize each other in the regulation of the cyclin-dependent kinase (CDK) Cdc2. Phosphorylation of Cdc2 on a conserved tyrosine by Wee1 inhibits its activity in initiating mitosis, whereas dephosphorylation by Cdc25 promotes Cdc2 activity and mitotic entry. This fine balance is easily disrupted by additional factors: too much inhibitory activity, and the cells cannot enter mitosis; too little, and they enter mitosis so prematurely that they are not big enough to divide, a phenotype called mitotic catastrophe (Fig. 2). The cyclin is an essential partner required for Cdc2 kinase activity, and its levels vary through the cell cycle, so periodic association with the cyclin provides additional regulation. Significantly, this basic mechanism was shown to be the same in all eukaryotes, providing the first evidence that fundamental events of the cell cycle are conserved.

The fission yeast cell uses a particularly streamlined version of this mechanism. Subsequent studies in metazoan cells identified additional cyclin-kinase pairs that act throughout the cell cycle, rather than a single kinase with different cyclin partners, as in *S. pombe*. In human cells, there are multiple additional factors that regulate CDK activity. These include transcription factors that affect the synthesis of the cyclin or small inhibitor molecules; other kinases and phosphatases that activate or inactive the CDK (including Wee1); inhibitor molecules that modulate CDK activity; and proteolytic machinery that regulates destruction of cyclins, inhibitors, and substrates. Thus, the CDK is the final stop for a network of regulatory molecules that control the rate of cell cycle progression. Changes in the network directly affect the rate of cell cycle progression, and genes identified in human cancers are often components of this network. The genetic, molecular, and biochemical data gleaned from the yeast system are thus directly relevant to the study of malignancy in human cells.

Further studies. The establishment of the CDK paradigm for cell cycle regulation spurred further investigation of the regulation of cell cycle events in fission yeast cells. Two hotly studied areas are particularly relevant: cell cycle order, and chromosome structure and replication.

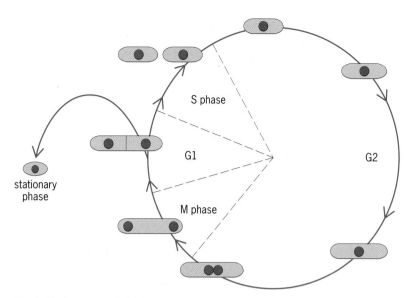

Fig. 1. Fission yeast cell division cycle. The solid ellipticals are nuclei.

Cell cycle order. How do cells maintain the order of events during the normal cell cycle and in response to perturbations that affect cell cycle progression? Normally, cells can undergo only a single S phase per cell cycle. Mutants were identified that allowed cells to repeat S phase multiple times without any intervening mitosis, leading to a re-replication phenotype (Fig. 2). Interestingly, most of these proved to be components of CDK regulation, demonstrating that in addition to promoting mitosis, Cdc2 is required to switch off S phase.

Additional studies identified a number of genes that are not required for normal cell division but become essential when the cells are damaged (for example, by radiation) or treated with drugs (for example, drugs that inhibit the mitotic spindle). Under these conditions, normal cells will activate a series of

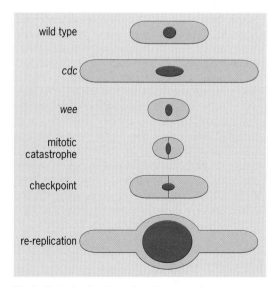

Fig. 2. Mutants affecting cell cycle progression or regulation show characteristic phenotypes. The solid ellipticals are nuclei.

"checkpoint" genes that function to delay the next step of the cell cycle until appropriate repair occurs. In the absence of the checkpoint, the cells do not delay the cell cycle in response to perturbation, and proceed through replication and mitosis as though they have not been damaged. This can be fatal to the yeast or cause damage or mutation (Fig. 2). These checkpoint genes are conserved in mammals, and thus are additional crucial regulators of the cell cycle. Mutations in these genes can make cells particularly prone to damage or genetic instability that promotes cancer.

Chromosome structure and replication. The haploid fission yeast has three chromosomes, each of which contain multiple origins of DNA replication. The activity of these origins depends upon conserved proteins found in all eukaryotes, and *S. pombe* lends itself readily to dissection of the details of protein function. Replication accuracy is particularly important because it is the accumulation of mutations in human cells that ultimately leads to sufficient defects to cause cancer. Studies of DNA repair and cellular response to damage are also important in *S. pombe*. Aspects of chromosome metabolism, replication origin and centromere structure, and the presence of conserved chromatin factors also are surprisingly similar between the fission yeast and larger eukaryotes.

Work in *S. pombe* has thus made significant contributions to understanding the regulation of normal human cell growth. Continued investigations of the mechanisms that maintain genome integrity and chromosome dynamics in this tractable model organism promise to provide additional insights into human cell biology in the years to come.

For background information *see* CELL (BIOLOGY); CELL CYCLE; CELL DIVISION; CHROMOSOME; MITOSIS; YEAST in the McGraw-Hill Encyclopedia of Science & Technology.   Susan L. Forsburg

Bibliography. S. L. Forsburg, The best yeast?, *Trends Genet.*, 15:340-344, 1999; L. H. Hartwell and M. B. Kastan, Cell cycle control and cancer, *Science*, 266: 1821-1828, 1994; J. Hayles and P. Nurse, Genetics of the fission yeast *Schizosaccharomyces pombe*, *Annu. Rev. Genet.*, 26:373-402, 1992; S. A. MacNeill and P. Nurse, Cell cycle control in fission yeast, in J. Pringle, J. Broach, and E. W. Jones (eds.), *The Molecular and Cellular Biology of the Yeast Saccharomyces*, vol. 3 of *Cell Cycle and Cell Biology*, Cold Spring Harbor Laboratory, New York, 1997; A. Murray and T. Hunt, *The Cell Cycle*, W. H. Freeman, New York, 1993; P. Nurse, A long twentieth century of the cell cycle and beyond, *Cell*, 100:71-78, 2000; M. Sipiczki, Where does fission yeast sit on the tree of life?, *Genome Biol. 2000*, 1:1011.1-1011.4, 2000.

Food web theory

Food webs describe the feeding, or trophic, relationships between species. They are road maps for the flow of energy and matter in an ecosystem (**illus. *a***).

As a result, food webs are an increasingly important tool for understanding how ecosystems should respond to perturbation as well as how controlled perturbations might be used to obtain desired ecosystem properties (for example, low occurrence of ecological pests). The information conveyed by a food web ranges from a static map of trophic connections (topological webs) to a more dynamic description of the relative importance of each trophic connection. In the latter case, relative importance can be represented in terms of either the impact of a consumer on resource abundance (interaction webs) or energy flow to a consumer (energetic webs). In empirical studies, food webs are often simplified to linear food chains that link important dynamic pathways between primary and secondary consumers and basal resources, or trophic levels (illus. *b*).

Historically, food web research has focused on three areas: (1) how food web topology determines ecosystem stability; (2) how species interactions determine the diversity of species in a food web; and (3) the relationship between the number of trophic levels, or trophic structure, and the relative abundance of organisms making up each trophic level.

Diversity and stability of model ecosystems. One of the most contentious issues in food web ecology is the relationship between species diversity and ecosystem stability. Many textbooks trace the origin of the debate back to the arguments of Charles Elton and Robert MacArthur. Elton observed that simple ecosystems—like those found in laboratory cultures, agroecosystems, and relatively species-poor island habitats—were more highly susceptible to invasion by exotic species or outbreaks of pests, and thus less likely to persist over time. Elton reasoned that more diverse systems, like the tropics, should be more resistant to these external disturbances. MacArthur added a mechanism to Elton's observations, suggesting that diversity and the number of interactions among the species pool should enhance ecosystem stability because the effects of changes in the abundance of one species would more rapidly dilute across a higher number of pathways in more speciose systems.

These early ideas were enormously influential on ecological thinking but were incomplete because few studies had linked stability to more complex systems. More importantly, this verbal model lacked a well-formulated quantitative theory. This deficiency was redressed by Robert May, who analyzed the stability of model ecosystems varying in complexity and diversity. May constructed artificial food webs with randomly drawn diversity and connectance (links between species in the web). He assigned interaction strength to each trophic link by drawing values from a uniform random distribution bounded by $+/-1$. This was repeated over a range of diversity, connectance, and interaction strength.

May found that stability could be bounded by a relationship between species diversity, connectance, and interaction strength—illustrated by the rule $b(SC)^{1/2} < 1$, where b is the average interaction strength of the system, S is the number of species (diversity), and C is the number of links between species (system connectance). Larger values of S increase the left-hand side of the inequality, resulting in decreased stability. (Note that increasing b or C also destabilizes a food web.) Therefore, May concluded that diversity begets instability, rather than stability—quite a different picture than Elton and MacArthur envisioned.

Some of the food web patterns predicted by May appeared to be confirmed by data from the field. For example, speciose webs tended to have lower connectance, suggesting a trade-off between S and C in stable real webs. This result led others to suggest that omnivory (feeding on more than 1 trophic level) should destabilize food webs, and thus be relatively rare in nature. However, numerous field studies showed that omnivory was not only common but prevalent in real food webs. This contrast between model predictions and empirical observations fueled a growing skepticism about the model assumptions and the data used to test the models.

Interaction strength in real food webs. One potential bias in May's approach was the construction of model food webs in which interaction strength varied randomly. This assumption ignored Robert Paine's earlier empirical results from the rocky intertidal, which suggested that strong interactions were rare. Paine found that the presence of a single species, the seastar *Pisaster ochraceus*, determined species diversity in sessile invertebrate communities. *Pisaster*'s effect was strong, not because *Pisaster* was a common predator in the system, but because it preferentially preyed upon the dominant space-holder and competitor in the system, the mussel *Mytilus californianus*. Reduction in space held by *Mytilus* allowed for successful colonization by sessile species that were otherwise excluded from the system. Paine dubbed *Pisaster* and other species with disproportionately strong effects on community diversity "keystone species."

Paine's pioneering work had two important implications for the plausibility of model ecosystems used by May to address the diversity-stability hypothesis. First, weak interactions ($b \approx 0$) appear to be more common in nature than strong ones ($b \approx +/-1$). Thus, it is not surprising that early theory based on a uniform distribution of interaction strengths (an equal chance of weak and strong interactions) predicted an inverse diversity-stability relationship. As species are added, the number of strong interactions increases faster than it would in nature, adding bias toward instability with increasing diversity. Second, Paine's work demonstrated that diversity (S) and interaction strength (b) are not necessarily independent. Single keystone species may in fact enhance (or in some cases diminish) species diversity. In some cases, these single interactions may summarize the stability properties of the system more adequately than the topology of the food web.

In addition to bias in the model ecosystems, several authors have pointed out potential biases in the

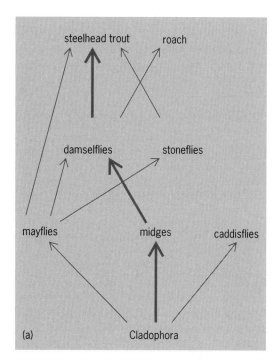 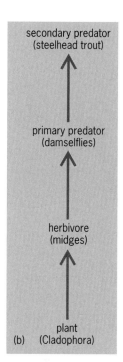

Examples from a northern California stream ecosystem: (a) food web and (b) food chain. (Modified from M. E. Power, M. S. Parker, and J. T. Wootton, Disturbance and food chain length, in G. A. Polis and K. O. Winemiller, eds., Food Webs: Integration of Patterns and Dynamics, Chapman & Hall, New York, 1996)

empirical data sets used for model evaluation. Although results from model food webs suggest that omnivory destabilizes food webs, omnivory has been shown to be the rule rather than the exception in a variety of natural ecosystems. For example, in desert ecosystems, diverse assemblages of arthropods are characterized by equally high connectance as a result of widespread omnivory. The results of several recent field experiments support the opposing view that omnivory stabilizes food webs. This suggests that the data used to evaluate May's predictions were potentially incomplete. Both Paine and Gary Polis argued that data in published food webs could be biased through a trade-off between accurate measurements of diversity and connectance. Effort spent on collecting species (b) in the field may preclude effort spent in the lab characterizing the diets of these species (C). Thus the commonly observed inverse relationship between diversity and connectance may be the simple result of using data collected to address only one of the variables needed to test the diversity-stability question.

Results from more recent theoretical studies suggest that weak interactions are an important component of ecosystem stability, such that diversity may beget stability, as predicted by Elton and MacArthur. For example, increasing numbers of weak or intermediate links stabilize otherwise unstable population cycles between strongly interacting predators and prey when models include populations that do not tend toward an equilibrium (undergo stable population cycles), and interaction strengths are drawn from distributions more similar to those observed in nature. Increasing diet generalization by predators (omnivory) reduces the dependence of predators on a single prey, thereby diminishing the coupling between the predator and a preferred prey. These results suggest that weak interactions are the "glue" that binds communities together, regardless of diversity and connectance.

Food chain dynamics and plant biomass regulation. A final tenet of food web ecology is that "the world is green" (covered with vegetation) because consumption of herbivores by carnivores releases plants from grazing pressure by the exploited herbivore populations. The number of trophic levels determines the effects of herbivores on edible plant tissues: The world should be vegetated when food chains have odd numbers of trophic levels, but barren (plants suppressed by herbivores) when food chains have even numbers of trophic levels. The idea that top predators can influence plants by depressing herbivore densities has been dubbed the trophic cascade because the effects of predators cascade down to lower trophic levels. Trophic cascades appear to be fairly common in lake and stream ecosystems, where fish can have dramatic indirect effects on the standing stocks of edible algae. In both systems, fish constitute the fourth and top trophic level. Experimental removal of these fish leads to release of secondary predators, a decline in herbivore abundance, and a vegetated stream bottom or lake water column (illus. *b*).

In contrast to the prevalence of trophic cascades in fresh-water ecosystems, cascades appear to be less common in terrestrial habitats. On land, the removal of a third trophic level may lead to an increase in the damage caused by herbivores to the edible tissues of plants, but not necessarily reduction in plant biomass. Several explanations have been offered for the difference. First, one of the earliest critiques of green world theory was that the world was green not because of trophic interactions but because not all vegetation can be consumed by herbivores. Herbivores in aquatic systems typically consume whole plants (phytoplankton or algal filaments). However, this is not true in most terrestrial systems. Wood, used by trees (but not most aquatic vegetation) to increase structural integrity, is less edible than softer vegetation in the leaves. As a result, herbivores are simply not able to consume whole plants in the terrestrial realm. Second, structural and chemical defenses of plants against herbivores may prevent these consumers from reducing plant biomass, and these defenses may be more characteristic of higher plants found in terrestrial settings. (However, chemically and structurally defended algae characteristic of highly productive lake ecosystems suggest that defense may be common in both terrestrial and aquatic systems.) Third, it has been argued that cascades occur in aquatic but not terrestrial food webs because aquatic food webs have much lower diversity. Aquatic systems more closely approximate linear food chains, where full exploitation of the dominant algal species at the bottom of the chain is not compensated by growth of unexploited species.

In terrestrial systems, higher species diversity may dilute the effects of top predators on plants; trophic cascades become trophic "trickles." This idea is similar to MacArthur's early ideas, where stability is achieved by increasing numbers of pathways (trophic interactions) by which disturbances may attenuate.

Applications. Diversity-stability theory and the keystone species concept have had a strong influence on environmental policy decisions, with the most direct application to the preservation of biodiversity and the design of ecological reserves. Threatened species can in some cases play keystone roles in ecosystems and have been used as "umbrella species" for the conservation of larger sets of species or whole ecosystems. For example, sea otters have keystone effects on subtidal kelp ecosystems through their predation on urchins that would otherwise clear kelp forests. These species have been used as an umbrella species to protect subtidal habitats in southern California and other areas along the Pacific coast. However, the observation that weak interactions stabilize food webs suggests that while the extinction of species with strong effects in ecosystems may lead to profound ecosystem-level effects, weak interactions may be important in keeping these strong interactions in check when keystone species are above endangered levels. *See also* SUSTAINABLE FOREST MANAGEMENT.

The trophic cascade concept has been applied to manipulate real food webs to produce desired ecosystem attributes in both aquatic and terrestrial ecosystems.

Aquatic ecosystems. Resource managers of freshwater lakes realized early on that fish could be used in lake ecosystems to control blooms of noxious species of algae. Specifically, even or odd numbers of trophic levels in lake food chains would lead to clear (barren) or algae-infested (green) water, respectively.

Terrestrial ecosystems. In an agricultural setting, a green world is most often the desirable state. Healthy plants produce stronger crop yields. One method for reducing crop damage by invertebrate pests is biocontrol—the planned release of crop pest enemies to control pest outbreaks and enhance crop yields. Successful examples of biocontrol include the use of parisitoids (insects that kill their hosts by laying their eggs in or on host bodies) to control the insect pests of citrus, olive, and walnut orchards.

One of the most debated issues in biocontrol is whether successful control agents are characterized by low stable densities of the host or unstable dynamics and extinction. If the latter is true, Elton's original observations may have real application—control of pests by predators in monoculture crop settings may be facilitated by the simplicity of the system.

For background information *see* BIODIVERSITY; ECOLOGICAL COMMUNITIES; ECOLOGICAL ENERGETICS; ECOSYSTEM; FOOD WEB; INSECT CONTROL, BIOLOGICAL; POPULATION ECOLOGY; TROPHIC ECOLOGY in the McGraw-Hill Encyclopedia of Science & Technology.   John Sabo

Bibliography. M. Begon, J. L. Harper, and C. R. Townsend, *Ecology: Individuals, Populations and Communities*, Blackwell Scientific, Oxford, 1996; S. R. Carpenter and J. F. Kitchell, *The Trophic Cascade in Lakes*, Cambridge University Press, 1993; C. B. Huffaker and P. S. Messenger (eds.), *Theory and Practice of Biological Control*, Academic Press, New York, 1976; R. M. May, *Stability and Complexity in Model Ecosystems*, Princeton University Press, 1973; S. L. Pimm, *Food Webs*, Chapman & Hall, New York, 1982; G. A. Polis and K. O. Winemiller, *Food Webs: Integration of Patterns and Dynamics*, Chapman & Hall, New York, 1996.

Forensic chemistry

Forensic chemistry is the application of analytical chemistry to solve disputes in legal proceedings. The legal proceedings may involve legislative, executive, or administrative decisions but usually involve judicial proceedings, such as criminal prosecutions or civil litigation. Forensic chemists are formally trained in analytical chemistry to examine physical evidence and to testify as expert witnesses before a judge or jury. The evidential materials commonly associated with forensic chemistry include residues from explosives, flammables, and discharge from firearms; drugs and poisons; blood and body fluids (including nuclear and mitochondrial DNA extracts); fingerprints; alcohol (including breath and body fluids); inks and pigments; and the general category of trace evidence, which includes fibers, glass, soil, hair, and paint.

The methods and techniques used to examine the physical evidence depend on the type of evidence and the resources available to the forensic chemist. Although microscopy continues to be used as an important tool of the forensic chemist, instrumental methods allow for very sensitive detection and identification of a number of analytes (compounds being analyzed). Analytical chemistry methods that have been applied to analytes of forensic interest include enzyme immunoassays for detection of drugs from body fluids; gas chromatography coupled to mass spectrometry (GC-MS) for the identification of volatile compounds, drugs, and explosives; high-performance liquid chromatography (HPLC) coupled to mass spectrometry for the identification of some drugs and explosives; electrophoretic methods (gel electrophoresis and capillary electrophoresis) for the separation of very small quantities of organic compounds and proteins (including the powerful DNA fingerprinting methods); and spectroscopy methods, such as infrared analysis for the identification of drugs or polymers from fibers and coatings, and microspectrophotometry for the analysis of paints and coatings.

Often, materials are characterized by their trace-element composition in order to distinguish between similar objects. For example, glass evidence is sometimes left behind at a scene of a violent episode such as a hit-and-run accident. The glass recovered from the crime scene can be linked to its source when a suspect vehicle is identified. The more sensitive the analytical methods applied, the better the source glass can be distinguished from a large number of potential sources. The more sensitive techniques achieve a closer association between the recovered material and a suspect source, resulting in better evidence. Inductively coupled plasma mass spectrometry (ICP-MS) is a sensitive method for the characterization of trace-element content when analyses of inorganic materials are required.

A common problem for forensic chemists is the sampling of the analytes of interest from complex matrices. For example, as in the case of a large bombing scene, the task of detecting and identifying very small quantities of explosive residues (on the order of nanograms of analytes) from a large amount of debris (kilotons of building material) can be very difficult. Advances in remote sensing, fast screening, and easy sampling can enable the detection of evidence that previously went unnoticed.

Recent advances. The types of evidential materials analyzed by forensic chemists have remained relatively unchanged over the last 100 years. What has changed is the amount of individualizing information obtained with today's sophisticated instrumentation. Where once a cigarette butt might have given investigators an indication of what brand a suspect smoked, DNA now can be extracted from saliva on the butt and a "genetic fingerprint" obtained and searched against a national database.

This increased application of more sophisticated analytical instrumentation has impacted many areas of forensic chemistry. Recent research in the application of nondestructive techniques has involved the use of instruments such as scanning electron microscopy (SEM) combined with energy dispersive x-ray analysis (EDX), total reflectance x-ray fluorescence (TXRF), and laser ablation inductively coupled plasma (ICP) combined with mass spectrometry in the examination of residues, stains, marks, and trace materials. Drugs, inks, and pigments are more successfully analyzed by capillary electrophoresis (CE) and high-performance liquid chromatography combined with mass spectrometry techniques. Glues, fibers, and coatings are being more fully characterized by pyrolysis gas chromatography and mass spectrometry. Biological fluids are now individualized by new and extremely sensitive techniques, including the polymerase chain reaction (PCR). The ever-increasing use of sophisticated instrumentation has greatly increased the amount of individualizing information available from evidential materials.

Case studies. The following case studies show how new developments in analytical chemistry have led to improved detection and analysis and better characterization of materials.

Poisoned water or contaminated vessel? In 1995, the Miami-Dade Police Department (MDPD) was summoned by the sergeant at arms of the Miami-Dade County Commission chambers because he suspected the possible poisoning of members of the elected county commissioners. A water company had delivered a 5-gallon jug of drinking water, and a foul odor was detected coming from the water. Poison was suspected, and the police department transported the water to the crime laboratory for analysis. The commonly used method to extract organic compounds, such as poisons, is to perform a solvent extraction of the water, concentrate the extract, and run a GC-MS analysis. Solvent extractions can consume a relatively large amount of sample and often produce high background interference from the concentration step. This case coincided with the application of a new forensic technique that the MDPD was working on with Florida International University (FIU). The technique, called solid-phase microextraction (SPME), is an improved method for extracting flammables and explosives from complex matrices while preserving the sample for further analysis. SPME was performed on the water, and diesel fuel was identified at a concentration of less than 500 parts per billion. It was later revealed that one of the water jugs had been used to store diesel fuel, and a small amount of diesel fuel had remained in the jug.

The SPME device (**Fig. 1**) was developed by Janusz Pawliszyn of the University of Waterloo, Canada, for sampling, preconcentrating, and extracting organic compounds from aqueous samples. The SPME device holds a fiber coated with an adsorbent (or absorbent) material that is exposed to the sample. The analytes of interest are preferentially adsorbed (or absorbed) on the sorbent, and the fiber can then be inserted in the injection port of a GC-MS to thermally desorb the analytes for analysis.

The application of this very sensitive technique by forensic chemists at the MDPD and FIU is an example of how a new analytical technique is introduced into the forensic laboratory. In order for a new technique to be accepted by the courts, it must undergo rigorous scrutiny, including general acceptance of the technique within the scientific community and its publication. Hundreds of scientific publications have illustrated the utility of SPME since 1995, and the American Society for Testing and Materials (ASTM) adopted SPME as a standard method of analysis for flammables in 2001.

Hit-and-run manslaughter. On June 1995, a hit-and-run accident occurred in Providenciales, a Caribbean island off the Turks and Caicos. The victim was apparently struck by a vehicle while walking home from work. A local constable carefully documented the crime scene and recovered nine large pieces of glass next to the body. A suspect was identified 11 days later and his vehicle was examined. The vehicle was clean of any body fluids but had recent damage to the driver's-side front fender. It also had a broken headlamp, and fragments of glass were recovered from inside the car bumper.

The common method for glass analysis is to compare the refractive index of the recovered glass with the suspected source. Automobile headlamps, however, all produce similar refractive indices, making it difficult to distinguish one headlamp from another. A more sensitive technique for comparing headlamps was developed around 1983 and then improved, making it a useful tool for distinguishing between all types of glass. The method involves ICP-atomic emission spectroscopy, and most recently ICP-MS. For analysis, a small quantity of the glass sample (~2 mg) is dissolved in acid and aspirated into a hot argon plasma where the metals are ionized and then extracted into a mass spectrometer for identification and quantitation (**Fig. 2**). The instrument can detect metals in the part per trillion to part per billion range, and many metals are analyzed at once. The match of refractive index was not nearly as convincing as the match of the elemental profile, making the evidence more valuable with the use of the sensitive ICP-MS method.

Future directions. In the near future there will likely be more rapid recognition of evidence and better detection of materials, including drugs, explosives, and accelerants at crime scenes via improved biological detection systems (including canine detection), which will continue to enhance the investigator's set of tools in the field. The development and application of more powerful instrumental analysis techniques will allow analysts to detect smaller amounts of chemical substances with greater reliability. Miniaturization will allow for more field analysis with immediate results to aid investigators. As the instrumental detection limits are lowered, they will approach the natural background levels of materials in the environment, and contamination issues will be increasingly important. This will necessitate studies of background levels at these new detection limits and the development of new databases. Statistical methods of analysis will be increasingly important as databanks become available on more evidentiary materials.

The combination of miniaturized instruments, computers, and expanded databases will allow for

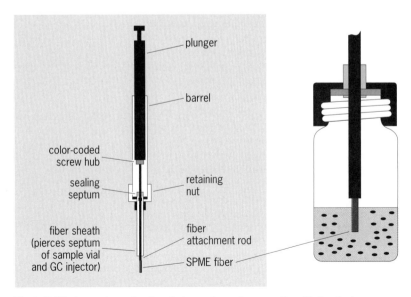

Fig. 1. Solid-phase microextraction device and sampling operation. (*Varian, Inc.*)

plunger

barrel

color-coded screw hub

sealing septum

retaining nut

fiber sheath (pierces septum of sample vial and GC injector)

fiber attachment rod

SPME fiber

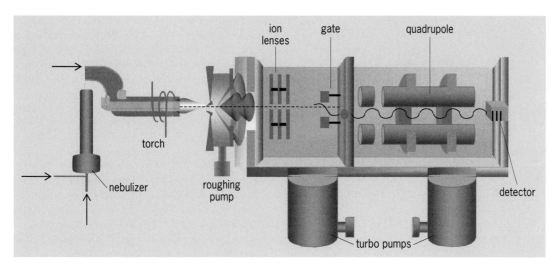

Fig. 2. Inductively coupled plasma mass spectrometer.

immediate on-site analysis, identification, and probability assessment, significantly improving investigations. For example, an on-site sampling of a suspect's hands by SPME-GC might reveal the presence of accelerants, confirming a canine alert or other information. Currently, fingerprints can be scanned directly into a computer system and searched against an Automated Fingerprint Identification System (AFIS). Improved analysis instrumentation may allow for the on-site development of difficult-to-resolve latent fingerprints and additional useful data, such as the age of the print and perhaps the DNA profile of the source. The number of forensic chemistry standard methods will continue to grow rapidly, improving the quality of the analyses.

For background information *see* ANALYTICAL CHEMISTRY; CHROMATOGRAPHY; EMISSION SPECTROCHEMICAL ANALYSIS; FORENSIC CHEMISTRY; GAS CHROMATOGRAPHY; MASS SPECTROMETRY in the McGraw-Hill Encyclopedia of Science & Technology.

José R. Almirall; Kenneth G. Furton

Bibliography. G. Davies (ed.), *Forensic Science*, American Chemical Society, Washington, DC, 1986; R. Saferstein (ed.), *Forensic Science Handbook*, vols. 1–3, Prentice Hall, Englewood Cliffs, NJ, 1982, 1988, 1993.

Forensic engineering

Forensic engineering, which is engineering applied toward the purposes of law, is a rapidly evolving forensics field. In forensic engineering, the majority of investigations are carried out in the context of civil litigation; for example, the cause of a plane crash is often due to defective design, which is a civil litigation consideration. However, the crash could be due to a bomb or terrorist activity in the cockpit, which is a criminal consideration. Often the cause of an accident is initially unknown or misattributed. For example, when TWA flight 800 exploded over Long Island

on July 17, 1996, the cause was first thought to be a terrorist bomb or a missile. After the fragmented aircraft was recovered and meticulously reconstructed, sophisticated laboratory and mathematical analysis indicated that the explosion was likely due to a center fuel tank design defect. However, Pan Am flight 103, which exploded over Lockerbie, Scotland, on December 21, 1988, was brought down by a terrorist bomb and not a design defect. Engineers were essential in analyzing both plane crashes.

Accident reconstruction. Accident reconstruction is the a posteriori scientific process of using the facts of an accident and the appropriate natural laws of science to determine the possible circumstances and causes, consistent with the available data. The quality and reliability of the reconstruction depend on the amount and quality of the available data, as well as the skill and knowledge of the investigator.

Consider, for example, a two-vehicle, right-angle, intersection collision (see **illus.**). The two vehicles had proceeded through an intersection, and then an impact occurred. Each vehicle had skidded prior to impact, and the final positions of the vehicles are indicated relative to the point of impact. For an analysis, the input data required (distances and angles as shown along with vehicle weights) would have to be obtained from a field investigation and from databases of vehicle properties. The principle of conservation of linear momentum, which follows from a straightforward application of Newton's laws, where all calculation details are given, enables the engineer to calculate the speeds of both vehicles at impact (v_1^i, v_2^i) and at the onset of pre-impact skidding (v_1, v_2), based solely on the data (independent of driver or witness statements). Depending on the actual input data and the calculated speeds, the results may be used in a variety of ways. Driver 2, for example, may claim that he stopped at the stop sign, and the accident occurred because driver 1 was speeding through the intersection. Based on the input data for the case illustrated, this claim cannot

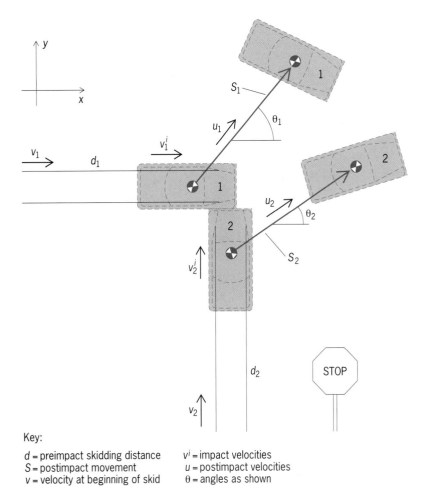

Key:

d = preimpact skidding distance
S = postimpact movement
v = velocity at beginning of skid

v^i = impact velocities
u = postimpact velocities
θ = angles as shown

Intersection collision.

be true: the accident occurred because driver 2 did not stop at the stop sign, and driver 1 was not speeding.

Product liability. In a product liability investigation, engineers are asked to determine whether a product is defective and whether the defect was causally related to any injuries that may have occurred. This is a challenging area of investigation and analysis. The three types of product defects recognized by law are (1) design defects (the product lacks those elements necessary for its safe and foreseeable uses), (2) manufacturing defects (the product was not made according to the manufacturer's specifications), and (3) failure to warn or instruct (the product was not accompanied by adequate warnings or instructions for its proper and safe use). In a lawsuit, if a forensic engineer has been retained by the plaintiff (the person bringing the lawsuit), the engineer's role is to objectively analyze the product for defects, which may or may not be found, and to causally relate the plaintiff's injuries to the product defect. If a forensic engineer has been retained by the defendant, the engineer's role will be to objectively defend the product, if it can be done, in view of the defect allegations made by the plaintiff. As part of the defense, the engineer may opine that the alleged defect does not exist or that the injuries sustained by the plaintiff are

not causally related to the product design. Product liability investigations can be very intense and can involve considerable analysis and calculations, testing, and computer modeling, often requiring the expenditure of large sums of money by both the plaintiff and defendant.

Design defects. As a simple but meaningful example of a design defect, consider cigarette lighters. Children are naturally inquisitive and may pick up lighters and try to light the flame. Some lighters have bright colors and shapes that make them particularly attractive to children. If the lighter does not have a child-proof or -resistant mechanism, which prevents children from lighting the flame, the lighter is defectively designed. The plaintiff's engineer in this case would show that at the time of manufacture it was practical and feasible to have provided a child-proof mechanism, and that the device may have even been in existence on other lighters, which would have prevented the accident.

Manufacturing defects. The fact that a person is injured using a product does not necessarily mean the product is defective. In addition, defective products do not cause injuries every time they are used. For example, a knife is designed to have a sharp edge in order to cut. If the user is cut, it does not necessarily follow that the knife is defective. However, suppose that a person is using a knife to carve a turkey and the blade fractures, causing the person to lose balance, to fall forward, and to be stabbed by the fractured blade. After the accident, a metallurgical analysis of the knife blade reveals that the steel was not properly heat-treated and that there were inclusions in the blade, which weakened it and caused the failure; this determination indicates that the knife was defectively manufactured.

Failure to warn. Car air bags can be life-saving devices if properly designed and used. When an air bag deploys in an accident, its front can move at up to 200 mi/h (320 km/h). A person sitting too close to the air bag can be seriously injured or killed by the impact to the head or upper torso. Small adults and children are particularly susceptible, as well as infants in rearward-facing child seats. Automobiles carry prominent warnings that occupants should not sit too close to air bags, that small children should sit in the rear, and that rearward-facing child seats should never be placed in the front seat of an air bag–equipped vehicle. Without such a warning, the vehicle would be defective by virtue of a failure to warn. The plaintiff's biomechanical engineer in this case would show by tests and/or calculations, or even using the automobile manufacturer's own data, that a person or child sitting too close to the deploying air bag would be subjected to injurious levels of impact forces.

Biomechanics of injury and death. Engineering science and methodology is being applied to injury and death investigation, a subject rooted in human biology. Biomechanical engineers working in the field of injury and death investigation are generally interested in applying engineering principles to the understanding of injury causation mechanisms in accidents, and in developing the criteria and conditions for when injury is likely to occur. For an accident, the kinematics of movement of the human body must be calculated and then correlated with any impacts and injuries that the body may have sustained. In an automobile accident, the movement of an occupant with respect to the vehicle (occupant kinematics) is often correlated with the acceleration-time history of the occupant, the occupant's change in velocity (delta-v), and the resulting injuries. This is a critical step if the occupant alleges an injury because of the accident or a vehicle defect. A very active research field in biomechanics is the determination of criteria that delineate the conditions in which injury may occur in a particular accident.

Computer forensics. This rapidly changing area of forensic engineering refers to methods for solving crimes committed using computer technology. Such crimes include extortion, violations of Internet security, hacking into supposedly secure Web sites and computers, computer theft of sensitive information and proprietary files (including national security data and financial data), and creation and propagation of computer viruses. Computer forensics uses techniques and procedures for recovering magnetic data, tracking hackers, and analyzing audio tapes, videotapes, and photographs for possible tampering and alteration.

Cause and origin. This area refers to investigations to determine the cause and origin of fires and explosions, and to determine if the cause was accidental (electrical shorting, product defect, cooking mishap, and so on) or intentional (arson or planting an explosive device). An understanding of heat transfer, burn patterns, combustion rates, pressure-wave propagation, and analysis of explosive residues is essential for cause and origin investigators.

Structural collapse, blast loading. This area refers to the catastrophic failure of structures either during construction or after the structure is in service. For example, the Hyatt Regency Skywalk collapse in Kansas City, Missouri, on July 17, 1981, was caused by a simple design error and resulted in the loss of 114 lives. The collapse was caused by an improperly designed connection, which did not meet Building Codes, where the vertical hanging rods passed through the box beams that supported the pedestrian walkways. Forensic civil engineers are also concerned with understanding the effect of blast and seismic loading on structures in order to ensure that future structural designs are safer and more resistant to dynamic loading arising as a result of explosions and earthquakes. *See also* EARTHQUAKE ENGINEERING.

Significant structural lessons were learned from the bombing of the Murrah Federal Building in Oklahoma City on July 19, 1995, and its resulting collapse; these lessons (outlined in the Proceedings of the 1st Congress of the American Society of Civil Engineers in 1997) will be used to mitigate blast loading effects on future reinforced concrete building designs.

Forensic engineers are becoming more actively involved in developing counterterrorism tactics and procedures, and this activity will likely grow as a separate area of forensic engineering.

For background information *see* BIOMECHANICS; COMPUTER SECURITY; INTERNET; NEWTON'S LAW OF MOTION; STRUCTURAL DESIGN in the McGraw-Hill Encyclopedia of Science & Technology.

Steven C. Batterman; Scott D. Batterman

Bibliography. S. C. Batterman, *Education of Forensic Engineers*, 20(6), American Academy of Forensic Sciences, 1990; S. C. Batterman and S. D. Batterman, Accident Investigation/Motor Vehicle (accident reconstruction and biomechanics of injuries), in J. A. Siegel et al. (eds.), *Encyclopedia of Forensic Sciences*, Academic Press, 2000; S. C. Batterman and S. D. Batterman, Forensic Engineering, chap. 12 in Y. H. Caplan and R. S. Frank (eds.), *Medicolegal Death Investigation: Treatises in the Forensic Sciences*, 2d ed., Forensic Science Foundation Press, 1999; F. P. Beer and E. R. Johnston, *Vector Mechanics for Engineers: Statics and Dynamics*, McGraw-Hill, 1977; D. T. Greenwood, *Classical Dynamics*, McGraw-Hill, 1977; K. L. Rens (ed.), *Forensic Engineering: Proceedings of the 1st Congress*, American Society of Civil Engineers, 1997; H. Yeh and J. I. Abrams, *Principles of Mechanics of Solids and Fluids: Particle and Rigid Body Mechanics*, McGraw-Hill, 1960.

Forensic entomology

Forensic entomology involves any interaction between the legal system and evidence derived from insects. It includes three subdisciplines: stored product entomology, structural entomology, and medicolegal entomology. Stored product entomology deals with damage to stored products, such as food materials or clothing, caused by insect activities. Structural entomology deals with damage to buildings and other structures by pest species such as termites or carpenter ants. Both these areas typically involve civil actions when there is a need to assess monetary damages resulting from insect activity. Medicolegal entomology deals with insects as evidence in criminal events, most frequently homicides. This is the area most commonly recognized by entomologists, law enforcement, and the public as forensic entomology. The application of entomological evidence to criminal investigations is not new, with the first recorded use coming from twelfth-century China. Beginning in the early 1980s, there has been a resurgence of interest in the field, and new applications of insect evidence are being exploited.

Insect evidence. Insects can serve as evidence in criminal investigations in several different ways. They can provide data for estimating the time since death (postmortem interval), detecting possible movement of the body following death and circumstances of the crime scene, and assessing antemortem versus postmortem wounds on the body.

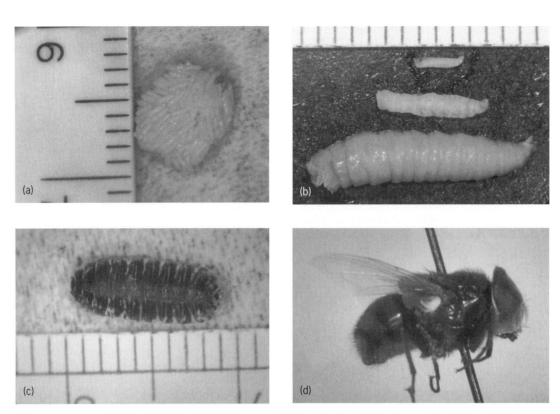

Chrysomya megacephala (family Calliphoridae). (*a*) Egg mass. (*b*) Larval instars: first (top), second (middle), and third (bottom). (*c*) Puparium. (*d*) Adult female.

In addition, insect evidence can provide alternative specimens for toxicological analyses in highly decomposed bodies, yield DNA material linking a suspect to a victim for ectoparasitic taxa, and help assess periods of abuse or neglect of children and the elderly. Of these, the most frequent application is the estimation of the postmortem interval.

A dead body provides a changing and ephemeral food source for a wide variety of organisms, ranging from bacteria to vertebrate scavengers. Among these organisms, insects are major factors because they arrive at decomposing remains in a predictable pattern and develop in known patterns. Insects arrive shortly following death, often within minutes. The first arrivals are most frequently flies in the families Calliphoridae (blowflies and bottle flies) and Sarcophagidae (flesh flies). The female flies arrive and begin to investigate the natural body openings of the head, anus, and genitals. Wounds present on the body provide another area for activity. The female flies take a meal of blood or other exudates from the body and then deposit their eggs or, in the case of the Sarcophagidae, larvae into these body openings. This action starts a biological clock which is stopped and interpreted by the forensic entomologist when the body is discovered and the insects are collected. The use of insects to estimate the postmortem interval requires an understanding of the insect's life cycle, the relationship of the insect to the remains, and the relationship of the remains to the habitat in which they are discovered.

Insects pass through a number of distinct life-cycle stages. Using the blowfly in the family Calliphoridae as an example, the female fly will arrive at the body and deposit eggs (**illus.** *a*) in the natural body openings or in wounds. These eggs will hatch into the larvae (maggots) which will feed on the decomposing tissues. There are three distinct larval stages (illus. *b*), called instars, with a molt between each stage. Once the maggot has reached complete development, it will cease feeding and move away from the remains to enter the pupal stage (illus. *c*), an inactive stage during which the larval tissues are reorganized to produce the adult fly (illus. *d*).

The insects encountered on a corpse in any given habitat will consist of some species unique to that particular habitat and some species having a wider distribution. The former group may be restricted to a particular geographic area or a particular habitat type within a given geographic area. For example, in Hawaii there are taxa that are restricted to a rainforest habitat, while others are specific to a more arid tropical habitat. Taxa having wider distributions are frequently encountered in several different habitat types and are typically highly mobile species, such as those taxa that are found both in rainforests and on arid beaches. Many of those taxa closely tied to carrion show this wider pattern of distribution. In estimating the postmortem interval, taxa from both groups may, under given circumstances, provide essential information concerning the history of the corpse.

Species analysis. Of those insect taxa having a direct relationship to the corpse, there are four generally recognized relationships:

1. Necrophagous species actually feed on the corpse. Included are many of the Diptera (Calliphoridae and Sarcophagidae) and Coleoptera (families Silphidae and Dermestidae). Species in this group may be the most significant isolatable taxa for use in postmortem interval estimates during the earlier stages of decomposition, defined here as days 1–14.

2. Parasites and predators of necrophagous species constitute the second most significant group of carrion-frequenting taxa. Many of the Coleoptera (Silphidae, Staphylinidae, and Histeridae), Diptera (Calliphoridae and Stratiomyidae), and Hymenoptera parasites of Diptera larvae and puparia are included. In some instances, Diptera larvae, which are necrophages during the early portions of their development, become predators during their later larval development.

3. Omnivorous species include taxa such as ants, wasps, and some beetles, which feed on both the corpse and associated arthropods. Large populations of these may severely retard the rate of carcass removal by depleting populations of necrophagous species.

4. Adventive species include those taxa which use the corpse as an extension of their own natural habitat, as in the case of the Collembola, spiders, and centipedes. Acari in the families Acaridae, Lardoglyphidae, and Winterschmidtiidae, which feed on molds and fungi growing on the corpse, may be included in this category. Of less certain association are the various Gamasida and Actinedida, including the Macrochelidae, Parasitidae, Parholaspidae, Cheyletidae, and Raphignathidae, which feed on other acarine groups and nematodes. In these investigations, what is being estimated is actually a period of insect activity rather than an actual time since death.

Interpretation. During the early stages of decomposition, the estimate of the postmortem interval is most frequently based on the development rates of individual species of Diptera, most frequently the flies in the family Calliphoridae. The most mature specimens collected from the body are preserved and identified. Given the particular species and stage of development represented, the entomologist can determine the period of time required to reach that stage. As the insects are dependent on environmental conditions, particularly temperature, for their rate of development, this must be factored into the estimate. Data obtained from detailed laboratory studies of life cycles conducted under controlled conditions can be correlated with conditions at a crime scene using weather data from National Oceanic and Atmospheric Administration (NOAA) stations and the concept of accumulated degree hours or accumulated degree days, a technique originally developed to predict pest outbreaks in agriculture. This calculated time period will represent the minimum postmortem interval. It is important to note that the time period is the period of insect activity and not the actual postmortem interval. Under most conditions, these will be quite similar, but there may be factors that delay the onset of insect activity, such as wrapping or concealment of the body.

After the first 2–3 weeks, depending on the location and environmental conditions, those flies in the initial wave of invaders will have completed their development and departed the body. At this point, the estimation of the postmortem interval is based on the ecological succession of insects onto the body. Those insects initially colonizing the body, change the body by their activities. These changes make the body attractive to another group of insects, which feed and change the nature of the body, thus making it attractive to yet another group of insects. This process continues until the resource of the body has been completely exhausted. By making complete collections of the insects and other organisms present on a decomposing body and comparing these taxa with results of detailed decomposition studies conducted in similar habitats and areas, periods of time during which the death most probably occurred can be determined. Care must be taken in applications of results from decomposition studies as even geographically proximate localities may support different arthropod populations. These differences in species composition and relative abundance may serve to alter the successional picture. Generally speaking, the closer to the time of death, the more accurate the estimated postmortem interval will be. The estimate begins in terms of hours and proceeds to days, months, and finally seasons of the year.

Entomological evidence is not present in all cases and, when present, may not ultimately prove to be of major significance. When insect evidence is present and properly interpreted, it may prove to be a powerful tool for the solution of the case.

For background information *see* ARTHROPODA; DEATH; DIPTERA; FORENSIC BIOLOGY; INSECTA in the McGraw-Hill Encyclopedia of Science & Technology.

M. Lee Goff

Bibliography. J. H. Byrd and J. L. Castner (eds.), *Forensic Entomology: The Utility of Arthropods in Legal Investigations*, CRC Press, Boca Raton, FL, 2000; E. P. Catts and N. H. Haskell (eds.), *Entomology and Death: A Procedural Manual*, Joyces Print Shop, Clemson, SC, 1990; M. L. Goff, Estimation of postmortem interval using arthropod development and successional patterns, *Forensic Sci. Rev.*, 5:81–94, 1993; F. Introna and C. P. Campobasso, *Entomolgia Forense: Il ruolo dei ditteri nelle indagini medico legali*, Essebiemme Editore, Italy, 1998.

Forensic physics

Forensic science has been dominated by the fields of chemistry (mainly analytical chemistry) and biology (serology and DNA profiling). Indirectly, physics has contributed to forensic science via the invention of the microscope, the electron microscope, the mass spectrometer, and optical spectrometers; but directly, its role in forensic science has been minimal. Forensic physics has traditionally involved the measurement of density (soil and glass examination), index of refraction, and birefringence (fiber analysis, glass examination). In the last 25 years, the use of the photoluminescence phenomenon for physical evidence examination has emerged, with latent fingerprint detection the most notable application. In criminalistics, fingerprint detection is important because it provides absolute identity, and does not suffer from the contamination problems to which DNA profiling is prone.

Photoluminescence. Light emission by substances can have various causes, such as heat (incandescence), electricity (electroluminescence), or a chemical reaction (chemiluminescence), as in the reaction of luminol (3-aminophthalhydrazide) with blood, a much-used criminalistics procedure. The most important origin of luminescence in forensic science is the prior absorption of light by the substance and subsequent emission of light at longer wavelength, called photoluminescence. The photoluminescence phenomenon was used in various science fields long before its quantum-mechanical underpinnings were understood. Early on, two types of photoluminescence were distinguished on the basis of photoluminescence lifetime, namely fluorescence (of short lifetime) and phosphorescence (of long lifetime). On practical grounds, this distinction suffices. Fluorescence techniques are presently more widely used in criminalistics than phosphorescence techniques, simply because intense fluorescence is generally encountered in nature more frequently than phosphorescence. Phosphorescence approaches, however, are beginning to show much promise, especially in instances in which fluorescence approaches fail.

Fluorescence techniques. In the past, a white surface would typically be dusted with a black powder to reveal a latent fingerprint. This mode of visualization involves absorption/reflectance phenomena. In the case of a weak print, to which only few of the absorbing black powder particles adhere, the print would reflect only slightly less than the surroundings. This inherently limits sensitivity because of the small difference between the two relatively large light signals. The photoluminescence approach detects a small signal, rather than a small difference between large signals and is therefore inherently more sensitive. Thus, photoluminescence-based techniques have generally supplanted absorption/reflectance techniques in most scientific fields. The principal application of photoluminescence in forensic science is fingerprint detection. Fingerprint residue left on an article typically has a mass of about a tenth of a milligram, and is composed primarily of water (about 98–99%), which soon evaporates, with the remaining residue made up about equally of inorganic material (mainly salts, largely useless for detection purposes) and a myriad of organic compounds. Thus, only nanograms of organic material are available to be made visible to the naked eye. Not only does the material have to be detected, but also its spatial distribution has to be visualized in fine detail. Thus, lasers, by virtue of their power (at the right wavelength to excite luminescence) were the first choice of illumination for purposes of photoluminescence detection of fingerprints (**Fig. 1**). Filtered lamps have since seen wide use, but sacrifice sensitivity in favor of price. Other photoluminescence applications in criminalistics include fiber analysis (especially in finding otherwise elusive fibers), document examination (detecting altered and erased writing), detection of body fluids (mostly semen and blood), trace evidence detection, authentication of currency and passports, explosives origin specification, and DNA profiling (with the radioactive tags of the past replaced by fluorescent tags).

Phosphorescence techniques. In fingerprint detection, one often encounters background fluorescence from the substrate on which the fingerprint is located, and this background fluorescence, which typically has a lifetime on the order of a nanosecond, cannot always be eliminated by optical filtering. Detection sensitivity is therefore reduced. This is a general problem with analytical fluorescence methods.

There has emerged in recent years a procedure whereby fingerprints are treated so that they exhibit phosphorescence, and then time-resolved imaging techniques are applied to suppress background fluorescence, to detect otherwise elusive prints. In such techniques, the imaging device is turned on and off in synchronization with the photoluminescence excitation source, which is also periodically turned on and off. Recording by the imaging device is delayed with respect to the excitation source turn-off, such that the background fluorescence has decayed by the time the imaging device turns on. Likewise, the imaging device turns off before the onset of the next illumination pulse (**Fig. 2**). Typically, a charge-coupled device (CCD) camera

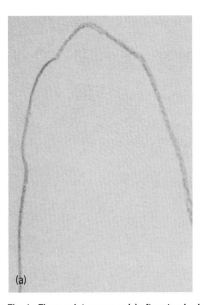
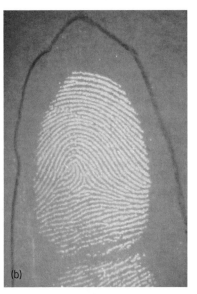

Fig. 1. Fingerprint on paper (a) after standard ninhydrin development and (b) after subsequent processing by zinc chloride and laser photoluminescence excitation.

equipped with a proximity-focused microchannel plate image intensifier is used as the imaging device. The time-resolved detection technique is not yet routinely used in crime laboratories, but operational prototype instruments have been in existence in United States and Canadian university forensic science laboratories since the mid-1990s.

A number of fingerprint treatments that lend themselves to time-resolved imaging, mostly based on lanthanide compounds that display millisecond luminescence lifetimes, have been developed. Time-resolved imaging techniques typically require phosphorescence lifetimes longer than microseconds. This lifetime requirement places considerable restrictions on the procedures that can be used to treat latent fingerprints.

Treatments that result in fingerprint luminescence lifetimes of 10–1000 nanoseconds can be used if phase-resolved imaging is implemented. Instead of turning the illumination source on and off (by means of a light chopper), the source intensity is modulated sinusoidally. This allows for much higher modulation frequencies than is possible with the abrupt on-off modulation of time-resolved techniques. In response to the sinusoidal luminescence excitation, the luminescence will also be sinusoidally modulated in intensity, but will be phase-shifted by an angle ϕ with respect to the excitation according to $\tan \phi = \omega\tau$, where ω is the modulation angular frequency and τ is the luminescence lifetime. If the gain of the imaging device (CCD camera equipped with a proximity-focused microchannel plate image intensifier) is modulated with the appropriate phase delay versus the modulation of the light source, a suppression of background fluorescence (small ϕ) with respect to the fingerprint luminescence is achieved (optimally ϕ is roughly $45°$ because phase shifts approaching the maximum of $90°$ lead to large demodulation of the luminescence intensity). The general scheme here is reminiscent of the lock-in amplifier, long used as a frequency and phase selective device. Phase-resolved imaging has been implemented in cell microscopy, for instance, but instruments have yet to be designed for forensic applications, perhaps because the corresponding fingerprint treatments have yet to reach maturity. There are, however, promising fingerprint treatments in the offing that appear suitable for phase-resolved imaging. The luminescence excitation sources needed for time- and phase-resolved imaging will have to be lasers.

Nanoparticles. Semiconductor material, when fabricated in size on the order of nanometers, may exhibit luminescence intermediate between what would be obtained from the bulk material and would be obtained from individual ions. The photoluminescence lifetimes of such nanoparticles tend to be 10–1000 ns. The color of the luminescence and the luminescence lifetime can be tailored by particle size. The broad absorption spectrum of the particle is typical of semiconductors, and thus amenable to optical luminescence excitation spanning a wide range of wavelengths; whereas the particle's luminescence

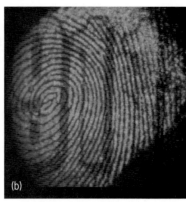

Fig. 2. Fingerprint (*a*) after standard photoluminescence development and (*b*) after time-resolved development. Note the suppression of the background fluorescence from the numbers 911.

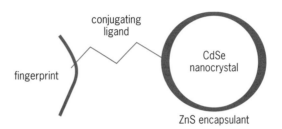

Fig. 3. General scheme for fingerprint detection with nanocrystals.

is sharply defined, akin to molecular fluorescences, and occurs at the (long-wavelength) absorption band edge. Nanoparticles are ideally suited for fingerprint work in terms of their spectroscopic properties, especially in their ability to suppress background fluorescence.

Presently, nanoparticles are used as selective taggants mostly in the biosciences, but the tagging chemistries should apply to fingerprint detection, making this an active area of current forensic research. Forensic application at present focuses on cadmium sulfide nanoparticles incorporated into tree-shaped polymers called dendrimers. The dendrimers have terminal functional chemical groups that selectively tag fingerprints. Cadmium selenide nanocrystals are also used. These crystals are typically encapsulated in zinc sulfide (or silica). The encapsulant protects the nanocrystal, such that its optical properties are not compromised, and functions as the attachment site for molecules (conjugating ligands), which chemically react with the material to be detected (**Fig. 3**).

Given that nanotechnology is a very active field today, with much bioscience application, it would seem to be only a matter of time before it finds its way to routine forensic science work. To date, financial considerations and the conservative nature of the forensic science community itself have been obstacles to its implementation.

For background information *see* CHARGE-COUPLED DEVICES; CRIMINALISTICS; FINGERPRINT; FLUORESCENCE; LUMINESCENCE; PHOSPHORESCENCE; PHOTO-

LUMINESCENCE in the McGraw-Hill Encyclopedia of Science & Technology.   E. Roland Menzel

Bibliography. H. C. Lee and R. E. Gaensslen, *Advances in Fingerprint Technology*, Elsevier, 1991; E. R. Menzel, *Fingerprint Detection with Lasers*, 2d ed., Marcel Dekker, 1999; E. R. Menzel, *Laser Spectroscopy: Techniques and Applications*, Marcel Dekker, 1995; E. R. Menzel, Photoluminescence detection of latent fingerprints with quantum dots for time-resolved imaging, *Fingerprint Whorld*, 26(101): 119–123, 2000; E. R. Menzel et al., Photoluminescent CdS/dendrimer nanocomposites for fingerprint detection, *J. Forensic Sci.*, 45(4):770–773, 2000; E. R. Menzel et al., Photoluminescent semiconductor nanocrystals for fingerprint detection, *J. Forensic Sci.*, 45(3):545–551, 2000; R. Saferstein (ed.), *Forensic Science Handbook*, Prentice Hall, 1982; R. Saferstein, *Criminalistics*, 6th ed., Prentice Hall, 1998; D. A. Tomalia et al., A new class of polymers: Starburst-dendritic macromolecules, *Polymer J.*, 17(1):117–132, 1985.

Functional magnetic resonance imaging (fMRI)

Functional magnetic resonance imaging is a noninvasive technique for measuring and mapping neural activity in the brain. In a typical fMRI experiment, volunteers lie inside an MRI scanner and perform a cognitive task. During the task, a series of three-dimensional images are taken of the participant's brain. Typically, these images have a spatial resolution of 3–4 mm and a temporal resolution of 1–4 s.

The images are designed to utilize the blood oxygen level–dependent (BOLD) effect, whereby the intensity of the MR signal in each portion of the image is affected by the amount of neural activity in that region. The BOLD effect is based on the fact that an increase in local neural activity produces an increase in local cerebral blood flow. This increase produces an increase in the local ratio of oxygenated relative to deoxygenated hemoglobin. Oxygenated and deoxygenated hemoglobin have different magnetic properties (deoxygenated hemoglobin is paramagnetic) such that the magnetic susceptibility of blood varies linearly with oxygenation over a broad range. The result is that an increase in local neural activity produces an increase in the local MR signal.

This increase in MR signal (often referred to as the hemodynamic response) is thought to be an approximately linear transformation of the underlying neural activity. However, the BOLD hemodynamic response to neural activity is both delayed and protracted over time, resembling the effect of a low-pass filter. For example, the hemodynamic response to a brief burst of neural activity rises to a peak approximately 6 s after the neural activity and does not return to baseline levels until 12–20 s following the neural activity (**Fig. 1**). It should be noted also that there is no inherent baseline for the MR signal. The absence of an absolute baseline (a specific MR signal value corresponding to zero neural activity) means that fMRI must be used as a contrastive or subtractive technique. That is, neural activity associated with one task condition or in response to one type of stimulus must always be assessed in contrast to the activity associated with a different task or in response to a different type of stimulus.

Experimental design. The delayed and protracted nature of the MR response to underlying neural activity complicates the design and analysis of fMRI experiments, as the responses to multiple closely spaced trials will overlap in typical cognitive experiments. In so-called blocked fMRI experimental designs, two task conditions or two types of stimuli (for example, familiar and unfamiliar stimuli) are presented in alternating blocks, each typically 20–40 s in duration. The hemodynamic responses to the trials within each block, when added together, provide a percent MR signal change that is greater than the MR signal change produced in a single trial. Typically, the fMRI data from each voxel (a three-dimensional unit in the image measuring perhaps 3 mm^3) are analyzed by correlating the MR signal in that voxel with a reference function based on the alternation between task conditions or stimulus types and adjusted for typical values of the hemodynamic response. Voxels with activity significantly correlated with the alternation in the blocked design are identified and typically color-coded according to the strength of the correlation and then overlaid on high-resolution structural MRI images (**Fig. 2**).

Two types of event-related fMRI designs can also be used. In spaced or nonoverlapping designs, stimuli are presented at sufficiently long intervals (typically at a rate of one stimulus per 16–20 s) such that the hemodynamic response to individual events does not overlap. The MR signal in response to different task conditions or stimulus types is then calculated by simple event-locked selective averaging of the fMRI data. In rapid or overlapping designs, stimuli are presented at the same rates as are used in typical cognitive experiments in the laboratory (for

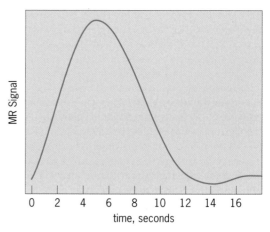

Fig. 1. Hemodynamic response time following neural activity. The response time is delayed and protracted.

example, one stimulus every 2 s). The analysis of rapid event-related designs relies on the observation that the hemodynamic response to multiple events sum together in an approximately linear fashion so that the response to different types of trials can be reconstructed with selective averaging or general linear modeling of the fMRI data.

Experimental findings. Since its initial development in 1992, fMRI has become a widely used tool in cognitive neuroscience (the study of cognition and the brain). It has been applied in areas ranging from visual perception to executive function and consciousness. For example, fMRI data have been used not only to identify occipito-parietal and occipito-temporal regions of the brain responsible for visual perception, but also to identify the different roles played by these regions. fMRI has also been used to demonstrate that attention, driven by a frontal-parietal network, has a modulatory effect on activity even in primary sensory regions (cortical areas that receive sensory input).

For the study of higher-level cognition, fMRI has been particularly useful for the study of the neural basis of learning and memory. Memory is considered to consist of three phases: learning (or encoding), storage, and retrieval. The learning phase refers to the initial acquisition and internal representation of the information that will constitute a memory. The storage phase refers to the period of time between learning and retrieval, during which the memory may become reorganized, weakened through forgetting, or strengthened through rehearsal. The retrieval phase refers to the subsequent accessing of memory (or some part of the memory) to guide thought or action.

Neuropsychological studies of the effects of brain lesions in both humans and experimental animals have provided the cornerstone for the present understanding of the neural basis of learning and memory. These studies have identified a system of structures within the medial portions of the temporal lobes (MTL) that are essential for the formation of long-term memory. Damage to these structures impairs the ability to learn new facts and events (also known as declarative memory). The advent of fMRI promises to greatly expand understanding of the mnemonic role of the MTL (that is, its role in learning, storage, and retrieval) and its anatomical components. In the few years that fMRI has been successfully used to study the neural basis of memory, it has been discovered not only that activity in the medial temporal lobe increases during the learning of new information, but also that this activity is lateralized based on the type of information being learned. During the learning of words, only the left MTL is engaged. During the learning of pictures of namable objects, both the left and the right MTL are engaged. During the learning of nonverbalizable visual objects (for example, nonsense objects), only the right MTL is engaged. Further, the amount of activity in the MTL during the learning of new information can predict how well that information will later be remembered.

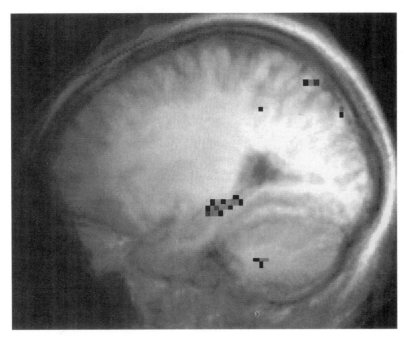

Fig. 2. Structural MRI image overlaid with fMRI-determined voxels of neural activity.

The MTL has also been observed to demonstrate increased fMRI activity during the retrieval phase of memory. In addition, this activity has been found to correlate with whether or not individuals can successfully remember having learned the information. That is, fMRI activity in the MTL appears to correlate more with retrieval success than with retrieval effort. Further, because of the relatively high resolution available in fMRI, these activities can be localized to individual structures within the MTL, providing information to inform and constrain theories of MTL function. For example, recent studies have found activity within the hippocampus related to recollective success during simple recognition memory tasks. The hippocampus is at the top of the anatomical hierarchy within the MTL and receives input from other structures within the system.

It is also of interest that the frontal lobes are typically quite active during learning and during retrieval, as well as when information is being held in mind for short periods of time (a process called working memory). Frontal lobe activity appears to be more related to retrieval effort than to retrieval success.

Although fMRI studies have been informative and compelling, the impact of fMRI on the study of cognition and the brain will be even greater in the coming years. In less than a decade since its discovery, fMRI has become a major tool in cognitive neuroscience. As the MRI hardware, pulse sequences, and data analysis techniques continue to develop and as more researchers apply fMRI to investigations of the brain, the promise and potential of fMRI will continue to grow. As with any new technique, the full promise of fMRI lies not in what it may reveal in isolation, but in the possibility of using information from fMRI in conjunction with other techniques in both humans and experimental animals.

For background information *see* BRAIN; COGNITION; ELECTROENCEPHALOGRAPHY; HEMISPHERIC LATERALITY; INFORMATION PROCESSING; LEARNING MECHANISMS; MEDICAL IMAGING; MEMORY; PARAMAGNETISM; PERCEPTION in the McGraw-Hill Encyclopedia of Science & Technology.

Craig E. L. Stark; Larry R. Squire

Bibliography. R. Cabeza and A. Kingstone (eds.), *Handbook of Functional Neuroimaging of Cognition*, MIT Press, Cambridge, MA, 2001; C. T. W. Moonen and P. A. Bandettini (eds.), *Functional MRI*, Springer, New York, 1999; D. L. Schacter and A. D. Wagner, Medial temporal lobe activations in fMRI and PET studies of episodic encoding and retrieval, *Hippocampus*, 9:7–24, 1999; B. R. Rosen, R. L. Buckner, and A. M. Dale, Event related fMRI: Past, present, and future, *Proc. Nat. Acad. Sci. USA*, 95:773–780, 1998.

Fungal genotyping

The dramatic rise in the incidence of fungal infections over the last two decades is related largely to an increase in the number of patients at risk for opportunistic fungal infections, due to the advent of newer technologies and therapies, such as bone marrow or solid-organ transplants, extensive surgery, and chemotherapeutic and broad-spectrum antibacterial agents. These practices as well as the AIDS epidemic have resulted in the rise of severely ill, immunocompromised, and long-term hospitalized patients. In addition to the escalating incidence, the emergence of new fungal pathogens (**Table 1**), the shifts in the distribution of *Candida* species, and the increased resistance to antifungal agents underscore the need to develop optimal methods for the detection, identification, and typing of fungal pathogens.

Genotyping (or molecular typing) of medically important fungi has become a subdiscipline of medical mycology, and it has been applied to a variety of epidemiologic studies. The goal of epidemiologic typing (DNA polymorphism analysis) is to determine whether two or more fungal isolates of a given species collected from one or several patients are "the same"—that is, they are "clonal identicals" and represent the same strain—or "different," belonging

to different clonal lineages. Various molecular methods have been developed or adapted to serve as epidemiologic tools to identify fungal strain heterogeneity and homogeneity, respectively. Since each method harbors its own set of benefits and limitations, no single molecular typing technique has evolved as a dominant "gold standard" method. The choice of technique depends on the aims for genotyping an isolate collection.

Typing systems and techniques. Since there is no ideal typing system, several performance criteria should be considered when evaluating and applying typing methods. Primarily, these criteria include typeability, reproducibility, and discriminatory power, followed by stability, flexibility, rapidity, accessibility, and cost (**Table 2**). In each case, however, the primary objective is to obtain a DNA (or,

TABLE 1. Examples of opportunistic fungal pathogens

Yeasts and yeastlike organisms	Filamentous fungi
Major pathogens	
Candida albicans	Aspergillus fumigatus
Cryptococcus neoformans var. neoformans	
Common and emerging pathogens	
Candida dubliniensis	Aspergillus flavus
Candida (Torulopsis) glabrata	Aspergillus niger
Candida guilliermondii	Aspergillus terreus
Candida kefyr	Fusarium spp.
Candida krusei	Penicillium marneffei
Candida lusitaniae	Rhizopus spp.
Candida parapsilosis	Scedosporium spp.
Candida tropicalis	
Cryptococcus neoformans var. gattii	
Exophiala spp. and other "black yeasts"	
Saccharomyces cerevisiae	
Rare pathogens	
Candida famata	Aspergillus nidulans
Candida holmii	Paecilomyces lilacinus
Candida inconspicua	Zygomycetes (for example, Mucor, Absidia, Mortierella)
Candida norvegensis	
Candida pelliculosa	
Candida rugosa	
Candida utilis	
Candida viswanathii	
Candida zeylanoides	
Geotrichum capitatum	
Prototheca wickerhamii	
Rhodotorula spp.	
Trichosporon spp.	

TABLE 2. Comparison of major characteristics of some genotyping systems

Methodology	Typeability	Reproducibility	Discriminatory power	Rapidity	Costs
Plasmid analysis	Most	Moderate	Poor to moderate	Good	Low
Ribotyping	All	Good	Moderate	Moderate	Low
Pulsed-field gel electrophoresis	All	Good	Good to high	Moderate	Moderate
Arbitrarily primed polymerase chain reaction	All	Moderate to good	Good to high	High	Low
Sequencing	All	High	High	Good	High

infrequently, RNA) profile of the pathogenic isolate. After extracting the nucleic acid from the microorganism, two general approaches are used in chromosomal and gene-fragment fingerprinting to obtain a characteristic pattern of multiple nucleic acid fragments. This aim is achieved by either digesting DNA with restriction enzymes (restriction fragment length polymorphism analysis) or amplifying random fragments of target DNA with the use of oligonucleotide primers (amplified fragment length polymorphism analysis), followed by analysis of the fragments by electrophoresis. With the advent of computer-assisted analysis and other techniques (for example, DNA sequence determination and microarray technology), larger collections of fungal isolates will be typeable and more data and complex patterns will emerge. *See also* PROTEOMICS.

Restriction fragment length polymorphism analysis. Restriction fragment length polymorphism analysis involves various techniques that can be used alone or in combination. Common techniques include pulsed-field gel electrophoresis and restriction endonuclease analysis.

Pulsed-field gel electrophoresis allows the separation of large DNA fragments, including entire chromosomes (electrophoretic karyotyping). Since the chromosomes of different fungal strains may vary in size, different strains are characterized by a particular karyotype (chromosomal constitution). After lysing fungal cells in situ, digestion of chromosomal DNA with rare cutting endonucleases (restriction enzymes that cleave DNA at specific sites), such as *Bss*H II, *Not* I, and *Sfi* I, generates strain-specific fingerprints. Due to its excellent discriminatory power and stable reproducibility, pulsed-field gel electrophoresis has been applied extensively to fingerprint *Candida albicans* and other *Candida* species (Table 1), *Cryptococcus neoformans*, *Pneumocystis carinii*, *Histoplasma capsulatum*, *Coccidioides immitis*, and several *Aspergillus* species.

Restriction endonuclease analysis involves the digestion of total genomic DNA with specific endonucleases (usually *Eco*R I and *Hin*f I). Although restriction endonuclease analysis is relatively easy to perform, the complexity of its fingerprints may hinder the interpretation of the clonal relationship. However, the combination of restriction endonuclease analysis with pulsed-field gel electrophoresis or hybridization with gene probes (labeled DNA) may offer some advantages, such as simplified banding patterns. With respect to standardization, labor time, and costs, restriction fragment length polymorphism based on hybridization of digested DNA with repetitive genetic elements (short tandemly repeated sequences) shows an excellent typing parameter. By using short oligonucleotide probes that are complementary to these sequences, informative DNA fingerprint profiles consisting of multiple hybridization bands can be obtained. However, restriction fragment length polymorphism methods are generally hampered by unwieldy and time-consuming sample preparation and electrophoretic procedures.

Amplified fragment length polymorphism analysis. Amplification-based fingerprinting is usually represented by modifications of the prototype of nucleic acid amplification, known as the polymerase chain reaction. Generally, these techniques are methodologically easier, less time-consuming, and more cost-effective than older genotyping methods, particularly pulsed-field gel electrophoresis. (Interestingly, the identity and spread of the current largest living organism, the fungus *Armillaria*, have been recently confirmed by polymerase chain reaction fingerprinting.)

The most popular method is the arbitrarily primed polymerase chain reaction, also called random amplification of polymorphic DNA, DNA-amplified fingerprinting, and polymerase chain reaction–mediated genotyping. In contrast to the conventional polymerase chain reaction, this method is based on using (mostly) one short arbitrary primer at nonstringent conditions (low annealing temperature), which anneals at multiple sites of both strands of the chromosomal and the plasmid DNA. This results in amplification of multiple DNA fragments of different length, which yields a fingerprint after separation in gel electrophoresis. However, a limitation of arbitrarily primed polymerase chain reaction is that its low-stringency conditions may result in poor reproducibility of typing results. A strict constancy of all assay parameters is imperative to obtain reproducible results.

Furthermore, the presence of repeat elements in microbial genomes, which may differ in their length and number, can be used for interrepeat spacer-length polymorphism analysis by modified polymerase chain reaction techniques. In general, genotyping methods targeting identified conserved sequences are more reproducible than random-based amplification techniques.

Genotyping of yeast. During the late 1980s, the first DNA-based procedures were described for yeasts, followed by numerous techniques that predominantly focused on the study of the epidemiology of the most pathogenic and frequent member of the *Candida* genus, *C. albicans*. Since other *Candida* species (Table 1) have been more frequently implicated in human mycoses, numerous fingerprinting techniques have been adapted for these species (**Table 3**). The application of genotyping to studies of *Candida* species has shown that DNA-based typing systems can reliably delineate strains among clinical *Candida* isolates (see **illus.**). Since individual patients are usually colonized by a single strain, infection due to *Candida* is most often of endogenous origin (from the patient's own flora). However, *Candida* strains can be passed from patient to patient, presumably on the hands of medical staff, resulting in nosocomial (hospital-acquired) infections. In addition, it is not uncommon to isolate multiple strains and species from single infections, indicating a high degree of complexity and dynamics of the *Candida* population. It has been shown that most non-*albicans* species are genetically very

TABLE 3. Examples of recent genotyping applications for yeasts and filamentous fungi

Typing system	Yeasts		Filamentous fungi	
	Authors	Application	Authors	Application
Restriction fragment length polymorphism				
Pulsed-field gel electrophoresis, electrophoretic karyotyping, restriction endonuclease analysis, and related techniques	Doi et al., 1992	*Candida albicans, C. (T.) glabrata, C. guilliermondii, C. kefyr, C. krusei, C.parapsilosis, C. stellatoidea, C. tropicalis*	Fekete et al., 1993	*Fusarium* spp.
			Verdoes et al., 1994	*Aspergillus niger*
	Willinger et al., 1994	*C. albicans, C. (T.) glabrata*	Lin et al., 1995	*A. fumigatus*
	King et al., 1995	*C. lusitaniae*	Geiser et al., 1996	*A. nidulans*
	Dib et al., 1996	*C. rugosa*	Wu et al., 1997	*Penicillium marneffei*
	Essayag et al., 1996	*C. inconspicua, C. krusei*	Rath et al., 1997	*A. fumigatus*
	Clemons et al., 1997	*C. albicans*	Tobin et al., 1997	*A. fumigatus*
	Hong and Leung, 1998	*C. albicans*	James et al., 2000	*A. flavus*
	Klepser and Pfaller, 1998	*Cryptococcus neoformans*		
	Di Francesco et al., 1999	*C. (T.) glabrata*		
	Deak et al., 2000	*C. lipolytica, C. zeylanoides*		
	Zancope-Oliveira et al., 2000	*C. parapsilosis*		
	Dassanayake et al., 2000	*C. krusei*		
Amplified fragment length polymorphism				
Arbitrarily primed polymerase chain reaction fingerprinting, random amplified polymorphic DNA, and related techniques	Lehmann et al., 1992	*Candida albicans, C. (T.) glabrata, C. haemulonii, C. lusitaniae, C. parapsilosis, C. tropicalis*	Loudon et al., 1996	*A. niger*
			Rath et al., 1997	*A. fumigatus*
			Brandt et al., 1998	*A. fumigatus*
	King et al., 1995	*C. lusitaniae*	Khan et al., 1998	*A. fumigatus*
	Howell et al., 1996	*C. albicans*	Leenders et al., 1999	*A. fumigatus, A. flavus*
	Cresti et al., 1999	*C. albicans, C. (T.) glabrata, C. krusei, C. tropicalis*	Rodriguez et al., 1999	*A. fumigatus*
			Symoens et al., 2000	*A. terreus*
	Di Francesco et al., 1999	*C. (T.) glabrata*	Birch and Denning, 2000	*A. terreus*
	Forche et al., 1999	*C. albicans*	Van Belkum et al., 1993	*A. fumigatus*
	Meyer et al., 1999	*C. neoformans var. neoformans*	Rath and Ansorg, 2000	*A. fumigatus, A. flavus, A. niger, A. terreus*
	Deak et al., 2000	*C. lipolytica, C. zeylanoides*		
	Becker et al., 2000	*C. (T.) glabrata*		
	Dassanayake et al., 2000	*C. krusei*		
	Zancope-Oliveira et al., 2000	*C. parapsilosis*		
	Cogliati et al., 2000	*C. neoformans var. neoformans*		
	Imai et al., 2000	*C. neoformans var. gattii*		
Interrepeat polymerase chain reaction	Barchiesi et al., 1997	*C. albicans*		
	Di Francesco et al., 1999	*C. (T.) glabrata*		
Polymerase chain reaction–single-strand conformational polymorphism analysis	Walsh et al., 1995	*C. albicans, C. parapsilosis, C. tropicalis*		
	Graser et al., 1996	*C. albicans*		
	Forche et al., 1999	*C. albicans*		
Polymerase chain reaction–ribotyping	Uijthof et al., 1994	*Exophiala dermatitidis*		
	Smole Mozina et al., 1997	*Saccharomyces* spp., *Torulaspora* spp.		
	McCullough et al., 1998	*Saccharomyces* spp.		
Restriction/amplified fragment length polymorphism combinations and other techniques				
Polymerase chain reaction–restriction fragment length polymorphism	Howell et al., 1996	*Candida albicams*		
	Barchiesi et al., 1997	*C. albicans*		
	Di Francesco et al., 1999	*C. (T.) glabrata*		
	McCullough et al., 1999	*C. albicans, C. dubliniensis, C. stellatoidea*		
	Xu et al., 1999	*C. albicans*		
	Cresti et al., 1999	*C. albicans, C. (T.) glabrata, C. krusei, C. tropicalis*		
Polymerase chain reaction sequencing	Niesters et al., 1993	*C. albicans, C. (T.) glabrata, C. krusei, C. tropicalis*		
	Forche et al., 1999	*C. albicans*		

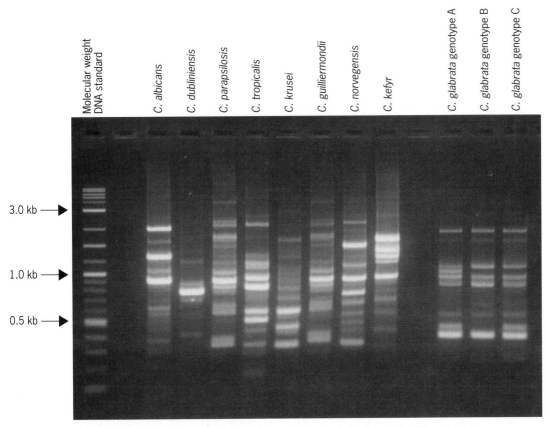

Agarose gel electrophoresis patterns demonstrating intra- and interspecies differences of *Candida* species by using arbitrarily primed polymerase chain reaction fingerprinting.

heterogeneous. In contrast, a lack of genetic differentiation between geographically diverse isolates has been revealed for *C. albicans* and *C. glabrata*. A slower evolution rate of some *Candida* clones has been proposed as an explanation for this observation. The demonstration of minor banding pattern differences of epidemiologically related isolates ("substrain shuffling") further complicates the question of clonal identity of different isolates. *See also* CANDIDA.

The use of modern genotyping methods revealed that cryptococcal infection of humans results from environmental sources and not from animals suffering from cryptococcosis. A correlation between the geographical distribution of the varieties of *Cryptococcus neoformans* in the environment and the clinical manifestation of cryptococcosis has been demonstrated.

Genotyping of filamentous fungi. To overcome the problems resulting from phenotypic variability, several genotyping methods have been applied to filamentous fungi, particularly species of the genus *Aspergillus* (Table 3). The first DNA-based typing method used for *Aspergillus* was restriction fragment length polymorphism analysis of ribosomal RNA genes by hybridization of *A. fumigatus* genomic DNA blots with a DNA fragment containing the ribosomal RNA intergenic spacer region from *A. nidulans*. Published molecular typing methods include several restriction fragment length polymorphism and aspergillus fragment length polymorphism methods, such as karyotyping, arbitrarily primed polymerase chain reaction/random amplified polymorphic DNA, microsatellite typing, and polymerase chain reaction of the ribosomal intergenic spacer region. The internal transcribed spacer region and intergenic spacer of the nuclear ribosomal RNA repeat units evolve the fastest and may vary among populations. A restriction fragment length polymorphism method using a species-specific retrotransposon-like sequence (part of the retroelement family, which is composed of transposable elements that move via an RNA intermediate) allows computer-aided analysis of large numbers of *Aspergillus* isolates.

Aspergilli are widely distributed in the environment (for example, dust, soil); and small outbreaks in medical facilities, sometimes in association with airborne populations of aspergilli, have been reported. Nevertheless, only a few typing studies are available comparing isolates from environmental conidia (asexual spores) and clinical specimens. In general, a wide range of different genotypes has been observed among clinical isolates. In a large-scale study, it was shown that patients with no clinical symptoms of aspergillosis appeared to carry several strains, whereas patients with proven pulmonary aspergillosis carried only one or two strains.

Furthermore, nosocomial transmission of a pulmonary infection occurred in more than one-third of the patients studied.

Currently, there is no evidence that a restricted group of strains of a given *Aspergillus* species or a single genotype is predominantly in clinical settings or is responsible for the majority of invasive aspergillosis cases. An extreme genetic diversity of strains has been found in hospital settings. In contrast, genotyping of environmental isolates from different continents has demonstrated indistinguishable types from dispersed locations. However, as with any typing scheme, indistinguishable isolates are not necessarily epidemiologically linked.

Despite the fact that colonization with *Aspergillus* species is a well-described risk factor for invasive disease in immunocompromised patients, only a few studies have investigated the relationship between colonization and invasive aspergillosis. Regarding patients with hematological malignancy, the data from this patient collective may be difficult to interpret, since clonal different strains from disseminated infection following bone marrow transplantation have been reported. For solid organ transplantation, only an apparently slight relationship between colonization and invasive disease has been found. In addition, involvement of multiple strains has been reported. Genotyping *Aspergillus* isolates of patients with chronic lung diseases (for example, cystic fibrosis) by several techniques demonstrated that many patients harbor several mold strains over a relatively short time period but some patients show prolonged colonization with a single strain. In general, the results of *Aspergillus* genotyping suggest a complex ecology of these molds.

For background information *see* FUNGAL INFECTIONS; FUNGI; GENE AMPLIFICATION; GENETIC MAPPING; MEDICAL MYCOLOGY; MOLECULAR PATHOLOGY; POLYMORPHISM (GENETICS); RESTRICTION ENZYME in the McGraw-Hill Encyclopedia of Science & Technology.   Karsten Becker

Bibliography. M. R. Micheli and R. Bova (eds.), *Fingerprinting Methods Based on Arbitrarily Primed PCR*, Springer, Berlin, 1997; A. P. Monaco (ed.), *Pulsed Field Gel Electrophoresis: A Practical Approach*, IRL Press, Oxford, 1995; M. A. Pfaller, Epidemiology of fungal infections: The promise of molecular typing, *Clin. Infect. Dis.*, 20:1535–1539, 1995; D. R. Soll, The ins and outs of DNA fingerprinting the infectious fungi, *Clin. Microbiol. Rev.*, 13:332–370, 2000; D. J. Sullivan et al., Molecular genetic approaches to identification, epidemiology and taxonomy of non-*albicans Candida* species, *J. Med. Microbiol.*, 44:399–408, 1996.

Gamma-ray astronomy

The *Compton Gamma-Ray Observatory* (*CGRO*), launched on April 4, 1991, was the second of the National Aeronautics and Space Administration's four great observatories (**Fig. 1**), following the Hubble Space Telescope and preceding the *Chandra X-ray Observatory* (*CXO*) and the planned *Space Infrared Telescope Facility* (*SIRTF*). For over 9 years, *CGRO* took high-quality images and spectra of a variety of high-energy sources, including gamma-ray bursts, active galaxies, pulsars, supernova remnants, and solar flares. Deorbited on June 4, 2000, *CGRO* leaves behind a legacy of discoveries and advances in high-energy astrophysics and the initiation of a new era of gamma-ray astronomy. *See* X-RAY ASTRONOMY.

CGRO carried four instruments, each designed to cover a portion of the 30 keV–30 GeV energy range. The Compton Telescope (COMPTEL) observed in the 1–30 MeV band, taking images and spectra of objects like active galactic nuclei and supernova remnants (SNRs). The Oriented Scintillation Spectrometer Experiment (OSSE) took spectra in the 50 keV–10 MeV range, looking for nuclear lines in solar flares and radioactive decay in supernova remnants. The Burst and Transient Source Experiment (BATSE), an all-sky monitor, studied gamma-ray bursts and other transient events in the energy range 20 keV to more than 1 MeV. Finally, the Energetic Gamma Ray Experiment Telescope (EGRET), the highest-energy instrument on board *CGRO*, observed phenomena such as pulsars and blazars in the 30 MeV–30 GeV band.

Gamma-ray bursts. About once per day a large flash of gamma radiation is seen from a random point in the sky. These flashes, gamma-ray bursts (GRBs), are more energetic than any other single events in the universe, surpassed only by the big bang. The gamma-ray flash can last from a few milliseconds to a minute or more. Although gamma-ray bursts were discovered in the late 1960s, their nature remains largely a mystery. With BATSE, an all-sky monitor designed to search the sky for gamma-ray bursts other transient events, combined with broadband gamma-ray coverage by OSSE, COMPTEL, and EGRET, *CGRO* provided the most comprehensive gamma-ray burst observations to date.

Before *CGRO*'s launch, most astronomers believed gamma-ray bursts were associated with neutron stars in the plane of the Milky Way. BATSE was expected to confirm this view. However, the gamma-ray bursts that BATSE discovered were isotropically distributed on the sky and had a deficiency of weak bursts compared to a volume-filling homogeneous population, proving that gamma-ray bursts are not confined to the galactic plane and suggesting a cosmological origin (**Fig. 2**).

During 1997, the Italian-Dutch satellite, *BeppoSAX*, discovered that gamma-ray bursts glow in x-rays long after the burst. In addition, *BeppoSAX* was able to pinpoint bursts to positions nearly 50 times more precise than those found by BATSE, allowing follow-up radio and optical observations. These measurements confirmed that long gamma-ray bursts are of cosmological origin. (A subclass of short-duration gamma-ray bursts is still of unknown origin.)

To facilitate multiple-wavelength observations of

gamma-ray-burst afterglows, BATSE and *BeppoSAX* reported coordinates of bursts in nearly real time to the astronomical community. On January 23, 1999, this quick communication of a blast's position from BATSE allowed a robotic camera on the ground to capture visible-light observations simultaneously with the gamma-ray flash for the first time.

In another observation, EGRET detected long-lived (more than an hour) emission of high-energy gamma rays from a gamma-ray burst. *See* GAMMA-RAY BURSTS.

Active galaxies. Quasars are the bright cores of distant galaxies and are thought to be powered by accreting supermassive black holes. The emission from quasars can be visible at all wavelengths from radio to gamma rays. Like gamma-ray bursts, quasars are among the most distant objects observed.

Before *CGRO* was launched, only one quasar, 3C 273, had been seen in gamma rays (by the European Space Agency's *COS-B* satellite). When *CGRO* looked at 3C 273, it caught another quasar, named 3C 279, flaring in the same field of view.

EGRET observed a total of 93 quasars, 66 identified with high confidence. Most were seen to be highly variable in their gamma-ray flux. The hypothesis is that the quasars detected by EGRET are those with relativistic jets viewed nearly along the set axis.

It is probable that several of the EGRET sources not yet identified with known sources are blazars. The observed isotropic, extragalactic high-energy gamma-ray background could be explained by a large number of unresolved blazars.

Neutron stars. A pulsar is a rotating neutron star with a high dipolar magnetic field. Strong, beamed

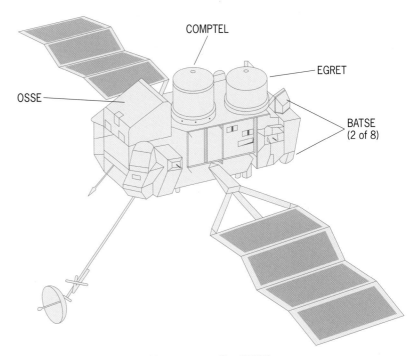

Fig. 1. *Compton Gamma-Ray Observatory* satellite. (*NASA*)

emission comes from the star's magnetosphere, which may not be co-aligned with its rotational poles. Thus, as the star rotates, the beams sweep through a stationary observer's line of sight, creating pulses.

Traditionally, pulsars lie in the domain of radio astronomy. However, pulsars are seen to pulse in all wavebands, typically synchronized in time with the radio pulses. Before *CGRO* only two pulsars, the Crab

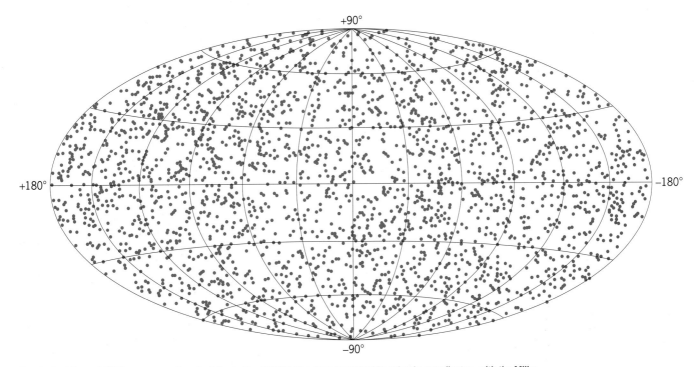

Fig. 2. Positions of 2704 gamma-ray bursts detected by BATSE. The map is plotted in galactic coordinates, with the Milky Way horizontally across the graph. (*NASA*)

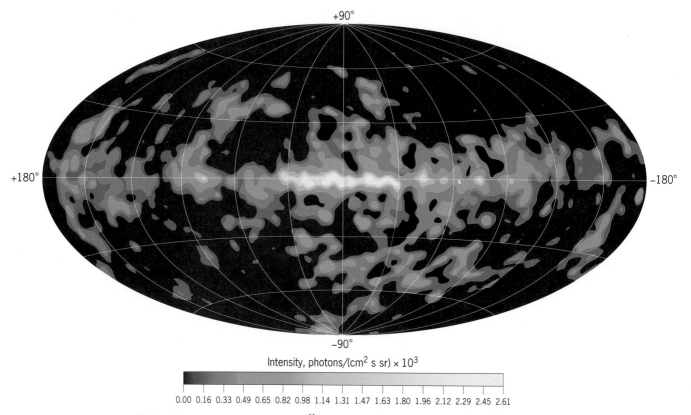

Intensity, photons/(cm^2 s sr) × 10^3

0.00 0.16 0.33 0.49 0.65 0.82 0.98 1.14 1.31 1.47 1.63 1.80 1.96 2.12 2.29 2.45 2.61

Fig. 3. Sky map of the concentration of ^{26}Al gamma rays from COMPTEL data. The ^{26}Al is concentrated in areas of recent star formation. The map is plotted in galactic coordinates. (*NASA*)

and Vela pulsars, had been known to emit gamma-ray pulses. *CGRO* discovered five more with gamma-ray pulses not necessarily synchronized with the radio emission.

One object, Geminga, had been an enigma for years. It was first detectable in gamma rays only as a bright steady source. In 1992 x-ray pulses were discovered from Geminga by the *ROSAT* x-ray observatory, proving it to be a nearby (470 light-years) pulsar. Using the data on timing of x-ray pulses, EGRET also found gamma-ray pulsations.

Soft gamma repeaters (SGRs) are another class of neutron stars that *CGRO* studied. Soft gamma repeaters are the brightest repeating bursters, and only three were known before *CGRO*. A fourth one, and possibly a fifth, have been added by BATSE. Three of the five soft gamma repeaters appear to be associated with supernova remnants. It is hypothesized that they are related to highly magnetized neutron stars called magnetars.

Nucleosynthesis. In the final stages of a massive star's life, the central core exhausts its nuclear fuel and can no longer support itself against gravitational collapse. The core then collapses in a fiery explosion, a supernova. This explosion releases a huge amount of energy and blows off the star's outer envelope.

Two products of a supernova blast are radioactive isotopes of titanium and aluminum. COMPTEL's detectors were able to observe gamma rays from the decay of ^{44}Ti and ^{26}Al, with half-lives of about 60 and

700,000 years, respectively. Studies of these isotopes aid in the discovery of both old and young supernova remnants. As an example, COMPTEL and *ROSAT* independently discovered SNR GRO/RX J0852. This supernova remnant lies at a distance of about 650 light-years and is about 700 years old, which makes astronomers today wonder why there was no record of the event from astronomers 700 years ago. *See* NUCLEAR ASTROPHYSICS.

COMPTEL's map of ^{26}Al emission shows concentrations in areas of recent star formation and supernova activity (**Fig. 3**). The map was used to determine that there are about 1 to 2 solar masses of ^{26}Al in the Milky Way Galaxy, in agreement with theories of element synthesis in supernovae.

Solar flares. A solar flare is an explosion of energetic particles and electromagnetic radiation in the outer atmosphere of the Sun. Studying flares in the nearby Sun can facilitate understanding of other astronomical events since particle acceleration processes involved in solar flares are also thought to occur in objects like pulsars, quasars, and black holes.

Although *CGRO* was launched just after the 1990 solar maximum (the last peak in solar activity), the Sun was active enough in the summer of 1991 for *CGRO* to observe several large flares. During a flare on June 4, 1991, OSSE detected gamma-ray emission lines from the elements iron, magnesium, neon, silicon, carbon, oxygen, and nitrogen. From these lines, astronomers inferred relative amounts of these and other elements in the Sun's coronal gas. On June 11, EGRET detected an afterglow from a solar flare in which no spectral cut-off was seen; therefore, photons of higher energies than EGRET could detect were presumably present. Both EGRET and COMPTEL observed gamma-ray emission lasting for several hours after the impulsive phase for two flares. These were surprising results, implying particle acceleration leading to gamma radiation long after the initial flare.

Since its detectors were able to distinguish between neutron and gamma-ray signals, COMPTEL was able to obtain a particle image of the Sun in neutrons during a flare on June 15, 1991. This was the first image of any astrophysical object in particles.

CGRO's impact. Before *CGRO*'s remarkable run, only 40 gamma-ray sources were known; *CGRO* increased that number by a factor of 10 (not including the gamma-ray burst observations). With these, new populations were discovered and new insight gained into high-energy astrophysical processes. Exotic objects were revealed. More than 200 of the sources remain unidentified. BATSE increased the number of gamma-ray bursts detected from about 300 to more than 2700. During *CGRO*'s last years, scientific journals published about 180 *CGRO*-specific articles per year.

Although *CGRO* has been deorbited, its discoveries have spawned the next generation of gamma-ray missions. Between late 2000 and 2006 there are no fewer than five missions launched or planned for launch, including the *High-Energy Transient Explorer-2* (*HETE-2*; launched in October 2000), the *High Energy Solar Spectroscopic Imager* (*HESSI*), the European Space Agency's *International Gamma-Ray Astronomy Laboratory* (*INTEGRAL*), NASA's *Swift* gamma-ray burst mission, and NASA's *GLAST* (*Gamma-ray Large Area Space Telescope*) high-energy gamma-ray emission.

For background information *see* GAMMA-RAY ASTRONOMY; NEUTRON STAR; NUCLEOSYNTHESIS; PULSAR; QUASAR; SUN; SUPERNOVA; X-RAY ASTRONOMY in the McGraw-Hill Encyclopedia of Science & Technology.   Neil Gehrels

Bibliography. G. J. Fishman and D. H. Hartmann, Gamma-ray bursts, *Sci. Amer.*, 277(1):34–39, July 1997; N. Gehrels and J. Paul, The new gamma ray astronomy, *Phys. Today*, 51(2):26–32, February 1998; P. J. T. Leonard and C. Wanjeck, Compton's legacy: Highlights from the Gamma Ray Observatory, *Sky Telesc.*, 100(1):48–54, July 2000.

Gamma-ray bursts

The *Vela* defense satellites were launched in the 1960s to monitor a treaty that forbade nuclear explosions in space. The *Vela* satellites did not detect any nuclear explosions. However, a serendipitous by-product of this mission was the discovery of celestial gamma-ray bursts (GRBs): short and intense pulses of gamma rays arriving from random directions in the sky. Several times a day a cosmic explosion takes place somewhere in the universe, and its echoes reach us as a burst of gamma rays. The burst appears suddenly, lasts for a few seconds, and then fades and disappears forever. The lack, until recently, of any long-lasting counterparts to these short bursts hindered their study, making them one of the most puzzling phenomena in modern astronomy.

During the last decade BATSE (Burst and Transient Source Experiment) on the National Aeronautics and Space Administration's *Compton Gamma-Ray Observatory* (*CGRO*) and the Dutch-Italian satellite *BeppoSAX* revolutionized the understanding of gamma-ray bursts. In the early 1990s BATSE discovered that gamma-ray bursts are so powerful that they are observed from across the universe. In a few seconds a gamma-ray burst releases more energy than a star, such as the Sun, does in its entire lifetime. During these few seconds it outshines the rest of the universe. Gamma-ray bursts are the most powerful explosions in the universe.

Fireball model. According to the fireball gamma-ray burst model, a black hole accelerates jets to relativistic velocities, at about 0.9999 of the speed of light. These jets carry a huge amount of kinetic energy. Internal collisions within the jets dissipate this energy and produce the gamma-ray emission. The fireball model has a clear prediction. Some of the kinetic energy of the jets should dissipate later when these jets encounter the surrounding interstellar material. At this stage the motion is slower (but still close to the speed of light); consequently the emission has lower energy (x-ray, optical, or radio) and it lasts longer. Thus, gamma-ray bursts must have a low-energy afterglow. On February 28, 1997, the wide-field camera of *BeppoSAX* triggered on a gamma-ray burst and measured its position with high precision. A sensitive x-ray detector on board *BeppoSAX* was pointed 8 hours later toward this position and discovered an x-ray afterglow. The long-sought, long-lasting counterparts of gamma-ray bursts were finally discovered. This x-ray afterglow could be seen for several days.

The accurate x-ray position enabled astronomers to point optical telescopes in this direction and to discover an optical afterglow, which lasted for several months. A radio afterglow was discovered later. These discoveries opened a new era in gamma-ray burst astronomy. They also confirmed the predictions of the fireball model (**Fig. 1**). Radio observations of the gamma-ray burst GRB970508, which was seen on May 8, 1997, established that the fireball reached a size of 10^{12} km (6×10^{11} mi) within a few weeks of the burst. It had to expand almost at the speed of light to reach this size. The explosion was moving faster and closer to the speed of light than any other celestial object. These bursts are the most relativistic objects presently known.

The fireball model makes another prediction. The interaction of the jet with the ambient matter begins while the internal collisions within the jet are still going on. This implies that the afterglow should

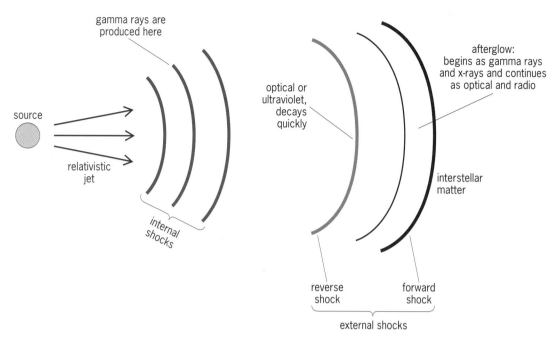

Fig. 1. Fireball model. Internal collisions within a relativistic jet produce the gamma rays. The x-ray, optical, and radio emission occurs later when the jet hits the ambient matter. A short-lived reverse shock (propagating backward into the jet) produces the prompt optical flash.

Fig. 2. Concept of a black hole surrounded by an accretion disk and emitting two jets. The gamma-ray burst is produced within these jets at a very large distance from the black hole. The afterglow is produced even farther out when the jets hit the ambient matter.

begin while gamma rays are still observed. This initial phase of the afterglow should radiate mostly in x-rays, but a significant emission should come out as an optical flash almost simultaneous with the gamma-ray burst. However, it is not easy to detect this flash because it lasts less than 1 minute and a telescope has to be pointed in the direction of the burst within this time. Luckily a system that transmitted within seconds BATSE's rough positions had been worked out. On January 23, 1999, the ground-based robotic telescope ROTSE (Robotic Optical Transient Search Experiment) responded to a BATSE trigger sent via the Internet and pointed itself toward the burst GRB990123 while the burst was still active. ROTSE took six snapshots revealing an extremely bright prompt optical flash that accompanied the burst. Even though this burst took place 10^{10} light-years away, the second snapshot recorded a gigantic 9th-magnitude flash (a 5th-magnitude star can be seen with the unaided eye). If the burst had happened in the Milky Way Galaxy, this signal would have had one-hundredth of the brightness of the Sun and would have been seen during daytime. These observations provided another spectacular confirmation of the fireball model.

Origin in black-hole formation. The successful fireball model explains how gamma-ray bursts work. However, it is still not clear how they form. During the last several years several clues to the nature of these processes have emerged.

The tremendous energy released in gamma-ray bursts gave the first indication that they are powered by black holes. So much energy can be generated so rapidly only in processes involving a compact object—more specifically, a black hole—by accretion of matter onto the black hole. The black hole swallows matter from an accretion disk that surrounds it (**Fig. 2**). A fraction of the energy of this matter is used to accelerate the relativistic jets. Approximately one-tenth of the solar mass has to be swallowed by the black hole to produce the required energy. Very rarely does a black hole engulf so much matter. This is most likely to happen when debris from the black-hole formation process still remains around the black hole. Gamma-ray bursts are the birth cries of newborn black holes.

Additional evidence for this picture arises from the positions of the observed afterglows. Deep optical searches by the Hubble Space Telescope and by the Keck telescope revealed faint host galaxies within which the bursts were located. In many cases the bursts occurred in star-forming regions within those galaxies. Massive stars live only a few million years (a very short time on astronomical scales), and therefore star formation is accompanied by star death. While star birth is a rather benign process, star death is violent, usually resulting in a gigantic supernova explosion. In some cases it is even more spectacular, and it results in a gamma-ray burst.

Progenitors. Several progenitors have been suggested as capable of producing a black-hole–accretion-disk system. Among those a collapsar—the death of a massive star—and the merger of binary neutron stars are the most promising candidates.

Collapsars. A collapsar occurs when a massive star dies. This happens when the star exhausts the nuclear fuel in its core. In most cases the core collapses forming a neutron star, a very dense and compact star (a typical neutron star contains 1.4 solar masses in a radius of 15 km or 9 mi). The stellar envelope explodes producing a supernova; the explosion can be seen for several months. Supernovae happen about once per hundred years per galaxy. Every once in a while, when the star is more massive than usual, its core collapses to a black hole. This can lead to a collapsar: Debris from the collapsing core surrounds the black hole, forming a disk. The accreting black hole produces two relativistic jets along the axis perpendicular to the disk's plane. These jets punch holes in the stellar envelope that now surrounds the black-hole–accretion-disk system. Once the jets emerge from the stellar envelope, they produce a gamma-ray burst. Sometimes the stellar envelope should explode, as in a normal supernova, and then it might be possible to see both a gamma-ray burst and an accompanying supernova explosion from the same source. There is some inconclusive evidence that a supernovalike signature has been seen superimposed on some afterglows.

Neutron-star mergers. An alternative route for forming a black-hole–accretion-disk system and a gamma-ray burst arises from the merger of a binary neutron star pair. In 1975, Russell A. Hulse and Joseph H. Taylor discovered the first binary pulsar: a binary system containing two neutron stars, one of which is a pulsar. Approximately half of all stars have companions, which combine to form a binary system. Some binary systems of massive stars form, after both stars die, binary neutron star pairs. Such a binary pair does not live forever. The neutron stars that rotate around each other emit gravitational radiation and spiral in toward each other due to the corresponding energy losses. Eventually the stars collide and merge. The binary pulsar discovered by Hulse and Taylor, for example, will merge with its companion some 300 million years from now. The merger of the two neutron stars produces a black hole surrounded by a disk of debris. The accretion of this disk onto the black hole generates the energy required to accelerate the relativistic jets and to produce a gamma-ray burst.

Prospects. Theory suggests that both processes, collapsars and neutron-star mergers, take place in nature. Collapsars produce the longer bursts while mergers generate the shorter ones. BATSE, the instrument that has detected more gamma-ray bursts than any before it, is now, with the rest of the *CGRO* satellite, at the bottom of the Pacific Ocean. In October 2000 NASA launched a new satellite, *High-Energy Transient Explorer-2 (HETE-2)*. It is smaller than BATSE, but it can pinpoint rapidly the positions of both short and long bursts and transmit them quickly to Earth for further exploration by other telescopes. It is up to *HETE-2* to confirm these ideas or to surprise us, as has happened may times with gamma-ray bursts, and send us back to the drawing board.

For background information *see* BINARY STAR; BLACK HOLE; GAMMA-RAY ASTRONOMY; NEUTRON STAR; PULSAR; RADIO ASTRONOMY; STELLAR EVOLUTION; SUPERNOVA; X-RAY ASTRONOMY in the McGraw-Hill Encyclopedia of Science & Technology.

Tsvi Piran

Bibliography. G. J. Fishman and D. Hartmann, Gamma-ray bursts, *Sci. Amer.*, 277(1):34–39, July 1997; T. Piran, Gamma-ray bursts and the fireball mode, *Phys. Rep.*, 314:575–667, 1999; J. van Paradijs, C. Kouveliotou, and R. A. M. J. Wijers, Gamma-ray burst afterglows, *Annu. Rev. Astron. Astrophys.*, 38: 379–425, 2000.

Gemini Observatory

Gemini Observatory consists of identical 8-m (319-in.) telescopes located at two of the best observing sites in the world. Gemini North (**Fig. 1**) at Mauna Kea, Hawaii (elevation 4214 m or 13,824 ft) and Gemini South at Cerro Pachon, Chile (elevation 2715 m or 8907 ft). Design-optimized for infrared observations, the telescopes use the latest materials-manufacturing techniques and computer control technology to achieve image quality that is unprecedented for ground-based astronomy. With adaptive optics systems that remove most of the atmosphere-produced distortion, Gemini can routinely produce high-resolution images (under 0.18 arcsecond in the K band infrared window at 2.2 μm wavelength).

Fig. 1. Exterior view of the Gemini North Observatory, showing the dome with vent gates and slit both open. Vent gates surrounding the dome allow air to flow throughout the dome, thereby minimizing distortion of images due to swirling. (*Neelon Crawford, Polar Fine Arts and Gemini Observatory*)

Gemini is a collaboration of seven partner countries (United States, United Kingdom, Canada, Australia, Chile, Argentina, and Brazil). The observatory's mission is to produce a large quantity of high-quality data using a queued observing schedule with a suite of scientific instruments dynamically matched to the current weather and atmospheric seeing conditions.

Mount. Gemini's telescope structure is an azimuth-elevation (AZ-EL) mount. It must hold all of the optical elements in precise alignment, while moving them to point at any object in the sky. The mount control system (MCS) software converts between the celestial coordinate system (right ascension, declination) and the telescope position coordinates.

The telescope structure (**Fig. 2**) is nearly seven stories tall (21.7 m or 71.2 ft) with a moving mass of 342 metric tons (377 U.S. tons) and rotates on cooled frictionless hydrostatic bearings. This behemoth can slew across the sky at $2°$ per second in azimuth and $0.75°$ per second in elevation to achieve a pointing accuracy of less than 2 arcseconds on the sky.

Primary mirror. The heart of the Gemini Telescope is the monolithic primary mirror (**Fig. 3**), weighing 22.22 metric tons (24.5 U.S. tons) and composed of ultralow-expansion glass. With a diameter of 8.1 m (26.58 ft or nearly 319 in.), it is very thin (20 cm or 7.87 in.). Both the primary and secondary mirrors were polished to a global root-mean-square smoothness accuracy of 15.6 nanometers.

Being so thin, the primary mirror (M1) is not rigid enough to hold the required concave hyperbolic shape. Most of its weight is supported by a passive air bag, with hydraulic actuators (120 axial and 60 lateral) continuously adjusting the mirror. The primary control system (PCS) can gently bend the mirror to less than 1 micrometer over the entire surface. Locally each actuator can achieve a resolution of 10 nm. When controlled by the wavefront sensor feedback loop, these actuators fine-tune the primary mirror in a mode called active optics.

Secondary mirror. Gemini uses a single Cassegrain focus located below the primary. The telescope upper ring houses the secondary mirror (M2) [1.023 m or 3.36 ft] weighing 45 kg (99.22 lb) with a convex hyperbolic shape yielding an effective focal ratio of $f/16$. It is composed of Schott Zerodur glass-ceramic composite and polished with a surface accuracy of 17 nm. The secondary support structure uses voice-coil actuators for precise three-axis mirror positioning (lateral XY for coma control and Z for focus control) with vibration canceling technology to minimize wind shake–induced structural resonances. A secondary control system (SCS) dynamically moves the mirror at up to 200 Hz in a fast tip-tilt mode to maximize image quality.

Infrared observations in the $5–30$-μm wavelength range require the secondary mirror to "chop" or oscillate at a slow ($\sim$10 Hz) rate. The detector alternately images the astronomical object (plus sky and telescope emissions) and then "empty sky" (which has only the sky and telescope emissions). The source can then be detected by subtracting the second image from the first since the background emissions eventually average to zero.

All of this is firmly held above the primary mirror by four ultrathin (10-mm or 0.39-in.) vanes to reduce obscuration.

Mirror coating. Instead of the traditional vacuum deposition process, magnetrons were used to sputter material (aluminum or silver) onto the mirror surface, achieving a better reflective coat. Electrodes attached to the conductive coating permit a current to be passed across the mirror, resulting in surface heating. The mirror must be very near the air temperature to produce the best images. Since it takes time to cool or heat a large mass like the entire mirror (an effect known as thermal inertia), heating just the surface is a fast and accurate way to match the mirror to the air temperature.

Cassegrain rotator. Mounted directly beneath the primary mirror, the Cassegrain rotator system precisely rotates the instruments in a direction and speed exactly opposite that which the AZ-EL telescope mount design imparts to the image. The Cassegrain rotator control system (CRCS) software controls the precise de-rotation velocity, which is a function of both the azimuth and elevation angles of the telescope. The rotator with the attached instrument support structure weighs 17.1 metric tons (18.9 U.S. tons) and can carry a total of 8.0 metric tons (8.8 U.S. tons) of scientific instruments. It has a range of $540°$ (1.5 circles) and a top speed of $2.0°$ per second.

Instrument support structure. The instrument support structure (ISS) attaches directly to the

Cassegrain rotator, and serves as the rigid framework to which the scientific instruments are mounted. Mounted on the ISS can be five different instruments at one time. All are to be facility instruments meant to exploit the full range of observing regimes allowed by the telescope design. Infrared cameras and spectrographs, optical multiobject spectrographs, an integral field unit spectrograph, and high-resolution cameras are some of the observing tools at the disposal of the astronomer. A facility calibration unit is available for all the instruments to use. Also planned is an adaptive optics system that will intercept the light before it reaches the instruments, dynamically correct the images, and then send the improved images onto the selected instruments.

Acquisition and guiding unit. The acquisition and guiding (A & G) unit is housed inside and above the instrument support structure. It contains a science fold mirror to direct the starlight to each of the instruments on the ISS. The A & G unit contains a charge-coupled device (CCD) camera for direct viewing during initial acquisition, a high-resolution wavefront sensor to accurately measure the primary mirror figure, and two off-axis CCD guide probes.

During observations, these guide probes are directed to stars out of the instrument field-of-view to enable auto-guiding, a closed feedback, to correct the raw telescope tracking. In addition to being auto-guiders, these probes are full Shack-Hartmann wavefront sensors that can measure the aberrations of the incoming off-axis light. By measuring the shape of these stellar wavefronts, a control matrix can then convert their aberrations to the proper correction for the on-axis wavefronts. This control loop of dynamic corrections relayed to the secondary mirror (focus and coma) and to the primary mirror (astigmatism, trefoil, and other higher-order aberrations) constitutes the active optics. Flexure-independent corrections and even more precise guiding are then accomplished by the on-instrument wavefront sensors, which are physically mounted close to the science detector in each instrument.

Enclosure. Protection against severe weather and the heat of the daytime sun is afforded by the dome. Instead of the traditional white, the steel of Gemini's enclosure has aluminum paint for thermal stability. This helps to minimize temperature changes and ensure that the outer surface (or skin) does not cool too fast at night via thermal radiation and become cooler than the night sky.

Studies have shown that the best approach at night is "no dome at all," mainly because of thermal effects. Temperature differences can cause the air inside the dome to undergo convective motions and swirl about, degrading the images by an effect known as dome seeing. To keep a steady flow of air through the dome and flush away such troublesome eddies, Gemini has two thermal vent gates which can open up to 10 m (32.8 ft) and stretch over 240° of the dome. Combined with the 9.3-m (30.1-ft) slit (through which the telescope looks out at the sky),

Fig. 2. Telescope structure and interior of the dome at the Gemini North Observatory. (*Gemini Observatory/AURA/NOAO/NSF; copyright Association of Universities for Research in Astronomy Inc [AURA], all rights reserved*)

the vent gates combine to give an open, breezy feel to the enclosure (Fig. 1).

The enclosure control system (ECS) rotates the 673 metric tons (741 U.S. tons) of steel on large train bogies so that the telescope is always coincident with the slit position when observing. While slewing to another target, the dome can complete a full rotation in about 2 minutes. At 36 m (118 ft) in diameter, the dome is just wide enough to enclose the telescope.

The enclosure control system also monitors the airflow through the dome and can activate large cooling fans and dehumidifiers for thermal stability during the day (or at night when the dome is closed). Heat management is important for the control electronics mounted on and near the telescope. All

Fig. 3. Primary mirror for the Gemini North Observatory being prepared for lifting into the telescope structure. The crucial lifting rig is above the mirror. (*Gemini Observatory/AURA/NOAO/NSF; copyright Association of Universities for Research in Astronomy Inc [AURA], all rights reserved*)

heat-producing electronics are mounted in a thermal enclosure cooled by a circulating chilled water system.

Telescope control system and console. Each of the aforementioned telescope systems can be compared to a virtuoso musician. The telescope control system (TCS) then takes those musicians and turns them into a finely tuned orchestra. It coordinates each subsystem and ensures that each is at the right place at the right time doing the right thing in the right order. Next in the hierarchy is the telescope control console (TCC), which provides a higher control layer with a user-friendly interface along with the graphical telescope system display.

Observatory control system. The penultimate control level is the observatory control system (OCS). This integrates all functions of the telescope with each instrument. It thus allows an instrument to control the specific telescope functions that the instrument needs to complete its observation. The observatory control system then coordinates with the data handling system (DHS) as the data are read out to on-site quick-look displays and then to distribution and archives. It also manages the observer queue list to prioritize the next observation sequence or program. It is the strategic overview manager.

For background information *see* ASTRONOMICAL COORDINATE SYSTEMS; ASTRONOMICAL OBSERVA-TORY; INFRARED ASTRONOMY; TELESCOPE in the McGraw-Hill Encyclopedia of Science & Technology.

John Hamilton

Bibliography. Big scope, *Sky and Space*, pp. 14–16, April/May 1999; D. Drollette, Extreme phototonics: Mega mirrors, *Photonics Spectra*, pp. 84–92, February 2000; Gemini North sees the light, *Sky Telesc.*, 98(3):18–19, September 1999; R. Graham, High-tech twin towers, *Astron. Mag.*, pp. 37–41, October 2000; M. Lemonic, Beyond Hubble, *TIME Mag.*, pp. 84–88, November 13, 2000; M. Lemmonic, Eyes on the skies, *TIME for Kids*, pp. 4–5, November 3, 2000; K. Sawyer, Unveiling the universe, *Nat. Geog. Mag.*, pp. 14–15, October 1999; G. Stix, A new eye opens on the cosmos, *Sci. Amer.*, 280(4):104–111, April 1999.

Gene therapy (cancer)

Cancer is the second leading cause of death in the United States. About 1,200,000 new cases will be diagnosed and over 550,000 Americans are expected to die of cancer in 2001. Despite developments in surgery, chemotherapy, and radiation therapy of cancer, new strategies for treatment are sorely needed. Advances in knowledge of the molecular basis for cancer have led to the understanding that cancer is a multistep process involving a variety of gene alterations. These genetic lesions accumulate until malignant cells are created and proliferate to form a malignant neoplasm, or tumor. To take advantage of these new understandings, gene therapy has emerged as a new treatment strategy. It involves introducing genes to either reverse or counteract the molecular defects or alterations associated with cancer cells. The potential advantage of gene therapy is that it can be directed to tumor cells with greater specificity than conventional chemotherapy. Thus it is less likely to affect normal cells or to cause undesired systemic effects.

Strategies. Several different strategies for gene therapy of cancer are undergoing clinical trials. The main approaches can be divided into four categories: (1) targeted activation of a treatment drug's metabolic ancestor, a so-called prodrug, in tumor cells; (2) inhibition of activated oncogenes through the use of antisense oligonucleotides; (3) enhancement of the host's immunological response to the tumor cells by cytokine genes or tumor cell vaccination; and (4) restoration of tumor suppressor gene function by gene replacement therapy.

In the first category, either a toxin gene or a gene that codes for an enzyme that converts a prodrug to an active agent is transferred into the tumor cell to kill it directly or sensitize it to the prodrug. For example, the gene coding for the enzyme cytosine deaminase can be delivered to tumor cells where it will convert the prodrug 5-fluorocytosine to the active chemotherapy agent 5-fluorouracil.

The second category takes advantage of the role of oncogenes in cancer. In many cancers, it is believed that the first tumor cell becomes malignant when dominant oncogenes are activated. The malignant progenitor cell then divides to form the tumor. In this category of gene therapy, DNA that prevents the oncogene from being copied correctly, called an antisense construct, is synthesized and delivered to the tumor cell to suppress or reverse its malignant phenotype.

The third category of gene therapy takes advantage of the fact that many types of tumor cells express antigens that can be recognized as foreign by the host, marking them for destruction by the immune system. Cells of the immune system can be isolated from the patient and transfected with genes coding for these tumor-specific antigens. The immune cells are then infused back into the patient, where they stimulate the immune response against the tumor cells.

The fourth category has a relatively large potential. When tumor suppressor genes are mutated or lost, a normal cell is more likely to become malignant. Thus, the strategy is to replace or restore the missing suppressor gene by transferring a good copy of the gene into the tumor cell. A relatively large number of genes could be explored in this category because more than 25 tumor suppressor genes have been identified.

Vectors. Gene therapy requires the use of a gene delivery vehicle (vector) to transfer the gene of interest into the cell in question. The gene therapy vectors currently in use can be classified as viral or nonviral. Viral vectors have received the most attention because viruses that infect mammalian cells already exist in nature. A number of different viruses have been investigated in this regard, including retrovirus, adenovirus, adeno-associated virus, herpes simplex virus, vaccinia, and lentivirus. All have advantages and disadvantages. The initial interest in retrovirus as a vector for gene therapy was based on the fact that it can integrate into the genome of dividing cells and, therefore, produce stable gene transduction. However, because most cancer gene therapy requires only transient gene expression, other vectors, notably adenoviruses that do not generally integrate their DNA into that of the host cell, are more suitable for this application. Adenoviruses also have advantages related to the high capacity for gene insertion and ease of vector production. Although adenoviruses have been widely used for gene therapy, their use is limited by a host immune response that can provoke inflammation.

Three basic approaches for delivering exogenous genes to target cells using nonviral vectors have been investigated. These involve the use of liposomes, naked DNA, or molecular conjugates. Of these, liposomes (charged lipid membranes capable of gene transfer through the formation of DNA-liposome complexes) appear to be most attractive, but they do not specifically target tumor cells. This shortcoming may be overcome with molecular conjugates, which consist of tissue-specific antibodies or ligands bound to a nucleic acid of the gene of interest. Molecular conjugates are limited by the fact that they are unstable in vivo. The simplest of the three nonviral vectors is naked DNA. However, this vector must be injected into the tissue of interest, and genes can be delivered only to cells near the injection site.

There is no single vector of either viral or nonviral origin that satisfies all criteria for a successful cancer gene therapy vector. Considerable effort is being expended on the search for and development of new vectors for gene therapy that will offer the possibility of safe systemic delivery with a high degree of target cell specificity and efficient gene transduction.

Gene replacement of p53 function. One of the first clinical trials for cancer gene therapy involved the restoration of p53 function. The tumor suppressor gene p53 is the most commonly mutated gene found in tumor cells. It is thought that p53 may be nonfunctional even in those tumors where it is not mutated, due to other mechanisms. Tumor suppressor genes such as p53 generally reside in molecular pathways that are critical for controlling the cell division cycle, regulating DNA repair mechanisms, or facilitating the mode of cell death known as apoptosis. Typically, tumor cells have multiple defects in tumor suppressor genes, abrogating the activity of all of these pathways. Therefore, it was initially thought that simply replacing one of these gene defects would not have a therapeutic impact. However, the clinical trials conducted to date indicate that correcting the function of certain tumor suppressor genes, such as p53, that seem to be central to the development of the malignant process may be sufficient to exert a therapeutic benefit.

The gene replacement strategy incorporating p53 evolved from laboratory studies to clinical trials in a very deliberate manner. Non-small cell lung cancer (NSCLC) was chosen as the target tumor type based on its poor outcome and high prevalence. Both in vitro and in vivo laboratory models of NSCLC were then investigated for restoration of cell-cycle regulation and apoptosis following transduction of an exogenous p53 gene. Based on the efficacy demonstrated in laboratory models, a clinical trial of gene replacement therapy of p53 in NSCLC was initiated using a retroviral vector. This early trial, while small, demonstrated that tumors could be safely injected with gene therapy vectors. Moreover, biopsy specimens taken after treatment showed that the vector DNA sequences were active in the tumor cells and levels of apoptosis had increased.

Since these promising results were first published, additional trials have been initiated using an adenoviral vector to deliver the p53 gene (Ad-p53) in patients with NSCLC. However, instead of using the Ad-p53 as a single agent, these new trials are examining the efficacy of this gene therapy vector in combination with conventional therapies, including chemotherapy and radiation therapy. This idea is based on studies in laboratory models in which these combinations together killed more NSCLC cells than the total of the two treatments alone. Similar strategies have also been initiated for the treatment of head and neck squamous cell carcinomas.

Future research. Cancer gene therapy is still in its infancy. Although the clinical trials have shown promising results, the number of tumor targets and genes examined has been limited. Nonetheless, the trials have clearly demonstrated that the method works, and have generated such excitement that research in the area has expanded rapidly. The challenge now is to build on this base of experience and test new strategies using second-generation vectors. These new vectors are engineered to blunt the parient's immune response against viral proteins, thereby reducing inflammation and prolonging therapeutic gene expression. Such vectors may be used systemically to target the spread of metastasis as well as the primary tumor. Moreover, as scientists continue to investigate the molecular basis for cancer, taking full advantage of the database provided by the Human Genome Project, they will uncover additional genes that may prove to be the basis for more effective gene therapies.

For background information *see* ADENOVIRIDAE; CANCER (MEDICINE); DEOXYRIBONUCLEIC ACID (DNA); GENE; GENE ACTION; MITOSIS; MUTATION; ONCOLOGY; RETROVIRUS; TUMOR VIRUSES in the McGraw-Hill Encyclopedia of Science & Technology.

Raymond E. Meyn

Bibliography. G. Kouraklis, Progress in cancer gene therapy, *Acta Oncologica*, 38(6):675–683, 1999; J. A. Roth et al., Adenovirus-mediated p53 gene transfer to tumors of patients with lung cancer, *Nat. Med.*, 2:985–991, 1996; J. A. Roth, S. G. Swisher, and R. E. Meyn, p53 tumor suppressor gene therapy for cancer, *Oncology*, 13(10)(supp. 5):148–154, October 1999; W. Walther and U. Stein, Therapeutic genes for cancer gene therapy, *Mol. Biotechnol.*, 13:21–28, 1999.

Genetically engineered crops

Genetic engineering is the use of molecular techniques to transfer genes from one organism to another, with the transfer usually occurring across species. This process allows scientists to create new crop varieties with novel combinations of traits. The resulting plants may confer enormous benefits or risks on both humanity and the environment at large. Worldwide, almost 70 million acres of farmland were sown with genetically engineered crops in 1998, and the acreage devoted to these "high-tech" crop varieties is continuing to expand.

For millennia, humans have manipulated the genetic code of crop plants. Farmers have improved crop yields and flavor by selecting, for example, the most productive or most tasty individual plants as the seed source for the following year's crop. Crop improvement is usually quite slow because the number of unique gene combinations in the crop population is often limited. To speed up the process and rapidly create many different gene combinations, plant breeders learned to bombard individual plants with mutagens, such as radiation or ultraviolet light. This process, however, is highly undirected, and the vast majority of mutations that result are agronomically worthless.

To gain greater control over the process of crop improvement, geneticists now transfer genes into a crop's genome that code for particularly desirable traits such as insect resistance, improved nutritional quality, herbicide resistance, or improved shelf life. In some ways, then, genetic engineering of crops is nothing new—people have a long history of manipulating crop genomes for their own benefit. On the other hand, some aspects of genetic engineering are entirely novel. One key difference is that genetic engineering allows the creation of gene combinations that were never previously possible, including the transfer of genes from an animal or a bacterium into a crop plant. For example, jellyfish genes transferred into crops make the recipient plants glow, and genes from the bacterium *Bacillus thuringiensis* produce a toxin that makes the crop resistant to certain insect pests. The novelty of the resulting gene combinations means that there is increased uncertainty about the effects of genetically engineered crops.

Clearly, there is great potential for either improvement or harm to result from these genetic manipulations; the difficulty is in predicting which outcome is more likely. Because large-scale planting of genetically engineered crops began only recently and the technology itself is fairly new, few studies have been performed to assess either the positive or negative effects of these crops. In addition, because it is difficult to foresee all of the potential ways in which genetically engineered crops could interact with other species, it is likely that some of the risks and benefits will become apparent only after these crops have been commercially produced for many years. Despite these difficulties, a number of recent studies have identified some possible risks and benefits of genetically engineered crops.

Benefits. Possible benefits of genetically engineered crops include improved human nutrition, reduction in pesticide use, and improved soil conservation. For example, a recently developed variety of rice has been genetically engineered to synthesize beta carotene, the precursor to vitamin A, and to accumulate increased quantities of iron. If this variety becomes widely adopted by farmers, it may greatly reduce the worldwide incidence of blindness and anemia.

Widespread use of crops genetically engineered to be resistant to insect pests could lead to reductions in pesticide use. The harmful effects of chemical pesticides on nontarget species are well documented. In particular, pesticides harm many invertebrate species that are beneficial to farmers, such as insects that feed on pest insects, and worms that aid in decomposition and nutrient cycling. In contrast, it may be possible to genetically engineer insect-resistant crops to target only the pest species, leaving the beneficial species unharmed. In fact, a study by the U.S. Department of Agriculture found that widespread planting of genetically engineered corn, cotton, and soybeans may have been responsible for a 1% reduction in pesticide use between 1997 and 1998.

Genetically engineered crops could also result in improved soil conservation if farmers choose to control weeds with herbicides (in combination with crops genetically engineered to tolerate herbicides) instead of using management practices that cause soil loss (such as excessive tilling).

Risks. Potential risks arising from genetically engineered crops include harmful effects on human health and the environment. Because genes can interact with one another in unanticipated ways, genetic engineering could cause plants to produce allergenic compounds. However, these risks are generally considered to be unlikely because the chemical changes produced by genetic engineering are specific enough to allow easy assessment of their allergenic properties. In contrast, the environmental effects of genetically engineered plants are less easily specified. Consequently, much debate has surrounded the potential environmental impacts of genetically engineered crops, including increased herbicide use, the creation of new weed problems, and harmful effects on nontarget organisms.

Increased herbicide use. The adoption of crops genetically engineered to tolerate herbicides could result in improved soil conservation. However, in order for this benefit to materialize, farmers must increasingly rely on chemical herbicides for weed control. Unfortunately, the use of chemical herbicides is not without its own set of costs. Herbicides vary in their toxicity, but some cause substantial harm to wildlife and human health. Of particular importance are herbicides that persist in the environment and can accumulate in water supplies.

Creation of new weed problems. Genetic engineering of crops could actually enhance weed problems or create new weed species. Genetic engineering is frequently used to create crops that can withstand herbicides or that are toxic to insect pests. However, if the transferred genes were to spread to related, uncultivated species, the population growth rate and hence the weedlike tendencies of noncrop varieties could be increased. In fact, 12 of the world's 13 most important food crops hybridize with wild relatives in some part of their distributions. Thus the opportunity for genes to move from genetically engineered crops to noncrop plants is substantial and could result in the formation of herbicide-resistant or insect-resistant weed populations that would be difficult to control with chemicals or insects.

Harmful effects on nontarget organisms. The potential risk associated with genetically engineered crops that has received the most attention is the harm to nontarget organisms. In particular, two recent studies have found that monarch butterflies may be harmed by genetically engineered "Bt corn"—a variety that contains genes from the bacterium *B. thuringiensis* and produces a toxin effective against the European corn borer. Monarch butterfly caterpillars do not feed on corn, but they do feed on the milkweed plants that are common in corn fields. Milkweeds can become coated with the copious pollen of the surrounding corn crop. Both studies found that monarch caterpillars experience substantially higher mortality when they feed on milkweed coated with the pollen of Bt corn. In contrast, a third study found that pollen from Bt corn is unlikely to harm black swallowtail butterflies. The opposing results of these studies illustrate that the effects of genetically engineered crops are likely to vary from species to species and can be difficult to predict.

However, the findings regarding the potential negative impact of certain types of Bt corn pollen on monarch butterflies prompted the Environmental Protection Agency (EPA) to take action. In December 1999, the EPA instituted a "data call-in" program requiring companies that have registered Bt corn fields to obtain and report information such as the proportion of potential monarch butterfly feeding habitat in and around the Bt corn fields, the lethal concentration of Bt corn pollen for monarch butterfly larvae, how long the lethal concentration of Bt corn pollen remains on milkweed, and whether monarch larvae are feeding on milkweed during pollen shed. Research protocols were due in March 2000 and data were due in March 2001. Many academic scientists have also initiated field experiments to address these data needs. The results of these studies will go a long way toward resolving the heated debate regarding the safety of Bt crops.

Indirect effects on nontarget organisms. The nontarget effects of genetically engineered crops might also be indirect. In other words, the effects of the crops could be propagated through the food web, eventually affecting species that do not directly interact with the genetically engineered crop. For example, genetically engineered crops could affect predatory insects that do not feed on the crop itself but do feed on the insect pests of the crop. If present in sufficient numbers, the predatory insects could be quite beneficial to farmers. However, a recent study found that the mortality rate of one such predator, the green lacewing, almost doubled when the predators were fed pest insects that had fed on Bt corn. Such effects are counterproductive for farmers. Ideally, a genetically engineered crop would kill the insect pests but leave the predatory insects unharmed.

Decrease in bird diversity. A recently published mathematical model demonstrated that adoption of genetically engineered crops could have consequences for the diversity of birds occurring in and around farmland. Specifically, the model examined the effects of herbicide-resistant sugarbeets and a concomitant increase in herbicide usage on weed populations. According to the model, a decrease in weed abundance could lead to a decrease in the abundance of bird species that rely on weed seeds as a food source.

Toxin persistence. One factor that may complicate the assessment of risks is that some by-products of genetically engineered crops may persist and accumulate in the environment. For example, Bt toxin that is genetically engineered into crop plants usually degrades quickly in the environment. However, a recent study found that under particular circumstances the Bt toxin can bind to clay particles in the soil and remain biologically active for at least 230 days. As a result of this soil-binding property, Bt toxin may accumulate over time and increase to much higher concentrations than previously anticipated. Such findings highlight the point that the effects of genetically engineered crops will likely vary greatly, depending on local conditions. In all likelihood, the risks and benefits of genetically engineered crops will have to be assessed on a case-by-case basis.

Future prospects: nationwide monitoring. In early 2001, the National Research Council sponsored a workshop that brought scientists, policymakers, farmers, economists, and biotechnology representatives together to discuss the importance of ecological monitoring of genetically modified crops. Many of the challenges associated with implementing a nationwide monitoring program were discussed. For example, it is not clear which and how many species should be monitored, or how many sites to monitor and at what frequency. Some of the workshop participants suggested that, given limited funding, the program might rely upon farmers to survey fields and report data. It is important that these issues be resolved soon because, as several workshop participants argued, sustained monitoring of environmental problems will be essential for continued public acceptance of transgenic crops.

For background information *see* AGRICULTURAL SCIENCE (PLANT); BIOTECHNOLOGY; BREEDING (PLANT); GENE; GENE ACTION; GENETIC ENGINEERING; HERBICIDE; PESTICIDE in the McGraw-Hill Encyclopedia of Science & Technology.   Michelle Marvier

Bibliography. N. C. Ellstrand, H. C. Prentice, and J. F. Hancock, Gene flow and introgression from domestic plants into their wild relatives, *Annu. Rev. Ecol. Systemat.*, 30:539–563, 1999; J. E. Losey, L. S. Rayor, and M. E. Carter, Transgenic pollen harms monarch larvae, *Nature*, 399:214, 1999; M. A. Marvier, Ecology of transgenic crops, *Amer. Sci.*, 89:160–167, 2001; R. Pool and J. Esnayra, *Ecological Monitoring of Genetically Modified Crops: A Workshop Summary*, National Academy Press, Washington, DC, 2001; D. Saxena, S. Flores, and G. Stotzky, Insecticidal toxin in root exudates from Bt corn, *Nature*, 402:480, 1999; A. R. Watkinson et al., Predictions of biodiversity response to genetically modified herbicide-tolerant crops, *Science*, 289:1554–1557, 2000; L. L. Wolfenbarger and P. R. Phifer, The ecological risks and benefits of genetically engineered plants, *Science*, 290:2088–2093, 2000.

Genome instability

The recent completion of the "working draft" sequence of the human genome has revealed exciting findings that will help scientists to understand the genetic factors that are unique to humans. One surprising finding is that humans have fewer genes than originally thought, perhaps only 30,000–40,000, a number only twofold greater than the worm *Caenorhabditis elegans* and the fruit fly *Drosophila melanogaster*. However, there are some obvious differences between the human genome and the genomes of lower eukaryotes, which may help explain the much greater complexity of humans. One difference is that human genes are more complex themselves, with mechanisms such as alternative splicing that yield many more protein products than the number of genes alone would predict. Another difference is that a much larger percentage of the human genome is derived from repetitive deoxyribonucleic acid (DNA) sequences (interspersed repeats). Interspersed repeats, consisting primarily of long interspersed nuclear elements and short interspersed nuclear elements, were long thought to be purely "junk DNA" or genetic parasites. It is increasingly clear, however, that these repeats contribute to genomic diversity over evolutionary time by increasing the fluidity and complexity of the genome. One type of long interspersed element, the L1 retrotransposon, has contributed more to the current knowledge of the human genome than any other interspersed repeat. L1 retrotransposons decrease genome stability via a variety of mechanisms, including insertion by retrotransposition, L1-mediated transduction, unequal homologous recombination, and retrotransposition of nonautonomous elements in trans.

Insertion by retrotransposition. L1 retrotransposons are a class of small, mobile DNA sequences that can retrotranspose, or move from one genomic location to another by producing ribonucleic acid (RNA) that is transcribed by reverse transcriptase into DNA that is then inserted at a new site. They are further classified as autonomous retrotransposons, meaning they encode the proteins required to promote their own mobilization. The mechanism of human L1 retrotransposition remains largely undetermined. However, the study of these elements in a cultured cell assay has provided some clues, and a proposed mechanism has been made based upon the study of similar retrotransposons from other organisms (see **illus.**).

L1 elements are transcribed from an internal RNA polymerase II (Pol II) promoter to produce a bicistronic RNA, a single transcript with two areas that can be translated into protein (known as open reading frames, or ORFs). The RNA is translated in the cytoplasm by an unknown mechanism to produce

the ORF1 and ORF2 proteins. The ORF1 protein is an RNA-binding protein, and the ORF2 protein has both endonuclease (it can cleave nucleic acids in the middle of the chain) and reverse transcriptase activity. ORF1 and ORF2 proteins tend to bind and promote retrotransposition of the RNA molecule that encoded them (called a cis preference). After gaining access to the nucleus, the transcript is reverse-transcribed and reintegrated into the genome, likely by a process called target-primed reverse transcription. During target-primed reverse transcription, the endonuclease domain of the ORF2 protein cleaves genomic DNA, creating a structure that the reverse transcriptase domain of the ORF2 protein can use to prime reverse transcription of the RNA. Reintegration often creates a 7- to 20-base-pair direct repeat of the endonuclease target site on each end of the inserted L1, called the target site duplication. The retrotransposition process results in a DNA copy of the L1 element that produced the RNA transcript; however, most of the L1 insertions truncate or invert during insertion, resulting in an inactive copy.

L1 insertion has had four important effects on the genome. First, L1 elements have expanded the genome. L1 elements move by a "copy and paste" mechanism and are relatively stable once inserted;

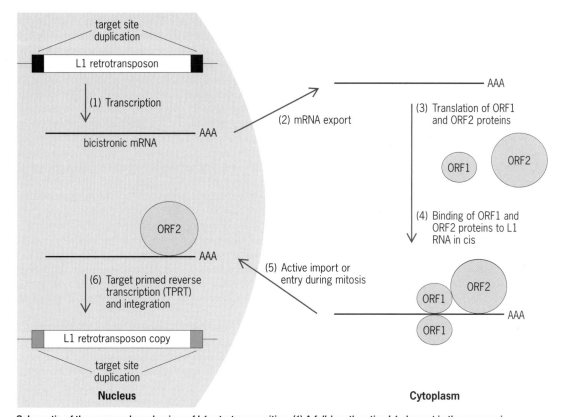

Schematic of the proposed mechanism of L1 retrotransposition. (1) A full-length active L1 element in the genome is transcribed using an internal promoter. (2) The L1 transcript is exported to the nucleus. (3) The ORF1 and ORF2 proteins are translated. (4) The ORF1 and ORF2 proteins preferentially bind the RNA molecule which encoded them (cis preference). (5) The L1 RNA and associated protein(s) return to the nucleus by active transport or entry during nuclear membrane breakdown at mitosis. (6) The L1 RNA is reverse-transcribed and integrated into the genome by target-primed reverse transcription (TPRT). The process depicted results in a DNA copy of the original L1 element at a new genomic location. Note that the target site duplications flanking the original L1 element will differ from the target site duplications flanking the L1 copy at a new genomic location. The new L1 copy may also differ from the original by undergoing 5′ truncation or 5′ inversion and truncation, processes that often occur during integration.

therefore they increase in copy number over evolutionary time. It appears that over 500,000 copies of L1 elements now populate the genome, constituting approximately 17% of the genome mass.

Second, L1 elements insert into genes and cause disease. Although most of the L1 elements in the genome have been inactivated by truncation, inversion, or accumulation of point mutations (mutations due to addition, loss, replacement, or change of sequence in one or more base pairs), there are an estimated 40 to 70 active L1 elements per diploid genome which remain active. L1 elements retrotranspose relatively randomly throughout the genome and are not precluded from inserting into genes. There are now 13 known recent or de novo L1 insertions that have resulted in independent cases of human disease. L1 elements insert into exons (coding sequences), causing disease by a variety of mechanisms including introduction of nonsense codons into the coding sequence, skipping of the disrupted exon during splicing, or insertion into the 5′ untranslated region, thereby affecting transcription or transcript stability. L1 elements can also cause disease when inserted into introns (noncoding sequences) by introducing alternative splice sites that result in an improperly spliced transcript, by decreasing transcription of the gene, or by decreasing the stability of the primary transcript.

Third, L1 elements can insert near a gene and alter its expression. For example, an ancient L1 insertion approximately 20 kilobases upstream of the apolipoprotein(a) gene has been demonstrated to contain a region that serves as an enhancer of apolipoprotein(a) expression. Another study has shown that the antisense L1 5′ untranslated region has promoter activity which can drive transcription of adjacent genes.

Fourth, L1 elements may actually be positively selected in some regions of the genome. For example, the X chromosome has a higher density of L1 elements than the autosomes, and it has been postulated that L1 elements may actually participate in X-chromosome inactivation.

L1-mediated transduction. As a consequence of their mechanism of retrotransposition, L1 elements occasionally carry 3′ end genomic sequences with them when they retrotranspose to a new genomic location (transduction). Therefore, L1 elements may serve as a vehicle to move exons, promoters, or other regulatory sequences throughout the genome. Studies have demonstrated that this is a relatively common process since 20–25% of recent L1 retrotransposition events are associated with a transduced sequence. The average length of the transduced sequence is about 200 nucleotides. When extrapolated to the entire genome, perhaps as much as 1% of the genome consists of sequences that have been transduced by L1 elements.

Unequal homologous recombination. Their high copy number, random distribution, and high sequence identity provides opportunities for L1 elements to participate in unequal homologous recombination events (DNA exchange between identical chromosome regions that are not precisely paired), resulting in deletion and duplication. There are three known unequal homologous recombination events mediated by L1 elements that have resulted in deletions causing human disease. It is likely that many more deletion events occur that are not detected because they create either a lethal phenotype or no phenotype. The concomitant duplication which is created during an unequal homologous recombination event may be a general mechanism for duplicating relatively large regions of the genome.

Retrotransposition of nonautonomous elements in trans. Although the L1-encoded retrotransposition machinery displays a strong preference for retrotransposing the L1 RNA which encoded it, some nonautonomous retrotransposons have apparently evolved a mechanism to undermine the strong cis preference. Alu elements (a family of short interspersed repeat elements), SVA elements (complex elements derived from endogenous retroviruses, variable number tandem repeats, and Alu elements), and processed pseudogenes (nonfunctional genes that are very similar to known genes) are thought to use the L1 retrotransposition machinery based upon the fact that recently inserted copies of these elements are flanked by target site duplications that resemble those created by the L1 endonuclease. These elements have had a similar impact on the genome as the L1 elements. First, they expand the genome. When these elements are taken into account, the L1-encoded proteins have been responsible for creating over one-third of the human genome. Second, they cause disease by insertional mutagenesis. There are over 18 independent cases of disease-causing Alu insertions and at least two cases of SVA elements causing disease. Third, they may serve some beneficial function. Recent evidence suggests that Alu elements may be positively selected in GC-rich (containing many guanine-cytosine base pairs), gene-dense regions of the genome, perhaps because their transcription is beneficial during times of stress. Fourth, they participate in unequal homologous recombination. A recent survey of disease-producing mutations caused by unequal homologous recombination of Alu elements reported 49 independent mutations. Nonautonomous retrotransposons have not yet been demonstrated to transduce downstream sequences, but it remains a distinct possibility.

Other functions. Most of the effects of L1 on the genome tend to increase the fluidity of the genome or decrease overall stability. Although many of the changes introduced to the genome as either a direct or indirect result of retrotransposition are neutral or negative, it is increasingly clear that retrotransposons can occasionally mediate positive changes. For example, it has been proposed that L1 may occasionally insert into double-stranded breaks and thereby repair them. This mechanism has been demonstrated experimentally for another mobile element in yeast, but it remains unknown if L1 elements may also serve this function. (There exists at least one reported human DNA rearrangement that may represent this activity.) It has also been hypothesized that both L1 elements and Alu elements may be positively selected in parts of the genome, where they could be providing a beneficial function.

For background information *see* DEOXYRIBO-NUCLEIC ACID (DNA); GENE; HUMAN GENOME PROJECT; PROTEIN; REVERSE TRANSCRIPTASE; RIBONUCLEIC ACID (RNA); RECOMBINATION (GENETICS); TRANSPO-SONS in the McGraw-Hill Encyclopedia of Science & Technology.   Haig H. Kazazian, Jr.; Eric M. Ostertag

Bibliography. R. J. Maraia, Alu elements as a source of genomic variation: Deleterious effects and evolutionary novelties, in R. J. Maraia (ed.), *The Impact of Short Interspersed Elements (SINEs) on the Host Genome*, pp. 1–24, Landes Bioscience, 1995; J. V. Moran and N. Gilbert, Mammalian LINE-1 retrotransposons and related elements, in N. Craig (ed.), *Mobile DNA*, ASM Press, 2001.

Geochronology

Time is the framework of geology, and the concept of geologic time is inextricably linked to the marine fossil record. This biostratigraphic record lies in sedimentary rocks; and radiometric techniques, which place direct time constraints on the sedimentary rock record, are critical for refining the geologic time scale. The most widely accepted techniques for dating the sedimentary rock record rely on analyzing minerals from igneous rocks, such as volcanic ashes, with established relationships to sedimentary rocks and then extrapolating to the fossil record to estimate the ages of period boundaries. Recent advances in analytical techniques for uranium-lead (U-Pb) zircon dating and the association of volcanic ashes at the Precambrian-Cambrian boundary in Siberia have allowed geologists to change the position of the boundary age from the widely accepted 570 million years ago (Ma) to 544 Ma. The Precambrian-Cambrian boundary marks the appearance of multicellular organisms, and this 26-million-year change in the boundary halved the Cambrian Period, significantly shortening the time available for the explosion of life. However, time resolution is poor for much of the sedimentary rock record because datable igneous rocks do not occur at every biostratigraphic boundary. A number of recent advances are allowing geologists to place direct time constraints on sedimentary rocks by dating minerals and organic matter that form at the time of sedimentation.

In order for a mineral, whatever its origin, to be a geochronometer, it must contain a radioactive isotope and it must remain a closed system with respect to the isotopic system used for dating; that is, there must be no addition or removal of parent or daughter isotope after the time the mineral forms. Radioactive isotopes may have long half-lives and have been around since the origin of the solar system (such as ^{238}U, ^{235}U, ^{232}Th, ^{87}Rb, ^{40}K). They may be the result of intermediate daughter isotopes in some of these decay chains (such as ^{234}U, ^{230}Th, ^{231}Pa) or may be produced in the atmosphere or on the surface of the Earth by cosmic bombardment (such as ^{14}C, ^{10}Be, ^{26}Al). The common isotope systems that are being used for sedimentary minerals are shown in the **table** with the half-lives of the parent isotopes. Isotope systems can be used to date rocks to about 7 half-lives, after which time they are virtually extinct. This, for example, limits the practical applicability of the carbon-14 (^{14}C) system to rocks no older than about 45,000 years (although advances in accelerator mass spectrometers are pushing this back significantly) while the Earth is considered to be 4.56 billion years old.

Uranium-lead dating. The U-Pb system is powerful for radiometric dating because there are two uranium isotopes that decay through independent decay schemes with very different half-lives to their stable lead daughters (see table), and this provides a cross-check for concordance (equal ages) of the two decay schemes. Uranium and lead behave very differently geochemically and can be separated from each other in nature. The U-Pb system is used to date igneous minerals such as the zircons, but more recently geologists have been exploring the potential for using this system to date sedimentary minerals, thus directly dating the sedimentary rock record. It is essential to select minerals that form at the time of sedimentation rather than minerals that are incorporated from preexisting rocks (detrital minerals) or minerals that form after sedimentation due to the inevitable passage of fluids through the system (diagenetic minerals). Several studies have shown that carbonate minerals such as dolomite [$MgCa(CO_3)_2$] and calcite ($CaCO_3$), which formed during soil formation, can have favorable U/Pb ratios for dating and

Common isotope systems for dating sedimentary rocks			
Radioactive isotope	Half-life*	Daughter	Types of materials that can be dated
^{14}C	5.7 ky	^{14}N (not measured)	Organic material, carbonate minerals
^{235}U	0.7 By	^{207}Pb	Carbonates, oil, phosphates, chert?
^{231}Pa[†]	32 ky	^{227}Ac (not measured)	Carbonates
^{238}U	4.5 By	^{206}Pb	Carbonates, oil, phosphates, chert?
^{234}U[†]	248 ky	^{230}Th	Carbonates
^{232}Th	14 By	^{208}Pb	Carbonates
^{87}Rb	48.6 By	^{87}Sr	Chert, clay minerals, glauconite
^{40}K	1.3 By	^{40}Ar	Clay minerals, glauconite, jarosite

*ky = 10^3 years, By = 10^9 years.
[†] Intermediate daughter of the preceding isotope but used as a dating scheme in its own right.

date the time of soil formation. The precision of the ages appears to relate to the duration of paleosol (fossil soil) development, and those paleosols from rapidly deposited sedimentary sections can be dated precisely. Based on the U-Pb ages of paleosols from the Permian-Carboniferous, the age uncertainty of this boundary has been improved by a factor of 10. Additionally, the sea-level changes resulting from the advance and retreat of glaciers, which are recorded in sedimentary cycles, have been shown to be similar in duration to that of the Pleistocene, suggesting that similar processes were pacing glaciation about 300 Ma as compared with the present. This is important because although the Earth is currently in an interglacial period, geologists believe that the glaciers will advance again. The Pleistocene record of glacial advances and retreats is only about 1.6 million years, while the Permian-Carboniferous glaciation lasted about 60 million years. By combining geochronologic studies with the records of climate change, geologists may be better able to predict future climate changes.

Because the half-lives of ^{235}U and ^{238}U are long (see table), using the U-Pb system has only recently received attention for dating the youngest part of the sedimentary record. Recent studies have shown great potential for dating cave deposits and lake deposits in the Cenozoic. These deposits can contain important information about the climate system at the time of their formation. Cave deposits have been shown to archive a detailed climate record in their layers for the Pleistocene, based on uranium-series studies. Using the U-Pb system in similar deposits will allow geologists to gain insight from an extended climate record. Because of these advances in dating, new types of sedimentary materials can be taken into the older sedimentary record.

Uranium-series dating. Both ^{235}U and ^{238}U decay to a number of radioactive intermediate daughter products on the way to the stable daughters of ^{207}Pb and ^{206}Pb. The intermediate daughters of the uranium systems have relatively short half-lives, but a few of them have long enough half-lives that they can be separated from each other during geologic processes. In secular equilibrium, all of the isotope ratios in a decay scheme have an activity ratio equal to one. The activity ratio is simply the concentration of the isotope multiplied by its decay constant. Separation of the intermediate daughter products leads to secular disequilibrium. In a closed system, the isotope system will return to secular equilibrium after about 7 half-lives of the intermediate daughter. Researchers take advantage of this disequilibrium to date geologic events to as old as about 350,000. Recent advances in mass spectrometry have made it possible to obtain very precise ages on much smaller samples than was previously possible with a technique called alpha counting. Alpha particles are two protons and two neutrons (a helium nucleus), and the energy spectra of the alpha decay for each isotope is unique. The alpha counting technique requires large samples because it actually counts the decays rather than measuring the ratios of isotopes as is done with a mass

spectrometer. Using these improvements, geologists have been able to date corals, which grow up to sea level, to constrain the timing of sea-level rise and fall due to the advance and retreat of Pleistocene glaciers. The results of some of these studies confirm models suggesting that these climate changes are the result of cycles of change in the Earth's orbit that increase and decrease the amount of solar radiation in the Northern Hemisphere. By dating continental cave deposits, researchers have also been able to compare the marine and terrestrial records for the Pleistocene to show that they record a similar climate history, even though some geologists predicted that the terrestrial record should lag behind the marine record.

Uranium-series dating has also been used to calibrate the ^{14}C record. The ^{14}C method depends on an assumption about the initial ^{14}C concentration in the sample at the time it formed. This initial concentration depends on the rate of production in the atmosphere, and the carbon cycle in the continent-ocean-atmosphere system. Without the cross-check to another system, one must assume a constant starting composition. This assumption has been demonstrated to be inadequate from measurements of tree rings and corals.

Carbon-14 dating. Radiocarbon dating takes advantage of the fact that ^{14}C is produced in the atmosphere by bombardment of ^{14}N with cosmic rays. Through photosynthesis, plants take in carbon isotopes in a predictable way; and as long as the plant is alive, it will continue to take in ^{14}C. If the cosmic rays enter the atmosphere at the same rate through time, the production rate of ^{14}C will not change. By measuring the ^{14}C of a dead plant or animal (animals have the same isotopic composition as the plants they eat) and knowing the half-life of ^{14}C, one can calculate the time of death. As discussed, the production rate of ^{14}C has changed through time, and this must be taken into account when attempting to date organic material. In addition to organic material, carbonate minerals that form from water that is in contact with the atmosphere (such as oceans or lakes) may also have the atmospheric composition of ^{14}C and can be used for dating. However, in addition to changes in the production rate of cosmogenic isotopes, such as ^{14}C, there are reservoir effects that must be understood in order to use this system to date the sedimentary rock record.

Developments in accelerator mass spectrometry and preparation techniques are reducing the sample sizes needed for ^{14}C dating and pushing the maximum age back to 70,000 years or more. However, the sample may have different carbon components. The issue of contamination of samples by near-modern carbon is a serious concern because the measurement is of the decaying isotope of ^{14}C, and modern carbon has the highest content of ^{14}C. Due to its approximately 5000-year half-life, only about 1/16 of the original ^{14}C remains in a sample that is 40,000 years old. A 70,000-year-old sample with a 0.1% contamination by modern carbon would appear to be 55,000 years old. However, it appears that progressive leaching methods may remove much or all of the contamination, which may be dominantly absorbed onto outer surfaces of datable materials. In order to evaluate the ages after pretreatment, the sample must be dated by another means such as uranium-series dating. Such studies on coral samples have demonstrated that progressive leaching of samples results in accurate ^{14}C dates.

For background information *see* CLIMATE HISTORY; DATING METHODS; GEOCHRONOMETRY; GEOLOGICAL TIME SCALE; ISOTOPE; LEAD ISOTOPES (GEOCHEMISTRY); MASS SPECTROMETRY; RADIOCARBON DATING; ROCK AGE DETERMINATION; SEDIMENTOLOGY in the McGraw-Hill Encyclopedia of Science & Technology.

E. T. Rasbury

Bibliography. A. P. Dickin, *Radiogenic Isotope Geology*, Cambridge University Press, 1995; W. B. Harland et al., *A Geologic Time Scale*, Cambridge University Press, 1989; J. S. Noller, J. M. Sowers, and W. R. Lettis, Quaternary Geochronology: Methods and Applications, *AGU Ref. Shelf Ser.*, vol. 4, 2000.

Glucose transporter

Facilitated glucose transport is the movement of glucose across cell membranes that is driven by the glucose concentration gradient but assisted (facilitated) by carrier proteins. It is energy-independent, and it is stereospecific in that only the D-glucose isomer is transported; the L-glucose isomer is excluded. This process occurs in all mammalian cells and is essential for the maintenance of whole-body glucose metabolism and energy balance. Currently, there are five established functional facilitative glucose transporters in mammalian cells, termed GLUT1, GLUT2, GLUT3, GLUT4, and GLUTx. Each of these transporters has distinct but overlapping tissue distributions, which underscore their specific physiologic function.

GLUT1 is generally expressed and is thought to be responsible for the basal (minimum) uptake of glucose. GLUT2 is predominantly expressed in the liver and pancreatic beta cells, where it functions as part of a sensor that mediates hepatic glucose output during states of fasting and insulin secretion in the postprandial (after a meal) absorption state. In contrast, neurons primarily express the relatively high-affinity GLUT3 isoform necessary to maintain high rates of glucose metabolism for energy production. GLUTx appears to provide important function during early embryogenesis, whereas GLUT4 is exclusively expressed in insulin-responsive tissues, adipose tissue, and striated (skeletal) muscle. These latter tissues provide the key functional elements responsible for the insulin stimulation of glucose uptake and are key targets for disregulation in states of insulin resistance and diabetes.

Intracellular trafficking of GLUT4. In the basal (postabsorptive) state, the GLUT4 protein is localized to poorly defined intracellular storage compartments that constitute components of the trans-Golgi network (responsible for protein sorting) and the endosome system (responsible for protein recycling). Under these conditions, GLUT4 slowly cycles to and from the plasma membrane so that only approximately 2–5% of the GLUT4 protein is found at the cell surface. However, following insulin stimulation, the rate of GLUT4 exocytosis increases approximately tenfold, with a concomitant decrease (two- to threefold) in the rate of GLUT4 endocytosis. This results in a large increase in the number of glucose transporters at the cell surface and thereby accounts for the increase in glucose uptake. In muscle, GLUT4 can also translocate in response to exercise; however, the molecular mediators are different from those of insulin. For example, exercise-induced GLUT4 translocation has been shown to require mobilization of intracellular calcium, but phosphatidylinositol 3-kinase enzyme activity, which is essential for insulinstimulated translocation, does not seem to be necessary.

Several studies have investigated the structural basis for the unique intracellular sequestration of the GLUT4 protein. GLUT4, like all the facilitative glucose transporters, is a multiple membrane protein composed of 12 membrane-spanning domains with both amino and carboxyl termini localized toward the cytoplasm, and a single extracellular glycosylation site between transmembrane domains 1 and 2. Several studies have identified critical internalization sequences at both the amino and the carboxyl regions. However, a specific contiguous intracellular sequestration sequence has yet to be established.

Nevertheless, the insulin-regulated trafficking of the GLUT4 protein has several similarities with the trafficking, docking, and fusion of synaptic vesicles in neurons responsible for neurotransmitter release. It has been well established that the interaction between vesicle receptors (v-SNAREs) and target membrane receptors (t-SNAREs) provide the necessary information for both specificity and induction of molecular fusion. In the case of GLUT4, specific plasma membrane t-SNAREs interact with intracellular GLUT4 compartment v-SNAREs.

Insulin-stimulated GLUT4 translocation. The insulin receptor is composed of two extracellular alpha (α) subunits and two transmembrane beta (β) subunits crosslinked by disulfide bonds into an $\alpha_2\beta_2$ heterotetrameric structure. Insulin binding to the α subunits results in the generation of an intramolecular transmembrane signal that activates the intracellular β-subunit tyrosine kinase domains. This activation event stimulates insulin receptor autophosphorylation and enhances the receptor tyrosine kinase toward several substrates, including the insulin receptor substrate family of proteins (IRS1-4) and two isoforms of the Cbl family (c-Cbl and Cbl-b). The tyrosine phosphorylation of these proteins generates recognition sites for a variety of SH2 domain-containing effector proteins. For example, the tyrosine phosphorylation of IRS1 induces the association/activation of the small adaptor proteins Grb2 and Nck, the Src family tyrosine kinase Fyn, the protein tyrosine phosphatase SHP2, and the type 1A phosphatidylinositol 3-kinase. This latter enzyme catalyzes the formation of phosphatidylinositol-3,4,5-triphosphate from phosphatidylinositol-4,5-bisphosphate in the plasma membrane. In turn, this lipid product activates the phosphatide-dependent protein kinase resulting in the phosphorylation and activation of protein kinase B and the atypical protein kinase C isoforms zeta and lambda. Thus, the insulin-dependent stimulation of a lipid kinase converts the insulin receptor tyrosine kinase signal into a serine/threonine kinase cascade. Although the specific roles of protein kinase B and/or protein kinase C in mediating insulin-stimulated GLUT4 translocation remains controversial, numerous studies have clearly documented that the activation of the phosphatidylinositol 3-kinase and subsequent generation of phosphatidylinositol-3,4,5-triphosphate are essential for this process.

Despite the necessity for activation of phosphatidylinositol 3-kinase, various studies have demonstrated that this pathway alone is not sufficient to mediate GLUT4 translocation. Recently, a novel insulin receptor signaling pathway has been elucidated that appears to function in concert with the phosphatidylinositol 3-kinase pathway. This pathway involves the tyrosine phosphorylation and recruitment of the Cbl protein to specialized microdomains of the plasma membrane enriched in glycosphingolipids and cholesterol, termed lipid raft domains. It appears that a portion of the insulin receptor is localized to these lipid raft domains through its interaction with a resident protein termed caveolin.

The Cbl protein is also recruited to these plasma membrane regions through interaction with an adaptor protein, CAP (Cbl-associated protein). CAP contains a carboxyl terminal SH3 domain that interacts with a proline-rich motif in Cbl, and an amino terminal domain that interacts with another lipid raft resident protein, flotillin. In this manner, CAP links Cbl with the plasma membrane lipid raft microdomains containing the insulin receptor. Importantly, disruption of these interactions results in a near-complete inhibition of insulin-stimulated GLUT4 translocation without any significant effect on the insulin activation of the phosphatidylinositol 3-kinase pathway.

Regulation of GLUT4 expression. In addition to the acute regulation of GLUT4 trafficking and subcellular localization, the long-term expression of the GLUT4 gene is also controlled by various hormonal, nutritional, and metabolic states. For example, during states of insulin-deficiency such as fasting and following destruction of the insulin-secreting β cells by the drug streptozotocin, thereby inducing diabetes, GLUT4 expression levels are severely reduced in adipose, heart, and skeletal muscle. Similarly, obesity, a high-fat diet, and muscle denervation result in decreased GLUT4 expression and subsequent insulin resistance. In contrast, GLUT4 is upregulated by exercise training, thyroid hormone treatment, and insulin therapy—treatments that increase insulin sensitivity.

Role of GLUT4 in type II diabetes. The dynamics of glucose uptake in patients with non-insulin-deficient diabetes mellitus (also known as type II diabetes)

is complex and may be due, in part, to a combination of alterations in GLUT4 trafficking and expression. Adipose tissue has been found to reflect a loss of GLUT4 protein and messenger RNA, but there is no significant alteration in skeletal muscle. Although a specific insulin-signaling defect in skeletal muscle has not been found, these patients display muscle insulin resistance despite a normal complement of GLUT4 protein. It is still uncertain whether disregulation of GLUT4 function is the primary cause or a consequence of insulin resistance. In any case, the presence of peripheral-tissue insulin resistance remains the best indicator for the prediction of subsequent diabetes and is currently the best target for interventional drug therapy.

For background information *see* ABSORPTION (BIOLOGY); CELL PERMEABILITY; DIABETES; ENDOCYTOSIS; GOLGI APPARATUS; INSULIN; SYNAPTIC TRANSMISSION in the McGraw-Hill Encyclopedia of Science & Technology.   Jeffrey Pessin; Silvia Mora

Bibliography. M. P. Czech and S. Corvera, Signalling mechanisms that regulate glucose transport, *J. Biol. Chem.*, 274:1865–1868, 1999; J. E. Pessin et al., Molecular basis of insulin-stimulated GLUT4 vesicle trafficking: Location! location! location!, *J. Biol. Chem.*, 274:2593–2596, 1999; E. A. Richter et al. (eds.), *Skeletal Muscle Metabolism in Exercise and Diabetes*, Plenum Press, New York, 1998; S. A. Summers et al., Signaling pathways mediating insulin-stimulated glucose transport, *Ann. N. Y. Acad. Sci.*, 892:169–186, 1999; A. Zorzano, C. Fandos, and M. Palacin, Role of plasma membrane transporters in muscle metabolism, *Biochem. J.*, 349(pt. 3):667–688, 2000.

Granular mixing and segregation

While the unique properties of granular materials were investigated by Michael Faraday in the 1830s, experimental studies of granular mixing were relatively meager thereafter. However, during the last decade there has been considerable interest, both experimental and theoretical, in granular materials and in their mixing and segregation. Starting with a homogeneous mixture of solid particles, the phenomenon of unmixing granules in stationary piles and under dynamic conditions is being intensely investigated. These investigations include noninvasive measurements of subsurface particle flow, numerical simulations that visually demonstrate the transient granule distributions, and attempts to develop a theory to understand the unmixing of granular mixtures causing particles to segregate or stratify. At present, theoretical models cannot predict whether a homogeneous granular mixture will degenerate to a heterogeneous state.

Nature of granular materials. The distinctive properties of granular materials suggest that they be considered a unique state of matter. A granule is a small particle that is usually visible to the unaided eye. Granular mixtures consist of a large number (of the order of 10^6) particles with at least two different qualities, such as diameter, density, shape, and surface properties. The assemblage of granules into a handleable mixture leads to a state of matter unlike solids, liquids, or gases. For example, granular flow in containers is characterized by slip flow, in which the granules adjacent to the wall are in motion, while liquid flow conditions at a wall are classically characterized by no-slip (the fluid velocity approaches zero near the wall). Also, the flow of granules in a heap or pile has both partly stationary layers and sections with high shear. Dry granule surfaces, free of water, are cohesionless. Thus, the dry granules exhibit only repulsive forces, and the shape of the granular mass is determined by the shape of the container and gravity. Granules, when mixed, undergo inelastic collisions. The kinetic temperature for the granular motion is zero. The behavior of granular mixtures can be treated independent of the interstitial gas (air). Furthermore, when different granular materials are mixed they tend to form a uniform batch; yet continued mixing can result in the granules separating or unmixing to form segregated mixtures and mixtures with stratified layers of like granules.

The mixing or unmixing of granules in a batch is related to differences in particle size, density, shape, and surface properties, as well as the angle of repose of the system. Particles in granular mixtures segregate when subjected to perturbations; for example, the tapping of containers of bread crumbs causes the larger particles to rise to the top. Due to sifting, the granules can separate in a matter of seconds. Without perturbations, granular mixtures can segregate when the mixture is being poured. As a pile becomes higher, larger granules flow down the outside as an "avalanche." Under certain conditions, a batch of powder in a rotating cylindrical mixer forms distinct slices of one size, a phenomenon that can be observed if different types of granules are colored differently.

Applications. The segregation of particles from solid mixtures is important in many industries, including ceramics, coal, food, pharmaceuticals, and the construction of houses and roads, as well as the preparation of paints. In the United States more than 10^{12} kg per year of granular materials are used in the bulk chemical, food, and pharmaceutical industries. In the shipping, handling, and storage of coal, segregation of particles is well known, and commercial operations have been designed to mitigate the effects of particle segregation.

Experiments. Experimental investigations of granular mixing and segregation have used magnetic resonance imaging (MRI) to discern subsurface particle flow in batches. The MRI technique has been used to observe the development of axial and radial segregations inside the bulk of granular batches in horizontal rotating drums or vibrated vertical containers. MRI images of granular materials display thin slices of the material (1–5 mm thick), parallel and perpendicular to the axis of drum rotation. Image intensity is directly proportional to MRI-sensitive particles present in the mixture. These images demonstrate that surface observations do not reveal the complexity of the system, which has hidden subsurface particulate flow. Sometimes there is no apparent particle movement at the surface, but interior patterns of segregation are evident in the form of undulating waves. Another experimental technique, gamma-ray tomography, has been used to conduct continuous, noninvasive observations of interior flow in granular mixtures discharging from hoppers.

Spontaneous granular segregation and stratification can be visually observed when a mixture of small and large particles is poured between two transparent slabs separated by a narrow gap. This system is called a Hele-Shaw cell; the gap size is about 5 mm (0.2 in.). Visual observations require only that the small particles be colored differently than the large ones, say red and white. In a Hele-Shaw cell experiment two spontaneous phenomena can be observed, self-segregation and self-stratification. In self-segregation the large particles tend to accumulate near the bottom of the cell, whereas the small particles are found near the top. In self-stratification the large and small particles tend to form alternating layers of large and small particles parallel to the surface of the pile.

The type of particle unmixing depends on the angle of repose of the particles. This angle is defined as the steepest angle of inclination of a surface at which a loose or fragmented collection of the particles can remain standing in a pile on the surface, rather than sliding or crumbling away. In the case of pouring, the surface in question is the incline of the particle pile, and the particle composition at the surface may well vary from that of the bulk material. The angle of repose depends on the particle density, shape, and surface properties.

Spontaneous self-segregation of particles (say, sand or glass beads) occurs when the large particles have a smaller angle of repose than the small ones. For self-stratification, the large particles have a larger angle of repose than the small particles. When particles roll down a pile, large-particle movement is easier on top of small particles than the flow of small particles on large ones. The small particles tend to be trapped in the spaces between large particles. On the other hand, large particles rolling down a surface of small particles experience the surface to be smoother, like a ball bearing. Thus, as particles of different size are poured from a point onto a plane, or Hele-Shaw cell gap, particle size separation can occur. At the base of the particle pile there is often an obvious difference in the particle size composition from that higher in the pile.

The stratification of particles occurs after an initial regime of segregation. If the incline angle of the pile is about equal to the angle of repose of small particles, and the large-particle angle of repose is greater than the small particles, then small particles can be trapped. This initiates stratification. Experiments indicate that a mixture of perfect spherical particles with the same angle of repose do not stratify.

The phenomena of segregation and stratification are of interest in explaining the origins of sedimentary deposits. These deposits are formed, in part, by the transport of particles of sands in the direction of the wind. The particles tend to crest on the upstream side of a dune. As the dune incline increases, the angle of inclination eventually exceeds the angle of repose, and the particles tend to roll down (an avalanche effect). This can cause the fine sands to deposit into layers as stratified sedimentary formations. Self-stratification may be able to explain rock slides, including why large boulders travel long distances during such an event. The travel of boulders can be related to the aforementioned ball bearing effect whereby large particles slide easily over small particles.

Theoretical models. Theoretical models have been proposed for the segregation of granular mixtures consisting of different types of particles, usually two, of different density, size, and shape. Unlike mathematical expressions for fluid flow in a continuous medium (the Navier-Stokes equations), the modeling of granular flow and behavior requires the representation of a large number of discrete granules that exhibit tendencies to behave unstably, with heterogeneous systems and hysteristic effects.

One model for granular flow with segregation considers a system to be a half-full cylinder with a major portion of the contained granules moving in solid-body rotation and a noncohesive layer (of powder) under continuous flow. This is the model starting point. Important special cases for avalanche flow can be derived. Equations are expressed for the volumetric flow rate, boundary-layer flow velocities, granule path lines, collision diffusion, drift velocities, segregation velocities, and other parameters. The movements of the granules are described from a lagrangian viewpoint. In order to compute results for the granule number density as a function of space and time, a grid is defined; the fraction of each specific type of granule in each bin in the grid is determined. These theoretical results are compared to experimental observations, and granular mixing and segregation phenomena can be simulated.

Chute flow. An important application of granular mixing and segregation is in powder flows out of chutes or hoppers. The powder flow rate and distribution are important variables for manufacturing processes and product quality. During the discharge of a hopper, granules tend to segregate the fine powders to the center of the flow duct. Particle-dynamic simulations model the powder mixture as one consisting of nearly elastic, inelastic, and frictional particles. Discharge flow rates and mass fluxes are due to pressure gradients. The pressure gradient in a column of granules is nonlinear. In the absence of granule segregation, the effective pressure difference will be constant as the solids flow. This is analogous to the constant flow out of an hourglass. For simple granular mixes, existing models predict powder flow out of a hopper reasonably well.

For background information *see* AVALANCHE; BULK-HANDLING MACHINES; CONVEYOR; FLOW OF SOLIDS in the McGraw-Hill Encyclopedia of Science & Technology.   Thomas R. Marrero

Bibliography. R. Behringer, H. Jaeger, and S. Nagel, Introduction to the focus issue on granular materials, *Chaos*, 9(3):509–510, 1999; K. M. Hill et al., Segregation-driven organization in chaotic granular flows, *Proc. Nat. Acad. Sci.*, 96(21):11701–11706, 1999; K. M. Hill, A. Caprihorn, and J. Kakalios, Bulk segregation in rotated granular material measured by magnetic resonance imaging, *Phys. Rev. Lett.*, 78(1):50–53, 1997; H. M. Jaeger, S. R. Nagel, and R. P. Behringer, Granular solids, liquids, and gases, *Rev. Mod. Phys.*, 68:1259–1273, 1996; H. A. Makse et al., Experimental studies of stratification in a granular Hele-Shaw cell, *Philos. Mag.*, B 77(5):1341–1351, 1998; H. A. Makse et al., Spontaneous stratification in granular mixtures, *Nature*, 386:379–382, March 27, 1997; T. Shinbrot and F. J. Muzzio, Nonequilibrium patterns in granular mixing and segregation, *Phys. Today*, 53(3):25–30, 2000.

Green chemistry

The pursuit and development of green (environmentally benign) technologies is one of the highest priorities for today's chemists. The United States chemical industry generates approximately 350 million tons of toxic waste per year, corresponding to more than 10 pounds per person per day. The safe disposal of this hazardous waste comes at the high price of $20 billion per year. A significant source of chemical waste that is often overlooked is the use of organic (or molecular) solvents. These solvents are deleterious to the atmosphere because they are volatile liquids and thus difficult to contain. Nonvolatile ionic liquids (or molten salts), composed exclusively of ions, are attractive substitutes for traditional molecular solvents. Sodium chloride (NaCl) melts above 800°C and could serve as a solvent for reactions proceeding at temperatures greater than 800°C. However, the number of feasible chemical reactions at such a high temperature is limited. Room-temperature ionic liquids (RTILs), however, are salts that melt at or below ambient temperature ($\sim$20°C or 70°F), a quality that renders them useful as reaction media.

Numerous low-melting ionic liquids are known. The majority consist of nitrogen-containing organic cations and inorganic anions (see **illus.**). Recently, imidazolium salts of carboranes (negatively charged cages composed of carbon and boron), which are among the weakest nucleophiles known, have been prepared. Even though research on the use of RTILs as solvents has prospered only in the past few years, ionic liquids are by no means new. The first RTIL, ethylammonium nitrate ($[CH_3CH_2NH_3][NO_3]$), was reported in 1914. Among the many attractive features that make ionic liquids interesting as solvents are that they (1) have no detectable vapor pressure; (2) display excellent thermal stability; (3) exhibit a large electrochemical window (>4 V); (4) are liquids over a wide temperature range, which allows for exceptional kinetic control; (5) are polar solvents, yet poorly coordinating; (6) in some cases, are hydrophobic; (7) are immiscible with a number of organic solvents such as diethyl ether and alkanes; (8) are highly conductive; and (9) dissolve a wide variety of organic and inorganic compounds.

Ionic liquids containing chlorocuprates, chloroaurate, and halogenoaluminates as anions are of limited usefulness because they are extremely sensitive to moisture and air. Hence, these ionic liquids must be handled under vacuum or in an inert atmosphere. Alkylammonium and imidazolium salts of tetraalkylborides, perchlorates, or nitrates are either difficult to prepare or weakly conducting. Furthermore, organic nitrates and perchlorates are potentially explosive, especially when dry. For these reasons, tetrafluoroborate, hexafluorophosphate, triflate, and to a lesser extent bis(triflyl)amide salts of N,N'-dialkylimidazolium cations are most commonly used, and hold the highest promise for application (see illus.).

Preparation. Many alkylammonium halides are available commercially at low cost. Alkylpyridinium and dialkylimidazolium halides can be prepared easily by the reaction of an appropriate halogenoalkane with pyridine or alkylimidazole [reaction (1)]. Prepa-

R = alkyl
X = Cl, Br, or I

ration of alkylammonium halides that are not commercially available can be achieved similarly. The desired ionic liquid is synthesized from the halide salt via metathesis with a silver, an alkali metal (group 1), or ammonium salt of the inorganic anion [reaction (2)]. An alternative method is acid-base neutralization [reaction (3)]. Ionic liquids of triflate salts

M = Ag, Li, Na, or K
Y = inorganic anion

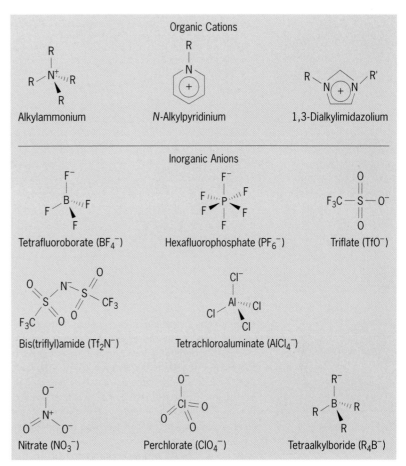

Organic cations combine with inorganic anions to produce room-temperature ionic liquids. R, R′ = alkyl group, usually methyl (CH_3), ethyl (—CH_2CH_3), or n-butyl (—$CH_2CH_2CH_2CH_3$).

are made in a single-step reaction in which the 1-alkylimidazole is reacted with a stoichiometric amount of methyl triflate [reaction (4)]. The halo-

genoaluminate ionic liquids are made simply by mixing the appropriate pyridinium or imidazolium halide salt directly with the aluminum(III) halide in the necessary ratios to give the desired composition.

The influences of alkyl substituents and anion size on the physical properties of imidazolium ionic liquids have been studied systematically. The results show that the desirable properties of an ionic liquid, including low melting point, low viscosity, and high conductivity, are favored by (1) small cation and anion sizes, (2) delocalization of charge, (3) low symmetry of the imidazolium cation, and (4) no alkylation of the C-2 on the imidazolium ring. The physical

Physical properties of some room-temperature ionic liquids[*]

Ionic liquid	mp, °C	d, g mL⁻¹	Viscosity, cP	Conductivity, mS cm⁻¹
[MeBuIm][AlCl₄]	65	1.10	—	—
[BuBuIm][AlCl₄]	55	1.01	—	—
[EtMeIm][BF₄]	15	1.18	38	14
[EtMeIm][TfO]	−9	1.39	45	8.6
[EtMeIm][Tf₂N]	−3	1.52	34	8.8
[BuMeIm][TfO]	16	1.29	90	3.7
[BuMeIm][Tf₂N]	−4	1.43	52	3.9

[*]Me = methyl (CH₃), Et = ethyl (—CH₂CH₃), Bu = *n*-butyl (—CH₂CH₂CH₂CH₃), Im = imidazolium. mp = melting point. d = density at 20°C or at the melting temperature when higher than 20°C. Viscosity refers to dynamic viscosity at 20°C, cP = 0.01 g cm⁻¹ s⁻¹. Conductivity refers to specific conductivity at 20°C.

properties for a few ionic liquids are shown in the **table**.

Despite their extreme moisture sensitivity, corrosiveness, and toxicity, halogenoaluminates are the most studied family of ionic liquids to date, probably because they were among the first RTILs prepared. Halogenoaluminates react with water to generate hydrochloric acid (HCl), which is itself a toxic gas. Nevertheless, dialkylimidazolium salts of tetrachloroaluminate ([RR'Im][AlCl₄]) have been successfully used in spectroscopic and electrochemical investigations of transition-metal halides. Several metal halides, such as hexachloromolybdenum(IV) (MoCl₆²⁻), that are only transients in molecular solvents were found to be stable in ionic media. The halogenoaluminates have also been used to deposit several aluminum–transition-metal alloys. In addition, the first catalytic reaction to be investigated in ionic liquids involved the use of the aluminates both as solvent and catalyst for electrophilic aromatic substitutions. Acidic compositions of chloroaluminates can give rise to superacidic protons, which stimulated much of the research in this area. Several patents have been issued for the use of chloroaluminate ionic liquids in oligomerization and polymerization of alkenes as well as in the preparation of branched oligomeric fatty acids from linear fatty acids.

Uses. Ionic liquids have found uses in four distinct areas: chemical synthesis, catalysis, separation science, and electrochemistry. Among the noncatalytic reactions that have been investigated in ionic liquids are cycloaddition of cyclopentadiene and methyl acrylate (Diels-Alder reaction), and alkylation of sodium β-naphthoxide. Catalytic science and technology will undoubtedly play a central role in providing a path to environmentally sustainable development. Thus, in the context of green chemistry, the most promising application for RTILs is in catalysis. While homogeneous catalysts offer high reactivity and selectivity, they suffer a major drawback, which is the difficulty of separating the catalyst from the products. When ionic liquids are employed as solvents, catalysts are easily recovered and recycled.

The net result is a combination of the efficiency, "atom economy," and selectivity of homogeneous catalysis with the ease of separation inherent in heterogeneous catalysis.

RTILs have been used successfully in the following catalytic transformations: (1) hydrogenation of carbon-carbon double bonds (C=C) and carbon monoxide (CO) with a variety of rhodium and ruthenium catalysts (the hydrogenation of carbon monoxide yields mixtures of ethylene glycol, methanol, and ethanol); (2) dimerization of butadiene catalyzed by palladium complexes; (3) hydroformylation of olefins catalyzed by a rhodium acetylacetonate complex; (4) coupling of an organoboron reagent with arylhalides (the Suzuki reaction) catalyzed by palladium phosphine complexes; (5) coupling of aryl halides or benzoic anhydride with alkenes (the Heck reaction) brought about by palladium catalysts; (6) epoxidation of alkenes and allylic alcohols with hydrogen peroxide (itself a green reagent) as oxidant and methyltrioxorhenium as catalyst; (7) asymmetric oxygen transfer to alkenes in methylene chloride/ionic liquid mixtures with a manganese catalyst; (8) copper-mediated polymerization of methyl acrylate to give narrow-dispersity polymers. All of the reported catalytic reactions that employ RTILs as solvents either exhibit selectivity and reaction rates comparable to those observed in molecular solvents or show pronounced enhancement in rates due to stabilization of the reactive form of the catalyst. For example, the Suzuki coupling reaction between bromobenzene and phenylboronic acid in organic solvents affords the biphenyl product with 88% yield in 6 hours, but in the ionic liquid dialkylimidazolium tetrafluoroborate the yield is 93%; more importantly, in the ionic liquid the reaction is completed in 10 minutes (36 times faster). All catalytic transformations in ionic liquids greatly improve product separation and catalyst recovery.

In separation technology, ionic liquids have been used to improve nuclear fuel processing, extract organic matter known as kerogen from oil shale, remove hydrogen sulfide (H₂S) and carbon dioxide

(CO$_2$) from contaminated natural gas, process biomass-derived renewable feedstock, and extract metal ions from aqueous media. In the field of electrochemistry, ionic liquids are promising electrolytes for batteries and solar cells.

Prospects. Although ionic liquids have the potential to be a major contributor in the development of green technologies, it is highly unlikely that they will provide all the answers. Alternative green reaction media include water, supercritical fluids such as supercritical carbon dioxide (CO$_2$), and fluorous phases. Some transformations might not require a solvent at all. In order to be successful in conquering the highly competitive world of existing chemical technologies, RTILs have to be less expensive, cleaner, and easier to purify. Purification could be of major concern if small amounts of halide contamination are detrimental to the application or process at hand. Most conventional preparation methods produce ionic liquids that contain a small amount of halide reactants. Nevertheless, ionic liquids will most likely soon be put to use in the fine-chemical sector.

Despite all the recent research attention given to RTILs, there is still a serious lack of physical scientific data. Comprehensive studies dealing with the toxicity, biodegradation, safety, and environmental risk and impact of ionic liquids are needed. In addition, analytical tools for assessing the purity of ionic liquids await development. However, the exploration of chemical reactions (catalytic and noncatalytic) in ionic liquids remains an exciting endeavor, since any reaction could produce an unexpected result. Pronounced differences in the behavior of many chemical reactions have already been observed between ionic liquids and molecular solvents, and many more are awaiting discovery.

For background information *see* CATALYSIS; CHEMICAL SEPARATION TECHNIQUES; ELECTROCHEMISTRY; FUSED-SALT SOLUTIONS; HETEROGENEOUS CATALYSIS; ORGANIC SYNTHESIS; SALT (CHEMISTRY); SOLVENT in the McGraw-Hill Encyclopedia of Science & Technology.   Mahdi M. Abu-Omar

Bibliography. P. T. Anastus and T. C. Williams (eds.), *Green Chemistry: Frontiers in Benign Chemical Syntheses and Processes*, Oxford University Press, New York, 1999; P. Bonhote et al., Hydrophobic, highly conductive ambient-temperature molten salts, *Inorg. Chem.*, 35:1168–1178, 1996; J. A. Cusumano, Environmentally sustainable growth in the 21st century: The role of catalytic science and technology, *J. Chem. Educ.*, 72:959–964, 1995; M. Freemantle, Eyes on ionic liquids, *Chem. Eng. News*, pp. 37–50, May 15, 2000; T. Welton, Room-temperature ionic liquids: Solvents for synthesis and catalysis, *Chem. Rev.*, 99:2071–2083, 1999.

Heat storage systems

Energy is used more efficiently when it is captured from heat that might otherwise be lost to the surroundings. This is true in industrial applications, and also on smaller scales such as the household use of solar energy. Heat storage systems reclaim this energy. In addition, heat storage materials and systems can be used in applications where the temperature needs to be kept within a certain range, such as prevention of damage to biological materials or electronic devices. The most important temperature range for heat storage materials for both industrial and consumer use is 0–100°C (32–212°F).

Heat storage materials. Heat storage systems rely on the materials within them to store thermal energy. These materials can be classified as sensible heat storage materials or phase-change materials (PCM).

Sensible heat storage. Every material stores energy within it as it is heated. This can be quantified by the heat capacity C, the temperature change ΔT (= final temperature − initial temperature), and the amount of additional heat stored ΔQ, such that $\Delta Q = C\Delta T$. The units for Q, C, and T are joules (J), J K^{-1}, and K, respectively (or Btu, Btu/°F, and °F). Clearly, other factors being equal, the higher the heat capacity (C) of a material, the greater will be the energy stored (ΔQ) for a given temperature rise (ΔT). Typical materials in a sensible heat storage device include rock, sand, wood, and water. An example of sensible heat storage is a night storage heater, which stores energy in bricks that are heated using nighttime electrical energy (available at reduced rates), and radiates this heat to the surroundings during the following day.

The energy density (defined as $\Delta Q/V$, where V is the volume of the system) of a sensible heat storage system depends on both the heat capacity and the density. For example, balsa wood has a specific heat capacity (heat capacity per unit mass) of 2.9 J K^{-1} g^{-1}, which is rather high compared with that of marble (0.88 J K^{-1} g^{-1}). However, balsa has a rather low density (0.16 g cm^{-3}) compared with marble (2.6 g cm^{-3}), which means that 1 cm^3 of marble can store 2.3 J for a 1°C (or, equivalently, 1 K) temperature rise, whereas 1 cm^3 of balsa can store only 0.46 J over the same temperature increment. Ideally, it would be best to have as small a system as possible to store heat energy. One way to improve the energy density of the heat storage system is to use phase-change materials.

Phase-change materials. Although all materials increase their heat content Q as the temperature is increased, a very large increase in Q occurs when materials change phase. For example, the heat content of water increases considerably as it goes through the phase change from ice to liquid; this is the familiar melting process. The step in Q at the phase change is the latent heat associated with the transition, usually represented as $\Delta_{trs}H$ (see **illus.**). The step in Q at the transition is in addition to the sensible heat storage capacity of the material.

A phase change can lead to a much larger quantity of energy stored, compared with sensible storage alone. The comparison for water is quite useful. Pure water has a heat capacity of 4.2 J K^{-1} g^{-1}, so for a 1°C temperature rise, 1 g of water can store 4.2 J. However, the latent heat associated with melting of

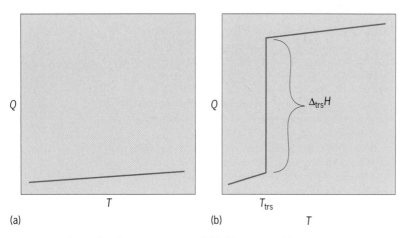

(a)   (b)

Heat content **Q** as a function of temperature **T**. (*a*) **Q** increases with increasing temperature, even if there is no phase transition, as in a sensible heat storage material. (*b*) When the material undergoes a phase transition at temperature T_{trs}, a dramatic increase in **Q** occurs; its jump corresponds to the value of the latent heat of the transition $\Delta_{trs}H$ as indicated on the diagram. This large increase in **Q** can be used to advantage in phase-change materials for heat storage.

ice is $330\,\mathrm{J\,g^{-1}}$. So taking 1 g of ice from just below its melting point to just above (with a total temperature difference of 1°C) absorbs 334 J (latent heat plus 4.2 J from sensible heat storage), about 80 times as much as the sensible heat storage capacity alone.

Although phase-change materials can be used to provide more energy storage, they have their limitations. The most critical limitation is that, in order to be most useful, they must have their phase transition in the temperature range at which the system will operate. For example, at ordinary pressures water has only two phase transitions, one associated with melting at 0°C (32°F) and one with boiling at 100°C (212°F). If a heat storage system were required to operate in the temperature range of 20–40°C (68–104°F), water could operate only as a sensible heat storage material, not as a phase-change material.

Glauber salts are commonly used phase-change materials. They are inorganic salt hydrates that undergo melting-phase transitions that absorb heat. While these materials are highly efficient at storing heat, their maximum storage capacity, again, is over a limited range (transition temperature from 30 to 48°C or 86 to 118°F). In addition, there are problems associated with phase separation on melting, and there is a need for supercooling in order to recover all the stored heat. While these materials are inexpensive, their corrosive nature can lead to high storage container costs. However, the energy storage capacity of Glauber salts and related materials is high, and they are used in commercial products.

Paraffin waxes are also used as phase-change materials. When a paraffin wax melts, it absorbs heat and stores energy. However, there are drawbacks associated with using paraffin waxes as heat storage materials such as oxidation, volatility, flammability, and the large volume change on melting (about 20% increase). Once again, containment costs can be high. Paraffins are often microencapsulated in commercial phase-change materials applications.

Pentaerythritol [$C(CH_2OH)_4$] and its derivatives absorb energy on going through phase transformations, but its phases (before and after the transition) are both solid. The attractive feature of using a solid-solid phase transition is that the mechanical properties of the material are much the same before and after the transition. In particular, the flow problems associated with melting transitions are overcome. The major problem with pentaerythritol is its relatively high transition temperature (188°C or 370°F), which severely limits its commercial use. The transition temperature can be lowered through chemical modification, but this also lowers the quantity of heat absorbed in the transition. Nevertheless, solid-solid phase transition phase-change materials offer considerable promise for heat storage systems.

Applications. Broadly speaking, applications of heat storage systems fall into two categories: scavenging of heat, and thermal control. Both rely on the principle that as energy is put into the system its temperature rises. Later, the energy is released as the temperature falls. In both cases, the goal is to achieve the maximum energy stored (and later released) for a given temperature change.

Scavenging of heat. These uses range from district heating or cooling to solar energy applications.

In district heating, heat storage systems are used to scavenge energy from what might otherwise be wasted sources, such as smokestacks associated with heavy industry. By appropriate heat exchange between the waste energy source and a fluid, this thermal energy can be moved to another location such as a steam heating system for housing and office buildings.

District cooling makes use of heat storage systems in a somewhat different way. For example, in Minato Mirai 21, a major commercial and residential development in Yokohama, Japan, an underground storage system makes use of off-peak nighttime electricity to cool phase-change materials to their low-temperature state. In the daytime, heat-exchange fluids carry the "cold" to the buildings in this district to provide air conditioning. When the fluids return to the phase-change materials system, their heat warms the material. The high energy density of the phase-change materials allows a large cooling capacity, while leveling out the diurnal electricity consumption.

In solar energy applications, the radiant energy of the Sun is used to increase the energy content of the heat storage system in the daytime. This energy is released from the system as the temperature decreases during the night, either through direct radiation or through exchange of energy with circulating fluids. As with the other heat-scavenging applications, the energy density is greatest if phase-change materials, optimized to the temperature cycle of the system, are employed.

Another example of a scavenging heat storage system is found in some automobiles. The heat storage system uses the engine's waste heat to store thermal energy that is later used to rapidly heat the air for the passenger compartment after a cold start.

Thermal control. Because a heat storage system aims to have a large energy storage for a small temperature change, these systems also have applications in thermal control.

Recently, ski boots have been introduced that make use of heat storage systems. Phase-change materials in the insole prevent major temperature drops. Any loss of energy because of the cold conditions is tempered by the presence of the material, which requires a large loss of energy (latent heat) before it cools into its low-temperature phase and then cools further. In a similar way, phase-change materials in insulated storage bags are used to keep pizza hot during delivery.

Preventing temperature extremes can be very important in applications involving biological materials or electronic devices. Here again, heat storage systems can be used to "buffer" against major temperature changes in the environment. If necessary, two phase-change materials can be used, with one providing a lower-temperature buffer and the other an upper-temperature buffer. If the temperature of the surroundings is too extreme, or out of range for too long, the storage capacity of the system can be overcome and the temperature can go outside the desired range.

Outlook. The convenience or actual savings associated with a heat storage system must outweigh its capital cost. One way to improve this situation is by the use of more efficient systems. Efficiency has two components: the energy recovered (which should be as high as possible) and the quality of the energy transferred (the fraction of the energy that is available for conversion to useful work, also called the exergy). New materials, such as high-efficiency phase-change materials with a wide range of phase-transition temperatures, would be very useful. New designs of the heat storage systems can further optimize efficiency. With further realization of the limitations of energy supplies, heat storage materials are likely to play a larger role in consumer and industrial energy distribution systems in the future.

For background information *see* CHEMICAL THERMODYNAMICS; ENERGY; ENERGY STORAGE; HEAT; HEAT CAPACITY; SOLAR ENERGY; THERMODYNAMIC PRINCIPLES; THERMODYNAMIC PROCESSES in the McGraw-Hill Encyclopedia of Science & Technology.

Mary Anne White

Bibliography. F. R. Kalhammer, Energy-storage systems, *Sci. Amer.*, 241(6):56–65, 1979; C. J. Weinberg and R. H. Williams, Energy form the Sun, *Sci. Amer.*, 263(3):146–155, 1990; M. A. White, *Properties of Materials*, Oxford University Press, New York, 1999.

Heat transfer

Heat transfer is vital in many processes, including electric power generation, automotive propulsion and climate-control systems, household heating and cooling equipment, control of body temperature through clothing, thermal management of electronic equipment, and Earth-atmosphere systems, to name a few. Whenever there is a temperature difference, there is heat transfer.

Heat Transfer Enhancement

The three modes of heat transfer are conduction, convection, and radiation. Convective heat transfer—specifically the convective, or fluid flow-induced, process—can be enhanced or improved.

Consider the basic transfer equation applied to a heat exchanger:

$$q = UA(T_a - T_b)$$

Here q is the rate of heat transfer, U the overall heat transfer coefficient, A the surface area for heat transfer, and $(T_a - T_b)$ the average temperature difference between the two fluids (or between the surface and adjacent fluid). The usual goals are to increase the heat transfer for a fixed area and temperature difference or to reduce the area (size) for a fixed heat transfer and temperature difference. Increasing the heat transfer coefficient, that is, enhancing the heat transfer, facilitates both goals. Passive enhancement techniques require no direct application of external power. Active enhancement techniques require external power.

Enhancement of convective heat transfer is one of the fastest-growing areas in heat transfer. Techniques for enhancing convective heat transfer include:

Passive techniques
   Treated surfaces
   Rough surfaces
   Extended surfaces
   Displaced enhancement devices
   Swirl-flow devices
   Coiled tubes
   Surface-tension devices
   Additives for fluids
Active techniques
   Mechanical aids
   Surface vibration
   Fluid vibration
   Electrostatic fields
   Suction or injection
   Jet impingement
Compound techniques
   Rough-surface tube with a twisted-tape insert, for example

Many other possibilities exist when two or more techniques are combined (compound enhancement).

Single-phase convection. Single-phase free convection is found in such diverse applications as cooling of household refrigerator coils and cooling of electronic components. Single-phase forced convection is the most common type of convective heat transfer, used in household environmental control systems and industrial processes.

Passive enhancement. Passive techniques focus on the flow inside channels. The most common passive enhancement technique is to add fins or surface

Fig. 1. Enhanced tubes for heat exchangers. (*a*) Corrugated or spirally indented tube. (*b*) Integral external fins. (*c*) Integral internal fins. (*d*) Deep spirally fluted tube. (*e*) Static mixer insert. (*f*) Wire-wound insert. (*G. F. Hewitt, G. L. Shires, and Y. V. Polezhaev, eds., International Encyclopedia of Heat & Mass Transfer, CRC Press, 1997*)

extensions. Interrupted fins are quite effective because they increase both U and A. Surface roughness has also been used extensively to enhance this type of heat transfer. Roughness and surface extension are usually combined, as most roughness also involves increasing the surface area. The real interest in extended surfaces is increasing heat transfer coefficients on the extended surface. Compact heat exchangers use several enhancement techniques such as offset strip fins, lanced fins, perforated fins, and corrugated fins. Heat transfer coefficients are several hundred percent above the smooth-tube values; however, the pressure drop is also substantially increased.

Internally finned (longitudinal or spiral) circular tubes are typically available in aluminum and copper alloys, but they also can be made in high-temperature materials such as silicon carbide. Correlations (for heat transfer coefficients and friction factors) are available for both straight and spiral fins. Improvements in computational techniques are expected, so that a wider range of geometries and fluids can be used. Examples of the wide variety of surfaces and inserts for tubes are shown in **Fig. 1**. Internal and external enhancement (double enhancement) is involved Fig. 1*a*, *b*, and *d*.

Many proprietary surface configurations have been produced by deforming the basic tube (Fig. 1*a*). A systematic survey of the single-tube performance of condenser tubes indicated up to 400% increase in the nominal inside heat transfer coefficient for water (based on diameter of a smooth tube of the same maximum inside diameter); however, pressure drops are as much as 20 times higher.

Displaced enhancement devices are typically in the form of tube inserts. In the tubes of a hot-gas-fired hot water heater, there is a trade-off between radiation and the turbulent convection. Wire-loop inserts have been used to enhance laminar and turbulent flow. Delta-wing and rectangular-wing promoters, creating vortices that co-rotate or counterrotate, have been studied.

Twisted-tape inserts have been widely used to improve heat transfer in both laminar and turbulent flow, and correlations are available. Several studies have considered the heat transfer enhancement of decaying swirl-flow, generated, say, by a short twisted-tape insert.

Active enhancement. Mechanically aided heat transfer in the form of rotation (stirring) or surface scraping can increase forced convection heat transfer. This is a standard technique in the chemical process and food industries when viscous liquids are involved.

Surface vibration has been demonstrated to improve heat transfer to both laminar and turbulent duct flow of liquids. Fluid vibration has been extensively studied for both air (loudspeakers and sirens) and liquids (flow interrupters, pulsators, and ultrasonic transducers). Although many studies have shown that heat transfer is improved when surfaces are vibrated (in some cases up to 10 times), this is not a popular technique because of possible equipment damage due to the intense vibrations. An alternative technique is used whereby vibrations are applied to the fluid and focused toward the heated surface. With proper transducer design, it is possible to improve heat transfer coefficients from simple heaters immersed in gases or liquids by several hundred percent.

A new development is a shape-memory-alloy coil that deploys at high temperature, when the heat transfer coefficient is low.

Some very impressive enhancements have been recorded with electrical fields, particularly in the laminar flow region. Electrostatic fields are particularly effective in increasing heat transfer coefficients in free convection up to 40 times, but with 100,000 volts. This technique has been studied extensively in the laboratory, and is used to cool cutting tools during machining. It is found that even with intense electrostatic fields the heat transfer enhancement is reduced as turbulent flow is approached in a circular tube.

Compound enhancement. Compound techniques hold particular promise for practical application because heat transfer coefficients can usually be increased more than is possible with any of the techniques acting alone. Some examples that have been studied for single-phase flow include rough tube wall with twisted-tape insert, rough cylinder with acoustic vibrations of the flow, internally finned tubes with twisted-tape inserts, externally finned tubes in fluidized beds, and externally finned tubes subjected to vibrations.

Boilers. Selected passive and active enhancement techniques have been shown to be effective for pool (static) and flow boiling.

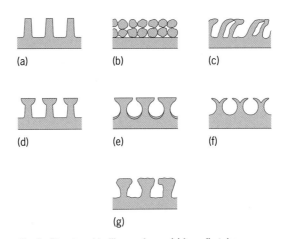

Fig. 2. Structured boiling surfaces. (*a*) Low-fin tube. (*b*) High-flux. (*c*) Thermoexcel-E. (*d*) GEWA-T. (*e*) GEWA-TX. (*f*) GEWA-TXY. (*g*) Turbo-B. (*After G. F. Hewitt, G. L. Shires, and Y. V. Polezhaev, eds., International Encyclopedia of Heat & Mass Transfer, CRC Press, 1997*)

Pool boiling. Surface material and finish have a strong effect on nucleate boiling (bubbles form at liquid-solid interface). However, reliable control of nucleation on plain surfaces is not easily accomplished. Accordingly, since the earliest days of boiling research, there have been attempts to relocate the boiling curve through relatively gross modifications of the surface. For many years, this was accomplished simply by area increase in the form of low helical fins (**Fig. 2a**). The subsequent tendency was to improve the nucleate boiling characteristics by fundamental changes in the boiling process. Many of these advanced surfaces, as depicted in Fig. 2, are being used in commercial boilers. Heat transfer coefficient increases of up to a factor of 10 are common with these surfaces. The advantage is not only a high nucleate boiling heat transfer coefficient, but also the fact that boiling can take place at very low temperature differences.

Active enhancement techniques include heated surface rotation, surface wiping, surface vibration, fluid vibration, electrostatic fields, and suction at the heated surface. Electrohydrodynamic enhancement was applied to a finned tube bundle, resulting in nearly a 200% increase in the average boiling heat transfer coefficient of the bundle, with a small power consumption for the field. Although active techniques are effective in improving heat transfer, the practical applications may be restricted, largely because of the difficulty of reliably providing the mechanical or electrical effect.

Compound enhancement has also been studied for pool boiling. Examples include fins and electric fields, electric fields in a bundle of treated tubes; radially grooved rotating disks; jet impingement on treated or rough surfaces; extended surfaces that are treated; and rough surfaces with additives.

Flow boiling. Helical repeated ribs, helically coiled wire inserts, and twisted-tape inserts have been used to increase in-tube vaporization coefficients in once-through boilers. Numerous tubes with internal fins, either integral or attached, are available for refrigerant evaporators. Original configurations were tightly

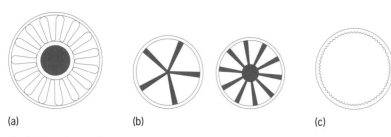

(a)   (b)   (c)

Fig. 3. Inner-fin tubes for refrigerant evaporators. (*a*) Strip-fin inserts. (*b*) Star-shaped inserts. (*c*) Micro-fin. (*After G. F. Hewitt, G. L. Shires, and Y. V. Polezhaev, eds., International Encyclopedia of Heat & Mass Transfer, CRC Press, 1997*)

packed, copper, offset strip fin inserts soldered to a copper tube, or aluminum star-shaped inserts secured by drawing the tube over the insert (**Fig. 3***a* and *b*). Average heat transfer coefficients (based on the surface area of a smooth tube of the same diameter) for typical evaporator conditions are increased by as much as 200%.

A cross-sectional view of a typical micro-fin tube, now widely used for convective vaporization, is shown in Fig. 3*c*. The average evaporation-boiling coefficient is increased 30–80%, and the pressure drop observed is frequently lower.

Rotating flow can be generated by inlet swirl generators, twisted-tape inserts, or tangential injectors along the test tube. Twisted-tape inserts are generally used to increase the limiting or burnout heat flux for subcooled boiling at high imposed heat fluxes (10^7–10^8 W/m^2 or 3×10^6 to 3×10^7 Btu/h ft^2), as might be encountered in the cooling of nuclear fusion reactor components. Examples of compound enhancement in flow boiling include rough surfaces and additives, jet impingement with electrostatic fields, and electrostatic fields with micro-fin tubes.

Condensers. Vapor-space condensation usually occurs with the condensate flowing as a continuous film. Surface extensions are widely employed for enhancement of film condensation. The integral low-fin tubing (Fig. 2*a*), used for kettle boilers, is also used for horizontal-tube condensers. With proper spacing of the fins to provide adequate condensate drainage, the average coefficients can be several times those of a plain tube with the same base diameter. These fins are normally used with refrigerants and other organic fluids that have low condensing coefficients, but which drain effectively because of low surface tension. The fin profile can be altered whereby condensation occurs mainly at the tops of convex ridges. Surface-tension forces then pull the condensate into convex grooves, where it runs off. The average heat transfer coefficient is greater than that for a uniform film thickness.

Vapor-space condensation has also been enhanced by compound techniques, including coiled and fluted tubes, rotating and finned tubes, radially grooved rotating disks, and fins with electrostatic fields.

The applications of convective, in-tube condensation include horizontal kettle reboilers, moisture-separator reheaters for nuclear power plants, and air-conditioner condensers. Internally grooved or knurled tubes, deep spirally fluted tubes, random roughness tubes, and conventional inner-fin tubes have been shown to be effective for condensation of steam and other fluids.

The micro-fin tubes (Fig. 3*c*) have been applied successfully to in-tube condensing. As in the case of the evaporation, the substantial heat transfer improvement is achieved at the expense of a lesser percentage increase in pressure drop. By testing a wide variety of tubes, it has been possible to suggest some guidelines for the geometry, such as more fins, longer fins, and sharper tips; however, general correlations are not yet available. For heat-pump operation, the tube that performs best for evaporation also performs best for condensation.

Twisted-tape inserts result in rather modest increases in heat transfer coefficient for complete condensation of either steam or refrigerants. The pressure-drop increases are large, however, due to large wetted surfaces. Coiled tubular condensers provide a modest improvement in average heat transfer coefficient. The only example of compound enhancement for convective condensation is electrostatic enhancement of a micro-fin tube.

Benefits. Many heat transfer enhancement techniques have made the transition from the laboratory to industrial practice, where they typically result in halving equipment size or doubling heat transfer rates. Successful application depends on enhancement of the lowest heat transfer coefficient making up the overall coefficient U, including due allowance for surface fouling or corrosion. Some examples of systems that will have reduced weight and volume due to enhanced heat transfer are laptop computers, wearable computers, and space suits.

Finally, life-cycle costs must be considered. While enhanced surfaces or devices cost more initially, they often have short payback periods, less than 2 years, for example. The incentive to use enhancement technology is much greater now that the price of energy is increasing.   Arthur E. Bergles

Interfacial Forces in Enhanced Phase-Change Heat Transfer

The use of interfacial forces to enhance heat transfer processes provides simple, passive, quiet cooling systems without mechanical pumps for small devices such as computers. These systems will be particularly useful in microgravity for the thermal control of satellites and space stations. In addition, the effective use of these forces leads to the optimization of many traditional processes.

Phase interfaces. Some characteristics of a liquid system, such as the vapor pressure at the interface, are a function of the intermolecular forces in the extremely thin boundary between the system and its surroundings. Due to the asymmetry of neighboring molecules, interfaces have intermolecular force fields different from those in a bulk solid, liquid, or vapor. Although the liquid-vapor interface is often shown as a two-dimensional surface without thickness, it is more accurately viewed as a diffuse dynamic interface with most changes occurring over a thickness of approximately 2 nanometers (8×10^{-8} in.) in which the molecules moving between the vapor and liquid have collisions and large velocities. A simple physical property that describes the average effect of this molecular asymmetry on the force field is surface tension. In a vapor, this interfacial tension (surface free energy) causes a small amount of liquid to form a spherical drop to minimize the free energy. On a solid substrate, sessile (attached) drops with different contact angles can form, depending on the surface free energies of the solid and liquid.

As the system becomes smaller, changes in the intermolecular forces at the interface become particularly important, and technologies are currently being developed that use these changes. In many cases, these additional changes can be related to the change in the surface shape.

Soap-bubble interface. The simple experiment of blowing a soap bubble shows that the pressure inside the bubble is greater than the pressure outside the bubble. Although this common system is more complicated than a pure liquid interface because it involves monolayers of soap molecules on both sides of a curved water film, the pressure jump can be shown to be a function of the radius of curvature of the bubble and its surface tension. The pressure on the concave side is greater than the pressure on the convex side and is inversely proportional to the radius of curvature. Very small bubbles give very large pressure differences because intermolecular forces are large. Measurement of the pressure jump and the radius of curvature gives the surface tension, which can be related to the intermolecular forcek field.

Length scales. Devices can be studied at the macro, micro, nano, molecular, or atomic scale. Based on the average volume occupied by a molecule in a bulk liquid, the average radius of a small molecule is approximately 0.3 nm (1.2×10^{-8} in.). The density of a liquid is approximately 1000 times that of a vapor at atmospheric pressure. Moving across the extremely thin interface between the liquid and vapor, the cohesive force per unit area due to intermolecular forces varies from the order of 3000 atmospheres in the liquid to essentially zero in the vapor. Since the larger part of this change in density occurs over a distance of approximately 2 nm (8×10^{-8} in.), large gradients in the intermolecular force field between molecules occur at interfaces. Large temperature gradients can also be generated within a small liquid system. Intermolecular forces are due to the fluctuating electrons in the various phases, and interfacial effects become asymptotically small at relatively large distances of the order of 0.1 micrometer (4×10^{-6} in.). On the other hand, they become very large at distances of the order of molecular dimensions.

Interfacial heat transfer. Some examples of heat transfer processes affected by interfacial forces are condensation or evaporation at a vapor-liquid interface such as the surface of an individual spherical

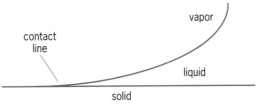

Fig. 4. Contact line region of a meniscus.

drop in a vapor or a sessile droplet on a substrate, and adsorption of an ultrathin film on a solid substrate. Since interfaces have properties unique from those of a bulk liquid or vapor, use of these properties can lead to the development of enhanced heat transfer devices or simply the description of common physical phenomena. For example, since the shape-dependent vapor pressure of a curved interface or an adsorbed film is different than that of the bulk liquid, this difference can be used to control an evaporation/condensation heat transfer process in a small device without the need for an auxiliary mechanical pump. These systems are called passive since the fluids are pumped by intermolecular forces which are a function of the imposed temperature field.

Meniscus intermolecular properties. An important common system that has a gradient in the interfacial force field is the contact line region of a meniscus on a substrate where the liquid-solid-vapor phases meet (**Fig. 4**). At the leading edge of the curved film where the liquid film approaches the thickness of a monolayer of adsorbed vapor (in the case of boiling of a completely wetting liquid), the adsorbed molecules see only solid molecules on one side (at the liquid-solid interface). Farther along the liquid-vapor interface where the meniscus becomes thicker than approximately 0.1 μm (4×20^{-6} in.), the interfacial molecules interact with only liquid molecules on one side of the interface and only vapor molecules on the other side of the interface. Except for the effect of curvature at this thickness, this force field is similar to the interface of a bulk liquid. Therefore, the force field in the thicker film region is significantly different from that in the monolayer. If the intermolecular forces of cohesion between liquid molecules are larger than the forces of adhesion between the liquid and solid, a system with a finite contact angle will form. If adhesion is stronger than cohesion, complete wetting will occur. Since the local vapor pressure is a function of the shape, temperature, composition, and surface tension, the

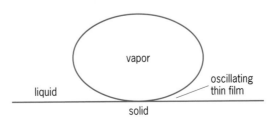

Fig. 5. Boiling bubble on a heated substrate.

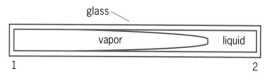

Fig. 6. Constrained vapor bubble heat exchanger with a square cross section and inside dimensions of 3 × 3 × 40 mm.

optimum use of these differences can lead to heat exchanger designs with enhanced performance. Use of intermolecular force concepts show that a superheat of the order of 40°C (72°F) is needed to remove the last monolayer of a simple fluid. Alternatively, theory shows that high rates of evaporation occur in the thicker region with negligible superheats at the liquid-vapor interface. In boiling, a large percent of the heat transfer occurs where the liquid between a vapor bubble and the solid substrate is thin (**Fig. 5**).

Heat transfer devices. For additional interfacial control, the vapor bubble can be constrained in a small container. A constrained vapor bubble (CVB) in a square glass cuvette (also called a wickless heat pipe) is presented in **Fig. 6**. Small-scale copper versions (micro heat pipes) are useful as passive heat exchangers for cooling computers. These systems are built by first pulling a vacuum and then partially filling the cuvette with liquid. The liquid preferentially fills the corners, forming a continuous meniscus in the axial direction. Energy input at the hotter end (1) vaporizes the liquid. The vapor then flows toward the colder end (2) and condenses along its length, where energy is removed due to heat loss. The shape-dependent interfacial pressure jump passively adjusts to give the necessary return flow of the condensate. Additional pumping capacity in a conventional heat pipe can be obtained by partially filling the cuvette with porous material which would give smaller radii of curvature. In the Earth's gravitational field, the shape of a large meniscus is distorted by gravity, and the usefulness of the device might decrease without the use of porous material. However, in microgravity the system can be larger without gravity affecting the performance. In fact, the viscous forces restricting the flow of liquid become less as the scale increases with a corresponding increase in the effective conductivity of the device. The passive evaporation/vapor flow/condensation mechanism in these devices can have an effective thermal conductivity 100–1000 times that of a bar of copper.

Ethanol vapor condensing on a glass substrate has an almost flat profile (that is, a small contact angle due to the strength of the intermolecular force field), making it easy to study. A not-to-scale schematic of the cross section of the CVB square glass channel with condensing ethanol is shown in **Fig. 7**. Looking at the condensation process through the wall of the container, using a microscope, shows a transient fringe pattern due to light interference (**Fig. 8**). The interference pattern gives the liquid thickness profile and, therefore, the force field. The uniformly bright region around the droplets indicates that only a very thin film of vapor is adsorbed in this region. The sessile droplets grow at a fast rate because the resistance to heat transfer is low in the regions where the liquid thickness is small. Once formed, the droplets are sucked into the corner of the CVB by capillary forces, thereby cleaning the surface for additional condensation. In small devices, the interfacial forces are larger than the gravitational force. Therefore, enhanced heat transfer can be obtained by having droplet condensation on a partially wetting surface instead of film condensation on a completely wetting surface. A finite contact angle can be obtained with ethanol by changing the molecular nature of the interface by heat-treating the glass. Proper use of these force fields lead to enhanced but simple passive heat transfer systems for high heat fluxes in small devices.

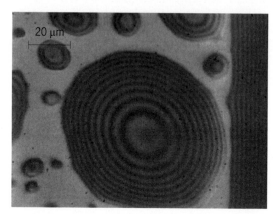

Fig. 8. Interference pattern due to corner meniscus and droplet condensation of ethanol near the corner.

(A significant portion of the recent research described herein was supported by the National Aeronautics and Space Administration under Grants NAG3-1834 and NAG3-2351. Any opinions, findings, and conclusions or recommendations expressed herein are those of the authors and do not necessarily reflect the view of NASA.)

For background information *see* BOILER; BOILING; CONVECTION (HEAT); FILM (CHEMISTRY); HEAT EXCHANGER; HEAT PIPE; HEAT TRANSFER; INTERMOLECULAR FORCES; INTERFACE OF PHASES; LAMINAR FLOW; SHAPE MEMORY ALLOYS; SURFACE TENSION;

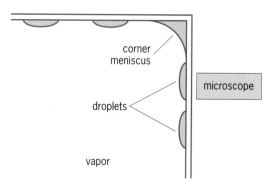

Fig. 7. Corner of a constrained vapor bubble heat exchanger with droplet condensation being viewed through a microscope.

TURBULENT FLOW; VAPOR CONDENSER in the McGraw-Hill Encyclopedia of Science & Technology.

Peter C. Wayner, Jr.

Bibliography. A. E. Bergles, Heat transfer enhancement: The encouragement and accommodation of high heat fluxes, *J. Heat Transfer*, 119:8–19, 1997; A. E. Bergles, Techniques to enhance heat transfer, *Handbook of Heat Transfer*, 3d ed., pp. 11.1–11.76, McGraw-Hill, New York, 1998; C. V. Boys, *Soap Bubbles and the Forces Which Mold Them*, Doubleday, Garden City, NY, 1959; S. DasGupta, J. L. Plawsky, and P. C. Wayner, Jr., Interfacial force field characterization in a constrained vapor bubble thermosyphon, *AIChE J.*, 41:2140–2149, 1995; A. Faghri, *Heat Pipe Science and Technology*, Taylor and Francis, Washington, DC, 1995; J. Israelachvili, *Intermolecular and Surface Forces*, 2d ed., Academic Press, New York, 1992; J. R. Thome, *Enhanced Boiling Heat Transfer*, Hemisphere, New York, 1990; P. C. Wayner, Jr., Interfacial forces in phase-change heat transfer: 1998 Kern Award Review, *AIChE J.*, 45:2055–2068, 1999; P. C. Wayner, Jr., Intermolecular and surface forces with applications in change-of-phase heat transfer, in R. T. Lahey, Jr. (ed.), *Boiling Heat Transfer*, pp. 569–614, Elsevier Science, 1992; R. L. Webb, *Principles of Enhanced Heat Transfer*, Wiley, New York, 1994.

Honeybee intelligence

A honeybee colony comprising some 25,000 individuals must collect sufficient nectar and pollen to keep its members alive and warm through the winter. An individual load is only about 40 mg, yet in one year a colony harvests about 80 kg of nectar and 20 kg of pollen. A colony gathers forage over an area that can extend more than 10 km in all directions from the hive. The locations of profitable flower patches within this area shift from day to day, and the colony can track and exploit these fluctuating resources. The ability to perform this feat involves both social and individual intelligence.

Social intelligence. Success is underpinned by sophisticated social organization, with different individuals performing specialized tasks and changing their jobs according to the current needs of the hive. Some bees (about 10% of the foraging force) act as scouts. They explore the terrain around the hive for patches of flowers and report back the locations of profitable sources. Other bees act on this information and collect nectar or pollen from these identified sources. Recruitment to good sites occurs through positive feedback: recruits report their success to the hive and in turn recruit more workers to the same site. Another group of workers unloads the foragers and stores the nectar and pollen that is brought to the hive. This division of labor allows bees recruited to a good site to devote all their skill and energy to exploiting it.

Waggle dance. Information brought from the field is transmitted to workers in the hive through the waggle dance. Karl von Frisch's classic work established that the information-bearing part of the dance is a repeated straight run on the vertical comb at the end of which the bee circles back to the start of the run. The direction of the food source relative to the sun is signaled by the angle of the straight run with respect to gravity, and the distance of the food from the nest is indicated by the duration of the straight run. To extract information from the waggle dance, a follower must determine the direction of a dancer among a jostling crowd of bees in the dark. Only followers positioned directly behind the dancer seem able to do so successfully.

Frisch found that the dance encodes not only the direction and distance of the source from the hive but also the source's profitability for allocating bees to good foraging sites. Tom D. Seeley's recent work emphasizes the crucial importance of communicating profitability. The bees' measure of this parameter combines both the richness of the nectar and the energy that the bee expends when traveling to harvest it: a rich distant site may have the same profitability as a poorer closer site. Bees thus report the net value of their find, and the way that they do this automatically distributes recruits among the different advertised sources according to the source's value.

Increasing profitability is signaled by increasing the duration of the dance. Potential recruits sample a small number of dance circuits from a particular dancer and have a fixed probability of flying out to the site indicated by its dance. Consequently, the more waggles performed for a site, the greater the number of bees recruited to it. This arrangement allows the colony to exploit many good sites simultaneously. In other tasks the colony needs to agree on a single best solution, such as when a swarm of bees selects their nest cavity. Scouts leave the swarm to find a hole with the right characteristics (size, height above ground, and exposure). They return to the swarm to signal the locations and suitability of the cavities that they have found, again using the waggle dance. Subtle differences exist between the dance behavior of bees signaling a food source and bees signaling a nest. In nest signaling, some bees eventually stop dancing, while some switch to the better sites advertised by others. Thus some sites drop out of the competition. Over a few days, all the scouts converge in signaling the same cavity, and the swarm flies off to find this agreed solution.

Tremble dance. The number of foragers fluctuates with the availability of forage, and the number of storer bees needs to keep pace. An insufficiency of storer bees means that foragers must wait to be unloaded. Foragers have adopted unloading time as a cue to the state of nectar flow into the hive. Increased unloading time tends to mean more inward nectar flow than the storers can handle. If foragers are kept waiting, they act to counter future delays by performing another dance, the "tremble," in which foragers run about in an irregular manner, shaking their body and turning alternately to the left and right. The effect of this dance is to reduce unloading time, which it does in two ways: it induces other workers to

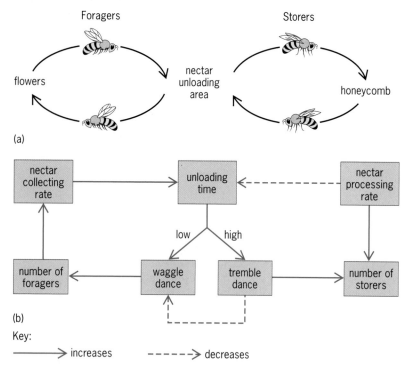

Foragers   Storers

flowers   honeycomb
                    nectar
                    unloading
                    area

(a)

nectar collecting rate → unloading time ← nectar processing rate

low / high

number of foragers ← waggle dance   tremble dance → number of storers

(b)

Key:

———→ increases   - - - - → decreases

Feedback loops in the social control of foraging. (a) Foragers and storers are involved in separate work cycles, with the output of the foraging cycle providing the input to the storing cycle. **(b)** The pace of the two cycles is kept balanced by feedback loops. The waggle dance is part of a positive feedback loop, increasing the number of foragers participating in the foraging cycle. The tremble dance, induced by a high unloading time, has the dual function of increasing the number of storers to keep pace with the incoming nectar flow and of reducing the number of foragers by suppressing the waggle dance. **(Based on work by T. D. Seeley)**

become food storers, and it causes active foragers to cease performing the waggle dance, preventing any further recruitment.

When foraging, the colony appears to act as a co-herent and purposive unit, with no managerial or central command structure. The system works en-tirely through individuals acquiring and transmitting information locally within the colony. The two inter-acting control systems that lead to coordinated forag-ing behavior are shown in the **illustration**. The first system matches the number of foraging bees to the available forage, whereas the second system matches the number of storer bees to the number of foragers.

Individual intelligence and learning. Another essen-tial component of efficient foraging is intelligence at the level of individual bees. The interacting units providing this intelligence are the 950,000 neurons within each bee's brain. The organization of insect brains differs in many ways from that of a colony of bees. The most important difference in this con-text is that individual bees are mobile and interact with a constantly changing population of other bees. Their communication with unknown hive-mates is through an inherited and limited code that remains fixed in order to be mutually comprehensible. Indi-viduals know much more than they can transmit to other bees. They cannot, for instance, inform other bees about landmarks to be encountered on the way to a foraging patch.

Neurons within the bee brain, unlike the bee it-self, are fixed in position and interact with a fixed cohort of neighbors. This stability allows an architec-tural richness that cannot be matched by a colony of mobile, undifferentiated units. Individual units in the brain can, for instance, come to act as complex filters by collecting information from a particular, selected set of sensory neurons, or they might control particu-lar actions by distributing information to a particular set of motor neurons. By increasing the strength of input connections from certain neighbors and de-creasing that from others, individual neurons can alter their properties and thus the information that they in turn pass on. Changing connectivity between fixed neurons allows individual bees to learn.

Navigation. Learning improves several aspects of navigation. It is needed to calibrate the bee's basic navigational instruments. Bees obtain compass direc-tion from the Sun. However, a major complication in relying on the Sun for compass information is that its azimuth changes over the course of a day, as the Sun rises in the east and sets in the west. The function that relates the direction of the Sun's az-imuth to time of day cannot be built in because it varies significantly with the bee's geographical lati-tude and the season of the year. A bee is born with a generic function that it adjusts through individual learning to make the learned function appropriate to its particular location in time and space.

Landmark recognition. Foragers can find a patch of flowers knowing only the direction and distance communicated by the waggle dance. But this infor-mation is imprecise, and it may take a bee a half-hour or more to find an advertised site. Once the site has been found, the bee shortens its journey time on subsequent trips. Part of the improvement comes from being able to make its own estimate of the site's distance and direction from the hive, one that is uncontaminated by errors introduced through reading the limited information in the waggle dance. But some errors remain, and an important compo-nent of the bee's increasing navigational accuracy is learning landmarks along the route. Bees learn the appearance of landmarks and attach motor responses to those landmarks. This acquired knowledge feeds back into social strategies. Bees are likely to remain faithful to a route they have learned, and are reluc-tant to waste their investment by changing to a new one.

Navigational learning is anticipatory and prepro-grammed. Bees leaving their hive or a feeding site for the first few times perform a complex stereo-typed flight that is specifically designed to pick up information about landmarks in the immediate vicin-ity of the goal. The first learning flights of naive foragers are limited to areas close to the hive, but with increasing experience the flights extend over longer distances. This broader geographical informa-tion allows bees displaced hundreds of meters from their hive to return home rapidly.

The learning of landmarks is an elaborately orga-nized process. Objects close to a goal are weighted most strongly in memory, presumably because these can provide precise guidance cues. However, more distant landmarks are not neglected. The spatial panorama surrounding a significant place is important in providing contextual cues that can aid in the recognition of local cues. Thus a small pebble may be crucial in leading a solitary bee to its nest, but it can only be recognized as the right pebble by virtue of the context in which it is embedded.

Conclusion. It seems that evolution arrived long ago at two distinct heuristics for different aspects of problem solving. The first is appropriate for mobile entities (such as bees) and has led to the engaging term "swarm intelligence." Individuals go out foraging for information, exploring the solution space in parallel. The swarm can then select good solutions from those discovered by individual members. The second heuristic is better suited to sedentary entities, such as neurons embedded in a network that can learn to acquire and sift information from their neighbors. Learning allows neural circuits, in some sense, to mirror the causal structure of a problem. Specific units (neurons) come to play specific roles, and the solution emerges through the operation of a complexly connected network of units that are familiar with each other.

It is often claimed that the complex social organization of insects can emerge despite the simplicity of the individual members that make up the society. This brief discussion shows that individual bees are by no means simple. Any simplicity in the interactions between individuals arises because of a limited channel of communication between them and because individuals do not interact consistently with a small group whose unique characteristics can be learned.

For background information *see* ANIMAL COMMUNICATION; BEE; HYMENOPTERA; SOCIAL INSECTS in the McGraw-Hill Encyclopedia of Science & Technology.

Thomas Collett

Bibliography. E. Bonabeau, M. Dorigo, and G. Theraulaz, *Swarm Intelligence: From Natural to Artificial Systems*, Oxford University Press, New York, 1999; T. S. Collett and J. Zeil, Places and landmarks: An arthropod perspective, in S. Healy (ed.), *Spatial Representation in Animals*, pp. 18-53, Oxford University Press, 1998; K. von Frisch, *The Dance Language and Orientation of Bees*, Oxford University Press, London, 1967; A. Michelsen, The dance language of honeybees: Recent findings and problems, in M. D. Hauser and M. Konishi (eds.), *The Design of Animal Communication*, pp. 111–131, MIT Press, Cambridge, 1999; T. D. Seeley, *The Wisdom of the Hive: The Social Physiology of Honey Bee Colonies*, Harvard, Cambridge, 1995.

HPLC–combined spectroscopic techniques

In the past, the components of mixtures of organic chemicals have been identified by isolating them in a pure form and then using spectroscopic techniques to characterize them. Developments in linking high-performance liquid chromatography (HPLC) with mass spectrometry (MS) and nuclear magnetic resonance (NMR) spectroscopy have greatly speeded up the identification of unknowns by removing the need for prior isolation. In the case of nuclear magnetic resonance, these developments have required practical and efficient solutions to the, apparently, intractable problems of efficient solvent suppression and (relatively) poor sensitivity. For mass spectrometry the difficulties associated with the use of chromatographic eluents and flow rates have also been solved, and HPLC-MS is now in routine use. However, when used singly neither HPLC-MS nor HPLC-NMR can always provide unequivocal identification of unknowns, and often both are required. Given the pressing need for rapid structure determination, it was inevitable that combined HPLC-NMR-MS systems would be constructed. The first example was reported in 1995 and was rapidly followed by applications to a range of analytes (pharmaceutical mixtures, natural products, drug metabolites, and so on).

Method. With current systems, it is possible to perform HPLC-NMR-MS experiments using reversed-phase (RP) chromatography with conventional columns and flow rates of 1 mL min^{-1}. There are, however, a number of factors that must be addressed in order to perform successful HPLC-NMR-MS, especially the use of mutually compatible solvent systems. This means avoiding nonvolatile buffer salts for mass spectrometry and minimizing the use of buffers, modifiers, and solvents that contribute unwanted signals in the nuclear magnetic resonance spectrum. If an acidic modifier is needed to adjust solvent pH, trifluoroacetic acid (TFA), with no interfering protons, would be ideal for nuclear magnetic resonance spectroscopy but in practice can suppress ionization in mass spectrometry, so formic acid is probably the best compromise. Ammonium formate buffers are generally compatible with both nuclear magnetic resonance and mass spectrometry, and these restrictions on the solvent systems used for HPLC-NMR-MS do not present insurmountable difficulties. Efficient solvent suppression methods that have been developed for nuclear magnetic resonance do not require the use of deuterated (D) solvents, but in practice mobile phases prepared from acetonitrile/D_2O mixtures makes the task of solvent suppression easier. The use of deuterated acetonitrile is not essential but allows the detection of peaks near the acetyl methyl resonance. While more expensive than using ordinary acetonitrile, the cost is often trivial compared with the cost of the experiment itself. The use of D_2O, while simplifying the acquisition of nuclear magnetic resonance data, can cause problems with mass spectrometry because of deuterium exchange. This difficulty can be eliminated by mixing the eluent with water or methanol prior to the mass spectrometer to exchange the protons. This phenomenon can actually be turned to advantage as, by recording mass spectra with and without deuterium exchange, the number of

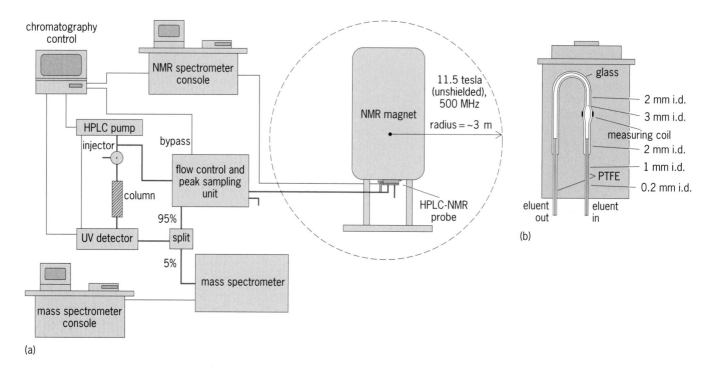

Fig. 1. HPLC-NMR-MS system. (*a*) Schematic. (*b*) Detailed NMR flow probe; i.d. = inside diameter.

exchangeable protons can be found, which is helpful in determining the number of hydroxyls, for example, present in an unknown.

A typical HPLC-NMR-MS system is shown in **Fig. 1***a*, comprising conventional high-performance liquid chromatography, a "sampling unit" in which peaks can be collected and stored for subsequent analysis, a UV detector, and the nuclear magnetic resonance and mass spectrometers (connected in parallel via a splitter at the outlet of the UV detector). The design of the nuclear magnetic resonance flow probe is shown in Fig. 1*b*. The nuclear magnetic resonance and mass spectrometers are not connected in-line, as placing the mass spectrometer after the nuclear magnetic resonance spectrometer results in the pressurization of the nuclear magnetic resonance flow probe. With current flow probe designs, this pressure can cause leaks as the probes were not meant to operate under such conditions. An advantage of the parallel operation of the two spectrometers is that mass spectrometry can be used to trigger stopped-flow (where the peak is held in the flow probe until a spectrum is obtained) nuclear magnetic resonance experiments on peaks of interest if stopped-flow nuclear magnetic resonance is required for reasons of sensitivity (or to perform complex nuclear magnetic resonance experiments). The stray magnetic field from the nuclear magnetic resonance spectrometer can interfere with the operation of the mass spectrometer which needs to be placed at a suitable distance from the magnet. However, the latest generation of actively shielded nuclear magnetic resonance magnets has greatly reduced this problem.

Given the high sensitivity of the mass spectrometer detector, the flow is usually split such that about

95% is directed into the flow probe of the nuclear magnetic resonance spectrometer and 5% to the mass spectrometer. The amount of sample needed to obtain a nuclear magnetic resonance spectrum depends upon a number of factors such as the residence time in the probe (that is, on-flow or stopped-flow), the design of the flow probe itself, field strength, nucleus (such as ^{1}H, ^{19}F), and the type of nuclear magnetic resonance experiment being performed (one- or two-dimensional). However, for a typical low-mass molecule and a 500-MHz nuclear magnetic resonance spectrometer, equipped with a 60-μL flow probe at flow rates of 0.5–1.0 mL min^{-1}, approximately 10–20 μg of sample would be expected to give a useful spectrum. If stopped-flow techniques are used, there is a corresponding reduction in the amount of material needed for a one-dimensional spectrum. Two-dimensional nuclear magnetic resonance experiments, such as ^{1}H-^{1}H TOCSY (total correlation spectroscopy), can be performed only in the stopped-flow mode.

Applications. Most applications of HPLC-NMR-MS have been in the area of pharmaceuticals and drug metabolite determination. In the first reported example, a mixture of fluconazole and two related triazoles was analyzed with RP-HPLC with mass spectrometry and nuclear magnetic resonance spectroscopy obtained at a flow of 1 mL min^{-1}. HPLC-NMR-MS was next applied to metabolites of acetaminophen (paracetamol) present in the solid-phase extracts of human urine. Acetaminophen provided a useful test as a major metabolite is a phenolic sulfate, which shows no diagnostic signal in the ^{1}H nuclear magnetic resonance spectrum. Thus, while the HPLC-NMR spectrum of the acetaminophen sulfate metabolite quite clearly showed all of the characteristics that could be expected for this metabolite, it could not be unequivocally identified—an easy feat with the mass spectrometry data. Other applications have included the identification of ibuprofen metabolites in urine extracts and studies on the metabolic fate of a range of aniline derivatives, including 2-trifluoromethyl-4-bromoaniline, the related 4-chloro substituted analog, and pentafluoroaniline. Using a slightly different approach, the metabolism of a nonnucleoside reverse transcriptase inhibitor was investigated, with the peaks of interest initially collected into the peak sampling unit. When all the relevant peaks had been collected in this way, they were first taken for nuclear magnetic resonance spectroscopy and then subjected to mass spectrometry (with or without back exchange using a protonated solvent). In another drug metabolism example, the HPLC-NMR-MS system incorporated a radioactivity monitor to detect the metabolites of a ^{14}C-labeled β-blocker present in urine following administration to the rat. Nuclear magnetic resonance and mass spectrometry data were then obtained for the high-performance liquid chromatography peaks corresponding to the drug-related components (in this case corresponding to the parent, a hydroxylated metabolite and its glucuronide conjugate). In addition to studying metabolism in mammalian systems, HPLC-NMR-MS has been used to study the fate of xenobiotics in plants. Thus, in maize, 5-trifluoromethyl pyridone was shown to form the *N*-glucoside and *O*-malonylglucoside conjugates of the parent pyridone.

Natural products are also a fertile area of application for the combination of HPLC-NMR-MS, and several examples are in the literature. Thus, the phytoecdysteroids (polyhydroxylated steroids) present in a plant extract (*Silene otites*) were identified using RP-HPLC via on-flow ^{1}H NMR-MS. This identified 20-hydroxyecdysone, 2-deoxy-20-hydroxyecdysone, and 2-deoxyecdysone as major components. The ecdysteroid integristerone A was also present as a minor component (identified by mass spectrometry and stopped-flow ^{1}H nuclear magnetic resonance spectroscopy). A typical chromatogram for this application, together with representative spectra for 2-deoxy-20-hydroxyecdysone, is shown in **Fig. 2**. In another study, RP-HPLC-NMR-MS was used on an extract of *Hypericum preforatum* to identify compounds that included the arabinoside and galacturonides of quercetin; and the rutinoside, glucoside, rhamnoside, and galactoside of quercetin, hypericin, protohypericin, pseudohypericin, pseudohypericin, and protopseudohypericin.

Another obvious use of this technology is combinatorial chemistry, and an example showing its use for a 10-compound model peptide library has been described.

HPLC-NMR-MS is a readily implemented technique for the identification of compounds present in complex mixtures, although not all samples require it for their solution. However, for suitable problems

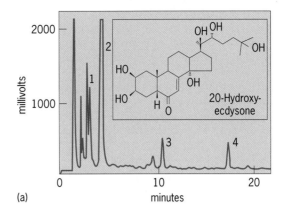

(a)

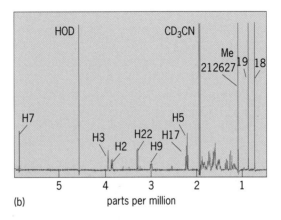

(b)

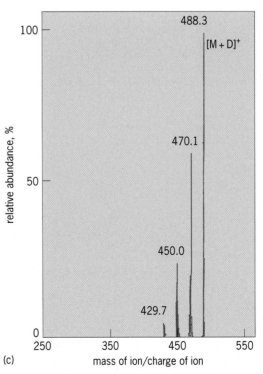

(c)

Fig. 2. Application of HPLC-NMR-MS to the identification of ecdysteroids present in a plant extract. (*a*) Chromatogram of the extract; peak 1 = integristerone A; peak 2 = 20-hydroxyecdysone; peak 3 = 2-deoxy-20-hydroxy-ecdysone; peak 4 = 2-deoxyecdysone. (*b*) NMR spectrum and (*c*) MS spectrum obtained for 20-hydroxyecdysone.

the combination of high-performance liquid chromatography with nuclear magnetic resonance and mass spectrometry represents a very powerful method for structure determination, and there is no doubt that the ease with which this can be achieved will result in such systems becoming much more widespread.

For background information *see* LIQUID CHROMATOGRAPHY; MASS SPECTROMETRY; NUCLEAR MAGNETIC RESONANCE (NMR) in the McGraw-Hill Encyclopedia of Science & Technology.   I. D. Wilson

Bibliography. J. C. Lindon, J. K. Nicholson, and I. D. Wilson, *J. Chromatogr. B*, 748:233–258, 2000; F. S. Pullen et al., *Rapid Comm. Mass Spectrom.*, 9:1003–1006, 1995; J. P. Shockcor et al., *Anal Chem.*, 68:4431–4435, 1997; I. D. Wilson, *J. Chromatogr. A*, 892:315–327, 2000; I. D. Wilson, J. C. Lindon, and J. K. Nicholson, *Anal. Chem.*, 72:534A–542A, 2000.

Hybrid silicon-molecular devices

For many years, scientists have imagined schemes for creating molecular electronic devices. In most instances, the intended capability of the molecular structure has been to switch an electron current on and off. Molecular building blocks, if linked together to form logic functions (AND, OR, NAND, and so on), would seem to lead to a computer of ultimate miniaturization. A simple estimate suggests it might be possible to make molecular-based computers about 1000 times smaller than silicon-based computers. Some key steps have recently been made toward achieving this goal. However, it is not yet known whether the molecular electronics scenario is ultimately reasonable.

The driving force for molecular-based technology is the same as for any other—the new approach must be better or cheaper. Replacing silicon will be a formidable task, because silicon has many near-ideal properties. For example, by adjusting the concentration of dopant atoms (such as boron or phosphorus) in a silicon crystal, the conductivity can be tuned over a very wide range. The oxide that grows on silicon has ideal insulating properties and, moreover, it does not trap charges. The latter feature allows electric fields to be directed through the oxide. Electric fields, in turn, can greatly alter the concentration of electrons (more generally, carriers, which can be electrons or holes) and therefore conductivity. It is via electric fields that silicon devices are turned on and off. These properties, combined with sophisticated lithographic techniques, allow very fine and numerous conducting, insulating, and semiconducting regions to be sculpted into a silicon crystal to create powerful devices.

Hybrid devices. A strategy for building molecular-based devices capable of supplanting silicon remains far off. For that reason, some researchers are considering hybrid silicon-molecular strategies. A hybrid approach helps to get around the extraordinarily challenging problem of electrically connecting an entirely molecular device. In addition, a hybrid strategy can be aimed at enhancing the capabilities of silicon rather than competing head-on with it.

There are natural niches for hybrid silicon-molecular devices to fill. Silicon is not well suited to light detection or emission—things that many molecules can do well. Also, silicon has no capacity to enter into the kind of subtle and discriminating interactions that molecules routinely share in, such as biochemical reactions. Compared to molecular electronics, the hybrid silicon-molecular approach is relatively broad in scope, and it is more appropriate to speak of molecular devices than molecular electronics.

One barrier to the creation of a hybrid, or any, molecular device has been the difficulty in knowing how to precisely position molecules. Such devices, which depend on the inherent properties of small collections of molecules, will in general be very intolerant of errors in placement. To make molecular-scale devices, it is necessary to link molecules to silicon with an extraordinarily detailed understanding of every atom and bond. Only if that can be achieved does it become possible to work toward

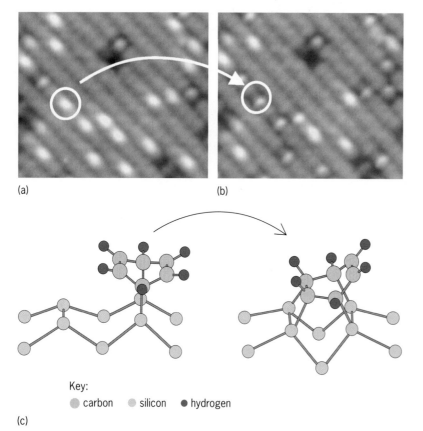

Key:
carbon   silicon   hydrogen

(c)

Fig. 1. Visualizing benzene chemistry and structure on a silicon surface. (*a*) This image was recorded several minutes after the molecules were adsorbed on the surface. (*b*) This image, recorded approximately 10 minutes later, reveals that a structural evolution occurred. In time, the bright features, which are most numerous in *a*, were replaced by less bright protrusions with an accompanying depression to one side as the molecule reached over and formed two additional C-Si bonds. (*c*) The diagrams show the benzene molecule in two- and four-bonded structures. Molecules in the two-bonded state appear relatively bright. The white circles and arrow follow one particular benzene molecule as it transforms to the relatively dark-looking four-bonded structure.

new molecular devices that serve, for example, as chemical sensors, optical devices, or biochemical monitors.

Molecular imaging and manipulation. The ability to study and engineer atom-scale structures has been made possible by a revolutionary tool called the scanning tunneling microscope (STM), for which the inventors were awarded a Nobel prize in 1986. The STM allows imaging and, in some cases, manipulation of individual atoms and molecules. Even though the STM provides a direct view of the position of a molecule, understanding a molecular image in detail is somewhat complicated. The manner in which matter is probed determines what structural aspects are revealed. From common experience, we know that x-rays and visible light, for example, provide very different views of our bodies. The STM uses tunneling electrons to probe valence electron energy levels, the same energy states involved in chemical bonding, which are spatially rather diffuse. The importance of this in the present context is that the STM tends to see only featureless bumps when it images a molecule—the individual atoms and their exact positions remain unresolved.

Silicon-organic interactions. Recently, advanced quantum-mechanical modeling techniques, together with powerful computers, have made it possible to unravel the STM images of molecules, effectively extending the resolution of the STM to see where the atoms are in a molecule and how exactly the molecule interacts with a silicon surface. An example is given in **Fig. 1**. The STM images show benzene (C_6H_6) molecules on a silicon surface. The barlike structures running diagonally across the image are rows of silicon dimers (paired surface silicon atoms). The white protrusions are due to individual benzene molecules. In benzene, each carbon has three neighboring atoms. Carbon atoms can share bonds with as many as four neighbors. Silicon atoms also prefer four neighbors. The free benzene molecule, which is planar, becomes highly distorted upon adsorbing. The two most distant carbon atoms within the benzene molecule rehybridize (that is, change electronic structure) to accommodate the new bonds to surface silicon atoms, as shown in the molecular model just above the image. Benzene adsorption is a stepwise process. Over time, each white protrusion is seen to suddenly be replaced by a less pronounced bright feature that is accompanied by a dark depression to one side. As explained by the molecular model, the change in the image results from a change in the molecule-surface structure, specifically, two additional C-Si bonds form and the benzene molecule further distorts.

Another example of a recently resolved silicon surface–organic molecule reaction is shown in **Fig. 2**. The molecule under study is 2-butene. The two central carbon atoms in 2-butene have only three neighbors. Upon reaction, each of those atoms forms a fourth bond. There are two distinct products that can form depending on which face of the molecule encounters the surface. The two product

molecules are distinct in the same way that right and left hands are. This property is known as chirality. R,R and S,S labels identify two molecules that have different "handedness" or chirality. Many biologically significant molecules are chiral. Furthermore, many biological processes involve only specific chiral species. It is therefore desirable that a detector of particular molecules be capable of discerning between different chiralities. As shown by this recent work, that is now possible on the single-molecule level.

Molecular assembly. The above molecule-silicon structural issues, as well as many others, were reviewed by R. A. Wolkow (1999). Many silicon-organic molecule systems remain to be explored, but it seems that the tools are now in place to understand those. Some attention is turning to a higher-order problem, that of controllably assembling multimolecule functional units. While scanned probe methods exist for moving atoms and molecules into place, such approaches are too limited in scope and are far too slow.

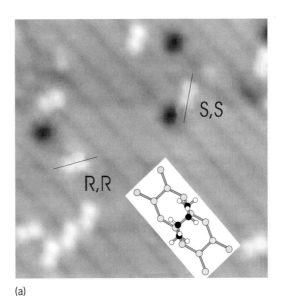

(a)

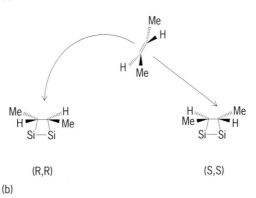

(b)

Fig. 2. *Trans-2-butene on silicon.* (*a*) STM image. The paired protrusions indicate the position of the methyl groups within the adsorbed molecule. (*b*) The schematic shows how reaction at each of the two faces of the molecule results in products of different chiralities. The (R,R) and (S,S), enantiomers, though indistinguishable using other techniques, are clearly identified in the STM image.

It is desirable to assemble many identical units in parallel, that is, simultaneously, in order to make device fabrication practical. A new approach has recently been discovered that uses scanned probe atom-level crafting, but only to a limited degree. The STM is used to create growth initiation sites by dislodging select hydrogen atoms from a hydrogen-terminated silicon surface. Molecular nanostructures are grown automatically by simply exposing the surface to reactant molecules. If large arrays of initiation sites are prepared, subsequent exposure will result in parallel fabrication. **Figure 3** shows a single-molecule-wide line being grown on a silicon surface.

Outlook. The capabilities discussed here were beyond reach just a few years ago. It might seem reasonable to expect that working molecular devices are just around the corner. Only time will tell. No matter the approach, it is clear that some substantial obstacles remain. The problem of electrically connecting molecular functional units must be overcome. Furthermore, if molecular detection is the goal, not only must discriminating chemical receptors (which might be synthetic or natural molecules) be incorporated into a device structure, but also a transduction strategy must be developed. In other words, some measurable change must be detected to signal the molecular detection event. Simple schemes exist that aim to control the gate of a field-effect transistor, but that approach is limited in scope. New methods of transduction will be needed to make the effort of building on the molecular scale worthwhile. Regarding light emission or light detection, the problems are similar. Metal-molecule and silicon-molecule interfaces are active research areas because of the importance of those issues in macroscopic organic light-emitting diodes. Approaches used there to transform electrical current to photon emission from a molecule may be transferable to the nanoscale. To create molecule-based light detectors, methods must be developed to coax the energy captured by a molecule to induce a measurable electrical signal. Possibly, research into natural and artificial light–harvesting molecules will merge with studies of molecular devices to approach this goal.

Much work remains to be done before hybrid silicon-molecular devices can be built, but it seems likely that fabrication procedures will be developed and applications will be found in a wide range of areas such as telecommunications, medical diagnostics, and prosthetic devices. *See also* BIOCOMPATIBLE SEMICONDUCTOR.

For background information *see* ADSORPTION; BIOELECTRONICS; INTEGRATED CIRCUITS; LOGIC CIRCUITS; MICROSENSOR; SCANNING TUNNELING MICROSCOPE; SEMICONDUCTOR; SILICON; STEREOCHEMISTRY in the McGraw-Hill Encyclopedia of Science & Technology.   Robert A. Wolkow

Bibliography. G. Binning et al., Surface studies by scanning tunneling microscopy, *Phys. Rev. Lett.*, 49:57–60, 1982; F. L. Carter, Molecular level fabrication techniques and molecular electronic devices, *J. Vac. Sci. Technol.*, B1:959–68, 1982; D. M. Eigler and E. K. Schweizer, Positioning single atoms with a scanning tunneling microscope, *Nature* 344:524, 1990; S. Hla et al., Inducing all steps of a chemical reaction with the scanning tunneling microscope tip: Towards single molecule engineering, *Phys. Rev. Lett.*, 85:2777–2780, 2000; S. N. Patitsas et al., Controlling organic reactions on silicon surfaces with an STM: Theoretical and experimental studies of resonance-mediated desorption, *Phys. Rev. Lett.*, 85: 2000; S. N. Patitsas et al., Current-induced organic-silicon bond breaking: Consequences for molecular devices, *Surf. Sci.*, 457:L425–L43, 2000; R. A. Wolkow, Controlled molecular adsorption on Si: Laying a foundation for molecular devices, *Annu. Rev. Phys. Chem.*, vol. 50, 1999.

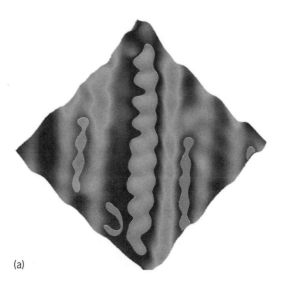

(a)

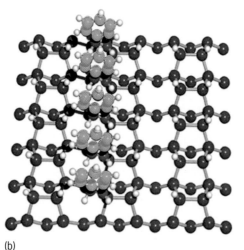

(b)

Fig. 3. A molecular line that has been induced to self-assemble on a hydrogen-terminated silicon surface. (*a*) STM image. (*b*) Schematic diagram. An initiation point was created by removing a hydrogen atom from the surface with the STM tip. Styrene molecules were introduced via the gas phase. A styrene molecule bonded to the reactive point on the surface and then pulled off a hydrogen atom from a neighboring site on the surface, thereby creating a new reaction site and the opportunity for a chain reaction.

Hydrometeorology, urban

Urban areas contain a large and ever-increasing proportion of the world's population, who use a disproportionate share of natural resources. Land surface and atmospheric alteration by urbanization leads to the development of distinct urban climates. The effect of these changes on rates of evapotranspiration is significant. Evapotranspiration can be defined as the loss of water from open water and from the soil by evaporation and by transpiration from the plants growing in the soil. It can occur only when water is available, and it requires that the humidity of the atmosphere be less than the evaporating surface. The evapotranspiration process also requires large amounts of energy (latent heat of vaporization) to convert liquid water into water vapor.

In the past, the widely held belief was that in cities evapotranspiration rates are negligible or nonexistent, and considerably less than those from the surrounding rural environments. This urban effect was attributed to the impervious properties of building materials (concrete and asphalt sidewalks, roads and parking lots, rooftops), the absence of transpiring vegetation, and the presence of engineering structures (drains and sewer systems) designed to remove water rapidly following rainfall. However, in many urban areas, trees and other vegetated surfaces cover a significant area (commonly up to 40% of city plan areas in North America). Moreover, in cities with little summer rainfall, irrigation is extensive and needed to maintain urban vegetation in a lush, green state. In the southwest United States, for example, 2.5-5 mm of irrigation per day is required to sustain grass and deciduous shrubs, and more is needed for ornamental flowers. Thus urban evapotranspiration can, in fact, be expected to be significant, and knowledge of rates and spatial variability is critical for understanding and predicting water demand and quality now and in the future. Urban evapotranspiration also has implications for issues such as groundwater recharge.

Measuring urban evapotranspiration rates. Evapotranspiration involves exchanges between the surface and the atmosphere of both water and energy (latent heat). It can be measured and analyzed in either a meteorological or hydrological framework (**Fig. 1**). For urban areas the water balance (an accounting of water inputs and outputs for a given area) is expressed in Eq. (1),

$$P + I + F = ET + R + S + A \quad [\text{mm h}^{-1}] \quad (1)$$

where P is precipitation, I is piped water supply, F is the water released due to anthropogenic activities, R is runoff (surface and ground water), S is the change in water storage in the period of interest, and A is the net moisture advection (water moved laterally through the area of interest). Similarly, the urban energy balance is expressed in Eq. (2),

$$Q^* + Q_F = Q_E + Q_H + \Delta Q_S + Q_A \quad [\text{W m}^{-2}] \quad (2)$$

where Q^* is net all-wave radiation (net balance of all solar and terrestrial radiation exchanges), Q_F the anthropogenic heat flux (energy from human activities), Q_E the latent heat flux, Q_H the turbulent sensible heat flux (energy that heats the air), ΔQ_S the change in heat storage (energy that heats the urban fabric), and Q_A the net heat advection (heat moving laterally through the area of interest). Q_E is related to evapotranspiration through the latent heat of vaporization (L_V), the energy required to bring about a phase change in water from a liquid to a gas.

At the micro-scale (the scale of individual lawns, parking lots, and so on), evapotranspiration can be estimated using lysimeters, devices in which a

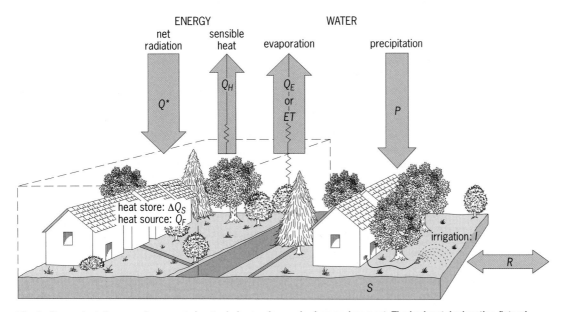

Fig. 1. Conceptual diagram of energy and water balances for a suburban environment. The horizontal advection (lateral movement) of heat and moisture is not shown.

(a)   (b)   (c)

Fig. 2. Three commonly used methods to measure evapotranspiration in urban environments. (*a*) Mini-lysimeters installed in lawns or parks. This photo shows a lysimeter being installed. The tape (which is removed before the system is operational) ensures that the monolith is oriented correctly when it is reinstalled. The lens cap provides a scale. Changes in mass of the monolith are used to compute evapotranspiration rates; the weighing system is located underneath the monolith. (*b*) Eddy covariance equipment. In the foreground is a three-dimensional sonic anemometer, then a krypton hygrometer and a one-dimensional sonic anemometer thermometer. The sonic anemometer measures the vertical wind speed (w'), and the krypton hygrometer measures fluctuations in humidity (q'). (*c*) Atmospheric profiling system (balloon and instrument package) to measure vertical properties of the atmosphere through time, from which evapotranspiration estimates can be made. Signals are sent from the instrument package back to the ground station, which is connected to a computer to archive data.

volume of the surface, commonly soil planted with vegetation, is hydrologically isolated and weighed continuously or periodically to estimate water losses by evapotranspiration (**Fig. 2***a*). Lysimeters have been used with some success to measure evapotranspiration of lawns and parks. However, issues of representativeness (there are so many different surfaces arranged in multiple ways in the urban environment that need to be measured) and appropriate spatial aggregation (how the readings from individual lysimeters get combined together) arise when these data need to be extended to larger areas.

The most direct way to measure evapotranspiration involves the eddy covariance approach (Fig. 2*b*). Evapotranspiration is measured as its energy equivalent (Q_E), using fast-response sensors that measure rapid fluctuations (5–10 Hz) in the vertical velocity (w') and moisture (q') properties of air parcels as they move toward or away from the surface. The covariance ($\overline{w'q'}$) averaged over an appropriate period (15–60 min) yields the latent heat flux. If the instruments are located at appropriate heights [more than two times the height of the building and trees; that is, at ~20–30 m (60–100 ft) in residential areas] and downwind of fairly homogeneous land cover, evapotranspiration data representative of areas the size of city blocks to neighborhoods (the local scale) can be obtained.

At the meso or regional scale, theoretically it is possible to estimate evapotranspiration by integrating between atmospheric profiles of the convective boundary layer (CBL) measured using radiosondes (balloon-borne instruments) or tethersondes (Fig. 2*c*). The convective boundary layer is the unstable boundary layer that forms at the Earth's surface and grows upward through the day as the ground is heated by the sun and convective currents transfer

heat upward into the atmosphere. However, profiles of humidity are often very complex, and assumptions about boundary layer conditions are very restrictive. To date, such convective boundary layer methods have met with limited success in urban areas.

Evapotranspiration rates. Evaporation rates for irrigated residential lawns and parks in the summertime have been shown to range up to 7 mm d^{-1}, with rates up to 15 mm d^{-1} from open water (for example, swimming pools in Tucson, AZ). Microscale advection (heat flow from warm built surfaces, such as roads and driveways, to cooler irrigated vegetated surfaces) with copious water from irrigation maintains these rates. Neighborhood-scale measurements, made using eddy covariance instruments, have shown values to range from much less than 1 mm d^{-1} in downtown and light industrial sites of Mexico City and Vancouver (given the small areal coverage of open water and vegetation and the lack of external water use) to 1–3 mm d^{-1} for vegetated residential areas across North America (Chicago, Los Angeles, Miami, Sacramento, Tucson, and Vancouver). Clear relations between evapotranspiration and the area vegetated, and even more so, the area irrigated, are evident.

The hydrological importance of evapotranspiration is best demonstrated by ratios of evapotranspiration to precipitation (P). Based on short summertime studies for residential neighborhoods in the North American cities listed above, this ratio ranges from 0.7 to 3.7 for those sites with precipitation. In Los Angeles and Sacramento, where there was effectively no rainfall during the study periods, *ET:P* tended to infinity, and evapotranspiration was sustained just by irrigation. Clearly these values will be expected to be different on a seasonal and year-to-year basis.

Impact. The data presented here show the importance of evapotranspiration in urban areas. In many instances, evapotranspiration exceeds precipitation, sustained by external water use from the piped urban supply. Independent support for the importance of urban vegetation and evapotranspiration is evident from thermal satellite images of urban areas, which show clearly the cooler temperatures associated with parks, golf courses, and other urban vegetation. In arid areas, irrigated cities may act like an oasis, with evapotranspiration rates in the urban areas greater than in the surrounding rural areas. Thus evapotranspiration should not be neglected in studies of urban hydrology. In fact, accurate knowledge of the magnitude and variability of evapotranspiration rates is critical for simulations of low flow conditions (important for water quality) and for predictions of long-term water demand. Given that the water used in urban environments often is piped in from large areas beyond the city boundaries, urban irrigation and evapotranspiration rates have significant impacts on regional-scale water resources and hydrology.

For background information *see* EVAPORATION; HUMIDITY; HYDROLOGY; HYDROMETEOROLOGY; PRECIPITATION (METEOROLOGY); URBAN CLIMATOLOGY in the McGraw-Hill Encyclopedia of Science & Technology.   C. S. B. Grimmond; C. Souch

Bibliography. H. A. Cleugh and C. S. B. Grimmond, Modeling regional scale surface energy exchanges and CBL growth in a heterogeneous, urban-rural landscape, *Boundary Layer Meteorol.*, 98:1–31, 2001; C. S. B. Grimmond and T. R. Oke, Rates of evaporation in urban areas, in Impacts of Urban Growth on Surface and Ground Waters, *Int. Ass. Hydrolog. Sci. Publ.*, no. 259, pp. 235–243, 1999; T. R. Oke, *Boundary Layer Climates*, 2d ed., University Press, London, 1987; R. A. Spronken-Smith, T. R. Oke, and W. P. Lowry, Advection and the surface energy balance of an irrigated urban park, *Int. J. Climatol.*, 20:1033–1047, 2001.

Hydrothermal ore deposits

Hydrothermal ore deposits are the predominant sources of scarce metals (copper, zinc, lead, tin, molybdenum, antimony, tungsten, mercury, bismuth, uranium, silver, and gold). Fascinating because of their variety, they are the subject of intense research aimed at understanding them better and finding them efficiently.

Formation. Hydrothermal ore deposits are inferred to have formed by mineral precipitation from hot hydrous fluids. This deposition occurred within the rocks through which these fluids traveled, or as the fluids entered bodies of water (ocean or lakes), or as they reached the atmosphere at the Earth's surface. These inferences are based on the following observations: (1) Most hydrothermal ores occur in places that favor the passage of fluids: fracture zones (veins), pipelike conduits (chimneys, pipes) in easily soluble rocks (such as limestone), sedimentary beds that are porous or susceptible to chemical replacement (mantos), fractured caps over intrusive cupolas (porphyries), and porous calcsilicates (skarns) at contacts of magmatic intrusives with limestone. Other hydrothermal ores occur where fracture zones reached lake or ocean waters (as along mid-ocean ridges) or the atmosphere (as in hot springs above presently active geothermal reservoirs). (2) Hydrothermal minerals often contain small fluid inclusions consisting mostly of water with some carbon dioxide; this water has significant concentrations of common salt and of several of the metals commonly found in hydrothermal deposits. These inclusions are interpreted as trapped remnants of the mineralizing fluids. Such fluid inclusions tend to homogenize between 50 and 500°C (120 and 930°F), suggesting that their host minerals formed at temperatures above the homogenization temperatures from one-phase fluids. (3) The original composition of hydrothermal ore host rocks was generally modified by the formation of water-bearing minerals (hydrothermal alteration). (4) Many hydrothermal deposits are in the vicinity of magmatic intrusive rocks, implying that they formed either from fluids emanating from a crystallizing magma or from fluids put in motion by magmatic heat.

Hydrothermal fluids may be generated in different ways: (1) as water expelled from a crystallizing magma; (2) as meteoric water that becomes ground water and is mobilized by gravity, tectonic activity (for example, tilting or folding), or magmatic heat and reacts with continental crustal rocks; (3) as seawater that infiltrates near mid-ocean ridges, is mobilized by ocean ridge magmatism, and reacts with oceanic crust; or (4) as water trapped in sediments (connate water) and released upon compaction and heating (diagenesis and metamorphism).

Composition of ore and host rocks. The ore minerals in hydrothermal deposits are mostly sulfides of copper, zinc, lead, tin, molybdenum, antimony, mercury, bismuth, and silver. However, some metals occur as oxides (tin and uranium), tellurides (gold), tungstates (tungsten), or native (gold). The accompanying nonvaluable gangue minerals consist mostly of sulfides (pyrite, pyrrhotite, orpiment, realgar), oxides (quartz), carbonates (calcite), sulfates (barite), fluorides (fluorite), and silicates (rhodonite).

Hydrothermal alteration. Hydrothermal alteration of the surrounding host rocks due to contact with the hot hydrous fluids may extend from a few meters to a few kilometers from an ore. The alteration generally consists of addition of silica and conversion of anhydrous silicates (orthoclase, plagioclase, olivine) to hydrous silicates (muscovite, sericite, kaolinite, biotite, epidote, chlorite) by a low-pH fluid. Where present, the sulfate mineral alunite indicates that the fluid was oxidized and acidic. Where magma intruded limestone, hydrothermal fluids promoted the formation of porous silicates (garnet, wollastonite)—the skarns that host hydrothermal ores of copper, zinc, gold, or tungsten.

Mineralization pulses and stages. Hydrothermal ores often contain specific minerals or mineral groups (paragenetic associations or assemblages) that either form successive layers or cut prior minerals, thus defining a paragenetic sequence of mineralization stages. Thin sections of transparent minerals and etching or electron microprobe analyses of opaque minerals reveal that each mineral consists of many hundreds of growth bands, proving that an ore deposit is made by hundreds or thousands of mineralization pulses.

Zoning. The chemical composition of hydrothermal ores often varies spatially within a mining district or within a given vein, manto, pipe, or disseminated deposit. Over a century ago, this led to a zonal theory, according to which certain minerals or elements were considered to be characteristic of the upper or peripheral parts (chalcedony, barite, fluorite, mercury, antimony, gold-silver), others of the central part (silver, lead, zinc, copper), and still others of the inner or lower part (pyrrhotite, arsenopyrite, tourmaline, cassiterite, gold, arsenic, bismuth, tungsten, tin). This was only a generalized scheme, because every mineral deposit and district has its own zoning sequence.

Hydrothermal zoning has traditionally been considered to result from progressive changes in the temperature, pressure, or composition of the hydrothermal fluids along their paths. However, field observations have so far failed to provide convincing evidence that a specific paragenetic stage (S-1) evolves chemically (from composition A to compositions B and C) from one part of a deposit to another (**illus. *a***). It seems more likely that successive stages (S-1, S-2, and S-3) are deposited by different fluids (A, B, C) at various points along the hydrothermal fluid paths, thus generating a zoned deposit (illus. *b*).

Oxidation, leaching, and supergene enrichment. The near-surface part of most hydrothermal ore deposits contains hydroxides and oxides of iron, copper, and manganese (limonite, hematite, cuprite, tenorite, pyrolusite, manganite), sulfate of lead (anglesite), carbonates of copper, lead, and zinc (malachite, azurite, brochantite, cerussite, smithsonite), and chloride of silver (cerargyrite). This oxide zone passes in depth into the underlying hypogene sulfides. Copper was often leached from the oxide zone, but precipitated again as a sulfide (chalcocite, covellite) between the sulfide and oxide zones to form a supergene enrichment zone or blanket. This sequence results from the near-surface atmospheric oxidation and dissolution of most of the original hypogene sulfides, which leads to the formation of acid ground waters. The downward-percolating ground waters are reduced at depth, just below the water table, leading to the precipitation of the supergene copper sulfides. Other factors influencing near-surface processes are the reactivity of the host rock to the acid ground water (for example, limestone is much more effective in neutralizing acid ground water than igneous, metamorphic, or other sedimentary rocks) and changes in the water table due to climatic or sea-level variations. Supergene enrichment of copper played an important role in making it economic to mine disseminated (porphyry) copper deposits in the southwestern United States, Mexico, and northern Chile. Residual enrichment in the oxide zone of relatively insoluble gold, and lead or zinc carbonates, may have played a role in the discovery of some deposits.

Ore deposition factors. Perhaps more important than the sources of the metals and of the hydrothermal fluids are the factors determining the uptake and transport of ore constituents in hydrothermal fluids, as well as the causes of hydrothermal ore

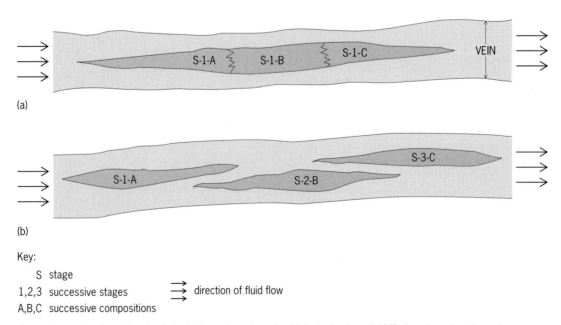

(a)

(b)

Key:

   S stage

1,2,3 successive stages   direction of fluid flow

A,B,C successive compositions

Alternative explanations of zoning in hydrothermal ore deposits. (*a*) A single-stage fluid (S-1) evolves along its path, depositing successively different mineral compositions or assemblages (A, B, and C). (*b*) Three different fluids flow through the vein at various times or stages (S-1, S-2, and S-3), depositing various mineral compositions or assemblages (A, B, and C).

deposition. It has long been known that elevated temperatures and acidity (low pH) promote increased metal solubility. Experimental studies also highlighted the importance of chloride, sulfide, organic, and other complexing agents in metal solubility. Thus, magmas contaminated by major evaporite sequences (which contain chloride in halite, and sulfide in gypsum or anhydrite) would produce more "fertile" hydrothermal fluids than uncontaminated magmas. Similarly, heated sedimentary basin brines or infiltrating seawater would be more likely to dissolve and transport significant amounts of metals than dilute ground waters. In contrast, hydrothermal ore deposition would require cooling or neutralization (increasing pH) of the metal-bearing fluids.

Temperature. In the early 1900s, the temperature of the hydrothermal fluids was considered to largely determine the mineralogical composition of an ore deposit. This led to a temperature classification of hydrothermal ore deposits (epithermal, leptothermal, mesothermal, hypothermal) based mainly on minerals deemed to be indicative of the temperature of ore deposition. However, subsequent laboratory studies of pertinent chemical systems demonstrated that supposedly "diagnostic" minerals are stable over wide ranges of temperature and are, therefore, generally unreliable as geothermobarometers. For example, pyrrhotite (FeS_{1-x}) had been thought to deposit at a higher temperature than pyrite (FeS_2), but the experimental studies demonstrated that both are stable over the general range of hydrothermal ore deposition (and that the deposition of one or the other depends more on the concentrations of sulfur and iron in the hydrothermal fluid). Unfortunately, reliable geothermometers, such as temperature ranges of different polymorphs of a mineral (for example, argentite and acanthite) or incompatible mineral assemblages (pyrite-magnetite versus pyrrhotite-hematite), only indicate large temperature ranges, are generally scarce, or are restricted to unusual ore deposition temperatures. Studies of fluid inclusion homogenization temperatures and of light-isotope fractionations have shown that hydrothermal ore deposition occurs at temperatures from several hundred to a few tens of degrees Celsius. In addition, they showed that in some ore deposits there was a fast initial increase and a subsequent gradual decrease in temperature, as evidenced by the depositional temperatures of successive paragenetic mineralization stages. These studies have so far failed to clearly demonstrate thermal gradients along hydrothermal solution paths for a single depositional stage. At present, the temperature gradient factor is supported mainly by pointing out the temperature gradients observed in geothermal fields and by noting mineral precipitation by cooling solutions in laboratory experiments.

Acidity. During the mid-1900s, it became evident that the hydrothermal alteration of rocks adjoining veins and disseminated ore results largely from attack by acid fluids. This has been reinforced by abundant laboratory evidence, and a pH increase due to the interaction of hydrothermal fluids with host rocks is now seen as a more general cause of hydrothermal ore deposition, especially for lead-zinc ores in limestone. In some geothermal fields (for example, Yellowstone), hot springs of substantially different pH occur in proximity, suggesting the possibility that fluid mixing may increase the pH of an acid metal-bearing fluid and cause mineral precipitation. Fluid mixing also may be responsible for mineral precipitation by bringing together ions that form insoluble compounds.

Pressure. Decreasing pressure gradients have been invoked as causes for mineral deposition because hydrothermal fluids tend to flow toward low-pressure regions. Fluid inclusion studies have shown that some minerals precipitated from boiling solutions (because different inclusions in a given mineral trapped varying proportions of liquid to water, thus giving an apparent range of homogenization temperatures). However, this appears to be an occasional rather than a predominant factor.

Most chemical equilibria vary with both temperature and pressure (that is, they are really geothermobarometers), so that reliable pressure gradients can be determined only if a pressure-independent geothermometer is available (such as a light-isotope fractionation).

Oxidation and reduction. Reduction of oxidized hydrothermal fluids is generally credited for the deposition of uranium, vanadium, copper, or silver ores in redbeds containing organic matter. Conversely, oxidation of hydrothermal fluids is also considered to be occasionally important in ore precipitation.

Fugacity. The decrease in emphasis on depositional temperature was supplanted by an increased awareness of the roles played by the fugacities of sulfur, oxygen, and carbon dioxide in hydrothermal solutions. (Fugacity is the partial pressure of a component in a perfect gas that is in equilibrium with a liquid or solid containing this gas in solution.) Eventually, this led to the current tendency to classify ore deposits as low-sulfidation, high-sulfidation, or acid-sulfate. But given that such expressions provide little information on the nature of a specific ore deposit, and given the difficulty involved in pinpointing the exact causes of ore deposition, there has been a parallel development of a hodge-podge terminology of ore deposit types, such as "Cordilleran vein, manto, pipe or breccia," "five-element (Ag-Ni-Co-Fe-As) vein," "disseminated or porphyry," "skarn," "volcanogenic massive sulfide," "redbed or sediment-hosted stratiform Cu-Ag," and "igneous-metamorphic."

For background information *see* CRATON; ELEMENTS, GEOCHEMICAL DISTRIBUTION OF; GEOCHEMICAL PROSPECTING; GEOLOGIC THERMOMETRY; IONIC EQUILIBRIUM; LIMESTONE; MAGMA; METAMORPHISM; MID-OCEANIC RIDGE; MINERALOGY; ORE AND MINERAL DEPOSITS; PORPHYRY; PRECIPITATION (CHEMISTRY); PROSPECTING; REDBEDS; SALINE EVAPORITES; SEDIMENTARY ROCKS; SEDIMENTOLOGY; SILICATE SYSTEMS; SKARN; VOLCANO in the McGraw-Hill Encyclopedia of Science & Technology.   Ulrich Petersen

Bibliography. H. L. Barnes, *Geochemistry of Hydrothermal Ore Deposits*, 3d ed., John Wiley, 1997;

J. M. Guilbert, Linkages among hydrothermal ore deposit types, *Pro-Explo 2001*, Lima, Peru, 2001; J. M. Guilbert and C. F. Park, Jr., *The Geology of Ore Deposits*, W. H. Freeman, 1986; H. D. Holland, Some applications of thermochemical data to problems of ore deposits, I. Stability relations among the oxides, sulfides, sulfates and carbonates of ore and gangue minerals, *Econ. Geol.*, 54:184–233, 1959; H. D. Holland, Some applications of thermochemical data to problems of ore deposits, II. Mineral assemblages and the composition of ore-forming fluids, *Econ. Geol.*, 60:1101–1166, 1965.

Immunoassays

Modern medical practice depends on sophisticated chemical analyses of body fluids—most commonly blood plasma, but also urine, cerebrospinal fluid, and many others. Substances measured range from the major plasma constituents, such as water, sodium chloride, and albumin, to various metabolites, proteins, and exogenous agents that may exert biological effects at concentrations of one part per trillion.

The detailed methodologies, known as assays, for measuring substances in body fluids have employed nearly all analytical technologies, from simple colorimetric measurements to ultrasophisticated mass spectrometry. Especially important in recent years has been a class of techniques known as immunoassay (IA) or, more broadly, ligand-binding assay (LBA). The key feature of these assays is molecular recognition: a macromolecular reagent is used that binds tightly and specifically to the substance of interest, and the extent of binding is then quantified. The binding reagent is still most commonly an antibody purified from an immunized animal. However, other types of binder can also be used. For example, nucleic acid sequences are recognized by a complementary nucleic acid. This article will specifically describe immunoassays, but most of the concepts have broader applicability.

Concepts and definitions. Essential characteristics of an assay include sensitivity (the ability to detect a small amount of substance); specificity (the ability to distinguish the substance of interest from others); precision (the ability to give reproducible results); and accuracy (agreement of results with some accepted standard of truth). The substance to be measured is an antigen (Ag), meaning that it is capable of eliciting an immune response. The protein produced by the immune system that binds the Ag is an antibody (Ab). The binding of Ag to Ab is described by the equation

$$K_{\mathrm{diss}} = \frac{[\mathrm{Ag}][\mathrm{Ab}]}{[\mathrm{Ag} \cdot \mathrm{Ab}]}$$

where the brackets denote concentration in moles per liter, and K_{diss}, the dissociation constant, expresses the strength of binding, with a smaller number indicating tighter binding. K_{diss} generally must be at least as low as the lowest concentration to be measured.

Naturally occurring Ab's are polyclonal; that is, they are a mixture of distinct proteins, each produced in the animal by a different clone of B-lymphocytes. In the laboratory, however, it is possible to extract lymphocytes from an animal and immortalize individual clones, thus producing monoclonal Ab's. A site on an Ag where Ab binds is called an epitope. Large Ag's such as proteins contain multiple epitopes, any of which can serve as a binding site for a monoclonal Ab.

Assay design. An immunoassay must generate a measurable signal that depends upon the degree of Ag-Ab complexation. This can be accomplished in many ways, making classification schemes difficult. Probably the most fundamental distinction is between assays that use the competitive binding approach and assays that use the immunometric approach (see **illus.**).

Competitive-binding immunoassay. The earlier, competitive-binding approach was pioneered in the assay for human insulin reported by R. S. Yalow and S. A. Berson. This assay became a prototype for many subsequent ones, collectively known as radioimmunoassays (RIAs), and the work was awarded a Nobel prize in 1977. In an RIA or other competitive assay, the

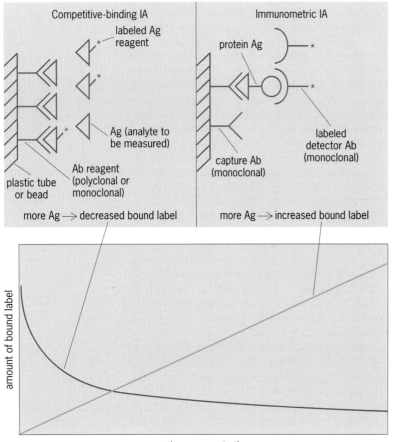

Schematic diagrams of the two major types of immunoassay. With competitive-binding assay, labeled Ag reagent and unlabeled Ag from the sample being analyzed compete for a limited number of binding sites. Hence more Ag results in less bound label, as shown in the calibration curve below. With immunometric assay, label is applied to the detector Ab, and more Ag results in more bound label.

binding Ab is present in limited amount, comparable to the amount of analyte. The second essential reagent is a labeled analog of the analyte—originally peptides were iodinated with the gamma-emitting isotope^{131}I. A fixed amount of binder and labeled reagent are combined with the sample to be assayed. Unlabeled analyte in the sample will then compete for binding sites with the labeled reagent—the more analyte in the sample, the less binding of labeled reagent that will take place. The concentration of unknowns is determined on the basis of a multipoint calibration curve (see illus.). Calibration may be done with a simple hand-drawn graph or by using a computer.

With a radioactive label such as ^{131}I, distinguishing bound label from unbound label requires a physical separation of the two, making this a heterogeneous assay. The separation can be conveniently accomplished by attaching the binder to a solid phase (see illus.), such as a plastic test tube or bead; unbound label is then simply decanted. Before the technology to bind Ab's to plastic was in wide use, the approach was to precipitate the Ab out of solution after binding took place, using either ammonium sulfate or a second antibody, then collect the precipitate via centrifugation and decant the supernatant. Some of the newer immunoassays have been made homogeneous by employing nonradioactive labels whose properties change in a measurable way upon binding to Ab.

Competitive-binding assays are very versatile, being applicable to small molecules, peptides, and proteins. For proteins at low concentration, however, there are drawbacks. Since the Ab concentration must be comparably low, the rate of binding is very slow, sometimes requiring days to approach completion. In addition, since different proteins often share certain epitopes, it can be difficult to obtain the desired specificity.

Immunometric immunoassay. Much-improved assay performance for proteins is usually achieved with an immunometric format. This assay is actually more straightforward than a competitive-binding assay, but its development essentially depended on the introduction of monoclonal Ab's. An immunometric assay generally employs two monoclonals: one (the capture Ab) linked to a solid phase, the other (the detector Ab) carrying the label. If these Ab's are directed against distant epitopes on the protein analyte, each molecule of analyte will cause a linking of the label to the solid phase (see illus.). Because the protein essentially becomes "sandwiched" by the two monoclonal Ab's, this is frequently referred to as a sandwich assay. This configuration allows excess Ab to be present, making it faster than the competitive-binding approach. It can also be much more specific, since formation of the complete complex requires recognition of two distinct epitopes. Finally, immunometric assay offers the advantage of a linear or near-linear calibration curve (see illus.).

Labeling schemes. Radioactive labels, as applied in the classical radioimmunoassays, behave in predictable fashion and offer high sensitivity, but they also have numerous drawbacks. Besides concerns about disposal and safety, their instability demands frequent assay calibration, and they are not well suited to homogeneous or automated assays. Newer labeling schemes are usually based on an optical measurement—light absorbance (colorimetry), fluorescence (emission of light following light absorption), or chemiluminescence (emission of light induced by a chemical reaction). Greater sensitivity is achieved measuring light emission (fluorescence and especially chemiluminescence) rather than absorbance. Sensitivity is also frequently enhanced by using as the actual label an enzyme, which can catalyze formation of a virtually unlimited amount of signal product. For example, certain substrates of the enzyme alkaline phosphatase develop, upon loss of a phosphate group, properties of color, fluorescence, or chemiluminescence. This allows for immunoassays with exquisite sensitivity.

A myriad of other labeling schemes have been devised. Of particular interest are labels suitable for homogeneous assays. The most popular to date have been enzymes (enzyme activity is inhibited when the label comes close to the binding Ab) and fluorescent molecules (the label's tumbling motion is slowed when binding to Ab takes place, which can be detected by measuring the rate of fluorescence depolarization). Homogeneous assay has been mainly applicable to competitive-binding assays where maximum sensitivity is not essential, but it may be extended to immunometric formats in the future. Also of great future interest are biosensor approaches, where ligand binding to an Ab-coated surface alters measurable properties of the surface, without the need for any actual label.

Binding molecules. The genius of the original radioimmunoassays was in harnessing, for analytic purposes, Ab molecules whose biological role depends on their outstanding affinity and specificity. The native polyclonal Ab's originally used for radioimmunoassays are not ideal reagents because they are inherently impure and irreproducible. Monoclonal Ab's, however, are pure reagents that can be produced in virtually limitless quantities. Monoclonal Ab's still have some disadvantages, since they are large, complex molecules with regions irrelevant to binding that may actually cause certain problems. Other biological proteins, such as transport proteins and receptors, can sometimes serve as binding reagents but can have similar disadvantages. At present, molecular technology allows fully in vitro production of Ab-like molecules, or entirely different types of polymers, that can have suitable binding properties. These should eventually become commonplace in clinical assays.

Modern immunoassays can be designed with sufficient sensitivity and linearity to measure nearly all known constituents of blood plasma within their physiological range. Assays in routine use range from relatively insensitive nephelometric techniques for proteins at 1 milligram per liter concentrations or higher; to competitive binding assays for low-molecular-weight hormones, such as estradiol, atk concentrations approaching 1 picomole (10^{-12} mole, or about 10^{11} molecules) per liter; to ultrasensitive immunometric assays for peptide hormones and tumor markers, occasionally at subpicomole per liter concentrations.

Automation. Even manually performed immunoassays are relatively labor-saving because their sensitivity and specificity are often so high that tedious purification steps are eliminated. Nevertheless, the increasing demand for endocrine, toxicology, and other tests in the clinical setting have been a strong impetus for automation. Automation of the early radioimmunoassay was cumbersome and not widely practiced. Modern immunoassays require less manipulation and have been successfully automated. Several manufacturers offer integrated hardware-software-reagent systems capable of performing several of the most popular immunoassays used in medical practice. These typically offer walkaway operation, where the instrument will identify the desired tests for each specimen based on a barcode, dispense sample and appropriate reagents to perform the assays, read the assay signal after an appropriate incubation, and compute the concentration result using a stored calibration curve. After the operator reviews the results, they can be directly uploaded to a clinical data management system.

Limitations. The major drawback of the immunoassay is that it is subject to interference—the assay signal can be affected by something other than the intended analyte. Interferences may come from closely related molecules, reflecting lack of perfect specificity of the binding reagent. Alternatively, interferences may come from molecules totally unrelated to the analyte. For example, blood plasma, the usual specimen analyzed, contains a very high concentration of diverse Ab's that can bind to either the analyte or the binding Ab, causing interference. Other times there may be a substance in plasma that mimics or affects properties of the label—for example, a fluorescence-based immunoassay may be affected if the patient is taking a fluorescent drug.

Automated immunoassay represents a demanding and relatively new technology and is subject to occasional malfunction. Reagents for immunoassay can be relatively expensive, due mainly to their high development cost.

The power and convenience of automated immunoassay has thrust it into a major role in many areas of diagnostics. Continued technological advances will enlarge the menu of analytes and make results more reliable and less expensive.

For background information *see* ANTIBODY; ANTIGEN; ANTIGEN-ANTIBODY REACTION; IMMUNOASSAY; IMMUNOFLUORESCENCE; RADIOIMMUNOASSAY in the McGraw-Hill Encyclopedia of Science & Technology.

Jay L. Bock

Bibliography. J. L. Bock, The new era of automated immunoassay, *Amer. J. Clin. Pathol.*, 113:628–646, 2000; D. W. Chan (ed.), *Immunoassay Automation: An Updated Guide to Systems*, Academic Press, New York, 1996; J. B. Henry (ed.), *Clinical Diagnosis and Management by Laboratory Methods*, 19th ed., Saunders, Philadelphia, 1996; C. P. Price and D. J. Newman (eds.), *Principles and Practice of Immunoassay*, 2d ed., Stockton Press, New York, 1997; R. S. Yalow and S. A. Berson, Immunoassay of endogenous plasma insulin in man, *J. Clin. Investig.*, 39:1157–1175, 1960.

Impulsive stimulated thermal scattering (ISTS)

Impulsive stimulated thermal scattering, also known as transient grating photoacoustics, is a purely optical, noncontacting method capable of characterizing the high-frequency acoustic behavior of surfaces, thin membranes, coatings, and multilayer assemblies. It has emerged as a generally useful analytical technique for thin-film materials research. Recent advances in experimental design have improved and simplified the ISTS measurement dramatically, resulting in straightforward, inexpensive setups and even a commercial ISTS instrument that requires no user adjustments of lasers or optics. This tool is now used routinely in the microelectronics industry for rapid (~1 s) and nondestructive measurements of metal film thickness with atomic-level precision and high spatial resolution.

Principles. In ISTS, picosecond pulses of light from an excitation laser stimulate motions which are

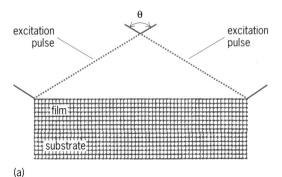

(a)

(b)

Fig. 1. ISTS experiment. (*a*) Crossed laser pulses induce coherent, monochromatic acoustic and thermal motions in a thin film. (*b*) Continuous-wave probing laser diffracts from the ripple on the surface of the sample. Measuring the intensity of the diffracted light with a fast detector and transient recorder reveals the time dependence of the motions.

then detected with a continuous-wave probing laser (**Fig. 1**). In the simplest case, a single pair of excitation pulses crosses at an angle θ at the surface of the sample. Optical interference between these pulses produces a sinusoidal variation in light intensity with a period Λ that is determined by the wavelength (color) of the excitation light and the crossing angle. Slight absorption by the sample generates mild impulsive heating at the intensity peaks of the interference pattern. The resulting thermal expansion launches three types of responses: (1) longitudinal acoustic wavepackets that propagate into the bulk of the sample and generate acoustic echoes upon reflection from subsurface interfaces; (2) coherent, monochromatic counterpropagating surface acoustic waves with wavelength Λ; and (3) thermally induced strain that relaxes via thermal diffusion. Each of these responses produces a small (size of the order of tenths of nanometers) amount of ripple on the surface of the sample. This ripple acts as a time-dependent diffraction grating for a probing laser beam which illuminates a portion of the surface that overlaps with the excited region of the sample. Measuring the intensity of the diffracted light with a fast detector and transient recorder reveals the complete time dependence of the ripple with each shot of the excitation laser. Current versions of commercial ISTS tools rely on the responses that occur on nanosecond or longer time scales: the in-plane acoustic and thermal modes. These motions can be excited and detected in real time with compact, inexpensive laser sources, fast detectors, and digitizing oscilloscopes. They provide information that enables thickness and other important physical properties (such as elastic constants, density, thermal diffusivity, and bonding integrity) to be measured in all types of metal films found in microelectronics.

Data collection. Figure 2 shows ISTS data collected using a commercial machine at a wavelength, Λ, of about 7 micrometers from an electroplated copper film on a silicon substrate. The onset of diffraction coincides with the arrival of the excitation pulses at $t = 0$. The oscillations are due to acoustic waves that propagate in the plane of the film. The gradual decay of signal that occurs during the first 25–50 nanoseconds is associated with thermal diffusion. The acoustic motions correspond to normal modes of the planar acoustic waveguide formed by the thin-film/substrate system. The power spectrum shows the frequencies of these modes. (The intensity of diffracted light is proportional to the square of the surface ripple. As a result, peaks in the power spectrum occur at the acoustic frequencies and at sums and differences of these frequencies.) The number and characteristics of the waveguide modes are determined by the elastic moduli, thickness, and density of the film, as well as the elastic moduli, thicknesses, and densities of other materials (for example, other films or the substrate) directly or indirectly mechanically coupled to it. **Figure 3** displays the computed distributions of displacements and the velocities associated with the two lowest modes in a

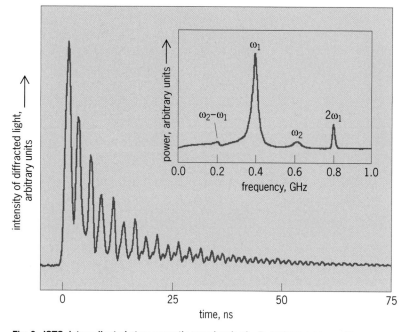

Fig. 2. ISTS data collected at an acoustic wavelengh of $\sim$7 μm from a copper film (thickness $\sim$1.8 μm) on a layer of silicon dioxide (thickness $\sim$0.1 μm) on a silicon substrate. The oscillations in the signal reveal the frequencies of two acoustic waveguide modes (ω_1 and ω_2) that are excited and monitored in this measurement. The inset shows the power spectrum.

waveguide formed by a layer of polymer on a silicon substrate. The response in Fig. 2 determines the velocities (products of the wavelength, Λ, and the response frequencies) of these two modes for the copper/silicon system. These and other waveguide modes are, in general, dispersive: their velocities vary with wavelength. The dispersion can be evaluated with ISTS through either (1) a series of measurements with different angles, θ, between a pair of excitation pulses to determine velocities at different wavelengths or (2) a single measurement that uses many crossed pulses to measure velocities at several wavelengths simultaneously.

Data interpretation. To obtain a qualitative picture of how the mode velocities and their dispersion vary with the thickness or mechanical properties of a film, it is useful to consider the simple case of the lowest-order mode of a single film layer on a substrate. In the limit where the acoustic wavelength greatly exceeds the film thickness, the acoustic wave is primarily contained in the substrate. Its velocity in this case approaches the Rayleigh velocity, v_{Rs}, for the substrate material, regardless of the film. The v_{Rs} is the velocity of the acoustic waveguide mode that is guided by a free surface. This mode corresponds to a surface-localized wave whose displacements decay exponentially into the depth of the substrate. In the opposite limit, where the acoustic wavelength is very small compared to the film thickness, the acoustic wave is entirely contained within the film. Its velocity then approaches the film material's Rayleigh velocity, v_{Rf}, regardless of the substrate. In between these two limits, the velocity varies continuously with the wavelength. Most metals are less

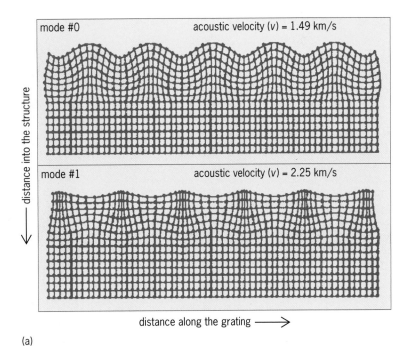

distance into the structure →

mode #0   acoustic velocity (v) = 1.49 km/s

mode #1   acoustic velocity (v) = 2.25 km/s

distance along the grating ──→

(a)

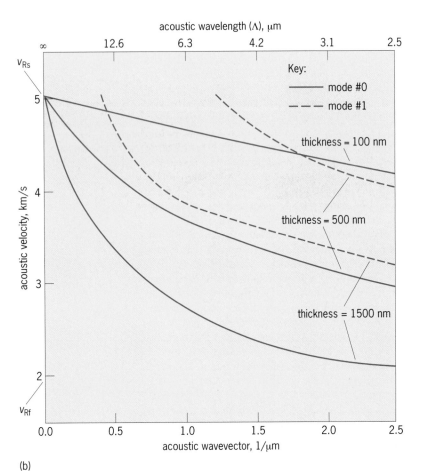

(b)

Fig. 3. Acoustic waveguide modes and their dispersion. (a) Computed displacement distributions for the lowest two modes in a polymer film supported by a silicon substrate. (b) Dispersion of similar modes in a waveguide formed by a layer of copper (thickness 100, 500, and 1500 nm) on a silicon substrate. The dependence of the velocities on thickness provides the basic for the ISTS thickness measurement. 1 km/s = 0.62 mi/s.

stiff than silicon and have, as a result, lower Rayleigh velocities. For a typical metal film on a silicon substrate, the surface acoustic wave velocity decreases as the wavelength is decreased or as the thickness is increased.

Acoustic waveguide theory allows the dispersion of guided modes in arbitrary layered structures to be calculated from equations of motion based on first principles. Figure 3 illustrates the computed dispersion for silicon-supported copper films with three different thicknesses: 100, 500, and 1500 nm. For this range of wavelengths, the 500- and 1500-nm films form waveguides that support two modes, while the 100-nm film supports only one. In each case the results for the lowest-order mode show the trends described above: the acoustic velocity decreases from the substrate limit at long wavelengths (small wavevectors $= 2\pi/\text{wavelength}$) to the film limit at short wavelengths. (The behavior of higher-order modes is similar, but unlike the lowest mode they have cutoff wavelengths beyond which they cease to be strictly guided.) This figure also illustrates how the velocities vary with film thickness, which is the foundation of the ISTS thickness measurement. At a given wavelength (that is, crossing angle θ), the velocity of the lowest-order mode progressively approaches the copper Rayleigh velocity as the copper thickness becomes large compared to the acoustic wavelength. Experimentally, this behavior would be observed as a decrease in the ISTS signal oscillation frequency as the thickness is increased.

Inversion algorithms based on the waveguide solutions can be used to determine the elastic constants, densities, or film thicknesses from the measured dispersion. For thickness evaluation, the elastic constants and densities are typically assumed to be known and are treated as fixed parameters in the inversion. For the simple case of evaluating the thickness of a single layer in a film stack whose other properties are known, the computational procedures reduce to a straightforward noniterative calculation that yields the thickness from a single measured phase velocity. The thicknesses of several films in a multilayer stack or the thickness and other properties (such as elastic constants and density) of a single layer can be determined by analyzing the dispersion or by using the velocities of multiple waveguide modes.

The accuracy of the thickness measurement is highest when the waveguide characteristics depend strongly on the thickness and when the properties of the other components of the structure are known accurately. For most materials on silicon substrates, the velocity of the lowest-order mode exhibits sufficient sensitivity when the acoustic wavelength is not small compared to the thickness. In many ISTS measurements of metal films for applications in microelectronics, the acoustic wavelength ranges from several to a few tens of micrometers and significantly exceeds the film thickness (typically 0.02–2.0 μm). In these cases, calibrated ISTS tools have thickness precision and accuracy that are both in the range of

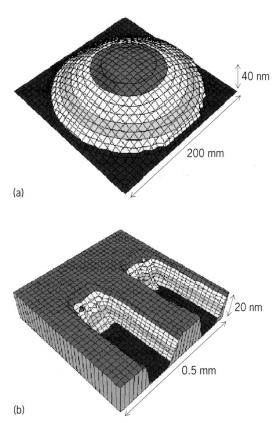

(a)

(b)

Fig. 4. ISTS thickness maps. (*a*) Full-wafer contour map of the thickness of a tungsten film across a silicon wafer coated with ~150 nm tungsten and 20 nm titanium nitride. The center point thickness of the tungsten is 145.3 nm. (*b*) Detailed thickness map of an 0.8 × 0.8 mm region of a sample of 200-nm copper/tantalum (patterned)/300-nm silicon dioxide/silicon substrate. In this case, the ISTS method determines the thickness of the buried tantalum. The tantalum thickness at the far right edge of this map is 20.4 nm.

a few tenths of a nanometer for measurement times of less than 1 second.

Application examples. The ISTS method is effective for rapid single-point evaluation. It can also generate maps of thickness or other properties with a spatial resolution that is comparable to the sizes of the excitation and probing spots (typically less than 100 μm). **Figure 4** shows two examples: a full-wafer map of the thickness of a plasma-vapor-deposited film of tungsten on titanium nitride on a 200-mm-diameter (8-in.) silicon wafer, and a high-resolution map of a patterned buried layer of tantalum in a multilayer stack of copper/tantalum/silicon dioxide/silicon substrate. In both cases, the precision is ~0.1 nm and the spatial resolution ~25 μm. The accuracy, as established through comparison to grazing incidence x-ray diffraction, is ~0.3–0.7 nm. Although these results and this article focus on metal films for microelectronics, ISTS is a general technique whose use is growing in biotechnology, data storage, photonics, and other areas of technology where thin films are important.

For background information *see* DIFFRACTION; INTERFERENCE OF WAVES; LASER; LATTICE VIBRATIONS; OPTICAL PULSES; SURFACE-ACOUSTIC-WAVE DEVICES in the McGraw-Hill Encyclopedia of Science & Technology.   John A. Rogers

Bibliography. H. J. Eichler, P. Gunter, and D. W. Pohl, *Laser-Induced Dynamic Gratings*, Springer, New York, 1986; G. W. Farnell and E. L. Adler, Elastic wave propagation in thin layers, in W. P. Mason and R. N. Thurston (eds.), *Physical Acoustics, Principles and Methods*, vol. 9, pp. 35–127, Academic, New York, 1972; M. Gotstein et al., Thin film metrology using impulsive stimulated thermal scattering (ISTS), in A. Diebold (ed.), *Handbook of Silicon Semiconductor Metrology*, Marcel Dekker, New York, 2000; J. A. Rogers et al., Optical generation and characterization of acoustic waves in thin films: Fundamentals and applications, *Annu. Rev. Mat. Sci.*, 30:117–157, 2000; J. A. Rogers and K. A. Nelson, Impulsive stimulated thermal scattering, in E. N. Kaufmann et al. (eds.), *Methods in Materials Research: A Current Protocols Publication*, p. 8a.3.a, Wiley, New York, 2000.

Information engineering (industrial)

In the next few years, an increasing number of companies separated by distance, function, and ownership will join to deliver products and solutions for the global marketplace. The trends toward virtual corporations and increasing global networking of economics will accelerate. Information systems, with the Internet as the primary communication and integration platform, will play an increasingly critical role in providing a competitive edge for organizations in the networked economy. Information engineering is the process of networking, collecting, analyzing, and reporting information, as well as controlling business, manufacturing, or service operations. The ongoing developments of information engineering in industrial operations address e-commerce concepts and software applications as well as business process and engineering technology in an integrated framework.

Intelligent manufacturing. Virtual enterprise systems such as (distributed) computer-integrated manufacturing, e-design, e-logistics, and e-business systems are typically large and complex. Fluctuations in market demands, technology evolution, and changing regulations require very flexible enterprise operations for reacting to changes. These reactions must be based on relevant and up-to-date information synthesized by decision support technology. Information engineering encompasses a wide range of ways for getting the needed information to decision-makers and improving business efficiency and effectiveness.

Production and manufacturing practices are undergoing revolutionary changes due to the acceptance of customer-centric business models that address global competition and Internet commerce. Products have shorter life cycles and are subject to frequent design changes. Processes have smaller lot sizes and limited in-process inventories. Many companies have used computer-aided design (CAD) and computer-aided manufacturing (CAM) to handle these challenges. Computer-integrated manufacturing (CIM) is now realizing the potential it offered many years ago as software systems have become widely available for product data management (PDM), manufacturing resource planning (MRP), and manufacturing execution systems (MES). Computer-integrated manufacturing systems promise to make production more agile and flexible in meeting customer demands and raw material and regulation restrictions.

To shorten product and process development cycles, designers and manufacturers are starting to use forms of groupware for coordinating design activities over the Internet with customers and suppliers. Some groupware supports virtual conferences for sharing CAD plans, material costs, and manufacturing schedules in real-time, and makes changes for exploring "what-if" scenarios. These e-design tools are connected to e-logistics and e-business systems for supporting the design-to-manufacturing and design-to-delivery activities.

Because of the popularity of e-commerce practices and the trend of outsourcing enterprise operations to alliance partners, efficient logistics support becomes more demanding and complicated. "Personalization" trends in business amplify these pressures. Thus, the logistics operations require an intelligent information management system (IIMS) to meet customer demands. The data warehouse (a database specifically structured for information access and reporting) plays a central role in the intelligent information management system. By networking data warehouse application to the logistics planning system (LPS), warehouse management system (WMS), and enterprise backbone systems, the intelligent information management system can manage limited resources effectively. For example, an intelligent transportation-planning module can facilitate resource allocation and execution to ensure that materials and finished goods are delivered at the right time, to the right place, according to schedule, and at minimal cost.

Due to dynamic changes and outsourcing trends in the business environment, many companies are building their e-business operations to meet customer demands and create competitive edges. Typically, e-business operations are supported by a collection of software, such as customer relationship management (CRM), supply chain management (SCM), enterprise resource planning (ERP), data warehousing, and decision support systems (DSS). For example, the supply chain management system (software) fulfills customer-specific needs for goods and value-added services in a timely, efficient, and cost-effective manner. The supply chain management software is composed of several modules for order planning, production, replenishment, and distribution management. For instance, the order-planning module generates and consolidates demand

forecasts from all business units in large corporations. The order commitment module allows vendors to accurately quote delivery dates to customers by providing real-time, detailed visibility into the entire fulfillment cycle from the availability of raw materials and inventory to production status and prioritization rules. Advanced scheduling and manufacturing planning modules provide detailed coordination of all manufacturing and supply efforts based on individual customer orders. Scheduling is based on real-time analysis of changing constraints throughout the process from equipment outages to supply interruptions. These software modules can be equipped with a range of statistical and forecasting tools to formulate the needed business intelligence.

Information systems. In terms of quality, information systems should be accurate (for example, no information distortion among preceding supply chains), secure, easy to use and access, scalable to handle more users and business functions, and reliable for meeting dynamic customer demands and information technology support service. It is very challenging to meet many of these requirements, considering the complexity of information technology infrastructure. Typical information technology infrastructure includes interface and communication devices [such as e-mail, electronic data interchange (EDI), groupware collaboration, and location tracking tools], databases (for example, legacy databases, relational databases, object databases, datamarts, groupware databases, and data warehouses located at various systems owned by different trading partners), and system architecture.

To make the enterprise information system useful to support decision-making activities, the design of information technology architecture should align well with the business vision and process, and consider people and technology issues. The computer-integrated manufacturing open-system architecture (CIMOSA) is commonly used in enterprise engineering to describe business, service, or manufacturing process functionality and behavior.

When designing an intelligent information management system architecture in an e-business operation, the heart of the design is the data warehouse. The data warehouse uses various tools such as data mining, online analytical processing, multidimensional data visualization, statistical analysis, metadata, Internet browsers, and report-generating systems to extract needed knowledge from the stored data. This warehouse is connected to customer relationship management and sales chain management systems to support marketing, sales, and customer service operations. The warehouse is also networked to the enterprise resource planning, manufacturing execution systems, supply chain management, logistics planning system, and other application software to support activities in material procurement, production, logistics, and partner alliance. The data warehouse also supports the financial, accounting, auditing, and human resource management systems to meet the expectations of stakeholders and employees. The Internet becomes the information and market exchange platform, and the Extensible Markup Language (XML) is the integration enabler. The selection of architecture, application software systems, integration platforms, and security software agents is the key challenge in e-business integration and e-collaborative operations (such as e-design, e-buying, and e-auction) for sharing information and resources with third parties seamlessly and securely. *See also* XML (EXTENSIBLE MARKUP LANGUAGE).

In building the data warehouse, structuring a set of well-defined principles will assist an organization in adapting to changes in business operations and technology development. Such principles should strive to ensure consistent data models; data cleaning and restructuring processes to guarantee that the information supports queries, reports, and analyses; as well as metadata, scalability, and warehouse management standards.

Information synthesis tools, such as data and text mining procedures, support the data warehouse. Commonly used data mining algorithms include association rules, decision trees, classification and regression procedures, artificial neural networks, instance-based learning, and many statistical methods. There are also various data segmentation, aggregation, and visualization tools available for helping users to search relationship, trend, or hidden patterns for solving problems. Many vendors provide data mining software and consulting support. Very often, in practice, decision-makers face the problem of "information overload." A challenge in data mining is processing large amounts of information in a short time window. One solution is to split the data intelligently into many pieces, process the data using parallel computing techniques, and then integrate all synthesized results in a coherent manner. Text mining is needed to extract useful knowledge from operational reports, business transaction documents, or research and development publications. Ultimately, the data and text mining tools will provide users business and manufacturing intelligence at different hierarchical levels of details.

Outlook. Society is moving from the industrial age to the information age. Information systems and decision-making technologies will play a central role in daily life and business activities. There are many opportunities for improving the quality and efficiency of the information systems, the underlying concepts of building the system, and the implementation process.

For background information *see* COMPUTER-AIDED DESIGN AND MANUFACTURING; COMPUTER-INTEGRATED MANUFACTURING; DATA COMMUNICATION; DATA MINING; DATABASE MANAGEMENT SYSTEMS; INDUSTRIAL ENGINEERING; INFORMATION MANAGEMENT; INFORMATION SYSTEMS ENGINEERING; INFORMATION TECHNOLOGY; INTERNET in the McGraw-Hill Encyclopedia of Science & Technology.
Jye-Chyi Lu

Bibliography. R. Kalakota, M. Robinson, and D. Tapscott, *e-Business: Roadmap for Success*, Addison-Wesley, 1999; J.-C. Lu, Methodology of mining massive data sets for improving manufacturing quality/efficiency, in D. Braha (ed.), *Data Mining for Design and Manufacturing: Methods and Applications*, Kluwer Academic, 2000; E. Sperley, *The Enterprise Data Warehouse: Planning, Building and Implementation*, Hewlett-Packard Professional Books, 1999.

Integrity (navigation)

Four parameters define the requirements for navigational aids that operate in civil airspace: accuracy, integrity, continuity, and availability. Accuracy refers to how well the system can provide correct position information in the absence of a failure. Integrity refers to the ability of the system to inform the user in a timely manner of a latent failure that may cause a hazardous condition. (A latent failure is an undetected degradation of the navigational aid operation.) Continuity refers to the ability of the system to let the user navigate without interruption. For example, radio-frequency interference (jamming) and blockage of radionavigation signals are potential causes of a loss of continuity. Availability is the probability that the system will function and provide required levels of accuracy, integrity, and continuity.

Ground-based navigational aids such as very high frequency (VHF) omnidirectional range (VOR), distance-measuring equipment (DME), and instrument landing systems (ILS) have been providing integrity information for most of the en route navigation and approach and landing phases of operation in the National Airspace System (NAS). Signals from these systems are transmitted from a known location, and an integrity monitor at the transmitter location can easily detect when performance of the broadcast signal is degraded. However, signals of the Global Positioning System (GPS) are not monitored for latent failures at the satellite. These latent GPS satellite failures may be caused by satellite clock errors, by erroneous uploads or changes in satellite ephemeris data, or by distorted signals. They create a more complex integrity monitoring problem in an effort to prevent GPS receivers from reporting hazardously erroneous navigation information.

Parameters associated with integrity. Requirements for integrity are usually stated in terms of three parameters: the maximum allowable rate of hazardously misleading information (HMI), the time-to-alert, and the alert limit.

Maximum allowable rate of HMI. This is the maximum allowable number of integrity failures per unit time. For example, the U.S. Federal Aviation Administration (FAA) has set this requirement at 10^{-7}/h for en route through nonprecision approach phases of flight, and 2×10^{-7} per approach of 150 seconds duration for Category I precision approaches. This rate is computed as the product of the rate of GPS satellite latent failures and the probability that a given failure will go undetected. With respect to a GPS satellite's latent failure, the FAA has used a rather conservative failure rate of 1×10^{-4}/h. This rate yields a missed detection probability requirement of 0.001 to meet 10^{-7}/h requirement for en route through nonprecision approach phases of flight. For a precision approach lasting for 150 s, the rate of 1×10^{-4}/h is equal to $(150/3600) * 10^{-4} = 4.2 \times 10^{-6}$ per approach. This rate, with a missed detection probability of 0.001, yields a 4.2×10^{-9} probability of an undetected GPS failure causing a hazard during a Category I precision approach. This probability is set well below the 2×10^{-7} per approach requirement because other integrity risks must be recognized.

Time-to-alert. The FAA requirements for the time-to-alert depend on the phase of navigation, namely, 1 min, 30 s, 10 s, and 10 s for the oceanic/remote, en route, terminal, and nonprecision approach operations, respectively. For precision approaches, it varies from 1 to 6 s, depending on the category of approach.

Alert limit. This is the maximum position error allowable for the given phase of flight without an alert being raised. The alert limits for the oceanic/remote, en route, terminal, and nonprecision approach phases of operation are 4 nmi, 2 nmi, 1 nmi, and 0.3 nmi (1 nmi = 1.85 km), respectively, where distance is measured in the local horizontal plane. For precision approaches, the alert limit is 5–12 m (16–50 ft) in the vertical direction, depending on the category of approach.

In addition to the alert limit, there is a parameter called the protection level which, with a very high probability, the system guarantees would not be

exceeded by the true position error. The user equipment estimates the protection level and compares it to the alert limit in real time. If the protection exceeds the alert limit for the given phase of operation, the integrity function is declared unavailable, and the operation at hand may be forced to transition to another phase of operation requiring a less stringent alert limit (for example, abort a landing approach and transition back to terminal operations).

Means of integrity for stand-alone GPS. When the U.S. Department of Defense originally developed the GPS for military purposes, this service did not provide a sufficient level of inherent integrity for acceptance by civil aviation authorities. For example, in case of a satellite anomaly, it could take as long as 90 min before the satellite signal was taken off the air. Although the timeliness of signal quality monitoring by the GPS system itself has improved significantly over the years, it cannot be relied on to meet civil aviation navigation requirements, which require alerts within seconds. Therefore, when the FAA first considered the use of GPS for civil aviation, it needed to find a way to provide integrity for the GPS signals.

Concepts. Three different concepts emerged.

1. Receiver autonomous integrity monitoring (RAIM) provides integrity within the user avionics by checking the consistency of the range measurements from redundant observations on GPS satellites. Use of integrity methods based on this concept was the first approach approved by the FAA. It was approved for use as a primary means of navigation for oceanic/remote airspace operations and as a supplemental means of navigation for en route through nonprecision approach flight operations.

2. A satellite-based augmentation system (SBAS) is designed to enable Category I or near Category I precision approaches in addition to serving as a primary means of navigation for en route through nonprecision approaches.

3. A ground-based augmentation system (GBAS) is designed to provide Category I, II, and III precision approaches at major airports. Unlike the first concept, the latter two concepts use ground-based integrity monitors (discussed later).

Range comparison method. The principle exploited by RAIM to detect the presence of a fault in the satellite navigation system is as follows. A GPS user typically can view a few more satellites than are needed to make a position fix. For the nominal 24-satellite GPS constellation, about 7 to 8 satellites are visible on average. When redundant satellites are available, the user equipment can perform a self-consistency check. How this is done can be seen by considering an example where five satellites are assumed to be visible (see **illus.**). In this case, one solves for position using each subset of four satellites at a time. By using this position estimate and the satellite position information broadcast by GPS, the predicted range to the fifth satellite can be calculated. If there is no failure on any of the satellites, the range residuals, that is, the differences between the predicted and the measured ranges to the fifth satellite, should be small,

Range comparison method for detecting the presence of a fault in a satellite navigation system.

and vice versa. Therefore, RAIM may declare the presence or absence of a failure, depending on the sizes of the range residuals. However, RAIM calculations as implemented are more sophisticated than this simple heuristic explanation implies because they must simultaneously satisfy requirements on both the minimum valid detection probability and the maximum false detection probability. Two leading RAIM methods are the chi-square method, where the sum of the squares of the range residuals is used as the test statistic, and the parity space method, in which the residuals are projected onto a "parity" space that facilitates detection and identification of the failed satellite.

FDE. The basic RAIM algorithm can provide integrity by detecting the presence of a fault. However, that may not be sufficient for a user who needs to continue navigation upon detection, as required when the system is to be used as a primary means of navigation. In this case, the user must also be able to identify and exclude the bad satellite in order to continue navigating using the remaining satellites. This capability, called fault detection and exclusion (FDE), can be provided by RAIM when enough satellites are in view to form a good geometry. However, fault detection and exclusion requires much more stringent geometric criteria in order for it to work properly (for example, fault detection requires at least five satellites, and fault detection and exclusion requires at least six), which can significantly reduce the service availability. The level of service that can be provided with stand-alone GPS using RAIM will depend on inherent GPS capabilities with regard to the constellation size, the GPS satellite integrity failure rate that can be guaranteed, and the receiver range measurement accuracy. With the current GPS receiver capability and for the nominal 24-satellite constellation, stand-alone GPS using RAIM can provide only a supplemental means of navigation for en route through nonprecision approach phases of flight. In the future, however, it may be possible for this system to provide a primary means of navigation for these phases of flight if planned improvements are made to the GPS service including increased constellation size, smaller GPS satellite guaranteed latent failure rate, and improved ranging accuracy. However, for precision approaches, RAIM performance with stand-alone GPS will not be sufficient because of the significantly smaller alert limits. Therefore, the user receiver will have to be augmented by integrity data generated from ground monitors.

Two GPS augmentation systems. A type of SBAS called a wide-area augmentation system (WAAS) and a type of GBAS called a local-area augmentation system (LAAS) are currently under development by the FAA.

The WAAS is planned to provide to the users over a wide area (for example, the United States) a precision approach capability by providing sufficient accuracy for vertical guidance as well as a primary means of navigation capability in the en route, terminal, and nonprecision approach flights. The WAAS will comprise at least 25 ground reference stations, redundant master stations, and geostationary satellites. The various reference stations provide the range measurements of the GPS signals, and the master stations collect the data and determine both the corrections to the GPS signals and integrity data. The data are then broadcast to the users over a large service volume via geostationary satellites.

The LAAS will employ redundant ground reference receivers that monitor the quality of the GPS signals and make range measurements. The corrections and integrity data are transmitted to the users within a terminal area over a VHF link. LAAS service is required out to a range of 23 nmi (43 km), from the terminal area. LAAS will service operations including terminal area navigation, surface navigation, and precision approach within this region.

The performance of a WAAS and a LAAS depends on how accurately and reliably the ground stations monitor the GPS signals. For this reason, both systems employ redundant receivers at each reference site and provide integrity of the ground monitoring by checking the consistency of the measurements between different receivers. Separate monitors cover any latent GPS signal errors. In each of these systems, the ground segment provides fault detection and exclusion of a reference receiver or a failed satellite. In a WAAS, two error sources can degrade the accuracy of postcorrection range measurements for the user as a function of the distance between the reference sites and the user location. One is imprecise satellite emphemeris corrections, and the other is the different amount of propagation delays of signals traveling different paths through the ionosphere to different locations on the Earth. In contrast, because a LAAS provides service only for the users in an area adjacent to the monitor receivers, this geographic decorrelation is minimized. Therefore, a LAAS can provide precision approach capability with much higher accuracy to the users in its coverage area than a WAAS. *See* DIFFERENTIAL GPS.

For background information *see* AIR NAVIGATION; AIR-TRAFFIC CONTROL; ELECTRONIC NAVIGATION SYSTEMS; SATELLITE NAVIGATION SYSTEMS in the McGraw-Hill Encyclopedia of Science & Technology.

Young C. Lee

Bibliography. R. Braff, Description of the FAA's local area augmentation system (LAAS), *Navigation* (*Journal of the Institute of Navigation*), 44(4):411–424, Winter 1997–1998; R. G. Brown, A baseline GPS RAIM scheme and a note on the equivalence of three RAIM methods, *Global Positioning System* (*Red Book*), vol. V, pp. 101–116, Institute of Navigation, 1998; Y. Lee, Analysis of range and position comparison methods as a means to provide GPS integrity in the user receiver, *Global Positioning System* (*Red Book*), vol. V, pp. 5–19, Institute of Navigation, 1998; R. Loh et al., The U.S. wide-area augmentation system (WAAS), *Global Positioning System* (*Red Book*), vol. VI (Selected Papers on Satellite Based Augmentation Systems), pp. 5–35, Institute of Navigation, 1999.

Interfacial analysis

Whether regarding a glass of water, a pond, or an ocean, at the liquid surface there is a boundary between the liquid water and the air. This boundary or interface is a transition region between the air (gas phase) and the water (liquid phase). A gradient with respect to the water density exists in this region and can sometimes be quite complex. There also exists a surface potential. As the water molecules move about in the bulk, there is randomness to their overall structure. The water molecules that are in the surface region have a net orientation with respect to the surface. The structural organization of an interface is unique because it lacks inversion symmetry about the interface. This lack of symmetry gives rise to an increase in the second-order nonlinear response of the interface to electromagnetic radiation and is the basis for second-harmonic, sum-frequency, and difference-frequency generation.

For more than a century, scientists contemplated the structure and chemistry of liquid interfaces, and although thermodynamic theory allowed them to understand the macroscopic properties of these surfaces, a molecular-level understanding remained elusive. This was particularly true for the surface of complex liquid mixtures.

In the 1960s, laser technology became available, and during that decade a nonlinear optical phenomenon was postulated and experimentally proven to exist in noncentrosymmetric (nonsymmetric with respect to a central point) media. Since that time, the field of nonlinear optics has intimately affected a broad range of sciences, including materials, biological, and environmental research.

Sum-frequency generation. The application of second-order nonlinear optics to the study of interfaces has been a major breakthrough for surface science. There are three types of second-order nonlinear phenomena, all of which are inherently surface-selective; second-harmonic generation (SHG), sum-frequency generation (SFG), and difference-frequency generation (DFG). Second-harmonic generation has been used to infer concentrations, acidity, molecular orientation, and structure at surfaces. However, second-harmonic generation lacks the molecular selectivity of sum-frequency generation and difference-frequency generation. Both of the latter have the ability to probe molecular vibrational modes to produce a spectrum that one can use to identify a molecule and a specific bonding structure from a complex liquid surface. Sum-frequency generation has the added convenience that the combination of visible and infrared (IR) light will give rise to sum-frequency light in the visible region, making it particularly convenient to detect. Difference-frequency generation gives rise to longer-wavelength light. The surface selectivity combined with molecular selectivity and detection in the visible wavelength spectral region makes sum-frequency generation a powerful surface science tool.

In the 1980s, surface vibrational sum-frequency generation (VSFG) emerged as a promising new area of surface science. This emergence was accompanied by improvements in laser technology and infrared optical parametric generation, which pushed VSFG to the forefront of surface science research. VSFG spectroscopy is used to probe the vibrational signature of surface species, allowing the identification of molecular species in the surface region of complex liquids. It is also used to probe other structures such as monolayers on solid surfaces in many environments, including high-pressure regimes. The analytical techniques that dominated surface science in past decades did not possess the surface selectivity inherent to second-order nonlinear optical techniques, and in many cases they did not possess the ability to probe in mid- to high-pressure regimes. The combined strengths of optical spectroscopy and the surface sensitivity have made VSFG uniquely suited to study a variety of interfaces under many environmental conditions, with molecular selectivity.

Analysis process. A vibrational spectrum of an interface using VSFG spectroscopy is obtained by spatially overlapping a visible laser beam and a tunable infrared laser beam at a surface of interest (**Fig. 1**). The beam angles with respect to the surface plane are optimized according to reflection and transmission properties (Fresnel coefficients) of the two boundaries that join to form the interface. Angles from the surface plane of 25–35° are typical when the beams initially travel through the lower-refractive-index medium, that is, if the two beams are incident on the air side of an air-water interface. The sum of the two frequencies is generated via the second-order nonlinear interaction of the photons with the surface molecules (**Fig. 2**). For monochromatic light, when a visible photon of wavelength 800 nm combines with an infrared photon of wavelength 4 μm (wave number of 2500 cm^{-1}), the sum of the two frequencies is calculated to be a 667-nm photon [that is, $1/(800 \text{ nm}) + 1/(4000 \text{ nm}) = 1/(667 \text{ nm})$]. Therefore, a 667-nm photon is generated from the 800-nm and 4000-nm photons and is detected spatially between the reflections of the two beams that had

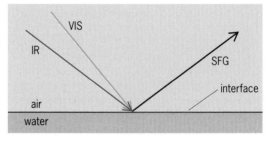

Fig. 1. Laser beam geometry for a sum-frequency generation (SFG) experiment in reflection mode. The infrared (IR) and the visible beams are overlapped on the interface of interest, temporally and spatially (~40° and 30° respectively from the surface). The SFG is detected at an angle that is determined by momentum conservation laws. This angle can be calculated from the frequencies and the incoming angles of the IR and visible beams, the SFG frequency, and the refractive index of the interfacial region.

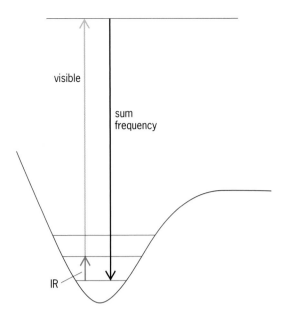

visible

sum
frequency

IR

Fig. 2. Schematic of sum-frequency generation. The curve represents the potential surface of a molecule. The lines within the curve represent the vibrational energy levels. The line at the top of the diagram represents a virtual energy state. The IR and visible probe beams are represented as vectors, and their lengths are proportional to their frequency. The sum of the two frequencies is generated, and an enhancement of the sum-frequency response is due to the interaction with a resonant IR pulse (the IR frequency is resonant with a vibration of the molecule being probed).

initiated the nonlinear sum-frequency generation response. Momentum conservation accounts for the spatial resolution of the sum-frequency photon. Detection is simplified by knowing the path of the photons and the sum-frequency generation wavelength. The sum-frequency generation photons are in the visible-wavelength region, which allows ease of detection since there are numerous detectors (such as photomultiplier tubes and charge-coupled devices) that are sensitive in this spectral range.

Since sum-frequency generation has a low probability of occurrence, short-pulsed laser beams are typically used to improve the efficiency of sum-frequency generation. Lasers provide large photon fluxes. More incident photons translate into a larger number of sum-frequency photons generated, resulting in improved signal-to-noise ratios. (The sum-frequency generation light, the signal, and the background light, the noise, are concurrently being detected.) Ultrafast laser pulses on the order of picoseconds (10^{-12} s) and shorter dramatically improve the nonlinear response efficiency due to the high peak powers associated with these short pulsed beams. (The power of the laser-produced photons is condensed into a shorter time, thus giving relatively high peak powers during the duration of the pulse.) However, longer pulses in the nanosecond (10^{-9} s) regime generate sum-frequency photons with only slightly longer averaging or integration times after detection.

Optical spectroscopies, including nonlinear optical techniques (such as VSFG), have several advantages over ultrahigh vacuum surface science technologies, including use in or out of vacuum environments. Optical spectroscopies are uniquely flexible with respect to environmental conditions. Vibrational sum-frequency generation spectroscopy is selective molecularly via vibrational identification of molecular bonds, and it also has advantages over infrared and Raman spectroscopy for use as a surface-selective analytical tool since it is inherently surface-selective. (One can calculate from quantum mechanics that the lack of inversion symmetry at any interface gives rise to a second-order nonlinear optical response.) Infrared spectroscopy and Raman spectroscopy, although excellent for use in many applications, are not well suited for studying liquid surfaces. Upon reflection, the evanescent wave of the incident light beam decays into the bulk and therefore penetrates the surface on the order of the wavelength of light used. This is a problem since infrared light is on the order of micrometers (10^{-6} m) and a surface monolayer is on the order of subnanometers ($<10^{-9}$ m). (Visible light, typically used for Raman studies, is on the order of 500–800 nm.) The advantage that vibrational sum-frequency generation spectroscopy has over its optical relatives is the surface-selectivity of VSFG. Sum-frequency generation intensity is produced only from the interface, which is typically much smaller than the depth probed. This property makes VSFG a valuable surface analytical tool, particularly for liquid surfaces, including liquid-liquid interfaces.

Application. Sum frequency, particularly vibrational sum-frequency generation spectroscopy, has propelled research in all areas of interfacial science. Impacts are currently being felt in many research areas as mentioned above. However, VSFG is far from being developed to its fullest capacity. In 1998, the introduction of broad-band infrared generation allowed researchers to probe a surface in one laser pulse. This new approach to VSFG referred to as broad-band SFG (BBSFG) allows researchers to probe molecular-level details of surface dynamics in order to understand a multitude of dynamic interfacial processes, such as orientational recovery, structural reorganization, and reaction mechanisms.

Researchers now have the ability to directly probe interfaces at the molecular level to answer questions about structure (**Fig. 3**), molecular orientation, and chemical reaction mechanisms at an interface. For example, sum-frequency generation studies have shown that the surface of liquid water has three oriented structures at the air-water interface: (1) hydrogen-bonded water molecules that are oriented primarily in the plane on the liquid side, (2) a lower density of partially hydrogen-bonded water molecules that straddle the interface, and (3) in the least dense region, gas-phase water molecules that are oriented with their hydrogen atoms on average pointed toward the bulk liquid. This new level of molecular-level understanding crosses many disciplines, which encompass biological, environmental, and industrial processes. Analysis of these

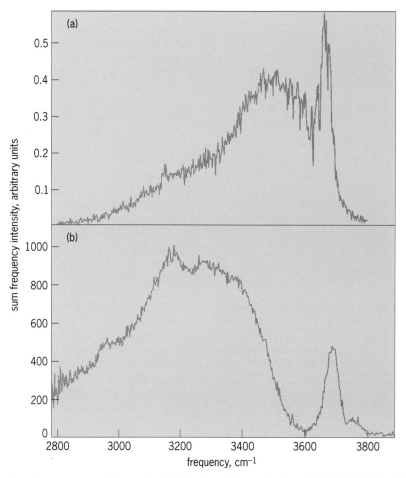

Fig. 3. Vibrational sum-frequency generation spectra. (*a*) Spectrum of the carbon tetrachloride (CCl_4)/water interface obtained using a nanosecond laser VSFG scanning system. The sum-frequency intensity peak at ~3665 cm^{-1} is assigned to the vibrational mode of dangling OH from water molecules that straddle the CCl_4/water interface (the OH protrudes into the CCl_4 phase). At lower energy, intensity is attributed to loosely hydrogen-bonded water molecules at the interface.
(*b*) Spectrum of the air/water interface using a picosecond laser VSFG scanning system. The sum-frequency intensity peak at ~3700 cm^{-1} is assigned to the dangling OH vibration of the water molecules that straddle the air/water interface (the OH protrudes into the air phase). The symmetric stretching vibrations of the water molecules on the liquid side of the air/water interface differ in frequency relative to the water molecules in the water phase of the CCl_4/water interface. The vibrations of the water molecules at the air/water interface are lower in energy and span a larger frequency range. This indicates that the interfacial water structure may be composed of several different types of intermolecular interactions. The interfacial structure of the low-energy water molecules at the air/water interface is sometimes referred to as icelike (from the intensity in the 3200 cm^{-1} region), inferring a high intermolecular symmetry of the symmetric stretching vibration of the water molecules. The observations from these two VSFG studies show that even simple interfaces can have complex interfacial structures.
(*Reprinted with permission from J. Phys. Chem. A, 104(45):10220–10226; copyright 2000 by American Chemical Society*)

interfacial systems has improved the fundamental understanding of biological membranes, industrial catalysis, nanostructures, polymer interfaces, soil-pollutant interactions, and heterogeneous atmospheric processes such as cloud droplet chemistry and aerosol growth.

For background information *see* CONSERVATION OF MOMENTUM; INTERFACE OF PHASES; LASER; LASER SPECTROSCOPY; NONLINEAR OPTICAL DEVICES; NONLINEAR OPTICS; OPTICAL PULSES; SURFACE AND INTERFACIAL CHEMISTRY in the McGraw-Hill Encyclopedia of Science & Technology.   Heather C. Allen

Bibliography. N. Bloembergen, Surface nonlinear optics: A historical overview, *Appl. Phys. B*, 68:289–293, 1999; M. G. Brown et al., The analysis of interference effects in the sum frequency spectra of water interfaces, *J. Phys. Chem. A*, V104(45):10220–10226, 2000; K. B. Eisenthal, Liquid interfaces

probed by second-harmonic and sum-frequency spectroscopy, *Chem. Rev.*, 96:1343–1360, 1996; E. L. Hommel and H. C. Allen, Broadband sum frequency generation with two regenerative amplifiers: Temporal overlap of femtosecond and picosecond light pulses, *Anal. Sci.*, 17:137–139, 2001; P. B. Miranda and Y. R. Shen, Liquid interfaces: A study by sum-frequency vibrational spectroscopy, *J. Phys. Chem. B*, V103:3292–3307, 1999.

Isoprostanes

The isoprostanes (iPs) are a class of natural products discovered recently. They are isomeric with another class of natural products, the prostaglandins. The prostaglandins are the result of enzymatic oxygenation of polyunsaturated fatty acids, in particular arachidonic acid. By contrast, the iPs are formed in vivo by nonenzymatic, free-radical oxygenation of arachidonic acid. This important distinction in the mode of formation is responsible for a more complex mixture of iPs being generated in vivo. For example, whereas the endoperoxide PGG_2 is formed by the cyclooxygenase enzymes (COX-1 and COX-2), four classes of iPs are formed as a result of the free-radical oxygenation of arachidonic acid, with each class containing 16 iPs for a total of 64 isoprostane molecules.

Two different mechanisms for the formation of iPs have been proposed based on in-depth regiochemical and stereochemical analysis of the peroxidation process (**Figs. 1** and **2**). Using the endoperoxide mechanism, a more detailed mechanistic view of the formation of group VI iPs is presented in **Fig. 3**. The four iPs shown in the box were prepared by total synthesis, and then used as markers in identifying these compounds in human urine (**Figs. 4** and **5**).

A comprehensive nomenclature for iPs that utilizes as much as possible the symbols in use in prostaglandin nomenclature has been recently introduced. The new classification (roman numerals) reflects the use of the ω end of the molecule instead of the COOH in the double-bond counting system—the same way that double bonds in polyunsaturated fatty acids are referred to as ω-3, ω-6, and so on.

Measurement of iPs. Gas chromatography/mass spectrometry methods have been developed to quantitate individual iPs, particularly group VI iPs, which appear to be most abundant in urine. Group VI was chosen as the main initial focus because the position of the OH group on carbon-5 of the upper side chain allows a lactone to form, which can easily be separated from the rest of the iPs, allowing for a more accurate and reliable determination in biological fluids (Fig. 4).

The gas chromatography/mass spectrometry method outlined in Fig. 4 has enabled measurement of in vitro and in vivo elevated levels of $iPF_{2\alpha}$-VI in a number of clinical settings, such as in ischemia-reperfusion syndromes, hypercholes-terolemia, Alzheimer's disease, and alcohol-induced liver disease. Elevated levels of $iPF_{2\alpha}$-VI have been measured in oxidized low-density lipoprotein (LDL) and in aortic atherosclerotic lesions. In all cases, the discovery and identification of iPs of different groups, including those shown in Fig. 4, have been made possible by the initial availability of authentic material prepared by total synthesis.

Liquid chromatography tandem mass spectrometry (LC/MS/MS) is another very promising method for the identification, separation, and quantitation of iPs. Neither derivatization nor prior purification is required. Figure 5 shows the chromatogram of group VI iPs from human urine. The chemical structure in the chromatogram identifies the individual iPs. The four most abundant iPs of the urinary group VI have the *cis* configuration of the side chain.

Biological implications. The importance of these free-radical-generated iPs encompasses three general areas.

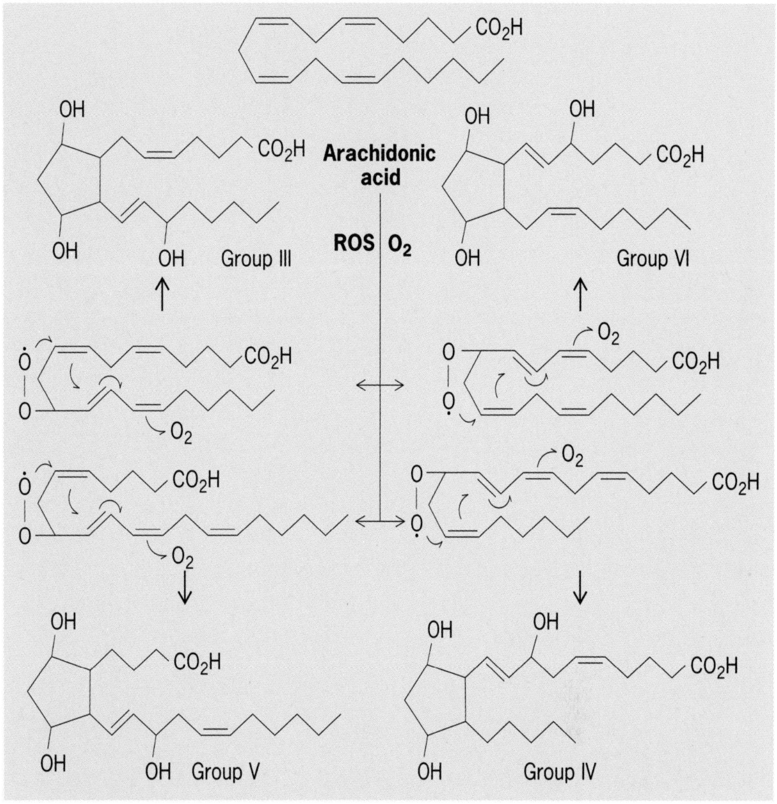

Fig. 1. Endoperoxide mechanism for the formation of isoprostanes. ROS = reactive oxygen species.

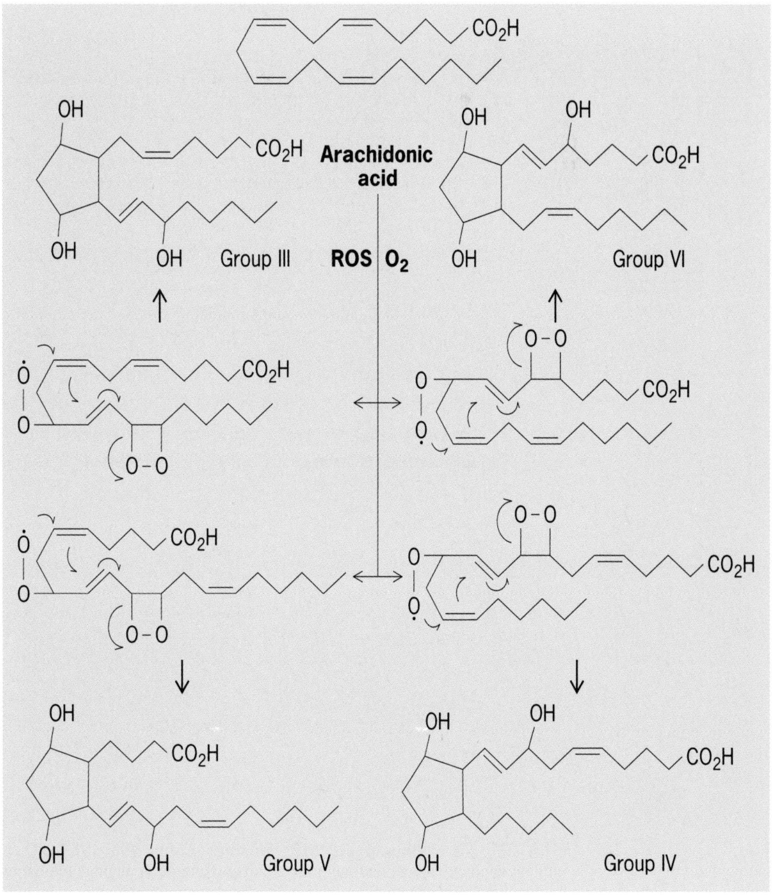

Fig. 2. Dioxetane mechanism for the formation of isoprostanes.

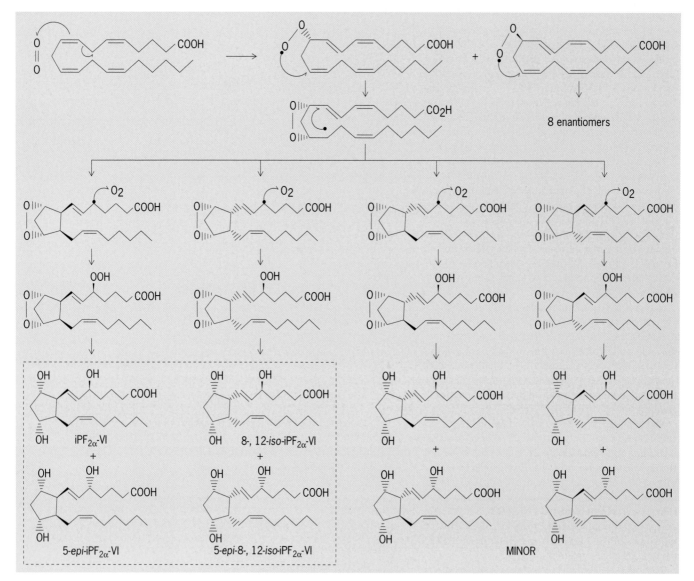

Fig. 3. Mechanism for the formation of group VI isoprostanes. For clarity, the eight enantiomers are not shown.

Damage to cell membranes. One consequence of oxidant injury is alteration in the physiochemical properties and function of cellular membranes. This has been attributed to an accumulation of products of lipid peroxidation in lipid bilayers. Phospholipids are essential constituents of the cell membrane. Radical oxygenation of the polyunsaturated fatty acids, located at the *sn*-2 position of the phospholipids, is bound to create disturbances in the cell membrane. The tightness of the fatty acid portion of the

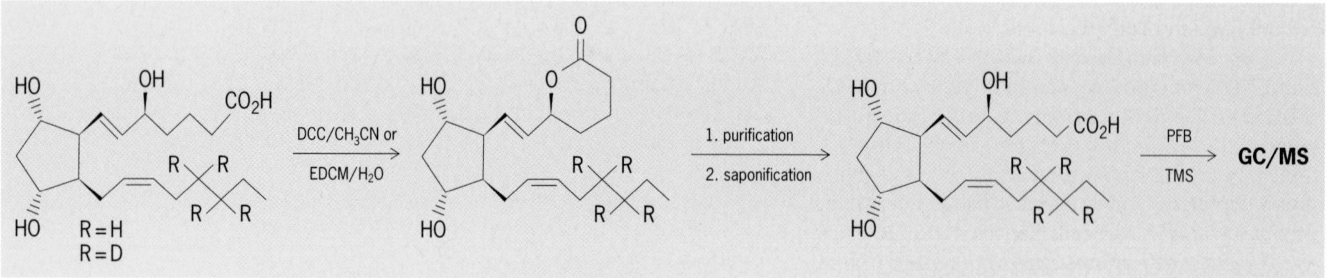

Fig. 4. Outline for GC/MS measurement of iPF$_{2\alpha}$-VI. DCC = dicyclohexylcarbodiimide; EDCM = 1-ethyl-3-(3-dimethyl-aminopropyl)carbodiimide methiodide; PFB = pentafluorobenzyl; TMS = trimethylsilyl; GC/MS = gas chromatography/mass spectrometry; H = hydrogen; D = deuterium.

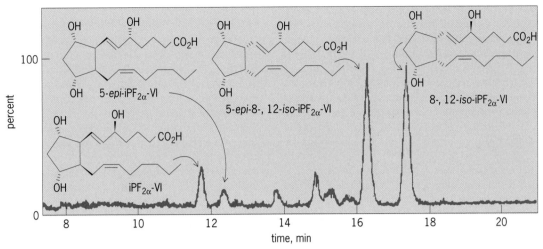

Fig. 5. LC/MS/MS chromatogram of group VI isoprostanes in urine.

phospholipids is what causes the cell membrane to be hydrophobic and guarantees the integrity of the cell membrane. Polyhydroxylation of the unsaturated fatty acid to yield, for example, iPs, which contain three OH groups, is bound to disrupt the tightness of the membrane and have profound effects on the fluidity and integrity of the cellular membrane. For example, a leak could develop in the membrane due to repulsion between the newly created OH groups and the adjacent lipophilic fatty acids, leading to cell death.

Potential physiological function of isoprostanes. Although the mechanism by which iPs are released from the cell membrane remains to be elucidated, phospholipid-containing iPs are subject to cleavage by phospholipases. These iPs are released into the bloodstream, circulate through the liver and kidney, and eventually are excreted in urine, although their metabolic behavior in vivo remains largely unknown. Several iPs have been shown to have biological effects in vitro. For example, $iPF_{2\alpha}$-III is a vasoconstrictor when infused into the rat renal artery and the rabbit pulmonary artery. In both instances, the effects are prevented by pharmacologic blockage of thromboxane receptors. Similarly, $iPF_{2\alpha}$-III modulates platelet function. Although it does not cause irreversible aggregation, it facilitates the ability of subthreshold concentrations of conventional agonists to induce this response. These effects require micromolar concentrations of $iPF_{2\alpha}$-III. Interestingly, much lower concentrations of this compound have been reported to modulate platelet adhesion.

Much less information is available about the potential biological activity of other iPs. For example, $iPE_{2\alpha}$-III behaves much like $iPF_{2\alpha}$-III on platelet aggregation. In contrast to these compounds, which appear to act as incidental ligands via the thromboxane receptor, 8,12-*iso*-$iPF_{2\alpha}$-III induces proliferative responses in NIH 3T3 cells via the $PGF_{2\alpha}$ receptor. This compound activates this prostanoid receptor selectively and in a saturable manner, but at concentrations roughly one order of magnitude greater than its natural ligand.

Given the difficulty in interpreting the biological potency of individual isoprostanes when multiple species are likely to be formed coincidentally in vivo, it is premature to decide whether the biological properties of these compounds might be relevant under conditions of oxidant stress in vivo. Nonetheless, the possibility of coordinate cellular activation by multiple species, their delivery to cell membranes in highly concentrated form, and the superiority of thromboxane receptor antagonism over cyclooxygenase inhibition in experimental settings, suggests a potential role for these compounds as mediators as well as indices of oxidant stress.

Isoprostanes as indices of oxidant stress. Perhaps the most interesting aspect of current isoprostane research is the possibility that measurement of these compounds in biological fluids might afford a quantitative index of lipid peroxidation in humans. Increased plasma and urinary concentrations of the $iPF_{2\alpha}$-VI and $iPF_{2\alpha}$-III have been reported in several syndromes thought to be associated with excessive generation of free radicals. These include coronary reperfusion after a period of ischemia, poisoning with paraquat and paracetemol, cigarette smoking, and alcohol-induced liver disease. Isoprostanes, which are increased in low-density lipoprotein when it is oxidized in vitro, have been found to be increased in the urine of asymptomatic patients with hypercholesterolemia, and are present in human atherosclerotic plaque. These observations, and their measurement in murine models of atherosclerosis, raise the likelihood that they may be useful in elucidating the role of low-density lipoprotein and cellular protein oxidation in atherogenesis. In perhaps the most striking evidence to date of the free-radical involvement in atherosclerosis, it has been shown that suppression of elevated isoprostane generation ($iPF_{2\alpha}$-VI) in vivo reduces the progress of atherosclerosis in apoE-deficient mice, even though the level of cholesterol remains high and unchanged.

Isoprostane analysis, particularly noninvasive analysis in urine, may be especially applicable to the selection not only of appropriate targets for interventional trials of antioxidants but also of rational doses of antioxidant drugs and vitamins. Presently, phase 3 clinical trials of antioxidant vitamins yield conflicting and confusing data. Oddly, there is little biochemical basis for the doses of antioxidants used in these trials, so it is not known if the confusion emanates from differences in trial design, dosing, or selection of patients. A reproducible, noninvasive index of oxidant stress would be particularly useful in addressing these issues. It is also unclear how alterations in endogenous antioxidant defenses might modulate such dose-response relationships.

Considerable progress is being made toward refining indices of free-radical-catalyzed modification of both protein and DNA. Analysis of iPs may complement these efforts as new insight is gained into oxidant stress as a mechanism of disease in humans.

[This work was supported by National Institutes of Health Grants HL54500, MOIRR00040, HL61364, HL62250 (GAF), and DK44730 (JR); and by National Science Foundation for AMX-360 NMR Instrument grant CHE9013145 (JR).]

For background information *see* ATHEROSCLEROSIS; CHOLESTEROL; CIRCULATION DISORDERS; EICOSANOIDS; GAS CHROMATOGRAPHY; MASS SPECTROMETRY in the McGraw-Hill Encyclopedia of Science & Technology.   Garret FitzGerald; Joshua Rokach

Bibliography. L. P. Audoly et al., Cardiovascular responses to the isoprostanes iPF(2α)-III and iPE(2)-III are mediated via the thromboxane A(2) receptor in vivo, *Circulation*, 101:2833–2840, 2000; A. Burke et al., Specific analysis in plasma and urine of 2,3-dinor-5,6-dihydro-isoprostane F$_{2\alpha}$-III, a metabolite of isoprostane F$_{2\alpha}$-III and an oxidation product of γ-linolenic acid, *J. Biol. Chem.*, 275:2499–2504, 2000; S. W. Hwang et al., Total synthesis of 8-*epi*-PGF$_{2\alpha}$: A novel strategy for the synthesis of isoprostanes, *J. Amer. Chem. Soc.*, 116:10829–10830, 1994; J. A. Lawson, J. Rokach, and G. A. FitzGerald, Isoprostanes: Formation, analysis and use as indices of lipid peroxidation in vivo, *J. Biol. Chem.*, 274:24441–24444, 1999; H. Li et al., Quantitative high performance liquid chromatography/tandem mass spectrometric analysis of the four classes of F$_2$-isoprostanes in human urine, *Proc. Nat. Acad. Sci. USA*, 96:13381–13386, 1999; D. Pratico et al., IPF$_{2\alpha}$-I: An index of lipid peroxidation in humans, *Proc. Nat. Acad. Sci. USA*, 95:3449–3454, 1998; D. Pratico et al., Vitamin E suppresses isoprostane generation in vivo and reduces atherosclerosis in apoE deficient mice, *Nature Med.*, 4:1189–1192, 1998; Z. Pudukulathan et al., Diels-Alder approach to isoprostanes total synthesis of iPF$_{2\alpha}$-V, *J. Amer. Chem. Soc.*, 120:11953–11961, 1998; L. J. Roberts and J. D. Morrow, Measurement of F(2)-isoprostanes as an index of oxidative stress in vivo, *Free Radical Biol. Med.*, 28:505–513, 2000; J. Rokach et al., Nomenclature of isoprostanes: A proposal, *Prostaglandins*, 54:853–873, 1997; E. A. Weinstein et al., Prothrombinase acceleration by oxidatively damaged phospholipids, *J. Biol. Chem.*, 275:22925–22930, 2000.

Laser-induced nuclear reactions

Lasers have a long history of use in inducing nuclear reactions in the context of thermonuclear fusion research. The energy carried by individual optical photons (1–4 eV in the infrared to ultraviolet range) is too small to directly excite transitions between quantum-mechanical states in nuclei, which are typically of the order of several kiloelectronvolts to several megaelectronvolts. However, the collective effect of coherent laser photons produces large oscillatory electric and magnetic fields, which couple to electrons. The interaction of these electrons, quivering in the laser field, with other electrons, ions, or collective waves in a plasma, can heat the plasma. Laser-heated plasmas are used to drive ablation shock waves to compress and heat deuterium or tritium fuel to fusion temperatures. The $d(d,n)^3$He fusion reaction occurs when the deuterium ion kinetic energy exceeds approximately 10 keV, at which point colliding deuterons overcome their Coulomb barrier and can fuse.

Ultrahigh-intensity lasers. The development of ultra-intense lasers in the last decade has opened new opportunities for inducing nuclear reactions with lasers. This has been made possible by use of the chirped pulse amplification (CPA) technique, which involves stretching a laser pulse in time by many orders of magnitude, amplifying it to high energy but below the critical intensity for the onset of nonlinear phenomena in the optical system, and then compressing the pulse in time to extremely high intensity. Lasers have been built which achieve power levels of up to 1000 terawatts (1 petawatt, or 10^{15} W), and many lasers having several terawatts or more are in routine use worldwide. (To give some idea of the size of these quantities, the net electric power capability of the United States in 1999 was 0.78 TW, and the total solar radiant energy flux incident on a perpendicular area of 691,000 km^2 or 266,800 mi^2, equal to that of Texas, is 0.945 PW.) If a petawatt laser pulse is focused to a near-diffraction-limited spot, optical intensities up to 10^{21} W/cm^2 are possible. At this intensity, the electric field **E** of the laser approaches 10^{14} V/m, and the force, **F**, on an electron charge, e, is given by the Lorentz force law, $\mathbf{F} = e\mathbf{E} + \mathbf{v} \times \mathbf{B}$, where the magnetic term (that is, $\mathbf{v} \times \mathbf{B}$, the vector product of electron velocity **v** and magnetic flux density **B**) is equal in magnitude to the electric field. The motion of electrons in the laser field is fully relativistic. In a single optical cycle, a quivering electron is accelerated to greater than 10 MeV, which is sufficient to initiate nuclear reactions. Collective interactions in a relativistic laser plasma, such as inverse bremsstrahlung, result in a broad spectrum of energies of electrons expelled from the laser focus, having an approximately exponential distribution with a logarithmic slope of up to 10 MeV.

Photonuclear reactions. High-energy electrons accelerated directly by ultra-intense lasers to energies above the threshold for nuclear excitation can interact with nuclei through the electromagnetic force

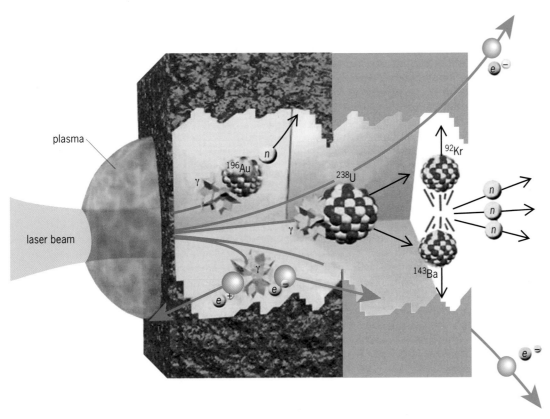

Fig. 1. Interaction of an ultra-intense laser beam with a solid target of gold and uranium. (*Courtesy of S. Wilks*)

via virtual or real photons. Most common is the interaction via real, bremsstrahlung photons, where the laser-generated electrons are first inelastically scattered in the field of a nucleus and emit a high-energy bremsstrahlung photon, which is then absorbed by a second nucleus, promoting it to an excited quantum-mechanical state. The nuclear de-excitation can proceed through many channels, including the emission of gamma rays, neutrons, or other nuclear particles, or by undergoing fission. Such photonuclear reactions have been widely studied using energetic electrons from conventional accelerators; however, ultra-intense laser matter interactions represent a new technique for production of photonuclear reactions, with somewhat different characteristics. For example, the pulse duration of an ultra-intense laser, and the corresponding duration of the electron and secondary bremsstrahlung radiation are similarly subpicosecond.

Figure 1 shows a schematic view of the laser interaction with a solid target of gold and uranium. The incident laser heats and produces a plasma, in which electrons are violently accelerated. As the electrons propagate through the material, hard bremsstrahlung photons are generated which liberate neutrons in (γ, n) reactions in gold, and initiate fission of the uranium nuclei. These processes, and many other photonuclear processes, have been observed experimentally. An example of the induced radioactivity in a laser target is shown in **Fig. 2**.

Ion acceleration. High-intensity lasers can also induce nuclear reactions by accelerating ions to high energy. These ions initiate reactions when they collide with other nuclei. This is analogous to the laser-plasma interaction acting as an ion accelerator. There are several mechanisms by which lasers can accelerate ions. Most are based upon electrostatic acceleration of the ions by an electric field, which is set up in the plasma by the laser displacing a large quantity of electrons from the initially stationary ions. On the time scale of the laser oscillation, electrons can be driven away from the neutralizing ions in the plasma. The resulting charge separation results in a strong electric field, whose magnitude can be as high as the average kinetic energy of the electrons, kT_e (where T_e is the temperature of the electrons and k is the Boltzmann constant) divided by the product of the electron charge e and the electrostatic shielding distance in the plasma. This distance is the Debye length, $\lambda_d = \sqrt{\varepsilon_0 kT_e/ne^2}$, where n is the number of electrons per unit volume and ε_0 is the permittivity of vacuum. For highly relativistic plasmas, $kT_e/e\lambda_d$ can be of order 10^{12} V/m or larger, and can be maintained for a sufficient time to accelerate the background ions to several megaelectronvolts per atomic mass unit (MeV/amu).

Figure 3 represents schematically some of the principal ion acceleration processes for an ultra-intense laser pulse incident on a thin plastic foil. A hot plasma is formed on the front surface of the foil due to electron and ion heating by the laser (region I). The Debye length of the hot electrons induces a charge-separation electric field at the surface of this plasma which drives its outward

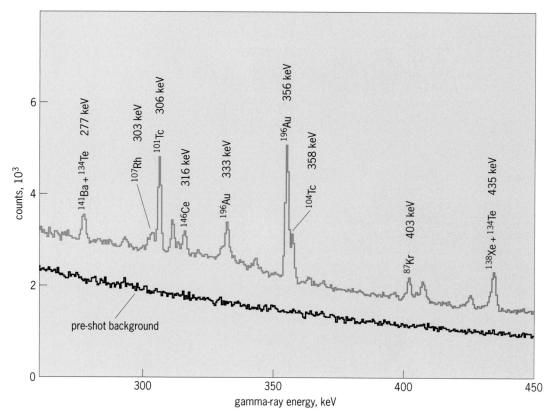

Fig. 2. Example of induced radioactivity (gamma radiation) in the target of an ultra-intense laser beam. Radioactive nuclides and gamma-ray energies are identified.

expansion, thus accelerating the plasma ions. At the critical density surface, where the plasma oscillation frequency equals the laser frequency, the laser light is absorbed. The laser drives electrons into the target plasma, creating a charge separation, which accelerates ions in the direction of the laser, into the target foil (region II). A third mechanism is unique to high-intensity laser interactions, in which the relativistic, several-megaelectronvolt electrons penetrate the target foil and set up an electron plasma sheath on the rear, nonirradiated surface of the target foil (region III). The strong electric field in the plasma

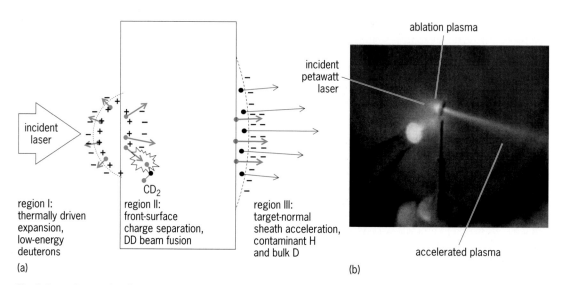

Fig. 3. Laser-ion acceleration mechanisms for a short-pulse laser striking a deuterated polyethylene (CD_2) foil. (*a*) Low-energy (a few MeV) deuterons and carbon ions are ejected with the front-side ablation plasma (region I). MeV deuterons are accelerated into the target, where they can undergo D-D beam fusion (region II). The rear-surface sheath field ionizes hydrogen-containing contaminants and accelerates protons to about 50 MeV, and deuterons from the bulk to lower energies (region III). (*b*) Petawatt laser shot on thin foil. Front-side ablation plasma and rear-side plasma beam are visible.

sheath, up to 10^{12} V/m, can ionize atoms from the rear surface of the foil, and accelerate these ions up to energies of several tens of MeV/amu. Protons of up to 50 MeV have been observed by this process in petawatt laser irradiation of gold and plastic foils.

Charged-particle nuclear reactions. These various ion acceleration mechanisms have been observed experimentally by the spectrum of ions ejected from the target and by charged-particle nuclear reactions induced by these particles. Several experiments have irradiated deuterated plastics (containing CD_2) and observed the acceleration of deuterons into the target material, where they undergo $d(d,n)^3$He fusion reactions with stationary deuterons. These so-called beam fusion reactions are distinct from thermonuclear fusion in which a collection of deuterium must be heated to ion temperatures above about 10 keV. Interactions of laser-accelerated protons, deuterons, and carbon ions with various targets have yielded a variety of nuclear reactions. For example, proton-induced reactions in titanium, producing long-lived isotopes of vanadium, ^{48}Ti$(p,n)^{48}$V, are used to diagnose the total yield and energy spectrum of protons accelerated in petawatt laser-matter interactions.

Coulomb explosion of clusters. Another recent advance has been in driving fusion reactions by the Coulomb explosion of small clusters of atoms. Clusters are small assemblies of atoms, for example, about 100 to 1000 atoms of deuterium, held together by van der Waals forces. If they are irradiated with high-intensity laser light, the electrons can be stripped away from the atoms making up each cluster very rapidly, leaving a positively charged sphere of naked ions. Their mutual Coulomb repulsion from the net positive charge distribution accelerates the ions radially outward, to energies of several tens or hundreds of kiloelectronvolts. If a mist of such clusters is irradiated, the Coulomb explosion of many nearby clusters results in an almost isotropic distribution of energetic ions. Ions from neighboring clusters can collide and undergo fusion reactions if their mutual kinetic energy is sufficient to exceed the repulsive $\sim$10-keV Coulomb barrier. Experiments on short-pulse laser irradiation of deuterium clusters have exhibited intense bursts of deuterium-deuterium fusion neutrons, which indicate a very efficient coupling of laser energy to the clusters. This is understood in terms of a resonant interaction of the laser light with the expanding cloud of electrons, when the electron density drops to three times the critical density of the laser. Cluster explosions offer the possibility of an efficient source of fusion neutrons, with the added feature that the source duration is very short, subnanosecond, which could allow time-resolved studies of neutron interactions with matter.

Prospects. The generation of intense beams of electrons, ions, and neutrons holds great promise for high-intensity laser interactions to contribute to many fields of science and technology that presently involve the use of particle accelerators. Laser particle and radiation sources can be simultaneously extremely intense and very short in duration, bringing new characteristics to the study of the interaction of particles with matter which may not be accessible with conventional accelerators. This would allow, for example, the time-resolved study of the dynamical processes involved in the interaction of ions with matter.

In addition, ion acceleration by the sheath mechanism holds great promise for the development of controlled beams of protons or ions. The quality of a proton beam accelerated in this way is comparable or better than that of conventional radio-frequency accelerators. In principle, if the target foil is appropriately shaped, the accelerated ion cloud could be highly collimated, or focused to extremely high intensity in a micrometer-scale spot. Whereas radio-frequency technology is quite mature, high-intensity laser technology is still undergoing tremendous development, and it is credible to assume that in several years much higher efficiency, diode-pumped solid-state lasers could be available, which would allow true "table-top" nuclear physics applications. These developments could provide more compact, more portable, cheaper, or brighter sources of nuclear particles and radiation for a host of applications, such as production of radiopharmaceuticals, portable neutron generators for industrial radiography, fundamental materials research, and research on inertial confinement fusion.

For background information *see* ATOM CLUSTER; BREMSSTRAHLUNG; COULOMB EXPLOSION; LASER; NUCLEAR FISSION; NUCLEAR FUSION; NUCLEAR REACTION; OPTION PULSES; PARTICLE ACCELERATOR; PLASMA (PHYSICS); RADIOACTIVITY in the McGraw-Hill Encyclopedia of Science & Technology.

Thomas E. Cowan

Bibliography. T. E. Cowan et al., Photonuclear fission from high energy electrons in ultra-intense laser-solid interactions, *Phys. Rev. Lett.*, 84:903, 2000; T. R. Ditmire et al., Table-top nuclear fusion with laser-heated exploding deuterium clusters, *Nature*, 398:489, 1999; K. W. D. Ledingham et al., Photonuclear physics when a multiterawatt laser pulse interacts with solid targets, *Phys. Rev. Lett.*, 84:899, 2000; M. D. Perry and G. Mourou, Terawatt to petawatt subpicosecond lasers, *Science*, 264:917, 1994; R. A. Snavely et al., Intense high energy proton beams from petawatt laser irradiation of solids, *Phys. Rev. Lett.*, 85:2945, 2000.

Lattice models

The most fundamental entity in nature is the point-like particle. All microscopic phenomena known today are explained by interactions among several kinds of particles. When the interactions are weak it is possible to make quantitative predictions by direct methods, but when they are strong it is necessary to revert to lattice models that can be simulated on the computer. Recently a way was discovered to incorporate chirality, a fundamental property of particles, into lattice models. This development extended vastly the reach of lattice models and may even expose the mechanism by which chirality is realized in nature.

Spin. Particles possess a fundamental property called spin. Spin is a quantum property: It is quantized, taking on only values that are an integral or a half-integral multiple of the universal constant $\hbar$ ($\hbar$ is Planck's constant divided by 2π; $\hbar = 1.05457 \times 10^{-34}$ J $\cdot$ s). All of matter is made out of particles whose spin equals $\hbar/2$ (or "spin $^1/_2$"). These matter particles interact with each other through the exchange of force-carrier particles. Almost all force carriers have spin $\hbar$ (or "spin 1").

Relativistic quantum fields. The laws governing the interactions of spin-0, $^{-1}/_2$, and -1 particles are restricted by the combination of the principles of special relativity and quantum mechanics. It is known that up to some energy Λ only a finite number of particle types exist. (We know, experimentally, that Λ is greater than 100 GeV, roughly 100 times the mass of the proton.) The structure of the equations governing the interactions between these particles is completely determined by some simple patterns into which the particles fit. Although these equations are not totally correct, their predictions are numerically very accurate, as long as the energies for which the predictions are made are small relative to Λ. These equations are partial differential equations for various fields. Each field is a function of space-time associated to a specific type of particle. The equations are relativistic: Observers moving relative to each other at constant speeds should use the same set of equations and would agree on the predictions for a jointly observed experiment.

Dirac and Weyl equations. Fields representing spin-$^1/_2$ particles obey the Dirac equation. Thus, all of matter is governed by the Dirac equation. Dirac fields are the single category of fields that enter all equations only linearly, never raised to a power higher than 1. Dirac's equation is in general able to describe both massive and massless spin-$^1/_2$ particles. The patterns into which spin-$^1/_2$ particles fall in nature, however, restrict this mass to be zero at the fundamental level. The property responsible for this restriction is called chirality.

The spin of a single, massless, spin-$^1/_2$ particle is a directional quantity that can point in either the same direction as the particle moves or in the opposite direction. A massless particle moves with the velocity of light, so all observers would agree on which way the spin is pointing. If the spin is parallel to the motion, the particle is called right-handed; if it is antiparallel, the particle is called left-handed. For massless spin-$^1/_2$ particles the Dirac equation breaks into two separate and simpler sets, sometimes referred to as Weyl equations. These more elementary equations, and the Weyl fields they act on, are the correct description of matter at the most fundamental level known today. Weyl fields are chiral; namely, they describe spin-$^1/_2$ particles of definite handedness. A

Dirac particle corresponds to a pair of Weyl particles of opposite handedness. A consequence of the chiral nature of the Weyl particles that describe our world is that there is no symmetry under parity: If we looked at our world as reflected in a mirror we could tell, just from the way various processes would proceed, that the world in the mirror is different from ours.

Dimensional analysis. The set of equations governing matter and forces make accurate predictions for energies E that are much smaller than Λ. When the number E/Λ is much smaller than 1 the theory is useful, albeit not perfect. When the energies are close to Λ the theory is no longer predictive. At these high energies a new theory would be needed, valid to another energy, Λ', which is much larger than Λ. When Λ increases and E is kept fixed (with Λ much larger than E), the theory should depend on Λ only weakly. The basic restriction that this implies is derived by dimensional analysis. This analysis is most convenient in a special system of units. In this system of units the velocity of light is set to 1 and $\hbar$ is also set to 1. As a result, length, time, inverse energy, inverse mass, and inverse momentum are all measured using the same unit. The typical theory has some free parameters, and it is possible to systematically expand in these free parameters if they are dimensionless. Some of the parameters could be built up from dimensional quantities. Such a quantity is a typical energy, E_0, and if E_0 is much smaller than Λ one can expand in the dimensionless combination $E_0/\Lambda \ll 1$.

Anomaly cancellation. The theory describing nature is unique, but many simplified models can be constructed, which contain only a subset of the full theory. The study of such models has revealed that it is, indeed, possible to maintain insensitivity to Λ in most cases. However, when Weyl fields are necessary, and do not pair up into Dirac fields, additional complications occur when quantum effects are taken into account. These complications resolve themselves only when a particular condition is met. When this condition holds, one says that anomalies cancel. Anomaly cancellation saves the theory to any finite number of steps in the expansion. However, it seemed impossible to ascertain that anomaly cancellation also saves the theory when the expansion is summed up. During the 1990s significant progress was made: It now seems likely that anomaly cancellation really saves the theory from a strong dependence on Λ. This significant progress involves developments in so-called lattice field theory.

Lattice field theory. Lattice field theory is the only reliable tool to deal with a theory as a whole rather than expand in its parameters. The basic approach is very direct: The space-time continuum is replaced by a discrete and finite grid. Then the full equations can be analyzed without any further approximations, using computers. No assumption about the size of the dimensionless parameters needs to be made. Being interested only in energies small relative to Λ is equivalent to only considering observations over time differences Δt and spatial separations Δx

much larger than the length $1/\Lambda$. The grid is made dense by choosing the lattice spacing, a, equal to $1/\Lambda$. Thus, any Δt, Δx of interest are guaranteed to be very large relative to a and the lattice model differs significantly from the true theory only at energies Λ, where the true theory itself no longer is valid.

Conceptual difficulties. However, the first attempts to define Weyl fields on the lattice met with an unpleasant surprise. No matter what was tried, each desired Weyl particle ended up being paired up with an undesired one, producing a Dirac particle. The chirality properties of the true theory could not be reproduced on the lattice. This was interpreted by a small number of physicists as an indication that something was fundamentally wrong with the chiral theories that are used to describe nature. However, the theory describing nature was working too well to be discarded, and the majority of elementary particle physicists chose to ignore the lattice difficulties. Still, it was just not logically acceptable that Weyl fields could not fit on the lattice. If the general philosophy was right, either the lattice had to have some way to accommodate Weyl fields or something was wrong with the basic logic of the entire approach to describing nature at short distances.

Lattice models with chirality. The solution is relatively simple but requires the addition of an unexpected ingredient. For every single Weyl field in nature, one introduces an infinite set of lattice Weyl fields. It is a subtle and important point to make the infinity an odd infinite number—this is perfectly well-defined mathematically. The construction is feasible because the Weyl fields enter the equations only linearly. The infinite set of fields acts as if it consists of one massless Weyl field together with an infinite number of pairs of Weyl fields. Each pair makes up a Dirac field and is given a mass of order Λ. So, strictly speaking, we have the right number of particles with masses small relative to Λ. But we cannot really eliminate the infinite number of extra heavy fields, except in a particularly simple limit, when parity is restored. This limit provides a lattice model for strong interactions, which is closer to nature than older models because it preserves chiral symmetries.

Many researchers are developing numerical and computational techniques for implementing the new chiral spin-$\frac{1}{2}$ particles on the lattice. Doing this successfully and efficiently would be useful to the large numerical project of solving the theory of strong interactions. In isolation, strong interactions can, in principle, also be solved without the new Weyl fields, but the task is technically daunting. The new developments on chiral particles, together with steady progress in computation, bring the goal of solving strong interactions closer than ever.

Implications. A tantalizing but speculative possibility is that the embracing theory provided by nature, just as the lattice model, also employs a truly infinite number of extra-heavy Dirac particles accompanying every type of Weyl particle. This new infinity could reflect one or more new dimensions, beyond the four we know. From our four-dimensional perspective, the Weyl fields representing particles that explore the entire higher-dimensional space appear as having an infinite multiplicity.

For background information *see* ELEMENTARY PARTICLE; PARITY (QUANTUM MECHANICS); QUANTUM CHRONODYNAMICS; QUANTUM FIELD THEORY; QUANTUM MECHANICS; RELATIVISTIC QUANTUM THEORY; RELATIVITY; SPIN (QUANTUM MECHANICS); STANDARD MODEL; SYMMETRY LAWS (PHYSICS); WAVE EQUATION in the McGraw-Hill Encyclopedia of Science & Technology.   Herbert Neuberger

Bibliography. M. Creutz, Aspects of chiral symmetry on the lattice, *Rev. Mod. Phys.*, 73:119–150, 2001; H. Neuberger, Exact chiral symmetry on the lattice, *Annu. Rev. Nucl. Part. Sci.*, in press; Proceedings of the 17th International Symposium on Lattice Field Theory, *Nucl. Phys. B (Proc. Suppl.)*, pp. 83–84, 2000.

Lava flow

The flow of lava across the surface of the Earth and other planetary bodies is governed by the rheology (flow behavior) of the melt and the surface conditions present during eruption. High-resolution images of Venus, Mars, and Io (satellite of Jupiter) acquired during the past decade clearly show that lava flow is perhaps the most important process in the creation of rocky planetary surfaces. Detailed studies of the processes involved in lava flow are needed to better understand the evolution of our neighboring planets and to better mitigate volcanic hazards on the Earth.

Lava flows are complex currents of melt, solids, and volatiles that span a wide range of flow regimes (such as laminar and turbulent flow). The delicate interplay of flow regime and heat transfer to the surrounding environment governs flow morphology at both large and small scales. Flow regime and heat loss are symbiotic partners that ultimately determine the length, width, and thickness of a flow, and the fine-scale appearance of the lava surface.

Lava rheology. Lava is a mixture of partially molten silicate minerals, solid minerals, dissolved and exsolved volatiles, and pieces of solid rock plucked from the surrounding environment called xenoliths. Each of these components has a significant effect on lava rheology.

The rheological behavior of the melt component of lava is dependent upon the silicate melt structure. Silica contents of typical lavas range from approximately 50% for basaltic (mafic) melts to nearly 70% for rhyolitic (silicic) melts. The degree to which silica polymerizes in a melt greatly affects the lava viscosity. Rhyolitic melts tend to have greater polymerization and much higher viscosities than their basaltic counterparts; the viscosity of rhyolitic melt with 2% water at a temperature of 1000°C (1832°F) is 3×10^6 pascal seconds, approximately four orders of magnitude (10^4) higher than a basaltic melt under the same conditions. (For comparison, water has a viscosity of approximately 10^{-3} Pa · s.)

Volatiles, both dissolved in the melt and exsolved into vesicles (gas bubbles), also affect rheology. Water is the most abundant volatile species in silicate melt, and when dissolved it destroys Si-O bonds, leaving the lava less polymerized (lower viscosity). A rhyolitic melt at 800°C (1472°F) with 0.1% water would have a viscosity of approximately 10^{10} Pa · s, whereas increasing the water content to 5% would cause a viscosity decrease of five orders of magnitude (10^5 Pa · s). Basalt exhibits a similar but lower-magnitude decrease in viscosity at higher water contents.

The exsolution of volatiles may also change the phase relations in a melt by increasing the liquidus temperature, resulting in a viscosity increase through the crystallization of minerals and formation of vesicles that retard movement of the fluid. The viscosity of lava is affected by the volume fraction of gas bubbles, crystals, xenoliths, or pieces of cooled surface crust entrained in the lava during turbulent flow, the effect of which is generally modeled by the Einstein-Roscoe relation $\eta/\eta_0 = (1 - R\phi)^{-2.5}$, where η is the viscosity with these items present, η_0 is the viscosity with no bubbles or solids present, ϕ is the volume fraction of bubbles or solids, and R is taken as 1.67 for lavas. For example, a void fraction of 2.2% ($\phi = 0.022$) will increase viscosity by 10%.

As lava is first erupted onto a planetary surface, radiation is the primary mechanism of heat loss to the atmosphere and into the rock substrate over which the lava flows. Rapid cooling results in the formation of upper and lower thermomechanical crusts that gradually thicken as heat moves, by conduction, outward from the flow interior. Heat then exits the flow through conduction to the rock below or to the upper surface where it dissipates through atmospheric convection. The lowering of internal lava flow temperatures, in turn, affects the viscosity of the flow through greater melt polymerization and as crystallization and volatile exsolution create a larger volume fraction of minerals and vesicles. The presence of a cooled crust complicates efforts to model lava rheology because such crust promotes flow by insulating the hot flow interior while providing a strong carapace that resists flow. Thus, the rheological behavior of lava as it flows across the surface is time-dependent and complex, presenting a number of challenges in modeling these systems.

Flow regimes and flow morphology. Because of their extremely high viscosities, silicic lava flows exhibit laminar flow, with Reynolds numbers (Re) in the 10^{-10} to 10^{-4} range (Re $= \rho Vd/\eta$, where ρ is density [M/L^3], V is flow velocity [L/T], d is flow depth [L], η is viscosity [$M/(LT)$], and M, L, and T stand for the dimensions of mass, length, and time). These flow conditions create lava domes, which are generally circular in planform, with thicknesses ranging from tens to hundreds of meters. Flows that create silicic lava domes move extremely slowly (<0.01 m/s or 0.03 ft/s), allowing ample time for cooling and the formation of thick, brittle surface crusts during emplacement. The brittle crusts are broken as flow proceeds, resulting in a blocky, glassy morphology that dominates the flow surfaces (**Fig. 1**).

Fig. 1. Volcanologists studying the roughness characteristics of Obsidian Dome, a 500-year-old silicic lava dome in eastern California. The extremely high viscosity of these flows, combined with a strong confining crust, produces flows with extremely low Reynolds numbers (10^{-10} to 10^{-4}). The blocky surface is typical of most silicic domes, and shows the influence of a thick, cool, brittle crust on the surface morphology. (*Photo by Steve Anderson*)

Fig. 2. Exploitation of an active aa basalt flow at Mount Etna, Italy. Molten lava is quickly removed and fashioned into ashtrays for the many tourists that visit this volcano. This surface morphology is influenced by the laminar flow regime, viscosity, and strain rate. (*Photo by Steve Anderson*)

The relatively low viscosity of basaltic lava results in higher Reynolds numbers and different flow morphologies. In Hawaii, the laminar (Re < 2000) flow of basaltic lava traveling at rates of 0.1–10 m/s (0.3–33 ft/s) and resulting Re of 1–100 yields several different morphologies, depending on the ratio of the lava viscosity to strain rate experienced during emplacement. Basalt with a slightly higher viscosity (perhaps due to cooling, crystallization, or vesiculation) combined with relatively high strain rates (such as those encountered as the lava traverses steep slopes) results in a rough, clinkery morphology called aa lava (**Fig. 2**). Lower-viscosity basalts combined with low strain rates produce a smooth, ropy texture referred to as pahoehoe lava (**Fig. 3**).

Creating basaltic lava flows that display transitional (Re = 500–2000) or turbulent (Re > 2000) behavior is problematic, given the viscosity of the liquid. Few examples of modern turbulent lava flows are available, although turbulent flow is suspected as the dominant flow regime during the emplacement of high-volume komatiite lava flows during the Archean Period (>2.5 billion years ago). Komatiites are similar to basalts in chemistry, but have a slightly lower silica content and were perhaps 200°C (390°F) hotter than present-day basalts. Therefore, they had very low viscosities (~1 Pa · s) which, when combined with their high eruption rates and resultant velocities (>10 m/s or 33 ft/s), produced Reynolds numbers as high as 10^6. This turbulence would produce a well-mixed flow that would not have the vertical viscosity and temperature gradients that dominate the flow behavior of modern basaltic and silicic flows. Thus, the modeling of long komatiite flows requires a slightly different approach than that taken in modeling other lava flows. Although komatiites have not erupted on the Earth for over 2 billion years, lava flows of similar composition and behavior may be abundant on the Moon, Venus, Mars, and Io. Modeling the processes at work during the emplacement of these low-viscosity fluids is critical in understanding the evolution of other planetary surfaces.

Bingham behavior of lava flows. Many lava flow studies are prompted by a desire to predict the distance that a lava flow might travel under a given set of conditions. An understanding of how lava rheology and cooling are influenced by the forces that cause lava movement is needed to formulate models that could allow prediction of flow length. Lava rheology is typically approximated by the Bingham flow law, $\sigma = \sigma_0 + \eta\varepsilon$, that relates shearing stress (σ) and strain rate (ε) as a function of plastic viscosity (η) and yield stress (σ_0). Numerous studies have shown that the assumption of Bingham behavior (linear viscous behavior above a specified yield stress) allows for a more in-depth understanding of flow morphology, although these successes are tempered by the realization that time-dependent processes, such as vesiculation, crystallization, and cooling, complicate rheologic behavior to such an extent that flow length still cannot be predicted with any reasonable certainty. Modeling of these systems is further complicated as recent laboratory studies of lava deformation show that many rheological parameters, such as viscosity and yield strength, are not only time- and cooling rate-dependent but also strain rate-dependent.

Wax modeling. In an attempt to better understand the relationship between flow regime and flow

Fig. 3. Active pahoehoe lava flow at Kilauea, Hawaii. Each bulbous pahoehoe "toe" is approximately 0.3 m (1 ft) wide and <0.1 m (0.3 ft) thick. This lava is subjected to laminar flow conditions, and has essentially the same chemistry as the material erupted at Mount Etna. However, the morphology of the surface is much different, owing to a slightly lower viscosity and lower strain rate. (*Photo by Steve Anderson*)

morphology, some volcanologists have turned their attention to the laboratory. Several recent studies involve the extrusion of polyethylene glycol wax (PEG) into a cold sucrose solution where rheology, flow regime, and cooling conditions can be carefully controlled. These experiments allow volcanologists to investigate a wide range of eruptive behaviors and flow regimes, which provide a mechanism for mimicking the effects of other planetary environments. In a study of the effect of cooling on lava flow morphology, J. H. Fink and R. W. Griffiths (1998) found that flow morphology could be successfully predicted for natural and laboratory flows by using a dimensionless parameter ψ_b, which is the ratio of the time necessary for Bingham materials (fresh lava or PEG) exposed at the surface to reach its solidification temperature t_s to the time needed for viscous flow to advance a distance equal to its thickness t_a. These studies open a new avenue that holds great promise for understanding the complexities of lava flow.

For background information *see* BASALT; CONTINENTS, EVOLUTION OF; LAMINAR FLOW; LAVA; NON-NEWTONIAN FLUID; REYNOLDS NUMBER; RHEOLOGY; RHYOLITE; SILICA; TURBULENT FLOW; VISCOSITY; VOLCANO; VOLCANOLOGY; XENOLITH in the McGraw-Hill Encyclopedia of Science & Technology.

Steven W. Anderson

Bibliography. S. W. Anderson et al., Block size distributions on silicic lava flow surfaces: Implications for emplacement conditions, *Geol. Soc. Amer. Bull.*, 110:1258–1267, 1998; J. H. Fink and S. W. Anderson, Lava domes and coulees, in H. Sigurdsson (ed.), *Encyclopedia of Volcanoes*, pp. 307–319, Academic Press, 1999; R. W. Griffiths, The dynamics of lava flows, *Annu. Rev. Fluid Mech.*, 32:477–518, 2000.

Lean manufacturing

Lean manufacturing is a unique linked-cell manufacturing system initiated in the 1960s by the Toyota Motor Company. Known also as the Toyota Production System (TPS), the Just-in-Time/Total Quality Control (JIT/TQC) system, or World Class Manufacturing (WCM) system, in 1990 it was given a name that would become universal, "lean production." This term was coined by John Krafcik, an engineer working in the International Motor Vehicle program at MIT with J. P. Womack, D. Roos, and D. T. Jones, who observed that this new system used less of the key resources needed to make goods. What is different about the system is its use of manufacturing cells linked together with a functionally integrated system for inventory and production control. The result is low cost (high efficiency), superior quality, and on-time delivery of unique products from a flexible system.

Mass versus lean production. The key proprietary aspects of lean production are U-shaped manufacturing and assembly cells, using walking workers and

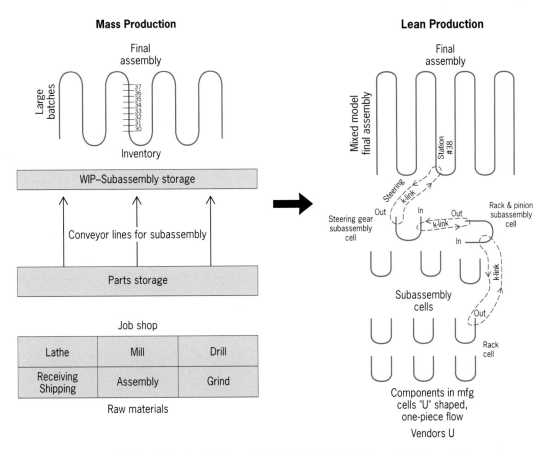

Mass Production

Final assembly

Large batches

37
36
35
34
33
32
31
30

Inventory

WIP–Subassembly storage

Conveyor lines for subassembly

Parts storage

Job shop

Lathe	Mill	Drill
Receiving Shipping	Assembly	Grind

Raw materials

Lean Production

Final assembly

Mixed model final assembly

Station #38

Steering
k-link

Out   In   Out
k-link

Steering gear subassembly cell

Rack & pinion subassembly cell

In

k-link

Subassembly cells

Out

Rack cell

Components in mfg cells "U" shaped, one-piece flow

Vendors U

Fig. 1. Restructuring of a job shop/flow shop final assembly design into a linked-cell manufacturing system.

are designed with system requirements in mind. Design decoupling allows the separation (decoupling) of processing times for individual machines from the cycle time for the cell as a whole, permitting the lead time to make a batch of parts to be independent of the processing times for individual machines. This takes all the variation out of the supply chain lead times, so scheduling of the supply system can be greatly simplified. The supply chain is controlled by a pull system of production control called Kanban. Using Kanban, the inventory levels can be dropped, which decreases the throughput time for the manufacturing system. *See also* SUPPLY CHAIN MANAGEMENT.

Figure 1 shows an entire mass production factory reconfigured into a lean manufacturing factory. The final assembly lines are converted to mixed model final assembly; this levels the demand for subassemblies and other components. The rate of production is determined by recalculating the monthly demand into a daily demand and trying to make the same product mix every day.

The subassembly lines are reconfigured into U-shaped cells. The daily output from these cells is balanced to match the demand from final assembly.

The traditional job shop generates a variety of unique products in low numbers with its functional design. It is redesigned into U-shaped manufactur-

ing cells that produce families of component parts (**Fig. 2**).

Lean manufacturers focus on sole-sourcing each component or subassembly (that is, they do not have multiple vendors supplying the same components), sharing their knowledge and experience in linked-cell manufacturing with their vendors on a one-to-one basis. For lean automobile manufacturers, the final assembly plant may have only 100 to 400 suppliers, with each supplier becoming a lean or JIT vendor to the company. In the future, the number of vendors supplying a lean manufacturer will decrease even more. The Mercedes-Benz plant in Vance, Alabama, has around 80 suppliers. The subassemblies contain more components as the vendors take on more responsibility for the on-time delivery of a larger portion of the auto.

Lean cell design. In a true lean manufacturing system, manufacturing processes and equipment are designed, built, tested, and implemented into the manufacturing cells. The machine tools and processes, the tooling (workholders, cutting tools), and the material-handling devices (decouplers) are designed specifically for cellular manufacturing. Simple, reliable equipment that can be easily maintained should be specified. In general, flexible, dedicated equipment can be built in-house better than it can be purchased and modified for the needs of the cell. However, many plants that lack the expertise to build

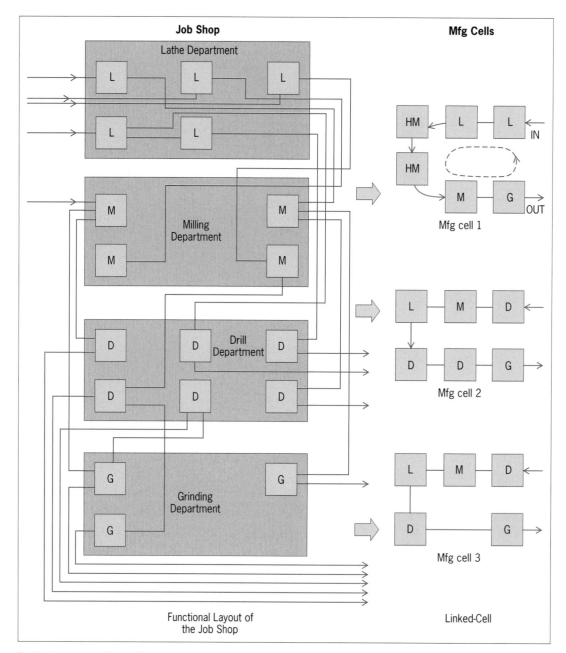

Fig. 2. Job shop portions of the factory require a system level conversion to redesign them into manufacturing cells.

machines from scratch do have the expertise to modify equipment to give it unique capabilities, and this interim cell approach can lead a company into true lean manufacturing cells.

Many companies understand that it is not good strategy simply to imitate manufacturing process technology from another company and then expect to make an exceptional product. When process technology is purchased from outside vendors, any unique aspects will be quickly lost. Companies must carry out research and development on manufacturing technologies and systems in order to produce effective and cost-efficient products, but the result makes such investment pay off. There are many advantages to this home-built equipment strategy.

Flexibility (process and tooling adaptable to many types of products). Flexibility requires rapid changeover of jigs, fixtures, and tooling for existing products and rapid modification for new designs. The processes have excess capacity; they can run faster if necessary, but they are designed for less than full-capacity operation.

Building to need. There are three aspects to this. First, there is no unused capability or options. Second, the machine can have unique capabilities that competitors do not have and cannot get access to through equipment vendors. For example, in the lean cell shown in **Fig. 3** is a broaching (cutting) machine for producing the gear teeth on a rack bar. The angle that these teeth make with the bar varies for

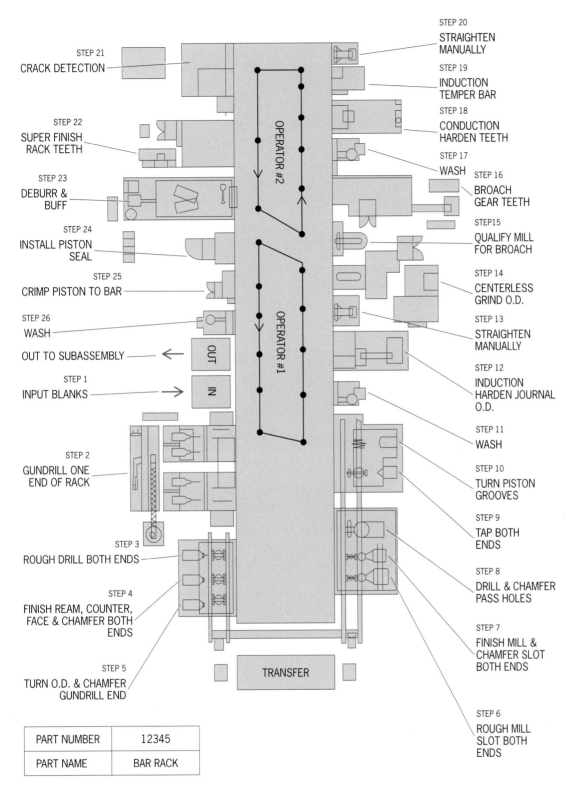

Fig. 3. Lean manufacturing cell for producing rack bars for a rack-and-pinion steering gear.

different types of racks. Job shop broaching machines are not acceptable for this cell because of their long changeover times, so a unique machine tool for broaching was designed and built using a proprietary process. Third, the equipment should allow the operator to stand and walk. Equipment should be the appropriate height to allow the operators to easily perform tasks standing up and then move to the next machine in a step or two (that is, the design should have a narrow footprint).

Built-in maintainability/reliability/durability. Equipment should be easy to maintain (to oil, clean, changeover, or replace worn parts, and to use standardized screws). Many of the cells at lean vendors are clones of each other. The vendor company, being the sole source, has the volume and the expertise to get business from many companies, making essentially the same components or subassemblies for many original equipment manufacturers (OEMs). The equipment can be interchanged from one cell to another in emergencies.

Machines, handling equipment, and tooling built for the needs of the cell and the system. Machines are typically single-cycle automatics but may have capacity for process delay. An example of process delay is a heat treatment process that takes 4 min in a cell with a 1 min cycle/time. If the heat treatment machine can hold four units, each getting 4 min of treatment, one unit is still produced per minute.

Safety. Equipment is designed to prevent accidents (fail-safe).

Equipment designed to be easy to operate, load, and unload. Research has shown that manufacturing cells have ergonomic benefits over the job shop. Toyota ergonomists recommend unloading with the left hand, loading with the right hand, while walking right to left. The wide variety of tasks over the cell cycle time keep the operators from developing cumulative trauma injuries.

Equipment designed to process single units, not batches. Small-area, low-cost equipment is the best. Machining or processing time (MT) should be modified so that it is less than the cycle time (CT), the time in which one unit must be produced. Equipment processing speed should be set in view of the cycle time, such that MT < CT. The MT is related to the machine parameters selected. For example, in a cutting operation this approach often permits the reduction of the cutting speed, thereby increasing the tool life and reducing downtime for tool changes. This approach also reduces equipment stoppages, lengthens the life of the equipment, and may improve quality.

Equipment that contains self-inspection devices (such as sensors and counters) to promote autonomation. Autonomation is the autonomous control of quantity and quality. Often the machine is equipped to count the number of items produced and the number of defects.

Equipment that is movable. Machines are equipped with casters or wheels, flexible pipes, and flexible wiring. There are no fixed conveyor lines.

Equipment that is self-cleaning. Equipment disposes of its own chips and trash.

Equipment that is profitable at any production volume. Equipment that needs millions of units to be profitable (R. J. Schonberger calls them supermachines) should be avoided, because if the production volume ever exceeds the maximum capacity that the first supermachine can build, it will be necessary to purchase another supermachine and the new one will not be profitable until it approaches full use. Schedulers of the equipment will typically divide the volume into two machines, so neither will be profitable.

Lean cell operation. The manufacturing cell shown in Fig. 3 typically uses two operators. These standing, walking workers move from machine to machine in counterclockwise (CCW) loops, each completing the loop in about 1 min. Operator 1 typically addresses 10 stations, and operator 2 addresses 11 stations. Most of the steps involve unloading a machine, loading another part into the machine, checking the part unloaded, and dropping the part into the decoupler (handling) elements between the machines. The stock-on-hand (SOH) in the decouplers and the machines helps to maintain the smooth flow of the parts through the machines. The decouplers also can be designed to perform inspections for part quality or necessary process delays while the parts heat up, cool down, cure, and so on. The stock-on-hand is kept as low as possible.

Sometimes the decoupler elements perform the inspection of the part, but mostly they serve to transport parts from one process to the next. The decoupler may also perform a secondary operation such as deburring or degaussing (removal of residual magnetic fields).

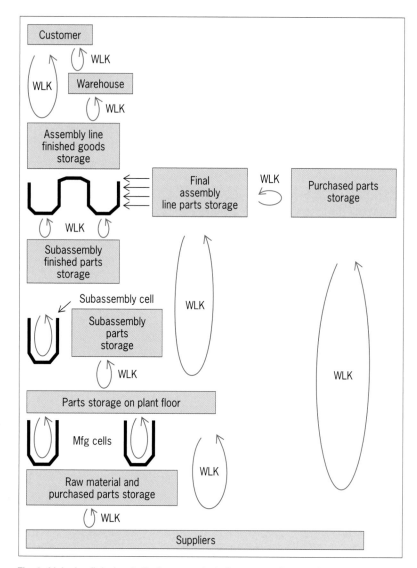

Fig. 4. Linked-cell design. In the lean manufacturing system, the manufacturing and assembly cells are linked to the final assembly area by Kanban (card) inventory links or loops. WLK indicates a withdrawal card.

By design, one operator controls both the input and output of the cell. One operator always controls the volume of material going through the cell. This keeps the stock-on-hand quantity constant and keeps the cell working in balance with the final or subassembly lines it is feeding.

At the interface between the two operators, either one can perform the necessary operations, depending on when they arrive and when the processes in the machines are finished. That is, the region where the two operators meet is not fixed, but changes or shifts depending upon the way parts are moving about the cell. This is called the relay zone. This added flexibility requires that the workers are cross-trained on all the processes in the cell.

Ergonomics of lean cells. Ergonomics deals with the mental, physical, and social requirements of the job, and how the job is designed (or modified) to accommodate human limitations. For example, the machines in the cell are designed to a common height to minimize lifting of parts, transfer devices are designed for slide on/slide off, and automatic steps equipped with interrupt signaling help workers monitor the process. When the job is primarily loading/unloading, ergonomic concerns regarding lifting and placing parts in machines and operating workholding devices must be addressed. Lean manufacturing cells are relatively free of cumulative trauma syndrome (CTS) problems because the operators' tasks and movements are so varied from machine to machine. Cell designers should try to incorporate ergonomic issues initially rather than trying to implement fixes later.

Human performance in detecting and correcting cell malfunctions will establish utilization and production efficiency. The design of machines for maintainability and diagnostics is critical. In manufacturing and assembly cells where workers operate machines, it is important that all the machines are ergonomically identical. Sewing machines in a cell are a good example: To the operator, all the machines should feel the same in terms of control.

Manufacturing system integration. Many believe that the only way in which manufacturing companies can compete is to automate. This approach, known as computer integrated manufacturing (CIM), was recently renamed agile manufacturing. The concept is to achieve integration through computerization and automation. This often results in trying to computerize, robotize, or automate very complex manufacturing and assembly processes, and it works only when there is little or no variety in the products. Lean manufacturing systems take a different approach—integrate the manufacturing system, then computerize and automate (IM, C). Experts on CIM now agree that lean manufacturing, especially development of manufacturing and assembly cells, must come before efforts to computerize the system. While costs of these systems are difficult to obtain, the early evidence suggests that the lean cell approach is significantly cheaper than the CIM approach.

Outlook. The lean factory is based on a different design for the manufacturing system in which the sources of variation in time are minimized and delays in the system are systematically removed. In the linked-cell manufacturing system, in which manufacturing and assembly cells are linked together with a pull system for material and information control (**Fig. 4**), downstream processes dictate upstream production rates. The linked-cell manufacturing system strategy simplifies the manufacturing system, integrates the critical control functions before applying technology (automation, robotization, and computerization), avoids risks, and makes automation easier to implement. This is the strategy that will predominate in the next generation of factories both in the United States and abroad.

For background information *see* AUTOMATION; COMPUTER-INTEGRATED MANUFACTURING; FLEXIBLE MANUFACTURING SYSTEM; INDUSTRIAL ENGINEERING; INVENTORY CONTROL; MANUFACTURING ENGINEERING; MATERIALS-HANDLING EQUIPMENT; PRODUCTION ENGINEERING; PRODUCTION METHODS; PRODUCTION PLANNING; QUALITY CONTROL; ROBOTICS in the McGraw-Hill Encyclopedia of Science & Technology.

J. T. Black

Bibliography. R. U. Ayres and D. C. Butcher, The flexible factory revisited, *Amer. Scientist*, 81:448-459, 1993; J. T. Black, Cell design for lean manufacturing, *Trans. NAMRI/SME*, 23:353-358, 2000; J. T. Black, C. C. Jiang, and G. J. Wiens, Design, analysis and control of manufacturing cells, *PED*, vol. 53, ASME, 1991; J. T. Black and B. J. Schroer, Simulation of an apparel assembly cell with walking workers and decouplers, *J. Manuf. Sys.*, 12(2):170-180, 1993; Y. Monden, *Toyota Production System*, Industrial Engineering and Management Press, IIE, 1983; S. Nakajima, *TPM, Introduction to TPM: Total Productive Maintenance*, Productivity Press, 1988; R. J. Schonberger, *Japanese Manufacturing Techniques: Nine Hidden Lessons in Simplicity*, Free Press, 1982; K. Sekine, *One-Piece Flow: Cell Design for Transforming the Production Process*, Productivity Press, 1990; S. Shingo, *A Study of The Toyota Production System*, Productivity Press, 1989; N. P. Suh, Design axioms and quality control, *Robotics and CIM*, 9(4/A):367, August–October 1992; J. P. Womack, D. T. Jones, and D. Roos, *The Machine That Changed The World*, Harper Perennial, 1991.

Leptin

Leptin is a 167-amino-acid protein that is important in the regulation of body weight, metabolism, and reproduction. It was discovered in 1994 when it was identified as the missing protein in mice with a spontaneous single-gene defect that caused obesity. These *ob/ob* mice were very obese, had many of the metabolic abnormalities associated with obesity, and were infertile. However, when leptin levels were examined in other animal models of obesity and in obese humans, the levels were found to be elevated and not low or absent (as in the *ob/ob* mice). Thus, general obesity (not due to leptin deficiency) has been termed a leptin-resistant state. Since these initial findings, the biology of leptin has proven to be more complex than originally thought.

Secretion. Leptin is secreted primarily from adipocytes (fat cells), although other cells in other locations in the body, including the placenta, stomach, and skeletal muscle, also make leptin. Leptin is secreted in a pulsatile fashion, and the average levels vary over the day (with the lowest levels in midmorning and the highest levels at night). The amount of leptin secreted by adipocytes is in proportion to the size of the cells. In animals, leptin levels increase soon after a meal and decrease with fasting. In humans, leptin levels do not change significantly with normal meals or with reasonable levels of exercise. However, with massive overfeeding, leptin levels will increase by up to 40%, and with fasts longer than 24 hours, leptin levels will decrease as much as 50%. Thus in humans, leptin appears to play a role in the more chronic regulation of body fat rather than the meal-to-meal changes in energy intake and energy expenditure.

Insulin requirement. Insulin is required for leptin secretion. In animals and in cell culture, insulin increases leptin expression. Likewise, leptin levels decrease when animals are made insulin-deficient. There is also a correlation between insulin levels and leptin levels in humans, but it is difficult to separate the confounding effect of increasing adipose tissue. When humans are deficient in insulin, leptin levels are low, but body fat is also decreased. However, when humans are given excess insulin for several hours (while keeping the blood glucose normal), most studies find no changes in leptin levels. Other hormones that increase leptin secretion include glucocorticoids (such as prednisone) and cytokines (such as tumor necrosis factor alpha and interleukin-1). Hormones that decrease leptin secretion include catecholamines (such as isoproterenol and epinephrine), testosterone, and perhaps thyroid hormone.

Gender difference. The leptin level in women is almost three times higher than the leptin level in men. One reason for this gender difference is that women have a higher percent of body fat than men. Additionally, women have more subcutaneous fat (they tend to be "pear-shaped") than men (they tend to have more abdominal fat and are "apple-shaped"). Subcutaneous fat produces more leptin than abdominal fat, thus contributing to this gender difference. Finally, testosterone, which is present in much greater concentrations in men, decreases leptin and further augments the gender difference.

Receptors. The primary role of leptin appears to be in decreasing food intake by way of its action in the brain. Leptin's actions are exerted by binding to specific receptors. There are at least six different splice variants (isoforms) of the receptor for leptin. All of these receptors contain a leptin-binding area, and all but one receptor span the cell membrane. However, only one receptor (the long-receptor iso-

form or Ob-Rb) has complete signaling function and is found predominantly in the hypothalamus (in areas known to control feeding behavior and hunger). The short-receptor isoforms are located mainly in peripheral tissues (such as the lung, kidney, liver, gonads) and the choroid plexus, but the current role of these short forms of the leptin receptor is still under investigation. They likely play a role in leptin transport in the blood and across the blood-brain barrier. At least part of the resistance to leptin associated with obesity might be due to decreased transport across the blood-brain barrier.

JAK-STAT signaling pathway. When leptin binds to the long form of the receptor, the receptor dimerizes (two receptors come together in combination with leptin) and several cascades are initiated by way of activation of the JAK (janus kinase)–STAT (signal transducers and activators of transcription) pathway (see **illus.**). This activation by JAK results in phosphorylation of tyrosine amino acids on the leptin receptor, which causes changes in gene expression (via STAT pathways that ultimately decrease appetite and increase energy expenditure), downstream phosphorylation and dephosphorylation of enzymes, alterations in ion channels, and other signal modulation (in part via SOCS-3, a suppressor of cytokine signaling). SOCS-3 can actually feed back to the leptin receptor and inhibit leptin signaling, and this may also play a role in the leptin resistance seen in obesity.

It has recently been shown that the effects of leptin on appetite occur through the STAT pathway, but the effects on maintaining reproductive hormones require only the JAK and not the STAT pathway. Additionally, activation of the leptin receptor can also activate pathways analogous to those of the insulin receptor, including the PI-3 kinase pathway (phosphatidyl inositol kinase pathway, which stimulates glucose uptake and other metabolic effects of insulin) and the MAP kinase pathway (mitogen-activated kinase pathway, which stimulates cell growth).

Regulation of body weight. Leptin is believed to regulate body weight through its effects on appetite and energy expenditure. When leptin is given to animals, there is a marked decrease in appetite and an increase in energy expenditure.

Neural pathways of appetite regulation. The decrease in appetite associated with leptin administration is due to the stimulation of neurons that inhibit feeding and the inhibition of neurons that promote feeding. The long form of the leptin receptor is found on two distinct types of neurons, the POMC/CART (proopiomelanocortin and cocaine- and amphetamine-regulated transcript) neurons and the NPY/AGRP (neuropeptide-Y and agouti-related peptide) neurons. The POMC/CART neurons decrease food intake when stimulated by neurotransmitters such as leptin and α-MSH (melanocyte-stimulating hormone, a product of POMC that is made in the hypothalamus and binds to the melanocortin receptors on these neurons). Leptin also binds directly to the NPY/AGRP neurons and decreases food intake by inhibiting the neurons'

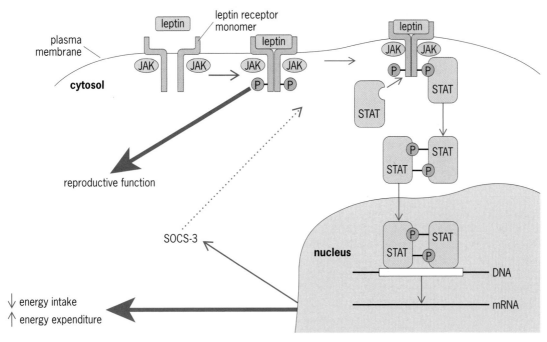

Leptin signal cascade. Leptin binds to the long form of the receptor, resulting in receptor dimerization and initiating several signaling cascades. The primary cascade signals by way of activation of the JAK-STAT pathway. The activation by JAK results in phosphorylation of tyrosine amino acids (shown as P in circle) on the leptin receptor. This in turn results in changes in gene expression via STAT pathways, which ultimately decrease appetite and increase energy expenditure. The maintenance of reproductive function requires only the JAK and not the STAT pathway. Signal modulation occurs in part via SOCS-3, a suppressor of cytokine signaling. SOCS-3 can feed back to the leptin receptor and inhibit leptin signaling. (*Adapted from D. L. Nelson and M. H. Cox, Lehninger Principles of Biochemistry, 3d ed, Worth Publishers, New York, 2000*)

ability to be stimulated. However, NPY and AGRP stimulate feeding. NPY stimulates food intake by binding directly to receptors on the NPY/AGRP neurons, whereas AGRP stimulates food intake indirectly by antagonizing or counteracting the effect of α-MSH on the melanocortin receptors.

Effects on energy expenditure. The increase in energy expenditure is due to stimulation of the sympathetic nervous system, which causes increases in oxygen consumption and metabolic rate. In addition, leptin has been shown to induce adipose cell death (apoptosis). As animals become more obese, the response to leptin given peripherally (into the abdomen) decreases, and they become leptin-resistant. Although the ability to respond to central leptin administration (leptin given directly into the brain) remains intact, with continued obesity there is also a decrease in the response to central leptin administration. This adipose cell death, along with an alteration of how the body handles nutrients following leptin administration, might be why there is a preservation of muscle with weight loss (in contrast to weight loss from caloric restriction alone, which is accompanied by a loss in muscle).

Regulation of reproductive hormones. When leptin-deficient mice were first discovered, they were found to be infertile as well as obese. When leptin was given back to these animals, fertility was restored. In addition, when leptin is given to animals prior to puberty, they undergo puberty earlier than control animals. Thus leptin appears to be part of the system that tells the brain there are adequate energy stores (adipose tissue) for reproduction and allows puberty to commence. Leptin has been shown to increase the secretion of the hormones that regulate reproductive function, including gonadotropin releasing hormone, luteinizing hormone, and follicle stimulating hormone.

Other potential roles. Although most of the attention of leptin has centered on its role in decreasing food intake and decreasing weight, it is not certain that this is its primary role. Until very recently, there has been little need for a hormone that inhibits an organism from gaining as much weight as possible, for more efficient weight gain appears to protect against the next famine. Therefore, leptin likely has other roles as a hormone.

Fat accumulation. One role that has received much attention lately is that of a hormone that prevents fat accumulation in nonadipose tissues, including muscle, the liver, and the pancreas. Excess fat accumulation has been shown to be detrimental to these organs and may play a significant role in the development of diabetes. Leptin may traffic fat in such a way that it does not accumulate in these tissues. In addition, leptin increases utilization of fat by producing heat and not energy. Thus leptin might enhance more appropriate and rapid disposal of fat that enters these tissues.

Fasting. Another potential role of leptin is in mediating the endocrine response to fasting. When a human or mouse is fasted, there is a decrease in levels of leptin, reproductive hormones, thyroid hormone, and insulin. However, there is an increase in levels of glucocorticoids, catecholamines, and glucagon. This results in a conservation of energy, an increase in the production of glucose for the brain, and an increase in the release of fatty acids from the adipose tissue for muscle. When mice were fasted but given leptin, there was a blunting in several of these responses, including the changes in levels of glucocorticoids, reproductive hormones, and thyroid hormone. Thus rather than acting as a signal of excess adipose tissue, leptin's primary role might be in signaling a decrease in energy, and initiating the stress response to this decrease.

Blood cell and bone formation. There is currently much debate on the extent of direct leptin action outside the brain. The gastrointestinal tract does have the long form of the leptin receptor, and signals from this receptor might play a role in nutrient absorption or appetite regulation. In addition, there is evidence that leptin has a direct role in the formation of bone red blood cells, white blood cells, and blood vessels.

Treatment of obesity. There are only five known people (from two families) who are deficient in leptin (like the *ob/ob* mice). They are very obese and seem to have reproductive abnormalities, but they do not appear to have other abnormalities such as increased glucose or insulin. Treatment with leptin has had a dramatic impact on their weight and has allowed puberty to occur. There are also three people (from one family) who do not make normal leptin receptors. These individuals have very high leptin levels, are very obese, do not have normal reproductive function, and have abnormal levels of thyroid hormone and growth hormone. Similar to the patients with leptin deficiency (and unlike the mouse models), they do not have high levels of blood glucose.

Leptin is also under investigation as a treatment for obesity in people with elevated leptin levels. Initial reports suggest that leptin administration results in weight loss in these subjects (with preservation of muscle), with the degree of weight loss similar to other pharmacologic treatments for obesity (about 8%). However, since leptin is a protein, it cannot be taken by mouth and must be injected, which sometimes results in pain, redness, inflammation, and bruising at the injection site. This mode of administration and the injection site reactions limit its use as a treatment. Whether administration of leptin or of a leptin-like compound will ever become a treatment for obesity remains to be seen. However, the discovery of leptin has opened a new area in the understanding of obesity and body weight regulation.

For background information *see* DIABETES; ENDOCRINE SYSTEM (VERTEBRATE); HUNGER; INSULIN; LIPID METABOLISM; METABOLIC DISORDERS; OBESITY in the McGraw-Hill Encyclopedia of Science & Technology.   William T. Donahoo; Robert H. Eckel

Bibliography. R. S. Ahima and J. S. Flier, Leptin, *Annu. Rev. Physiol.*, 62:413–37, 2000; C. A. Baile, M. A. Della-Fera, and R. J. Martin, Regulation of metabolism and body fat mass by leptin, *Annu. Rev. Nutrit.*, 20:105–27, 2000; R. B. Harris, Leptin—much more than a satiety signal, *Annu. Rev. Nutrit.*, 20:45–75, 2000; B. Jeanrenaud and F. Rohner-Jeanrenaud, Effects of neuropeptides and leptin on nutrient partitioning: Dysregulations in obesity, *Annu. Rev. Med.*, 52:339–351, 2001; D. L. Nelson and M. M. Cox, *Lehninger Principles of Biochemistry*, 3d ed, Worth Publishers, New York, 2000; R. H. Unger, Leptin physiology: A second look, *Regulatory Peptides*, 92(1–3):87–95, 2000.

Light pollution

Light pollution, or sky glow, is the excessive brightness of the sky (especially the night sky) caused by artificial light sources on the ground. In addition to obscuring naked-eye viewing of the stars from Earth, light pollution has a significant adverse effect on observational ground-based astronomy, which relies on discerning extremely distant, faint light sources against a background of lesser brightness.

The combination of population growth and the attendant residential and commercial developments, coupled with the growth in lighting technology, has led to a greatly increased use of outdoor lighting. Some new lighting installations, well designed and correctly installed, offer excellent nighttime visibility at reasonable lighting levels and provide excellent ambience for their settings. Many, however, produce glare, overlighting, and light trespass and add greatly to urban sky glow, as well as wasting energy.

Origin of sky glow. The Earth's atmosphere causes the light coming from sources in urban areas to scatter, creating the halo of light visible over cities even from a great distance. Even a single bright source in an otherwise dark locale can be a source of local sky glow.

Light emitted directly into the sky, as well as light that is reflected from the ground, buildings, or other objects, is scattered by molecules and aerosols (solid or liquid particles) present in the air, which both scatter and absorb light. A study by Roy Garstang of the University of Colorado showed that an empirically derived relation, published by Merle Walker of the University of California, is a good approximation for sky glow: $\Delta I = 0.01Pr^{-2.5}$, where ΔI is the increase in sky glow level at a vertical angle of $45°$ toward the city, P is the population of the city, r is the distance from the city to the observing site in kilometers, and 0.01 is a constant typical for most cities with average mixture of outdoor lighting. A ΔI value of 0.2 means a sky glow increase of 20% over the natural background level. The equation seems to fit best to situations where the average luminous flux per person is between 500 and 1000 lumens. Air pollution exacerbates the problem of light scattering. While air molecules scatter light even in the cleanest atmosphere, adding human-made particulates increases the effect significantly. Cloudy skies, although not an issue for astronomers, also increase sky glow.

Enacting quality lighting measures does help to ameliorate sky glow. Many communities have now adopted outdoor lighting ordinances, the first comprehensive one being in Tucson, Arizona, in the early 1970s. These ordinances help darken night skies while improving visibility and energy efficiency for residents. All such ordinances require that the light shine down, where it is needed, rather than up, and many recent ones limit the amount of light that can be used, with different amounts allowed for different "environment zones" such as city centers, suburbs, and rural areas. The city of Tucson has grown greatly from the time that its first outdoor lighting control ordinances went into effect in 1970, and now has close to 800,000 people. Yet the sky glow at the observatories about 70 km (43 mi) from the city have not increased much over that period.

Effect on telescopes. Many astronomical observatories have already suffered a loss of effective telescope aperture and value due to the increasing urban sky glow. For many types of research, especially work on the faintest objects, one measure of the value of a telescope is its light-collecting area, specifically the square of the aperture of its main light-collecting area. The **table** shows the calculated loss of value for a 4-m-aperture (157-in.) telescope, where X is the increased sky glow level above the natural background and 1.0 denotes the natural background level. A value of 1.2 means a 20% increase above the natural background, 2.0 is double the natural background, and so on. The equivalent or effective aperture is defined to be (aperture squared/X)$^{1/2}$, and the value of the facility is assumed to scale as (aperture/4)$^{2.7}$. Clearly, the loss to observatories due to any significant amount of urban sky glow is very large, and potentially growing.

Effect on quality of life. There are four quality-of-life problems associated with outdoor lighting. All are energy-inefficient and adversely impact the outdoor environment, and none actually improves visibility.

1. Glare is the sensation produced by luminance within the visual field that is sufficiently greater than the luminance to which the eyes are adapted, causing annoyance, discomfort, or loss of visual ability. Good outdoor lighting design should minimize glare.

2. Light trespass is light that is shining where or when it is not wanted or needed. Street lighting, for example, should light the streets and not the interior of houses or rooftops. Light trespass occurs whenever the light shines beyond its intended target.

3. Uplight is light that goes directly up into the night sky, is lost into space, and serves no useful purpose. Uplight often results from the same sort of inefficient fixtures that produce glare and light trespass.

4. Excessive light results when light levels exceed that needed for the task. Many installations greatly exceed the recommended lighting levels proposed by national and international professional lighting organizations, such as the Illuminating Engineering Society of North America and CIE (International Commission on Illumination). Examples include gasoline service stations and convenience marts, some of which use lighting levels 10–15 times the recommended levels; buildings in which the entire façade is washed with light much above recommended levels, to make the building stand out from others in the area; and rural highways that are lit by installations more appropriate for city centers.

Good lighting practices. The following recommendations help reduce sky glow: Do not light an area if it is not necessary; use the correct lighting level; turn the lights off at times when they are not needed; eliminate light above the horizontal plane; minimize; light output at high vertical angles (70–90°); eliminate light trespass by selecting a luminaire (light fixture or other complete lighting unit) of appropriate light intensity and directionality; and conceal the lamp source and bright reflector sections from direct view.

Cutoff fixtures, which are designed not to output direct light above the horizontal, are most effective at minimizing sky glow. In areas near observatories, the use of low-pressure sodium (LPS) lighting sources is highly recommended, and in fact outdoor lighting ordinances in many such locales require its use. It is a monochromatic light source, which renders no colors, but it ensures that the sky remains dark at all wavelengths (colors) except that of the sodium resonance lines. It is also a very energy efficient lighting source and is used elsewhere for that reason alone.

For background information *see* AIRGLOW; ASTRONOMICAL OBSERVATORY; ASTRONOMY; ILLUMINATION; OPTICAL TELESCOPE; SCATTERING OF ELECTROMAGNETIC RADIATION; TELESCOPE in the McGraw-Hill Encyclopedia of Science & Technology.

David L. Crawford

Bibliography. R. H. Garstang, The status and prospects for ground-based observatory sites, *Annu. Rev. Astron. Astrophys.*, 27:19–40, 1989; R. H. Garstang, *Publications of the Astronomical Society of the Pacific* (*PASP*), 103:1109, 1991; R. H. Garstang, *PASP*, 101:306, 1989; R. H. Garstang, *PASP*, 98:364, 1986; *Illuminating Engineering Society of North America Handbook*, 2000.

Loss of value for a 4-m-aperture (157-in.) telescope due to increased urban sky glow

	Equivalent aperture		
X	m	in.	Percent of original value
1.00	4.00	157	100
1.10	3.81	150	88
1.20	3.65	144	78
1.25	3.58	141	74
1.50	3.27	129	58
2.00	2.83	111	39
3.00	2.31	91	23
4.00	2.00	79	15
5.00	1.79	70	11

Lotus

The more than 175 legume species that constitute the tribe Loteae are found worldwide, except for the very cold arctic regions and the lowland tropical areas of Southeast Asia and South and Central America. Approximately three-quarters of all species in the genus *Lotus* are Old World, yet only four Old World species receive major attention: *L. corniculatus* (broadleafed birdsfoot trefoil), *L. glaber* (narrowleafed birdsfoot trefoil), *L. uliginosus* (big trefoil), and *L. japonicus*. Several species of *Lotus* are used as forage legumes to provide a high-quality feed source for grazing livestock. Some *Lotus* species have utility as tools for advancing molecular biology. *Lotus corniculatus* has the most agricultural importance and the widest distribution. Its native range extends from Scandinavia across lowland and high alpine regions of central Europe to Asia Minor and Africa, and it occurs as naturalized populations in North and South America.

Ecology. The forage legumes and grasses that are used to produce the majority of the world's grazing livestock coevolved in mixtures. The value of legumes in grassland systems is derived from their ability to grow symbiotically with nitrogen-fixing bacteria (*Rhizobium* and *Bradyrhizobium* spp.). The fixed nitrogen used by the growing legume gives its foliage a high protein and nutritional value for livestock. In turn, the nitrogen retained in the feces and urine of grazing animals becomes bioavailable to the grasses and forbs that contribute to pasture productivity.

While most field crops have become distinctly different from their wild and weedy counterparts through a process of domestication, *Lotus* has changed little from its wild ancestry. Locally adapted landraces of *Lotus* evolved from wild populations found in natural pastures managed by generations of indigenous farmers. The landraces were carried by immigrants to new lands and used as the genetic resources for new plantings. *Lotus* was not purposefully sown in mixtures used to seed meadows until the midnineteenth century, yet *Lotus* can make a major contribution to livestock production in improved, naturalized, and native pastures. It is often grown in environments that other legumes cannot tolerate—in soils characterized as being acidic, of low fertility, or having poor drainage—and in locations that are difficult to crop. Much of its wide range of adaptation is due to its highly variable genetic diversity. The broad range of phenotypes is believed to have developed as a result of environmental adaptation and through a continuum of intraspecific hybridization. *Lotus* is polymorphic for growth habit, degree of plant pubescence, leaf color, flower number per umbel (inflorescence), reproductive compatibility, root morphology, chromosome karyotypes, forage quality characteristics, insect resistance, and biochemical constituents.

Reproductive behavior. *Lotus* species generally are long-day plants having a critical day length for flowering. Flowering is retarded at day lengths less than 15 hours, limiting *Lotus* to latitudes above 30°N or below 30°S. The importance of flowering in perennial *Lotus* species is for seed production to aid the persistence of a population where individual plants may live only 2 or 3 years. The persistence of *Lotus* species is reliant on seedlings to replace individual plants lost to mortality.

Lotus plants are highly susceptible to crown- and root-rotting bacteria and fungi. Opportunities to control the crown and root diseases by chemical treatment are not practical, necessitating population management through agricultural practices. Harvesting or grazing *Lotus* at intervals helps maintain an open plant architecture and creates a dry leaf canopy, unsuitable to the growth of pathogens. Stand persistence results from balancing the grazing or harvesting management to reduce the disease potential with the need for seed production and replacement seedlings. *Lotus corniculatus* and *L. uliginosus* also produce rhizomes, lateral underground stems, adding opportunities for spread in favorable environments.

Secondary metabolites. *Lotus* species contain secondary metabolites of biochemically related flavonoid, isoflavonoid, and polyphenolic end products, and of cyanogenic glycosides. Condensed tannin and cyanogenic glycosides are metabolites implicated in providing defense mechanisms against predation, competition, and disease. The nutritional value of *Lotus* is ranked high among forage species, but considerable attention has also been paid to the presence of condensed tannins in the herbage of the cultivated *Lotus* species.

Condensed tannins in *Lotus* species confer bloat resistance and increased nitrogen utilization to ruminants. Bloating is a potentially lethal condition caused by excess formation of gases in the rumen, resulting in its expansion to the point of exerting damaging pressure on the heart. Condensed tannins bind with soluble proteins, making them less available in the rumen and reducing the conditions that induce bloat. Some of the condensed tannin-bound protein is digested later in the lower intestine, providing rumen bypass nutrition. Bypass protein can improve utilization of the digested *Lotus* forage, leading to improved animal health, condition, and gain. However, high levels of condensed tannin can also reduce forage intake and digestibility. A negative association between condensed tannin content and cyanogenic glycosides is found among several *Lotus* species, indicating the possibility that they play overlapping roles in these species. The major role of cyanogenic glycosides in *Lotus* is as a feeding inhibitor. Cyanogenic glycosides release hydrogen cyanide after the tissues have been damaged by grazing or by insect or fungal attack. Like cyanogenic glycosides, condensed tannins may be a protection factor from insect attack and grazing.

New World species. Several native North American *Lotus* species can be useful for range improvement and land reclamation. *Lotus crassifolius* has been used to provide ground cover, improve soil nitrogen, and provide forage for big game animals in clearcut timberland in the Pacific Northwest and the Central Sierra Nevada. Other North American species which may be useful for wildlife forage and revegetation are *L. scoparius, L. micranthus, L. nevadensis,* and *L. oblongifolius.* Urban development and the expansion of and changes to agricultural systems threaten the distribution and abundance of native *Lotus* species.

Biotechnology. *Lotus* species act as hosts to two bacterial pathogens which result in crowngall (*Agrobacterium tumefaciens*) and hairy root (*A. rhizogenes*) diseases. On infection, these bacteria transfer some of their genes into the *Lotus* plant's cells, where they cause an uncontrolled proliferation of cells or roots. Cell biologists use these agrobacteria as gene vectors to transfer other genes into the plant. In contrast to many other legumes, *Lotus* species regenerate shoots from callus at high frequency. A variety of tissues of four species—*L. glaber, L. uliginosus, L. japonicus,* and *L. corniculatus*—have been shown to regenerate shoots efficiently. Stable gene transformation systems using *Agrobacterium* have been developed sufficiently in *L. corniculatus* and *L. japonicus* to use them as genetic exploration tools. *Lotus japonicus* is a preferred model system for genetic and molecular biology research because it is a diploid ($2n = 2x = 12$) perennial self-pollinizer and has a relatively short sexual regeneration time of 3 months; it has a relatively small genome size; and it is susceptible to *Agrobacterium. Lotus corniculatus* is a more complex genetic tool because it is a tetraploid ($2n = 4x = 24$) and cross-pollinating. Exploitation of a broader range of genetic resources will be possible as technologies are developed to efficiently transfer useful alleles from exotic germplasm into cultivated species.

For background information *see* ANIMAL FEEDS; BIOTECHNOLOGY; FORAGE CROPS; GENETIC ENGINEERING; LEGUME FORAGES in the McGraw-Hill Encyclopedia of Science & Technology.

Paul R. Beuselinck

Bibliography. R. F. Barnes, D. A. Miller, and C. J. Nelson (eds.), *Forages,* 5th ed., vol. 1: *An Introduction to Grassland Agriculture,* Iowa State University Press, Ames, 1995; P. R. Beuselinck (ed.), *Trefoil: The Science and Technology of Lotus, Crop Sci. Soc. Amer. Spec. Publ.,* no. 28, 1999; D. T. Fairey and J. G. Hampton (eds.), *Forage Seed Production,* vol. I: *Temperate Species,* CAB International Monograph, Wallingford, U.K., 1998.

Magnetic interactions, ligand-mediated

The ability to control magnetic interactions between metal centers across a bridging ligand is of fundamental importance in inorganic and materials chemistry. Magnetic exchange interactions between metals in dinuclear (or larger) complexes are dependent on the nature of the pathway linking the metal ions. If the metal-based magnetic orbitals are sufficiently close to overlap directly, the nature of the magnetic interaction depends on their relative symmetry (Goodenough-Kanamori rules). This has been exploited in the preparation of complexes with predictable magnetic properties. If, however, the magnetic orbitals are too far apart to overlap directly and require the participation of bridging ligand orbitals to mediate the interaction (a superexchange process), the properties of the bridging ligand become extremely important. This principle has received relatively little systematic attention for metal complexes, in contrast to the extensive work on the magnetic properties of organic polyradicals as a function of structure and topology. This may seem surprising considering the obvious advantages of metal-based radicals over organic radicals for use in magnetic materials, such as their higher spin density (in many cases), chemical stability, and possible redox activity for switching purposes. The ability to design desirable properties into magnetic materials by controlling the bridging pathways at the construction stage is therefore important.

A demonstration of the principles of ligand mediation of magnetic properties is most effectively provided by a study of two different but closely related types of dinuclear complexes of molybdenum. These two types are based on (1) a molybdenum nitrosyl core, $[Mo(NO)]^{2+}$, in which the Mo is usually in oxidation state $+1$ (d^5 configuration, assuming that the nitrosyl group is formally coordinated as NO^+), and (2) an oxo-molybdenum core, $[Mo(O)]^{3+}$, in which the Mo is in oxidation state $+5$ (d^1 configuration). The molybdenum nitrosyl core is stabilized by the tripodal ligand hydrotris(3,5-dimethylpyrazolyl)borate, $Tp^{Me,Me}$ (structure **1**), forming neutral para-

(1)

(2)

magnetic 17-electron Mo(I) complexes with pyridine (py) as the sixth ligand, $[Mo(NO)Tp^{Me,Me}Cl(py)]$. The oxo-molybdenum(V) core is similarly stabilized by $Tp^{Me,Me}$ (Tp*), forming neutral paramagnetic complexes with phenolates (OAr$^-$) as the sixth ligand, [Mo(O)Tp^{Me,Me}Cl(OAr)]. By use of appropriate bridging ligands containing two (or more) pyridyl or phenolate termini, dinuclear complexes incorporating these metal fragments (structure **2**) when Z = NO or Z = O are prepared, such as [{Mo(NO)Tp^{Me,Me}Cl}$_2$-(4,4'-dipyridyl)] (**7**) and [{Mo(O)Tp^{Me,Me}Cl}$_2$(1,4-OC$_6$H$_4$O)] (**12**). Each metal atom in these species has one unpaired electron.

Despite the difference in the formal oxidation states of the {Mo^I(NO)Tp*Cl} and {Mo^V(O)Tp*Cl} fragments, their magnetic properties may be regarded as comparable, for two reasons. First, the unpaired electron in each fragment is in the d_{xy} orbital. In the former case the NO ligand is a strong π-acceptor. Taking the Mo-NO axis as the z axis, the two empty NO π^* orbitals overlap with the d_{xz} and d_{yz} metal orbitals, thereby lowering them, but leaving d_{xy} unchanged. The Mo(I) electron configuration is therefore $d_{xz}^2 d_{yz}^2 d_{xy}^2$ (**3**). In the latter case the oxo ligand is a strong π-donor. The filled oxygen p_x and p_y orbitals overlap with the metal d_{xy} and d_{yz} orbitals, thereby raising them, but leaving d_{xy} unchanged. The Mo(V) electron configuration is therefore $d_{xy}^1 d_{xz}^0 d_{yz}^0$ (**4**). In each case, the mag-

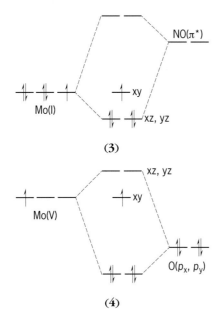

(3)

(4)

neticorbital is of the correct symmetry for $d(\pi)$-$p(\pi)$ overlap with the bridging ligand. Second, the strongly electron-withdrawing nature of the NO ligand attached to Mo(I), and the strongly electron-donating nature of the oxo-ligand attached to Mo(V), means that the actual electron densities at the metal centers are comparable, which has been confirmed by photoelectron spectroscopy. The favorable interaction between the d_{xy} orbital—which is the magnetic orbital—with the bridging ligand π-system means that the ligand-mediated interactions between the metal centers are exceptionally strong.

Nitrosyl-molybdenum(I) complexes. The electron paramagnetic resonance (EPR) spectra of di-, tri-, and

tetranuclear complexes are consistent with magnetic exchange between the unpaired spin on each metal center. However, the EPR spectra give no direct information about the nature of the magnetic interaction [ferromagnetic (spin values aligned parallel) or antiferromagnetic (spin values aligned antiparallel)], or its strength, as measured by J, the exchange coupling constant between pairs of metal atoms. The magnitude of J is obtained from solid-state susceptibility measurements, by fitting the variable-temperature magnetic susceptibility data in the temperature range 1.6–250 K. The data are interpreted using the exchange spin hamiltonian in the form H = $-J(S_1 \cdot S_2)$, with positive J indicating ferromagnetism and negative J indicating antiferromagnetism.

The most significant result is the alternation in the sign of J as the substitution pattern of the bridging ligand changes from 4,4' to 3,4' to 3,3' for the bipyridine ligand series (**7, 8, 9**) where $J = -33$, $+0.8$, and -1.5 cm^{-1}, respectively. This suggests a spin-polarization mechanism for propagation of the exchange interaction (**5** and **6**). The unpaired elec-

(5)

(6)

tron on one Mo center polarizes the spin of the electron cloud of the adjacent nitrogen atom in the opposite sense; the next atoms in the sequence will be spin-polarized in the opposite direction to the N atom, and so on around the bridging ligand. With 4,4'-bipyridine as the bridging ligand, the second Mo spin will be antiferromagnetically coupled to the first one. As the bridging pathway changes in length by one atom to 3,4'-bipyridine (**8**), this model predicts that the exchange interaction must become ferromagnetic, which is observed. Changing the path length by an additional atom, as in (**9**), restores antiferromagnetic exchange.

The spin-polarization mechanism arises from the Longuet-Higgins molecular orbital model for

(7)

(8)

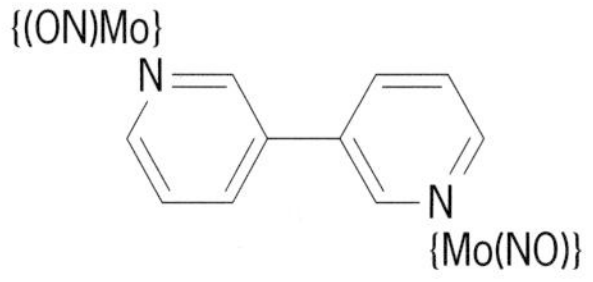

(9)

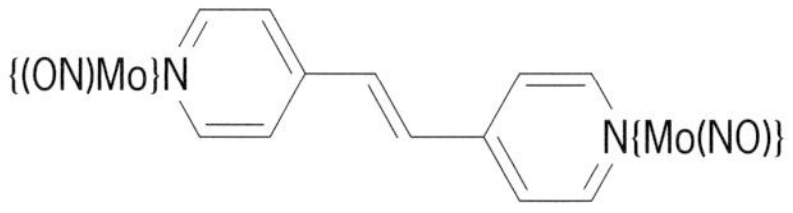

(10)

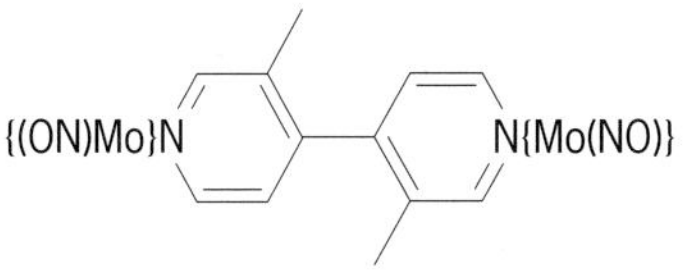

(11)

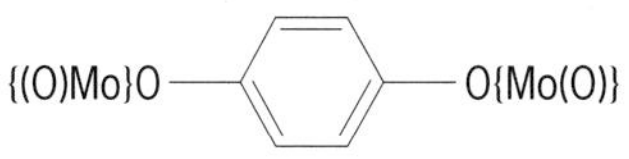

(12)

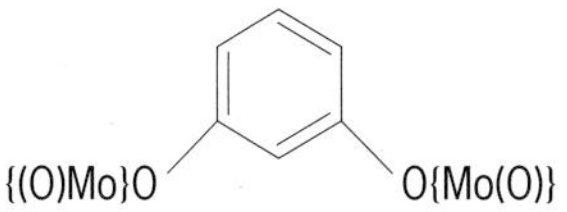

(13)

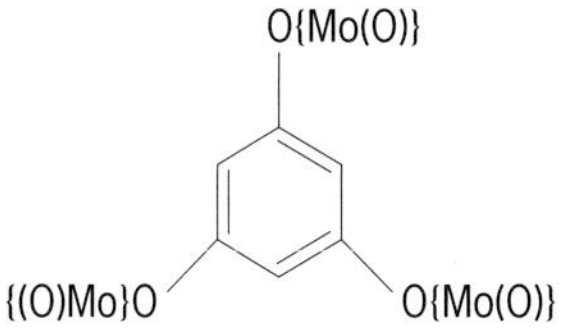

(14)

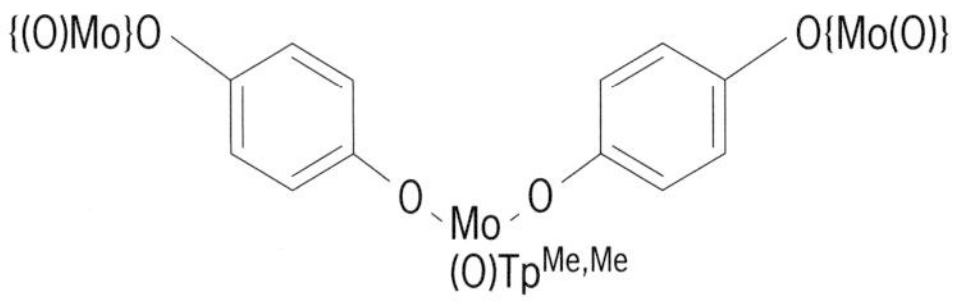

(15)

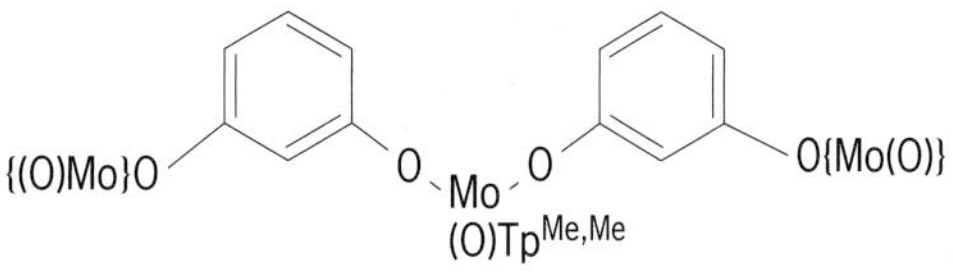

(16)

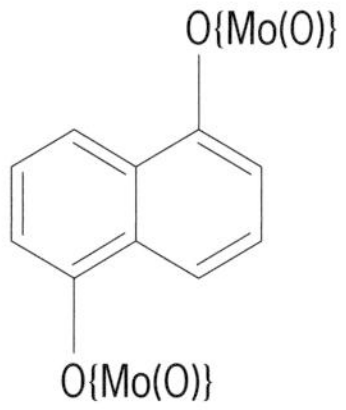

(17)

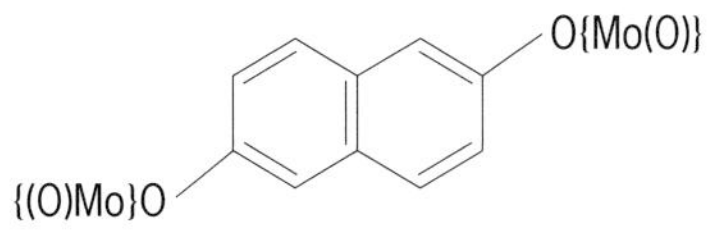

(18)

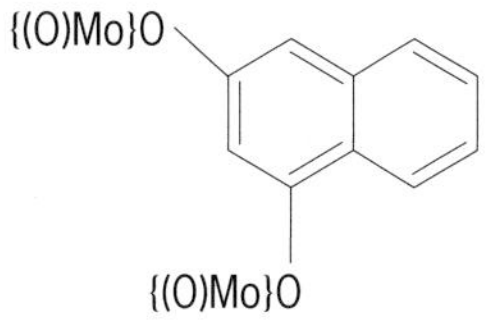

(19)

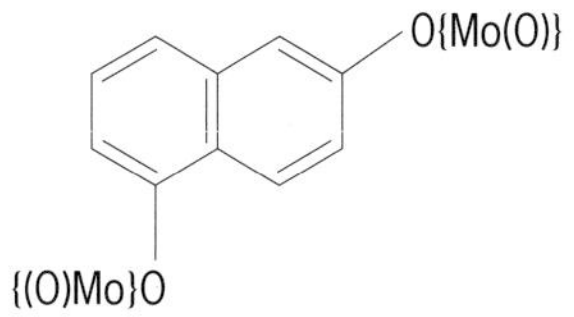

(20)

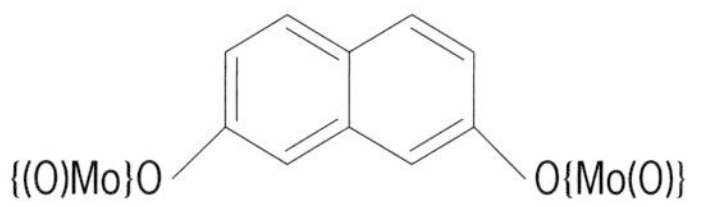

(21)

conjugated alternant hydrocarbons, which results in ferromagnetic coupling between two radicals separated by an *m*-phenylene bridge. This spin-polarization effect may also be applied to organic diradicals so that a 3,4'-biphenyl bridge promotes ferromagnetic exchange between two unpaired spins, whereas a 3,3'-biphenyl bridge promotes antiferromagnetic exchange.

Oxo-molybdenum(V) complexes. The same patterns of magnetic behavior appear in the dinuclear oxo-Mo(V) complexes with bis-phenolate ('OO') bridging ligands. First, between structure **12** (*para*-substituted bridge) and **13** (*meta*-substituted bridge) the

exchange interaction changes sign from antiferromagnetic to ferromagnetic and $J = -80$ and $+9.8$ cm^{-1}, respectively.

The spin-polarization principle also works for the trinuclear complexes (**14**, **15**, **16**). In **14** the bridging ligand, 1,3,5-benzenetriolate, results in a *meta* geometric relationship between each pair of oxo-Mo(V) centers. Complexes **15** and **16** are linear chains, with 1,4- and 1,3-benzenediolate bridges between adjacent pairs of metals. Based on spin-polarization, **14** should have a quartet ($S = 3/2$) ground state with all metal centers ferromagnetically coupled, and this was confirmed by magnetic susceptibility measurements ($J = +14.4$ cm^{-1}). For **15** and **16**, application of the spin-polarization principle predicts ground states of $S = 1/2$ (spin arrangement ↑↓↑) and $S = 3/2$ (spin arrangement ↑↑↑) arising from antiferromagnetic and ferromagnetic coupling respectively between adjacent spins, and this was also confirmed with $J = -33$ and $+4.5$ cm^{-1}, respectively.

The principles described above can also be applied to more complex polycyclic aromatic bridging ligands that are topologically similar to the simpler monocyclic *meta*- and *para*-substituted analogs. Complexes **17** and **18** are topologically similar to **12**, and the magnetic interaction is, as predicted, antiferromagnetic, with $J = -9.6$ and -28.5 cm^{-1}, respectively, coupling which is lower than that observed for **12** (-80.0 cm^{-1}), in part because the metal atoms are attached to different aromatic rings. The J values for the ferromagnetically coupled **19**, **20**, and **21**, topologically equivalent to **9**, are $+28.7$, $+17.4$ and $+12$ cm^{-1}, respectively, which may be compared with the smaller value for **9**, $+9.8$ cm^{-1}.

The strength of magnetic coupling attenuates with increasing distance between the magnetic centers. Thus, J values for **10** and **22** are substantially smaller

(22)

(23)

than the comparable **7** and **12** (-18 and -13.2 cm^{-1}, respectively). In addition, twists within the bridging ligand can also cause reductions in J, which is -3.5 cm^{-1} in **11** and -2.8 cm^{-1} in **23**, both an order of magnitude smaller than in **7** and **10**. These results indicate that the strength of the interaction between the unpaired spins is conditioned by the effectiveness of the overlap of the metal's magnetic orbital with the π-orbitals of the bridging ligand.

Applications. The nitrosyl- and oxo-molybdenum complexes described in this article are ideal for studing the long-distance electronic and magnetic interactions across conjugated bridging ligands (molecular wires). The interactions between two or three metal fragments are exceptionally strong because of the near-ideal matching in both symmetry and energy in terms of the relevant metal orbitals with those of the bridging ligands, and the complexes are amenable to study by a wide range of techniques [voltammetry, electron paramagnetic resonance (EPR) spectroscopy and infrared (IR) spectroscopy, and magnetic susceptibility], which complement each other and allow electrochemical and magnetochemical properties to be studied in tandem. The electronic and magnetic interactions between two or more metal centers are determined by the nature of the bridging ligand, and are of fundamental interest for the design of future multinuclear compounds in the areas of molecular electronics and magnetic materials.

For background information *see* ANTIFERROMAGNETISM; ATOMIC STRUCTURE AND SPECTRA; ELECTRON PARAMAGNETIC RESONANCE (EPR) SPECTROSCOPY; ELECTRON SPIN; FERROMAGNETISM; LIGAND; MAGNETOCHEMISTRY; MOLYBDENUM; PARAMAGNETISM in the McGraw-Hill Encyclopedia of Science & Technology.   Jon A. McCleverty; Michael D. Ward

Bibliography. S. Bayly et al., *Inorg. Chem.*, 39:1288–1293, 2000; J. B. Goodenough, Theory of the role of covalence in the perovskite-type manganites, *Phys. Rev.*, 100:564–573, 1955; O. Kahn, *Molecular Magnetism*, VCH Publishers, New York, 1993; J. Kanamori, Superexchange interaction and symmetry properties of electron orbitals, *J. Phys. Chem. Solids*, 10:87–98, 1959; J. A. McCleverty and M. D. Ward, *Acc. Chem. Res.*, 31:842–851, 1998.

Magnetoelectronics

Magnetoelectronics, also known as spin electronics or spintronics, has developed rapidly since the discovery of giant magnetoresistance and tunneling magnetoresistance in the late 1980s. Both effects are based on the spin of the electron, and its use in electronic devices.

Magnetoelectronic Devices

In addition to charge, electrons have the property of spin. As for any rotating charge, the spin gives rise to a magnetic moment for the electron. This magnetic moment is quantized. The most obvious effect of this magnetic moment can be seen in ferromagnetic materials, where a macroscopic number of electrons have their spins pointing into one direction, causing a macroscopic magnetic moment. Until 1987, this magnetic moment was the only effect of the spin that was used in applications such as data storage in hard disks and magnetic tapes. The only access to the magnetically stored information was via the magnetic field which was generated by magnetized particles in the material and was typically picked up by an inductive read head (transducer) using a small magnetic coil.

Spin-polarized transport and giant magnetoresistance. In 1988, a new spin-related effect was discovered. In a ferromagnetic metal layer, the conduction electrons are mainly spin-polarized. A. Fert and P. Grünberg independently showed that electrons in a ferromagnetic metal layer could travel through a thin nonmagnetic metal layer without losing their spin polarization. If a second ferromagnetic metal layer is in contact with the nonmagnetic metal, the electrons entering the second magnet will exhibit a scattering which is dependent on the relative magnetization of the magnets. For parallel alignment, the scattering will be lower than for antiparallel

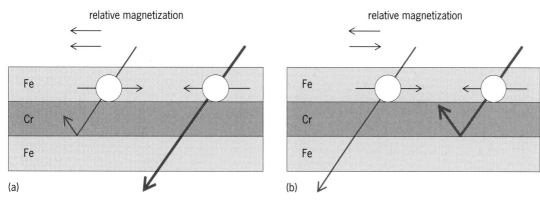

Fig. 1. GMR structure in the (*a*) low-resistance and (*b*) high-resistance state. In the high-resistance state, electrons coming from the upper layer exhibit stronger scattering in the low-resistance state.

magnetization. This effect can be seen in sandwiched metal layers with two ferromagnetic layers and a thin (a few nanometers) nonmagnetic metal layer in between, such as Fe/Cr/Fe or Co/Cu/Co. The magnetization-dependent scattering results in an effect known as giant magnetoresistance (GMR) [**Fig. 1**]. Today the effect is used by most of the hard disk manufacturers for new magnetoelectronic read heads, resulting in an enormous increase in storage density. It is also widely used in magnetic sensors and sometimes in special magnetic random access memories.

Tunneling magnetoresistance. Another spin-related effect is tunneling magnetoresistance. If two conductive layers are coupled by an insulating barrier of very low thickness, the wave functions of the electrons on both sides can couple through the barrier, resulting in a finite transmission and thus a finite conductivity. This quantum-mechanical effect is called tunneling and depends strongly on the density of states of the electrons on both sides of the barrier. In a ferromagnet, the density of states for the spin-up electrons and the spin-down electrons differs considerably. It can thus be expected that the tunneling resistance of a tunnel barrier in between two ferromagnets would depend on the relative magnetization of the two magnets. The effect, called tunneling magnetoresistance (TMR), was demonstrated by J. S. Moodera in the late 1980s and subsequently was investigated for possible memory applications. **Figure 2** shows a typical geometry for a TMR memory cell. If a voltage is applied from point *a* to *b*, the

current through the TMR structure gives a measure for the relative magnetization of the two magnetic layers, thus reading out the information. The writing process is more complex. A current is sent from *a* to *c* in such a way that the induced field is slightly smaller than the opposing field of the top magnetic layer, but is too small to switch its magnetization. A second current of the same magnitude is sent from *b* to *d*. At the cross point of the two lines, the magnetic fields add up to a magnitude that is large enough to switch the top magnetic layer of the TMR structure underneath. Memory cells fabricated this way are fast, nonvolatile (retain information when power is turned off), and very small.

Spintronics in semiconductors. Although several magnetoelectronic devices based on metallic structures are already in use, there are good reasons to use magnetoelectronics in semiconductors. First, all magnetoelectronic memory structures need an interface to typical semiconductor integrated circuits. This interface can be realized more easily if the impedance of the magnetic random access memory (MRAM) devices matches the impedance of the semiconductor devices, namely metal-oxide semiconductor field-effect transistors (MOSFETs). For TMR structures this can be done, but only with a very thin tunnel barrier (0.5–1 nanometer) and low yield. For GMR structures, which are all metal, the impedance is always much lower than for semiconductor structures. Moreover, much in contrast to metal structures, semiconductors allow for a control of the electron density by doping or field effect, which was the main reason for the development of the transistor and the MOSFET and which is the basis for all semiconductor electronics. Recently, it was demonstrated that the spin scattering in semiconductors takes place on a scale of micrometers or greater. While typical scales for the effects in GMR or TMR devices are in the range of a few nanometers, they would be extended by several orders of magnitude if spin-polarized transport could be used in a semiconductor. This would give rise to new semiconductor devices that manipulate both charge and spin, or which could directly integrate GMR or TMR effects into a classical transistor or FET. Such devices may

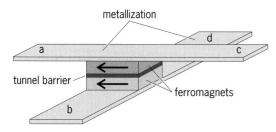

Fig. 2. TMR memory cell. The magnetization of the bottom magnetic layer is pinned in a particular orientation (made insusceptible to switching by an applied magnetic field).

lead to new high-speed, low-power electronics, or more complex electronic devices, for example, by adding nonvolatile high-speed programmability to a standard logical gate in a field programmable gate without extending its dimensions.

The main property of semiconductor spintronics is the injection of spin-polarized electrons into semiconductors. Research groups have tried to achieve this goal based on the experience of GMR just by adding ferromagnetic metal contacts to semiconductor devices. After a series of failing experiments, a theoretical model was developed, showing that the mismatch in conductivities of the metal contacts and the semiconductor prevents spin injection in these devices, and that there is no way to circumvent this effect without using additional barriers. However, typical ferromagnetic metals exhibit a spin polarization of only 60–70%. The model also predicts that using materials with a spin polarization very close to 100%, instead of the metallic ferromagnets, can lead to high-efficiency spin polarization.

One class of materials that can exhibit such a high spin polarization are the dilute magnetic II–VI semiconductors. Some of these paramagnetic semiconductors such as ZnMnSe at low temperatures and moderate magnetic fields (4.2 K; 1–2 tesla) exhibit a spin polarization of 99.9% or more. In 1999, the first successful demonstration of spin injection into GaAs was published. In the experiment, a layer of a dilute magnetic semiconductor (ZnBeMnSe) was grown on top of a GaAs/AlGaAs *pin* light-emitting diode (**Fig. 3**). If such a diode is fed from one side with unpolarized holes and from the *n*-side with spin-polarized electrons, the emitted light of the diode will show a circular optical polarization whose degree is proportional to the degree of spin polarization of the electrons. The results of the experiment indicated a spin polarization of the injected electrons as high as 90%. While this was the first demonstration of high-efficiency spin injection into semiconduc-

tors, it was also the first example of a spin-based semiconductor device. It can easily be seen that the concept may also be transferred to lasers, and that a laser with a tunable polarization would at least double the transfer rate of fiber optics.   Georg Schmidt

Magnetoelectronic Materials

Recent hybridization of semiconductors and ferromagnetic materials has yielded novel functional transport, optical, and magnetic devices. Such materials promise new functionality that potentially may find application in spin-based quantum computation and communication, using quantum-mechanical coherence (coupling) of spins.

Diluted magnetic semiconductors. In semiconductor devices, electron spin has long been neglected since most semiconductors in use are nonmagnetic. As the interest in spin-polarized transport has increased, much effort has been devoted to the development of a new class of semiconductors that exploit semiconductor-based spintronics—ferromagnetic semiconductors.

Just as doping of ionized impurities (donors or acceptors) makes intrinsic semiconductors conductive, doping of magnetic elements, such as manganese (Mn), can make nonmagnetic semiconductors magnetic. Among the various alloys of nonmagnetic semiconductors and magnetic elements, called diluted magnetic semiconductors (DMS), Mn-doped II–VI compounds such as zinc selenide (ZnSe) and cadmium telluride (CdTe) and III–V compounds such as indium arsenide (InAs) and gallium arsenide (GaAs) have been extensively studied.

In II–VI compounds, the divalent cation (Zn or Cd) can be easily replaced by divalent Mn. However, magnetic interaction among the Mn spins is antiferromagnetic, resulting in paramagnetic, antiferromagnetic, or spin-glass behavior in II–VI dilute magnetic semiconductors.

In III–V compounds, substitution of Mn for the trivalent cation (Ga or In) induces holes, making the material *p*-type. In III–V compounds, the low solubility of magnetic elements has been the main obstacle to enhancing the magnetic properties of these materials. A breakthrough, however, was made by using low-temperature molecular beam epitaxy, a thin-film growth technique in vacuum under highly nonequilibrium conditions. In 1989, epitaxial growth of uniform indium manganese arsenide (InMnAs) films on GaAs substrates was reported, and in 1992 partial ferromagnetic order was found in this alloy. In 1996, uniform gallium manganese arsenide (GaMnAs) was grown on a GaAs substrate with Mn concentrations up to about 7%, which was found to be ferromagnetic with a Curie temperature T_C of ~110 K. The discovery of this GaAs-based ferromagnetic semiconductor spurred experimental studies of semiconductor ferromagnet/nonmagnet heterostructures.

Hole-induced ferromagnetism. The direct magnetic interaction between the Mn atoms in III–V diluted magnetic semiconductors was shown to be also

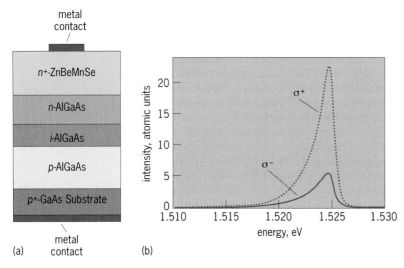

Fig. 3. Light-emitting diode with spin injection contact. (*a*) Schematic structure. (*b*) Emission spectrum for left (σ^-) and right (σ^+) circular optical polarization. The difference in intensity indicates the degree of spin polarization of the driving current.

antiferromagnetic in *n*-type InMnAs and in semi-insulating GaMnAs, indicating that the ferromagnetism is induced by holes doped in these materials. Indeed, ferromagnetism is observed in II-VI diluted magnetic semiconductors when holes are induced. Two approaches have been proposed to explain hole-induced ferromagnetism. The mean field model, which is based on exchange interactions mediated by delocalized holes in the ensemble of localized spins, accounts for the experimental findings such as the Curie temperature (T_C) of GaMnAs quite well. The theory was applied to calculate the T_C of various diluted magnetic semiconductors as functions of hole and Mn concentrations, and it predicted a high T_C above room temperature for such materials as zinc oxide (ZnO) and gallium nitride (GaN). The other approach explains carrier-induced ferromagnetism in terms of the double exchange interaction, in which *d*-electrons jump between localized Mn atoms, bringing about ferromagnetic ordering of their magnetic moments. Based on these theories, material design and synthesis of new diluted magnetic semiconductor with room-temperature ferromagnetism are in progress.

Spin-dependent transport in all-semiconductor structures. The use of diluted magnetic semiconductor instead of metals offers an advantage in that the properties of the heterointerface, such as lattice matching and band offset, are well understood and controllable, using mature semiconductor technologies.

Spin-dependent transport, such as giant magnetoresistance and tunneling magnetoresistance effects, have been observed in all-semiconductor structures based on ferromagnetic GaMnAs and nonmagnetic gallium aluminium arsenide [Ga(Al)As]. In a GaMnAs/AlAs/GaMnAs trilayer structure, where aluminium arsenide (AlAs) serves as a tunneling barrier for spin-polarized holes, a tunneling magnetoresistance effect of 70% resistance change has been observed at 8 K.

Spin injection. Spin injection, that is, applying spin-polarized electrical current to nonmagnetic semiconductors, is the basis of future semiconductor spintronics. For this purpose, numerous experiments were done using contacts of ferromagnetic metals with semiconductors, but no clear evidence of effective spin injection has been obtained yet. Instead, by using diluted magnetic semiconductor as a source of spin-polarized carriers, direct spin injection has been realized.

Figure 4a shows an example structure in which ferromagnetic GaMnAs was used as a source of spin-polarized carriers (holes). To measure the spin injection, a light-emitting-diode structure was used. By analyzing the polarization of the light emitted from the structure, one can evaluate the efficiency of spin injection since it is related to the spin polarization of the recombined carriers by the selection rule for optical transitions. Figure 4b shows the observed hysteretic loop of luminescence polarization, indicating that the current injected from GaMnAs is partly spin-polarized.

The effects of spin-polarized transport have been observed in ferromagnetic/nonmagnetic semiconductor heterostructures at low temperatures because of the lower T_C of ferromagnetic semiconductors. However, these results proved the ability of semiconductor-based spintronics devices, and promoted further development in the understanding of spin-polarized transport in semiconductors and the creation of new ferromagnetic semiconductors with higher T_C.

Outlook. Since the late 1990s, quantum computing communication has attracted great interest. It promises an exponential speed increase for such a problem as factorization due to quantum-mechanical parallelism. To implement quantum computers, quantum states (quantum bits) are needed with sufficiently long spin coherence times. In several proposal solid-state quantum computers, electron or

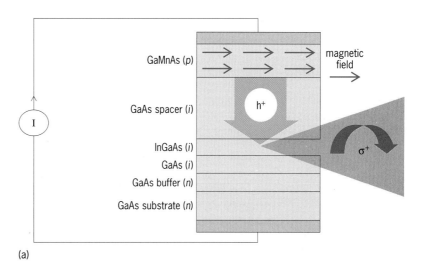

(a)

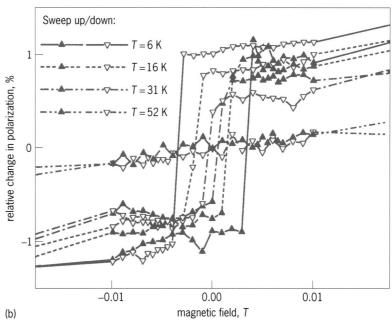

(b)

Fig. 4. Spin injection light-emitting-diode structure. (*a*) Cross-sectional view. (*b*) In-plane magnetic field dependence of the luminescence polarization measured at 6–52 K.

nuclear spins are the most promising candidates for quantum bits, where spin injection might be a useful tool to manipulate nuclear spins and detect the spin state of an electron or nuclei after a calculation. Although there are many hurdles to overcome, the observation of long spin coherence times and the realization of electrical spin injection in semiconductors advance the development of semiconductor quantum computers. *See* QUANTUM COMPUTATION.

For background information *see* ANTIFERROMAGNETISM; BAND THEORY OF SOLIDS; COMPUTER STORAGE TECHNOLOGY; CURIE TEMPERATURE; ELECTRICAL CONDUCTIVITY OF METALS; ELECTRON SPIN; FERROMAGNETISM; FREE-ELECTRON THEORY OF METALS; HOLE STATES IN SOLIDS; INTEGRATED CIRCUITS; MAGNETIC MATERIALS; MAGNETIC RECORDING; MAGNETORESISTANCE; PARAMAGNETISM; SEMICONDUCTOR; SEMICONDUCTOR HETEROSTRUCTURES; SEMICONDUCTOR MEMORIES; SPIN GLASS; TUNNELING IN SOLIDS in the McGraw-Hill Encyclopedia of Science & Technology.   Yuzo Ohno

Bibliography. M. N. Baibich et al., Giant magnetoresistance of (001)Fe/(001)Cr magnetic superlattices, *Phys. Rev. Lett.*, 61:2472–2475, 1988; G. Binasch et al., Enhanced magnetoresistance in layered magnetic structures with antiferromagnetic interlayer exchange, *Phys. Rev. B*, 39:4828–4830, 1989; S. Das Sarma et al., Theoretical perspectives on spintronics and spin-polarized transport, *IEEE Trans. Magn.*, 36:2821–2826, 2000; R. Fiederling et al., Injection and detection of a spin-polarized current in a light-emitting diode, *Nature*, 402:787, 1999; R. Fitzgerald, Magnetic semiconductors enable efficient electrical spin injection, *Phys. Today*, 53(4):21–22, April 2000; H. Ohno, Making non-magnetic semiconductor magnetic, *Science*, 289:951–956, 1998; G. A. Prinz, Magnetoelectronics, *Science*, 282:1660–1663, 1998; G. Schmidt et al., Fundamental obstacle for electrical spin injection from a ferromagnetic metal into a diffusive semiconductor, *Phys. Rev. B*, 62:R4790-R4793, 2000.

Maternal serum screening

Until recently, prenatal screening for common fetal disorders such as Down syndrome and neural-tube defects was based solely on clinical examination of the pregnant woman and a review of her medical and family history. Pregnancies identified as at risk were then followed up with more accurate diagnostic testing methods such as amniocentesis or chorionic villus sampling. In general, only about 1% of these at-risk pregnancies were found to have a significant problem, and the diagnostic testing methods had approximately a 0.5% risk of causing the loss of a normal pregnancy. Since the means of identifying at-risk pregnancies were so inaccurate and the diagnostic testing carried a risk of the loss of a normal pregnancy, most women could not be offered an effective method of advising them of the risk of problems in the pregnancy. As a result, more than 80% of all pregnancies with Down syndrome or neural-tube defects were not found until birth.

Recently, however, it has been demonstrated that reliable information about fetal health can be obtained by measuring substances released by the fetus into the maternal blood. This new technique has made it possible to offer all pregnant women a blood-screening test to assess fetal health.

Alpha fetoprotein. One of the first fetal substances that was found in the mother's blood (maternal serum) was alpha fetoprotein (AFP). The United Kingdom Collaborative Study published in 1977 clearly demonstrated that alpha fetoprotein could be measured in the maternal blood sample and would give reliable information regarding the health of the fetus. The study researchers had access to maternal sera from women who were known to have pregnancies both with and without neural-tube defects. By pooling the data from this population, it was firmly established that an elevated level of alpha fetoprotein in the maternal blood was very strongly correlated with the presence of a neural-tube defect in the pregnancy. This is probably because most neural-tube defects are associated with a break in the skin that covers the back, causing the fetus's blood to come in closer contact with the amniotic sac and alpha fetoprotein to be transferred at higher levels into the mother's circulation.

Standardized measurement methods have been developed to accurately and reliably measure alpha fetoprotein levels in a pregnant woman's circulation according to her week of gestation (from approximately 14 through 22 weeks). Using relatively standardized mathematical modeling (likelihood ratios), the level of alpha fetoprotein in the mother's circulation can be calculated into a risk of the fetus having a neural-tube defect. Pregnancies identified as being at higher risk can then have appropriate diagnostic testing (either amniocentesis or detailed fetal ultrasound) to provide a more accurate assessment of the health of the fetus. By setting cutoff levels of alpha fetoprotein based on known population occurrence of neural-tube defects, all women can be offered screening for these conditions. In most screening programs it has been found that by advising 1% of women that their alpha fetoprotein is in the at-risk range, about 70% of fetuses with neural-tube defects can be identified.

Multiple-marker pregnancy screening. Following the recognition that substances in the maternal circulation can be used in the detection of fetal disease, which was clearly proven with the association of alpha fetoprotein and neural-tube defects, numerous biochemical markers (including alpha fetoprotein, estriol, human chorionic gonadotropin, inhibin, and pregnancy-associated protein A) in the maternal blood were identified that can be used for the detection of Down syndrome. These (and future) markers are studied using the same population methods used to study alpha fetoprotein (by comparing marker levels in pregnant populations with and without a particular disease). Depending on the marker's positive rates (the proportion of pregnancies identified as at risk) and detection rates (the proportion of pregnancies from the positive population that are found to have a problem), its use in screening can be determined. In pregnancies with Down syndrome, maternal serum levels of alpha fetoprotein, estriol, and pregnancy-associated protein A tend to be low, whereas levels of human chorionic gonadotropin and inhibin tend to be high. Various combinations of these markers can be used to provide risk estimates for pregnancies with Down syndrome. This use of multiple-marker testing has led to more complicated mathematical modeling, which has been made practical as computers have become available in the laboratory-testing sector. This method of maternal serum screening for fetal disease is now relatively common in most areas of the world. Although local methods of screening (for example, diseases screened or laboratory methods used) vary from locale to locale, most programs at least screen for Down syndrome and neural-tube defects.

Multiple-marker screening programs can include combinations of different information regarding the woman's age, the exact gestation of the pregnancy as determined by ultrasound, physical changes identified by ultrasound, as well as a combination of serum markers. Depending on which combination of markers is used, conditions such as neural-tube defects, Down syndrome, trisomy 18, and gastroschisis (a rare condition in which the intestines develop outside the stomach) can be reliably screened for.

Screening has become more complicated as more biochemical markers have been detected in maternal circulation, and there are now markers (alpha fetoprotein, estriol, human chorionic gonadotropin, inhibin, and pregnancy-associated protein A) that can be used as early as 11 weeks of gestation and as late as 19 weeks. Used in combination with medical history and ultrasound dating, these maternal serum markers can provide increasingly accurate information regarding the health status of the fetus.

Screening considerations. Because of all the variables involved, maternal serum screening must be started with an informed-consent philosophy in place. That is, a pregnant woman should clearly understand her options and the intended results of screening before any testing is begun. For example, the advantages of early (first trimester) screening in pregnancy are that it generally occurs before the woman is physically showing signs of pregnancy and usually allows for short outpatient treatment, rather than hospitalization, if a decision is made to terminate the pregnancy. However, the major disadvantage is that about one-third of the pregnancies detected with problems in the first trimester would miscarry spontaneously without any screening. Therefore, about one-third of the patients who require treatment as a result of first-trimester screening would not require treatment if they waited until the second trimester.

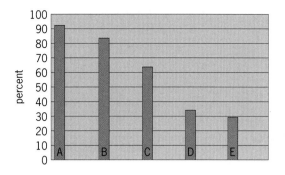

Key:

A Age, ultrasound, and four serum markers

B Age, ultrasound, and two serum markers

C Age, no ultrasound, and four serum markers

D Age, no ultrasound, and one serum marker

E Age alone

Comparison of the effectiveness of different multiple-marker pregnancy screening programs for Down syndrome. Estimates of detection rates with a 5% positive rate (5 of 100 woman are advised that they have at-risk pregnancies). A: age, ultrasound, and four serum markers. B: age, ultrasound, and two serum markers. C: age, no ultrasound, and four serum markers. D: age, no ultrasound, and one serum marker. E: age alone.

Some women elect for early pregnancy testing based on information from only one blood test, whereas others elect for a series of tests and the single result available by integrating all the information available by various tests. There are now numerous options available for pregnancy screening—all with various strengths and weaknesses that should be fully understood as part of an organized screening program. Each combination of markers provides a different degree of detection and positive rates (see **illus.**). The selection of the best method of screening is determined by many factors, such as cost, desired level of detection, and access to follow-up diagnostic services.

For background information *see* ALPHA FETOPROTEIN; CONGENITAL ANOMALIES; DOWN SYNDROME; PREGNANCY; PRENATAL DIAGNOSIS in the McGraw-Hill Encyclopedia of Science & Technology.

Philip Wyatt

Bibliography. W. J. Krzanowski, *Principles of Multivariate Analysis: A User's Perspective*, Oxford University Press, 1990; L. S. Penrose and G. F. Smith, *Down's Anomaly*, Churchill, 1966; N. Wald and I. Leck, *Antenatal and Neonatal Screening*, 2d ed., Oxford University Press, 2000.

Metamorphism (geology)

Metamorphism is the process whereby the mineralogy, microstructure, and chemistry of rocks are changed due to changing conditions of pressure and temperature. The information contained in metamorphosed rocks thus sheds light on the burial and thermal history of Earth's crust and underlying mantle

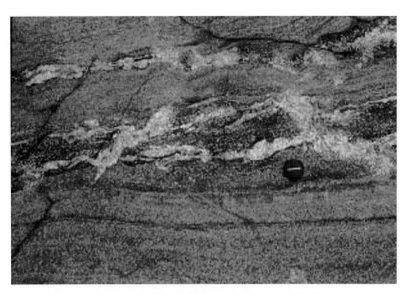

Fig. 1. Migmatite at outcrop. Light material is granite that records the location of melt. From the ancient Acadian mountain belt of west-central Maine.

in response to plate tectonics. Geologists who study metamorphism are concerned with processes at scales that range from atomic (in minerals) to global (interactions between tectonic plates). These processes occur at (1) convergent plate boundaries, where subduction (such as produced the Andes) or continental collision after a phase of subduction (such as produced the Himalayas) takes rocks deep into Earth's mantle during mountain building (orogeny); (2) at plate boundaries that involve predominantly lateral displacement (such as produced the San Andreas fault system); and (3) at divergent plate boundaries, where oceanic crust is metamorphosed immediately after formation by hydrothermal circulation of seawater at high temperature. These diverse tectonic environments generate different, perhaps unique metamorphic signatures in rocks that span a range in temperature from several hundred to more than a thousand degrees Celsius and in depth from near-surface to several hundred kilometers.

Exciting new discoveries in metamorphism have come in part from the development of improved instrumentation and the introduction of new techniques to investigate the composition and structure of Earth materials, and in part from better methods to quantify metamorphic conditions and time scales and the rates and kinetics of processes. Measured time scales have become shorter as the precision in age determination has increased. As a result, rates are now accepted that would have been thought far too fast only a few years ago. On the other hand, the slow kinetics of many metamorphic processes, especially along the decreasing temperature part of the metamorphic cycle during exhumation, which also involves decreasing depth (and, therefore, pressure), means that equilibrium commonly is not achieved. Disequilibrium features can be used to calculate the speed of tectonic processes, using

diffusion rates that are calibrated based on laboratory data. The whole metamorphic cycle involving burial (increasing pressure and temperature) and exhumation is called a pressure-temperature-time-deformation (*P-T-t*-d) path.

During the last decade, much of the attention in metamorphic petrology was focused on the extremes of pressure and temperature apparently recorded by crustal rocks involved in mountain building. The fields of ultrahigh-pressure (UHP) and ultrahigh-temperature (UHT) metamorphism have evolved in the past 15 years. UHP and UHT metamorphism were identified as a direct result of the improved ability to quantify the conditions of metamorphism. UHP metamorphism is related to the subduction of continental materials to great depths during collisional mountain building [greater than 100 km (62 mi) depth, with some arguments to suggest that crustal material may have been buried to >300 km (186 mi) depth in some places]. UHT metamorphism is related to the exhumation from depth of overthickened continental crust that commonly has been invaded extensively by magmas to generate temperatures in excess of 1000°C (1832°F) at lower crustal depths (35–40 km; 22–25 mi). Because of an improved understanding of the extent to which lower crustal rocks have been affected by metamorphism, it is now believed that the deep crust has contained melt for a significant part of the history associated with mountain building and erosion. The presence of melt has been postulated in both the Andes and the Himalayas to explain geophysical observations, and is evidenced in the exposed deep levels of ancient mountain belts as migmatites (**Fig. 1**) and residual granulites, rocks that preserve the physical evidence of melt flow pathways.

Alpe Arami enigma. In 1996, a controversial proposal was made that the Alpe Arami peridotite in Switzerland contains mineralogical evidence—in the form of abundant titanate ($FeTiO_3$) rods embedded within olivine crystals—implying an origin at a depth >300 km (186 mi). The original depth estimate was based on the crystallography of these titanate rods, and this has been supported by experiments that demonstrate an increased solubility of titanium dioxide (TiO_2) in olivine with pressure. Other corroborating evidence for UHP conditions includes exsolution lamellae of clinoenstatite (a low-calcium pyroxene) within diopside (a calcic pyroxene) crystals in the same rock as the olivine with titanate rods. Of the five different polymorphs known for low-calcium pyroxenes, the only precursor for the exsolution lamellae that is consistent with all of the crystallographic and geologic evidence is high-pressure clinoenstatite (HPclen). The conditions necessary for exsolution of HPclen provide independent evidence of a minimum depth of origin of the Alpe Arami peridotite of 250 km (155 mi). In spite of the increasing amount of apparent corroborating evidence to support the original postulate for a deep origin, others have argued against such great depth, although all agree that the peridotite formed at >100 km (62 mi).

Examples are known from other continental collision zones that imply very great depth of exhumation of mantle rocks, or subduction of crustal rocks to such depths. These include the Sulu terrane of eastern China, the Western Gneiss region of Norway, and the Erzgebirge of Germany. Thus, deep subduction and exhumation has perhaps occurred multiple times during the Phanerozoic, which would indicate that this phenomenon is a normal part of subduction and collisional mountain building. Whether such exhumation is a single- or multiple-step process is an important question for future research.

UHT metamorphism. It used to be thought that the temperatures necessary for melting crustal rocks in the absence of a free water-rich metamorphic volatile phase were unlikely to be achieved. However, scientists now know differently, and the temperatures recorded by the mineral assemblages in some crustal rocks are 300–400°C (570–750°F) above the beginning of melting due to the breakdown of hydrate minerals (micas and amphiboles). In particular, the aluminum content in solution in the mineral orthopyroxene increases with temperature, and a high aluminum content in orthopyroxene coexisting with garnet, cordierite, sillimanite, or sapphirine indicates UHT metamorphic conditions. Recently, peak temperatures of at least 1120°C (2050°F) have been calculated for a granulite from the Tula Mountains of Enderby Land, East Antarctica, based on the aluminum content of orthopyroxene coexisting with sapphirine and quartz.

Successively overprinted coronitic [successive rings of new material products replacing the outer part of a large grain of a reacting (unstable) mineral] and intergrowth reaction microstructures in granulites have allowed them to be used to deduce reaction histories and, from these, to infer *P-T-t*-d paths of UHT granulite terranes (**Fig. 2**). However, since crustal rocks under UHT metamorphic conditions generally are melt-bearing, misreading the record of microstructural features due to back reaction with coexisting melt must be avoided. After the peak temperature of metamorphism has been achieved, both close-to-constant pressure (isobaric) cooling and close-to-constant temperature (isothermal) decompression are documented in different UHT granulites. An important but commonly ambiguous issue concerns the increasing temperature path during burial and heating to the peak metamorphic conditions, which may follow a clockwise path in *P-T* space (Fig. 2) or, much less likely, a counterclockwise path involving heating before substantial burial. Different *P-T-t*-d paths help to constrain the tectonic environment within which metamorphism proceeds. For example, three different *P-T* paths are illustrated in Fig. 2: (I) from a deep contact metamorphic zone around a body of hot magma, (II) from a continental collision zone, and (III) from a UHT terrane. Composite isothermal decompression–isobaric cooling–

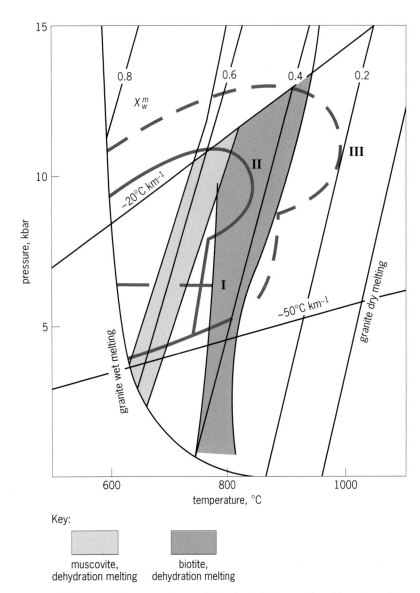

Key:

muscovite, dehydration melting   biotite, dehydration melting

Fig. 2. *P-T* diagram of the anatectic zone, that region in *P-T* space above the wet granite melting curve in which melt may be present in many common crustal rocks. The symbol X_w^m is used to denote the mole fraction of H_2O in the melt and is considered to equal the activity of H_2O in the melt. Schematic *P-T* paths: (I) Isobaric heating—cooling path characteristic of deep contact metamorphism (for example, around deep granites of the Cascades of Washington state). (II) Stepped clockwise path, characteristic of collisional metamorphism (for example, the Himalayas or the ancient Variscan belt of Europe). (III) Stepped clockwise path at UHT (for example, ancient rocks in peninsular India).

isothermal decompression postpeak temperature *P-T* paths may record periods of exhumation separated by a period of stability.

Geochronology. Although establishing rates and time scales of processes such as heating has recently come within scientists' grasp, determining the age of peak *P-T* conditions of metamorphism in ancient mountain belts remains elusive, although it can be achieved. During the past decade, techniques have been developed, principally using the common metamorphic mineral garnet, both to date close-to-peak *P-T* and to constrain the period of heating by knowing the time scale for growth of garnet. There are several advantages to using garnet as a

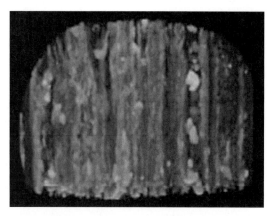

Fig. 3. Projection of three-dimensional image of layered migmatite, derived from a stack of two-dimensional representations of high-resolution x-ray computed tomography scans, created using VoxBlast.

chronometer, including its common occurrence in metamorphosed sedimentary rocks and the fact that garnet chemistry, when garnet grows in equilibrium with appropriate other minerals, can be used to obtain a good measure of the *P-T* conditions of metamorphism (using a technique called thermobarometry). Information on the time scale of metamorphism during burial and heating and exhumation and cooling also can be gained by careful use of multiple isotope systems and different methods on accessory phases, such as the minerals zircon, monazite, rutile, titanite, and apatite, although there remains the issue of being sure of what we are dating when we do not understand fully the exact crystallization history of some of these minor minerals.

Ideally, in situ dating is to be preferred in *P-T-t*-d studies so that all metamorphic and deformation information is related to a specific interval of time during the evolution. An important development in relation to structural and metamorphic studies is the

rapid chemical dating of monazite, a thorium-rich accessory mineral common in metasedimentary rocks (rocks originally deposited at the Earth's surface as sediments before burial during mountain building). This rapid dating method has been enabled by the development of in situ mapping of thorium, uranium, and lead concentrations using the electron-probe microanalyzer (an instrument more commonly used for analyzing major element concentrations in minerals and only rarely used to analyze elements present in trace quantities). For metasedimentary rocks, which are commonly used to track the burial and thermal history of ancient mountain belts, this technique offers a rapid method to obtain age information as part of routine petrology. This is particularly useful in Precambrian belts where the lower precision on ages obtained in comparison with conventional isotope techniques is sufficient to distinguish between major episodes of mountain-building activity. However, rapid chemical dating of monazite using the electron-probe microanalyzer does not eliminate the need for more time-consuming but higher-resolution and higher-precision geochronological studies (such as using isotope dilution mass spectrometry, secondary ion mass spectrometry, and laser ablation–inductively coupled mass spectrometry).

Postpeak thermal histories are better known than prograde thermal histories because there are a number of minerals from which information can be retrieved along the whole exhumation *P-T* path. These data are also commonly used to constrain models for the tectonic evolution of ancient mountain belts. It has become clear during the past decade that the time-integrated rates of cooling of orogens vary considerably. Thus, exhumation may vary from very slow, for example $\sim 1.5^{\circ}$C Ma^{-1} for at least 150 million years following the last phase of high-grade metamorphism, to extremely rapid, with rates of up to 100°C Ma^{-1} during 1–10 million years. This variation suggests a range of behaviors between limited vertical tectonic displacement and approximate isostatic equilibrium, and significant vertical tectonic displacement probably reflecting tectonic exhumation. Further, rates of cooling change during the period of exhumation.

New tools in petrology. Most new tools relate to microscopy or in situ analysis, although the increase in computer power and its availability has allowed more complex tectonic models of mountain belts to be developed. Advances in image analysis have also opened up new research directions. The development of techniques for mapping in situ distributions of elements within minerals has allowed advances in understanding processes reflected by differences in element distributions. For example, serial sectioning and three-dimensional reconstruction of compositional zoning from electron backscatter images and quantitative x-ray elemental maps can be used to examine the three-dimensional growth history of large crystals grown during metamorphism. In the mineral garnet, there is a coupling between major

Fig. 4. Oblique view of outcrop of layered migmatite with sheetlike bodies of granite that record the magma extraction pathways through the crust. From the ancient Acadian mountain belt of west-central Maine.

and accessory phases during reaction progress, and because trace elements are sensitive to changes in accessory mineral assemblage or fluid composition, these features can be used to calibrate trace-element thermobarometers, identify changes in reacting assemblage, and reveal information not recorded by the major elements.

High-resolution x-ray computed tomography (HR x-ray CT) has revolutionized the analysis of the three-dimensional spatial relationships among features in rocks without destroying them, as would occur in serial sectioning. In hand samples of migmatite, for example, HR x-ray CT illustrates well the three-dimensional distribution of the melt flow network represented by granite due to the mass density contrast between it and the residual matrix (**Fig. 3**); this distribution is seen to be similar to meter-scale magma transfer sheets observed at outcrops (**Fig. 4**). In partially melted rocks, scanning electron microscope cathodoluminescence (SEM-CL) allows identification of textural features related to melting, crystallization, and melt movement that are not resolvable with the petrographic microscope. These two novel techniques offer exciting potential to characterize the nature and distribution of petrologic features, especially as related to evolution of melt-bearing rocks.

Outlook. Metamorphism has benefited from dramatic advances in both analytical capabilities and the development of quantitative methods to determine the depth-time evolution of rocks in orogens. *P-T-t*-d paths are the link between petrology and tectonics, and between the small-scale interactions between deformation and metamorphism, and mountain building. Thus, although metamorphic petrology is an essential tool available to the geologist, it is a valuable tool only if it forms part of an integrated study linked with structural geology and chronology. Indeed, the regional scale availability of *P-T-t*-d information in an orogen will enable us to constrain this history and thus discriminate among different tectonic models for the evolution of that orogen. Ultimately, the *P-T-t*-d path of UHP and UHT rocks will enable us to discriminate between hypotheses for the formation and preservation of these rocks. In the case of UHP metamorphism it is the process of exhumation that remains enigmatic, whereas in the case of UHT metamorphism it is the source of the extreme heat at modest depths of crustal thickening that must be explained.

For background information *see* DATING METHODS; GEOCHRONOMETRY; GNEISS; METAMORPHISM; MIGMATITE; OROGENY; PETROLOGY; SOLID SOLUTION in the McGraw-Hill Encyclopedia of of Science & Technology.   Michael Brown

Bibliography. M. Brown, From microscope to mountain belt: 150 years of petrology and its contribution to understanding geodynamics, particularly the tectonics of orogens, *J. Geodyn.*, vol. 32, 2001; M. Brown, The generation, segregation, ascent and emplacement of granite magma: The migmatite-to-crustally-derived granite connection in thickened orogens, *Earth Sci. Rev.*, 36:83–130, 1994; M. Brown and G. S. Solar, The mechanism of ascent and emplacement of granite magma during transpression: A syntectonic granite paradigm, *Tectonophysics*, 312:1–33, 1999; W. G. Ernst and J. G. Liou, Overview of UHP metamorphism and tectonics in well-studied collisional orogens, *Int. Geol. Rev.*, 41:477–493, 1999; S. L. Harley, The occurrence and characterization of ultrahigh-temperature (UHT) crustal metamorphism, in P. J. Treloar and P. O'Brien (eds.), What Drives Metamorphism and Metamorphic Reactions?, *Geol. Soc. Spec. Publ.*, no. 138, 1998; S. L. Harley and D. A. Carswell, Ultra-deep crustal metamorphism—A prospective view, *J. Geophys. Res.—Solid Earth*, 100:8367–8380, 1995.

Meteoric inclusions

It is believed that the vast majority of meteorites come from asteroids, and the fascinating range of minerals and mineral textures displayed by meteorites is a reflection of the diversity of these asteroids and their geological histories. Some carbonaceous and ordinary chondrite (stony) meteorites, long touted as primordial material relatively unchanged since the formation of their primitive parent asteroids, have been profoundly altered by interactions with liquid water within the first 10 million years after asteroid formation. The evidence for these reactions include the widespread occurrence in meteorites of water-bearing clay minerals, as well as many other minerals, such as carbonates and sulfates, that must have formed from aqueous solutions. However, the location and timing of the aqueous alteration and the nature of the aqueous fluids themselves have remained unknown. Related questions include: How abundant was liquid water in the early solar system? How did water arrive at the Earth? Where are the current reservoirs of water—places that could harbor life—in the solar system?

Researchers have attempted to model and understand these aqueous processes through analysis of hydrated minerals present in the meteorites and through computer simulations of the alteration process. Until recently, the major impediment to understanding the aqueous alteration has been the absence of actual aqueous fluid samples in meteorites. In 1998 aqueous fluid (water) inclusions from within halite (NaCl) were discovered and characterized in two ordinary chondrite meteorites.

Monahans and Zag meteorites. On March 22, 1998, an ordinary chondrite (called Monahans 1998) fell in Monahans, Texas. The meteorite was broken open in a filtered-air, clean room facility less than 48 hours after the fall, effectively eliminating aqueous alteration or other contamination of its interior. Upon breaking open the first sample, it was noted that the gray matrix contained locally abundant aggregates of purple halite crystals, measuring up to 3 mm in diameter (**Fig. 1**). Megascopic halite had not been previously seen in any extraterrestrial sample, although

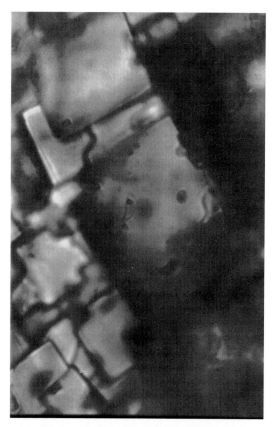

Fig. 1. Transmitted light image of halite crystals in the Monahans meteorite. The view measures 1 mm across. A small elliptical aqueous fluid inclusion can be seen near the center of the image.

it had been reported in minute quantities. Crystals of sylvite (KCl) were present within the halite crystals, similar to their occurrence in terrestrial evaporites. The purple color of the halite was probably caused by exposure to beta-decaying potassium isotopes (^{40}K) in the sylvite. The presence of halite and sylvite only within the gray matrix indicates that the halite may have formed at the surface of the asteroid, before final aggregation of the meteorite from primordial bits of dust, mineral grains, rock fragments, and chondrules. However, its deposition must have postdated the most severe thermal metamorphism and shock, because halite is such a brittle material that it could not have survived these processes. In August 1998, a similar meteorite (called Zag) fell in Morocco. Inspection of pristine samples of this meteorite also showed the presence of dark blue halite crystals. This suggests that halite may be relatively abundant in certain chondrite meteorites but is routinely destroyed during residence on the ground or during sample handling. The halite from Monahans and Zag was dated by three isotopic techniques (Rb/Sr, $^{39}Ar/^{40}Ar$, I/Xe), and all gave dates at the dawn of the solar system, some 4.5 billion years ago.

Findings. Fluid inclusions are microsamples of fluid that are trapped as isolated inclusions in a crystal during growth (primary inclusions) or at some later time along a healed fracture in the mineral (secondary inclusions). Both primary and secondary fluid inclusions were found in the halite from both Monahans and Zag. The presence of secondary inclusions in the halite indicates that aqueous fluids were locally present following halite deposition, suggesting that the aqueous activity could have been episodic.

The inclusions ranged up to 15 micrometers in their longest dimension (**Fig. 2**). At room temperature, a few of the inclusions (~25%) contained bubbles that were in constant motion, indicating that the inclusions contained both a low-viscosity liquid and vapor. Raman microspectroscopy showed that the fluid was an aqueous solution. The freezing point of these aqueous fluids indicates that the inclusions likely contained divalent cations such as Fe^{2+}, Ca^{2+}, or Mg^{2+}, in addition to Na^+ and K^+; and dissolved iron and calcium are the most likely source of the lowered freezing point. The rarity of vapor bubbles in the fluid inclusions in Monahans and Zag halite suggests a low formation temperature (<100°C).

The most important finding was that an actual sample of the aqueous fluid presumably responsible for aqueous alteration of early solar system material was identified and characterized as a brine (salt solution). The two possible origins for the brine are indigenous fluids flowing within the asteroid, and exogenous fluids delivered to the asteroid surface from a salt-containing icy object such as a comet. In either case, the inclusions provide information about the nature of water in the early solar system.

The halite in these meteorites dates to within a few million years of the birth of the solar system, as dated by the most primitive meteorites. The water

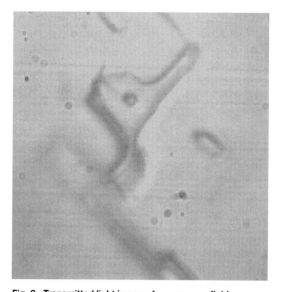

Fig. 2. Transmitted light image of an aqueous fluid inclusion in a halite crystal in the Monahans meteorite. The fluid inclusion measures 15 micrometers across. A round vapor bubble can be seen inside the fluid inclusion.

enclosed within the halite must be at least as old, and possibly far older. It may be preserved interstellar water, which would tell a great deal about the chemistry of interstellar clouds. The finding of preserved liquid water in these meteorites indicates that it was far more abundant at the very dawn of the solar system than realized, and that the conditions necessary for the development of life may have been present at this ancient time. It is possible that all the Earth's water arrived within chondrite meteorites very much like Monahans and Zag, as well as more carbon-rich carbonaceous chondrites. Further analysis of the water locked up in the halite in meteorites may finally answer these critical questions.

For background information *see* ASTEROID; COSMOCHEMISTRY; ELEMENTS, COSMIC ABUNDANCE OF; HALITE; ISOTOPE; METEORITE; ROCK AGE DETERMINATION; SOLAR SYSTEM; WATER in the McGraw-Hill Encyclopedia of Science & Technology.

Michael Zolensky

Bibliography. D. Barber, *Geochimica et Cosmochimica Acta*, 45:945, 1981; J. Berkley, J. Taylor, and K. Keil, *Geophys. Res. Lett.*, 5:1075, 1979; H. Y. McSween, *Meteorites and Their Parent Planets*, 2d ed., Cambridge University Press, 1999; M. E. Zolensky et al., *Science*, 285:1377–1379, 1999; M. E. Zolensky and H. Y. McSween, in J. Kerridge and M. Matthews (eds.), *Meteorites and the Early Solar System*, University of Arizona Press, 1988.

Microfabrication (nerve repair)

If their axons are severed, nerves in the peripheral nervous system retract from the injury site and sprout growth cones from the tips of the proximal axons (that is, the segment of the axon that remains attached to the neuron cell body). These growth cones sense the local environment and try to lead the axon to grow toward an appropriate target using various mechanisms (**Fig. 1**). The distal segment of the

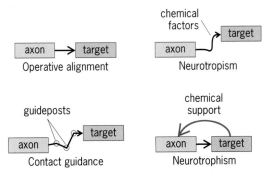

Fig. 1. Mechanism for growth cone guidance. With operative alignment, the axon must be physically aligned with the target. For neurotropism, the growth cone follows a chemotactic gradient of neurotransmitters secreted by the target tissue. With contact guidance, the axons follow physical guideposts to reach their target. In neurotrophism, growth cones will reach a target and receive chemical support if it is the correct one.

nerve—the part separated from the neuron cell body—begins to degenerate, and the Schwann cells there (the cells that support and insulate the neurons) secrete neurotrophic factors to attract the growth cones and stimulate nerve regeneration. The proximal axons eventually will cross the injury gap and reach the distal segment. Once there, they will enter the endoneural tubes that remain from the degenerated nerve segment and extend to make new synaptic connections. Neurons that fail to reach their target within a certain amount of time will lose function or die. In these cases, recovery would require the surgical grafting of new neurons from donor tissue.

A promising technique to enhance peripheral nerve regeneration is the artificial nerve graft, a porous or resorbable tube containing matrix material that may lead axons to grow in the desired direction. Successful axonal regeneration in the peripheral nervous system has been achieved for short graft lengths in nonhuman primates. However, there has been little or no success with longer grafts. These conduits control nerve growth only on a macroscopic scale, while the ability to control nerve growth at the micrometer or cellular level is crucial in improving the nerve regeneration process. Directional nerve growth is important for the axons on the proximal end of a severed nerve to meet with the distal stump to enable regeneration to occur. For successful regeneration, the nerve growth cone needs both physical guidance and chemical signals from one or more neurotrophic compounds, such as laminin, collagen, nerve growth factor (NGF), basic fibroblast growth factor, and fibronectin.

Effect of physical and chemical cues. Nerve cells recognize the three-dimensional geometrical configurations of substrates, and their growth can be guided and controlled at the cellular level by fabricating microgrooves on artificial substrates. Outgrowth of neurites (thin, hairlike neuronal projections that may go on to form axons), occurs along these microgrooves, and deeper microgrooves show better guiding ability. The influence of chemical cues on neurite outgrowth has been demonstrated by using micropatterned laminin regions on glass coverslips to guide nerve growth. The neurons also exhibit sensitivity to the dimensions of the chemical micropatterns. Neurons cultured in glass dishes without grooves form simple neural networks, and it is difficult to recognize connections among them. Mathematical models have been developed to understand neurite growth-cone directional changes in response to physical and chemical cues.

Micro- and nanofabrication techniques used in the computer industry for making microprocessors and other integrated circuits can be used to create materials to provide physical or chemical guidance cues for directional axon extension. Such biomimetic (mimicking life) approaches can lead to improvements in peripheral nerve regeneration to obtain desired growth rates and regain functionality.

Physical micro/nanopatterning techniques. Photolithography, a photochemical patterning technique, can be used to fabricate microgrooves on substrates. A substrate, such as silicon or quartz, is coated with a photoresist (a light-sensitive coating) and exposed to ultraviolet light through a photomask (a clear substrate imaged with the desired pattern for exposure). Next, a developer solution removes either the exposed regions of a positive photoresist or unexposed regions of a negative photoresist. The remaining photoresist serves as a mask for selective etching of the substrate. Deep channels or grooves are cut into the substrate's surface by reactive ion etching (RIE) with ion plasma. The groove depth is controlled by the etching time, and the patterns can be etched to a resolution of 1.5 μm.

Microscale grooves can be directly etched onto a substrate using laser ablation—energetically ejecting atoms from the substrate's surface. The amount of laser-ejected material depends on the acceleration voltage, number of pulses, and pulse duration.

Atomic force microscopy (AFM) can also be used to form grooves in a surface. An atomic force microscope consists of a probe tip that scans the top of a surface. During normal use, the tip is deflected by the molecular forces of the surface atoms, and a laser is used to monitor these deflections. If a force is applied to the tip of the atomic force microscope, the molecular forces between the tip and surface atoms cause the atoms to shift out of place, leaving nanometer-sized grooves on the substrate's surface. The depth of the grooves is dependent on the force applied to the tip. The surface must be made as smooth as possible to prevent uneven patterning.

These techniques can be used only with certain materials. Photolithography requires harsh chemicals and high temperatures to create the patterns. As a result, patterns are usually made on silicon or quartz surfaces. For similar reasons, reactive ion etching also typically uses these materials.

Though studies on oriented nerve growth can be done in vitro on silicon and quartz substrates, creating patterns on the surface of a biodegradable polymer such as poly-(D,L-lactide) is desirable from a clinical standpoint, because after the nerve is regenerated the polymer will degrade, eliminating the need for a secondary surgery to remove it. For physical guidance, grooves can be directly etched into a polymer surface using laser ablation or atomic force microscopy (**Fig. 2**). Reactive ion etching can be used indirectly by creating a silicon or quartz die and then spin-coating a polymer solution onto the surface of the die and drying. In this instance, groove depth is limited to approximately 4 μm. Compression molding is used to transfer the grooves from the die to the biodegradable polymer film, but this typically leads to accelerated degradation of the polymer.

Chemical patterning techniques. Besides creating physical patterns to control the direction of nerve growth, bioactive molecules can be selectively attached to substrate surfaces to guide the regener-

Fig. 2. Micropatterned polymer substrate with microgrooves that are 10 μm wide and 3.4 μm deep with 10 μm spacing between grooves.

ating neurites. Photolithography can be combined with other techniques for selective chemical deposition and cell adhesion. For example, a photoresist can be coated on top of a substrate, patterned, and developed. The entire substrate is then coated with a cell adhesion protein such as laminin. The remaining photoresist is removed, leaving a pattern of laminin on the surface. This technique can be used to make patterns as small as 5 μm.

Another method for patterning proteins onto flat surfaces is microstamping, or microcontact printing. This method accommodates biodegradable polymers readily, unlike the previous method. With this technique, a master die is created using standard photolithographic techniques. The die can then be used to make elastomeric (rubberlike) stamps. The stamps can be coated with a protein that will be transferred to the substrate. Microstamping can be used to create repeating biomolecular patterns on the substrate. It also can stamp patterns of different molecules on the same substrate. However, the flexibility of the elastomer can lead to lower resolution.

Conclusions. Much work has been accomplished using microfabrication techniques to guide the growth of neurons and Schwann cells in vitro. Recent studies have shown that microstamping techniques can be used to create microscale patterns for nerve growth. Dies formed by reactive ion etching, coupled with solvent casting of polymer films with adsorption of laminin in the microgrooves, have been used to create biodegradable substrates with microgrooves. Close to 100% alignment of axons was achieved by culturing neurons on these substrates. These linear systems have been incorporated in the nonlinear environment of an animal model and found to accelerate peripheral nerve regeneration.

For background information *see* BIOMEDICAL CHEMICAL ENGINEERING; INTEGRATED CIRCUIT; NERVE; NERVOUS SYSTEM (VERTEBRATE); NEUROBIOLOGY; NEURON; POLYMER in the McGraw-Hill Encyclopedia of Science & Technology.

Gregory Rutkowski; Surya K. Mallapragada

Bibliography. H. C. Hoch et al. (eds.), *Nanofabrication and Biosystems: Integrating Materials Science, Engineering and Biology*, Cambridge University Press, 1996; C. D. James et al., Aligned microcontact printing of micrometer-scale poly-L-lysine structures for controlled growth of cultured neurons on planar microelectrode arrays, *IEEE Trans. Biomed. Eng.*, 47:17–21, 2000; C. A. Miller et al., Micropatterned polymer substrates for guided Schwann cell growth in vitro, *Biomaterials*, 2001; H. C. Tai and H. M. Buettner, Neurite outgrowth and growth cone morphology on micropatterned surfaces, *Biotechnol. Prog.*, 14:364–370, 1998.

Microwave organic synthesis

In the electromagnetic spectrum, microwaves (0.3–300 GHz) lie between the radiowave frequency (RF) and infrared (IR) frequency and have relatively large wavelengths. Microwave radiation is non-ionizing and, therefore, incapable of breaking bonds. Microwaves (a form of energy and not heat) are manifested as heat through their interaction with a medium or material. They can be reflected (by metals), transmitted (by good insulators, which will not heat), or absorbed (decreasing the available microwave energy and rapidly heating the sample). Microwave energy transfer occurs by ionic conduction, either to ions in solution or to compounds with a dipole, such as water molecules, trying to align with the electric field. The amount of energy transferred is a function of both the dipole moment and dielectric constant. Even though Percy Spencer invented the underlying principles of the microwave (MW) oven about a half century ago, until recently, the application of microwave heating has been limited in the chemical industry.

Applications. The use of microwave radiation is gaining popularity for rapidly and selectively heating polar molecules. For the most part, the exploratory research in organic chemistry was done with household microwave ovens, designed to heat water using inexpensive microwave generators (magnetrons), which produced 600–1200 W of 2450-MHz microwave beams. Early laboratory microwave ovens were used to accelerate the digestion (dissolution) of envi-

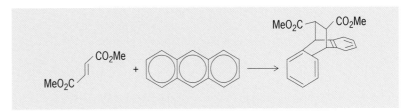

Fig. 1. Diels-Alder reaction of anthracene with dimethyl fumarate.

ronmental or mineral samples in sealed systems. Microwave applications now include waste treatment, polymer technology, ceramics, hydrolysis of proteins and peptides, and a wide range of organic reactions wherein the microwave energy is selectively absorbed by polar molecules. (Nonpolar molecules are inert to microwave dielectric loss.) For example, the polymerization of epoxy resins, used as the organic matrices in composite material, has been accomplished using microwaves.

Since the appearance of the first article on the use of microwaves for chemical synthesis in 1986, the approach has become useful for a variety of applications in organic synthesis. Diels-Alder, Claisen, and Ene reactions have been demonstrated to proceed rapidly using microwaves. The Diels-Alder reaction of anthracene with dimethyl fumarate in *p*-xylene occurred in 10 min in a microwave oven, whereas it took 4 h using conventional heating conditions (**Fig. 1**).

Hydrogenolysis of various bonds and hydrogenation of double bonds can be rapidly accomplished by microwave-assisted catalytic transfer hydrogenation, which substitutes for the hydrogen gas tank and associated pressure reaction vessel normally used (**Fig. 2**).

The initial microwave organic reactions used high dielectric solvents, such as dimethyl sulfoxide and dimethylformamide, in sealed glass or Teflon vessels with claims of a special "microwave effect." However, the rate enhancements in these reactions are now attributed to the rapid superheating of the polar solvents. These solution-phase microwave reactions had some limitations, such as the development of high pressures, the requirement of specialized containers, and a few incidents of explosions in domestic microwave ovens. To overcome these difficulties, special microwave ovens, equipped with variable frequency or focused systems, have been developed

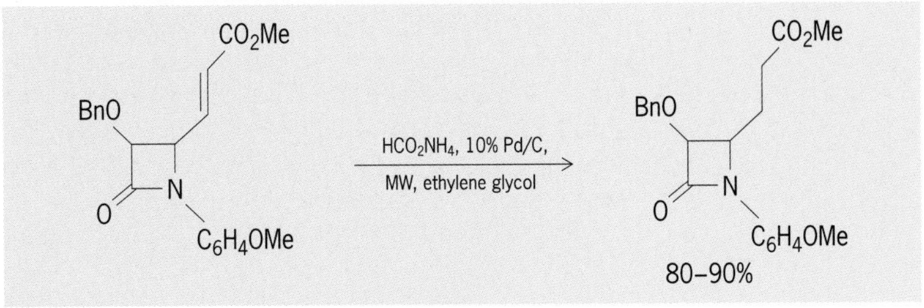

Fig. 2. Catalytic transfer hydrogenation reaction.

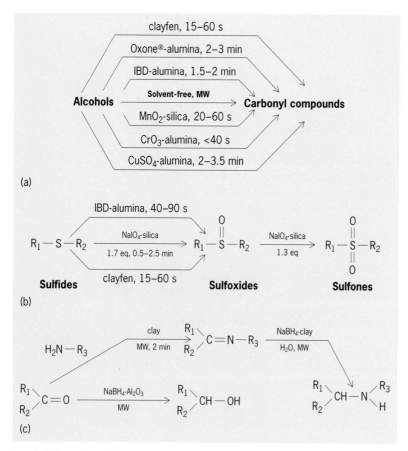

(a)

(b)

(c)

Fig. 3. Solvent-free oxidation and reduction reactions: (*a*) oxidation of alcohols to carbonyl compounds; (*b*) oxidation of sulfides to sulfoxides and sulfones; (*c*) borohydride reduction of carbonyl compounds and reductive amination reaction.

or minimizing the use of organic solvents. This approach capitalizes on the utility of heterogeneous organic reactions using the surface of inorganic oxides, such as alumina, silica, and clay or "doped" supports, which have already proven useful to chemists in the laboratory as well as in industry. These reactions are enhanced by the immobilization of the reagents on the porous solid supports, and have advantages over the conventional solution-phase reactions because of their good dispersion of active reagent sites, associated selectivity, and easier workup procedures. Some popular reagent systems that have been used in solventless oxidation and reduction reactions (**Fig. 3**) and in the synthesis of heterocycles (**Fig. 4**) are iron(III) nitrate–clay (clayfen), copper(II) nitrate–clay (claycop), hydroxylamine hydrochloride–clay, iodobenzene diacetate (IBD)–alumina, sodium periodate–silica, chromium trioxide–alumina, active manganese dioxide–silica, and sodium borohydride–clay.

The microwave-assisted synthesis of a wide variety of industrially important compounds and intermediates, namely enones, imines, enamines, and nitroalkenes, can be rapidly accomplished in the absence of solvents since the resulting water molecules in these reactions are selectively heated and eliminated. In conventional reactions, a Deans-Stark apparatus is used which requires an excess of aromatic hydrocarbons, such as benzene or toluene, for azeotropic water removal, contributing to chemical pollution.

The syntheses of radiolabeled (tritiated) organic molecules by nucleophilic aromatic and aliphatic substitution reactions, esterification, and complexation reactions have been accomplished using microwaves. Hydrogen exchange reactions that normally require elevated temperatures and extended reaction times are the primary beneficiary of the approach. The excellent regioselectivity and high purity of the labeled materials using microwaves have circumvented the traditional disadvantages

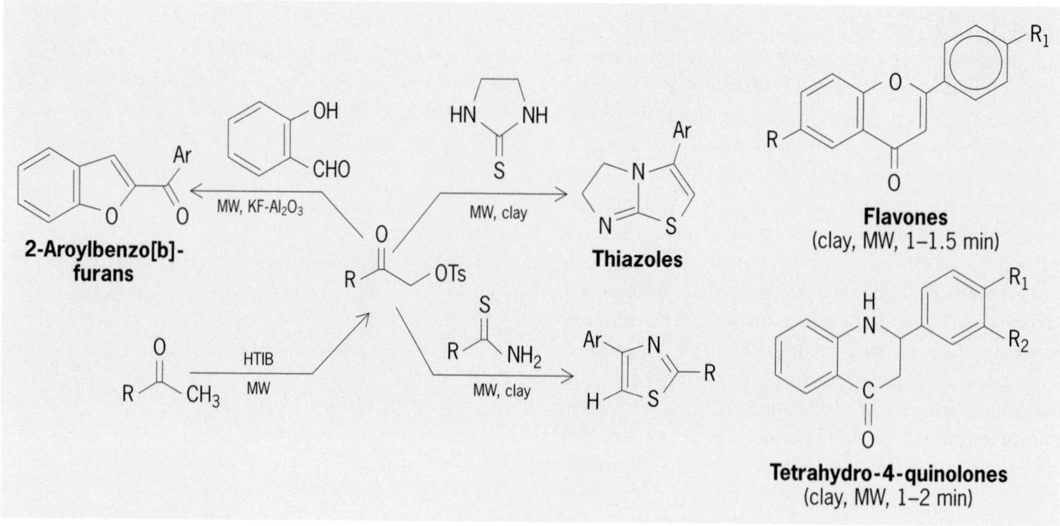

that offer reliability, efficiency, safety, and control features.

Recent developments. Recent research has focused on solvent-free synthetic methods wherein the neat reactants, often in the presence of mineral oxides or supported catalysts, undergo facile reactions to provide high yields of pure products, thus eliminating

Fig. 4. Synthesis of thiazoles and aroylbenzofurans via α-tosyloxyketones.

encountered in labeling techniques where radioactive waste generation is a problem.

Solvent-free microwave reactions occur at relatively low bulk temperature, although higher localized temperatures may be reached. The recyclability of solid supports makes these processes into truly eco-friendly ("green") protocols. Further, such microwave-assisted reactions provide an opportunity to work with open vessels, thus avoiding the risk of building up high pressures and in turn increasing the commercial scale-up potential of such reactions. *See also* GREEN CHEMISTRY.

Conclusion. From a practical standpoint, microwave reactions are easily adapted for routine laboratory methods, since they can be performed in open glass containers (test tubes, beakers, and round-bottomed flasks) using neat reactants under solvent-free conditions in an unmodified household microwave oven. The experimental procedure simply involves mixing the reactants with the catalyst or adsorbing them on mineral or doped supports, followed by a brief exposure to microwaves. This methodology is exemplified by a concise synthesis of a number of pharmacologically significant heterocycles, such as flavones, tetrahydro-4-quinolones, 2-aroylbenzofurans, and thiazole derivatives. As an example, thiazole derivatives are synthesized by the reaction of thioamides with in situ generated α-tosyloxyketones (Fig. 4) which, in turn, are obtained in solid-state reactions from readily available acetophenones and [hydroxy(tosyloxy)iodo]benzene.

An added advantage of microwave-assisted chemistry is that solventless multicomponent reactions can be adapted for the rapid generation of a library of molecules by parallel synthesis, thus providing a valuable tool for the emerging area of combinatorial chemistry. The synthesis of imidazo[1,2-a]annulated pyridines, pyrazines, and pyrimidines involving a three-component Ugi reaction is one example (**Fig. 5**).

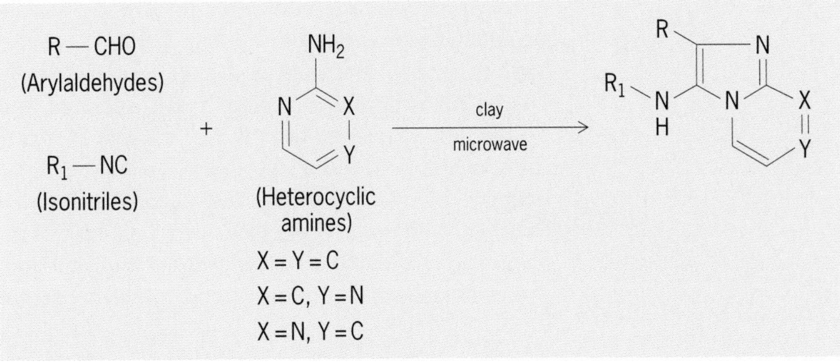

Microwave systems provide useful enabling technology for chemical laboratories, which may result in increased productivity. The engineering and scale-up prospects of the microwave-enhanced clean chemical process are actively being considered, with a handful of major industrial applications already in place. These include applications in chlorination, preparation of hydrogen cyanide, drying of pharmaceutical powders, and pasteurization of food products. The increased efficiency afforded by chemo-, regio-, and stereoselective synthesis of high-value chemicals may encourage the chemical industry to pursue large-scale microwave preparations, in view of the resulting solvent and chemical conservation. However, the successful design of bigger microwave reactors to harness the true potential of this clean technology ultimately rests on the close-knit participation of multidisciplinary teams of chemists and chemical and electrical engineers.

For background information *see* CATALYSIS; COMBINATORIAL CHEMISTRY; HETEROCYCLIC COMPOUND; HETEROGENEOUS CATALYSIS; HYDROGENATION; OR-

Fig. 5. Multicomponent microwave reactions.

GANIC CHEMISTRY; ORGANIC SYNTHESIS; OXIDATION-REDUCTION; STEREOCHEMISTRY in the McGraw-Hill Encyclopedia of Science & Technology.

Rajender S. Varma

Bibliography. R. S. Varma, Environmentally benign organic transformations using microwave irradiation under solvent-free conditions, in P. Tundo and P. T. Anastas (eds.), *Green Chemistry: Challenging Perspectives*, pp. 221–244, Oxford University Press, 2000; R. S. Varma, Expeditious solvent-free organic syntheses using microwave irradiation, in P. T. Anastas et al. (eds.), *ACS Symp. Ser.*, no. 767: Green Chemical Syntheses and Processes, pp. 292–313, American Chemical Society, Washington, DC, 2000.

Mine mechanization

The global mining industry is experiencing substantial technological advancement as well as productivity gains. Reported gains and expected benefits include increased safety, increased equipment use, improved mine cash flow, more effective maintenance, more effective time management, and improved throughput times.

Long term, it is not enough to just invest in new equipment or facilities. The mining industry must acquire agility and invest in a well-trained work force and new mining technologies, such as automation and telerobotics. Important technology development issues for mine mechanization include mining process design, machine intelligence and intelligent mine planning, and production control.

Teleoperation and automation are challenging. They relate not only to technology transfer but also to issues such as systems safety, reliability, traffic control, human factors, and the management of change. Mining automation, teleoperation, telerobotics, smart monitoring and positioning systems, and intelligent software systems are examples of technologies that will offer the potential to the industry to boost its productivity and enhance its ability to compete and prosper in the global marketplace.

Current status. During the last 15 years, mine mechanization has gained momentum. Unlike manufacturing automation, where mobile machines operate under virtually constant environmental and safety conditions within well-controlled and -monitored factory processes, the underground mine represents a challenging environment involving uncertain information and logistics, and harsh operating environments. For example, a haulage truck operates in a production environment that may change radically from area to area within the mine. In addition, the hauling activity is not always optimized or well tuned to other operational processes due to the lack of reliable and affordable communications and real-time positioning systems. Taking into account the nature of vertical transport between levels and the frequent isolation between production areas, the coordination and monitoring of individual and total mine operating systems on a daily basis are very demanding.

In the last decade, significant advances were made in underground mine communications and teleoperation. Today it is possible to operate underground mobile machines from the surface, significantly reducing the need for human involvement in underground mining operations. For instance, Inco Limited in Canada has demonstrated the teleoperation of more than one mobile underground loader at different levels within a mine from the surface—at a location 400 km (240 mi) away—using ground-based fibers optics. In the context of the evolving Internet, teleoperation may have application for the future exploitation of ore bodies in remote parts of the world. Other examples include teleoperation from the surface of drill rigs and teleoperation with automatic guidance of underground loaders (scooptrams by Inco Limited), as well as Noranda's (Canada) automatic guidance of a 26-ton truck with speeds in excess of 22 km/h (14 mi/h).

In underground soft rock mining, where rock masses are appreciably weaker than in metal mines, operations approaching continuous mining systems can be employed. The history of past mechanization has enabled these mines to contemplate full automation more readily. An example of advance mechanization in soft rock mining is provided by the Potash Corporation, mainly in its Rocanville operation in Saskatchewan, Canada.

Surface mines today have greater capability to define and control operating conditions. Recent advances in surface mine scheduling, machine monitoring, and global positioning systems (GPS) have contributed to higher productivity gains and better production control (see **illus.**).

Mines, at large, are approaching a milestone—the implementation of a comprehensive communications infrastructure that has proven to be sufficiently reliable and economically feasible to support the automation of routine mine production by using either teleoperated or autonomous machines. As a result, there is a deeper commitment on the part of industry to real-time monitoring and management.

Issues in underground mine mechanization. The future highly mechanized, teleoperated/automated underground mine will require a new set of design criteria for what will be a radically different underground environment. Machines will initially be teleoperated from the surface, although eventually these will gain in autonomy as machine intelligence is developed. The absence of human beings underground will change the needs for ventilation, heating or cooling of air, ground support, and infrastructure. Machine design will not be constrained by the need to accommodate onboard operators. Excavation sizes, layouts, and profiles will be more adaptable to ground conditions and local geology. Access tunnels will likely be smaller with reduced development costs and increased rates. In addition, productivity will be enhanced by prolonged operation of machinery, limited primarily by maintenance and service requirements.

The prospect of a highly mechanized and automated underground mine raises four key issues: (1) the suitability of conventional mining methods and equipment systems for the future teleoperated/automated mine; (2) mining performance and risk in terms of productivity, efficiency, and quality to be expected from the combinations of machinery systems and extraction methods; (3) new design developments required for both mining methods and machines to optimize the future teleoperated/automated mine; (4) automated mining performance in difficult conditions associated with increased depth, narrow ore bodies, complex geology, and low-grade distribution.

The evolution of automation in the mining industry is expected to have radical implications for future improvements in safety, productivity, and economics. The teleoperated/automated mine should facilitate lean mining, which aims to minimize throughput time, stockpiles, wastage, and rework, and improve workers' safety.

Transitional approach to high-tech mining. The global mining industry, and in particular the North American industry, is expected to face a technology breakthrough. The question is not whether the technological revolution will happen, but when will it occur. Envisioning future breakthroughs is critical to the survivability of a company or an entire sector. A smooth transition from a conventionally

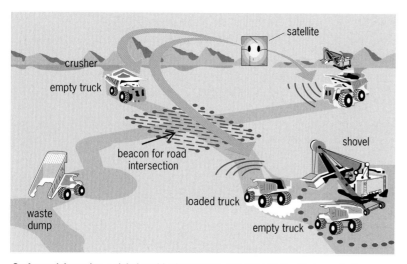

Surface mining using a global positioning system. (*Modular Mining Systems, Inc.*)

based resource industry to a dynamic and innovative global industry will require a multidisciplinary effort, involving the industry itself; technology developers and equipment manufacturers; educators and researchers together with other research organizations, centers, and institutions; governments; and international mining organizations and societies.

Such a multidisciplinary effort will create a business environment where new technologies can enjoy economical justification and support. The coordinated efforts of the mines, technology developers and manufacturers, and academic and research organizations will foster a technological environment where research and development is well focused, with a high degree of adaptability to innovation, competitiveness, and the productivity plans of the mines. At the same time, academia will educate and train the future mining work force. Working together with government will balance legislation encouraging mining innovation and application with issues related to the environment and human resources.

For background information *see* AUTOMATION; HUMAN-MACHINE SYSTEMS; INTERNET; MINING; REMOTE-CONTROL SYSTEMS; ROBOTICS; SATELLITE NAVIGATION SYSTEMS; SURFACE MINING; UNDERGROUND MINING in the McGraw-Hill Encyclopedia of Science & Technology.   Nick Vagenas

Bibliography. G. R. Baiden and E. Henderson, LHD Tele-operation and guidance—Proven productivity improvement tools, *CIM Bull.*, pp. 47–51, October 1994; R. Burgelman, M. Maidique, and S. Wheelwright, *Strategic Management of Technology and Innovation*, 2d ed., Irwin Publishers, 1996; Mining adds $524 billion to the US economy, *Min. Eng.*, pp. 16, September 1997; M. Scoble, Canadian underground mine automation: Progress and issues, *CIM Bull.*, pp. 29–32, January 1996; M. Sopko, Mining's impact on Canadian economy, *13th Mine Operators Conference*, Sudbury, Ontario, Canada, February 16–19, 1997; N. Vagenas, M. Scoble, and G. Baiden, A review of the first 25 years of mobile machine automation in underground hard rock mines, *CIM Bull.*, pp. 57–62, January 1997; J. White and L. Zoschke, Automating surface mines, *Min. Eng.*, pp. 510–511, June 1994.

Molecular cooling

The ability to cool and trap atoms has led to a renaissance in atomic physics. Workers in the field of chemical physics are looking for ways to apply atom-cooling techniques to molecules. Molecular cooling will make it possible to study molecular collisions in a regime where the wave properties of the molecules are dominant as described by quantum chemistry. Molecular cooling may also allow the testing of fundamental theories of physics.

Temperature. The temperature of a gas is related to the velocity distribution of the molecules that make up the gas. By lowering the temperature of the gas, the velocity spread is decreased. In a conservative system, the number-density of a group of particles and its temperature are coupled through Liouville's theorem; lowering of the temperature can be accomplished only by an accompanying decrease in density. The number of molecules per position interval (density) and velocity interval (temperature) is called the phase-space density. This is the quantity defining the number of cold collisions in a gas and determines, for instance, the onset of Bose-Einstein condensation. Techniques that lower the temperature of a gas and lead to an increase of the phase-space density are referred to as real cooling to distinguish these techniques from those that lower the temperature at the expense of density only. One very powerful and general method to cool a gas is by letting it expand through a nozzle into a vacuum, resulting in a dramatic decrease in the rotational, vibrational, and translational temperature. In a pulsed supersonic expansion, a beam can be produced with a high density (typically 10^{13} molecules per cubic centimeter) and a very narrow velocity distribution at a low associated temperature (typically around 1 kelvin). This expansion results, however, in a large absolute velocity and thus a high temperature in the laboratory frame. By using time-varying electric fields, the average velocity of the molecules in the laboratory frame can be lowered, thus making the low temperatures obtained in a molecular beam available in the laboratory frame.

Molecular beam deceleration. Although neutral as a whole, polar molecules do interact with electric fields as the charge is not distributed homogeneously over the molecules (one end of the molecules is more positively charged and the other end is more negatively charged). This charge separation leads to a dipole moment. The interaction of the dipole moment with an external electric field is known as the Stark effect. Depending on the orientation of the molecule with respect to the electric field, which is an inherent property of its quantum state, a molecule is attracted (a so-called high-field seeker) to or repelled (a so-called low-field seeker) from a high electric field. Static electric fields have been used to deflect and focus polar molecules since the 1920s. In the 1950s, it was realized that time-varying electric fields could be used to change the longitudinal velocity of polar molecules. A molecule in a low-field-seeking state will decelerate upon entering an electric field. If the electric field is greatly reduced before the molecule has left the electric field, it will not be accelerated when exiting the field, keeping its lower velocity. This process may be repeated by letting molecules pass through multiple pulsed electric fields. Molecules can thus be slowed down and eventually brought to a total standstill. This can be considered as the neutral analog of a charged particle accelerator. In this process, the phase-space density remains constant; the cooling has taken place in the supersonic expansion.

Comparison with other methods. In the 1990s, much progress was made in gaining control over the motion of neutral atoms, as demonstrated by the realization of Bose-Einstein condensation and the atom laser. These developments were made possible by the use of laser cooling, a technique in which atoms are cooled by many consecutive absorption-emission cycles, leading to a significant momentum transfer from the laser to the atom. Laser light acts as a friction force, damping the motion of the atoms, thereby increasing the phase-space density. Unfortunately, laser cooling requires a "simple" energy level structure, excluding most atoms and all molecules. Therefore, molecules have long been disregarded in cooling and trapping experiments. In 1998, thirteen years after the first atoms were trapped, two different methods were demonstrated to cool and trap molecules. The first method uses collisions with a cold helium buffer gas to thermalize the molecules, after which they are trapped in a magnetic trap. In preliminary experiments, 10^8 calcium monohydride molecules were trapped at a temperature of 400 mK. If the buffer gas can be removed sufficiently fast, it will be possible to cool the molecules further by evaporation. This method is generally applicable to any paramagnetic molecule or atom. In the second method, the molecules are directly formed from laser-cooled atoms, thus avoiding the problems associated with molecular cooling altogether. The so-formed dimers can be trapped in the focus of an intense laser beam. An exciting new development is to apply this laser-association technique to an atomic Bose-Einstein condensate. These two methods are restricted to paramagnetic molecules and (mixed) alkali-dimers, respectively. Deceleration of molecular beams using time-varying electric fields can be used to cool and trap polar molecules.

Applications. Lowering the velocity of molecules increases the time scale on which the molecules can be studied, and thus the accuracy of the measurement. Ultimately, all spectroscopy is limited by the time that the molecules spend in the measuring apparatus. Therefore, cooling molecules to the millikelvin regime increases the measurement accuracy by orders of magnitude. Besides studying molecular structure in detail, this might also be useful for studying fundamental physics. Molecules are currently being used to study time reversal, chirality, and other fundamental physics theories. Another goal is to study the collisional properties of cold molecules. At very low temperatures, the associated de Broglie wavelength of the molecules becomes larger than the classical length of the intermolecular bonds. Therefore, the classical picture of colliding marbles must be replaced by a total quantum-mechanical description of the collision. Studying cold collisions may therefore open up an entire new field in chemistry. Another interesting topic is the study of the mutually interacting electric dipoles. At low temperature, the translational energy is comparable to the dipole-dipole interaction energy. It may be possible to observe the formation of a dipolar crystal analog to the occurrence of ionic crystals in ion traps.

Decelerating and trapping polar molecules. A pulsed molecular beam is formed by expanding a mixture of ammonia and xenon in a vacuum (**Fig. 1**). Due to rotational and vibrational cooling, only the lowest quantum states of ammonia are populated in the beam. In a second vacuum chamber, the ammonia molecules are slowed down from 260 m/s to 13 m/s (853 ft/s to 43 ft/s) using a Stark decelerator consisting of 63 deceleration stages. The deceleration stages are formed by two 3-mm-diameter (0.12-in.) rods spaced 2 mm (0.079 in.) apart. The electrodes are connected to four switchable power supplies delivering a voltage of either 10 or −10 kV (**Fig. 2**).

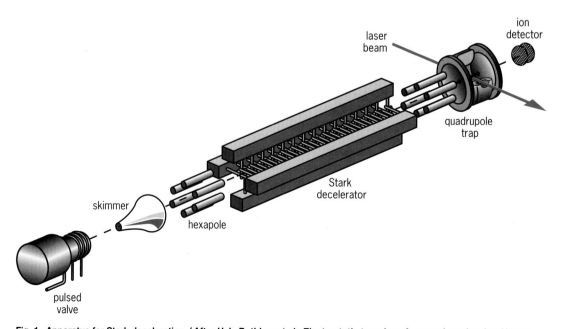

Fig. 1. **Apparatus for Stark deceleration.** (*After H. L. Bethlem et al., Electrostatic trapping of ammonia molecules, Nature, 406:491–494, 2000*)

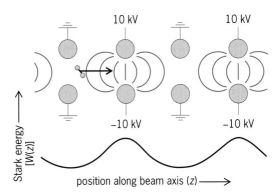

Fig. 2. Scheme of Stark deceleration. The electrode configuration is shown together with the potential (Stark) energy of a molecule as a function of its position along the molecular beam axis. (*After H. L. Bethlem et al., Trapping neutral molecules in a traveling potential well, Phys. Rev. Lett., 84:5744–5747, 2000*)

TEMPERATURE PHYSICS; MOLECULAR BEAMS; PARTICLE TRAP; STARK EFFECT; STATISTICAL MECHANICS in the McGraw-Hill Encyclopedia of Science & Technology.   Hendrick L. Bethlem; Giel Berden; Rienk T. Jongma; Gerard Meijer

Bibliography. J. T. Bahns, P. L. Gould, and W. C. Stwalley, Formation of cold ($T \leq 1$ K) molecules, *Adv. Atom. Mol. Opt. Phys.*, 42:171–224, 2000; H. L. Bethlem et al., Electrostatic trapping of ammonia molecules, *Nature*, 406:491–494, 2000; B. Goss Levi, Hot prospects for ultracold molecules, *Phys. Today*, pp. 46–50, September 2000; J. A. Maddi, T. P. Dinneen, and H. Gould, Slowing and cooling molecules and neutral atoms by time-varying electric-field gradients, *Phys. Rev. A*, 60:3882–3891, 1999; J. D. Weinstein et al., Magnetic trapping of calcium monohydride molecules at millikelvin temperatures, *Nature*, 406:148–150, 1998.

Molecular rheology

The friction between two moving surfaces, such as a piston in an automobile engine and the cylinder surrounding it, is routinely reduced by a lubricant between the surfaces. The viscosity of a lubricating fluid is determined by measuring the force (per unit area) of the surface required to maintain the surfaces moving at a constant relative speed: the higher this force, the higher the viscosity (**Fig. 1**). Another way to determine viscosity is to measure the terminal speed with which a metal sphere falls through a column of the fluid: the lower the terminal speed, the higher the fluid's viscosity. Viscosity thus measures a fluid's ability to resist flow. Clearly, water is much less viscous than motor oil, which in turn is much less viscous than a typical polymer melt (such as polyethylene). Moreover, the viscosity of a typical polymer decreases as the shear rate (the ratio of the relative velocity of the two moving surfaces to the distance between them) increases, which is known as shear thinning. Shear thinning behavior is an example of non-newtonian behavior. Ultimately, the differences in the viscosity of these fluids must reflect differences in the shape of the molecules in the fluids and the forces between the molecules. For example, polymer molecules are long and can become entangled, thus

The computer-controlled pulsers switch the electric fields synchronously with the bunch of decelerated molecules. This decelerated bunch is focused using a hexapole lens into a electrostatic quadrupole trap formed by a ring electrode and two end-cap electrodes. By initially applying an asymmetric voltage to the end caps, the molecules are decelerated upon entering the trap. When the molecules are brought to a standstill near the center of the trap, the voltages are switched to create a symmetric electric field with a minimum at the center of the trap. The molecules are attracted to this minimum and have insufficient energy to escape from it. The depth of the potential well in which the ammonia molecules are then confined is 350 mK. The trapped molecules are monitored by shining in a pulsed laser beam. Densities of trapped ammonia molecules of 10^6/cm^3 at temperatures below 350 mK have been observed, with higher densities and lower temperatures anticipated for the near future—the ultimate limitation being only the phase-space density of the initial beam. At present, the ammonia molecules can be observed up to 1 second after loading the trap, limited by collisions with background molecules. This method will work for any polar molecule, including molecules that are highly relevant from a chemical perspective, such as H_2O and OH. Once these molecules are trapped at sufficiently high density, they can be further cooled by evaporation, opening the way to study them at unprecedentedly low temperatures. Cold collisions can also be studied using a crossed molecular beam setup. The ability to tune the molecular beam velocity allows collision cross sections to be determined as a function of the collision energy. Recent theoretical work has predicted that this cross section will have rather narrow resonance at low energy, caused by long-lived transition states in the collision complex. Studies of this resonance will give information complementary to that provided by conventional spectroscopy of these complexes.

For background information *see* BOSE-EINSTEIN STATISTICS; DIPOLE MOMENT; LASER COOLING; LOW-

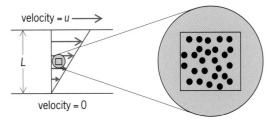

Fig. 1. Schematic of shearing motion. The lubricating fluid is sheared between the upper surface moving at velocity *u* relative to the lower surface. The surfaces are separated by a distance *L*. The shear rate is *u/L*. The force per unit area of the surface required to maintain this steady state is called the shear stress, and the viscosity is the ratio of the shear stress to the shear rate.

yielding a high viscosity; as they are sheared at high rates, they elongate and become unentangled, resulting in non-newtonian shear thinning behavior. The field of molecular rheology aims to understand and predict these rheological properties, particularly viscosity, based on molecular forces and architecture, using theory and/or molecular simulation.

Molecular simulation methods. In recent years, as a result of algorithmic advances and the continuing exponential growth in computational power, molecular simulation techniques have emerged as one of the most important tools in molecular rheology. Molecular simulation techniques for studying transport properties, including diffusivity and thermal conductivity as well as viscosity, can be broadly classified into equilibrium molecular dynamics (EMD) and nonequilibrium molecular dynamics (NEMD) methods. In an EMD simulation, Newton's equations, or a variant thereof, describing the motion of the atoms in the model system are solved as a function of time. Transport properties can be measured as integrals of autocorrelation functions determined during the simulation. For example, the viscosity η is given by

$$\eta = \frac{V}{k_B T} \int_0^\infty \langle P_{xy}(t) P_{xy}(0) \rangle dt$$

where V is the volume of the system, k_B is Boltzmann's constant, T is temperature, and t is time. The quantity $P_{xy}(t)$ is the value of the xy component of the stress tensor at time t, and so $P_{xy}(t)P_{xy}(0)$ is the stress-stress autocorrelation function and $\langle P_{xy}(t)P_{xy}(0) \rangle$ is its ensemble average (indicated by $\langle \ldots \rangle$) measured during the course of the simulation. While EMD has been used extensively for calculating transport properties, in recent years the NEMD method has become popular. In particular NEMD is a useful technique for studying rheological properties since it can also be used to probe the non-newtonian regime, common to polymers and other high-molecular-weight systems, in which the transport properties are nonlinear in the applied field.

Molecular models. As computational power has increased, so has the sophistication of the models used to describe the internal structure of molecules and the interactions between them. At the most detailed level molecules can be modeled explicitly, where each atom in the molecule is represented explicitly in the model as a single interaction site. Such explicit atom (EA) models have been developed for simple systems, such as alkanes, though for large chains they represent a significant increase in computational time over the simpler united atom (UA) model description. In a UA model the interactions between two molecules are simplified by adding together interactions between functional groups rather than atoms. For example, in a UA representation of the alkanes, the CH_2 and CH_3 groups are represented as single interaction sites. Using a UA potential instead of an explicit atom potential reduces the cost of computation significantly for larger systems. For a C_{30} alkane, for example, the computational time is reduced by a factor of about 50, due to the reduced number of force calculations and being able to ignore the fast stretching motion of the C-H bond. In order to simulate very large molecules, such as polymers, more coarse-grained models are often used which neglect much of the detail of the polymer while retaining as much of the general architecture as possible, such as connectivity. For polymeric fluids, this is typically achieved by using a "bead-spring" model where a polymer is represented as a chain of beads connected by springs. The monomers interact via a Lennard-Jones (LJ) potential (a semiempirical approximation to the potential of the force between two molecules), and nearest-neighbor monomers on the same chain interact with an additional finitely extensible nonlinear elastic (FENE) "spring" potential. In such a model, the monomers represent effective groups of many atoms. This methodology has been used to study the structure and dynamics of a wide variety of complex systems, including glass-forming liquids, polymer melts, and immiscible polymer blends.

Classification of rheological behavior. The types of systems for which molecular simulation has been used to predict transport properties can be divided into two broad classes: low-molecular-weight fluids, which do not exhibit non-newtonian behavior over any range of shear rate accessible to experiment; and high-molecular-weight fluids, such as polymer melts, colloidal suspensions, and polymer solutions, which do exhibit non-newtonian behavior over experimentally accessible shear rates. From NEMD simulations it is known that all fluids become non-newtonian at high enough shear rates, with the shear rate at which non-newtonian behavior begins generally increasing with decreasing molecular weight. As might be expected, this division into low- and high-molecular-weight fluids is not precise: for example, alkanes in the C_{20}-C_{40} range would traditionally be considered as low-molecular-weight fluids. However, in practical lubrication applications, the shear rates experienced by these fluids (such as when confined to nanoscopic gaps between surface asperities) can be sufficiently high (10^8 s^{-1} and beyond) that the fluid will become non-newtonian. At such high shear rates, it is a challenge to measure rheological properties under controlled experimental conditions; hence the value of molecular simulation in probing this regime.

Newtonian fluids. An extensive review of the application of the EMD and NEMD methods for predicting transport properties of low-molecular-weight fluids is provided by P. T. Cummings and D. J. Evans (1992). In general, molecular simulation methods have proved quite successful in predicting transport properties, even when the force fields used were not designed to predict them. Since that review, there has been considerable activity in transport properties prediction, particularly focusing on the viscosity of alkanes. The interest in alkanes has centered on molecules in the C_{20}-C_{40} range, spurred by their application in lubrication, as alkanes in this mass range form the major constituents of lubricant base stocks. In addition to the prediction of the viscosity and

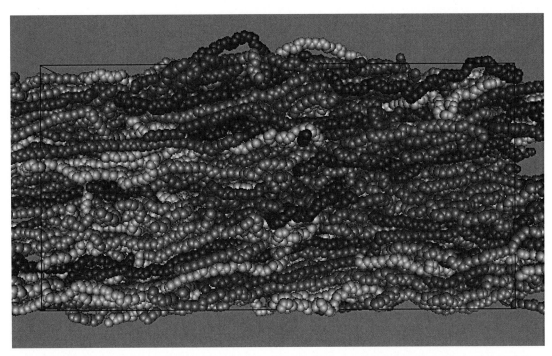

Fig. 2. Configuration of $C_{100}H_{202}$ melt undergoing shear at a high rate (large enough to induce shear-thinning behavior). In this united atom model, the molecules are shaded differently to make it easier to distinguish among them. Each interior sphere represents a CH_2 group, and the CH_3 end groups of each molecule are black. Notice how the molecules are stretched out in the direction of the shearing motion. (*From J. D. Moore et al., Molecular dynamics study of a short-chain polyethylene melt, II. Transient response upon onset of shear, J. Non-Newtonian Fluid Mech., 93:101–116, 2000*)

diffusion coefficient, recent publications have focused on predicting more sophisticated measures of alkane lubricant performance, such as the viscosity index, viscosity number, and pressure-viscosity coefficient. The ability of molecular simulation methods to efficiently predict these measures of lubricant performance is building the foundation for computer-aided lubricant design.

Non-newtonian fluids. In the realm of high-molecular-weight systems, current computing capabilities have limited the systems that can be modeled explicitly with either explicit atom or united atom models to polyethylene with a molecular weight of around 1500 (specifically, $C_{100}H_{202}$ has been the system on which both explicit atom and united atom EMD and united atom NEMD have been performed over simulation time scales long enough to probe transport properties, typically 10–100 nanoseconds) [**Fig. 2**]. It is expected that simulations of fluids containing longer polyethylene molecules and other polymers will be accomplished as computing capabilities continue their exponential increase in speed. Part of the difficulty with long-chain systems is that as chain length increases, the relaxation times increase as the square of the chain length, thus requiring systems to be simulated for much longer times. In the meantime, both EMD and NEMD simulations of high-molecular-weight systems have largely been of bead-spring models. These models permit testing of the predictions of molecular rheology theories but are not suitable for direct comparison with experiment. Nonetheless, the ability of these methods to provide direct unequivocal tests of molecular theories is ex-ceptionally valuable and has contributed directly to the development of more sophisticated molecular rheology theories.

Outlook. The future is bright for molecular rheology and the computational simulations which now lie at the core of this field, as improvements in algorithms are coupled to continuing improvements in computational speed. The increasing use of EMD and NEMD as tools in molecular rheology is an example of the inroads that molecular modeling is making into many fields of basic science and technology.

For background information *see* ALKANE; FLUIDS; LUBRICANT; MODEL THEORY; NON-NEWTONIAN FLUID FLOW; POLYMER in the McGraw-Hill Encyclopedia of Science & Technology.

Peter T. Cummings; Clare McCabe

Bibliography. M. P. Allen and D. J. Tildesley, *Computer Simulation of Liquids*, Oxford University Press, 1987; R. B. Bird, R. C. Armstrong, and O. Hassager, *Dynamics of Polymeric Liquids: Fluid Mechanics*, vol. 1, 2d ed., John Wiley, New York, 1987; P. T. Cummings and D. J. Evans, Molecular approaches to transport properties and non-newtonian rheology, *Ind. Eng. Chem. Res.*, 31:1237–1252, 1992; M. Doi and S. F. Edwards, *The Theory of Polymer Dynamics*, Oxford University Press, 1986; D. J. Evans and G. P. Morriss, *Statistical Mechanics of Nonequilibrium Liquids*, Academic Press, New York, 1990; K. Kremer and G. S. Grest, Simulations for structural and dynamic properties of dense polymer systems, *J. Chem. Soc. Faraday Trans.*, 88:1707–1717, 1992; C. McCabe et al., Examining the rheology of 9-octylheptadecane to giga-pascal pressures, *J. Chem. Phys.*, 14:1887–1891, 2001; J. D. Moore et al., Lubricant characterization by molecular simulation, *AICHE J.*, 43:3260–3263, 1997; J. D. Moore et al., Molecular dynamics study of a short-chain polyethylene melt, II. Transient response upon onset of shear, *J. Non-Newtonian Fluid Mech.*, 93: 101–116, 2000; J. D. Moore et al., Rheology of lubricant basestocks: A molecular dynamics study of C_{30} isomers, *J. Chem. Phys.*, 113:8833–8840, 2000; W. Paul, G. D. Smith, and D. Y. Yoon, Static and dynamic properties of a n-$C_{100}H_{202}$ melt from molecular dynamics simulations, *Macromolecules*, 30:7772–7780, 1997.

Mutation

In the late 1980s, the universality of the accepted paradigm for how spontaneous gene mutations form was challenged. The paradigm of S. E. Luria and M. Delbrück (1943) described spontaneous mutation as a process akin to random thermal noise—that is, spontaneous mutation occurred before cells encounter an environment in which the mutation might prove beneficial, in a cell generation-dependent manner, and more or less randomly in the genome. This was in harmony with neo-Darwinian evolutionary thought, in which rare, randomly formed mutants gain the ability to survive by accidentally acquiring genotypes more favorable for their environment than their competitors' genotypes. In this view, the environment and its role in selection of favorable genotypes is not linked to the generation of new genotypes.

In 1988, John Cairns and colleagues (and others concurrently) described a different process. They reported observing adaptive mutations—mutations that arise *after* cells are exposed to a selective environment, that occur in nongrowing or slowly growing cells, and that confer an advantage in that environment. The mere possibility of adaptive mutational processes provoked controversy because of the implicit tinge of Lamarckian evolution—adaptations to an environment that then become heritable. Adaptive mutation has now been reported in many different assay systems in bacteria and yeast and, in at least one of these systems, is known to be "real." That is, the adaptive mutations are demonstrably not normal growth-dependent mutations because they form via a novel molecular mechanism.

Recombination-dependent adaptive genetic change. The first adaptive mutation responses demonstrated to occur via molecular mechanisms different from spontaneous growth-dependent mutation were in *Escherichia coli* cells carrying a lactose catabolism (*lac*) +1 frameshift mutation on an F′ sex plasmid, and a large deletion of the chromosomal *lac* genes. These cells cannot utilize lactose. When such cells are spread on medium with lactose as the sole carbon source, Lac⁺ mutant colonies (able to utilize lactose) arise over several days. The first-appearing colonies

(day 2) represent spontaneous growth-dependent mutants that formed in the cell population before exposure to lactose. These are similar to Luria-Delbrück spontaneous mutants.

It is now known that two independent mechanisms of genetic change occur after the cells are exposed to lactose, both of which generate Lac⁺ colonies that appear later, over the next week: adaptive point mutation and adaptive gene amplification. The point mutations are frameshift reversions that are nearly all −1 (single-base) deletions in small mononucleotide repeats (similar to simple repeat instability seen in some cancers). The amplifications are direct repeats of 30 or so copies of the weakly functional *lac* frameshift allele (reminiscent of chromosomal instability seen in some cancers). The multiple copies confer enough (weak) *lac* gene function to allow growth on lactose without acquisition of a reversion mutation. Both kinds of genetic change occur after cells encounter the selective environment (starvation on lactose) and so represent inducible or adaptive genetic instability mechanisms.

Recombination-dependent adaptive point mutation. Adaptive point mutation requires proteins of the double-strand break-repair system of *E. coli*, the RecBCD system of homologous recombination, whereas growth-dependent Lac reversion does not. Because RecBCD operates only on linear DNA, double-strand breaks or ends are implicated as intermediates. [The origin of the DNA double-strand breaks is not yet known, nor is the DNA homology used in recombination, although sister molecules are possible, even in starving cells. On the F′ sex plasmid, breakage may be aided by the F-encoded transfer proteins that catalyze single-strand cleavage of the plasmid's sexual replication or transfer origin (*oriT*).] Recombination-dependent adaptive mutations are postulated to result from DNA polymerase errors occurring during DNA replication primed by recombinational strand invasion in acts of double-strand break repair. The adaptive mutation sequences, nearly all −1 base deletions in small mononucleotide repeats, are different from the more heterogeneous growth-dependent reversions, and are characteristic of DNA polymerase errors. In growing cells, polymerase-generated small frameshifts are usually corrected by the cellular mismatch repair system. However, mismatch repair activity becomes limiting transiently in cells undergoing adaptive point mutation. This differentiation could occur because of excessive polymerase errors requiring correction, or by a separate regulatory process—these possibilities have not been distinguished.

Adaptive point mutation also requires induction of the SOS response, a complex inducible response to DNA damage that is the bacterial equivalent of eukaryotic cell cycle checkpoint control: cell division and DNA replication are blocked, and DNA repair, recombination, and mutation genes are induced. One of the three DNA polymerases that are upregulated by SOS, DNA polymerase (pol) IV, is required for adaptive but not growth-dependent *lac* frameshift mutation. This intriguing role for pol IV in inducible mutability may suggest similar functions for its homologs in eukaryotes. The human homolog of this protein is highly expressed in spleen and germ-line cells and so far has no known functions.

Important to the Lamarck versus Darwin question is the finding that the adaptive mutations in this system are not directed specifically to *lac* genes. Cells containing Lac$^+$ adaptive point mutations are strewn with other mutations throughout their genomes, having 100 times or more unselected mutations than their Lac$^-$ neighbors on the lactose plates. This implies that genome-wide hypermutation takes place, but only in a subpopulation of the starving cells, and generates both adaptive Lac$^+$ and other unselected mutations. The chromosomal mutations are not uniformly distributed. There are hot (more likely to mutate) and cold regions in the *E. coli* chromosome, suggesting an intriguing *cis*-acting control that is not yet understood. Proximity to double-strand break–accessible regions in stationary-phase chromatin is a possible basis for hot regions.

Adaptive amplification. In the same experiments that detect adaptive point mutation, amplified colonies also appear and become a greater proportion of the total colonies after several days on lactose plates. The amplifications are adaptive (generated after exposure to lactose medium), involve at least one nonhomologous recombination event to form the novel junctions in their 7–40-kilobase direct DNA repeats, and also appear to require RecA, RecBCD, and Ruv recombination genes. Thus, a double-strand break intermediate is implied in amplification too. A flow diagram suggesting possible mechanisms of adaptive amplification and point mutation in the *lac* system is given in the **illustration**.

Other adaptive mutation mechanisms. Adaptive mutations have also been reported in the yeast *Saccharomyces* and in many other assay systems in *E. coli*, *Pseudomonas*, and *Salmonella* bacteria. However, in most of these, little or nothing is known about the mechanism of formation of the postselection mutations, such that it is not yet possible to tell whether these are really different from spontaneous growth-dependent mutations. One fascinating system is transcription-associated mutation in *E. coli*, in which mutant cells unable to synthesize an amino acid up-regulate the transcription of the defective amino acid biosynthesis genes when starved for that amino acid, then preferentially accumulate mutations in the highly transcribing genes. This process has the potential to produce truly directed adaptive mutations. Transcription-associated mutation does not require RecA recombination protein, and many of the other assays mentioned here also do not. If all of these systems really do display new mutation mechanisms induced in response to a selective environment (adaptive mutation), the suggestion is that adaptive mutation is a collection of mechanisms (a strategy), not a single mechanism.

Possible analogies in multicellular organisms. Whether processes like adaptive mutation in microbes also occur in multicellular organisms is an interesting question. Certainly, special differentiated states of enhanced mutability occur in cells of the immune system. Somatic hypermutation of immunoglobulin genes is an example that may share mechanistic similarities to stationary-phase and adaptive mutation mechanisms in bacteria. In somatic hypermutation, the variable region (V) genes for antibodies are targeted for multiple rounds of point mutation (mostly substitutions) that, coupled with

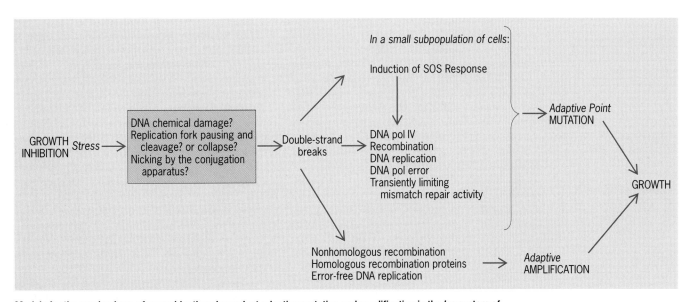

Models for the mechanisms of recombination-dependent adaptive mutation and amplification in the *lac* system of *Escherichia coli*. Adaptive point mutations reflect genetic instability akin to microsatellite (simple repeat) instability in humans, whereas adaptive amplification is similar to larger-scale chromosomal instability seen in some cancers. Because both processes require RecBCD, an enzyme specific to the processing of DNA double-strand ends (DSEs), it is inferred that amplification and mutation are alternative fates of processing DSEs. (*After S.M. Rosenberg, Evolving responsively: Adaptive mutation, Nat. Rev. Genet., 2001*)

selection for cells producing higher-affinity antibodies, ultimately result in antibodies that bind antigen better. DNA breaks have been discovered recently to occur near the mutation hot spots in the V genes. A process like adaptive point mutation at *lac* seems probable, in which breaks provoke error-prone synthesis upon their repair. However, the putative DNA polymerase(s) involved are not yet known.

In 2000, a large new superfamily of error-prone DNA polymerases was discovered. The DinB/ UmuDC superfamily (named for the genes encoding its two *E. coli* representatives, the two SOS mutator polymerases pol IV and pol V) is very well conserved in all three domains of life (archaebacteria, eubacteria, and eukaryotes). Whether DinB1, the ortholog of pol IV that is responsible for much of the adaptive mutation at *lac*, is also required in somatic hypermutation is a subject of active investigation. As the molecular mechanisms of adaptive genetic change become better understood in bacteria, these will guide dissections of the apparently similar and possibly homologous processes in other organisms, in addition to providing a picture of the underlying nature of evolution. Far from being a barely controlled aspect of entropy, genetic change may be highly orchestrated to occur when and where it is useful.

(Supported by National Institutes of Health grants R01-GM53158, and R01-AI43917.)

For background information *see* BACTERIAL GENETICS; DEOXYRIBONUCLEIC ACID (DNA); ESCHERICHIA; GENE; MUTATION; ORGANIC EVOLUTION; RECOMBINATION (GENETICS) in the McGraw-Hill Encyclopedia of Science & Technology.

Susan M. Rosenberg

Bibliography. J. Cairns et al., The origin of mutants, *Nature*, 335:142–145, 1988; R. S. Harris et al., Somatic hypermutation and the three R's: Repair, replication, and recombination, *Mutat. Res. Rev.*, 436:157–178, 1999; P. J. Hastings et al., Adaptive amplification: An inducible chromosomal instability mechanism, *Cell*, 1031:723–731, 2000; S. E. Luria and M. Delbrück, Mutations of bacteria from virus sensitivity to virus resistance, *Genetics*, 28:491–511, 1943; G. J. McKenzie et al., SOS mutator DNA polymerase IV functions in adaptive mutation and not adaptive amplification, *Mol. Cell.*, 7:571–579, 2001; M. J. Prival and T. A. Cebula, Adaptive mutation and slow-growing revertants of an *Escherichia coli lacZ* amber mutant, *Genetics*, 144:1337–141, 1996; S. M. Rosenberg, Evolving responsively: Adaptive mutation, *Nat. Rev. Genet.*, 2001; B. E. Wright et al., Hypermutation in derepressed operons of *Escherichia coli* K12, *Proc. Nat. Acad. Sci. USA*, 96:5089–5094, 1999.

Natural fission reactors

Natural fission reactors (NFR) are high-grade uranium deposits in which self-sustained fission chain reactions took place approximately 2 billion years ago. Found only in southeast Gabon, Africa, they are unique physical phenomena in the Earth's crust and the only place where some minerals are composed of elements with nonprimordial isotopic abundance. Despite great efforts, evidence for large-scale nuclear reactions has not been found anywhere else. Natural fission reactors have recently been studied as "natural analogs" for the disposal of high-level radioactive waste in the lithosphere, because they provide the exceptional opportunity to study geochemical behavior of natural fission products that do not normally occur in any significant quantities.

Uranium sources. Fifteen NFR have been recognized since their discovery in 1972, most in the Oklo-Okelobondo uranium deposit 60 km northwest of Franceville, Gabon (**Fig. 1**). The latest reactor was found at Oklo in 1991. The natural fission reactors occur within the 2.1-billion-year-old sediments in the Franceville basin. Their primary source of uranium was uraniferous thorite weathered from the nearby 2.9–2.6-billion-year-old granite and deposited in conglomerate, now at the bottom of the sedimentary formation. Uranium was removed from the thorite by oxidizing fluids, possibly derived from rainwater descending along fractures and faults during uplift of the Franceville basin. The uranium-bearing conglomerate is covered by sandstone. The natural fission reactors are confined to the uppermost layer of sandstone, in contact with an overlying layer of black shale (Fig. 1) that is rich in organic matter formed from algae. During burial of sediments at depths up to 4 km (2.5 mi), the organic matter turned into petroleum. The uranium deposits formed when oxidized uranium-bearing fluids, circulating upward, from the conglomerate encountered reducing hydrocarbon-bearing fluids migrating from the black shale into the sandstone. Uranium precipitated as uraninite in the fractures and pores of the sandstone.

The conditions that caused the uranium deposits to become nuclear reactors included the occurrence of uranium enriched in the fissionable isotope uranium-235, in highly concentrated uranium minerals that formed at least 50-cm-thick (20-in.) seams. Two billion years ago the amount of uranium-235 was 3.5%, similar to the enriched uranium used for nuclear power reactors. The availability of a neutron moderator (water), the presence of neutron reflectors (quartz), and low concentrations of neutron absorbers (boron, lithium, manganese, vanadium, and heavy rare-earth elements) were equally essential for the operation of NFR. The major difference between conditions in NFR and those in nuclear power reactors is the role of water. Unlike in power reactors, hot water in NFR directly contacted the uranium minerals during nuclear reactions.

Evidence of fission. Natural fission reactors consist of massive uraninite (the reactor core) enveloped by clay minerals (Fig. 1). The uranium concentrations in the reactor cores range from about 20 to 80%. The primary indicator of neutron-induced natural fission reactions is the depletion in uranium-235 (about 5000 kg, or 11,000 lb, consumed). The ratio of uranium-235 to the most abundant uranium isotope, uranium-238, in some NFR is as low as 0.0029, while the normal terrestrial ratio is 0.0072. The concentration of uranium-235 varies among reactors and within a single reactor. Usually the central portions of reactor cores are more deficient in uranium-235 than the outer parts.

The reactors operated intermittently for periods ranging from 24,000 to 270,000 years. They produced significant amounts of the same elements formed in nuclear power reactors, including plutonium, neptunium, and technetium. For instance, the total mass of plutonium in an extremely rich uranium deposit elsewhere is about 10^{-12} g, whereas several kilograms of plutonium were produced by each NFR. A significant portion of the present-day uranium-235 in NFR was formed by alpha decay of plutonium-239. In some places, that led to a uranium-235/uranium-238 ratio as high as 0.00804.

Associated minerals. The major minerals in NFR are uraninite, clay minerals (illite and chlorites), coffinite (uranium silicate), and galena. Uraninite occurs as massive aggregates or single crystals dispersed within both clay minerals and solid bitumen. Some crystals display zoning caused by alteration

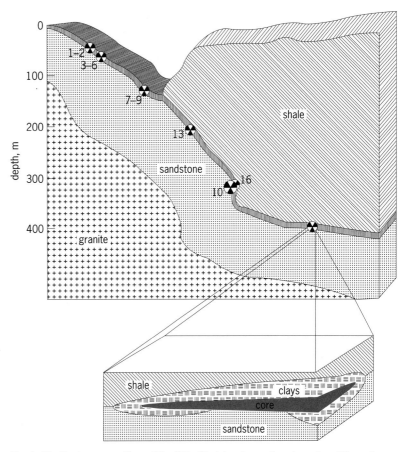

Fig. 1. Idealized cross sections of the Oklo-Okelobondo uranium deposit and through a single reactor. Most reactors were numbered following their discovery at Oklo; a reactor without a number is at Okelobondo. The lengths of reactors vary from 10 to 40 m (33 to 131 ft). The thickness of the reactors cores range from a few centimeters to 1 m (3 ft).

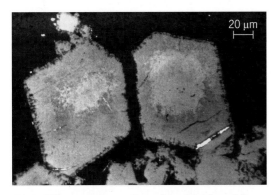

Fig. 2. Microscopic image of zoned crystals of uraninite from a natural fission reactor at Oklo. The bright centers of the crystals are original uraninite rich in lead (up to 19% by weight), which is the final product of radioactive decay of uranium. The centers are surrounded by an altered zone from which lead was removed. The crystals occurred within a clay (chlorite) matrix.

processes (**Fig. 2**). The age of the original (nonaltered) uraninite was determined to be 1968 ± 50 million years. Despite the long-term neutron irradiation (neutron fluxes were comparable to those in research reactors) and high doses of alpha decay experienced by uraninite in NFR, the uraninite remained crystalline due to rapid self-annealing of radiation-induced defects. Polycrystalline aggregates of nanometer- to micrometer-sized crystallites within single grains of uraninite, seen by transmission electron microscopy, suggest high radiation damage on a micro scale. Uraninite in NFR is a natural spent nuclear fuel; however, its chemical composition is significantly different from irradiated uranium dioxide (UO_2) in spent fuel from nuclear power reactors (see **table**).

Products of nuclear reactions in NFR were incorporated into numerous minerals. These minerals have isotopic signatures distinct from other terrestrial materials, although their chemical and physical properties are the same as their counterparts with normal isotopic composition (**Fig. 3**). Minerals containing depleted uranium and those which incorporated major amounts of fission products include uraninite, coffinite, ruthenium minerals, and rare-earth aluminous phosphates (crandallite and florencite). Apatite incorporated some fission products

Chemical compositions (in weight percent) of light water reactor (LWR) spent fuel and uraninite from two natural fission reactors (NFR)

Oxides*	LWR	NFR-16	NFR-10
UO_2	95.4	90.5	77.4
PuO_2	0.9	Traces	Traces
AcO_2	0.1	—	—
FP	3.6	?	0.3–0.5
PbO	—	5.6	18.7
ThO_2	—	0.1	0.1
Impurities	—	2.0	0.7
U^{6+}/U^{4+}	0.001	?	0.10–0.19

*Ac = neptunium and americium; FP = fission products; impurities = calcium, iron, and silicon.

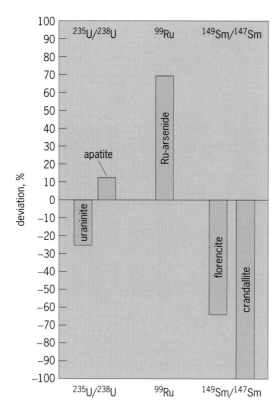

Fig. 3. Deviation of isotopic compositions from normal terrestrial values in minerals from natural fission reactors.

and plutonium as trace elements. Of particular interest are arsenides and sulfarsenides of ruthenium, rhodium, and palladium. In commercial spent nuclear fuel, those metals, together with molybdenum and technetium, form an alloy that precipitates along grain boundaries in UO_2. In NFR, ruthenium-bearing minerals occur within or adjacent to uraninite and coffinite as grains up to 4 micrometers in diameter. These minerals consist of up to 43% ruthenium-99, a decay product of now-extinct technetium-99. Technetium-99 is a fission product of major concern in radioactive-waste disposal because, due to its half-life of 210,000 years, it remains highly radioactive for periods longer than those proposed for underground repositories for high-level waste (10,000 years). Technetium-99 produced in NFR decayed to ruthenium-99 within a million years after the end of fission reactions. Ruthenium-99 is, therefore, an indicator of technetium-99 behavior. Isotopic abundance analysis of uraninite showed that a significant amount of technetium was released from the NFR, transported by hot fluids, and deposited approximately 10 m (33 ft) above the reactors. The tellurium found in association with ruthenium minerals was probably also a fission product.

The reactors experienced alteration by hot fluids, which caused the dissolution of uraninite and its replacement by coffinite. Heat generated by fission reactions turned ground water into high-temperature (250–450°C; 480–840°F) and moderately saline fluids. During the 0.6–1-million-year cooling period of

the reactors, the fluids changed to low-temperature (100°C; 212°F) and highly saline water solutions. The reactor clays that formed at that time replaced the dissolved sandstone. Circulation of the fluids was local because heat from nuclear reactions dissipated in surrounding rocks within 20 m (66 ft) of the reactors. Another stage of alteration of NFR was caused by the intrusion of dolerite (basalt) dikes into the Franceville basin some 780 million years ago. Hot fluids (260°C; 500°F) caused the release of lead and some fission products from uraninite by diffusion-driven processes. The loss of lead from uraninite was almost complete in most NFR. Lead was precipitated as galena and, locally, as metallic lead. Despite the actions of the fluids, more than 90% of uranium has remained in NFR due to their reducing environment. The discovery of free hydrogen and oxygen in the fluid inclusions of quartz from reactor clays suggests locally oxidizing conditions due to the decomposition (radiolysis) of water in a highly radioactive environment.

Actinides and fission products that were mobilized during various stages of uraninite dissolution and alteration either precipitated within or migrated out of the reactor zone. As a result, NFR are surrounded by halos of dispersed actinides and fission products that were released from the uraninite. Noble gases and volatile elements (iodine, cesium, strontium, and rubidium) were almost completely lost from the reactors. Other fission products and actinides were confined within 10 m (33 ft) of the NFR. Sorption and mineral precipitation were the most important retardation processes, limiting the mobility of dissolved actinides and fission products. The most efficient sorbents were chlorite, titanium oxide (anatase), and goethite. Dissolved uranium and plutonium precipitated on mineral surfaces as secondary uraninite.

Reactors near the surface have additionally been affected by tropical weathering for at least 63,500 years. Oxidizing rainwater has percolated through the reactor cores, causing dissolution of uraninite and coffinite, which at ambient temperatures are stable only in a reducing environment. Dissolved uranium precipitated as uranyl minerals (minerals with hexavalent uranium) or was trapped by goethite. Present-day leaching of uranium in shallow reactors is negligible.

Implications. From the perspective of radioactive waste disposal, NFR occurred in an unfavorable environment. They were vulnerable to the direct action of hot fluids and oxidized water solutions for long periods. Yet, many biologically hazardous fission products and actinides have remained in place or migrated only short distances out of NFR due to the effectiveness of various minerals that have acted as barriers, limiting the migration of radionuclides.

Natural fission reactors are not only interesting physical and geochemical phenomena in the Earth's crust but also provide a rare opportunity to study processes relevant to the disposal of radioactive waste in the lithosphere. Unfortunately, due to the termination of mining operations at Oklo in late 1997, these unique natural phenomena are no longer available for ongoing research.

For background information *see* ACTINIDE ELEMENTS; GEOCHEMISTRY; NUCLEAR FISSION; NUCLEAR FUELS; NUCLEAR POWER; NUCLEAR REACTOR; RADIOACTIVE MINERALS; RADIOACTIVE WASTE MANAGEMENT; URANINITE; URANIUM in the McGraw-Hill Encyclopedia of Science & Technology.

Janusz Janeczek

Bibliography. F. Gauthier-Lafaye, P. Holliger, and P.-L. Blanc, Natural fission reactors in the Franceville basin, Gabon: A review of the conditions and results of "critical event" in a geologic system, *Geochim. Cosmochim. Acta*, 60:4831–4852, 1996; H. Hidaka and F. Gauthier-Lafaye, Redistribution of fissiogenic and non-fissiogenic REE, Th and U in and around natural fission reactors at Oklo and Bangombe, Gabon, *Geochim. Cosmochim. Acta*, 64:2093–2108, 2000; J. Janeczek, Mineralogy and Geochemistry of Natural Fission Reactors in Gabon, *Reviews in Mineralogy*, Mineralogical Society of America, Washington, DC, 1999; R. Naudet, *Des Réacteurs Nucléaires Fossiles: Étude Physique*, Eyrolles, Paris, 1991.

Neuromorphic engineering

Neuromorphic engineering is a new field that uses the functional principles of biological nervous systems to inspire the design and fabrication of artificial neural systems, such as vision chips and roving robots. In turn, these artificial neural systems are used to test biological principles, such as learning and sensorimotor control.

Neuromorphic engineering began with Carver Mead's 1989 monograph, *Analog VLSI and Neural Systems*. Mead explained the technology of analog VLSI (very large scale integration) chips and their potential as devices to model neurons. Prior use of transistors had increasingly focused on their ability to generate clear-cut digital pulses that were rapid and well above a detection threshold. However, Mead's use of transistors operated well below this high level of pulsed output. The advantage of using the analog operating range for modeling neurons was that these subthreshold transistors produced tiny low-power devices with a large range of possible values. Encompassed within their limited confines were many values, rather than a single binary 1/0 value, all-or-none response. In the biological world, communication from one neuron to another is via axons that can extend over long distances. The signals sent along these axons are digital in the form of all-or-none action potentials of roughly equal size. However, when a signal reaches the target neuron and is processed within the neuron, the processing is graded in amplitude and more accurately described as analog. At the heart of Mead's vision was the recognition that evolution had been a powerful force in the realm of biology, efficiently engineering biological compo-

nents to survive against competition far more intense than that of corporate pressure. Thus, the use of functional principles derived from biology could generate considerably more efficient machines than those inspired only by the imaginations of their designers. This principle has been applied to the design of sensory devices such as retina chips, neural circuit models, and robotic implementations of walking and flying machines.

Retina chip. Mead's book captured the imagination of many young people, especially a group of his graduate students at Caltech. Among the early implementations developed by the group was a retina chip, followed by the first silicon analog neuron designed by Misha Mahowald. The field has since seen the development of more complex and sophisticated devices for a range of uses, some commercial and some for basic research. Now there are also hybrid systems that employ analog and digital elements to obtain the optimal efficiency of both. For example, rapid long-range communication is best achieved, as it is by neurons, in the digital domain, while analog computation can be used to better approximate internal neuronal processes and slower short-range interactions among neurons.

The first retina chip described in Mead's book captured some of the important features of the mammalian retina. Each pixel (the smallest element of an electronically coded picture image) serves as a light-sensitive photoreceptor with circuitry that models the early stages of biological retina visual processing. Biological photoreceptors are recreated in the retina chip by a dense hexagonal array of pixels that together perform visual processing in the focal plane; the pixels are oriented to match the surface topography of the retina, thereby guaranteeing a correct representation of space. The retina chip represents the network formed by biological retinal cells that spatially and temporally averages the photoreceptor output in real time.

Despite its complexity and early success, the first retina chip was far simpler than its biological counterpart. Recent models have incorporated more of the important characteristics of the mammalian retina (compare the left and right circuits in **Fig. 1a**). These and future implementations promise to provide better visual decoding of the world with so much of the functionality of the mammalian retina that they could one day serve as important prosthetic devices for the visually impaired. Other implementations have included adaptation in the receptors that allows operation over several orders of magnitude of light intensity, and some have modeled the extremely efficient motion detection system of the fly retina. Mead has gone on to build a new digital camera with 4000×4000 analog VLSI pixels. The camera will provide dramatic photographic clarity beyond anything presently available.

Neuron chip. Mahowald's neuron chip has formed the backbone of numerous circuits. The basic circuit uses analog transistors in a variety of arrangements to implement the functional characteristics

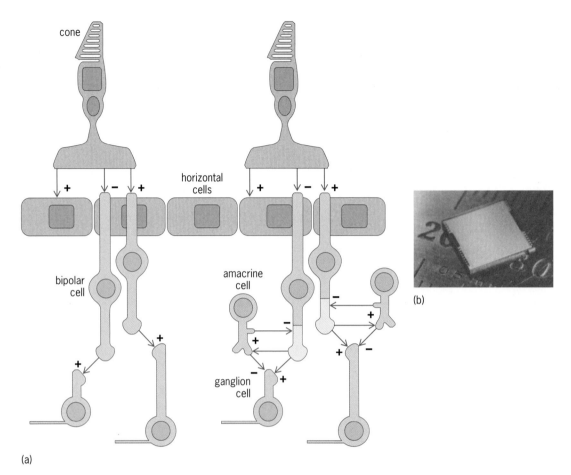

Fig. 1. Retinal circuitry implemented in a VLSI vision chip. (*a*) The left photoreceptor has the circuitry used in M. Mahowald's early retinal receptors. The right photoreceptor has the more complex circuitry used in K. A. Boahen's retina chip, which incorporates a variety of circuit configurations. (*b*) The recently fabricated retinal chip has 104 × 96 pixels with 104 × 96 photoreceptors and 4 × 52 × 48 ganglion cells.

of real nerve cells, to efficiently emulate the major voltage-sensitive ion channels (**Fig. 2***a*). In their biological counterpart, these channels permit ions to cross the impermeable neuron membrane when there is a suprathreshold voltage change across the membrane. The analog circuits operate in real time. Their low power consumption and small size allow many circuits to be fabricated within a remarkably small space (Fig. 2*b*). Other implementations have modeled more complex channels such as the multipurpose excitatory glutamate channel. Recently, complex analog VLSI neurons have been combined with digital elements to model cortical circuitry that can process sensory inputs. The channel circuits have also been used as elements in neuron implementations by R. Calabrese and S. DeWeerth that have complex behavior patterns such as bursting similar to that of the leech nerve cord controlling heart pumping.

Synapses can also be implemented, using floating gate transistors in the analog range. Floating gates employ continuously adjustable stored charge; thus the synapses permit nonvolatile memory storage for adaptive processing systems. The floating gate synapse transistors incorporate long-term memory and local computation that is quite sophisticated and can

operate over a range of time scales, making it possible to approximate a host of biological synaptic interactions, including learning and other slow-time-scale adaptive responses. The synapse transistors can be used as links between neuron chips or to change the temporal characteristics of any silicon implementation of neuronal devices.

Robotics. Robotics has benefited both directly and indirectly from the new vision of neuromorphic engineering. Direct benefit is exemplified by a circuit to control limb movement designed by a group of engineers, roboticists, and biologists. A small analog chip is used to generate the swinging motion of a limb or a pair of limbs (**Fig. 3**). The chip is designed to provide oscillatory output to a motor that propels the limb. Feedback from sensors at the hip of the limb adaptively controls the oscillation to keep the swing centered and in the correct amplitude range. Such a design has potential for considerably more complexity. The chip can be fitted with floating gates to slowly change the parameters of the system; the feedback can be increased and used to change the frequency and amplitude of the swing and the position of the limb. The feedback can also be used to correct for perturbations encountered by the limb and for changes in slope and

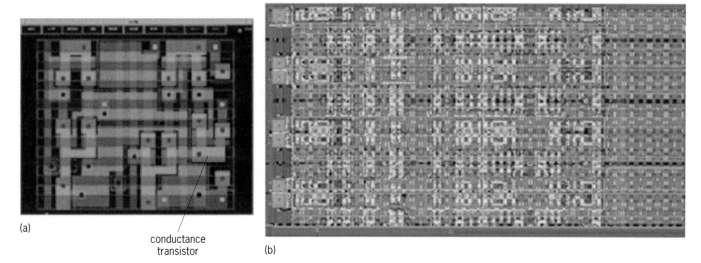

(a)

conductance
transistor

(b)

Fig. 2. Analog VLSI neuron chips. (*a*) Schematic layout for a delayed rectifier potassium circuit, one of the channels of a neuron circuit. The highlighted regions correspond to transistor channels and thus denote the presence of a transistor. The conductance transistor is the very short one on the far right. The area of the circuit is quite small: a chip area of 4 mm^2 can contain roughly 650 such circuits. (*b*) Seven complete silicon neurons back to back. The left-most transistors (highlighted areas) are used for injecting direct current into the neuron's membrane capacitor. The total chip area is roughly 2 mm × 2 mm. However, most of the chip area is empty.

type of terrain. The small size and low power consumption means the chip can be placed directly on the robots to permit autonomous on-board adaptive control.

Vision chip. Vision chips have been put onto autonomous robots that rove about on wheels or fly like insects, negating the need for workstations or other special-purpose digital signal processors (DSPs). The chips—with their potential for fairly complex visual processing, motion detection, and adaptation; small size; and low power consumption—can be good on-board visual guidance and navigational devices. They allow the robot to track high-contrast lines or specified images, to move toward or away from light sources, and to find designated objects in its environment, among other things. They also can be used for optomotor control of moving head devices to localize objects in space.

Cochlea chip. There has also been considerable progress toward a clinically useful cochlea chip. The device has already been used for hearing prostheses. These devices can be combined with visual processing chips on roving robots to allow more efficient localization of sound sources. As knowledge about perception grows, the visual and auditory devices and their uses in robots will become even more efficient, sophisticated, and complex.

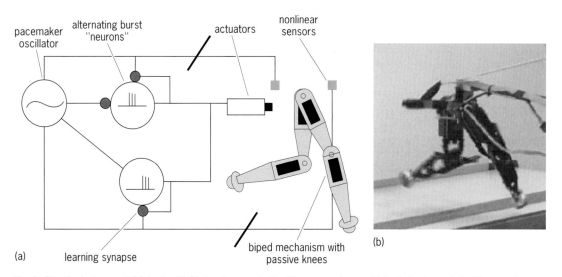

Fig. 3. Biped robot control. (*a*) Analog VLSI circuit as controller. The pacemaker oscillator is implemented with an analog VLSI circuit that drives bursting neurons whose output is integrated and used to control the actuator for the leg. The oscillator, burst neurons, and integrator are implemented on a single chip 2 mm on a side, which draws less than 1 microwatt of power. (*b*) Image of the legs, designed by Iguana Robotics, Inc.

Form versus function. Robotics has benefited indirectly from the idea of using evolutionary processes to guide design, rather than just morphology. While some researchers are mimicking biology and building robots that emulate biological forms and their respective locomotor patterns (for example, humans, fish, lobster), others use robots to test the principles of biology. Herein lies an important distinction. Biomimetic robots do not necessarily incorporate biological principles, but use biology for their morphological inspiration (for example, a fish robot that looks like a swimming fish but does not use the same principles of movement generation). Biomorphic or neuromorphic devices do not try to emulate the form, but strive to emulate or model the function of their biological counterparts.

Biomorphic or neuromorphic devices can be used to test theories of learning, sensorimotor control, and other difficult biological motor control principles. It is, for example, possible to examine the interactions between sensory input and motor output and how the input from sensory structures can potentially be used to improve the performance of a robot. One example of this approach is the work of Hiroshi Kimura and his colleagues in Tokyo. They have been able to implement several sensory mechanisms on a quadrupedal robot controlled with a central pattern generator (a neural circuit that generates a complex rhythmic motor pattern) that can successfully navigate over moderately irregular terrain. Randall Beer and his colleagues have taken evolution and its insights in a different direction by using genetic algorithms to find the most efficient implementations for a hexapod robot that they built.

Clearly, neuromorphic engineering benefits both engineering and neuroscience by using insights from one to enlighten and improve understanding of the other.

For background information see EYE (VERTEBRATE); LOGIC CIRCUITS; NERVOUS SYSTEM (VERTEBRATE); ROBOTICS; TRANSISTOR; VISION in the McGraw-Hill Encyclopedia of Science & Technology.

Avis H. Cohen

Bibliography. R. Douglas, M. Mahowald, and C. Mead, Neuromorphic analogue VLSI, *Annu. Rev. Neurosci.*, 18:255–281, 1995; G. Indiveri and R. Douglas, Robotic vision: Neuromorphic vision sensors, *Science*, 288:1189–1190, 2000; C. A. Mead, *Analog VLSI and Neural Systems*, Addison-Wesley, Reading, MA, 1989.

Nobel prizes

The Nobel prizes for 2000 included the following awards for scientific disciplines.

Chemistry. The chemistry prize was awarded to Alan J. Heeger, Alan G. MacDiarmid, and Hideki Shirakawa for the discovery and development of electrically conductive polymers in the late 1970s. Since then, conductive polymer research has expanded into a multidisciplinary field and produced a range of metallic and semiconducting polymers that can be coated or cast as films (low-cost processing advantages). These materials now find uses as rechargeable batteries, sensors, antistatic protective coatings, and light-emitting diodes, to name a few, and show promise in the emerging field of molecular electronics.

In 1974, Shirakawa and coworkers at the Tokyo Institute of Technology accidentally added too much catalyst while working on the polymerization of acetylene. Instead of the expected black powder, they observed a silver-colored polyacetylene $(CH)_n$ film lining the reaction vessel.

Around the same time, Heeger and MacDiarmid were studying nonmetallic electrical conductors. When MacDiarmid heard about the metallic-appearing polyacetylene film, he thought it might be conductive, and invited Shirakawa to work with his group at the University of Pennsylvania. Shirakawa there refined the polymerization process for polyacetylene. After characterization, it turned out that polyacetylene did not have metallic conductivity, but was a semiconductor.

Conductivity—the ability of a material to carry electric current, expressed in units of siemens per meter (S/m)—is the ratio of the amount of charge passing through the material per unit area (perpendicular to the current direction) per second, divided by the electric field intensity (the force on a unit charge).

In metals and semiconductors (such as silicon) the charges responsible for current are free electrons and holes (missing electrons that act like positive charges). These free electrons or holes are not bound to any particular atom and can move freely in the field. Conductivity due to free electrons is called *n*-type, and conductivity due to holes is called *p*-type.

In 1977, MacDiarmid treated polyacetylene with halogen vapor, an oxidation process called *n*-doping. Heeger, who had been characterizing polyacetylene's electrical properties, measured conductivity as high as 10^5 S/m (metallic conductivity) for doped polyacetylene. For comparison, silver (the highest-conductivity metal) has a conductivity of 10^8 S/m. Semiconducting silicon and *trans*-polyacetylene have conductivities of 10^{-4} and 10^{-3} S/m, respectively. Teflon (polytetrafluoroethylene), an insulator, has a conductivity of 10^{-16} S/m.

Polyacetylene and all conductive polymers have conjugated double bonds (alternating single and double bonds). In conjugated polymers, the π-bond electrons are delocalized over the entire polymer, but conjugation alone does not make these polymers conductive. Added dopant removes electrons from the polymer, and the resulting holes migrate along the polymer backbone, making it conductive. In practice, however, conductive polymers must be highly doped for this mechanism to work.

By applying an electric field perpendicular to a doped-polymer film (a salt), the anions (such as I_3^-) are made to diffuse into or from the polymer structure, causing the doping reaction to proceed forward or backward (that is, switch the conductivity on or off).

Doped polyacetylene is highly crystalline (brittle), insoluble, and unstable (air- and humidity-sensitive). Although polyacetylene was the model for polymer conductivity, it was never commercialized. Other conductive polymers have been synthesized that exhibit greater flexibility, such as doped polyaniline, and that have side-chain substitution to improve solvent solubility for film application, such as poly-(ethylenedioxythiophene).

For background information *see* CONDUCTION (ELECTRICITY); HOLE STATES IN SOLIDS; ORGANIC CONDUCTOR; SEMICONDUCTOR in the McGraw-Hill Encyclopedia of Science and Technology.

Physics. The physics prize was awarded in two parts, with half shared by Herbert Kroemer and Zhores I. Alferov for developing semiconductor heterostructures used in high-speed- and opto-electronics, and half going to Jack S. Kilby for his role in the invention of the integrated circuit.

Kroemer, at RCA in Princeton, New Jersey, in 1957, put a high-band-gap semiconductor together with a low-band-gap semiconductor such that one material gradually transitions into the other. The transition region is called the heterojunction. (The band gap is the energy difference between two ranges of allowed energies in the semiconductor.) The gradual variation in material implies an electrical energy gradient, which imposes a force similar to that due to an electric field. He referred to such forces as quasi-electric fields, which act on negatively charged electrons and positively charged holes differently. This led to the development of semiconductor heterostructures [layered semiconducting structures such as gallium arsenide (GaAs) and aluminum gallium arsenide (AlGaAs)] for use in transistors, such as heterostructure bipolar transistors (HBTs).

In 1963, Kroemer began researching light-emitting semiconductors at Varian Associates in Palo Alto, California. He proposed a crystal with two heterojunctions containing two wide-band-gap semiconductors flanking a lower-band-gap material. When an electric field is applied, the electrons and holes flow in opposite directions and are trapped in the middle region. The gathered electrons and holes recombine (in a coherent fashion at room temperature and under continuous operation), with the excess energy emitted as light particles (photons). This discovery led to the development of lasers and light-emitting diodes (LEDs). That same year, Alferov of the Ioffe Physico-Technical Institute in St. Petersburg (formerly Leningrad) independently reported the same principle in a patent application. In May 1970, Alferov's group reported a technical breakthrough, the first continuous operating room-temperature heterostructure laser. Such devices have made fiber-optic communication possible.

Both Alferov and Kroemer invented and developed fast opto- and micro-electronic components based on semiconductor heterostructures. Using heterostructure technology, fast transistors are used in radio link satellites and the base stations of mobile telephones. The same technology is used to make laser diodes (used in bar-code readers, compact-disk players, and fiber-optic cables) and LEDs (used in automobile brake lights, traffic lights, and warning lights).

Kilby, while working at Texas Instruments in 1958, built the first integrated circuit, an oscillator, by soldering all its components together, using only silicon as a base material. Later that year, he fabricated a complete circuit on a piece of germanium, the most common semiconducting material in use at that time.

In 1959, Kilby filed a patent for "miniaturized electronic circuits." As the inventor of the integrated circuit, he laid down the various components on a substrate and connected them with gold wires. Through this invention, microelectronics has grown to become the foundation of all modern technology, for example, powerful computers and processors, which control a wide range of products from washing machines to space probes. His insights and contributions were instrumental in the invention of the microchip. The microchip has also led to the development of small electronic apparatuses, from calculators to personal computers.

For background information *see* INTEGRATED CIRCUITS; SEMICONDUCTOR HETEROSTRUCTURES in the McGraw-Hill Encyclopedia of Science & Technology.

Physiology or medicine. Arvid Carlsson of Göteborg University in Sweden, Paul Greengard of The Rockefeller University in New York, and Eric Kandel of Columbia University in New York were jointly awarded the Nobel Prize for Physiology or Medicine for their pioneering research on signal transduction in the nervous system.

Carlsson discovered that dopamine is a key chemical transmitter in the brain and very important in the control of movement. In the late 1950s, Carlsson developed a high-sensitivity assay to measure dopamine levels in tissue and found that dopamine was concentrated in different areas of the brain compared with norepinephrine, indicating that dopamine was not a precursor of norepinephrine, as was previously believed, but a transmitter in and of itself. He found that dopamine was particularly concentrated in the basal ganglia, an area of the brain involved in motor control. Carlsson went on to discover that levels of L-dopa, a precursor of dopamine, are abnormally low in the basal ganglia of Parkinson's disease patients. Consequently, L-dopa was developed into a drug to treat Parkinson's disease and today remains the most important treatment for this disorder. In addition, Carlsson discovered that antipsychotic drugs affect synaptic transmission by blocking dopamine receptors.

Carlsson's work has also contributed to the treatment of depression. In the late 1960s, Carlsson and his research team created zimelidine, the first clinically active inhibitor of serotonin reuptake for depression. This research contributed to the development of selective serotonin reuptake inhibitors (SSRIs) such as Prozac, one of the most widely used drugs for the treatment of depression.

Building on Carlsson's research, Greengard elucidated the mechanism by which dopamine and other chemical transmitters (such as norepinephrine and serotonin) affect the nervous system. Greengard discovered that the transmitters alter the function of nerve cells through a process known as slow synaptic transmission, which controls functions such as mood and alertness. When dopamine and other transmitters bind to a receptor on the surface of a neuron (nerve cell), they increase cellular levels of a secondary messenger known as cyclic adenosine monophosphate (AMP). Cyclic AMP then activates an enzyme called protein kinase A, which phosphorylates (adds a phosphate group to) key cellular proteins, altering their form and function and ultimately leading to changes in neuron function. (For example, phosphorylation of ion channel proteins can alter a neuron's excitability and ability to transmit electrical impulses.) Greengard went on to discover that in some neurons dopamine and other chemical transmitters cause a cascade of phosphorylations and dephosphorylations (removal of a phosphate group) through the activation of a regulatory protein known as dopamine and cyclic AMP-regulated phosphoprotein-32 (DARPP-32). Activation of DARPP-32 indirectly changes the function of a number of other proteins and ultimately leads to alterations in signal transmission at particular fast synapses, which control functions such as speech, movement, and sensory perception. Greengard's research has also elucidated the mode of action of several drugs that work by affecting phosphorylation of proteins in different nerve cells.

Kandel's prize-winning research revealed that protein phosphorylation also plays an important role in the molecular mechanisms underlying learning and memory formation. Using the simple nervous system of the sea slug *Aplysia* as a model, Kandel showed that memory formation occurs at the synapse (site of nerve cell interaction). Short-term memories are formed when a weak stimulus activates ion channels via protein phosphorylation, increasing the amount of transmitter released into the synapse. In contrast, long-term memory formation requires a stronger and longer-lasting stimulus and the synthesis of new proteins. In long-term memory formation, the increases in the levels of cyclic AMP and protein kinase A trigger phosphorylation of different proteins that signal the cell nucleus to synthesize new proteins. This may change the shape of the synapse, enabling more transmitter to be released. The end result is a long-lasting increase in synaptic function. Kandel showed that in *Aplysia* amplification of the protective reflex (a form of learning) results when certain stimuli enhance the function of the synapses connecting sensory nerve cells to the nerve cells that activate the muscle groups that produce the reflex. Kandel has since demonstrated that these fundamental mechanisms apply not only to sea slugs but also to humans.

This increased understanding of the molecular mechanisms that underlie memory formation brings scientists closer to developing new drugs to treat memory loss in patients with various types of dementia.

The combined work of these three Nobel laureates has elucidated the molecular mechanisms underlying slow synaptic transmission, which has been vital for understanding normal signal transduction in the brain and how transmission disturbances can engender neurological and psychiatric diseases. Their research has led to the development of drugs to treat disorders such as Parkinson's disease and depression, elucidated how many antipsychotic drugs affect the nervous system, and provided fundamental knowledge that will in the future enable scientists to create drugs to improve memory function.

For background information *see* DOPAMINE; LEARNING MECHANISMS; NERVOUS SYSTEM (INVERTEBRATE); NERVOUS SYSTEM (VERTEBRATE); PARKINSON'S DISEASE; SEROTONIN; SIGNAL TRANSDUCTION; SYNAPTIC TRANSMISSION in the McGraw-Hill Encyclopedia of Science & Technology.

Nomenclature, phylogenetic

In the past decade, knowledge of the phylogeny (evolutionary history) of life has improved dramatically as a result of advances in molecular biology and computer technology. Computer analysis of genetic and phenotypic data has permitted biologists to delimit a great many clades (branches of the tree of life), and this process of discovery continues to accelerate. Scientific communication about phylogeny requires that clades be named, but the traditional system of biological nomenclature is poorly suited to do so. The inability of traditional nomenclature to provide stable, unambiguous names for the branches of the tree of life motivated the development of a new, suitable system called phylogenetic nomenclature. The fundamentals of this system were published by Kevin de Queiroz and Jacques Gauthier in 1990 and 1992. A formal set of rules (the PhyloCode) has been drafted but not yet implemented.

Phylogenetic definitions. The objective of phylogenetic nomenclature is to give clades (and eventually species) names that are unique, unchanging, and unambiguous. The clade to which a name applies is specified by means of a phylogenetic definition.

Node-based. For example, the name Mammalia might be defined as the clade comprising the most recent common ancestor of *Ornithorynchus anatinus* (platypus), *Didelphis marsupialis* (opossum), and *Homo sapiens* (human), and all of the descendants of this ancestral species. This definition specifies the entire clade of extant mammals because all three subclades (monotremes, marsupials, and placentals) are represented, but the definition excludes extinct animals that predated the divergence of the three extant groups and that might be classified as mammals by some biologists.

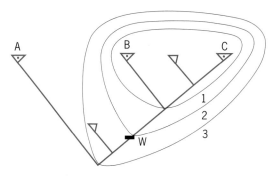

Three kinds of phylogenetic definitions. The nested bubbles designate named clades 1, 2, and 3. The triangles are other (subsidiary) clades. The bar designates the evolution of character W. The dots indicate species A, B, and C (each contained within a subsidiary clade). The name applied to clade 1 would have a node-based definition (for example, the clade comprising the most recent common ancestor of species B and C and all of its descendants). The name applied to clade 2 would have an apomorphy-based definition (for example, the clade stemming from the first species to possess character W homologous with that in species C). The name applied to clade 3 would have a stem-based definition (for example, species C and all organisms that share a more recent common ancestor with C than with species A). (*After K. de Queiroz and J. Gauthier, Phylogenetic taxonomy, Annu. Rev. Ecol. Syst., 23:449–480, 1992*)

An alternative wording that specifies the same clade is the least inclusive clade containing *Ornithorynchus anatinus*, *Didelphis marsupialis*, and *Homo sapiens*. Both of these are called node-based definitions because they specify a particular node (branching point) on the tree of life (see clade 1 in the **illus.**).

Stem-based. A stem-based definition (for example, clade 3 in the illus.) specifies the clade comprising all organisms that are more closely related to species C than to species A. For example, Bryophyta (mosses) might be defined as the clade comprising *Bryum argenteum* and all organisms that share a more recent common ancestor with this species than with *Marchantia polymorpha* (a liverwort), *Anthoceros punctatus* (a hornwort), or *Polypodium vulgare* (a vascular plant).

Apomorphy-based. An apomorphy-based definition (for example, clade 2 in the illus.) specifies the clade stemming from the first species to possess a particular apomorphy (derived feature). For example, Aves (birds) might be defined as the clade stemming from the first species with wings bearing feathers homologous (sharing the same evolutionary origin) with those in *Passer domesticus* (house sparrow). The inclusion of the word "homologous" in the definition clarifies the meaning of "feathers" if featherlike structures evolved more than once in the history of life.

Phylogenetic versus rank-based nomenclature. The explicit linkage of a name to a clade in phylogenetic nomenclature is radically different from the way that names are defined in traditional (rank-based) nomenclature. (The latter is often called Linnaean nomenclature because it is based on the Linnaean hierarchy of ranked categories; however, several key

aspects of the current form of rank-based nomenclature were developed after Linnaeus's time.) In rank-based nomenclature, taxa are named by reference to a rank (for example, genus, family, order, and so on) and a type (a specimen or included taxon to which the name of a taxon is permanently attached). For example, according to the International Code of Botanical Nomenclature, Asclepiadaceae is the family that contains the type species of *Asclepias* (the milkweed genus). If this group were classified as a subfamily rather than a family, as it is by some botanists, it would have to be called Asclepiadoideae. In contrast, the phylogenetic definition of Asclepiadaceae might be the clade stemming from the first species with pollinia (masses of fused or adhering pollen) homologous with those in *Asclepias syriaca* (common milkweed).

The Asclepiadaceae example illustrates two differences between phylogenetic and rank-based nomenclature: (1) Because the rank-based definition of the name Asclepiadaceae makes no reference to phylogeny, the name may or may not apply to a clade. Indeed, a name may be applied to a clade by some people and to a subjectively circumscribed group of species by others, and there is no way for users of the name to know which is the case without delving into the literature. (2) When rank-based nomenclature is applied to clades, the name of the clade will depend on its rank (for example, Asclepiadaceae versus Asclepiadoideae). If a clade is reclassified at a different rank, which can occur for a variety of reasons, its name is usually changed. (This rule is not mandatory at all ranks in current codes of rank-based nomenclature, but it is generally followed even when not mandatory.) Such name changes do not occur in phylogenetic nomenclature because ranking is not required and does not affect the name.

In the interest of clarity, both rank-based and phylogenetic nomenclature mandate that each name correctly applies to only one taxonomic group and that each group has only one correct name. When competing names exist, the date of publication is generally used to determine which name is correct. However, this principle of priority takes a different form in the two systems. Under rank-based nomenclature, when two or more names have been applied to the same taxonomic group, the name that was published earliest at the rank at which it is currently classified usually has priority over the others, which are considered synonyms. If the same group were classified at a different rank, a different name would take precedence. In phylogenetic nomenclature, the rules of precedence ignore rank. If two or more names have been applied to the same clade, the first one to have had a phylogenetic definition published (after implementation of the PhyloCode) has precedence, regardless of rank.

In the rank-based system, the process of naming clades is inseparable from the process of arranging them into a hierarchy of ranked categories (classification). In phylogenetic nomenclature, these processes are separated, and the use of ranked categories is optional, making it much easier to name clades. With this sytem, it is possible to name clades one at a time as they are discovered, without developing a new classification or changing the names of other clades. In contrast, naming a newly discovered clade with rank-based nomenclature may require developing a new classification, because the clade must be given a rank in order to name it, and its rank will depend on and affect the ranks of other taxa in the classification. It may also cause a cascade of name changes at lower levels in the hierarchy as clades within the newly discovered clade shift downward in rank and must therefore be renamed.

PhyloCode. A formal set of rules governing phylogenetic nomenclature, called the PhyloCode, is under development. Once implemented, it will standardize and regulate the application of phylogenetic nomenclature, including the use of publication date to select the correct name for a clade when there are competing names. It will thus serve a function comparable to that of the existing codes of rank-based nomenclature and will operate in parallel with them. Decisions about the scope and content of the PhyloCode were made at an initial workshop at Harvard University in August 1998. Many of the participants remained involved as an advisory group during the subsequent writing of the code, a draft of which is available on the Internet.

Species nomenclature under the PhyloCode. The tree of life includes species as well as clades. A complete system of biological nomenclature should include rules for naming both kinds of entities. The Linnaean method of naming species, which is universally used today, is incompatible with phylogenetic nomenclature. This is because Linnaean binomials require the existence of a genus category, whereas a basic tenet of phylogenetic nomenclature is abandonment of mandatory ranked categories such as genus. Linnaean binomials are also relatively unstable; because the genus name is part of the species name, the species name must change if the species is reclassified as belonging to a different genus. This problem does not exist in phylogenetic nomenclature or at other levels of the taxonomic hierarchy in rank-based nomenclature; transferring a genus to a different family does not require that its name be changed.

Many alternative methods for naming species in the context of phylogenetic nomenclature have been proposed, each with advantages and disadvantages, and there is currently no consensus as to which is preferable. As a result, a decision was made to restrict the first version of the PhyloCode to clade names and delay coverage of species names until a later version is developed, permitting further discussion of the issues involved. Although it has not yet been decided what form species names will take under the PhyloCode, two generalizations can be made: Under most naming methods that have been proposed to date, species names (1) would not look dramatically different from how they look now but (2) would be less prone to change than species names currently are because naming would be uncoupled from classification. For example, under one proposed naming method, the human species name would be converted from *Homo sapiens* to *sapiens*. The name *sapiens* could optionally be combined with a clade name (for example, *Homo*) when necessary for clarity of communication, but the clade name would not be part of the species name in the way that the genus name is part of the species name in Linnaean nomenclature. Since many specific epithets (the second part of a Linnaean binomial—for example, *sapiens*) are not unique, the species name under the PhyloCode will be complemented by a unique registration number which may be cited when desirable for clarity.

Under a different proposed method, *Homo sapiens* would be converted to *Homo-sapiens*. This hyphenated name would be unique and completely stable; it would remain unchanged even if it were later determined that this species did not belong to a clade called *Homo*. With this method, the name has two parts, but the first part does not imply relationship as it does in Linnaean nomenclature. These hyphenated species names would thus have less information content than Linnaean binomials, but they would be more stable. The permanence of species names under both of these proposed methods would have distinct advantages for database management and for communication in general.

For background information *see* CLASSIFICATION, BIOLOGICAL; PHYLOGENY; TAXONOMIC CATEGORIES; TAXONOMY; ZOOLOGICAL NOMENCLATURE in the McGraw-Hill Encyclopedia of Science & Technology.

Philip D. Cantino

Bibliography. P. D. Cantino, Phylogenetic nomenclature: Addressing some concerns, *Taxon*, 49:85–93, 2000; P. D. Cantino et al., Species names in phylogenetic nomenclature, *Syst. Biol.*, 48:790–807, 1999; K. de Queiroz and J. Gauthier, Phylogenetic taxonomy, *Annu. Rev. Ecol. Syst.*, 23:449–480, 1992; K. de Queiroz and J. Gauthier, Toward a phylogenetic system of biological nomenclature, *Trends Ecol. Evol.*, 9:27–31, 1994.

Nuclear astrophysics

Recent advances in the understanding of stellar nucleosynthesis, the formation of elements in the interior of stars, feature processes at both extremes of the chart of the nuclides. The r-process, which may occur in the hot neutrino wind that results from collapse of a massive star to a neutron star, passes through neutron-rich nuclides that lie roughly 20 neutrons beyond the neutron-rich side of stability, and reach the heaviest nuclei that nature can produce. A process known as alpha-rich freezeout precedes the r-process; it involves both stable nuclei and unstable nuclei close to the stable nuclides. The path of the high-temperature rp-process, which can occur in a hydrogen-rich high-temperature environment produced by accretion onto a neutron star or black hole, passes through the most proton-rich nuclei that exist in nature. Studying the nuclei involved in these processes, as well as the nuclear reactions, has proved to be extremely difficult for scientists (although nature seems to have no difficulty producing and using them for the *r*- and *rp*-processes).

The r-process. Recent developments in the understanding of the *r*-process involve not only advances in experimental and theoretical understanding of the nuclei through which the *r*-process passes, but work on neutrino astrophysics as well.

Effects of closed neutron shells. In the first context, work is proceeding to produce and study the nuclides along the *r*-process path. The *r*-process synthesizes heavier nuclei by adding neutrons to the lighter nuclei. Because addition of only a few neutrons to a stable nucleus will make it unstable, all the nuclei along the *r*-process path are unstable, usually with very short half-lives. As these nuclei increase in mass, they occasionally reach waiting points, nuclei at which further progression along the *r*-process path is inhibited by either a long (compared to the time required for the *r*-process to synthesize the heaviest nuclei observed in nature; the order of seconds) half-life or an inability to capture another neutron. Both of these features occur at the neutron closed shells. Since abundance will build up at these waiting-point nuclei, they will become the progenitors of the abundant *r*-process nuclei when the conditions of the *r*-process end and the progenitors decay back to stability. Thus the closed-neutron-shell nuclides are crucial to the understanding of the *r*-process.

A number of measurements of both the structure and half-lives of such nuclides have been made recently by the ISOLDE group at CERN, an accelerator facility in Switzerland, particularly of progenitor nuclei of the stable nuclei around the *r*-process peak at a mass of 130 atomic mass units. Future work will determine if the shell closures in very neutron rich nuclei occur at the same neutron number as in the stable nuclei. Should the shell closures change or weaken for the very neutron rich nuclides, the effects on the predictions of *r*-process models could be profound. Additional experiments will push toward the progenitors of the nuclei around the *r*-process abundance peak at mass 195 u. In addition, recent theoretical work has suggested that the inclusion of the details of the possible beta decays of the excited states of the nuclei along the *r*-process path may also impact its nucleosynthesis; further work on this aspect of the *r*-process should also advance understanding.

Alpha-rich freezeout. Experimental work on understanding alpha-rich freezeout, which is thought to precede the *r*-process, has also involved studies of unstable nuclei. This scenario seems consistent with the basic constraints on the *r*-process. Alpha-rich freezeout is the process of nuclear assembly from neutrons and alpha particles via nuclear statistical equilibrium. At the temperature at which alpha-rich freezeout begins, tens of billions of Kelvin, all preexisting nuclei are destroyed. Then as the temperature decreases, nuclei are reformed; that is, they freeze out. Nuclear statistical equilibrium depends exponentially on the nuclear binding energies and so tends to make stable nuclei. However, unstable nuclei are made in appreciable abundance also.

One nucleus that is synthesized in alpha-rich freezeout is titanium-44 (^{44}Ti). A group at Argonne National Laboratory is studying reactions involving ^{44}Ti that can affect the amount of it that is produced in a core collapse supernova. The ^{44}Ti nucleus is sufficiently long-lived that it can be expelled by such a supernova to provide a diagnostic of the core collapse process. Indeed, its decays have been observed in the nebula of two supernovae by orbiting gamma-ray observatories. These observations produced a potential crisis. Past measurements had produced two values of the half-life of ^{44}Ti. The larger value produced tenuous agreement between theory and experiment in explaining the amount of ^{44}Ti observed, whereas the smaller value produced a serious conflict. Several remeasurements of the half-life produced a confirmation of the larger half-life, averting the crisis, but only marginally. Work is continuing to try to understand the implications associated with ^{44}Ti production and expulsion into the interstellar medium. *See* GAMMA-RAY ASTRONOMY.

Effect of neutrinos. Another aspect of recent *r*-process research has involved the effect of neutrinos on *r*-process nucleosynthesis. The neutrinos are produced in enormous numbers in a core-collapse supernova, and carry away 99% of the energy produced by the collapse of the star. The *r*-process must occur in an extremely neutron-rich region, so that the many neutron captures necessary to promote the seed nuclei to the heavy nuclei that must be made in the *r*-process can occur. However, the neutrinos may be hazardous to the *r*-process. They can interact with the helium-4 (^{4}He) nuclei in the *r*-process region to produce hydrogen-3 (^{3}H) or helium-3 (^{3}He), which can then be captured on other ^{4}He nuclei to make lithium-7 (^{7}Li) or beryllium-7 (^{7}Be). Once these nuclei are made, heavier nuclei can be created by subsequent capture reactions. These reactions would greatly accelerate the rate at which nuclei could be promoted beyond carbon-12 (^{12}C) during alpha-rich freezeout, at which point they would capture neutrons to become *r*-process seed nuclei. Unfortunately for the *r*-process, these seed nuclei would capture so many neutrons that the *r*-process site would run out of neutrons before the heavy nuclei that the *r*-process must synthesize could be made.

Another hazard to the *r*-process involves the risk that the neutrinos could convert many of the neutrons that exist in the region just outside the core of the collapsed star to protons, which would reduce the net neutron density to a value far below that necessary to drive the *r*-process. However, neutrinos are known, as a result of recent observations from the neutrino detector Super-Kamiokande, to oscillate; that is, they can change from one type to another. Thus a neutrino that might convert a neutron to a proton could oscillate to a type of neutrino that could not do so. These oscillations may be absolutely essential in the core of a star for the *r*-process to occur at all. Terrestrial experiments on neutrinos may be able to confirm some of these features of neutrinos, but observations of the neutrinos from a supernova would do vastly more to expand the knowledge of the properties of these elusive particles.

The rp-process. Progress has also been made in understanding the *rp*-process, especially in the matter accreted onto the surface of a neutron star. The accretion can produce x-ray bursts, which have been observed by x-ray satellites. Understanding the energy production and nucleosynthesis of these events is the goal of the research being conducted. The important effects are thought to occur in several layers that lie just above the surface of the neutron star, and are separated by distances of order 10 m (30 ft). The infalling matter heats up as it falls into the deep gravitational well of the neutron star, achieving a temperature of 1.5×10^9 K. At this temperature, the protons from the infalling matter are quickly captured on the seed nuclei, promoting them to the most proton-rich nuclides that can exist. The *rp*-process passes through proton-rich waiting-point nuclei, which subsequently become the progenitors of the abundance peaks of the *rp*-process. *See* X-RAY ASTRONOMY.

Explanation of Mo and Ru isotope abundances. Particularly challenging for the theory of the *rp*-process is an explanation of the synthesis of the very lightest molybdenum (Mo) and ruthenium (Ru) isotopes. These are among a class of nuclides, the *p*-nuclei, that cannot be made by the two processes of nucleosynthesis that synthesize most of the nuclides heavier than iron. Because these *p*-nuclides can be made only by complex processes, they generally have low abundances, typically less than 1% of those of their respective elements. The molybdenum and ruthenium *p*-nuclides are anomalous, however, because they have very large abundances. Only recently has any description of a nucleosynthesis process come close to successfully predicting the abundances of the molybdenum and ruthenium *p*-nuclides.

One such effort involved pushing the parameters of the *rp*-process—its density, temperature, and processing time—to their extremes to see what nucleosynthesis effects would result. Striking effects were found; large abundance enhancements were achieved for the molybdenum and ruthenium *p*-nuclides. However, some questions still exist. First, how can the nuclei produced in this form of nucleosynthesis escape from the gravitational field of the neutron star and into the interstellar medium? Second, is it possible to reconcile the fact that some of the nuclides synthesized in large abundance are also made in other nucleosynthesis processes, thereby producing too much of that isotope? In response to the first question, a tiny fraction of the synthesized material would be expected to escape from the neutron star, and the molybdenum and ruthenium *p*-nuclei are produced in enormous quantities in this production site. The second question may disappear when all the data needed to describe the nuclear reactions and half-lives of the nuclei involved are known.

Termination point. One interesting feature of this form of the *rp*-process is that it has a termination point. The nuclei just beyond tin-100 (^{100}Sn) are known to decay by emission of an alpha particle. This means that when successive proton captures reach these nuclides, they will emit an alpha particle, producing a cycle much like the carbon-nitrogen-oxygen (CNO) cycle and prohibiting progression beyond them. Thus abundance will build up in those nuclei, the progenitors of the stable nuclei around palladium-104 (^{104}Pd). This might produce an observable signature in, for example, x-ray emission of these nuclei, assuming they could be emitted from the neutron star.

The (rp)²-process. A variant of this *rp*-process would exist if *rp*-processing could occur in many successive bursts; this has been termed the $(rp)^2$-process. The $(rp)^2$-process requires the same temperature as the *rp*-process, but the density and processing time can be relaxed considerably from the *rp*-process values required to synthesize the molybdenum and ruthenium *p*-nuclides. The path of the $(rp)^2$-process would differ somewhat from that of the *rp*-process described above, as the short processing times would not push the preexisting nuclides as close to the extremes of stability as they would in the *rp*-process. The $(rp)^2$-process could also circumvent the mass-104 termination point of the high-temperature *rp*-process, as the nuclei that decay back to stability after each processing burst would progress beyond any waiting points imposed by the previous processing pulse.

Prospects. In both the *r*- and *rp*-processes, the nuclei through which the processes pass are so far from stability that they have often never been produced in the laboratory. Future work using radioactive beam facilities will certainly produce information that will be crucial to the understanding of both the *r*- and *rp*-processes. And future theoretical work that incorporates these data will almost certainly provide new testable predictions.

For background information *see* BINARY STAR; BLACK HOLE; CARBON-NITROGEN-OXYGEN CYCLES; GAMMA-RAY ASTRONOMY; NEUTRINO; NEUTRON STAR; NUCLEOSYNTHESIS; NUCLEAR REACTOR; NUCLEAR STRUCTURE; RADIOACTIVITY; SUPERNOVA; X-RAY ASTRONOMY in the McGraw-Hill Encyclopedia of Science & Technology.   Richard N. Boyd

Bibliography. L. Bildsten and T. Strohmayer, New views of neutron stars, *Phys. Today*, 52(2):40, February 1999; R. N. Boyd, The effect of beta-delayed proton emission on the path of the *rp*-process, Proc. PROCON99, edited by J. Batchelder, *AIP Conf. Proc.* (American Institute of Physics), no. 518; p. 239; D. O. Caldwell, G. M. Fuller, and Y.-Z. Qian, Sterile neutrinos and supernova nucleosynthesis, *Phys. Rev.*, D6:123005(12 pages), 2000; B. S. Meyer, Neutrino reactions on ^{4}He and the *r*-process, *Astrophys. J.*, 449:L55, 1995; H. Schatz et al., *rp*-process nucleosynthesis at extreme temperature and density conditions, *Phys. Rep.*, 294:167–263, 1998; G. Wallerstein et al., Synthesis of elements in stars: Forty years of progress, *Rev. Mod. Phys.*, 69:995–1084, 1997.

Ocean warming

Among all liquids, water has nearly the greatest capacity to store heat while undergoing relatively small temperature changes. In fact, the upper 2 m (6.2 ft) of the ocean can store as much heat as the entire overlying atmosphere, which has much less mass and much less ability to store heat. The meteorologist-climatologist Carl Rossby and other scientists before him drew attention to the fact that the great mass of the world ocean not only could store large amounts of heat but also could remove this heat by direct contact with the atmosphere for long periods of time, from years to millennia. This could be accomplished through the downward movement of the water that is observed to occur in certain regions of the world ocean such as the Labrador Sea. Relatively dense surface water can sink and displace less dense subsurface water. Even if the temperature of the surface water is warmer than the subsurface water it displaces, the surface water can still be denser than the subsurface water if its salinity is greater than the subsurface water. At subsurface depths this heat can be transported via ocean currents over large distances in the horizontal as well as the vertical before returning to the sea surface in modified form due to mixing of water masses of different temperatures and salinities. This mechanism is called the thermohaline circulation ("thermo" refers to temperature and "haline" to salinity) and is known popularly as the ocean conveyor belt.

The role of the ocean is so important in climate variability that the Intergovernmental Program on Climate Change, the World Climate Research CLIVAR program, and the U.S. National Research Council have identified understanding the ocean as being critical to climate system research. The Earth's climate system comprises the world ocean, the global atmosphere, and all forms of ice, including continental and mountain glaciers as well as sea ice.

Temperature data. Although it has been long hypothesized that the world ocean plays a major role in maintaining the Earth's climate system, until recently little scientific work had been done in systematically identifying ocean subsurface temperature variability on ocean basin and global scales. This was due in large part to the lack of data. In the 1970s, the international scientific community initiated Ship-of-Opportunity Programs (SOOP) to make routine observations of the upper ocean temperature from merchant vessels using expendable instruments. The scientific community has also conducted international observations programs, such as the World Ocean Circulation Experiment and the Joint Ocean Global Ocean Flux Study, that have made high-quality measurements of temperature from the ocean surface to the bottom, along with measurements of other aspects such as salinity and nutrients. During the past several years, international projects to locate and digitize historical subsurface temperature data from the world ocean have resulted in the addition of more than 2 million of these profiles to the

electronic, CD-ROM database known as *World Ocean Database 1998* (WOD98).

Using a total of more than 5 million temperature profiles for the 1948–1996 period from WOD98, S. Levitus and coworkers (2000) estimated that the upper 300 m (984 ft) of the world ocean has warmed on average by approximately $0.31°C$ ($0.5°F$), and the upper 3000 m (9842 ft) has warmed on average by approximately $0.06°C$ ($0.11°F$). Although some parts of the world ocean, such as the subarctic of the North Atlantic, have cooled during this period, the total volume of seawater in the upper 3000 m of each major ocean basin has exhibited a net warming. It is important to note that the data have been analyzed in such a way as to minimize the creation of any temperature change during analysis; that is, the analyses are biased to produce zero temperature change. The results suggest that the observed warming is not due to sampling biases or lack of data in some regions. Although the temperature differences described may seem small, they represent very large changes in the heat content of the ocean and the Earth's climate system. These results have great significance in the current debate about the causes of the observed warming of the Earth's surface (surface air and sea surface temperature) since the mid-1970s.

Figure 1 shows the variability of the yearly, mean temperature anomalies of the upper 300 m (984 ft) of the world ocean for 1948–1998. Starting in the mid-1960s, the world oceans shifted from a relatively cool to a relatively warm state. However, the warming of the world ocean did not occur steadily over time, and the beginning of world ocean warming preceded the observed warming of surface temperatures that has been attributed to the increases of greenhouse gases in the atmosphere due to fossil fuel burning. The Pacific and Indian oceans cooled during the early 1980s before warming again, and cooled yet again in the early 1990s before warming. The Atlantic Ocean exhibited similar behavior but to much less extent.

Figure 2 shows the variability of the yearly, mean temperature anomalies of the upper 3000 m (9842 ft)

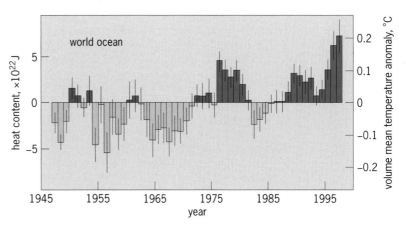

Fig. 1. Time series for 1948–1998 of mean ocean temperature anomaly and ocean heat content in the upper 300 m of the world ocean. Note that 1.5×10^{22} J equals 1 watt year m^{-2} (averaged over the entire surface of the Earth). Vertical lines through each yearly estimate represent plus and minus one standard error of the estimate of heat content.

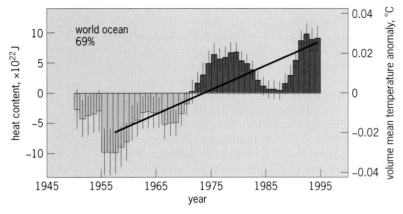

Fig. 2. Time series of 5-year running composites of mean ocean temperature anomaly and heat content in the upper 3000 m for the world ocean. Note that 1.5×10^{22} J equals 1 watt year m^{-2} (averaged over the entire surface of the Earth). Vertical lines represent plus and minus one standard error of the 5-year mean estimate of heat content. The linear trend is estimated for the time series for the period 1955–1996, which corresponds to the period of best data coverage. The trend is plotted as a diagonal line. The variance accounted for by this trend is 69%.

of the world ocean for 1948–1995. To have enough data to make meaningful estimates of the variability of the temperature of the deep ocean requires compositing all the data for 5-year periods as a "running" average. This procedure has the effect of producing smoothed results. Prior to the mid-1950s, there were relatively few deep-sea observations, and the results are tentative for this early part of the record. The average temperature of the upper 3000 m of the world ocean increased from the mid-1950s until the early 1980s, whereupon it decreased for several years before beginning to warm again. Approximately half of the warming of the world ocean during the past 50 years occurred at depths greater than 300 m (984 ft), which verifies the long-held idea, articulated by Rossby and other scientists, that the world ocean, and particularly the deep ocean, constitutes a major part of the Earth's climate system.

Modeling studies. These results have implications for climate system research in general, and more specifically for the debate that has taken place during the past 10 years regarding the causes of the warming of the Earth's surface since the mid-1970s. It is not possible to partition the observed warming between an anthropogenic component and a component associated with natural variability based on observations alone. Complex modeling studies and/or estimates of climate forcing by volcanoes, solar variability, and other phenomena are required to be able even to attempt such a partition.

One of the criticisms of the atmospheric general circulation models that were initially used to quantify the effect of increasing greenhouse gases on the Earth's atmospheric climate is that the models were too sensitive to the effect of increasing greenhouse gases in the model atmosphere. To try to determine the effect of greenhouse gases on the atmosphere, the models were started with a set of initial conditions of the state of the model atmosphere and forced over time with the observed increase in atmospheric greenhouse gases, which have been

added to the atmosphere in increasing amounts over the past 100 years. The models were criticized because their atmospheres warmed much more than the real atmosphere has warmed during the past 100 years. The climate modeling community responded by suggesting that since their models did not have an ocean "attached" to absorb the additional heat trapped in the Earth's climate system, the models were, in fact, not too sensitive. A leading climate modeler, James Hansen, computed the amount of heat that could be expected to be found in the ocean based not on model results but on the forced heating of the Earth's climate system expected from observations of volcanic aerosols, stratospheric ozone depletion, greenhouse gases, and solar variability. Hansen and coworkers concluded that a planetary radiative disequilibrium of about 0.5–0.7 W m^{-2} existed for the 1979–1996 period (with the Earth system gaining heat) and suggested that the excess heat must primarily be accumulating in the ocean. The heating rate for the world ocean based on the curve shown in Fig. 2 is 0.3 W m^{-2}. It is possible that the present estimate of the increase in temperature of the ocean is an underestimate. The results strongly support the explanation by the climate modeling community as to why atmosphere-ocean general circulation models forced with the observed increase of greenhouse gases produced more warming of the model atmosphere than has been observed in the real atmosphere. With the advent of more powerful computers, the climate modeling community is now routinely simulating not only the atmospheric response to increasing greenhouse gases but also the response of more of the Earth's climate system, by including ocean models coupled to the atmospheric models, as well as sea-ice models. This is one of the most active climate research areas, and results will be published in the near future.

Implications. Global sea surface temperature time series for the past 100 years show two distinct warming periods. The first occurred during the 1920–1940 period and was followed by a period of cooling; the second warming began during the 1970s. It is now believed that the early warming may have been due to natural variability of the Earth's climate system, but the second period of warming is widely believed by the scientific community to be of anthropogenic origin. Analysis of independent data, so-called proxy climate data such as temperature reconstructions from tree rings, corals, and lake sediments, indicate that the warming of the last 100 years is real and is unprecedented for the past 1000 years. The observed temperature increase at the Earth's surface is also highly correlated with the increase of greenhouse gases observed in the Earth's atmosphere for the past 100 years. All these lines of evidence point to an anthropogenic explanation of the recent warming of the Earth's surface. The results presented for the subsurface ocean represent a shorter record than the surface time series but are critical to substantiating an anthropogenic explanation of the recent warming. This is because the capacity of the world ocean to store heat means that the recent warming of the Earth's surface could be due to internal variability of the Earth's climate system, with heat stored in the ocean many years ago now warming the Earth's surface. The fact that the entire world ocean has exhibited a warming in the amount predicted by independent analysis of the forcing of the Earth's climate system is strong evidence that the warming is human-induced.

The most important consequence of the results described is that the present warming trend at the Earth's surface is expected to continue. The warming, however, may not be steady. For example, volcanic eruptions, such as El Chichón in 1981 and Mount Pinatubo in 1991, that transfer large amounts of sulfur dioxide to the Earth's stratosphere result in a temporary (2–4 years) cooling of the Earth's surface of approximately $0.5°C$ ($1.0°F$). These eruptions most likely are responsible for the decrease in heat content of the world ocean observed after the eruptions.

To reduce uncertainly about the results, substantial amounts of historical oceanographic data existing only in manuscript form need to be digitized and, along with other data, placed into integrated ocean profile databases for use by the international research community. In addition, the planned Global Ocean Observing System (GOOS) must be implemented immediately so that the scientific community has access to a real-time global oceanographic database to monitor the state of the world ocean, that is, the temperature and salinity of the ocean as a function of depth. The international oceanographic community has developed the technology to implement such a system, and it is critical that deployment begin now. Such a system will make available the data to improve seasonal-to-interannual forecasts of phenomena such as El Niño and to perhaps allow for the forecasting of decadal climate anomalies in a statistical sense.

For background information *see* CLIMATE HISTORY; CLIMATE MODELING; CLIMATE MODIFICATION; CLIMATIC PREDICTION; HEAT BALANCE, TERRESTRIAL ATMOSPHERIC; OCEAN CIRCULATION; OCEANOGRAPHY in the McGraw-Hill Encyclopedia of Science & Technology.   Sydney Levitus

Bibliography. J. Hansen et al., *J. Geophys. Res.*, 102: 25679, 1997; S. Levitus et al., Warming of the world ocean, *Science*, 287:2225–2229, 2000; S. Levitus et al., *World Ocean Database 1998a*, vol. 1: *Introduction*, NOAA Atlas NESDIS 18, 1998.

Optical activity

In order to explain optical activity, it is helpful first to discuss linearly and circularly polarized light. Light can be thought of as a wave or a particle (wave-particle duality). In wave theory, light is an electromagnetic wave propagating with a velocity v ($c = v$ in vacuum). The wave consists of an electric field E oscillating in a plane perpendicular to the direction of propagation, and a mutually perpendicular magnetic field H. Both are oscillating with a frequency v, and the distance between successive maxima of E or H is the wavelength λ, such that $\lambda = v/v$. If E oscillates in a specific plane relative to the propagation direction (plane of polarization), the light is said to be linearly polarized. If two mutually perpendicular linearly polarized electromagnetic waves are combined, having the same frequency and direction of propagation but with phases differing by one-quarter wavelength, the resultant E vector rotates about the propagation direction. If the direction of rotation is clockwise as viewed in the direction of the light source, the light is said to be right-circularly polarized (RCPL), and it is left-circularly polarized (LCPL) if the rotation is in the opposite direction. In particle theory, a quantum of light is called a photon, with energy equal to Planck's constant h times the frequency, or hv. Photons also have unit intrinsic angular momentum of $+\hbar$ (equal to $h/2\pi$) along the direction of propagation (RCPL) or $-\hbar$ against the direction of propagation (LCPL). Linearly polarized light is an equal mixture of these two photon beams.

Optical rotation. The French scientist D. F. J. Arago (1786–1853) discovered that the plane of polarization of a beam of linearly polarized light rotates upon passing through certain transparent crystals such as quartz. Jean Baptiste Biot (1774–1862) noted the same optical rotation in some liquids and vapors such as terpentine. The phenomenon of rotation of the E vector of plane-polarized light is called optical activity, and materials exhibiting this characteristic are described as being optically active. Biot also noted that some materials rotate the electric vector of light (E vector) to the right or to the left as viewed looking toward the direction of the source of the light beam. Substances which rotate the E vector of light to the right (or in a clockwise sense) are called dextrorotatory or *d*-rotatory. Substances which rotate the E vector to the left (or counterclockwise) are called levorotatory or *l*-rotatory.

In 1848, Louis Pasteur made the monumental discovery that individual molecules could exhibit optical activity. While examining crystals being formed during the making of wine, he noted that both *l*- and *d*-rotatory salts of tartaric acid were being formed. Upon dissolving these crystals in water, the two solutions containing the molecular subunits of the two crystals were also seen to be optically active. The rotations of the plane-polarized light for the two solutions were of the opposite sense, one *l*-rotatory and the other *d*-rotatory. He concluded that the molecules making up these crystals were "right-handed" or "left-handed." Although the two molecules were chemically the same, they were mirror images of each other but somehow different in their ability to rotate the polarization of light. Pasteur had discovered optical stereoisomers.

Pasteur was not satisfied just to discover a new branch of chemistry, now called stereochemistry—he demonstrated that the chemistry of living material also exhibits a preferred handedness. Upon mixing equal amounts of the two optically active tartaric acid

Enantiomorphs of bromochlorofluoroiodomethane:
(*a*) original molecule, (*b*) mirror image molecule.

solutions, he produced a mixture that was optically inactive; that is, there was no rotation of the plane of polarization of light. He attributed this to the fact that the solution had equal numbers of molecules of opposite handedness (called a racemic mixture). However, Pasteur observed that as time elapsed, the solution once again became optically active. Noting that molds had grown in the dish, he postulated that the growing mold had ingested only molecules of one handedness, leaving the other-handed molecules seemingly untouched.

Chirality. A molecule whose mirror image cannot be superimposed upon itself is called chiral. One of the simplest chiral molecules is the all-halogenated methane molecule (see **illus.**). Chiral molecules that are nonsuperimposable mirror images of one another are said to be enantiomorphs of each other, and each is called an enantiomer. When one enantiomer exists to the exclusion of the other in a system, the system is said to be homochiral.

Chiral molecules may exhibit optical activity. Optical activity is observed as both optical rotation as discussed above and circular dichroism, the difference in extinction for RCPL and LCPL passing through the chiral medium. Since linearly polarized light is a coherent superposition of RCPL and LCPL (or equal numbers of right- and left-handed photons), the two phenomena are related through the Kramers-Kronig relations. Knowledge of one can be used to calculate (in principle) the other. The optical rotation angle, θ (in radians), can be written as $\theta = (\pi/\lambda)(n_L - n_R)\ell$, where $n_L - n_R$ is the difference in the index of refraction for LCPL and RCPL at the wavelength λ and ℓ is the path length. The precise angle through which the polarization is rotated depends on the concentration of the chiral molecules, path length, wavelength, and temperature. The specific rotation, $[\alpha]$, has been adopted as a useful quantitative measure of these effects: $[\alpha]_\lambda^T = \theta/\ell c$, where c is the concentration of chiral molecules in (g mL^{-1}) and ℓ is the path length in decimeters, T is the temperature at which the experiment is run, and λ is the wavelength of the linearly polarized light.

It is now known that most of the molecules that make up living things are of one specific chirality. The amino acids that polymerize to form proteins in cells exist almost exclusively in one enantiomeric form. All but one (glycine, the simplest of the 20 biologically important amino acids) are chiral and can exist in both the L and D absolute configurations. The symbols L and D are not to be confused with the des-

ignation *l*-rotatory or *d*-rotatory, which refers to the direction of rotation of linearly polarized visible (yellow) light rather than the structure of the molecule itself. Amino acids which make up the proteins found in cells are almost exclusively L-enantiomers. This is true regardless of the organism from which the protein is extracted. There are but a few exceptions to this preference; some involve a group of antibiotics such as penicillin. Also, only D sugars (D-ribose and 2-deoxy-D-ribose) are found in the polymers of nucleic acids that make up the backbone of DNA and RNA. Thus, specific homochirality of biomolecules represents a fundamental characteristic of life. In fact, homochirality appears to be an accepted tenet of any theory of the origins of life. Understanding how this specific handedness or homochirality evolved is fundamental to the origins of life. Two general theories have been evoked: (1) a random or chance event that selects the L-amino acids and D sugars and (2) the existence of an internal/external chiral force which could favor a prebiotic asymmetric selection of the building blocks of life.

Asymmetric synthesis in nature. It is well known that exposing a racemic mixture of molecules to circularly polarized light can lead to an excess of one enantiomer over another, called an enantiomeric excess. This can occur by converting one enantiomer into its mirror image or by preferential destruction of one enantiomer. Thus, circularly polarized light in nature could provide an advantage for one enantiomer over another, a type of neo-Darwinism. A number of sources of circularly polarized light (CPL) on Earth and in the universe have been identified. Sunlight passing through the atmosphere is slightly elliptically polarized (not quite circular) in the morning (LCPL) and evening (RCPL). Supernovas are also known to emit CPL at right angle to the normal linearly polarized bremsstrahlung (electromagnetic radiation emitted by electrons when they pass through matter) generated from electrons rotating in a circle (like synchrotron radiation). CPL is emitted perpendicular to the plane of bremsstrahlung, RCPL on one side and LCPL on the other. Thus in the cases of the sunrise/sunset and supernova, both right- and left-CPL are produced so that any advantage would again occur by a chance encounter with either form of CPL. Another source of asymmetry comes from radioactive beta decay. Beta decay, however, exhibits a specific chirality.

All matter interacts through four known forces (in order of strength): nuclear, electromagnetic, weak, and gravity. Of these forces, only the weak interaction is fundamentally chiral (can distinguish between left and right). The weak interaction is responsible for radioactive beta decay, the spontaneous emission of electrons from metastable nuclei. Electrons possess an intrinsic one-half unit of spin angular momentum, or $\frac{1}{2}\hbar$. An electron in motion can have its spin angular momentum vector along or against its direction of propagation (that is spin-polarized), just like circularly polarized light. Thus the electron is chiral (nonsuperimposable mirror image), and its spin describes a right- or left-hand helicity as it moves through space. In 1957 it was predicted and observed that the weak nuclear force does not conserve parity. This means that beta (β) particles emitted from the nucleus are highly spin-polarized in one preferred helicity. It was observed that parity was being violated and that beta rays are primarily left-handed. Positron emission gives rise to right-handed antielectrons (positively charged electrons). It is often said that we live in a left-handed universe since we are made of matter rather than antimatter. Shortly after the fall of parity, predictions of the effect of the chiral weak interaction on atomic physics were confirmed. Since the nucleus of an atom contains a chiral force (weak interaction), all atoms are thereby chiral and should exhibit an optical rotation and circular dichroism. This force will also mix states of opposite parity, giving rise to the observation of "forbidden transitions" in atoms. Both of these minute atomic effects have been seen and are taken as support of the theory of the electroweak interaction. A bigger question remains, "Can the weak interaction play a role in chemistry?"

Weak interaction effects. There are at least two ways that the weak interaction could influence a chemical reaction: (1) a direct parity-violating energy difference in the energy levels of enantiomers, and (2) the interaction of left-handed electrons with matter following beta decay.

It has been pointed out that parity nonconservation leads to an energy difference between mirror image molecules. One enantiomer will go up in energy and the other will go down in energy in comparison to the exact energy degeneracy expected in normal quantum electrodynamics. This is called the parity-violating energy difference, or PVED. PVED is calculated to be exceedingly small for most biomolecules containing low atomic number (Z) atoms (carbon, oxygen, nitrogen, hydrogen), and this energy difference has not been measured. For amino acids the PVED is expressed as a difference in energy divided by the total energy, $\Delta E/E$, and has been calculated to be on the order of 10^{-18}. Although all of the L-amino acids (except for glycine which is achiral) lie lower in energy than the D-amino acids, this energy is generally believed to be too small to account for the dominance of L-amino acids in biology. However, the PVED is proportional to the atomic number of the atoms making up the molecule raised approximately to the sixth power, Z^6. Thus, for larger molecules containing many atoms with higher atomic number, this energy difference is expected to become experimentally observable with current techniques such as nuclear magnetic resonance (NMR), Mössbauer (nuclear gamma resonance fluorescence), and infrared (IR) spectroscopy. The first measurement of an energy difference between enantiomers in a complex iron molecule (crystal) was recently reported using the Mössbauer technique. Due to the fact that this was done for a crystal, it is too soon to attribute it to PVED for a chiral molecule, although the difference in energy

($\sim 10^{-10}$ eV) is close to that predicted by the scaling with atomic number (Z^6).

It was suggested in the 1950s that circularly polarized electrons generated in beta decay may be responsible for the homochirality of biological molecules. When the energetic electrons collide with matter, they generate circularly polarized gamma rays that could induce asymmetry in a subsequent photochemical reaction. This could occur in an initially racemic reactant in which one enantiomer reacts faster than the other, thus leaving an enantiomeric excess in recovered starting material. Photoproducts of the reaction could also be optically active. Such photochemistry is known using circularly polarized ultraviolet photons, but in general enantiomeric excess in recovered reactant or photoproduct is very low. An asymmetry induced by the direct reaction of the spin-polarized (lefty) electron with the substrate is also possible.

Experimental evidence. Several attempts were made in the last several decades to see if this hypothesis is plausible. F. Vester and T. L. V. Ulbricht were the first researchers to test the suggestion. They carried out three dozen organic reactions initiated with several different beta sources, and obtained no positive results. Subsequently A. S. Garay subjected separate alkaline solutions of the enantiomeric amino acids, D-tyrosine and L-tyrosine, containing ^{90}SrCl$_2$ (strontium chloride), a weak source of beta particles. After 18 months Garay found the D-enantiomer to have reacted faster that the L. Unfortunately, the method of analysis was flawed. W. Darge later also obtained positive results. He observed that racemic DL-tryptophan, another amino acid, in frozen solution containing ^{32}PO$_4^{-3}$ (phosphate), another beta emitter, became significantly enriched in the D-enantiomer after 12 weeks. In a set of exhaustive experiments W. A. Bonner demonstrated that the claims of Garay and Darge were untrue. Bonner also attempted two other types of experiments. In the first, a set of several amino acids labeled with carbon-14 (^{14}C), which also undergoes beta decay, was allowed to undergo internal decay; no enantiomeric excess was found in any of the recovered amino acids. In the second, Bonner was more successful. Here solid samples of DL-leucine, a still different amino acid, were subjected to 120-keV antiparallel spin and parallel spin polarized electrons generated in a linear accelerator. The antiparallel longitudinally polarized electrons afforded a slight excess of L-leucine in recovered amino acid, while the parallel electron afforded a slight excess of the D-enantiomer.

More success has recently been obtained in a different type of experiment involving beta particles. Sodium chlorate (NaClO$_3$) is an achiral molecule consisting of achiral spherical sodium cations and achiral chlorate anions of C$_{3v}$ symmetry (pyramidal shape). When the ionic compound is crystallized from water, the resulting crystals are chiral because they crystallize in a chiral space group. Because the crystallization process yields the d (+) and l (−) crystals randomly, a random distribution of d and l crystals is obtained if enough crystals from a large number of crystallization experiments are examined. When aqueous solutions containing sodium chlorate were evaporated slowly in front of an energetic ^{90}Sr source, considerably more *d* than *l* crystals (about two to one) were obtained. Interestingly, when the crystallizations were carried out in front of a sodium-22 source ([22]Na, a positron emitter), a two-to-one excess of *l* crystals was obtained. Thus beta particles with left-handed helicity afforded an excess of *d* crystals, while positrons with right-handed helicity afforded an excess of *l* crystals. Is it possible that the asymmetric synthesis of a crystal under the influence of polarized electrons or positrons was somehow involved in the origin of homochirality in the biological world? Many more experiments of this type are need in order to establish such a connection.

For background information *see* BREMSSTRAHLUNG; DICHROISM; ELECTROMAGNETIC RADIATION; ELECTROWEAK INTERACTION; MÖSSBAUER EFFECT; OPTICAL ACTIVITY; PARITY (QUANTUM MECHANICS); POLARIZED LIGHT; RACEMIZATION; STEREOCHEMISTRY; STEREOISOMER in the McGraw-Hill Encyclopedia of Science & Technology.

R. N. Compton; R. M. Pagni

Bibliography. W. A. Bonner, Experiments on the Origin and Amplification of Optical Activity, Chap. 1 in D. C. Walker (ed), *Origins of Optical Activity in Nature*, Elsevier, Amsterdam, 1979; M. Gardner, *The New Ambidextrous Universe*, W. H. Freeman, New York, 1990; R. A. Hegstrom and D. K. Kondepudi, The handedness of the universe, *Sci. Amer.*, no. 108, January 1990; A. Lahamer et al., Search for a parity-violating energy difference between enantiomers of a chiral iron complex, *Phys. Rev. Lett.*, 85:4470, 2000; S. Mahurin et al., The effect of beta radiation on the crystallization of sodium chlorate in water: A new type of synthesis, *Chirality*, vol. 13, 2001.

Optical data storage

This article discusses potential optical data storage systems that may be developed in the next 5–10 years. Each system would increase data capacity compared to present-day systems. Some systems could also significantly increase the rate at which information is stored and retrieved. Both high capacity and high data rate are important in future high-throughput applications.

At present, commercially significant optical data storage is disk-based. This format is convenient from both user and manufacturing perspectives. To the user, the disk format is easy to handle, stacks well in libraries, and is more rugged than the phonograph records and tape products it replaces. To the manufacturer, plastic disks are easy and inexpensive to fabricate. The success of optical storage is largely due to the fact that the entertainment industry embraced the compact disk (CD) at an early stage.

However, a serious limitation exists with disk-based optical data storage. As the data rate increases, the playing time for a fixed capacity decreases. Applications that require long playing times (and correspondingly high capacities) necessitate compression, which is lossless only for compression factors of 2 to 4. For example, a compact-disk read-only memory (CD-ROM) drive operating at 50 million bits per second (50 Mbps) takes only 102 seconds to read the entire disk. Correspondingly, a hypothetical digital-versatile-disk read-only memory (DVD-ROM) drive operating at 400 Mbps (a similar speed multiplier compared to the fast CD drive) takes less than 100 seconds to read a 4.7-gigabyte (GB) disk.

Technical principles. In optical recording (**Fig. 1**), light from a laser passes through a beam splitter and is focused into the recording layers by an objective lens. The recording layers are on a disk that spins under this lens. They contain spiral tracks of mark patterns that differ in reflectivity from the area between marks. As the focused laser beam passes over a mark, the reflected light level changes. These changes are sensed by using the beam splitter to direct a portion of the reflected light onto a silicon detector. The detector current, which is a representation of the mark pattern, is decoded to produce digital information.

The fidelity of the detector signal determines the amount of data per unit length of track, usually referred to as the linear density, that can be decoded with high reliability. The linear density along with the radial track spacing determines the areal density, which is usually reported in gigabits per square inch (Gb/in.^{2}). The areal density multiplied by the effective recording area is the disk capacity.

Several factors influence the fidelity of the detector signal. The most important factor for closely spaced marks is the focused spot size *s*. Large *s* blurs the reflected light signal. Conversely, if *s* is small, changes in the reflected signal are sharp as the marks traverse the spot. Therefore, as *s* decreases, more rapid changes in the mark pattern can be detected, and areal density can increase.

Unfortunately, *s* cannot be made arbitrarily small. Due to the physics of diffraction, the minimum spot

TABLE 1. Performance data of disk-based products					
	CD-1X	CD-40X	DVD-1X	DVD-40X	DVR-1X
Capacity, GB	0.64	0.64	4.7	4.7	20
Data rate, Mbps	1.2	48	10	400	25
CRP	0.77	30.7	47	18,800	500
Retrieval time, min	70	1.7	62.7	1.6	106.7

size s is a function of the wavelength of the laser λ, the focusing properties of the objective lens, system aberrations, and the thin-film structure used as the recording layer. A simple relationship for estimating the full-width-at-$1/e^2$ spot size for conventional gaussian illumination is given by the equation below,

$$s = \frac{\lambda}{\sin \theta_m}$$

where λ is the wavelength in air and θ_m is the marginal ray angle (Fig. 1). The marginal ray passes just at the edge of the stop, which is the limiting aperture of the system. The value of $\sin \theta_m$ is the numerical aperture (NA) of the focusing objective. As the NA increases or λ decreases, the spot size s gets smaller, and areal density can increase.

A useful figure of merit is the capacity-rate product (CRP), which is the product of the capacity in gigabytes and the data rate in megabits per second. The CRP and other performance characteristics of disk-based products are given in **Table 1**. The data-rate speedup factor is shown as 1X or 40X, where 1X refers to the data rate of products when they were first introduced, like the CD-ROM in 1991. 40X refers to a data rate that is 40 times faster that the 1X rate. Also included in Table 1 are preliminary data concerning the digital video recorder (DVR), which is under development.

Future demand. An important factor affecting the growth in demand for digital storage is the emergence of digital television. Given the capacity offered by the storage medium, the video data rate determines the playing time offered by the system (**Table 2**). Analog video of standard quality can be compressed using MPEG-2 to average data rates of 5–6 Mbps. Standard-definition digital broadcasts will be 9–10 Mbps, which is the DVD standard in the United States and Europe. Two high-definition broadcast standards are HD-DVB at 19 Mbps in the United States and Europe, and DS4B at 24 Mbps in Japan.

The playing time for video recorders should be at least 4–8 hours (the time provided by current VCRs). A capacity of 45 GB is needed to satisfy this requirement for HD-DVB and DS4B (Table 2), which is well beyond predictions of products listed in Table 1, and will most likely not be available in a disk format

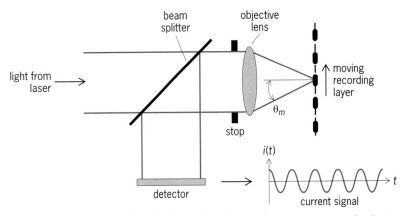

Fig. 1. Optical recording device. Light from a laser beam is focused onto a recording layer on a spinning disk with tracks of data marks. The reflected light is directed by the beam splitter onto a detector, which provides a current signal that is modulated by light reflected from the data marks.

until after 2005. Capacity can be increased by reducing s and using the same total recording area, or by keeping s constant and increasing the recording area.

Solid immersion lens system. By augmenting the illumination optics with a solid immersion lens (SIL) [**Fig. 2**], the spot size can be reduced significantly. SIL optics provide $s = \lambda/(n \sin \theta_m)$, where n is the refractive index of the lens. With $\lambda = 0.65$ micrometer, $\sin \theta_m = 0.6$, and $n = 2.0$, $s = 0.54$ μm. Disk capacity for SIL systems could easily reach 20 GB for each recording surface. The difficulty with such systems is that the lens must be very close (on the order of 0.1 μm) to the recording layer. Figure 2 shows one possible implementation, where the lens is mounted in a slider, like those found in magnetic hard-disk drives. As the disk spins, an air bearing is produced under the slider that keeps the lens a fixed distance above the surface. The distance between the lens and the top of the recording layers is called the air gap, h.

Since the air gap is so small, contamination on the disk surface is an important consideration for SIL systems. In CD and DVD disks, the focused laser light first passes through the substrate of the disk that provides some degree of coverplate contamination protection, but SIL systems lack protection from the substrate. The recoding layers are exposed to the

TABLE 2. CRP and play time (in parentheses) as a function of video format and capacity					
	Capacity, GB				
Format	4.7	10	20	45	100
MPEG-2 variable bit rate (5 Mbps)	23.5 (2h 05)	50 (4h 26)	100 (8h 52)	225 (20h 00)	500 (44h 00)
MPEG-2 fixed bit rate (10 Mbps)	47 (1h 03)	100 (2h 13)	200 (4h 26)	450 (10h 00)	1000 (22h 00)
HD-DVB (19 Mbps)	89.3 (0h 33)	190 (1h 10)	380 (2h 20)	855 (5h 16)	1900 (11h 42)
DS4B (24 Mbps)	112.8 (0h 26)	240 (0h 56)	480 (1h 51)	1080 (4h 10)	2400 (9h 16)

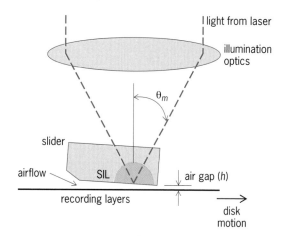

Fig. 2. Solid immersion lens (SIL) system. The SIL reduces spot size and increases data density. Since the SIL must be in proximity to the recording layers, an air-bearing slider is used to control the air-gap height as the disk rotates.

environment. A major challenge for SIL system development is engineering effective contamination control.

A five-beam SIL system operating with $\lambda = 0.40$ μm, $\sin \theta_m = 0.85$, and $n = 2.9$ could conceivably provide a 150-GB disk with CRP $\sim 1.5 \times 10^4$.

Optical tape storage. The effective recording area can be increased by using an optical tape system. The tape can be wound on a spool to create high volumetric efficiency. The medium is nearly flat and does not have a moving coverplate, like disk-based systems, so optical design of the components allows a relatively large amount of surface area to be examined for each position of the tape. Like magnetic tape systems, a large number of parallel channels can dramatically increase the data rate. Unlike magnetic tape systems, optical tape systems have no head wear, because the readout heads do not contact the medium. Also, the readout spots on the medium can be extremely close together, unlike the limitations imposed on magnetic tape recording. Optical tape systems currently under development can store 1000 GB of user data. With multiple laser beams, very high data rates (800 Mbps) can be expected. The CRP of optical tape devices is very high, with advanced prototypes exhibiting CRP $\sim 10^6$.

Two-photon volumetric storage. Volumetric storage also increases the effective recording area. One promising volumetric technology is two-photon storage, where a very intense light beam is used to record data with an effective wavelength that is half the wavelength of the laser source. That is, a 0.65-μm laser interacts with the medium as if it is a 0.325-μm laser. This property can be achieved only in the region directly around the focus point due to the high light level there. If the medium is designed to respond only to the two-photon energy (at 0.325 μm), the vast majority of the medium in the path of the focus beam does not interact with the laser. Therefore, the laser can be focused through very thick media with many data layers and interacts only with the layer closest to focus. In one possible implementation of two-photon technology, with a 1-in.-diameter (25-mm) disk format (**Fig. 3**), 500 layers at 1 GB each are read out at a data rate of 500 Mbps with multiple beams. The resulting CRP is $\sim 2.5 \times 10^5$.

Holographic storage. Holographic storage is also a volumetric technology, but the manner in which data are stored is different than in the technologies discussed above. Bits of information are not stored as discrete marks in the medium, but are distributed throughout the volume and recorded interferometrically. To accomplish this, a large page of digital information is assembled on a light valve, such as a liquid crystal display (**Fig. 4a**). Laser light is passed through this page and caused to interfere with another beam of light, which is called a reference beam, from the same laser. A page may contain as many as 1024×1024 bits, or 1 Mbit, of information. Additional pages, each containing different information, may be recorded in the same volume by using other reference beams that are uniquely different from each other in some attribute, such as direction or wavelength. An individual page is recovered by illuminating the recording volume with the corresponding reference beam and directing the image of the reconstructed page onto a photodetector array (Fig. 4b). Because the entire page illuminates the detector array at once, the bits appear in parallel, and exceptionally high data rates can be achieved. Data rates as high as 10 Gbps (10,000 Mbps) are not

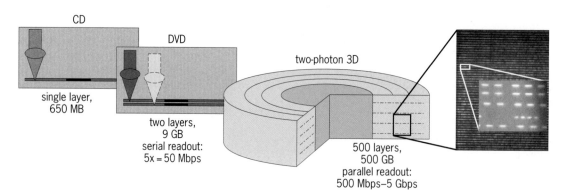

Fig. 3. Two-photon storage. A 1-in.-diameter (25-mm) two-photon disk has the potential to record up to 500 GB of user data, which is much greater than the conventional CD or DVD.

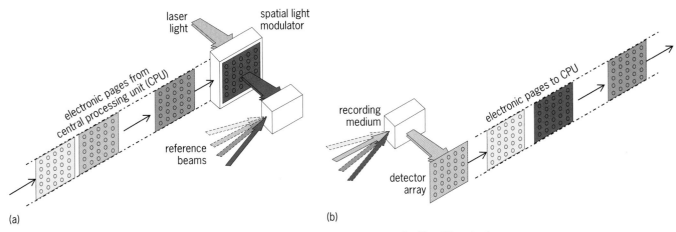

Fig. 4. Holographic storage. (a) Recording. Reference beams record one page at a time, each with a different reference beam. (b) Retrieval. One page is retrieved from many by using the corresponding reference beam.

uncommon. Several thousand pages can be stored in a common volume and, by using adjacent volumes, capacities of 125 GB can be realized. The resulting CRP is 1.25×10^6.

In order to record multiple pages in a common volume, a process called multiplexing, the recording medium must be thick, typically 200 μm-1 mm. Two primary candidates are inorganic photorefractive crystals and organic photopolymers. Photorefractive crystals, such as iron-doped lithium niobate, can be grown in large sizes (2–3 cm is not uncommon), and allow real-time recording of erasable holograms. In the recording process, electrons are excited in the bright areas of the interfering light beams, and migrate to dark regions of the pattern. They become trapped and create internal electric fields that change the local index of refraction. The pattern of refractive index changes is the hologram. The material must be temporarily fixed, typically by heat, or else the stored information will be erased upon readout. To reuse the material, the crystal is uniformly illuminated at an elevated temperature.

Photopolymer materials record holograms by converting monomers to long-chain polymers. As monomers are converted in the bright areas of the interfering light beams, the local monomer concentration decreases, and new monomer material diffuses in from the dark areas. This causes an unbalance of polymerized material and results in a modulation of the refractive index that mimics the exposing interference pattern, that is, the hologram. The process is completed by uniformly exposing the entire medium to polymerize any remaining monomer. Photopolymetric holograms cannot be erased, and represent WORM (write-once-read-many) storage.

For background information see COMPACT DISK; COMPUTER STORAGE TECHNOLOGY; DIFFRACTION; HOLOGRAPHY; NONLINEAR OPTICAL MATERIALS; NONLINEAR OPTICS; OPTICAL INFORMATION SYSTEMS; OPTICAL RECORDING in the McGraw-Hill Encyclopedia of Science & Technology.

Tom D. Milster; Glenn Sincerbox

Bibliography. A. B. Marchant, *Optical Recording: A Technical Overview*, Addison-Wesley, Reading, MA, 1990; T. W. McDaniel and R. Victoria, *Handbook of Magneto-Optical Data Recording: Materials, Subsystems, Techniques*, Noyes Publications, Park Ridge, NJ, 1997; T. D. Milster, Near-field optics: A new tool for data storage, *Proc. IEEE*, 88(9):1480–1490, 2000; D. Psaltis, D. G. Stinson, and G. S. Kino (introduction by T. D. Milster), Optical data storage: Three perspectives, *Optics & Photonics News*, pp. 35–39, November 1997.

Osteoprotegerin

Osteoprotegerin is a protein that plays a central role in regulating bone mass. Throughout life, bone is constantly being broken down and rebuilt in a process that replaces old bone with new bone. These events are mediated by osteoclasts, the cells that resorb bone, and osteoblasts, the cells that form bone. Maintaining bone mass is critical for the integrity of the skeleton, which provides protection for soft tissues, support for locomotion, a scaffold for muscle attachment, and a reservoir for mineral ions (calcium, phosphorus, magnesium).

Abbreviated terms are used in this subject matter:

IL	Interleukin
JNK	*c-jun* N-terminal directed protein kinase
KO mice	Knockout mice
M-CSF	Macrophage colony stimulating factor
mRNA	Messenger ribonucleic acid
NF	Nuclear factor
OCIF	Osteoclast inhibitory factor
OPG	Osteoprotegerin
OPGL	Osteoprotegerin ligand
PGE$_2$	Prostaglandin E$_2$
PTH/PTHrP	Parathyroid hormone/parathyroid hormone–related protein

RANK	Receptor activator of nuclear factor-κB
RANKL	Receptor activator of nuclear factor-κB ligand
TGFβ	Transforming growth factor beta
TNF	Tumor necrosis factor
TRAF	Tumor necrosis factor receptor–associated factor

Osteoblasts. Osteoprotegerin is secreted by osteoblasts, which originate from mesenchymal stem cells within bone marrow stroma. Osteoblasts have receptors for systemic hormones (such as the calcium-regulating hormones, parathyroid hormone/parathyroid hormone–related protein, and vitamin D), and for growth factors and cytokines that circulate in the blood or are made in the local bone microenvironment. Osteoblasts synthesize an extracellular matrix (made mostly of type I collagen and several noncollagenous proteins) within which mineral ions (calcium and phosphate), originating from the osteoblast and also present in the extracellular fluid, crystallize. Osteoblasts eventually become osteocytes, entrapped within their mineralized extracellular matrix, or lose their PTH/PTHrP receptors to become dormant bone-lining cells.

Osteoclasts. Osteoclasts are mononuclear or multinucleated cells originating from hematopoietic (blood-forming) monocyte/phagocyte lineage cells. Osteoclast precursors require a proliferation and survival factor, macrophage-colony stimulating factor, and contact with osteoblast or bone marrow stromal cells to develop. Although the hematopoietic stem cell has PTH/PTHrP and vitamin D receptors, osteoclast lineage cells do not. At its point of contact with bone, the osteoclast forms an acidic sealed compartment where osteoclast-derived hydrogen ions and degradative enzymes dissolve the mineral and digest the collagenous matrix, respectively. As a result, calcium and growth factors stored in the extracellular matrix are released into the local environment and blood.

RANK ligand. Bone mass is maintained by the coordinated activities of osteoblasts and osteoclasts in a highly regulated process that involves systemic hormones (parathyroid hormone, vitamin D, glucocorticoids, sex steroids, thyroid hormone, calcitonin) and local factors (which include growth factors, cytokines, and prostaglandins). Local factors may be autocrine (acting on nearby cells of the same type) and/or paracrine (acting on nearby cells of a different type). RANK ligand (also known as osteoprotegerin ligand) is a local paracrine factor that originates from osteoblasts and mediates the effects of most, if not all, agents that are known to impact osteoclast development within bone. In response to hormones and local factors, RANKL binds to its receptor (receptor activator of nuclear factor-κB) on osteoclast precursors, stimulating osteoclast differentiation and activation. Both processes are inhibited by OPG (also

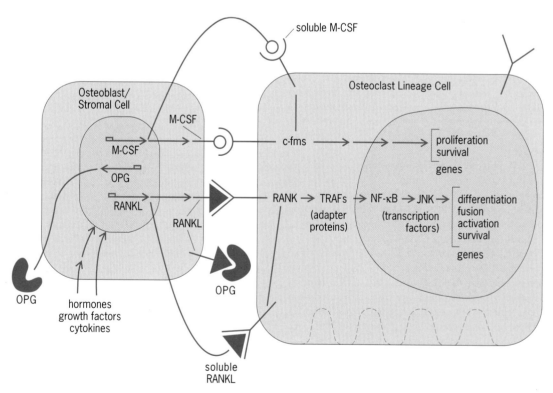

Fig. 1. Regulation of RANKL/RANK/OPG expression within the bone microenvironment. Hormones, growth factors, and cytokines direct the synthesis of membrane-bound (and secreted) RANKL and secreted OPG in osteoclasts/stromal cells. RANKL binds RANK on osteoclast lineage cells to stimulate osteoclast differentiation and activation. OPG can either inhibit or regulate both processes by binding to RANKL (depending on the RANKL to OPG ratio). M-CSF, macrophage colony stimulating factor; TRAFs, tumor necrosis factor receptor-associated factors; JNK, *c-jun* N-terminal directed protein kinase.

known as osteoclast inhibitory factor), the soluble decoy receptor for RANKL (**Fig. 1**).

Remodeling. At any one time, microscopic packets of bone (cells, extracellular matrix, and mineral) throughout the skeleton are dug out by osteoclasts (a process taking 10 days) and then replaced by new bone produced by invading osteoblasts (a process taking 3–4 months) [**Fig. 2**]. This cycle of bone resorption followed by bone formation is called remodeling. Normally, remodeling serves to meet metabolic demands for minerals and to replace bone that accumulates microscopic fractures resulting from stresses and strains of normal living. In young adults who have completed growth, the amount of bone replaced during remodeling is equal to that removed. Hormones and local factors made by osteoblasts/stromal cells, osteoclasts, and/or other cells within adjacent bone marrow control remodeling by regulating the relative levels of RANKL and OPG made by osteoblasts/stromal cells. This in turn impacts the differentiation, activation, and survival of osteoclasts (Fig. 1). Uncontrollable osteoclast numbers and activity contribute to the increased number and depth of excavation sites that occur with aging or in diseases such as postmenopausal osteoporosis, glucocorticoid-induced osteoporosis, cancer-induced bone diseases, hyperparathyroidism, rheumatoid arthritis, and chronic alcohol ingestion. More bone is removed than replaced, compromising bone strength and increasing fracture risk. Lack of remodeling, due to deficiencies in osteoclast formation or function, characterizes a spectrum of less common diseases (osteopetroses) that result in increased bone mass.

RANKL/RANK/OPG interactions. Mechanisms associated with RANKL/RANK/OPG in mediating the effects of agents known to increase osteoclast-mediated bone loss (proresorptive and proinflammatory agents) have been examined in several osteoblast/stromal cell culture systems and in animals. In osteoblast/stromal cell cultures, RANKL mRNA levels were increased by proresorptive hormones (PTH, vitamin D, and glucocorticoids), cytokines (IL-1, IL-11, and tumor necrosis factor), and prostaglandin E_2, and were suppressed by the antiresorptive factors (estrogen and transforming growth factor beta).

RANKL. RANKL is active in a membrane-bound or soluble form. The membrane-bound form is more important in bone because osteoclast precursors (from bone marrow, spleen, or blood) could not differentiate into osteoclasts unless they contacted osteoblast/stromal cells. Only an engineered soluble form of RANKL could replace osteoblast/stromal cells in these systems, indicating that the essential factor for osteoclastogenesis was RANKL. RANKL-injected animals had larger and more active osteoclasts, elevated blood calcium levels, and decreased bone mass. Knockout mice with a disrupted RANKL gene were osteopetrotic (lacking osteoclasts) with increased bone mass. Osteoblasts/stromal cells isolated from RANKL-KO mice could not stimulate normal

hematopoietic precursors to form osteoclasts, confirming that RANKL was necessary for osteoclastogenesis within bone in these animals.

RANK. RANK was found on the surface of all cells of the osteoclast lineage (and some immune cells).

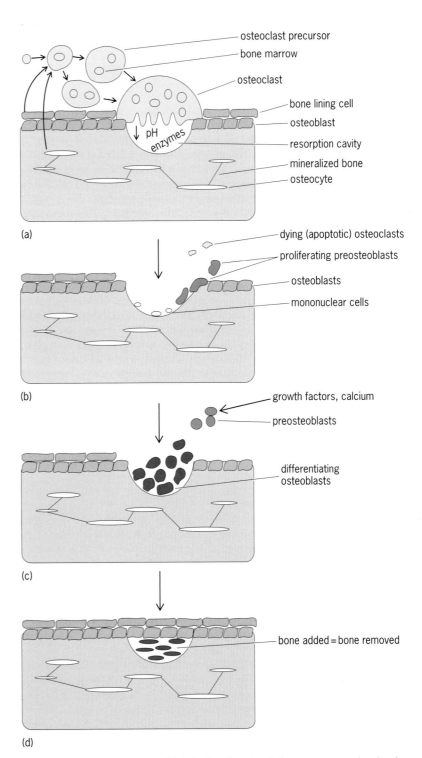

(a)

(b)

(c)

(d)

Fig. 2. Remodeling cycle in bone. (*a*) Activation: Hematopoietic precursors receive signals to proliferate, differentiate, fuse, and resorb bone. Resorption releases growth factors and calcium stored in the extracellular matrix. (*b*) Reversal: Osteoclasts die by apoptosis induced by calcium and cytokines. Calcium and growth factors attract preosteoblasts into cavity. (*c*) Formation: Growth factors, cytokines, and calcium stimulate osteoblast proliferation and differentiation. (*d*) Resting: Osteoblasts stop synthetic activity when cavity is filled.

In vitro, RANKL was shown to bind RANK on osteoclast precursors, and a soluble engineered version of RANK competed with this cell surface version for RANKL, inhibiting osteoclastogenesis. Animals injected with soluble RANK exhibited osteopetrosis (with abrogated osteoclastogenesis and increased bone mass). RANK signals intracellularly through adapter proteins such as tumor necrosis factor receptor-associated factors and transcription factors (NF-κB and *c-jun* N-terminal directed protein kinase) to upregulate genes expressed in mature osteoclasts (for example, calcitonin receptors and tartrate-resistant acid phosphatase). Mice with disrupted TRAF-6 (one of several TRAF family members associated with RANK signaling), NF-κB1 and NF-κB2, or *c-fos* (which associates with *c-jun* and is involved in JNK protein kinase signaling) genes were each osteopetrotic with increased bone mass, confirming their involvement in RANK signaling.

OPG. OPG, which is secreted by osteoblasts/stromal cells in bone, inhibited excessive osteoclast activity stimulated by proresorptive agents administered to animals (PTH, vitamin D, PTHrP, IL-1β, TNFα, RANKL) or in culture (PTH, vitamin D, IL-11, IL-1β, RANKL). In general, OPG is regulated in reverse fashion by hormones and cytokines that regulate RANKL expression: In vitro, OPG mRNA levels in osteoblast/stromal cells were downregulated by factors (PGE$_2$ and glucocorticoids) that upregulate RANKL. Although some proresorptive agents (IL-1β, TNFα, vitamin D) may increase OPG expression, the RANKL/OPG ratio remains conducive for osteoclast generation and activation. Antiresorptive agents (estrogen and TGFβ) that decrease osteoclast formation, activation, and survival increased OPG mRNA in osteoblasts/stromal cells. Transgenic mice overexpressing OPG were osteopetrotic with increased bone mass, whereas OPG-deficient mice developed osteoporotic bone fractures and spine deformities.

Cycle initiation. Although the stimulus for osteoclast precursors to initiate the remodeling cycle is unknown, it is thought to involve RANKL upregulation in response to signals from osteocytes sensing mechanical changes, such as microfractures. The involvement of hormones in remodeling is complex. Bone is one of the target organs of PTH to maintain mineral homeostasis. Normally, lowered levels of ionized blood calcium trigger the secretion of PTH and PTH-induced vitamin D synthesis. Both vitamin D and PTH upregulate RANKL/OPG in osteoblasts/stromal cells, resulting in osteoclastic bone resorption and release of calcium from bone.

Therapeutic potential of OPG. The results of in vivo and in vitro studies indicate that OPG may have therapeutic potential for diseases characterized by excessive bone loss.

Hyperparathyroidism. Sustained elevation of PTH or PTHrP in mouse models that were administered these hormones led to bone destruction and hypercalcemia, which OPG treatment prevented (likely by decreasing PTH- or PTHrP-induced RANKL).

Osteoporosis. In osteoblasts and osteoclasts, estrogen treatment increases TGFβ. This estrogen-induced TGFβ in bone likely increases OPG to regulate RANKL. Also, estrogen directly upregulates OPG in osteoblast/stromal cells. With aging (more in women after menopause with a decline in estrogen levels), bone replaced during bone remodeling is always lacking: Estrogen deficiency results in lower levels of TGFβ, OPG, and the decoy receptor for IL-1, and unleashes proresorptive agents (PGE$_2$, M-CSF) and inflammatory cytokines (IL-1β, TNFα, IL-6) to increase RANKL. Supporting this theory, OPG treatment prevented bone loss in animals administered IL-1β or TNFα or made estrogen-deficient by removing the ovaries.

Glucocorticoid-induced osteoporosis. In glucocorticoid-treated patients, serum OPG levels were decreased, suggesting that glucocorticoids enhance bone resorption by decreasing OPG levels. Glucocorticoids may decrease OPG levels by a direct effect on osteoblasts/stromal cells and, indirectly, by elevating levels of PTH. Both avenues increase the RANKL/OPG ratio.

Cancer-induced bone diseases. Because OPG prevented bone destruction in a mouse tumor-bearing model, PTHrP and other cytokines produced by cancer cells (metastatic to bone or not) likely stimulate osteoclastogenesis by upregulating the production of RANKL by osteoblast/stromal cells in bone.

Rheumatoid arthritis. In the inflamed rheumatoid joint, many inhibitors and stimulators of osteoclasts (IL-1, IL-6, IL-11, IL-13, IL-17, PGE$_2$, PTHrP) converge on the RANKL/RANK/OPG pathway in osteoblasts/stromal cells and immune cells. Synoviocytes and T-cells in these joints also produce RANKL; and in a T-cell-dependent model of rat arthritis, OPG treatment prevented bone and cartilage destruction.

These in vivo and in vitro studies support the convergence hypothesis whereby proresorptive and proinflammatory agents regulate bone mass through their effects on RANKL and OPG. This makes OPG and/or drugs designed to interfere with the RANKL/RANK pathway potentially potent therapies for bone destruction associated with these diseases.

For background information *see* ADRENAL CORTEX; ARTHRITIS; BONE; CONNECTIVE TISSUE; CYTOKINE; OSTEOPOROSIS; PARATHYROID GLAND DISORDERS; SKELETAL SYSTEM; SKELETAL SYSTEM DISORDERS in the McGraw-Hill Encyclopedia of Science & Technology.   Victoria Shalhoub

Bibliography. N. Bucay et al., Osteoprotegerin-deficient mice develop early onset osteoporosis and arterial calcification, *Genes Dev.*, 12:1260–1268, 1998; Y.-Y. Kong et al., RANKL is a key regulator of osteoclastogenesis, lymphocyte development and lymph-node organogenesis, *Nature*, 397:315–323, 1999; D. L. Lacey et al., Osteoprotegerin (OPG) ligand is a cytokine that regulates osteoclast differentiation and activation, *Cell*, 93:165–176, 1998; J. Li et al., RANK is the intrinsic hematopoietic cell surface receptor that controls osteoclastogenesis and regulation of bone mass and calcium metabolism,

97:1566–1571, 2000; W. S. Simonet Amgen EST Program et al., Osteoprotegerin: A novel secreted protein involved in the regulation of bone density, *Cell*, 89:309–319, 1997; H. Yasuda et al., Osteoclast differentiation factor is a ligand for osteoprotegerin/osteoclastogenesis-inhibitory factor and is identical to TRANCE/RANKL, *Proc. Nat. Acad. Sci. USA*, 95:3597–3602, 1998.

Panic disorder

Panic disorder is a severe and disabling psychiatric condition that often goes undiagnosed, misdiagnosed, or untreated. By definition, panic disorder has two components: there must be the occurrence of sudden, intense attacks of anxiety (known as panic attacks), and the attacks must be followed by persistent preoccupation, distress, or behavioral change related to these attacks. A panic attack is defined in the American Psychiatric Association's *Diagnostic and Statistical Manual of Mental Disorders*, fourth edition (DSM-IV), as a "discrete period of intense fear or discomfort" in which greater than three symptoms that characterize an attack "developed abruptly and reached a peak within 10 minutes." It should be emphasized that the fear can be overwhelming, as if death were imminent. The accompanying symptoms include a stronger or faster heart beat, shortness of breath or feeling of suffocation, nausea, trembling, dizziness or lightheadedness, chills, sweating, feeling of unreality or detachment (from oneself), and immense fear of losing control, going crazy, or dying. Simply having a panic attack does not mean a person has developed panic disorder, as panic attacks are known to occur in other clinical situations such as depression and social anxiety disorder. For a diagnosis of panic disorder, according to the DSM-IV, at least one of the attacks must have been followed by a month or more of persistent concern about having additional attacks, worry about the medical or psychological implications of the attack, or a significant change in behavior related to the attacks (such as avoidance of situations in which an attack previously occurred).

Agoraphobia is a related condition that is defined as anxiety about being in places or situations from which escape may be difficult or embarrassing. Examples include being in a crowd, being on a bridge, or traveling in a car, train, or plane. A person can have panic disorder either with or without agoraphobia. A person suffering from agoraphobia avoids such situations or endures them with marked distress or anxiety about having panic attack symptoms. Individuals with agoraphobia may require a companion in order to be able to tolerate such situations.

Studies of community samples from several countries, including the United States, suggest that 1.5–3% of the population suffer from panic disorder. This is a subset of individuals who have simply had a panic attack at some point (which is quite common and may approach 10% of the population). The typical age of onset of panic disorder is in early to middle adulthood, and the disorder is more common in women than in men. Panic disorder is associated not only with high rates of agoraphobia but also with clinical depression.

Etiology. Although research over the past two decades has been narrowing in on the underlying brain mechanisms involved in panic disorder, the exact mechanisms are still unknown. In general, when considering the cause of a psychiatric disorder such as panic disorder, the sum of research suggests the old concept of "nature versus nurture" should be replaced with "nature and nurture." For example, studies of identical and fraternal twins have demonstrated that there is a strong genetic component to panic disorder. There are also data from retrospective studies in humans indicating that stresses in childhood, such as loss of a parent or divorce, may increase vulnerability to panic disorder in adulthood. Animal research on the impact of stress during the nurturing period indicates the profound influence of stressful events on both brain anatomy and physiology and the vulnerability to abnormal responses in brain circuits that are hypothesized to be involved in panic disorder symptoms. Current hypotheses of the etiology focus on abnormal regulation of specific brain alarm systems. For example, it has been proposed that the body's natural suffocation alarm system becomes activated in panic disorder despite the absence of actual suffocation.

Other investigators have focused on the brain's "fight or flight" defense response and Pavlovian fear conditioning pathways as possibly playing a critical role in the elaboration of symptoms during a panic attack and anxious and avoidant behavior between attacks. For example, the area of the brain known as the amygdala, which mediates conditioned fear, is being investigated as a potential site of abnormal brain activity in panic disorder. Psychologists and neuroscientists alike have also proposed a role for heightened sensitivity to sensations in the body among people with panic disorder. The rapid advances of neuroimaging are now allowing the first looks at the actual brain areas involved in panic disorder symptoms. It is hoped that such active investigation at these many levels will soon reveal the neural circuitry responsible for the disorder.

Course and prognosis. The consequences of panic disorder are devastating. The first panic attack, because of its similarity to a heart attack, is often misinterpreted as a life-threatening event. The person having the attack will commonly go to an emergency room. While hearing that the symptoms are not due to a heart attack may be somewhat reassuring, often neither the cause of the devastating symptoms nor the appropriate treatments are adequately explained. Instead, the person may be told, "It is not serious" or "It is all in your mind," and no recommendations for treatment are given. This can lead sufferers to believe they have a serious medical condition that has yet to be uncovered, and may result in many subsequent trips to the emergency room or to health-care professionals in the hope that the cause will be revealed and addressed. Appropriate treatment must be initiated by a clinician who is aware of both the devastating toll that these symptoms have on the sufferer and the types of treatments that are available and effective. Left untreated, the sufferer may soon be crippled by the fear of future panic attacks and the avoidance of situations that are perceived to bring on attacks. For some, there may be a severe impact on the ability to work, socialize, or even travel anywhere outside the home, leading to severe restriction in activities. Such loss of livelihood and social activities is likely involved in the development of clinical depression that occurs so commonly in people with panic disorder.

Treatment. The two main treatments of panic disorder are psychotherapy and medication. A particular form of psychotherapy, known as cognitive-behavioral therapy, has been specifically adapted for the treatment of panic disorder, and clinical studies have proven its effectiveness at addressing many of the symptoms. Cognitive-behavioral therapy targeted for panic disorder generally includes three components. First, the therapist helps the sufferer identify thinking patterns that lead to misinterpretation of the significance of body sensations and to the false assumption that something catastrophic is occurring. Second, the therapist teaches breathing exercises that may help calm the sufferer during heightened anxiety, thereby preventing breathing patterns, such as hyperventilation or overbreathing, which may worsen anxiety symptoms. Finally, the therapist assists the sufferer with the gradual exposure to both feared body sensations and places and situations that are being avoided, slowly desensitizing the sufferer to such stimuli.

Medications that are effective for panic disorder are from two major classes, the antidepressants and the antianxiety/sedatives. First-line treatments include use of the selective serotonin reuptake inhibitors (SSRIs) as well as the higher-potency medications of the general class known as benzodiazepines. Benzodiazepines have effects on a class of receptors on neurons known as the GABA-A receptors. The result of benzodiazepines binding to such receptors is that the major inhibitory neurons in the brain, expressing the neurotransmitter GABA, have greater inhibitory effects on neurons expressing the GABA-A receptor. It is believed that the anti-anxiety effects of the benzodiazepines are due to the enhancement of the effects of these inhibitory neurons in circuits mediating anxiety. Alprazolam and clonazepam are examples of benzodiazepines that have known effectiveness in the treatment of panic disorder.

The SSRIs have a lag time of several weeks before they have an impact on symptoms. For some suffering from panic disorder, SSRIs may make the anxiety symptoms somewhat worse during the lag period. The benzodiazepines begin working within an hour of ingestion, but side effects such as sleepiness, poor concentration, and difficulty with balance may be problematic. Benzodiazepines are also potentially addictive. Recent evidence indicates a combination of cognitive-behavioral therapy and medication for panic disorder may provide the best outcome, although availability of skilled therapists and the cost of treatment may limit treatment options.

For background information *see* AFFECTIVE DISORDERS; ANXIETY DISORDERS; BEHAVIOR GENETICS; BRAIN; NEUROBIOLOGY; NEUROTIC DISORDERS; PHOBIA; PSYCHOPHARMACOLOGY; PSYCHOTHERAPY; SEROTONIN in the McGraw-Hill Encyclopedia of Science & Technology.   Gregory M. Sullivan

Bibliography. M. M. Antony and R. P. Swinson, *Phobic Disorders and Panic in Adults: A Guide to Assessment and Treatment*, American Psychological Association, Washington, DC, 2000; D. Barlow and M. Craske, *Mastery of Your Anxiety and Panic II*, Graywind Publications, Albany, NY, 1994; *Diagnostic and Statistical Manual of Mental Disorders*, 4th ed., American Psychiatric Association, Washington, DC, 2000; B. J. Sadock and V. A. Sadock, *Kaplan & Sadock's Comprehensive Textbook of Psychiatry*, 7th ed., Lippincott Williams & Wilkins, 2000.

Parallel processing (petroleum engineering)

Parallel processing, the simultaneous use of more than one processor of a digital computer to solve a problem, has found profound application in reservoir engineering, a subdiscipline of petroleum engineering where fluid flow inside porous rocks is simulated. Naturally occurring hydrocarbons, such as oil and gas, are located thousands of feet belowground. Oil and gas reside within the small pores of rocks, such as limestone or sandstone, called reservoirs. Oil reservoirs are located at an average depth of 1–2 km (0.6–1.2 mi), while gas reservoirs are located typically at 4 km (2.5 mi) depth or more. The fluids inside the reservoir rock are under tremendous pressure due to the weight of the rocks and earth above them [approximately 150–500 atm (15,200–50,600 kilopascals) pressure], and are produced from the reservoirs by vertical or horizontal wells drilled from the surface. The drilled wells are typically small in diameter, ranging 5–10 cm (2–4 in.) for oil and gas wells. Oil is brought to the surface either by the overlying pressure on the reservoir or by pumps.

Reservoir size depends on the geographical area. In the Middle East, where most of world's known oil and gas reservoirs are located, a typical oil reservoir has a surface area of 150–9000 km² (58–3500 mi²). In North America, the surface area of the reservoirs ranges 10–200 km² (3.8–77 mi²). The thickness of reservoir rock containing hydrocarbons also varies geographically. Typically, the thickness is 50 m (164 ft) in the Middle East and only 3–10 m (10–33 ft) in North America. Such large volumes of oil, water, and gas under hundreds of atmospheres of pressure will naturally flow through the porous and permeable rock into the wells. Under this natural depletion, however, only about 25% of the oil-in-place from these reservoirs can be produced. Most of the oil remains in the reservoir because the microscopic pores and heterogeneities in the rock restrict oil movement. In order to recover additional oil, water or gas is injected into the reservoir to sweep the remaining oil to the wells and provide sufficient energy to move the oil. Even then, about half of the initial oil remains within the reservoir. To produce the remaining oil, steam, carbon dioxide, and chemicals are injected. This process, called secondary oil recovery, is costly.

Reservoir simulation. The main concern in reservoir engineering is how long a reservoir can produce. In addition, reservoir engineers need to determine how many wells are needed, where the wells should be drilled, when the wells should be drilled and connected to surface facilities, and how much oil each well should produce. In short, oil companies need a means to manage oil and gas reservoirs most efficiently. Since the reservoirs are located thousands of meters below the ground and there is no way to see them or conduct experiments, their behavior must be studied through simulation.

Mathematical reservoir simulators are a collection of several partial differential equations describing fluid flow in oil and gas reservoirs. The equations, describing oil, water, and gas flow in porous media, are well established. The fundamental equation is Darcy's law, which states that the rate fluid flow in porous media is directly proportional to the permeability (conductivity) and the pressure differential between the inlet and outlet, and inversely proportional to the viscosity of the fluid. This equation is also called the momentum equation.

Darcy's equation combined with the continuity equations (mass conservation) for oil, gas, and water yields three partial differential equations in space and time. These equations are complemented by the capillary pressure and thermodynamic phase equilibrium in porous media. Using these auxiliary relations together with volume constraints, one obtains the complete system of equations describing the flow of oil, water, and gas in porous media. Production and injection wells provide the boundary conditions required to calculate the velocities of oil, water, and gas in the reservoir.

Numerical methods. It is impossible to solve these coupled and complicated partial differential equations using analytical methods. For this reason, numerical methods have been developed. Numerical methods require subdividing a reservoir into elements. These elements usually have rectangular shapes and resemble bricks used in constructing homes. Each brick is called a grid block. The size of the blocks varies, depending on the reservoir. Usually they are 50–250 m (164–820 ft) rectangular shapes, and thicknesses range 2–6 m (6–20 ft). **Figure 1** shows a typical reservoir grid in three dimensions.

These methods use high-speed computers. For large systems, oil industry engineers use supercomputers and high-end workstations. Recently, due to the rapid advancement in personal computers (PCs),

small-to-medium-size reservoirs are simulated using PCs. For large systems and for improving the speed of the computations, parallel computers or a cluster of PCs are used to solve these problems in a practical time frame.

Simulation process. A typical reservoir simulator model requires 100,000 grid blocks. A typical simulation study consists of two parts: history matching of the reservoir performance, and future forecasting. In the first part, a simulator model is matched to the reservoir's past production history. This means that the observed well pressure history, as well as production of oil, water, and gas from every well, is matched (duplicated) by the simulator. To achieve a good match, a simulation engineer usually adjusts formation properties that are unknown between the wells (usually 1 km or 0.6 mi apart). A reasonable match is necessary to have confidence in the simulation model as a predictive tool. Generally, the history match phase for a typical Middle East field spans 25–50 years of production history, and future forecasting (prediction) is done for 5–20 years.

Magnitude of computations. To make a full 50-year history match and to forecast 5–20 years of production with a reservoir simulator that has only 100,000 cells (grid blocks), it is necessary to compute three unknowns (pressure and amount of oil and water) in every grid block for each time step. For a 50-year simulation, the simulator uses at least 600 time steps. For each time step, there are at least two nonlinear (outer) iterations. In this case, for each iteration one needs to solve linear equations comprising 300,000 unknowns, that is, to solve 300,000 equations in 300,000 unknowns.

A 50-year history requires solving 1200 times a linear system of equations composed of 300,000

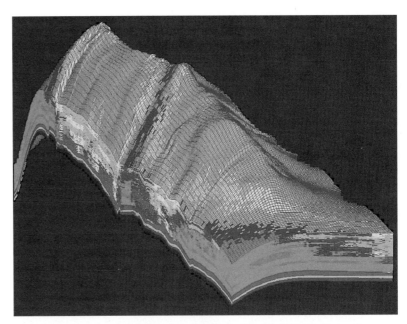

Fig. 1. Typical computational grid for an oil reservoir. (*From A. H. Dogru, Megacell reservoir simulation, J. Petrol. Technol., May 2000; copyright Society of Petroleum Engineers*)

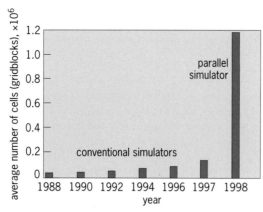

Fig. 2. Model sizes, 1988–1998. (*From A. H. Dogru, Megacell reservoir simulation, J. Petrol. Technol., May 2000; copyright Society of Petroleum Engineers*)

computer, simulators use Fortran 77 or Fortran 90, whereas parallel simulators running on multiple CPUs use specialized languages such as High Performance Fortran, that is, Fortran 90 with Open MP (Multi Processing) and MPI (Message-Passing Interface) programming languages.

Parallel computers. Parallel computers can be classified in two basic categories: shared memory and distributed memory. Shared memory computers, such as Cray, IBM Night Hawks, and SGI Origin series, are composed of multiple CPUs sharing a common memory. Each CPU is usually very fast and may use vector processing or very fast scalar computations. Distributed memory machines have different architecture: each processor (CPU) has its own memory and communicates to the other nodes (CPUs) through a high-speed network. While the number of CPUs in shared memory cannot be increased beyond a specific number (usually 4–16), distributed memory machines have no such limitations. Distributed memory machines (sometimes called massively parallel processors) consist of hundreds and thousands of CPUs (each CPU can be very cheap). However, the communication network and programming language become a major challenge in this type of architecture. Massively parallel computers are manufactured by some major hardware companies as well as some small ones. A collection (cluster) of workstations or a collection of PCs with a high-speed switch shows significant cost savings and provide attractive performance.

Parallel computations in reservoir engineering. Parallel computers in petroleum reservoir engineering were first used in 1980, and the first commercial simulators appeared in 1991 and 1998. Reservoir simulators consisting of millions of cells can be constructed on both shared and massively parallel computers. By

equations and 300,000 unknowns. Such a simulation study, using high-speed single-CPU (central processing unit) computers, such as an IBM SP-2 or a Cray machine, can usually be done within 6 hours.

For large models involving millions of cells, single-CPU computers are not sufficient because they do not have enough memory to fit the million-cell data into primary memory, and even if they did, the computations would take an impractically long processing time. For example, a million-cell model for the above reservoir has 3 million equations in 3 million unknowns per iteration. For the entire history match, one has to solve 600 times of the same large system of equations. For this reason, parallel computers are used.

To use parallel computers, the reservoir simulator has to be rewritten. The difference mainly comes from solving the set of linear equations and the programming language. Typically for a single-CPU

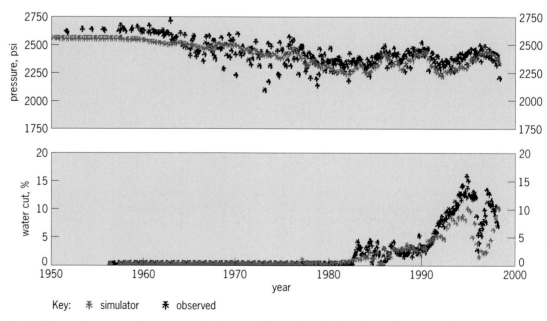

Key: ✳ simulator   ✳ observed

Fig. 3. Offshore oil reservoir with pressure and water cut match, as shown in the early runs in a megacell simulation. (*From A. H. Dogru, Megacell reservoir simulation, J. Petrol. Technol., May 2000; copyright Society of Petroleum Engineers*)

using such large simulation models, 30–50 years of history is run within hours.

Benefits of parallel simulation. The data collected for reservoir studies are stored in geocellular models. These models usually contain 200–300 vertical layers and a 250 × 250 m grid area, totaling millions of cells. To use standard nonparallel simulation technology, such a large system is collapsed to 100,000 cells or less. This process is called upscaling. Naturally, much important heterogeneity in the rock properties is lost during this process, and thus the upscaled reservoir model seldom represents the actual reservoir details. By using parallel simulation technology, such difficulties are eliminated or minimized. In theory, the original geocellular model can be run on a parallel computer as is. Naturally, it is expected that the results of such a detailed model would represent the field performance better (requiring little or no adjustment to the reservoir properties).

Figure 2 shows the 10-fold increase in model size achieved by using parallel technologies. **Figure 3** shows a good fit to field performance by a parallel simulator. Typically, when comparing the results of a parallel simulator with a nonparallel one, the parallel simulator will show oil trapped behind the advancing water front, while the nonparallel simulator shows no oil left behind. The trapped oil can be subsequently confirmed by use of a well log.

[The author thanks Saudi Aramco management for permitting publication of this paper.]

For background information *see* DIGITAL COMPUTER; MODEL THEORY; MULTIPROCESSING; PETROLEUM ENGINEERING; PETROLEUM ENHANCED RECOVERY; PETROLEUM RESERVOIR ENGINEERING; PROGRAMMING LANGUAGES; SIMULATION; SUPERCOMPUTER in the McGraw-Hill Encyclopedia of Science & Technology.   A. H. Dogru

Bibliography. K. Aziz and A. Settari, *Petroleum Reservoir Simulation*, Applied Science Publishing, London; K. Hwang and Z. Xu, *Scalable Parallel Computing: Technology*, Architecture WCB, McGraw-Hill, Boston, 1998; J. E. Killough and R. Bhogeswara, Simulation of compositional reservoir phenomena on a distributed memory parallel computer, *J. Petrol. Technol.*, November 1991; G. S. Shiralkar et al., *Falcon: A Production Quality Distributed Memory Reservoir Simulator*, Society of Petroleum Engineers and Reservoir Engineers, October 1998; A. H. Dogru, Megacell reservoir simulation, *J. Petrol. Technol.*, May 2000.

Peroxynitrite chemistry

Peroxynitrite chemistry, a century-old area of research, has experienced a renaissance with the annual number of publications on the subject increasing from only one in 1990 to nearly 200 in 1998–2000. This renewed interest was stimulated by the discovery of the biological roles of nitric oxide, distinguished by the 1998 Nobel prize, and the recog-

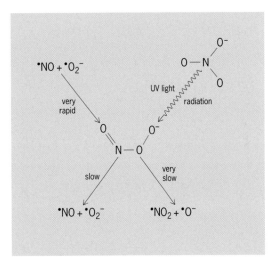

Fig. 1. Peroxynitrite formation and decomposition reactions. The dot denotes radical species.

nition that the conversion of nitric oxide into peroxynitrite may play a major role in human diseases associated with oxidative stress and in cellular defense against invading pathogens.

Occurrence. Peroxynitrite ($ONOO^-$) is a structural isomer of nitrate (NO_3^-) that contains a peroxo bond (**Fig. 1**). The physiological route to $ONOO^-$ is provided by the combination of a nitric oxide radical ($\cdot NO$) with a superoxide radical ($\cdot O_2^-$), an extremely rapid reaction occurring upon every encounter of these radicals. Both $\cdot NO$ and $\cdot O_2^-$ are the oxygen metabolic products simultaneously generated in a number of cell types within the human body. Compared to its precursors, peroxynitrite is a much stronger oxidant capable of oxidizing proteins, nucleic acids, and lipids.

In the environment, peroxynitrite can be produced by ultraviolet (UV) or ionizing irradiation of the nitrate ion, which induces its isomerization (Fig. 1). Because NO_3^- is one of the three most abundant anions present in the cloud water of the Earth's atmosphere, photochemical $ONOO^-$ generation can be significant, particularly in stratospheric aerosols, where the flux of solar ultraviolet light is high. Extensive radiation-induced generation of $ONOO^-$ can occur within the nuclear waste storage tanks at Hanford, Washington, which contain about 16,000 m^3 (60 million gallons) of highly radioactive, nitrate-saturated liquids and solids generated during nuclear weapon production.

Reactivity. Although almost 180 kJ/mol higher in energy than NO_3^- and, therefore, inherently unstable, the peroxynitrite anion does not isomerize through a concerted bond rearrangement. In the absence of acids, peroxynitrite solutions decompose with the half-life of approximately 20 hours to nitrite (NO_2^-), oxygen, and NO_3^- in a complex set of radical reactions initiated by the slow dissociations of the $ON\text{-}OO^-$ and $ONO\text{-}O^-$ bonds (Fig. 1).

The $ONOO^-$ anion is a moderately strong Lewis base (electron pair donor), which is the most

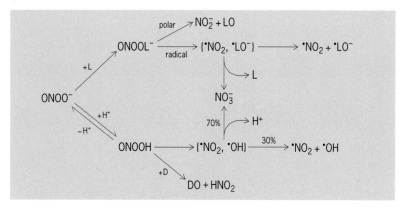

Fig. 2. Major peroxynitrite reactivity pathways. L = Lewis acid; D = electron donor; braces denote a geminate (contact) radical pair.

prominent characteristic determining its reactivity. Accordingly, both major pathways through which peroxynitrite performs its oxidative chemistry begin with the neutralization by an acid (**Fig. 2**). The combination of ONOO⁻ with a Lewis acid (electron pair acceptor, L) creates the ONOOL⁻ adduct, which then decomposes either through homolytic (radical) or heterolytic (polar) O-O bond cleavage. In the latter case, an O atom is transferred; that is, the two-electron oxidation of L occurs, such as the oxidation of trivalent chromium in alkali shown in reaction (1).

$$ONOO^- + Cr^{III} \longrightarrow NO_2^- + Cr^V{=}O \qquad (1)$$

Similar reactions are presumed to occur for some trivalent arsenic and antimony compounds and for several ketones (R_1R_2CO).

The homolytic decomposition of ONOOL⁻ transfers ·O⁻; that is, the one-electron oxidation of L occurs, creating a geminate radical pair within a solvent cage (Fig. 2). The caged radicals then either

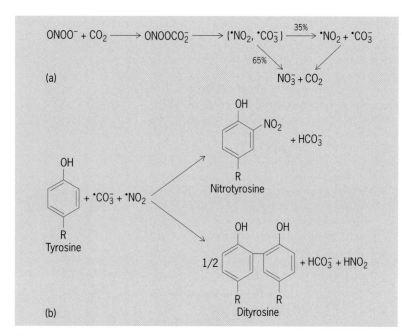

(a)

(b)

Fig. 3. Reactions showing (a) carbon dioxide–catalyzed peroxynitrite decomposition and (b) tyrosine oxidation. $R = CH_2{-}CH(CO_2^-)(NH_3^+)$.

recombine to give NO_3^- and regenerate L or diffuse apart to become the free radicals. Carbon dioxide (CO_2) is the most important, unexpected, and well-studied Lewis acid reacting via these pathways. It rapidly binds to ONOO⁻ forming an unstable adduct, which almost instantly breaks up to afford the carbonate (·CO_3^-) and nitrogen dioxide (·NO_2) radicals (**Fig. 3a**). In the absence of reductants, these radicals react with each other regenerating CO_2—hence the catalysis of ONOO⁻ isomerization to NO_3^-. Because carbon dioxide is ubiquitous in biological fluids, being present at a concentration level of 1 millimole per liter or greater, it is projected that nearly all of the ONOO⁻ that might be generated in these environments will, within 0.02–0.03 second, react preferentially with CO_2. Since both ·CO_3^- and ·NO_2 are potent oxidants, a bystander compound, otherwise unreactive toward ONOO⁻, can become oxidized in the presence of CO_2. For example, CO_2 promotes the nitration of guanine in deoxyribonucleic acid (DNA) and the oxidation of phenols, notably, tyrosine residues of proteins (Fig. 3b). In vitro experiments have shown that the nitration of only one tyrosine residue or, in some cases, a few tyrosine residues of an enzyme can seriously compromise its biological function. Because the yields of the major oxidation products, nitrotyrosine and dityrosine (Fig. 3b), depend upon the medium's pH and its tyrosine, CO_2, and ONOO⁻ contents, both the extent and the type of CO_2-mediated protein damage by ONOO⁻ may vary appreciably in different tissues. Reportedly, the dietary polyphenols consumed with vegetables, tea, chocolate, or wine can act as protectors by scavenging the peroxynitrite-derived radicals. Aldehydes (RHCO), which can be viewed as partially reduced CO_2, are weaker Lewis acids; they react with ONOO⁻ in the same manner, but more slowly.

Several porphyrins (P) of trivalent manganese (PMn^{III}) have been reported to scavenge ONOO⁻ through the Lewis acid homolytic pathway in Fig. 2, with the specific rates up to a thousand times greater than that for CO_2. In this case, the tetravalent manganese porphyrin ($PMn^{IV}{=}O$) is the reaction product, along with ·NO_2. Being weaker oxidants than ·CO_3^-, the $Mn^{IV}P{=}O$ species react more selectively. Although also capable of oxidizing tyrosine, they preferentially react with natural antioxidants, such as ascorbic and uric acids. Analogous porphyrins of trivalent iron are also extremely efficient ONOO⁻ scavengers. Further progress in this direction may lead to the development of drugs protecting cellular proteins and DNA against peroxynitrite by redirecting its reactivity toward less critical targets.

A somewhat different reactivity arises when ONOO⁻ accepts a hydrogen ion (H^+) to become peroxynitrous acid (ONOOH; Fig. 2). Unlike most hydroperoxides, ONOOH is a relatively strong acid; its pK value is 6.6. As a result, very rapidly interconverting ONOO⁻ and ONOOH coexist at comparable amounts in neutral (pH 7) solution, a situation that has occasionally resulted in ambiguity in the assignment of observed reactivity to one of these species.

Peroxynitrous acid is the only known hydroperoxide capable of spontaneous O-O bond scission at ambient temperature; within a few seconds, it decomposes via the two competing pathways producing both NO_3^- and a hydroxyl ($\cdot$OH) and $\cdot NO_2$ radical pair (Fig. 2). The highly reactive $\cdot$OH radical can almost indiscriminately oxidize a wide variety of organic and inorganic compounds. Accordingly, the mutagenic DNA lesions and strand breaks, oxidation of amino acids, inactivation of enzymes, and initiation of lipid peroxidation have all been observed upon incubation with peroxynitrite in vitro. However, the $\cdot$OH-mediated reactions of ONOOH will be of little or no significance in vivo, because of the competition from the more rapid CO_2-catalyzed pathway and from direct oxidations by ONOOH.

As with all hydroperoxides, ONOOH can engage in an oxygen atom transfer to electron donors (D; Fig. 2), such as the oxidation of dimethyl sulfide to dimethyl sulfoxide [reaction (2)]. The two sulfur-

$$ONOOH + (CH_3)_2S \longrightarrow HNO_2 + (CH_3)_2SO \qquad (2)$$

containing amino acids, methionine ($RSCH_3$) and cysteine (RSH), and their naturally occurring selenium analogs react in a similar fashion. Their reactivity is sufficiently high to compete with the spontaneous decomposition of ONOOH in most, and even with CO_2-catalyzed reactions in some, cellular environments. Unlike tyrosine oxidation, the oxidation of these amino acids can be reversed, sometimes rapidly, by the cellular enzymatic reduction systems. For example, the catalytic destructions of peroxynitrite by a bacterial enzyme, peroxiredoxin, and by a mammalian enzyme, glutathione peroxidase, are thought to be carried out by the oxidation-reduction cycling of their respective cysteine and selenocysteine residues.

Another class of biological molecules that are capable of rapid peroxynitrite scavenging is represented by heme proteins, which contain iron porphyrin active sites. These include a number of peroxidases, cytochromes, and hemoglobin. Depending, apparently, upon the oxidation state of heme iron and the protein structure, some of them react with $ONOO^-$ through the Lewis acid route, while others take the donor pathway and scavenge ONOOH (Fig. 2). In the cellular and subcellular compartments, such as red and white blood cells or mitochondria, with high content of heme proteins, they have the capacity to rival CO_2-directed peroxynitrite reactivity. This, however, does not always constitute protection, because the nascent products of peroxynitrite reactions with hemes are, in many cases, strong oxidants in their own right.

Environment. The oxidation of natural and anthropogenic sulfur dioxide (SO_2) by hydrogen peroxide (H_2O_2) in atmospheric fog and clouds is a major contributor to acid rain. Under typical cloud water conditions, the analogous reaction of SO_2 with ONOOH is about a hundred times more rapid than with H_2O_2. Accordingly, the presence of even relatively small amounts of $ONOO^-$ has the capacity to adversely affect the environment. In stratospheric clouds, the $\cdot$OH radical created during ONOOH decomposition is capable of initiating a chain reaction of ozone destruction in the presence of halogen-containing compounds. Although the contributions of peroxynitrite to atmospheric chemistry have not yet been completely accounted for, a correlation between the conditions favoring $ONOO^-$ formation and the depletion of ozone over the Earth's polar regions has been observed.

Biology. Peroxynitrite has been implicated as a causative agent in a number of human diseases, including neurodegenerative disorders, atherosclerosis, ischemic reperfusion injury, inflammation, and sepsis. The connection is made primarily based on (1) the likelihood of simultaneous (in place and time) generation of large fluxes of $\cdot$NO and $\cdot O_2^-$; (2) the in vitro observation of the exceptionally high peroxynitrite cytotoxicity and its adverse effects upon critical cellular components; and (3) immunohistological assays using nitrotyrosine-specific antibodies that reveal copious tyrosine nitration in the affected tissues (**Fig. 4**).

While deleterious in normal cells, the massive oxidative damage becomes beneficial when inflicted upon invading bacteria and parasites. A preponderance of circumstantial evidence has led to the suggestion that peroxynitrite may be one of the key bactericidal agents generated by several types of phagocytes, the specialized cells dedicated to combating microbial infection. Upon activation, these cells have been shown to induce nitration of added phenols and nitrotyrosine lesions in bacteria, consistent with the involvement of peroxynitrite.

Finally, peroxynitrite formation may play a regulatory role by controlling the biological activity of its nitric oxide and superoxide precursors through both their consumption and the oxidative inactivation of

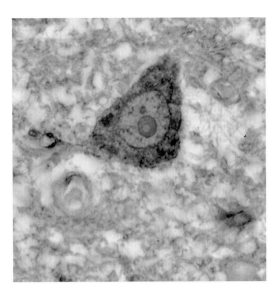

Fig. 4. Immunostaining for nitrotyrosine (dark areas) in a degenerating spinal motor neuron (large triangular cell) of a patient suffering from amyotrophic lateral sclerosis (Lou Gehrig's disease). Magnification 180×. (*Liliana Viera and Joseph Beckman, University of Alabama at Birmingham*)

the enzymes involved in their metabolism. Furthermore, there appears to be evidence of peroxynitrite involvement in cellular signaling events and in the modulation of immune response.

All the physiological roles of peroxynitrite are predicated on the basis that it is produced in biological systems in significant quantities, which remains a contentious point, mainly because the short lifetime of peroxynitrite under physiological conditions has prevented its accumulation and detection by direct means. Although nitrotyrosine found in cells (Fig. 4) is widely accepted as a biomarker of peroxynitrite chemistry, it is also recognized that, at least in certain cells and tissues, there exist other pathways of the ˙NO metabolism that may yield nitrotyrosine without the intermediacy of peroxynitrite. Clearly, the unraveling of the extremely complex biological roles of peroxynitrite is still at an early stage. As with any rapidly evolving field, controversies abound, and more research will have to be done before a comprehensive understanding emerges.

For background information *see* ACID AND BASE; AMINO ACIDS; ATMOSPHERIC CHEMISTRY; FREE RADICAL; NITRIC OXIDE; NITRO AND NITROSO COMPOUNDS; OXIDATION PROCESS; OXIDATION-REDUCTION; OXIDIZING AGENT; PEROXIDE; PEROXYNITRITE; PK; SUPEROXIDE CHEMISTRY in the McGraw-Hill Encyclopedia of Science & Technology.

Sergei V. Lymar

Bibliography. R. Bryk, P. Griffin, and C. Nathan, Peroxynitrite reductase activity of bacterial peroxiredoxins, *Nature*, 407:211–215, 2000; J. O. Edwards and R. C. Plumb, The chemistry of peroxonitrites, *Prog. Inorg. Chem.*, 41:599–635, 1994; J. T. Groves, Peroxynitrite: Reactive, invasive and enigmatic, *Curr. Opin. Chem. Biol.*, 3:226–235, 1999; J. K. Hurst and S. V. Lymar, Cellularly generated inorganic oxidants as natural microbicidal agents, *Acc. Chem. Res.*, 32:520–528, 1999; L. J. Ignarro (ed.), *Nitric Oxide: Biology and Pathobiology*, Academic Press, San Diego, 2000.

Petri nets

In many fields, such as physics or astronomy, where there are phenomena or systems that cannot be studied or observed directly, it is necessary to use models. The models, generally mathematical representations of the phenomena or systems, are intended to capture essential properties that are relevant for the study of the structure and dynamic behavior of the modeled system. The information that is obtained from the model can then be used to evaluate the accuracy of the model or to adjust the model. A Petri net is an abstract formal model of information flow. The major use of Petri nets has been as a graphical language for modeling systems with interacting concurrent components. Most of the models implemented with Petri nets consist of independent components that may interact with each other in synchronized manner or may carry out their activities simultane-

ously (concurrently) with other components of the system. The nets developed from the doctoral dissertation of Carl Adam Petri, *Kommunikation mit Automaten* (Communication with Automata), at the University of Bonn during the early 1960s.

Definition. In mathematical terms, a Petri net can be defined as a structure with four parts or components (a four-tuple), denoted by $C = (P, T, I, O)$. The components of this structure are a finite set of places, $P = \{p_1, p_2, \ldots p_n\}$, a finite set of transitions, $T = \{t_1, t_2, \ldots t_k\}$, an input function ($I$), and an output function (O). The input function I is a mapping from a transition t_j to a collection or bag of places known as the input places of the transition. The set of input places is denoted by $I(t_j)$. Likewise, the output function O maps a transition t_j to a collection or bag of places $O(t_j)$ known as the output places of the transition. A collection or bag is a set in which an element may appear more than once. The use of bags allows a place to be a multiple input or output of a transition. However, in Petri nets it is required that the set of places (P) and transitions (T) do not have any common element. That is, they are disjoint. In set theory notation, this can be written $P \cap T = \emptyset$, where $\emptyset$ denotes the empty set and $\cap$ denotes the intersection (the set of common elements) of these two sets.

A Petri net can be defined as a graph with two types of nodes: circles "◯" that represent places, and bars "|" that represent transitions. Here the circles and bars will be called places and transitions, respectively. In a Petri net there are directed arcs (arrows) that go from places to transitions and arcs that go from transitions to places. Directed arcs from places to transitions define input places (of a transition). Arcs from transitions to places define output places. Since there can be more than one arc emanating from a place or from a transition, the graph is called a multigraph. In addition, since the nodes of the Petri net can be partitioned into two sets (places and transitions) and they are such that each directed arc is from an element of one set to an element of the other set, the graph is a bipartite directed multigraph. If $P = \{p_1, p_2, \ldots p_n\}$ is the set of places and $T = \{t_1, t_2, \ldots t_k\}$ is the set of transitions, then V, the set of nodes of the graphs, is $V = P \cup T$ with $P \cap T = \emptyset$ (where $\emptyset$ is the empty set). The arcs of the graph form a bag $A = \{a_1, a_2, \ldots a_m\}$, where each element a_i is of the form (p_i, t_j) or (t_k, p_l). That is, an arc goes from a transition to a place or vice versa.

From a practical point of view, Petri nets can be defined using a formal or graphical representation. Both definitions are equivalent. **Figure 1** is the graphical representation of a Petri net defined by its input and output functions, I and O, given in Eqs. (1) and (2).

$$
\begin{aligned}
I(t_1) &= \{p_1\} & I(t_2) &= \{p_2, p_3, p_5\} \\
I(t_3) &= \{p_3\} & I(t_4) &= \{p_4\}
\end{aligned}
\tag{1}
$$

$$
\begin{aligned}
O(t_1) &= \{p_2, p_3, p_5\} & O(t_2) &= \{p_5\} \\
O(t_3) &= \{p_4\} & O(t_4) &= \{p_2, p_3\}
\end{aligned}
\tag{2}
$$

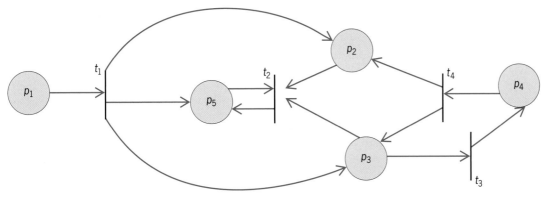

Fig. 1. Graphical representation of a Petri net. (*After J. L. Peterson, Petri Net Theory and Modeling of Systems, Prentice Hall © 1981. Reprinted by permission of Pearson Education, Inc., Upper Saddle River, NJ.*)

It is not necessary to write explicitly the set of places and transitions since the input and output functions define these two sets. The notation $I(t_1) = \{p_1\}$ indicates that p_1 is an input place and therefore there is a direct arc (arrow) from p_1 to transition t_1. Likewise, the notation $O(t_3) = \{p_4\}$ indicates that p_4 is an output place and therefore there is a direct arc (arrow) from t_3 to p_4. As indicated before, each node of a Petri net is either a place or a transition; therefore, there can be no isolated nodes in the graph. In consequence, the input and output functions define completely the Petri net.

Dynamic properties. To consider the dynamic properties of a Petri net, the basic definition of this structure needs to be expanded with the notion of markings, tokens, and rules of execution. Tokens are considered primitive concepts of Petri nets. Tokens are assigned to the places of a Petri net, and a mark-ing is an assignment of tokens to these places. In graphical notation a token is represented as a dot "•" within a circle. Tokens are used to execute a Petri net. A Petri net executes by firing transitions. To fire a transition implies removing tokens from its input places and creating new tokens that are sent to its output places. For a transition to "fire" it must be en-abled. A transition is said to be enabled if each of its input places has at least as many tokens in it as arcs from the place to the transition. Using the definition of an enabled transition, the concept of firing a tran-sition can be redefined. In this case, a transition can be said to fire by removing all of its enabling tokens from its input places and putting them into each of its output places—one token for each arc from the transition to the place.

With the notion of executing a Petri net it is possi-ble to design, validate, and evaluate automated

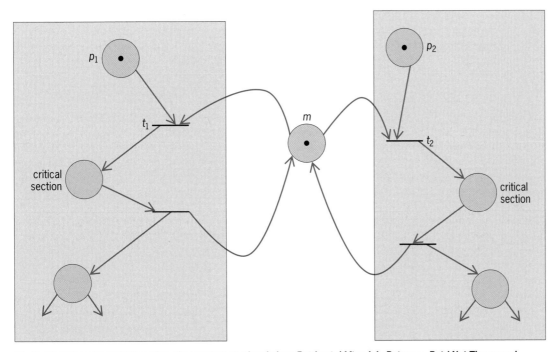

Fig. 2. Modeling of a critical section of a program or of code by a Petri net. (*After J. L. Peterson, Petri Net Theory and Modeling of Systems, Prentice Hall © 1981. Reprinted by permission of Pearson Education, Inc., Upper Saddle River, NJ.*)

manufacturing systems, parallel and distributed computer systems, and telecommunication networks. It is also possible to model the behavioral characteristics of computer software or hardware. The expressive power of Petri nets makes it possible to capture the semantics of concurrency, interleaving, and partial ordering formalisms. In other words, Petri nets are capable of modeling true concurrency. With Petri nets it is possible to use modeling strategies developed through stepwise refinement (top-down modeling) or modular composition (bottom-up modeling).

Application to parallel computation. An example of the expressive power of Petri nets is seen in their ability to model synchronization mechanisms that prevent race conditions at the operating-system level of a computer. Race conditions result from the explicit or implicit sharing of data or resources among two or more processes. A program in execution, especially one that supports interleaved execution of multiple programs, will be called a process or task. A race condition occurs when the scheduling of two processes is such that the various orders of scheduling them result in different outcomes. The key to prevent race conditions is to make sure that if one process is using a shared resource the other process is excluded from using the same resource at the same time. Since this concept applies to any of the participating processes, the processes should mutually exclude each other to avoid race conditions. The processes are said to be mutually exclusive; that is, only one of the two processes can use the shared resource at any one time.

That part of a program or of code that accesses a shared resource and that needs protection from interference by any other process is called a critical section. If there were a way to ensure that no two processes were ever in their critical sections at the same time, then race conditions could be avoided. **Figure 2** shows how Petri nets can model a critical section. The place m represents the permission to enter the critical section. If process p_1 wants to enter the critical section, it must have a token in p_1 and there must be a token in m granting the permission to enter. A similar condition applies to process p_2 if it desires to enter the critical section. If both processes want to enter the critical section at the same time, a conflict exists between transitions t_1 and t_2. Only one transition can fire. For example, if transition t_1 fires it will disable transition t_2 and will put process P_2 "to sleep." Process P_2 will continue sleeping until process p_1 leaves the critical section and puts a token back in place m.

For background information *see* CONCURRENT PROCESSING; DATAFLOW SYSTEMS; GRAPH THEORY; OPERATING SYSTEM; SET THEORY in the McGraw-Hill Encyclopedia of Science & Technology.

Ramon A. Mata-Toledo

Bibliography. K. Jensen, *Coloured Petri Nets*, vol. 2: *Analysis Methods*, Springer-Verlag, 1992; J. L. Peterson, *Petri Net Theory and Modeling of Systems*, Prentice Hall, 1981.

Photonics

Photonics deals with the practical generation, manipulation, analysis, transmission, and reception of photons (the quanta of electromagnetic energy) in the visible and nonvisible [infrared (IR) and ultraviolet (UV)] portions of the light spectrum. Photonics is contributing to a wide range of fields, including astronomy, biomedicine, data communication and storage, fiber optics, imaging, optical computing, optoelectronics, sensing, and telecommunication. As of 2001, photonics products made an impressive list, including:

Bandgap devices
Biomedical sensors
Birefringence measurement system
Compact spectrometer
Digital and analog transceivers
Dynamic polarization controller
Fiber laser
Fiber saver dense wavelength division multiplexer
Infrared detectors
Image sensor
Interferometer
Laser
Liquid crystal beam steerer
Miniature uncooled infrared camera
Nonlinear optical devices
Optical fiber
Optical micro-electromechanical systems
Scanning laser ophthalmoscope
Surface measurement instrument
Spectrophotometer
Spectroscopic ellipsometer
Terahertz spectrometry system
3-D measurement sensor
Tunable 2.5-Gb/s long-haul transmitter
Ultrafast photodetectors and modulators
UV polarization analyzer
Vision measurement workstation
Wavelength/power meter

Sensing. Photonic sensing is a large field with many subdivisions. For example, the Hubble space telescope with its sensitive photonic arrays detects photons in space and converts them into electrical signals that are beamed to Earth as radio waves and reconstructed in a visual format. This type of sensing is also considered imaging.

Atoms and molecules absorb and radiate photons, thus generating "spectral fingerprints." Energized or excited atomic and molecular species radiate light in precise quanta of characteristic wavelengths that immediately broadcast their identities, even from billions of miles away. Helium was first discovered on the Sun, not the Earth, when a photonic method (spectroscopy) identified the spectral wavelengths of the element. Molecules in a normal energy state absorb specific wavelengths, and the resulting attenuation is used to identify them and determine their concentrations. Gases, liquids, and solids can all be measured by their photonic absorption using spectroscopy.

Sensors are critical components used in a vast array of products, instruments, vehicles, systems, and manufacturing processes. Photonic sensors can detect physical characteristics, including temperature, position, color, size, texture, shape, and chemical composition, and translate the data into useful information for computers and humans. Photonic sensors are typically faster, more accurate, reliable, efficient, and less expensive than those based on other technologies. Explosion suppressors in a ship's hold (cargo deck) employ fast-responding UV sensors to trigger suppressant release to stop the pre-explosion "flash." Sensor arrays and scanners often provide the only practical method for certain physical measurements, especially when direct contact is impractical. For example, they can detect, measure, and analyze gases in the atmosphere with extreme accuracy. Photonics instruments detected the reduction of normal gas ratios that led to the discovery of the hole in the ozone layer.

Photonics is used to measure all kinds of phenomena beyond analyzing composition. One valuable feature of photon sensing is that it is a nondestructive method and can be done without physically contacting the sample. While spectroscopy has long been used to identify elements and complex molecules, mechanical and sensitive physical measurements are now done routinely using photonics. For example, silicon computer chips are sensitive to stress that can result during assembly into electronic packages. Laser scanners and interferometers can measure even the slightest curvature of a tiny chip, which serves as a stress indicator.

Imaging. Photonic imaging is also used in many disciplines, including biology, physics, chemistry, medicine, and even criminology. Seemingly invisible latent fingerprints are visualized, or "imaged," using nonvisible wavelengths via photoluminesence techniques. Photonics personal verifiers "see" the operator's fingerprint to authenticate the operator.

Infrared imagers that detect heat and provide real-time or instantaneous "heat pictures" are saving firemen's lives. While humans cannot directly see infrared, a myriad of imaging equipment can easily do this. Portable infrared imagers are used to "see" heat before entering rooms and opening doors. An imager showing a hot door or doorknob tells the crew that there is a dangerous fire situation on the other side. The infrared imagers can even look through dense smoke, making them doubly valuable.

Infrared imaging technology works equally well at longer range. Sensitive equipment can easily detect body heat. The Coast Guard and other rescue units can readily spot a person in the water, and this ability to see warm-blooded creatures in the dark has saved countless lives. Infrared units are also used by military and law enforcement. The viewers are so sensitive that a prison escapee, buried in dirt except for his nose, was detected from a helicopter.

Modern infrared and other imagers use semiconductor photodetector arrays similar to those in video cameras but made sensitive to the desired wavelength. These units can be mass-produced at reasonable cost. Consumers can now purchase the once-expensive night vision glasses, where light levels are boosted up to thousands of times, that were developed for the military.

Medical imaging is another critical, life-saving field of photonics where improvements continue to be made. Advanced medical instrumentation permits minimally invasive surgery based on photonic imaging technologies where tools can see inside the body. Medicine uses various wavelengths, like x-ray, to see inside, and atomic particles, like positrons, to also locate activity areas. The ability to convert invisible phenomena into high-resolution digital displays and hard-copy images is a key link. The computerized tomography (CT; also CAT for computer axial tomography) scan is perhaps one of the most publicized and amazing three-dimensional photonics imagers.

Communications. Photonics is now the most important technology for efficient terrestrial and transoceanic voice, video, and data communications. It has been adapted to high-speed Internet data transfer to make broadband or high-rate transmission possible. Web downloads would be impossibly slow without photons.

The photon, or light quantum, is nearly weightless compared to the electron and has no electrical charge. This makes the photon the ideal messenger that travels at the speed of light—186,000 mi/s (300,000,000 m/s). Nothing is faster than photons, and the low mass with electrical neutrality allows them to pass through many substances without interference or significant mutual interaction. Glass made from common silica (SiO_2) is an ideal conduit when formed into very pure fiber with the right dimensions. Internet messages travel along thousands of miles of optical glass fiber (fiber optics) in milliseconds. Today, most of the long-haul, or backbone, links employ fiber optics using advanced photonics. Additional capacity and bandwidth is now achieved by adding more "colors of light" (additional wavelengths) through the same optical fibers. This new technology, called wave division multiplexing (WDM), continues to add more wavelengths to deliver almost limitless bandwidth.

More recently, "free space" photonics data links have been deployed in metropolitan areas as an alternative to copper wire, fiber optics, and even wireless radio systems. Wavelengths within the infrared spectrum are beamed from building tops. Subscribers use small photo-receivers attached to the inside of windows to obtain securely encoded information. Very high data rates are possible. Multiple transmitters ensure that the data get through even in rain and fog. Messaging is handed off to a closer site if the error rate increases, just as the cellular telephone network. Systems have been set up and proven in Seattle, Washington, one the rainiest and foggiest metropolitan areas.

Data storage. Photonics has emerged as an advanced technology for storing digital data. Today, many computers come equipped with read/write disks that use photonics devices to store, erase, and retrieve data. The information is not actually stored as photons; rather, the photons alter recording media. Optical data storage offers significant performance advantages over other techniques, especially durability, high density, and low cost. Continuous improvements in such products as compact disks, higher-density DVDs, and rewritable magnetic/optical drives will provide the technology for multimedia software, digital libraries, and paperless offices. Photonic books will soon be available, with digital works from the Internet and digital libraries.

Computing. Many scientists believe that a photonic computer will be the ultimate computing machine, replacing the electronic design of today. Others point out that the photon's low mass and lack of charge make computing and storage difficult. But photons do possess valuable attributes, such as low interaction, that would allow the photonic computer to have much higher speed and power. However, most agree that any commercially useful machine is probably decades away.

For background information *see* COMPUTERIZED TOMOGRAPHY; ELECTROMAGNETIC RADIATION; LASER; LIGHT; LIGHT-EMITTING DIODE; NONLINEAR OPTICAL DEVICES; OPTICAL COMMUNICATIONS; OPTICAL DETECTORS; OPTICAL FIBERS; OPTICAL GUIDED WAVES; OPTICAL INFORMATION SYSTEMS; PHOTON; SPECTROSCOPY in the McGraw-Hill Encyclopedia of Science & Technology.   Ken Gilleo

Bibliography. V. Burkis, *Photonics: The New Science of Light*, Akran, 1986; G. Gilder, *Telecosm*, Free Press, New York, 2000; F. Graham-Smith and T. A. King, *Optics & Photonics: An Introduction*, John Wiley, 2000.

Phylogenetic bracketing

Determination of the biology of extinct organisms is hampered by the fact that, except in very unusual situations, only their hard parts (teeth, skeleton, shells) fossilize. In other words, a vertebrate paleontologist studying a particular dinosaur is almost always limited to the skeleton, and often only part of that skeleton. Information about the remainder of the animal's anatomy, its behavior, and other attributes must be inferred. This information is required in order to visualize the fossil as a living organism and to consider its paleoecology.

Unpreserved features of extinct organisms can sometimes be inferred directly from the characteristics of their fossils. Diet can be inferred from the shape of the teeth, and scars on bones can provide evidence regarding the position and size of muscles. However, this information can be misleading. Whereas the teeth of the giant panda suggest an omnivorous diet, this species feeds almost exclusively on bamboo, and muscle size does not always correlate with the size of muscle scars.

Phylogenetic inference. Unpreserved attributes of extinct organisms can also be inferred based on their phylogenetic relationships with living species: features found in living species that are considered to be closely related to the extinct species are inferred to also have been present in the extinct species. Some of these types of inferences are not problematic; it can be inferred with confidence that fossil species that share the common ancestry delimited by living mammals will probably share features, such as hair, mammary glands, and live birth, that are characteristic of this clade. These inferences are robust because the ancestry of the fossil species of interest is bracketed by living species that have the feature inferred to be present in the fossil. This "extant phylogenetic bracket" consists of two or more extant relatives of the fossil species whose most recent common ancestry includes the fossil (**Fig. 1**). Using this method, the closer the phylogenetic relationship between the fossil species and the bracket species, the larger the number of features that can probably be inferred in the fossil species.

Although it could be argued that the inferences in the above examples are obvious, other examples would be less straightforward, and the bracket approach provides a precise framework for determining exactly which phylogenetic inferences are valid. These inferences are conservative in that only features found in at least two related taxa can be unequivocally inferred in the fossil.

Phylogenetic inferences that do not involve a bracket approach are problematic. Traditionally

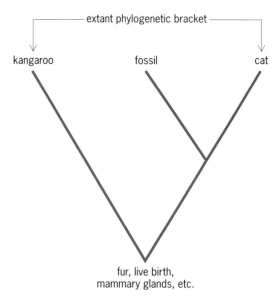

Fig. 1. Extant phylogenetic bracket. The presence of features such as fur, live birth, and mammary glands in the kangaroo and the cat (and other mammals) suggests that these features were probably present in the most recent common ancestor of these living species (basal node on cladogram). Given no evidence to the contrary, it is reasonable to assume that these features were also present in fossils that share this ancestry.

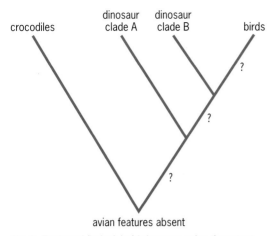

Fig. 2. Equivocal (indecisive) inferences using the extant phylogenetic bracket. Birds and crocodiles form the extant phylogenetic bracket for the two clades of dinosaurs. Using this method, avian features such as endothermy cannot be inferred to have existed in dinosaurs because these features are absent in crocodiles. The most recent common ancestor of crocodiles and birds almost certainly lacked endothermy; it must therefore have evolved along one of the three internodes leading to birds (marked as "?"). Because the appropriate internode is unknown, it cannot be inferred with confidence that this feature existed in either of the dinosaur lineages shown.

dinosaurs were considered reptiles, and living members of this group were used as models for dinosaurs. As a result, dinosaurs were interpreted as sluggish, ectothermic, dimwitted beasts. With the increasing evidence that birds are actually the closest living relatives to dinosaurs, there has been a recent tendency to use birds as living models for dinosaurs, and these extinct animals have been reinterpreted as more active, endothermic, and often relatively brainy. However, both sets of inferences are flawed because they rely on assumptions as to when the features of interest evolved (**Fig. 2**). Valid phylogenetic inferences about dinosaurs must involve at least their two most closely related living relatives, crocodiles and birds. Because both of these living groups have a four-chambered heart, for example, it can be inferred that their most recent common ancestor also had this feature. Because these two groups bracket dinosaurs phylogenetically, dinosaurs share that same ancestor; therefore, dinosaurs probably also had four-chambered hearts.

On rare occasions, fossils can take on the role of living species in the inference of unpreserved features. Although feathers are not normally preserved in fossils, they occur in *Archaeopteryx* (considered by most as the earliest known bird). *Archaeopteryx* and living birds form an "extant" phylogenetic bracket that allows the inference of feathers in fossil species that share the same most recent common ancestry. Using the same reasoning, the recent discovery of Chinese dinosaurs with feathers can be used to infer feathers in other dinosaurs that are bracketed by birds and these feathered Chinese dinosaurs. This method can also be used to infer unknown, but fossilizable, features in extinct species. In these instances, other fossils may form the bracket.

Osteological correlates. The phylogenetic bracket method can be used alone, as discussed above, or in combination with osteological correlates for soft anatomy or other attributes not normally preserved in fossils. Osteological correlates are skeletal features that can be causally associated with particular aspects of the soft anatomy or other nonskeletal features. Skeletal tissues tend not to self-differentiate; instead, they are the products of developmental induction involving nonosteological tissues. For example, the developing brain induces the ossification of the dermal skull roof, and the brain and cerebrospinal fluid determine the ultimate form of these bones. Because of this causal relationship, the dermal skull roof can be considered an osteological correlate for the brain. Once these causal relationships have been established, the presence of the osteological correlate can be used to infer the presence of the soft tissue. This information can then be incorporated into the phylogenetic bracket approach to provide strong inferences regarding the presence of unpreserved attributes in extinct species.

Once a causal relationship between a particular soft tissue and its osteological correlate has been established in two or more living groups, it can be inferred that the same causal relationship was also present in their most recent common ancestor. It can then be hypothesized that this causal relationship would have also been present in any fossil taxa that share that ancestry (in other words, that are bracketed phylogenetically by those living taxa). If the osteological correlate in question occurs in those fossils, there is a strong basis for inferring that the associated soft tissue was also present. In this situation, in which both of the bracketing taxa have the soft tissue and its osteological correlate, the inference of the soft tissue in the fossil species based on the occurrence of the osteological correlate is both positive and unequivocal (decisive). For example, mammals have a fossa on the lateral surface of the lower jaw that is an attachment site for the masseter muscle. The strong causal relationship between the fossa and this muscle provides the basis for an unequivocal inference of the presence of the masseter muscle in fossil mammals that are bracketed phylogenetically by other mammals and have that fossa on their lower jaw.

In some instances, the osteological correlate and the associated soft tissue occur in the closest living relative of the fossil species, but do not occur in the other bracketing taxon. In this situation, the presence of the osteological correlate and its associated soft anatomical feature in the most recent common ancestor of the fossil species and its closest living relative is uncertain (as shown in Fig. 2). As a result, the inference of the soft feature in the extinct species is equivocal (indecisive). The situation can sometimes be resolved by referring to more distant living relatives; if the soft anatomical feature and the osteological correlate occur in those relatives, it may be reasonable to infer that these features also occurred in that ancestor. Nonetheless, if the osteological correlate is present in the fossil, this evidence may be deemed sufficient in some cases to infer the presence of the soft anatomical feature. Inference of the high metabolic rates of living mammals in closely related nonmammalian fossil species is equivocal because the ancestors of the lineage on the other side of the extant phylogenetic bracket lacked this feature. High metabolic rates have been correlated with the occurrence of particular turbinate bones in the nasal cavity. If these bones are valid osteological correlates for high metabolic rates, the latter could be inferred to have existed in close nonmammalian fossil relatives of living mammals that have these bony structures.

Another possible situation is that a particular unpreserved feature and any osteological correlates may be absent in both of the living taxa that bracket the fossil taxon. In this instance, the most recent common ancestor of these living and fossil taxa is inferred to have lacked this feature, and the phylogenetic inference in the fossil species is negative and unequivocal. Given the absence of phylogenetic evidence for that feature, if an osteological correlate occurs in the fossil, a judgment must be made as to whether the causal relationship between the osteological correlate and the unpreserved feature is strong enough to infer the presence of that feature in the extinct species with any confidence. For example, various fossil reptile groups are found exclusively in marine sediments, but their closest living relatives are terrestrial animals, providing a decisive negative phylogenetic inference of aquatic adaptations in these fossil species. However, many of these fossils display convincing osteological correlates of these adaptations (for example, limbs modified into flippers) that, together with the sedimentological evidence, argue persuasively for these adaptations in these extinct species.

In the above examples, osteological correlates and the extant phylogenetic bracket were used in combination to infer the presence of unpreserved features in fossil species. Although this approach can provide very strong evidence for the presence of unpreserved attributes, it is limited to features that have osteological correlates. Also, when the two methods are employed separately, the results of one method can be used to test the results of the other.

Paleontologists seek to breathe life into the fossil species they study. In this process, they must infer the presence of features of these organisms that are not preserved in the fossil record. These reconstructions, therefore, include some inferences for which the evidence is poor or which are based on total speculation. However, by employing sound scientific methodologies, such as the extant phylogenetic bracket approach, whenever possible, valid scientific inferences can be separated from inferences based solely on speculation.

For background information *see* ANIMAL SYSTEMATICS; FOSSIL; LIVING FOSSILS in the McGraw-Hill Encyclopedia of Science & Technology.

Harold N. Bryant

Bibliography. H. N. Bryant and A. P. Russell, The role of phylogenetic analysis in the inference of unpreserved attributes of extinct taxa, *Phil. Trans. Roy. Soc. Lond. B*, 337:405–418, 1992; L. M. Witmer, The extant phylogenetic bracket and the importance of reconstructing soft tissues in fossils, in J. J. Thomason (ed.), *Functional Morphology in Vertebrate Paleontology*, pp. 19–33, 1995.

Pipeline design

Liquid and gaseous hydrocarbons remain the primary energy sources of the industrialized economies; in the United States, more than 700 million gallons of oil products and 51 billion cubic feet of natural gas are used every day. Pipelining is the primary means of delivering these hydrocarbons to the consumer. While less developed economies still rely heavily on batch mode such as trucks and rails as the primary means of distributing energy, the more developed economies rely more on the continuous mode of transport afforded by pipelines. In addition to being economic and very efficient, an underground pipeline network is an environmental and safety imperative. Burying pipelines preserves the esthetics of the environment while protecting the pipelines from accidental damage and sabotage.

In the United States, there are more than about 1.5 million miles of natural gas transmission and distribution pipelines and 170,000 miles of oil and oil products pipelines. **Figure 1** shows the steady growth in gas pipeline mileage in the United States since 1960. Although accurate data on the total pipeline mileage in the world are hard to come by, the mileage is growing very fast in the new industrial economies springing up around the world, as seen by the latest data on pipeline construction (**Fig. 2**).

Design. The scientific underpinnings of pipeline design have advanced significantly in the past 100 years. The theory of fluid flow in pipelines is well developed with the works of Reynolds, Stokes, Bernoulli, and others. Flow of fluids, such as oil or gas, in pipelines is fundamentally governed by the Navier-Stokes equations, the same equations that have been used to design modern airplanes. The equations result from mathematical accounting of the fluid's mass, momentum, and energy as the fluid moves down the pipeline. These fundamental equations describe the dynamic behavior of the transported fluid and its interaction with its environment, that is, the wall of the conduit (pipeline) in which it is flowing. Pumps in the case of liquids, and compressors in the case of gaseous fluids, provide the motive force for fluid flow. The motive force helps the fluid to overcome the frictional resistance to flow by the pipe wall and the internal friction within the fluid itself. In addition, the fluid must overcome the work of gravity. Depending on the terrain, gravity may either aid or inhibit flow.

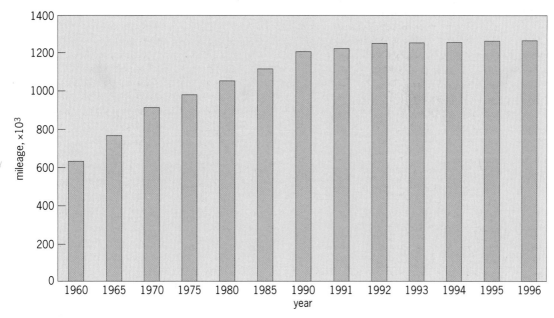

Fig. 1. Growth in gas pipeline mileage in the United States since 1960. (*After Oil Gas J., October 30, 2000*)

Solving these equations is a major challenge, and only very elementary forms of them are solved, using too many assumptions to make the solution realistic for practical pipeline flow. For example, neglecting the kinetic energy loss arising from changes in fluid flow velocity along the pipeline eliminates some of the nonlinearity in the governing equations and renders it easier to integrate to become a simplified design equation. Traditionally, this problem has been bypassed by resorting to a semiempirical approach in which all the unknown quantities are lumped into a friction term, which is obtained empirically. Modern computational methods, known as computational fluid dynamics (CFD), have helped alleviate this constraint significantly. Even then, many assumptions are still made so that the problem is managable. Computational fluid dynamics has helped pipeline scientists and engineers develop a better understanding of the fluid flow phenomena and quantify the errors that the semiempirical

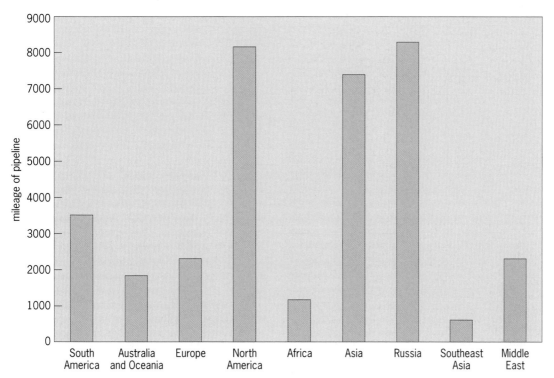

Fig. 2. Pipeline construction worldwide as of October 2000. (*After Oil Gas J., October 30, 2000*)

method, the standard for pipeline design, could introduce, and factor the errors into the design. Because of the level of mathematical complexity involved in CFD analyses, its use is still limited. CFD is growing systematically, and it is only a matter of time before it becomes routine for pipeline engineers to design actual pipelines with CFD packages.

The design principle is simple and straightforward. For a new pipeline, engineers need to size the pipeline and peripheral facilities (pumps, valves) to deliver the desired throughput of the fluid to the intended destination. For an existing pipeline, engineers need to determine what is operationally possible given the existing parameters. For example, given the pipeline size, internal roughness, and length, the maximum throughput of a given fluid is determined. The semiempirical equations used to solve these problems are simple in form. They usually relate flow rate to pressure loss. For liquid there is usually a direct proportionality between flow rate and pressure loss, whereas for gas the relationship is usually nonlinear and more complicated due to the compressible nature of gases. Both forms of the equations can be derived from the same starting point while factoring fluid properties into the derivation process, usually by some sort of an equation of state that relates the volumetric properties of the fluid to the prevailing pressure and temperature.

Pipeline design for single-phase flow (gas or oil) is a mature technology, whereas designing for multiphase flow (simultaneous flow of oil and gas) still needs more research and new technological development. Other issues pertain to the hostile environments in which new petroleum sources have been located, such as deep offshore locations and permafrost regions. In addition, formation and deposition of solid or semisolid materials in the pipelines, such as hydrates, paraffin, asphaltenes, and waxes, could lead to flow assurance problems.

Management and construction. In the United States, the pipeline systems have become like toll roads where users pay fees to transport their products. This is particularly true for natural gas transmission pipelines. Pipelines had been privately owned and used until about a decade ago when FERC Order 636 gave open access to all pipelines in the United States, so that natural gas owners could use the pipelines for a fee. FERC Order 636 essentially unbundled gas transportation from sales, forcing a greater level of competition in the natural gas industry.

Because of demand, each year hundreds of miles of pipelines are built. Building pipelines is a capital-intensive undertaking. However, if properly maintained, pipelines will last several decades.

From an engineering point of view, the construction of pipelines depends on the characteristics and the quantity of the fluid to be transported, the terrain, and the climate. The greater the quantity of fluid to be transported per unit time, the larger the pipe diameter required. Constructing pipelines for hostile environments, such as permafrost regions and deep

sea, remains a serious challenge. The building of the Alaskan pipeline is regarded as one of the greatest technological feats of the twentieth century.

As the quest for energy reaches the ocean deeps, as is now taking place in the Gulf of Mexico, on the West African coast, and at other places around the world, and extremely cold regions, such as Alaska and Siberia, pipeline design and operation face yet greater challenges. These environments are sensitive, and hence a greater degree of care and safety is required. Pipeline engineers are constantly searching for means of designing safer systems for these hostile environments and exploring new materials for pipelines that are better able to cope with the added requirements of the environments.

Safety. By most accounts, the petroleum pipeline industry has an excellent safety record, considering the complexity of the engineering and the operations involved in oil and gas pipelining. This is attributable to the diligence of trade organizations such as the American Gas Association (AGA), the American Petroleum Institute (API), the American Society of Mechanical Engineers (ASME), and the standards organizations such as American National Standards Institute (ANSI) and International Standards Organization (ISO). Over the years, standards have been established that ensure safe design, safe operational protocols and maintenance regimes, and sound practices. In addition, technological advances for detecting leaks and potential for leaks have made pipeline networks quite safe.

Environment. Over the past decade, the level of environmental awareness of the public has increased dramatically, and government's role has assumed a greater level of vigilance with tougher regulations coming out in rapid succession. Most pipeline projects now include a comprehensive environmental impact assessment at the design stage to have a credible plan in place to prevent or minimize environmental impact. For example, at river crossings, pipelines are buried below the river bed in order to prevent accidental pipe rupture.

Oil and gas will likely continue to be the dominant sources of energy in the twenty-first century. The challenge will be to continue to find new ways of preventing pollution and minimizing the environment impact, such as oil spillages and pipe leakages. The issue here is mainly quick and efficient response, rather than totally eliminating the possibility of spillage and leakage, because accidents will surely occur. For example, undetected gas leak is a fire hazard with sparks arising from any number of sources such as passing cars and lightning. The pipelining technology has sufficiently matured that with care and probity it will continue to bring energy to consumers without significant environmental degradation.

For background information *see* COMPUTATIONAL FLUID DYNAMICS; FLUID FLOW; FLUID-FLOW PRINCIPLES; GAS DYNAMICS; NAVIER-STOKES EQUATIONS; PIPE FLOW; PIPELINE in the McGraw-Hill Encyclopedia of Science & Technology.   Michael A. Adewumi

Bibliography. M. A. Adewumi, Natural gas transportation issues, *J. Petrol. Technol.*, pp. 139–143, February 1997; J. J. McKetta (ed.), *Piping Design Handbook*, Marcel Dekker, 1992; D. Yergin, *The Prize: The Epic Quest for Oil, Money, and Power*, 1993.

Polyglot programming

The first computer systems were built by English-language speakers and supported text in English. However, speakers and writers of other languages soon demanded support for their writing systems, and over the years computer systems have evolved to meet that demand. Because this process was evolutionary and because some languages are easier to support than others, the mechanisms used by programmers vary considerably.

Writing systems. A wide variety of writing systems for text are in use.

The Latin family of writing systems uses, more or less, the English alphabet. Some of these languages make use of diacritical marks to modify some letters, such as é. Others use modified versions of some letters, such as Ð. Some, such as Cyrillic and Greek, use letter forms that are unfamiliar to English speakers. Still, all of these systems have some important characteristics in common: Text is written from left to right, and divided into words separated by white space; the number of distinct symbols used in any one of these languages is less that 256, even counting all of the accented or modified letters; and there is a distinction between uppercase and lowercase letters.

Middle-Eastern languages, such as Arabic and Hebrew, are significantly different from the Latin family. They are written from right to left. The forms of some of the letters vary depending on whether they occur in the beginning, middle, or end of the word.

Many of the languages spoken in Southeast Asia use writing systems that are particularly complex. Examples include the languages of India, Tibet, and Cambodia. These writing systems include some of the same features as the Middle-Eastern languages, in which letter forms vary depending on context. However, the contextual changes are more complex. In some cases, vowels are written around their preceding consonant. In Tibet, characters are stacked in vertical clusters.

Chinese, Japanese, and Korean use a very large set of characters that originated in China. These characters are called Hanzi in China, Kanji in Japan, and Hanja in Korean. In contrast to the 26 symbols used in English, there are thousands of Chinese characters. These characters are often described as ideographic, in that a single character can represent an entire word or concept rather than a sound.

The Japanese writing system uses three sets of symbols: Kanji, and two phonetic scripts called Kana. Korean uses a phonetic writing system called Hangul in addition to the Hanja. In modern Korean, Hangul is the primary alphabet.

Text display and input. Considering all these writing systems together, it is clear that a computer system designed to support English might have considerable difficult handling the rest of the world. Simply displaying some of the more complex scripts is far more difficult than displaying Latin text. Programmers, however, rarely have to concern themselves with this function, since it is handled by operating systems and libraries.

Input is another problem area. It is not possible to build a keyboard with thousands of different symbols. Instead, systems provide input method editors (IME) that allow the end user to select a character by typing in a phonetic or other approximation, and then selecting from a list of candidates. As with the display, the average programmer does not have to create an IME or even write code that exercises much control over it.

Internal representation of text. Where programmers must interact with international text is in the representation of the text in memory. Computer languages such as C, C++, and COBOL represent text with a single 8-bit byte for each character. This follows the representation of text in the early versions of operating systems. Using 8 bits for each character is an effective procedure for the various Latin languages and even for the languages of the Middle East, but it is completely inadequate for Asian languages with thousands of characters.

Character sets. Over time, system developers have invented several different schemes for representing text in many languages in computers. These schemes are called character sets, coded character sets, or character encodings.

For the languages with comparatively few symbols, computer systems use single-byte character sets (SBCS). Because there are only 256 different symbols in one of these character sets, each one of them supports only one or a small number of different languages.

Single-byte character sets are inadequate for languages with many characters. This posed a problem since many computer languages, systems, and applications are designed to work with one byte of text at a time. System designers responded to this problem in different ways. Some systems added new data types that used more bits for each character. Initially, only mainframe systems took this path. Eventually, Microsoft adopted Unicode, a 16-bit-per-character representation as a fundamental feature of Windows NT and Windows 2000 allowing a character set with 65,536 symbols. Characters sets that use 16 bits for each character are called double-byte character sets (DBCS).

In most systems, however, the designers chose an intermediate approach that allowed them to represent a large number of symbols without requiring all the source code to be modified to support larger characters. This approach is called multibyte character sets (MBCS). In an MBCS representation, the data are still a stream of 8-bit bytes. Different characters, however, are represented by different numbers of bytes. Several different, incompatible schemes are in common use.

Examples. The following two examples involve a sentence that is English except for one Japanese character, 鰻 (meaning "eel"). The sentence is:

Your Hovercraft is full of 鰻

In the examples, the sequence "\xNN" is a way of showing a character which does not correspond to any letter in English.

Example (1) is in a character set called ISO-2022-JP, which is one of a family of ISO-2022 encodings for different languages. This character set is commonly used for electronic mail.

Your Hovercraft is full of

$$\x1b\$B\x31\x37\x1b(B \quad (1)$$

In ISO-2022-JP, escape sequences control the interpretation of the data. As a program scans this string from left to right, it starts out in Latin. When it encounters the sequence ESC-$-B (\x1b$B), the program must change its interpretation. The bytes following the escape sequence are interpreted in pairs. Each pair of bytes specifies a single Japanese Kanji character.

Example (2) is in a character set called ShiftJIS. This character set is commonly used on Windows systems.

$$\text{Your Hovercraft is full of } \x85\x56 \quad (2)$$

In ShiftJIS, there are no escape sequences. Instead, a special value, called a lead byte, introduces a two-byte sequence.

Advantages and drawbacks. Both schemes have advantages. In both, the system can handle Latin text without modifications, and it can pass more complex text through the system without changing the data type. However, there are some severe problems:

1. Many common programming practices corrupt text in these representations. If a program tries to scan a string, byte by byte, it can misinterpret the data by treating the bytes that are part of multibyte sequences as if they were ordinary Latin characters.

2. The large number of different character sets leads to complexity and confusion. Three different character sets are in common use on the Internet for Japanese, and many more are used on mainframes.

3. It is very difficult to build programs that work with text in more than one language at a time. Each of these character sets serves a single, or at most a handful, of languages. If a program needs to work with, for example, names of people in many countries, it must use a complex scheme in which different pieces of information are stored in different character sets.

Unicode. To address these problems, several major computer-related companies joined with experts on languages and writing systems to form the Unicode Consortium. This organization designed a character set called Unicode. Text represented in Unicode can include just about all the characters in all of the writing systems in current or historical use. The Unicode Standard evolves over time, and is progressively filling in the small gaps.

Windows NT uses Unicode for its 16-bit-per-character representation. Similarly, the Java programming language uses Unicode as its representation for text. Increasingly, computer systems and applications are moving to Unicode to allow them to support many languages. In some areas of the world, such as India, programming practice is moving rapidly to the Unicode as the normal encoding even for text in the local language.

For background information *see* ELECTRONIC MAIL; HUMAN-COMPUTER INTERACTION; WORD PROCESSING in the McGraw-Hill Encyclopedia of Science & Technology.   Benson Margulies

Bibliography. A. Deitsch and D. Czarnecki, *Java Internationalization*, O'Reilly & Associates, 2001; K. Lunde, *CJKV Information Processing*, O'Reilly & Associates, 1999; Unicode Consortium, *The Unicode Standard, Version 3.0*, Addison-Wesley, 2000.

Primate social organization

Primate social organization refers to the size and composition of primate groups, the social behavior and relationships of individuals living in those groups, and the ways in which those features are influenced by demographic and ecological conditions. Primates show an enormous diversity in social organization, more than most other groups of vertebrates: in some species individuals are largely solitary, while in other species animals live in groups that can contain several hundred individuals. The basic types of primate social organization are differentiated primarily by the temporal and spatial aspects of associations between adult males and females (**Fig. 1**).

In the noyau system, males and females range solitarily but come together occasionally to mate. The ranges of males generally encompass the ranges of more than one female, making the mating pattern effectively polygynous. This form of social organization is thought to be characteristic of ancestral primates and is common in nocturnal strepsirrhines and in orangutans.

Roughly 15% of primate species are characterized as monogamous, including gibbons, titi monkeys, sakis, owl monkeys, several strepsirrhines, and many callitrichids. Monogamy involves a more or less permanent spatial association between a single adult male and female, which commonly defend a territory against neighboring pairs. Typically, monogamous primates exhibit some sort of affiliative pair-bonding behavior, such as coordinated vocal duets or visual displays. Some callitrichids can live also in cooperative breeding groups. These groups generally contain a single breeding female and one or more breeding adult males, plus, on occasion, other reproductive-age females which do not reproduce but help care for the young of the breeders. Nonreproductive females are typically offspring of the breeding female, and the latter may suppress their reproduction behaviorally and/or physiologically.

Many species of primates live in one-male groups, in which a single breeding adult male is spatially associated with several adult females. These females may either be related to one another (as in guenons, patas monkeys, capuchins, and some colobines) or unrelated (as in gorillas, red howler monkeys, and some red colobus monkeys). In some species, bachelor males kept out of reproductive units may associate with one another in all-male groups. A similar form of social organization is known as an age-graded harem, in which several males (which are generally related to one another) reside in a group with several adult females but mating and reproduction are monopolized by the dominant, oldest male.

Some of the most familiar primates, including squirrel monkeys, ring-tailed lemurs, vervets, and macaques, live in bisexual multimale-multifemale groups, which are permanent associations of several breeding individuals of each sex. A few primates, notably chimpanzees, spider monkeys, and some muriquis, show a different type of multimale-multifemale social organization known as a fission-fusion society, in which a community of associating individuals splits into smaller parties of variable composition for daily foraging and travel. In this system, party size and composition are continually reorganized based on the availability of food and on the social and kin relationships among community members. Related males form the core of the community and together defend access to an area containing the ranges of several females, which are largely solitary.

Finally, some papionin primates, particularly geladas and hamadryas baboons, live in multilevel societies. In these species, bands consist of several one-male units, each of which includes a single adult male and several breeding females. Within bands, females are largely unrelated in hamadryas baboons but are matrilineal kin in geladas. In contrast, males within a band are related to one another in hamadryas baboons but unrelated in geladas. In these societies, bands may form temporary associations with varying degrees of regularity, sometimes leading to formation of troops containing hundreds of individuals. Temporary associations of multiple social groups are also seen in other primates.

Determinants. Part of the variety in primate social organization can be explained as the result of ecological factors, particularly the risk of predation and the need to maintain access to food resources for survival. The fact that most nocturnal primates are largely solitary has been interpreted as a strategy to avoid being detected by predators. For diurnal primates, sociality may reduce predation risk because predators are detected more effectively when additional animals are on the lookout (the vigilance effect), because an animal's chance of being the target of a predator's attack decreases as group size increases (the dilution effect), or because group

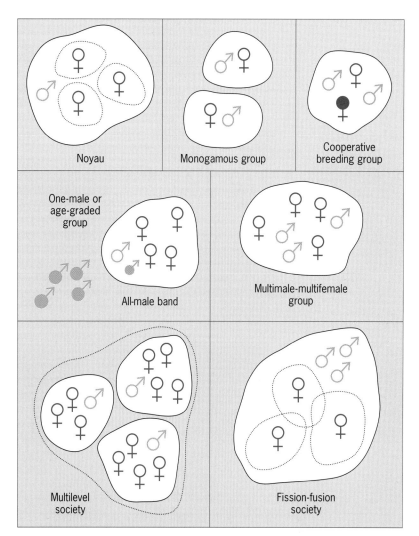

Fig. 1. Basic types of primate social organization. Only adult individuals are shown. Solid symbols represent nonreproductive animals. (*After J. G. Fleagle, Primate Adaptation and Evolution, 2d ed., Academic Press, San Diego, 1999*)

members might participate in cooperative defense against a predator. In terms of access to food, group living is generally considered to be costly to individuals because of intragroup feeding competition. If food resources are limited and must be divided among group members, an individual's share of a given resource decreases as group size increases. Consequently, animals living in larger groups may need to travel farther or spend more time feeding to meet their resource needs. On the other hand, living in a group can enhance an individual's access to food if groups can locate novel resources more readily than individuals, if group members share information about the locations of food sources, or if larger groups can displace smaller groups from desired food patches.

Socioecological theory maintains that other features of primate social organization, such as sex biases in dispersal patterns, the sexual composition of social groups, and the patterning of social relationships within groups, are also ultimately shaped by ecological factors—particularly by the way in which

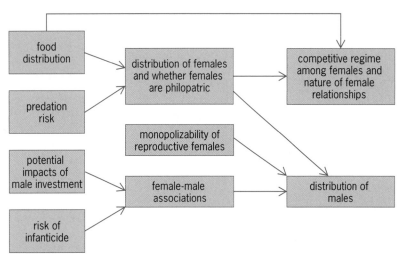

Fig. 2. Ecological and social influences on aspects of primate social organization. (*After N. B. Davies, Mating systems, in J. R. Krebs and N. B. Davies, eds., Behavioral Ecology: An Evolutionary Approach, 3d ed., Blackwell Scientific Publications, Oxford, 1991; and E. H. M. Sterck et al., The evolution of female social relationships in nonhuman primates, Behav. Ecol. Sociobiol., 41:291–309, 1997*)

the risk of predation and the distribution of food resources jointly determine the dispersion of females in the landscape and the nature of social relationships among females, and by how female dispersion subsequently influences the dispersion of males (**Fig. 2**). The primacy of females in this model stems from the fact that, for most mammals, females invest far more in offspring (through pregnancy, lactation, and other postpartum care) than do males. A female's fitness—the number of offspring she can raise in a lifetime—is closely tied to and limited by her access to food resources, and thus the distribution of resources determines the distribution of females. Females are expected to aggregate where resources are clumped and to be more evenly spaced where resource distribution is more uniform. Predation risk also influences female distribution in that, when high, it is an additional pressure favoring female groupings. Ecological factors also determine sex biases in dispersal. Where females benefit more than males in terms of foraging efficiency or protection from predators by remaining in a familiar natal range, females should be philopatric (remain in their natal groups for life) and males should disperse.

The nature of female social relationships within groups is also related to food distribution through its impact on the competitive regime over food. For species where females are philopatric, where resources are distributed such that they can be monopolized by dominant individuals or coalitions, female social relationships tend to be strongly hierarchical and characterized by nepotistic patterns of affiliation and cooperation. In contrast, where resources are less clumped and defensible, female relationships tend to be more egalitarian and cooperation between females less common.

As male fitness depends less on access to food resources than on access to females, male distribution patterns are determined by the distribution of females. Where multiple females, or the resources needed by those females, are easily monopolized by a single male, polygynous social systems (noyau, one-male units) can arise. Similarly, where lone males cannot defend access to multiple females, either monogamous or multimale-multifemale groups may arise, depending on whether females are solitary or grouped.

This ecological model does not address why, in many primates, the sexes should associate permanently. In particular, why should female primates tolerate the presence of males which in most instances are unrelated and may compete with them for food? Some researchers have argued that the benefits of male investment in offspring provided the impetus for females to group permanently with males which provided such paternal care. This hypothesis may help to explain the evolution of some monogamous and cooperatively breeding social systems, but is less satisfactory for the majority of primates, in which direct male care of offspring is negligible. More recently, researchers have argued that the evolution of permanent male-female associations is due mainly to the risk of infanticide. Female primates may benefit from permanent associations with males, if those males in some way help protect infants from other, potentially infanticidal males. These influences on primate social organization are summarized in Fig. 2.

While ecological factors are obviously important in shaping primate social systems, there is also a clear phylogenetic component to primate social organization. For example, among cercopithecine primates many species show similarities in group structure and intragroup social behavior despite having varied ecologies. These similarities include male-biased dispersal and female philopatry; the existence of stable, linear dominance hierarchies among females; and affiliative female social behavior that is directed along matrilineal kin lines. These social organization traits appear to be part of a suite of behavioral features that evolved in the ancestor of the Old World monkeys and are retained in most descendant taxa.

Genetics. Molecular techniques such as restriction analysis, mitochondrial and nuclear DNA sequencing, DNA fingerprinting, and nuclear marker analysis are increasingly used as tools to examine the genetic underpinnings and consequences of primate social organization. For example, DNA fingerprinting and multilocus genotyping have been employed to examine the breeding systems of several primates, including chimpanzees, marmosets, red howler monkeys, and a number of macaques. Multilocus genotyping has also been used to determine patterns of intragroup relatedness for comparison with behavioral data to investigate the role of kinship in structuring social relationships. For example, a molecular study of chimpanzees in Gombe National Park, Tanzania, revealed that males were more closely related to one another than females, supporting the suggestion that the social core of the community consists of close male kin.

Some of the genetic consequences of primate social structure have been examined by comparing genetic variability in both maternally inherited mitochondrial DNA and nuclear DNA within and between populations. In macaques, mitochondrial DNA shows little variation within local populations but substantial variation between populations. In contrast, most of the overall variation in nuclear DNA is found within rather than between groups. This result is consistent with field observations of male-biased dispersal and female philopatry in these species. Primates in which females disperse, such as chimpanzees and spider monkeys, should show far less of a difference between mitochondrial and nuclear DNA variability within groups.

The development of the polymerase chain reaction, which can be used to amplify minuscule amounts of genetic material obtained from feces, shed hairs, or noninvasively collected tissue samples, along with the growing availability of easily screened neutral genetic markers such as microsatellites, promises to facilitate more widespread application of genetic techniques to the study of primate mating systems, dispersal patterns, and intragroup relatedness.

For background information *see* BEHAVIORAL ECOLOGY; PRIMATES; SOCIAL MAMMALS; SOCIOBIOLOGY in the McGraw-Hill Encyclopedia of Science & Technology.   Anthony Di Fiore

Bibliography. E. Delson et al. (eds.), *Encyclopedia of Human Evolution and Prehistory*, 2d ed., Garland Publishing, New York, 2000; R. I. M. Dunbar, *Primate Social Systems*, Cornell University Press, Ithaca, 1988; J. G. Fleagle, *Primate Adaptation and Evolution*, 2d ed., Academic Press, San Diego, 1999; D. J. Melnick and G. A. Hoelzer, The population genetic consequences of macaque social organization and behaviour, in J. E. Fa and D. G. Lindburg (eds.), *Evolution and Ecology of Macaque Societies*, Cambridge University Press, 1996; P. A. Morin et al., Kin selection, social structure, gene flow, and the evolution of chimpanzees, *Science*, 265:1193–1201, 1994; B. B. Smuts et al. (eds.), *Primate Societies*, University of Chicago Press, 1987; E. H. M. Sterck, D. P. Watts, and C. P. van Schaik, The evolution of female social relationships in nonhuman primates, *Behav. Ecol. Sociobiol.*, 41:291–309, 1997.

Programmatic risk analysis

Traditional risk modeling tools focus on the quantification of either the technical risk or the management (cost or schedule) risk of a project. Yet, a comprehensive programmatic risk analysis needs to consider both because they may be linked. Technical failures generally occur during operations when a project does not perform its functions. The causes of these failures are often system design flaws that were not detected before operations. The failure may have been caused by human and organizational factors. Probabilistic risk analysis (PRA) methods are useful tools for analyzing a system, measuring quantitatively its technical failure risk, and supporting design decisions to minimize this risk. Probabilistic risk analysis was originally developed for the commercial nuclear power industry in the 1970s, and has been applied successfully to many other sectors, including aerospace, marine oil platforms, and chemical processing. The purpose of a technical probabilistic risk analysis is to examine the system for all potential damage states and the probability of each state. This process, however, does not include management failures such as project cancellations due to budget overruns. Management risks focus on factors that affect both the schedule and the budget of a project. Many projects have failed (that is, were canceled) because of large cost or schedule overruns. Monte Carlo simulation in conjunction with budgets and work-breakdown structures is often used to examine the risks of cost escalation or schedule slippage for a project. While beneficial, these tools (probabilistic risk analysis and simulation of cost and schedule risks) are too often used separately. Because it is difficult to balance simultaneously project cost, schedule, and performance given the dependencies between the types of risks, an advanced programmatic risk analysis framework known as APRAM has been developed to support decisions involving both management and technical risk factors.

Programmatic risk analysis is a tool that allows accounting for two critical issues in the management of programs of multiple projects: the trade-off between the probabilities of technical and managerial failures and the dependencies among projects. It was developed for the management of "faster, better, cheaper" unmanned space missions of the NASA Jet Propulsion Laboratory. Experienced managers generally have an intuition for what a project's financial reserves should be and the engineering knowledge to allocate the resources. However, new systems with complex, multiple dependencies can benefit from a rigorous programmatic risk analysis approach.

APRAM framework. The APRAM framework permits explicit quantification of technical and management risks for specific systems while considering interdependencies among projects in programs. The results of the process determine the best overall design alternative, the optimal budget allocation, and the initial budget reserves that minimize the overall failure risk. At the core of the model are (1) allocation of the overall budget among projects in a program and (2) for each individual project, allocation of the budget between technical reinforcement of the system and management reserves needed to resolve development problems. The framework is based on a sequence of optimization steps.

Independent-project programs. Consider first a single project independent of others in a program. The first step is to identify all financially feasible and technically acceptable configurations and the minimal cost among them given available design options. For example, one possible configuration could be a single-string system, while others may involve varying levels of redundancy. The difference between the lowest-cost alternative and the total project budget is the budget residual X. For each feasible configuration, the budget residual can be allocated between technical system improvements T to increase reliability beyond the minimal-cost alternative and initial project reserves R to solve unforeseen problems or errors. Therefore, $X = T + R$. The allocation of the budget residual among these components reflects the project manager's trade-offs between the two types of risk. The framework for a single project is thus a two-stage optimization, that is, an optimal allocation of the total residual budget between T and R and within T and R to minimize the overall risk of project failure.

Optimal allocation of technical improvement budget (T). A probabilistic risk analysis model is needed for each alternative design to quantify the probability of system failure as a function of the probabilities of component failures. The probabilistic risk analysis model includes functional block diagrams and fault trees for each design, probabilities of component (or subsystem) failures for the minimal cost design for each configuration, and functional relationships between the probabilities of failure of each component and the investments in its reinforcement. The probability of technical failure for each alternative is then modeled as a function of the probabilities of the system's failure modes and the effects of investments in the reinforcement of the various system components.

A nonlinear optimization algorithm based on the Karush-Kuhn-Tucker method (a nonlinear programming method) is used to minimize this probability of technical failure as computed by the probabilistic risk analysis model subject to the investment constraint defined by T. The optimization is repeated for all possible levels of T ($0 \leq T \leq X$). The results from this set of optimizations determine, for each design alternative and level of T, the investment in each subsystem that minimizes an objective function such as the expected costs of technical failures (both total and partial).

Optimal allocation of management reserve budget (R). In parallel to the technical optimization process, the analysis requires identification (by the project manager) of the possible development problems and mitigation alternatives associated with each design. For each possible level of budget reserve R, decision trees are used to determine the optimal use of these budget reserves to solve development problem scenarios and minimize the probability of management failure. The project results in a management failure if the resolution of a scenario requires resources that exceed the available budget reserves. The results of this sequence of optimizations determine the use of the available budget reserves that minimizes an objective function such as the expected cost of management failures (both total and partial) for each design alternative and level of reserves R ($0 \leq R \leq X$).

Overall optimal allocation. So far, the APRAM model has addressed separately the optimal mitigation of technical and management risks for all possible levels of T (investments in technical robustness) and R (project reserves), based on the budget residual constraint X. The final step is to select the design configuration and optimal initial budget allocation that minimizes the overall risk of mission failure, expressed either as the expected failure costs or as some other objective function. At the optimum, if some resources are to be invested in both T and R, the change in the objective function relative to T is equal to the equivalent change relative to R. The result of this process is the identification of the optimal system design and development risk mitigation strategy, and the best allocation of the budget residual between initial reserves (R) and system hardware and development (T).

Dependent-projects programs. The framework described earlier focuses on a single project and assumes that projects are developed independently. Greater risk, however, may be acceptable for an independent project than for a project upon whose success several future projects depend. Therefore, the manager of multiple projects may be more risk-averse in his or her decisions for each project. APRAM uses dynamic programming to include multiple projects in the model. Since recovering from an early project's failure often involves additional development costs for a later one, the probability of management failure for the second becomes a function of the possible development problem scenarios given the outcome of the previous project. The process is to optimize the second project conditional on the possible failure states of the first, and then to optimize the first project given the additional penalty costs incurred by the second for the different outcomes of the first. This procedure allows adapting the management of the first project's development given its potential effects on the success of later projects.

Progress and prospects. As early as the Apollo program, NASA seemed to have accepted the notion that quantitative risk analysis could be useful for decision support, but the failure probabilities computed for some missions of the Apollo program were largely overestimated because they were based on conservative estimates of subsystem failure risks. Because the results were so pessimistic at that time and showed such a small probability of mission success, NASA turned away from quantitative risk assessment methods. Following the *Challenger* accident, high-level commissions that investigated it concluded that the perceptions of shuttle failure risks had been overly optimistic and that NASA's risk assessment methods needed improvement. Since that time the use of probabilistic risk analysis at NASA has increased significantly, not just for the space shuttle but also for some unmanned space missions and for the space station. In the long term, this decision will improve the consistency and the efficiency of the management of NASA's space systems.

For background information *see* MONTE CARLO SIMULATION; NONLINEAR PROGRAMMING; RISK ANALYSIS; SYSTEMS ENGINEERING in the McGraw-Hill Encyclopedia of Science & Technology.

Robin L. Dillon; M. Elisabeth Paté-Cornell

Bibliography. C. B. Chapman and S. C. Ward, *Project Risk Management: Processes, Techniques, and Insights*, John Wiley, Chichester, 1997; R. L. Dillon and M. E. Paté-Cornell, An advanced programmatic risk analysis method, *Int. J. Technol. Policy Manag.*, vol. 1, no. 1, February 2001; E. Henley and H. Kumamoto, *Probabilistic Risk Assessment: Reliability Engineering, Design, and Analysis*, IEEE Press, New York, 1992; F. S. Hillier and G. J. Lieberman, *Introduction to Operations Research*, McGraw-Hill, New York, 1990; T. Williams, A classified bibliography of recent research relating to project risk managements, *Eur. J. Operational Res.*, 85:18–38, 1995.

Prospecting

Mineral assemblages surrounding ore deposits typically have unique characteristics that create zoning patterns (changes in composition with depth, lateral distance, or both) in a variety of ore deposit environments. These patterns result from the interaction of hydrothermal fluids with the surrounding host rock and may be used to help locate mineralization. Mapping alteration minerals—that is, determining the type and distribution of alteration minerals in a given area—is therefore extremely useful in identifying and outlining mineral deposits. It is a routine part of exploration for hydrothermal mineral deposits and aids in the assessment of exploration properties and the construction of deposit models. Short-wave infrared (SWIR) spectroscopy is used to identify these alteration minerals and focus exploration programs.

Spectral analysis. Short-wave infrared spectroscopy detects the energy generated by vibrations within molecular bonds. These bonds have bending and stretching modes within the 1.3–2.5-micrometer region of the electromagnetic spectrum. The observed absorption features, which are manifestations of energy absorption by molecules within the crystal lattices, represent the first and second overtones and combination tones of fundamental modes that occur in the mid-infrared region. The positions of the features in the spectrum and their characteristic shapes are a function of the molecular bonds present in the mineral. A typical spectrum consists of several characteristic features, including the hull (background), wavelength position, and feature depth and width. Variations in chemical composition may be detected as the wavelength positions of features shift consistently with elemental substitution. The SWIR range is particularly responsive to a variety of molecules and radicals typical of alteration minerals, including hydroxyl (OH), water (H_2O), ammonia (NH_4), and carbonate (CO_3), and cation-OH bonds such as Al—OH, Mg—OH, and Fe—OH.

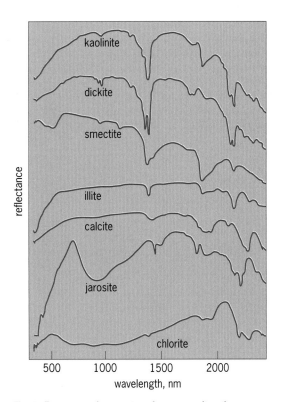

Fig. 1. Representative spectra of common alteration minerals. Analyses were collected with field-portable spectrometer covering the NIR and SWIR range. Reflectance values are offset for clarity.

SWIR spectroscopy detects minerals such as phyllosilicates, clays, carbonates, and hydrous sulfates (**Fig. 1**). Mineral identification is based on feature positions, the intensity and shape of absorption troughs, and the overall shape of the entire spectrum. The SWIR wavelength region is not suitable for most anhydrous silicates.

Mineral identification is based on the use of reference data sets that are empirical records of each mineral's characteristic spectra. Visual observation of a group of mineral spectra will quickly show variations based on numerous factors, including mineral chemistry, temperature, mode of formation, and other subtle changes. High-quality reference databases are critical to accurate interpretations of these variations. Reference samples should represent a wide variety of deposit environments and occurrences.

Automatic identification may be helpful when working with large data sets on well-defined areas. In order to achieve high-quality results, variations at the deposit scale must be observed and recorded by the user, using reference data sets created for that deposit. Deposit- or region-specific data sets appear to be critical in obtaining reliable results from attempts at unmixing (identifying mineral mixtures) using algorithms. Identification of complex mixtures requires geological context, user experience, and establishment of reference samples with supporting mineralogical information from other analytical techniques.

Fig. 2. Field-portable instrument in use at Goldfield, Nevada.

In recent decades, geologists in the remote-sensing community have spurred the development of field-portable SWIR spectrometers (**Fig. 2**). Remote-sensing geologists use a variety of bands within the electromagnetic spectrum, including the visible-near infrared (VNIR), short-wave infrared (SWIR), and mid-infrared (MIR). Field-portable instruments detect in the SWIR region, which is sensitive to molecular changes, and also in the VNIR, where color variations and changes in elemental oxidation states (such as iron and chromium) are observed.

Alteration mapping. Alteration maps typically are based on macroscopic field observations, but in many deposits the alteration minerals are fine-grained and difficult to identify in the field. Many current exploration programs use field-portable SWIR spectrometers to facilitate field identification of important alteration minerals. The use of SWIR spectrometers at a field base allows the mineralogy to be mapped or placed on cross sections. The resultant interpretation can be applied in real time to guide drilling and exploration. Extensive mapping programs using field observations and spectral analysis should also be supported by limited petrographic or x-ray diffraction studies. The combined data can ultimately be integrated with other data to develop targets, models, and regional guides.

All alteration mapping first requires mineral identification, followed by grouping of the minerals based on their associations. The purpose of the mapping is to construct a model of the deposit in space and time. Field observations must be made in a careful and systematic manner. The recording of basic observations, such as color, texture, modes of occurrence, and weathering state, is important for the correct interpretation of the mineralogy. The relationship among minerals must be determined with care prior to assigning them to a single assemblage or interpreting their relationship to other minerals. In addition, sample descriptions need to include several characteristics that directly affect the quality of spectral data, including grain size, transparency, sulfide content, water content, heavy-element content, contaminants (such as oil and organic material), and orientation of minerals (such as micas).

A series of logical steps should be followed in order to make realistic interpretations of the observed hydrothermal alteration. These steps include (1) detailed field and sample descriptions, (2) mapping distribution of minerals at several scales of observations, (3) consistent use of SWIR analysis to supplement mapping, (4) selected use of petrography and x-ray diffraction to provide references, and (5) continual reevaluation of the interpretation and integration with other data sets.

SWIR analysis aids exploration from property to regional scales. Although useful for identifying minerals in individual samples, field spectroscopy is most effective when data are collected in a large-scale, systematic manner. Large-scale surveys carried out on grid or other systematic patterns provide valuable information. For example, in complex zoned intrusive systems the alteration mineralogy determined during routine mapping helps to define the vertical and horizontal zoning and related ore environments. Within each environment the alteration mineralogy can define local zoning, providing vectors to mineralization. Spectral analysis may also provide information on subtle mineralogical and chemical changes even where the dominant mineralogy is obvious to the field geologists. Data processed and evaluated concurrent with mapping can have a direct impact on an exploration program.

Case study. SWIR analysis was used for prospecting for gold mineralization in an epithermal environment in Nevada. The analyses were collected with a field-portable SWIR spectrometer on rock chips from reverse-circulation drilling and were analyzed using references from a database of mineral spectra. A variety of minerals were identified, including K- and NH_4-alunite, buddingtonite (NH_4-feldspar), NH_4-bearing illite, dickite, kaolinite, and montmorillonite. A correlation was noted between alunite and elevated gold values, but more importantly the mineralogy near the bottom of the hole was found to reflect higher temperatures, indicating a higher potential for mineralization at depth. The drill log illustrates the detail in which mineralogy may be mapped, providing important information for exploration (**Fig. 3**).

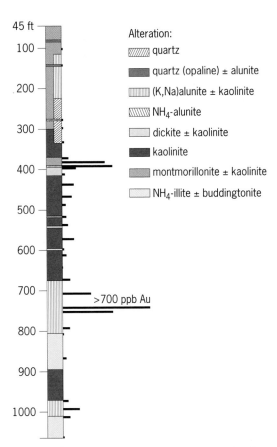

45 ft

100

200

300

400

500

600

700

>700 ppb Au

800

900

1000

Alteration:

////// quartz

quartz (opaline) ± alunite

(K,Na)alunite ± kaolinite

NH₄-alunite

dickite ± kaolinite

kaolinite

montmorillonite ± kaolinite

NH₄-illite ± buddingtonite

Fig. 3. Drill log for epithermal prospect in Nevada showing alteration mineralogy determined by SWIR analysis and Au values.

Summary. Field mapping of alteration mineral assemblages is critical to successful mineral exploration. Rapid and accurate mapping of mineralogy in the field enhances the understanding of the property and provides critical information which, when used concurrently with data from other analytical techniques, is important for the development of deposit models and regional exploration programs. The use of SWIR spectrometers also allows patterns over large areas to be delineated, creating comprehensive databases that are invaluable for future exploration.

Currently used extensively in the gold, copper, and uranium-mining industries, SWIR analysis is likely to be used in an ever-increasing number of geotechnical and environmental applications in the future, such as mineral processing, evaluation of tailings, and analysis of cement.

For background information *see* GEOCHEMICAL PROSPECTING; INFRARED SPECTROSCOPY; MINERAL; ORE AND MINERAL DEPOSITS; PROSPECTING; SPECTROSCOPY in the McGraw-Hill Encyclopedia of Science & Technology.   Anne J. B. Thompson

Bibliography. R. V. Kirkham et al. (eds.), Mineral Deposit Modeling, *Geol. Ass. Canada Spec. Pap.*, no. 40, 1993; A. J. B. Thompson, P. L. Hauff, and A. Robitaille, Alteration Mapping in Exploration: Application of Short-Wave Infrared (SWIR) Spectroscopy, *Soc. Econ. Geol. Newsl.*, vol. 39, 1999; A. J. B. Thompson and J. F. H. Thompson, *Atlas of Alteration: A Field and Petrographic Guide to Hydrothermal Alteration Minerals*, Geological Association of Canada, Mineral Deposits Division, 1996.

Proteasome

Intracellular proteins are continuously synthesized and degraded, and their levels in cells reflect the fine balance between these two processes. The rate of breakdown of individual proteins inside the cell varies widely and can be altered according to changes in the cellular environment. In eukaryotic cells, the site for degradation of most intracellular proteins is a large proteolytic particle termed the proteasome. Proteasomes are a major cell constituent, constituting up to 2% of cellular protein, and are essential for viability. They are found in the cytoplasm and nucleus of all eukaryotic cells. Simpler but homologous forms of the proteasome are also present in archaea and bacteria. Much of the knowledge about the structure and function of eukaryotic proteasomes has been gained from studies of these simpler systems.

The form responsible for the most intracellular protein breakdown is termed the 26S proteasome, which functions as a component of the ubiquitin-proteasome pathway (**Fig. 1**). This system catalyzes the rapid turnover of many critical regulatory proteins (for example, transcription factors, cell-cycle regulators, oncogenes); thus the proteasome plays a pivotal role in controlling a wide variety of cellular processes, ranging from cell division to circadian rhythms, gene transcription, and immune responses. Most normal long-lived proteins (which make up the bulk of proteins in cells) and damaged, misfolded, or mutated proteins (which could cause disease if they accumulated in cells) are also degraded by this pathway.

Proteasome versus protease. The proteasome differs from a typical proteolytic enzyme in many important respects. The typical protease is a single subunit enzyme of 20,000 to 40,000 daltons (Da). By contrast, the proteasome is up to 100-fold greater in size (2.5 million daltons) and contains about 50 proteins with multiple enzymatic functions, some of which require adenosine triphosphate (ATP). Unlike typical protease pathways, the ubiquitin-proteasome pathway is ATP-dependent and uses the energy stored in ATP to mark, unfold, and transfer the protein substrate into the degradative chamber of the proteasome. Traditional proteases simply cleave a protein and release the partially digested fragments, whereas the proteasome binds and cuts the protein substrate into small peptides ranging from 3 to 25 residues in length, ensuring that partially digested proteins do not accumulate within cells. Peptides released by the proteasome are rapidly hydrolyzed to amino acids by peptidases in the cytosol, which are then reutilized for new protein synthesis.

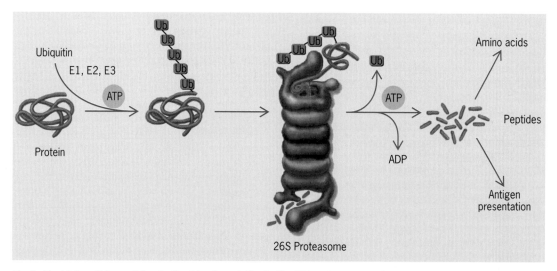

Fig. 1. Most intracellular proteins destined for degradation by the 26S proteasome are bound to multiple ubiquitin molecules through the action of three enzymes: a ubiquitin-activating enzyme (E1), a ubiquitin-conjugating enzyme (E2), and a ubiquitin-protein ligase (E3). Degradation of ubiquitinated proteins occurs within the central chamber of the proteasome. Ubiquitin conjugation, unfolding of a substrate, and its translocation into the inner cavities of the proteasome require energy obtained by ATP hydrolysis to ADP. The great majority of peptides released by the proteasome are further digested into amino acids by other peptidases in the cytosol, but a small fraction of peptides escapes destruction and serves in MHC class I antigen presentation.

Ubiquitin conjugation. Ubiquitin is a small (76 amino acids) protein found in all eukaryotic cells but not in bacteria and archaea. Most proteins degraded by the 26S proteasome are first marked by the covalent linkage to multiple ubiquitin molecules (Fig. 1). Ubiquitin conjugation represents a means of providing selectivity and specificity to the degradation process. Long chains of ubiquitin molecules are attached to proteins through the action of three enzymes called E1, E2, and E3. The E1 enzyme activates ubiquitin in an ATP-dependent manner and transfers it to one of the cell's approximately 15 E2's, which function as ubiquitin carrier proteins. The cell has hundreds of distinct E3 enzymes, each of which binds a specific group of proteins destined for degradation and catalyzes the transfer of the activated ubiquitin molecules from the E2 to a lysine side chain in the substrate protein. Proteins tagged with ubiquitin chains are then rapidly bound and destroyed by the 26S proteasome.

Structure and function. The 26S proteasome is a complex composed of two very different components—the core 20S particle and one or two 19S regulatory particles (**Fig. 2**). Proteins are degraded within the 20S (720-kDa) particle, a 28-subunit barrel-shaped structure composed of four stacked rings, each containing seven distinct but homologous subunits surrounding a central cavity. The two inner rings (formed by the β-subunits) contain the proteolytic active sites and together form a central chamber in which proteins are degraded. The two outer rings (formed by the α-subunits) together with the β-rings form two antechambers through which protein substrates must pass. The proteolytic active sites of the β-subunits face the central cavity so that cleavage of proteins can occur only within the cavity. Only three of the seven β-type subunits possess catalytic activity. Each is specific for a different type of amino acid sequence: two cleave preferentially after basic amino acids, two after large hydrophobic amino acids, and two after acidic residues. Thus, the 20S particle can cleave nearly all the peptide bonds in proteins.

Substrates can enter the 20S particle only at the center of the α-ring. The N-terminal sequences of the α-subunits form a gate that is normally maintained in a closed position. Proteins or peptides cannot,

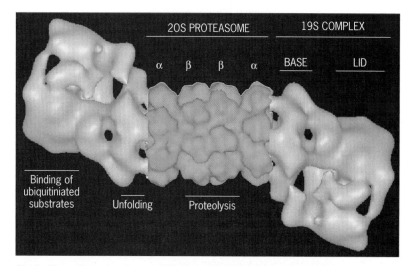

Fig. 2. Three-dimensional computerized reconstruction from electron microscopic images of the 26S proteasome. This large proteolytic complex is composed of about 50 different subunits that form subparticles, the 20S proteasome and the 19S regulatory particle. The lid of the 19S particle binds polyubiquitinated protein substrates and removes ubiquitin chains from the substrate. The substrate is then unfolded by the six ATPases located in the base of the 19S particle and translocated into the inner cavities of the 20S proteasome. Degradation occurs inside the central chamber of the 20S proteasome, formed by the two adjacent β-rings which contain six proteolytic sites. (*After W. Baumeister et al., The proteasome: Paradigm of self-compartmentalizing protease, Cell, 92:367–380, 1998*)

therefore, freely enter the inner cavity of the 20S proteasome from the cytosol. This gate must be opened by a 19S regulatory particle for protein substrates to gain access to the catalytic chamber of the proteasome. The small diameter of the entry channel—even in the opened state—requires polypeptides to be unfolded in order to enter the chamber and be degraded. The 19S particle not only binds ubiquitinated proteins but also unfolds them and promotes their translocation into the 20S particle. The architecture of the particle, in which proteolysis is isolated to a separate chamber and access is tightly regulated, prevents inappropriate destruction of cell proteins.

The 19S (890-kDa) regulatory particle associates with one or both ends of the 20S proteasome and is composed of at least 17 different subunits. It consists of two different functional entities, the lid and the base (Fig. 2). The lid provides specificity to proteolysis and contains the binding site for ubiquitin chains and enzymes that break down the polyubiquitin chains so that the ubiquitin can be recycled for use in subsequent rounds of proteolysis. The base, which touches the α-rings of the 20S core particle, contains six different ATPases (a type of enzyme that hydrolyzes ATP and uses the energy released to drive biological processes). These ATPases form a ring that uses ATP to unfold protein substrates. They also have the capacity to open the gate in the α-ring of the 20S proteasome. In doing so, they promote translocation of substrates into the inner chambers of 20S particle, where they are degraded. These structural features of the proteasome, the need for ubiquitination prior to degradation, and the energy requirement for proteolysis clearly evolved to provide a remarkable degree of selectivity and regulation to the degradative process and to safeguard against the nonspecific digestion of essential cellular proteins.

This complex system evolved from earlier proteolytic complexes found in archaea and bacteria, which lack ubiquitin and 26S proteasomes. Archaea, for example, contain 20S proteasomes that are arranged as in eukaryotes but appear to function in protein breakdown in association with an ATPase ring complex called PAN (proteasome-activating nucleotidase). PAN stimulates the ATP-dependent degradation of proteins by archaeal proteasomes, but it does not stimulate hydrolysis of small peptides that appear to freely enter into these particles. In prokaryotes, proteins bind directly to PAN. This hexameric-ring structure is homologous to the six ATPases found in the 19S base of the eukaryotic proteasome and has the capacity to unfold globular proteins and promote their entry into the 20S particle. Many features suggest that this PAN complex was the evolutionary precursor of the 19S regulator base of the eukaryotic 26S proteasome. The lid of the 19S regulator evolved from a distinct enzyme complex that allows degradation of ubiquitinated proteins. Thus, with evolution, the ATP-dependent function of the eukaryotic proteasome became linked to ubiquitin conjugation to provide greater selectivity and regulation, and lid subunits became necessary to bind the ubiquitinated protein substrates.

Catalytic mechanism. Proteolytic enzymes are normally classified according to which catalytic amino acids are in their active sites; proteins undergo a different nucleophilic attack depending on whether a serine, cysteine, aspartic, or metal group is in the active site of the enzyme. The sequences of the proteasome's catalytic β-subunits are not homologous to those of known proteases, and the pattern of sensitivity to various inhibitors differs from that of any known protease family. X-ray diffraction studies and mutagenesis of different amino acids in the proteasome have uncovered a new type of proteolytic mechanism. The active-site nucleophile of the proteasome is the hydroxyl group of a threonine at the amino terminus of the β-subunit. Because no other protease of this type is known, it has been possible to synthesize or isolate inhibitors that specifically block proteasomes, without affecting other cellular enzymes. These inhibitors have been valuable tools in clarifying the proteasome's mechanism and determining its role in cells.

Antigen presentation. In mammals, a small portion of peptides generated by proteasomes escapes destruction in the cytosol and is transported into the endoplasmic reticulum (ER). In the ER, peptides of defined lengths (between 8 and 11 amino acids) are bound by specialized membrane-bound proteins, known as major histocompatibility (MHC) class I molecules, and are subsequently transported to the cell surface. The immune system is continually surveying the surface of all cells for nonself peptides originating from viral or mutated proteins. If cells carrying such nonnative peptides are encountered, they are rapidly destroyed by cytotoxic T lymphocytes. Specialized proteasomes, called immunoproteasomes, play an important role in the generation of most MHC class I peptides. They are normally found in lymphoid tissues (for example, the spleen and thymus) but are induced in other cells of the body during infection or in response to the cytokine interferon-γ. Immunoproteasomes contain three alternative proteolytic β-type subunits that replace constitutive β-subunits. These alternative subunits change the way that peptide bonds are cleaved within proteins in order to favor the production of peptides that can preferentially bind MHC class I molecules.

20S proteasomes can also associate with an activator called PA28, or the 11S regulator. PA28 is a heptameric ring composed of two types of subunits that can be attached to one or both ends of the 20S proteasome. PA28 greatly enhances the proteasome's degradation of small peptides (but not of intact or ubiquitinated proteins) by opening up the channel in the base of the particle. Like the alternative β-subunits of the immunoproteasome, PA28 is induced by the cytokine interferon-γ. This cytokine also enhances the formation of a hybrid proteasome particle that consists of a 20S core particle capped with the 19S regulator on one side and with PA28 on the other. It is thought that within these hybrid particles ubiquitinated substrates initially bind to the 19S regulator and peptide products are released through PA28. These hybrid particles appear to play a major role in the generation of MHC class I antigens.

Role in disease. The novel catalytic mechanism of the proteasome has permitted the synthesis of selective inhibitors that can be used to study the function of proteasomes in intact cells. These studies resulted in many of the present insights about the role of the ubiquitin-proteasome pathway in various biological processes. Besides being involved in the rapid clearance of mutated and abnormal proteins which accumulate in various inherited diseases, this pathway is implicated in regulated degradation of transcription factors, oncogenes, tumor-suppressor genes, and cell-cycle regulators. Therefore it is not surprising that several prevalent diseases are caused by the gain or loss of function of this proteolytic system. For example, carcinomas arising from cells lining the uterine cervix are caused by certain strains of human papilloma virus (HPV). Upon infecting cells, these viruses produce a protein (E6) that in concert with an E3 (called E6-AP) causes rapid ubiquitination and proteasome degradation of the major tumor suppressor p53. The lack of p53 dramatically increases the tendency of cells to become malignant. Mutation of another E3, the von Hippel-Lindau (VHL) tumor suppressor, is responsible for some types of kidney tumors. Among its many functions, VHL regulates the levels of a specific transcription factor that controls the growth of blood vessels into tissues. This protein is normally maintained at a low level by rapid ubiquitination and degradation by proteasomes; however, when VHL is mutated, the transcription factor accumulates in cells, and tumors are able to form new blood vessels and grow rapidly.

The ubiquitin-proteasome system is also very important in the onset of inflammation. The transcription factor known as nuclear factor–kappa B (NF-κB) is the key regulator of expression of many inflammatory mediators. Normally, it is localized to the cytoplasm through binding of an inhibitory protein, termed IκB, which prevents its translocation to the nucleus. However, bacterial surfaces, oxidative stress, cytokines, and a cytokine tumor necrosis factor-α cause rapid phosphorylation, ubiquitination, and degradation of IκB, allowing NF-κB to enter the nucleus, bind to DNA, and activate inflammatory gene expression.

A general acceleration of the ubiquitin-proteasome pathway is responsible for the excessive breakdown of cell proteins in muscle during various disease states characterized by muscle wasting, such as cancer cachexia, diabetes mellitus, sepsis, inflammatory diseases (for example, AIDS), chronic renal failure, and nerve injury or disuse atrophy. There is appreciable interest in using proteasome inhibitors to treat various pathological states because of the key role of the proteasome in human disease. Promising results have been reported in animal models of various diseases, including cancer, rheumatoid arthritis,

and stroke. Human trials with these agents are in progress.

For background information *see* ADENOSINE TRIPHOSPHATE (ATP); AMINO ACIDS; CYTOLYSIS; ENZYME; HISTOCOMPATIBILITY; NUCLEOPROTEIN; PROTEIN; TUMOR VIRUSES in the McGraw-Hill Encyclopedia of Science & Technology.

Tomo Saric; Stewart Lecker; Alfred L. Goldberg

Bibliography. A. Ciechanover, A. Orian, and A. L. Schwartz, Ubiquitin-mediated proteolysis: Biological regulation via destruction, *Bioessays*, 22(5):442–451, 2000; D. H. Lee and A. L. Goldberg, Proteasome inhibitors: Valuable new tools for cell biologists, *Trends Cell Biol.*, 8:397–403, 1998; J. M. Peters, J. R. Harris, and D. Finley (eds.), *Ubiquitin and the Biology of the Cell*, Plenum Press, New York, 1998; K. I. Rock and A. L. Goldberg, Degradation of cell proteins and the generation of MHC class I-presented peptides, *Annu. Rev. Immunol.*, 17:739–779, 1999; A. L. Schwartz and A. Ciechanover, The ubiquitin-proteasome pathway and pathogenesis of human disease, *Annu. Rev. Med.*, 50:57–74, 1999; D. Voges, P. Zwickl, and W. Baumeister, The 26S proteasome: A molecular machine designed for controlled proteolysis, *Annu. Rev. Biochem.*, 68:1015–1068, 1999.

Proteomics

The completion of the human genome and the genomes of several model organisms is fundamentally changing how biological discovery takes place. More than any specific benefit, it has already changed how biological questions can be addressed. This is best illustrated in the application of deoxyribonucleic acid (DNA) microarray technology, by which thousands of gene segments are deposited on a glass slide or chip and then used to simultaneously analyze the expression levels of the corresponding genes. Thus, instead of testing the changes in expression of only a few genes at a time, it is now possible to measure changes in tens of thousands of genes in a single experiment. Having the complete blueprint of an organism in its DNA sequence is opening up possibilities at almost every level, including protein function.

Just as the genome corresponds to the complete sequence of DNA contained in the cells of an organism, the proteome refers to the complete set of proteins present in the various cells of an organism. The study of the proteome, known as proteomics, is changing the scope of questions that can be answered about proteins in a manner analogous to how microarrays have changed the study of gene expression. This new approach to the study of proteins encompasses a variety of techniques and strategies that make it possible to analyze collections of proteins simultaneously.

Studying protein function. The goal of proteomics is to create a model of cellular function by enhancing the current understanding of protein function and regulation. Because proteins perform most of the important functions within cells, and their function is so complexly regulated, an understanding of protein activity is necessary in order to proceed from a sequenced genome to an understanding of how all of that stored information is utilized and coordinated into a functioning cell.

At the protein level, there are several measurable properties that scientists have studied for years in order to elucidate the function and regulation of proteins. Generally, these include protein abundance, localization, posttranslational modification, and interactions.

Protein abundance is simply the amount of protein that is present within the cell at any one point in time. It is a steady-state readout of the balance between the rate of protein production and protein degradation. Protein localization refers to the subcellular organelle or structure where a given protein is found within a cell. (It may also refer to the organ, tissue, or cell type in which the protein is found.) This is of considerable importance, since for most proteins their localization at a specific site is essential for their function. Posttranslational modification (alteration of the chemical structure of a protein or polypeptide after it has been synthesized) is one of the primary means by which a cell regulates protein activity. There are over 200 different posttranslational modifications that can be made to a protein, and many of these are reversible and can be rapidly affected or changed through the action of enzymes. (An important and extensively studied modification is the addition of a phosphate to specific amino acids that is used by many cellular signaling cascades to propagate their signals.) Knowing the other intracellular proteins with which a given protein interacts, or protein-protein interactions, is also important for understanding its function. Proteins often function as part of multiprotein catalytic machines, such as the ribosome that synthesizes proteins or the ribonucleic acid (RNA) polymerase holoenzyme that transcribes genes to produce the encoded RNA. In these instances, it is impossible to understand the function of these proteins outside the context of their interactions with other proteins of the larger structure. However, even if a protein can perform its catalytic function alone, its activity, targeting, and localization can be affected by its interactions with other proteins.

Identifying proteins: mass spectrometry. A challenging technical hurdle in every proteomic strategy is identification of small quantities of protein material. Mass spectrometry is particularly suited to address this issue because it can analyze proteins far more rapidly and at far lower levels than traditional protein analysis methods.

Mass spectrometers are conceptually simple; all have an ion source, a mass analyzer, and a detector (**Fig. 1**). In proteomics, the two most common types of ionization sources are matrix-assisted laser desorption ionization (MALDI) and electrospray ionization (ESI). MALDI utilizes a laser pulse to ionize a mixture of the sample and small energy-absorbing molecules

(matrix). The matrix helps to transfer the energy of the laser to the sample. ESI uses a voltage potential difference to facilitate ionization of the liquid mixture.

For protein analysis, a protein is first digested with a proteolytic enzyme of known specificity, such as trypsin (Fig. 1a). The resulting mixture of smaller sections of the protein (peptides) can then be analyzed. Masses of these peptides, which are determined with great accuracy, are used as a fingerprint or mass map for that protein. These mass maps can then be compared with predicted digestion patterns of database proteins to identify the protein being analyzed. The databases of protein sequences are most often derived from databases of DNA sequences that are obtained by the Human Genome Project through analysis of chromosomal DNA or of collections (libraries) of DNA copies of messenger RNA molecules

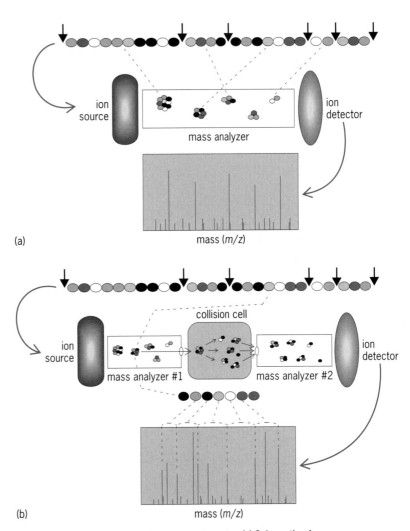

(a)

(b)

Fig. 1. Single-stage and tandem mass spectrometry. (a) Schematic of a mass spectrometer. The protein (ovals) is digested with a proteolytic enzyme which cuts at specific sites (arrows). The masses of the resulting peptides are then determined. (b) Schematic of a tandem mass spectrometer. The protein is digested with a proteolytic enzyme. The masses of the resulting peptides are measured, a specific peptide is isolated, and then it is passed to the collision cell, where it is fragmented. The masses of those fragment ions are measured in the second mass analyzer. As fragmentation occurs most often along the peptide backbone, the amino acid sequence can be inferred from mass differences between peaks.

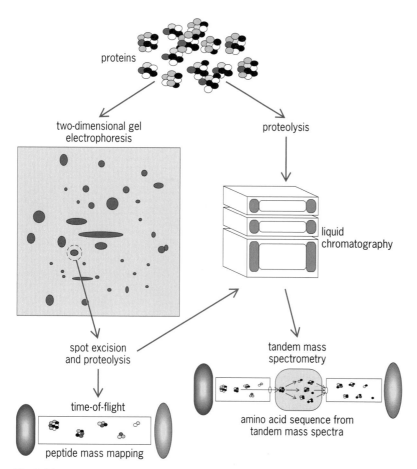

proteins

two-dimensional gel
electrophoresis

proteolysis

liquid
chromatography

spot excision
and proteolysis

tandem mass
spectrometry

time-of-flight

amino acid sequence from
tandem mass spectra

peptide mass mapping

Fig. 2. Identification of protein mixtures by mass spectrometry. A mixture of proteins can be resolved by two-dimensional gel electrophoresis and visualized by staining. Individual spots are excised and subjected to proteolytic digestion. The resulting peptide mixture can then be analyzed either by time-of-flight to determine the masses of the individual peptides (a mass map), or by liquid chromatography coupled with tandem mass spectrometery to determine the amino acid sequence of individual peptides. Alternatively, a mixture of proteins can be directly digested and analyzed by liquid chromatography coupled with tandem mass spectroscopy. (*Reproduced with permission from W. H. McDonald and J. R. Yates, Proteomic tools for cell biology, Traffic, 1:747–754, 2000*)

amino acids, the actual amino acid sequence can be deduced by the relative masses of these fragments. Tandem mass spectrometry on proteins is most often accomplished using electrospray ionization but, depending on the configuration of the instrument, can also be done using MALDI. Electrospray ionization has extended the power of the tandem mass spectrometer by first allowing the separation of peptides using liquid chromatography before they enter the mass spectrometer (Fig. 2). By coupling automated data acquisition to computer-based searches against protein databases, very complex mixtures of proteins can be analyzed.

Currently, MALDI-TOF mass mapping of proteins is the most rapid and easily automated strategy. However, this method does have weaknesses: high-quality databases with information about full-length proteins are required; it is difficult to identify multiple proteins from a single gel spot; and the amount of protein required for visualization on the preparative gel is greater than is needed for detection (since proteins must first be separated by two-dimensional gel electrophoresis). However, because of its speed, MALDI-TOF mass mapping will likely remain the method of choice for identifying proteins from two-dimensional gel electrophoresis. Electrospray tandem mass spectrometry, especially liquid chromatography coupled with tandem mass spectrometry (LC-MS/MS), avoids many of the weaknesses of mass mapping. However, the disadvantages of LC-MS/MS to most proteomic applications are the time required for analysis and the greater challenges to automation. It is currently the strategy of choice to identify proteins in complex mixtures without first having to resolve them by gel electrophoresis (well over 1000 proteins are possible). With these balancing considerations, however, it seems likely that both general strategies will be important tools for studying the proteome.

Applications. These technologies are helping to answer questions about proteins in a more global way by improving the study of the measurable properties integral to protein function and regulation: protein abundance, protein localization, posttranslational modification, and protein-protein interactions.

Protein abundance and posttranslational modification. Scientists have the ability to measure which proteins and how much of each protein are found within a cell under a certain condition. These types of analyses are the protein versions of microarray experiments. The most common strategy for such an experiment is to extract proteins from a cell and then separate them out into individual "spots" by two-dimensional gel electrophoresis. After staining to visualize these protein spots, it is possible to determine which proteins are present in one sample but not the other, and also which proteins change in relative abundance. The protein is then identified, usually by mass mapping analysis using a MALDI-TOF mass spectrometer. Again, while this is a powerful and easily automated strategy, it can often fail to identify the protein from a given spot; therefore tandem mass spectrometry needs to be employed for unequivocal identification.

that encode individual polypeptides. A commonly used mass spectrometer for such an analysis couples MALDI ionization with a time-of-flight (TOF) mass analyzer and is known as MALDI-TOF. For this type of analysis, a mixture of proteins must first be separated into individual proteins. More complex protein mixtures require two-dimensional gel electrophoresis to separate proteins into individual "spots" prior to mass mapping (**Fig. 2**).

A tandem mass spectrometer has two mass analyzers and a collision cell. These mass spectrometers allow additional information to be gleaned in a single experiment. Most importantly, when analyzing a peptide mixture, it is possible to determine not only the mass of a peptide but also the amino acid sequence of an individual peptide. This is accomplished by measuring the masses of the peptide mixture in the first mass analyzer, selecting a specific peptide to be fragmented in the collision cell, and then measuring the masses of the fragments from that peptide (Fig. 1*b*). Since fragmentation occurs most often along the peptide backbone between

Given the fact that many posttranslational modifications cause changes in a protein's position within the two-dimensional gel, some changes that can be observed using this strategy will reflect changes in the modification of a protein. An example of an application of this strategy is identification of proteins important in disease. For instance, the proteins expressed in cancer cells at various stages could be compared with proteins from noncancer cell controls. Determining which proteins differ between different types of cells can lead to a better understanding of the mechanisms that allow a cell to become cancerous. By looking directly at proteins, it is possible to detect many posttranslational differences that would be missed using a microarray strategy to look at gene expression differences. Particular proteins or specific posttranslational modifications found to be important in disease become potential targets for either developing treatment drugs or medical tests that are more sensitive and yield important information for physicians.

Protein localization. It is important to know not only the types of cells where a protein is present but also its location within an individual cell. Typically, cell biology labs have used various microscopic techniques to localize a protein to certain subcellular structures. A technique that is beginning to be employed on a genome-wide scale is "tagging" the protein of interest with green fluorescent protein and viewing the cells expressing the tagged protein by fluorescence microscopy. An alternative strategy is to isolate a particular subcellular structure (organelle) and identify the protein components of it. Larger membrane-bound organelles that have clearly defined purification protocols, such as mitochondria or Golgi apparatus, work best for this type of analysis.

Protein-protein interactions. Another challenge of proteomics is to identify protein-protein interactions. A common technique for this purpose is the yeast two-hybrid system, which takes advantage of the power of yeast molecular genetics to detect an interaction between two proteins that have been engineered to be expressed as "fusions" with yeast proteins. Protein purification has been used for years to isolate proteins that interact with one another and may, therefore, form protein complexes. Purification can be accomplished through "classic" biochemical means, in which enzymatic activity is followed through various purification steps to purify the specific protein(s) responsible for that activity. The major limitation of any protein purification strategy is the amount of material and time that is required to provide a positive identification. Another strategy is affinity purification, in which a known protein is used to "fish out" other proteins with which it interacts. This is markedly facilitated by using recombinant DNA methods to introduce into cells a tagged form of the protein that can easily be recovered (with its binding partners) by reagents that bind specifically to the tag. Often, as with measuring protein levels, components of a protein complex are first resolved by polyacrylamide gel electrophoresis and the individual proteins identified by mass mapping. Even greater speed and sensitivity can be achieved by directly analyzing digested protein complexes using LC-MS/MS.

For background information *see* GEL ELECTROPHORESIS; GENE; HUMAN GENOME PROJECT; LIQUID CHROMATOGRAPHY; MASS SPECTROSCOPY; PEPTIDE; PROTEIN in the McGraw-Hill Encyclopedia of Science & Technology.   Hayes McDonald; J. R. Yates III

Bibliography. J. S. Anderson and M. Mann, Functional genomics by mass spectrometry, *FEBS Lett.*, 480:25–31, 2000; S. Broder and J. C. Venter, Whole genomes: The foundation of new biology and medicine, *Curr. Opin. Biotech.*, 11(6):581–585, December 2000; S. Fields, Proteomics in genomeland, *Science*, 291:1221–1229, 2001; W. H. McDonald and J. R. Yates, Proteomic tools for cell biology, *Traffic*, 1:747–754, 2000; R. E. Rosamonde et al., Proteomics: New perspectives, new biomedical opportunities, *Lancet*, 356:1749–1756, 2000.

Proterozoic

The Proterozoic is a period of geological time that extends from about 2.5 billion years ago (Ga) to about 543 million years ago (Ma)—the beginning of the Paleozoic Era and Phanerozoic Eon (**Fig. 1**). It includes almost half of the history of the Earth (since ~2 Ga). The Proterozoic witnessed some of the most significant events in Earth history, such as the amalgamation and breakup of huge continental masses known as supercontinents, the most severe glaciations that the Earth has known, oxygenation of the atmosphere, and dramatic advances in the evolution of life. It is divided into three eras, the Paleo-, Meso- and Neoproterozoic, separated at 1600 Ma and 1000 Ma (Fig. 1).

Paleoproterozoic. The oldest Proterozoic era is represented by rock sequences such as the Huronian Supergroup of the north shore of Lake Huron, Canada. The Huronian is a thick (~12 km; 7 mi) accumulation of sedimentary and minor volcanic rocks. The lowest (oldest) sediments contain particles of minerals, such as pyrite and uraninite, that survived transportation and deposition in an ancient river system. Because such minerals would be destroyed by reaction with oxygen in the Earth's atmosphere today, it has been inferred that the early atmosphere contained little or no free oxygen. On the other hand, the upper part of the Huronian succession contains sedimentary rocks that are red, due to the presence of a highly oxidized iron mineral (hematite) that is considered to indicate the presence of at least some free oxygen in the Earth's atmosphere. The mechanism responsible for the inferred compositional change in the Earth's atmosphere is not fully understood, but there is a balance between oxygen-producing processes (mostly metabolic activities of photosynthetic organisms) and sinks for oxygen (mostly elements or compounds that react with oxygen). It was originally suggested that the rise of

Major subdivisions of geologic time		
International Union of Geological Sciences		Hofmann (1992)
Eon	Era	Geon
PHANER-OZOIC	CENOZOIC — 65 Ma —	0
	MESOZOIC — 245 Ma —	1
		2
	PALEOZOIC	3
		4
	— 543 Ma —	5
PROTEROZOIC	NEOPROTEROZOIC	6
		7
		8
	— 1000 Ma —	9
		10
		11
	MESOPROTEROZOIC	12
		13
		14
	— 1600 Ma —	15
		16
		17
	PALEOPROTEROZOIC	18
		19
		20
		21
		22
		23
	— 2500 Ma —	24
ARCHEAN		compressed interval
	— 3800 Ma —	37
HADEAN		compressed interval
	~4600 Ma —	45

Fig. 1. Schemes for subdivision of geological time. (*After K. A. Plumb, New Precambrian time scale, Episodes, 14:139–40, International Union of Geological Sciences 1991; and H. J. Hofmann, New Precambrian time scales: Comments, Episodes, 15:122–123, 1992*)

oxygen in the Paleoproterozoic atmosphere was related to increased photosynthesis when microorganisms, such as cyanobacteria, colonized shelf environments on the thicker and more widespread continental crust formed at the end of the Archean. These microorganisms built up carbonate-rich layers and reeflike mounds known as stromatolites (**Fig. 2**). The study of carbon isotopes preserved in the geological record indicates, however, that photosynthesis was important in the oceans even in the early part of Earth history, so that the release of free oxygen to the atmosphere may be more closely linked to changes in the efficacy of oxygen sinks in the oceans. For example, production of continental crust and cooling of the Earth's interior may have resulted in diminished hydrothermal circulation at mid-ocean ridges. It has also been suggested that oxygenation of the Earth's

atmosphere was due to a change in the redox state of volcanic gases related to mantle overturn or a high degree of mantle plume activity.

Widespread glaciation in the Paleoproterozoic is recorded by special kinds of sedimentary rocks such as diamictite—conglomerate comprising large rock fragments "floating" in a fine-grained matrix, and laminated sediments with large isolated rock fragments, called dropstones, thought to have been dropped from floating icebergs (**Fig. 3**). Some ancient glaciers left a scratched or striated substrate, as well as rock fragments bearing scars attesting to their rough passage in the ice. Such deposits are best known from North America but have also been reported on several other continents. The cause of this period of frigid climate is not known, but this early portion of Earth history was characterized by low solar luminosity, the cooling effects of which were probably mitigated by abundant carbon dioxide (and other greenhouse gases) in the atmosphere. The Paleoproterozoic glaciations could have been initiated by the removal of large amounts of carbon dioxide during weathering of the first supercontinents, which formed toward the end of the Archean, or by enhanced biological removal of carbon dioxide by photosynthesis and burial of organic matter in sediments.

Mesoproterozoic. The Mesoproterozoic Era (1.6 to 1.0 Ga) contains no clear evidence of dramatic climate change. There was significant addition of crustal material during this period, especially in the southwestern part of North America. This period is also characterized by the emplacement of significant

Fig. 2. Paleoproterozoic stromatolite from South Africa. These structures are built by the metabolic activities of cyanobacteria which trap or secrete carbonate minerals (light material) from water. The dark layers are silicified. These structures attest to the widespread nature of these photosynthetic microorganisms during the Proterozoic. Hammer head is 12 cm (5 in.) wide.

amounts of igneous rocks, such as anorthosites and granites, that were exposed in huge areas of the continental crust. The significance of these large-scale igneous intrusions is not known, but they have been considered by some to be the result of aborted attempts at continental rifting.

Toward the end of the Mesoproterozoic, widespread, continental collisions resulted in the production of a supercontinent known as Rodinia. Associated widespread mountain building ended at about 1.0 Ga, which is taken as the beginning of the third great Proterozoic era—the Neoproterozoic.

Neoproterozoic. The last major era of the Proterozoic was a time of dramatic climatic change and the evolution of complex life forms, including the first metazoans (organisms with complex cell structure and differentiated cells). Widespread glaciation in the Neoproterozoic was suggested in the 1960s, and its evidence has now been recognized on all the continents. The widespread nature of these glacial deposits and the discovery that many were formed at low latitudes, near the ancient equator, has led to revival of the idea that the entire planet may have been frozen (the snowball Earth hypothesis). A new aspect of these investigations is the discovery of exceptionally great fluctuations in the distribution of carbon isotopes in carbonate rocks. Many photosynthetic organisms sequester the lighter carbon isotope ^{12}C, so that during organic blooms ocean waters are enriched with the heavier isotope ^{13}C. Thus during periods of high organic activity the ratio of the heavy to light carbon isotope (expressed in a more complex formula as $\delta^{13}C$) increases. Exceptionally low $\delta^{13}C$ values (near-cessation of organic productivity) have been recorded in carbonate rocks deposited immediately below and above some Neoproterozoic glacial deposits, and the glaciations appear to have been followed by periods of exceptionally high organic productivity (high $\delta^{13}C$ values). Some have suggested that these periods of exceptionally high organic productivity may have triggered the glaciations, for the carbon dioxide content of the atmosphere is reduced by photosynthetic activity as carbon is used to build the plant and oxygen is released as a by-product. No adequate explanation has been offered for the initiation of such organic blooms. Alternatively, the glaciations may have been triggered by removal of atmospheric carbon dioxide by enhanced weathering on the Grenville supercontinent (Rodinia), and the observed variations in carbon isotopes may indicate the response of photosynthetic microorganisms to dramatic climatic variations, rather than being their cause.

Interpretation of the Neoproterozoic glaciations has been thrown into disarray by the discovery that many glacial deposits were formed at low latitudes. This has been interpreted to mean that the entire Earth (including tropical areas) was glaciated (snowball Earth hypothesis), but an alternative theory proposes that the tilt of the Earth's spin axis (obliquity of the ecliptic) was much greater in the geologic past. Accordingly, if the Earth had entered

Fig. 3. Granite dropstone in laminated mudstone of the Paleoproterozoic Gowganda Formation, from the Huronian Supergroup on the north shore of Lake Huron, Canada. Such isolated clasts in finely bedded sedimentary rocks suggest emplacement by floating glacier ice. The alternating fine and coarse layers (varves) are interpreted as deposits from seasonally freezing and thawing lakes. Coin is ~2 cm (0.8 in.) in diameter.

a glacial period, ice should have preferentially developed in tropical latitudes because these regions would have received less solar energy in a given year. Rocks associated with some low-latitude glacial deposits contain evidence of strong seasonal temperature variations. These include varved deposits, which are interbedded coarse and fine sediment layers that formed as a result of deposition from alternately freezing and thawing lakes (Fig. 3), and ice-wedge structures, which are also known to form by seasonally alternating freeze-thaw cycles. Such seasonal variations are difficult to explain in low latitudes (tropical regions have fairly uniform temperatures throughout the year) on today's Earth, but may be explained in tropical regions of an Earth that had a significantly increased obliquity (a tilt of $>54°$).

It is in the rocks, representing emergence of the planet from this period of climatic extremes, that the first evidence of complex organisms (metazoans) is found. In many parts of the world, fossil remains of the first abundant metazoans occur above the second of two major Neoproterozoic glaciations (the Varanger glaciation, at about 620 Ma). These are mostly preserved as imprints of the organisms in fine sandstones. They are known as the Ediacaran fauna, from a locality in South Australia where they were discovered. The affinities of these fossil organisms are the subject of ongoing debate. Some think they are ancestors of the prolific biota that appeared on Earth during the Cambrian "explosion" of life and gave rise to all modern forms, including humans. Others place them in a separate grouping (the Vendobionta) and regard them as a widely developed but failed experiment. It has been suggested that emergence of these complex life forms may have been an evolutionary response to the preceding climatic crises. Emergence of the metazoa is linked to the buildup of atmospheric oxygen which, together with development of a protective ozone layer, was a prerequisite for the metabolic activity of these complex life forms and colonization of terrestrial environments.

The Proterozoic Eon represents about half of the Earth's history. It is bracketed at the beginning and end by dramatic climatic fluctuations that may be linked to plate tectonics, the evolution of the atmosphere, and the emergence of complex life forms (metazoans) whose progeny (including humans) inherited the Earth.

For background information *see* ATMOSPHERE, EVOLUTION OF; CHEMOSTRATIGRAPHY; CLIMATE HISTORY; ECLIPTIC; GLACIAL EPOCH; METAZOA; PROTEROZOIC; SUPERCONTINENT in the McGraw-Hill Encyclopedia of Science & Technology.   Grant Young

Bibliography. P. E. Cloud, *Oasis in Space*, W. W. Norton, New York, 1988; A. J. Kaufman, A. H. Knoll, and G. M. Narbonne, Isotopes, ice ages, and terminal Proterozoic earth history, *Proc. Nat. Acad. Sci. USA*, 94:6600–6605, 1997; P. F. Hoffman et al., A Neoproterozoic snowball Earth, *Science*, 281:1342–1346, 1998; G. E. Williams, History of Earth's obliquity, *Earth. Sci. Rev.*, 34:1–45, 1993; G. M. Young, The geologic record of glaciation: Relevance to the climatic history of Earth, *Geosci. Canada*, 18:100–108, 1991.

Pump flow

The pump is the most widely used turbomachine and is the subject of intensive research and development. It is believed to be the second most frequently built machine of modern times, next to the electric motor. Pumps reach efficiency levels above 90% and are built with powers ranging from a few watts to megawatts. As pumps have evolved to handle a wider range of liquids at higher pressures and temperatures, whole industries have become dependent on them. The complex flow in pumps makes them among the most complicated fluid machines ever built. Although the most rapid development of pumps has occurred since around 1940, the pump has a long history that parallels the rise of modern engineering itself.

Pump classification. Pumps may generally be classified as intermittent-flow or continuous-flow. Intermittent-flow pumps may be divided into reciprocating or rotary; and continuous-flow pumps, into ejector or dynamic. The dynamic pumps are subdivided into axial-flow, centrifugal, and mixed-flow pumps. Another common classification of pumps is on the basis of flow coefficient, ϕ, or specific speed (**Fig. 1**).

The most widely used pumps are the reciprocating, axial-flow, centrifugal, and mixed-flow pumps. The centrifugal pump is by far the most used, and is estimated to encompass about 75% of all pump applications.

History of centrifugal pump flow. Credit for the invention of the centrifugal pump is in dispute. Leonardo da Vinci suggested the idea of using centrifugal force to lift liquid, and Johann Jordan is reported to have discussed a similar theory around 1680. Most place the origin of the centrifugal pump with the French physicist and inventor Denis Papin in 1689.

The theory of hydrodynamics and aerodynamics of pumps developed slowly in the twentieth century. The major initial task was to establish the relationships among the performance parameters of the pump's impeller, the form of the flow, and the form of the fluid passages. One of the most important relationships was that between the torque transmitted from the impeller to the fluid and the change of the properties of the fluid as it passes through the impeller.

The first intense research into centrifugal pump flow occurred between 1910 and 1930. A theoretical analysis of the flow was attempted, using the ideal assumption of two-dimensional irrotational flow of an inviscid and incompressible fluid in the impeller. At the same time, experiments were conducted to gain an understanding of the flow mechanism in centrifugal pumps. Secondary flows (fluid flows within the main flow but in a different direction) could be observed but, owing to the lack of adequate measuring and analytical tools, quantitative measurements of their structure were not possible.

K. Fischer and D. Thoma conducted the first comprehensive study of centrifugal pump flow. They showed that practically all the flow conditions for a real fluid were different from those theoretically derived for an ideal frictionless fluid. The velocity distribution along circles concentric with the axis of rotation of the impeller was not the same as generally assumed.

Until the end of the 1950s, due to the lack of knowledge of impeller and volute flow physics, the radial thrust in volute pumps and compressors was not well understood. Thus, designers and users faced serious problems of rapid wear in the bearings, leakage from the glands, and failure of the shaft due to fatigue.

Basic pump flow. A dynamic pump (axial, centrifugal, or mixed-flow) is a device which adds energy to

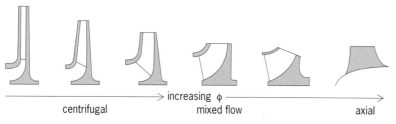

Fig. 1. Classification of pumps on the basis of flow coefficient ϕ.

centrifugal   increasing ϕ axial
                     mixed flow

a fluid by means of a fast-moving blade. In a cylindrical coordinate system, the energy transfer through pumps is governed by the rate of change of the angular momentum. A typical dynamic pump is made up of four basic components: a stationary inlet casing, a rotor, a stationary diffuser, and the collector or volute. The fluid is drawn in through the inlet casing into the eye of the rotor parallel to the axis of rotation. In order to add angular momentum, the rotor whirls the fluid. Thus, the energy level is increased, resulting in both higher pressure and velocity. The purpose of the diffuser is to convert some of the kinetic energy of the fluid into static pressure. Outside the diffuser is a scroll or volute whose function is to collect the flow from the diffuser and deliver it to the discharge pipe.

The hydrodynamic problem is to efficiently accomplish large fluid deflections and diffusion at high flow velocity, with the added difficulty that the cross-sectional areas of the flow passages must be very small in order to achieve good efficiency and high head. The individual components of the pump are capable of achieving high efficiency, but it is the efficiency of the whole pump that is of primary importance. The rotor is the rotating component of the pump, where the energy transfer occurs. The specific energy transfer can be derived from the velocity triangle, a vector diagram with the fluid velocities at the inlet and the outlet of the rotor. The rate of change of angular momentum will equal the sum of the moments of the external forces.

The most widely practiced method of analyzing flow in pumps and designing pumps has been to use very simple, one-dimensional inviscid flow techniques. While this is a convenient and straightforward design methodology, the flow within pumps is by no means one-dimensional and simple. **Figure 2** shows the actual complex flow in a centrifugal pump.

Advances in flow analysis and design. The emergence of computational fluid dynamics (CFD) since the 1970s has provided a major impetus to solve the Euler and Navier-Stokes equations, which govern the flow fields in turbomachines. This has been possible mainly due to advances in grid generation, turbulence modeling, formulation of boundary conditions, pre- and postdata processing, and computer architecture. Most of the techniques used for the solution of the Navier-Stokes equations can be classified as finite difference, finite area/volume, finite element, and spectral methods. Only the first two techniques are widely used in turbomachinery such as the dynamic pump.

The design of pumps has reached the stage where improvements can be achieved only through a detailed understanding of the internal flow. The prediction of the flow in dynamic pumps is very complicated due to the three-dimensionality of the flow and the highly curved passages in the rotating blades. Furthermore, the flow has an unsteady behavior, especially under off-design conditions, as a result of

the interaction between rotor and collector or stator. Because of these complexities, computer simulations of the flow are increasingly important.

Accuracy and productivity. Pump designers and analysts require CFD software to provide accurate solutions to their applications in a cost-effective and timely manner. With the widespread availability of affordable workstation-class software, the main issues are therefore accuracy and productivity. The information provided by CFD must be of sufficient accuracy to allow the designer to make appropriate decisions based on the available information, and it must be available rapidly enough to fit within the time scales of the design cycle. For the industrial designer, "accuracy" means providing (1) reliable qualitative information that correctly reproduces the important flow features, such as swirl, boundary layers, shocks, wakes, separation zones, stagnation points, and mixing layers; and (2) reliable quantitative predictions of such parameters as efficiency, work input, pressure rise, blade profiles, component loss coefficients, flow distortion parameters, incidence, and deviation or slip.

While it is desirable that such features be perfectly reproduced, this expectation is unrealistic given limitations due to mesh size, turbulence modeling, and other modeling assumptions such as steady-state flow. To be used confidently, it is therefore of

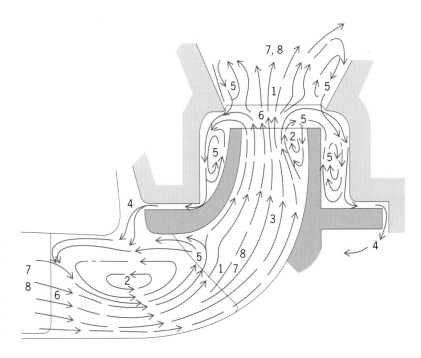

Key:

1 stall	5 unsteady flow fluctuation
2 recirculation (secondary flow)	6 wake (vortices)
3 circulation	7 turbulence
4 leakage	8 cavitation

Fig. 2. Complex flow in an actual pump. (*After Electric Power Research Institute, Centrifugal Pump Hydraulic Instability, CS-1445 Research Project 1266-18, Palo Alto, CA, 1980*)

importance that CFD provide (1) a "sufficient" level of accuracy and (2) repeatability and consistency.

The two factors that influence accuracy most for pump applications are the discretization accuracy of the flow solver and the computational mesh. Only after these two are adequately treated do other issues such as the turbulence model become of primary significance. In particular, the turbulence model is the usual scapegoat for poor CFD predictions, while in reality other causes are often more significant. Hence it is often the case that numerical errors exceed model errors.

Methods. There are two principal methods for the design and flow analysis of dynamic pumps. One method uses a computer-aided design (CAD) system to generate the pump geometry, which is then used as input in CFD analysis codes. The other method uses a self-contained program that generates the geometry of the pump and its components and sets up the CFD parameters automatically.

CAD system approach. In this approach, the design process generally flows from a CAD-generated solid model that is exported to a mesh generation package. The mesh generation package either can be contained within a commercial CFD system or can be an independent program. The mesh generation system must define the mesh spacing and topology. This work also involves fixing errors in the geometry that result from importing files. Critical areas of the mesh are examined and refined before applying the boundary conditions, and CFD processing then commences. Upon successful convergence of the calculated values, the results are examined, appropriate modifications are made to the geometry, and the entire process is repeated.

The one major drawback of the CAD system approach is its high total cost of analysis. This results from the extensive time required to repeatedly iterate on the geometry until an acceptable design is achieved. Typically, the time required to generate a three-dimensional (3D) solid model to export to CFD processing is of the order of 80–120 h. This time is a function of part complexity and ease of use of the CAD system. In order to reduce the complete process time, often the pump designer must know how to use and manipulate a 3D CAD system. This can be difficult and time-consuming, considering the complex pump geometry. A designer fluent in 3D CAD systems may also be available, but will be employed at the expense of increased interaction time between the hydrodynamicist and the designer.

Design system approach. In this approach, geometry generation, mesh node spacing, and boundary conditions are completely defined within an independent program that does not have a dependence upon CAD. Thus, the time from generating pump geometry to performing CFD can be significantly reduced, to the order of a few minutes. The time to complete CFD analysis depends upon the processor, mesh density, and convergence criteria. Iterations in geometry can be made within minutes prior to running CFD analysis. Thus, the design system approach allows the user to spend more time analyzing the results rather than setting up the geometry and mesh. Additionally, the reduced design time allows for quicker time to market by a pump manufacturer. Once an acceptable design has been achieved, the geometry is exported to the CAD system, where a drawing suitable for part procurement is produced. In this methodology, CAD is used as final documentation tool, not as an interactive design system. While this approach may significantly reduce iteration design time, it requires the development of a geometry generator. The geometry generator development is heavily dependent upon cartesian mathematics and convergence schemes.

For background information *see* CENTRIFUGAL PUMP; COMPUTATIONAL FLUID DYNAMICS; COMPUTER-AIDED DESIGN AND MANUFACTURING; PUMP; TURBULENT FLOW in the McGraw-Hill Encyclopedia of Science & Technology.　Abraham Engeda

Bibliography. I. J. Karassik et al. (eds.), *Pump Handbook*, 3d ed., McGraw-Hill, 2001.

Quantum computation

Quantum computers hold the promise to perform classically intractable tasks—computations which require astronomical time or astronomical hardware resources on the fastest classical computer—in minuscule time with minuscule resources. The building blocks of a quantum computer are atomic particles which, following the laws of quantum physics, behave like waves and exhibit interference phenomena as if they were in several locations at the same time. By equating different locations—for example, an electron in the lowest orbit or in an excited orbit of an atom—to binary digits 0 or 1, one may interpret the time-evolving state of the particles as executing several computations at the same time. One set of locations at a given time describes the result of one computation. Thus one atom can do two computations at once; two atoms can do four; three atoms can do eight. The challenge is to coerce the atoms to follow trajectories that amount to meaningful computations and to read out a definite result from the multitude of computations occurring in parallel. This article describes how this can be accomplished, summarizes the presently known quantum algorithms, explains how they can be run in a stable manner, and describes the current experimental state of the art and the hardware requirements for any quantum computer.

Principles. A classical computer manipulates strings of N classical bits, $(n_1, \ldots, n_N)$ with $n_j = 0$ or 1 ($j = 1, \ldots, N$), in such a way that intermediate states of the computation are also strings of classical bits. A quantum computer manipulates states of N two-level atoms, nuclear spins, or other entities, $|n_1, \ldots, n_N\rangle$, with $n_j = 0$ if the j-th atom is in the ground state and $n_j = 1$ if it is in the excited state, in such a way that intermediate states are superpositions of the states $|n_1, \ldots, n_N\rangle$. The 2^N states $|n_1, \ldots, n_N\rangle$ (the computational basis) are product states in which each atom is in either the ground state or excited state, and n_j is called the value of the j-th qubit (quantum bit); they represent the strings of classical bits. The superpositions include states in which an atom no longer has a sharp value of n_j (it has indefinite bit value), and states in which an atom no longer exists in a state separate from the other atoms (entangled states); both have no classical counterpart. A quantum computation starts with a product state $|n_1, \ldots, n_N\rangle$, allows the state to evolve according to the Schrödinger equation (1), with initial condition

$$i\hbar \frac{d}{dt}|\psi(t)\rangle = H(t)|\psi(t)\rangle \qquad (1)$$

$|\psi(0)\rangle = |n_1, \ldots, n_N\rangle$ and time-dependent hamiltonian $H(t)$ (energy operator) driving the coupled atoms, and ends with the measurement of the values of the qubits of the state $|\psi(t)\rangle$. The hamiltonian generates the unitary time evolution operator $U(t)$ which takes the initial state into the final state according to Eq. (2), with T the time-ordering operator. The

$$|\psi(t)\rangle = T \exp\left(-(i/\hbar)\int_0^t H(s)\,ds\right)|\psi(0)\rangle$$

$$= U(t)|\psi(0)\rangle \qquad (2)$$

measurement transforms $|\psi(t)\rangle$ into the output state $|n_1', \ldots, n_N'\rangle$ with probability $|\langle n_1', \ldots, n_N'|\psi(t)\rangle|^2$. The output is probabilistic because quantum measurements are so. The computation $|n_1, \ldots, n_N\rangle \mapsto |\psi(t)\rangle$ is a unitary transformation, hence reversible; the readout $|\psi(t)\rangle \mapsto |n_1', \ldots, n_N'\rangle$ is a projection (called the collapse of the wave function), hence irreversible. Thus, to perform a specific computation, one must drive the atoms with a specific hamiltonian; to read out the result, one must send the atoms through a series of state detectors.

Power of quantum computation. A quantum computer is more powerful than a classical computer for two reasons:

1. The quantum state space is much larger than the classical state space: N qubits can be in an infinite number of different states (any point on the unit sphere of the complex Hilbert space spanned by the 2^N basis vectors $|n_1, \ldots, n_N\rangle$); N classical bits can be only in 2^N different states (the points where the 2^N coordinate axes intersect the unit sphere). If states on the unit sphere can be resolved to within accuracy ε ($\varepsilon \ll 1$), the sphere hosts $O(\varepsilon^{-(2^N-1)})$ distinct states. Thus a quantum computer can store and access an exponentially large number of states compared to a classical computer.

2. The quantum computer operates in a massively parallel way: If the initial state is the uniform superposition of all basis states, Eq. (3), the time evolution

$$|\psi(0)\rangle = 2^{-N/2} \sum_{n_1, \ldots, n_N = 0, 1} |n_1, \ldots, n_N\rangle \qquad (3)$$

computes simultaneously $U(t)|n_1, \ldots, n_N\rangle$ for all 2^N

possible inputs $|n_1, \ldots, n_N\rangle$ by linearity of $U(t)$. The matrix element $\langle n_1', \ldots, n_N'|U(t)|n_1, \ldots, n_N\rangle$ is the probability amplitude that the computation converts the input $|n_1, \ldots, n_N\rangle$ into the output $|n_1', \ldots, n_N'\rangle$ along all possible classical computational paths in parallel (as given by Feynman's path integral). A classical computation can follow only a single path. The aim is to choose $U(t)$ such that paths corresponding to computations of no interest cancel each other by destructive interference, and paths of interest add constructively (that is, to select relevant computations by quantum interference).

Logic gates. Any computation $U(t)$, also called a quantum circuit, can be approximated by sequential application of a finite set of unitary transformations that operate only on one or two qubits. An example for such a universal set of quantum logic gates are the unitary transformations given in Eqs. (4) [Hadamard gate], (5) [T gate], and (6) [controlled-

$$H_j = (|0_j\rangle\langle 0_j| + |0_j\rangle\langle 1_j| + |1_j\rangle\langle 0_j| - |1_j\rangle\langle 1_j|)/\sqrt{2} \qquad (4)$$

$$T_j = |0_j\rangle\langle 0_j| + e^{i\pi/4}|1_j\rangle\langle 1_j| \qquad (5)$$

$$C_{jk} = |0_j 0_k\rangle\langle 0_j 0_k| + |0_j 1_k\rangle\langle 0_j 1_k| \\ + |1_j 0_k\rangle\langle 1_j 1_k| + |1_j 1_k\rangle\langle 1_j 0_k| \qquad (6)$$

not gate], where $|n_j\rangle\langle n_j'|$ acts only on qubit j, and $|n_j n_k\rangle\langle n_j' n_k'|$ only on qubits j and k. They correspond to the logic gates in a classical computer, but are reversible, unlike the classical "and" and "exclusive or" gates. The Hadamard gate transforms the states $|0_j\rangle$ and $|1_j\rangle$ into the superpositions $(|0_j\rangle \pm |1_j\rangle)/\sqrt{2}$; the T gate shifts the phase of the excited state relative to the ground state by $\pi/4$; and the controlled-not gate flips the target qubit k if and only if the control qubit j is in the excited state. The three gates are the analog of an optical beam splitter, phase shifter (refractive medium), and conditional mirror. Only the Hadamard gate creates multiple computational paths; the other two transform a single basis state into a single basis state. To approximate a general $U(t)$ to within accuracy ε requires $O(4^N N^2 [\ln(4^N N^2/\varepsilon)]^\alpha)$ gates, where $\alpha \approx 2$ (this is the Solovay-Kitaev theorem); the leading factor 4^N is set by the number of matrix elements of $U(t)$. For special computations, however, often a much smaller number of gates, independent of ε, suffices.

Fast quantum algorithms. A widely used class of encryption systems for public transmission of sensitive data derive their security from the difficulty of factoring a large, publicly transmitted integer. The fastest known classical algorithm factors an N-bit number in time $O(2^{\text{const} \times N^{1/3}(\ln N)^{2/3}})$. Peter Shor's quantum algorithm for the same task requires only time $O(N^2(\ln N)(\ln \ln N))$. The key task in Shor's algorithm is to find the period of a function related to the number to be factored, and the algorithm does this using the Fourier transform: The values at which the transform does not vanish yield the period, similarly to the way in which Bragg peaks in x-ray diffraction yield the crystal periodicity. The quantum fast Fourier transform, with no approximation, requires $O(N^2)$ operations, which is much less than the $O(2^N N)$ operations in the classical fast Fourier transform and much less than the Solovay-Kitaev bound. The period is obtained from $O(\ln N)$ repeated measurements of the qubits of the state representing the Fourier transform.

Other quantum algorithms that outperform classical algorithms by orders of magnitude are Lov K. Grover's algorithm for "finding a needle in a haystack" [search of an item in a database of 2^N items in $O(2^{N/2})$ instead of $O(2^N)$ steps]; estimation of the median and mean of 2^N items to precision ε in $O(N/\varepsilon)$ instead of $O(N/\varepsilon^2)$ steps; search of the minimum of a function sampled at 2^N points in $O(2^{N/2})$ instead of $O(2^N)$ steps; search of two distinct pre-images giving the same image of a two-to-one function sampled at 2^N points, in $O(2^{N/3})$ instead of $O(2^{N/2})$ steps; the Deutsch-Jozsa algorithm to determine if 2^N numbers are either all 0 (a constant function), or half are 0 and half are 1 (a balanced function), in one instead of up to $2^{N-1} + 1$ steps; and various allocation tasks and game-theoretic strategies.

Quantum error correction. Noise from imperfect computer operation poses no fundamental barrier to large-scale computations. A quantum error-correction code encodes the N logical qubits into N' carrier qubits ($N' > N$), runs the carrier qubits through a group of accordingly encoded logic gates, transforms the noisy state by appropriate projection operators (syndrome measurements) and unitary operators (recovery of the original carrier qubits) into an error-corrected state, and feeds the state into the next group of gates. Such periodic error correction prevents accumulation of errors in the state (similar to the quantum Zeno effect). At the end of the computation, the carrier qubits are decoded. The encoding spreads the state of the N logical qubits over all N' carrier qubits so that, when the syndromes are measured, no information about the state of the logical qubits is revealed: The projections preserve superpositions of the logical qubits, and the state of the carrier qubits is highly entangled even if the logical-qubit state is not. Remarkably, a discrete set of corrections can correct a continuum of errors. For a code to correct any error on any M carrier qubits, a necessary condition is $N' \geq 4M + N$ (the Knill-Laflamme bound), and a sufficient condition for large N is given by Eq. (7) [the Gilbert-Varshamov bound]. For exam-

$$N/N' < 1 - 2[-x \log_2 x - (1 - x) \log_2(1 - x)]_{x = 2M/N'} \tag{7}$$

ple, a code exists which encodes one logical qubit ($N = 1$) into 5 carrier qubits ($N' = 5$) and corrects any error on any one carrier qubit ($M = 1$) with 16 pairs of syndrome measurements and recovery operations (equality in the Knill-Laflamme bound).

Different codes require different encoding of gates, and of interest are encodings for which an error in an input carrier qubit or in the gate operation propagates only to a small number of output carrier qubits. Specifically, an encoded gate is called fault-tolerant if a failure with probability p of any single component (for example, one of the 10 "wires" feeding two logical qubits, each encoded by 5 carrier qubits, into a controlled-not gate) introduces an error in two or more carrier qubits in any logical output qubit with probability cp^2 at most, with c a constant and p small. Such a gate, when followed by error correction with $M = 1$, yields an error-free output with probability $1 - cp^2$, that is, reduces the error probability from p to cp^2 if $p < 1/c$. Fault-tolerant Hadamard, T, and controlled-not gates exist with $c \approx 10^4$ and $d \approx 10^2$, where d is the number of operations on carrier qubits used to encode and error-correct the gate. By hierarchical fault-tolerant encoding of all gates, a computation involving L gates can be carried out to within accuracy ε using only $O(L[\ln(L/\varepsilon)]^{\log_2 d})$ operations on carrier qubits, if $p < 1/c$ (threshold theorem for quantum computation). Thus, if the noise in individual carrier qubits is low enough, $p < 1/c \approx 10^{-4}$, arbitrarily large computations can be performed because the overhead for error correction grows only polynomial-logarithmically with the size of the computation, L.

Experimental state of the art. A remarkable array of experimental realizations of quantum computing devices and implementations of algorithms have been achieved. Current records with respect to the number of qubits that can be controlled and prepared in well-defined states are a 7-qubit nuclear magnetic resonance device (NMR, three ^{1}H and four ^{13}C nuclei in *trans*-crotonic acid, each in its spin up or down state, with a total of twelve 2-qubit gates driven by radio-frequency pulses; see **illus.**); a 4-qubit ion-trap device (IT, four ^{9}Be$^+$ ions in a linear electromagnetic trap, each in one of two hyperfine Zeeman levels, driven by Raman transitions); and a 3-qubit device based on cavity quantum electrodynamics (CQED, three Rb atoms, each in one of two Rydberg states and coupled to a cavity mode with zero or one photon, driven by microwave pulses). Successful computations carried out include Grover's algorithm on a 2-qubit NMR device; simulation of the dynamics of quantum harmonic and anharmonic oscillators on a 2-qubit NMR device; the Deutsch-Jozsa algorithm on a 5-qubit NMR device; finding the order of a permutation on a 5-qubit NMR device; finding the period of a function, using the quantum Fourier transform, on a 3-qubit NMR device; correction of any one-qubit error on a 5-qubit NMR device; and protection of an ion-trap qubit from decoherence (two ^{9}Be$^+$ ions encoding one qubit).

General hardware requirements. Any experimental realization faces three challenges:

1. The device must be able to control the state of each qubit separately, while allowing neighboring qubits to interact with each other during the operation of two-qubit gates. Control of individual qubits is achieved by different chemical shifts in the NMR device, by laser beams driving ions spatially

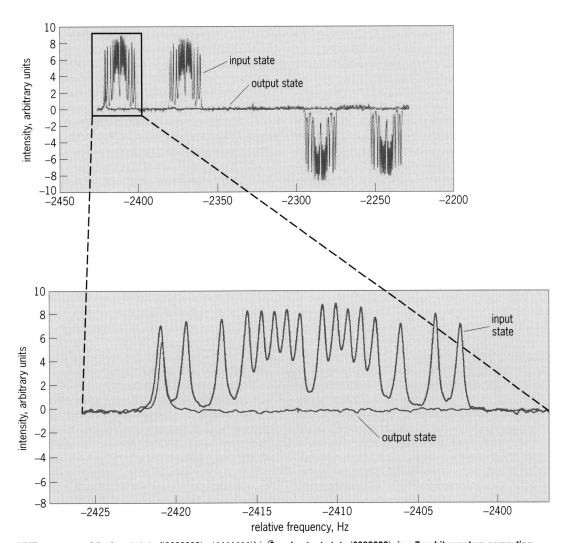

NMR spectrum of the input state $(|0000000\rangle + |1111111\rangle)/\sqrt{2}$ and output state $|0000000\rangle$, in a 7-qubit quantum computing experiment. (*After E. Knill et al., An algorithmic benchmark for quantum information processing, Nature, 404 (6776):368–370, 2000.*)

separated by tens of micrometers in the IT device, and by pulses addressing successive atoms traveling at spatial separation of several centimeters in the CQED device. Qubits interact via spin-spin coupling in the NMR device, via a shared phonon state of the ions (excitation of the center-of-mass mode) in the IT device, and via electric dipole coupling between each atom and the cavity mode in the CQED device.

2. The system must be switchable, so that interactions executing a prescribed sequence of gate operations [the hamiltonian $H(t)$] can be turned on and off by external control. Switching is done by magnetic field pulses in the NMR device, and by laser pulses in the IT and CQED devices.

3. The computer must be well isolated from the environment so that the decoherence time t_d, the time at which the computer and the environment depart significantly from a product state (entanglement of the computer with the environment) occurs, is long compared to t_g, the time it takes to operate a single gate. At time t_d, a generic qubit state $a|0_j\rangle + b|1_j\rangle$ will have degraded into the mixture

with the density matrix operator $|a|^2|0_j\rangle\langle 0_j| + |b|^2|1_j\rangle\langle 1_j|$, which no longer contains the interference terms necessary for quantum computation. Good isolation is provided by long spin-spin and spin-lattice relaxation times in the NMR device ($t_d = 10^{-2}$ to 10^8 s, $t_g = 10^{-6}$ to 10^{-3} s), long-lived hyperfine levels and stable trap and laser operation in the IT device ($t_d = 10^{-1}$ to 10^0 s, $t_g = 10^{-7}$ to 10^{-5} s), and the low spontaneous emission rate of Rydberg states and low photon escape rate from the cavity in the CQED device ($t_d = 10^{-3}$ to 10^0 s, $t_g = 10^{-5}$ to 10^{-4} s).

The condition $p < 1/c$ in the threshold theorem for quantum computation is much more demanding than $t_g < t_d$: Let $|\psi_i(t)\rangle$ be the state of the computer and environment interacting with each other, starting from a product state at $t = 0$; let $|\psi_n(t)\rangle$ be the same state, but noninteracting (product state at all times); and assume the overlap $F(t) = |\langle\psi_i(t)|\psi_n(t)\rangle|$ (called the fidelity) decays exponentially with time. The decay constant is the decoherence time, $F(t) = e^{-t/t_d}$, and the threshold condition requires $1 - e^{-2t_g/t_d} < 1/c$ [because $p = 1 - F^2(t_g)$], giving

$t_g < t_d/(2c)$ for large c. Thus the computer should remain isolated from the environment for about 10^4 gate operations for sustained computations. Present experimental devices are far from this goal.

Outlook. Numerous other approaches have been proposed and are under way for quantum computers with much longer decoherence times and control over many more qubits than currently available (a long-term goal is 200 qubits). These include solid-state NMR, electron spin resonance of donor atoms in semiconductors, electron spins in quantum dots (spintronics), electronic excitation in quantum dots, tunneling of single Cooper pairs across Josephson junctions, macroscopic persistent-current states (many Cooper pairs) in superconductors, magnetic resonance force microscopy, ion traps with hot ions, electrons trapped on superfluid helium films, neutral atoms trapped in an optical lattice, and linear optics (beam splitters, phase shifters, and single-photon sources). With so many new approaches under exploration, quantum computing as a technological tool holds a promising future.

For background information *see* BIT; CAVITY RESONATOR; CRYPTOGRAPHY; DENSITY MATRIX; DIGITAL COMPUTER; ELECTRON SPIN; FOURIER SERIES AND TRANSFORMS; INFORMATION THEORY; JOSEPHSON EFFECT; LOGIC CIRCUITS; NONRELATIVISTIC QUANTUM THEORY; NUCLEAR MAGNETIC RESONANCE; PARTICLE TRAP; QUANTIZED ELECTRONIC STRUCTURE (QUEST); QUANTUM ELECTRODYNAMICS; QUANTUM MECHANICS; RYDBERG ATOM; SUPERCOMPUTER; SUPERCONDUCTIVITY in the McGraw-Hill Encyclopedia of Science & Technology.   Peter Pfeifer

Bibliography. G. P. Berman et al., *Introduction to Quantum Computers*, World Scientific, Singapore, 1998; J. Gruska, *Quantum Computing*, McGraw-Hill, New York, 1999; C. Macchiavello, G. M. Palma, and A. Zeilinger (eds.), *Quantum Computation and Quantum Information Theory*, World Scientific, Singapore, 2001; M. A. Nielsen and I. L. Chuang, *Quantum Computation and Quantum Information*, Cambridge University Press, 2000; A. O. Pittenger, *An Introduction to Quantum Computing Algorithms*, Birkhäuser, Boston, 2000; C. P. Williams and S. H. Clearwater, *Explorations in Quantum Computing*, Springer, New York, 1997.

Quark-gluon matter

According to the established theory of the strong interaction, quantum chromodynamics (QCD), quarks and gluons are the fundamental building blocks from which atomic nuclei are assembled. The quark-gluon deep structure of nuclear matter is well concealed, however. Quarks and gluons are tightly packaged into protons and neutrons, which form the more conspicuous units. But theory predicts that under extreme conditions of high temperature or high density the physical nature of this matter changes drastically: Protons and neutrons dissolve. Quarks and gluons then become the obvious as well as the fundamental units of description. Physicists therefore speak of "quark-gluon matter."

The extremes of temperature and density, at which quark-gluon matter is encountered, are of considerable physical, as well as theoretical, interest. High-temperature quark-gluon matter was the dominant component of the universe during the early stages of the big bang. There is a vigorous international program to reproduce and study these conditions, on a small scale of course ("little bangs"), by colliding fast-moving heavy nuclei. High-density quark-gluon matter is important to another branch of astrophysics as well. During the collapse of massive stars, and deep in the interior of the neutron stars they often leave behind, tremendous pressures and densities occur. It is very likely that quark-gluon matter is produced under these conditions.

It is important to be clear about a possible source of confusion. In the context of quark-hadron matter, in speaking of high density a high baryon-number density is meant, which is to say a large excess of quarks over antiquarks per unit volume. This is what is obtained by compressing nuclear matter while keeping it cool. On the other hand, injecting a lot of energy into nuclear matter without compressing it results in a very large energy density and many quarks and antiquarks per unit volume, but only the initial (modest) excess of quarks over antiquarks. This matter is at high temperature, but not high density in our sense.

In general it is very difficult to solve the equations of quantum chromodynamics with useful accuracy. Fortunately, however, a reasonably simple, yet rigorous description of quark-gluon matter can be achieved by exploiting the asymptotic freedom property of quantum chromodynamics. Asymptotic freedom implies that events where quarks or gluons radiate or exchange large amounts of energy and momentum are rare. This property leads, in somewhat different ways, to crucial simplifications at high temperature and at high density.

High temperature. The simplest possible starting point for describing quark-gluon matter at high temperature is to imagine an ideal gas of these particles, ignoring their interactions altogether. Asymptotic freedom implies that this simplest possible description is not far from the truth. Indeed, at high temperatures the typical quarks, antiquarks, and gluons will be highly energetic, moving with large energy and momentum. When two such particles interact, either they exchange a large amount of energy-momentum—but this is rare, according to asymptotic freedom—or they are barely deflected. So including interactions does not drastically modify the free-particle distributions.

The consequences are dramatic. At relatively low temperatures (up to about 100 MeV, or 10^{12} K), hadronic matter (that is, matter consisting of hadrons, the strongly interacting particles) with zero baryon number density is well described as a dilute gas of π mesons, since these are the only particles whose mass is small enough to allow a significant

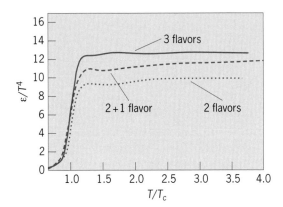

Results from the numerical integration of the equations of quantum chromodynamics at high temperature. The axes represent the energy density ε divided by the fourth power of temperature T, a quantity which is constant for weakly interacting massless particles in equilibrium; and the temperature. The rapid crossover from a small value of ε/T^4 to a very large one indicates the emergence of quark-gluon plasma at a critical temperature T_c, approximately 150 MeV. The curves indicate whether the strange quark mass is taken very heavy (2 flavors), very light (3 flavors), or intermediate (2 + 1 flavor) compared to T_c. (After F. Karsch, Lattice results on QCD thermodynamics, Los Alamos National Laboratory e-Print Archive (http://xxx.lanl.gov), hep-ph/0103314, 2001)

abundance. These represent three degrees of freedom, since there are three spin-0 particles (π^+, π^0, and π^-). But at high temperatures (about 200 MeV or more) a proper description involves quarks, antiquarks, and gluons. Since there are light u, d, and s quarks, each of which comes with two spin polarizations and three colors, and their antiquarks, plus eight gluons, each with two polarizations, the number of degrees of freedom explodes to $3 \times 2 \times 3 \times 2 + 8 \times 2 = 52$. This means that the specific heat becomes large: Energy pumped into the quark-gluon matter will be invested in creating new particles rather than increasing the temperature.

This signature of the profound reorganization of matter, from hadrons into quarks and gluons, has emerged clearly from numerical simulation of the theory (see **illus.**). Simple estimates lead to the expectation that fireballs produced at the Relativistic Heavy Ion Collider (RHIC) at Brookhaven National Laboratories on Long Island, and eventually at the Large Hadron Collider (LHC) at CERN, near Geneva, Switzerland, will achieve temperatures sufficient to produce quark-gluon matter. *See* LATTICE MODELS.

High density. The simplest possible, and until recently the traditional, starting point for describing quark-gluon matter at high density (and low temperature) is to ignore interactions. Recent work, however, has demonstrated that this is not correct.

A key insight is to recognize the profound analogy between a high density of quarks and a high density of electrons, as occurs in metals. Much is known about the latter, both theoretically and experimentally. At low temperatures, many metals exhibit the phenomenon of superconductivity. This represents a drastic change in their properties. Superconductivity would not occur in the absence of interactions, but

according to the accepted theory of John Bardeen, Leon N. Cooper, and John R. Schrieffer (BCS), it can be triggered by arbitrarily weak attractive interactions.

By adapting the methods of BCS theory to quantum chromodynamics, quark-gluon matter is found to undergo an analogous phase transition, called color superconductivity or color-flavor locking. In the new phase, quarks do not move independently, but are correlated with one another. Roughly speaking, they form a mobile superfluid of diquark pairs. This superfluid quickly responds to, and effectively neutralizes, all slow, low-energy-momentum color exchanges. What is left are only the high-energy-momentum color exchanges, which are rare according to asymptotic freedom. Thus many aspects of the physical behavior of quark-gluon matter at high density, in the color superconducting phase, can be predicted accurately from first principles, directly from the fundamental equations of quantum chromodynamics.

Again, the consequences are dramatic. For example, it is predicted that at high enough density matter will go into the so-called color-flavor-locked phase. In this phase it is a transparent insulator and expels all electrons. Prior to this analysis, it might have seemed highly counterintuitive that such ultradense matter, perhaps 10^{16} times denser than ordinary matter, would let light pass right through it.

Of course, astronomers are unlikely to get a chance to examine neutron stars optically any time soon. A frontier of current research is to deduce potentially observable astrophysical consequences from the new understanding of high-density quark-gluon matter.

In any case, the discovery of a new, highly nontrivial limit in which the equations of quantum chromodynamics can be solved has substantially advanced the understanding of this difficult theory. For example, the phenomena of quark confinement and of chiral symmetry breaking, which are notoriously difficult to prove at low density, are immediate consequences of the rigorous high-density theory.

For background information *see* GLUONS; NEUTRON STAR; PARTICLE ACCELERATOR; QUANTUM CHROMODYNAMICS; QUARK-GLUON PLASMA; QUARKS; SPECIFIC HEAT; SUPERCONDUCTIVITY in the McGraw-Hill Encyclopedia of Science & Technology.

Frank Wilczek

Bibliography. T.-P. Cheng and L.-F. Li, *Gauge Theory of Elementary Particle Physics*, Oxford, 1984; F. Wilczek, QCD made simple, *Phys. Today*, 53(8): 22–28, August 2000; F. Wilczek, Quantum field theory, *Rev. Mod. Phys.*, 71:S85–S95, 1999.

Retinoid receptors

Certain receptors mediate the actions of retinoids (natural and synthetic analogs of vitamin A) by regulating the transcription of retinoid-responsive genes in the cell nucleus. The rates of transcription of more than 500 genes are influenced by the binding of retinoids to retinoid receptors. These actions on gene transcription thereby confer on retinoid receptors the ability to regulate cell proliferation and differentiation and cell death. The retinoid receptors play an important role in the maintenance of normal growth and development; in the immune response; in male and female reproduction; in blood cell development; in the maintenance of healthy skin and bones; and ultimately in the general good health of the organism.

The biochemical details of how retinoid receptors regulate gene transcription have been worked out over the last decade and are well understood. The retinoid receptors recognize specific DNA sequences, called retinoid response elements, within the regulatory regions of genes. Through interactions with these response elements, and with coactivator or corepressor proteins present in the nucleus, the transcription of a responsive gene is either increased or diminished. In general, retinoid binding facilitates gene transcription by making the gene more accessible to enzymes and other factors responsible for the synthesis of RNA from the gene, whereas the absence of such binding lessens RNA synthesis.

Because retinoids are such potent regulators of cellular processes, there has been considerable interest in the use of retinoids to treat chronic diseases. This interest centers especially on disease states involving aberrant cell proliferation and differentiation.

Retinoids. The term "retinoid" refers to the class of chemical compounds, either natural or synthetic, and with or without the biological activity shown by vitamin A, that structurally resemble all-*trans*-retinol (vitamin A) [**Fig. 1a**]. The naturally occurring retinoids consist of vitamin A and its metabolites. The most abundant of the physiologically important vitamin A metabolites are all-*trans*-, 11-*cis*- and 9-*cis*-retinol, all-*trans*- and 11-*cis*-retinaldehyde, and all-*trans*- and 11-*cis*-retinyl esters. Less abundant, but still important physiologically, are the all-*trans*- (Fig. 1b), 13-*cis*-, and 9-*cis*- (Fig. 1c) isomers of retinoic acid. Many other less abundant naturally occurring retinoid metabolites have been identified in living cells or animals. Over 10,000 retinoids have been synthesized in the hope of developing effective pharmacological agents for use in the treatment of disease.

Receptor activation and distribution. The retinoid receptors act as ligand-dependent transcription factors. They mediate most retinoid-dependent actions (apart from the visual cycle). Upon binding to retinoic acid or other transcriptionally active retinoids, the nuclear receptors become activated and, in this form, they act to regulate the rate of transcription of retinoid-responsive genes. All-*trans*- and 9-*cis*-retinoic acid, stereoisomers of retinoic acid, have physiologically relevant roles in activating the retinoid receptors. Six different retinoid nuclear receptors have been identified, and each is the product of its own individual gene. Three of these, based on similarities in their protein structure, are classified as

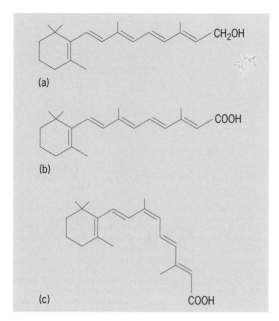

Fig. 1. Chemical structures for (*a*) all-*trans*-retinol (vitamin A), (*b*) all-*trans*-retinoic acid, and (*c*) 9-*cis*-retinoic acid. All retinoids have a chemical structure similar to that of all-*trans*-retinol. Usually this chemical structure consists of a cyclic tail (a cyclohexyl group in the case of natural retinoids), a polyene chain (either in *trans* or *cis* configurations), and a polar head group (an alcohol group in the case of retinols and a carboxylic acid in the case of retinoic acids).

retinoic acid receptors (RARs) and three are classified as retinoid X receptors (RXRs). The three distinct RAR subtypes are termed RAR-α, RAR-β, and RAR-γ; and the RXR subtypes are termed RXR-α, RXR-β, and RXR-γ. All-*trans*-retinoic acid activates members of the RAR family of nuclear receptors, and 9-*cis*-retinoic acid activates members of both the RAR and RXR family. Thus, different retinoid forms can activate differentially the RAR and RXR retinoid receptor families.

When activated, the retinoid receptors bind to response elements in retinoid-responsive genes, either to increase or to decrease the rate of gene transcription. The retinoid receptors bind these response elements as dimers, either as homodimers or more frequently as RAR-RXR heterodimers.

The different RAR and RXR species are distributed in distinct temporal and spatial patterns within tissues. Thus, not all tissues have the same complement of RARs or RXRs; different combinations of these RARs and RXRs are found in tissues and cells. It is not fully understood what significance these differences in RAR and RXR distribution have for mediating retinoid actions in the body. It is clear that almost all tissues and cells within the human body have one or more of the RARs or RXRs. This implies that retinoids play an important and diverse role in maintaining the health of all tissues within the body.

Regulation mechanism. The retinoid receptors are members of the steroid/thyroid/retinoid superfamily of ligand-dependent transcription factors. Each of the proteins that compose this superfamily

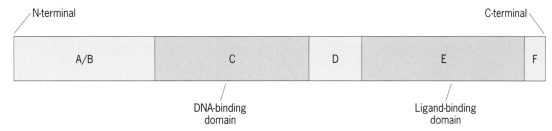

Fig. 2. Schematic of the structural and functional organization of retinoid receptors. The C-region of the receptor interacts with response element DNA [DNA-binding domain (DBD)], and the E-region, with the retinoid ligand [ligand-binding domain (LBD)]; both are highly conserved evolutionarily. All members of the steroid/thyroid/retinoid superfamily of nuclear receptors have the same structure-function organization as illustrated here. This conserved structure-function relationship is central for identifying and classifying nuclear receptors to this superfamily. The A/B-, D-, and F-regions are divergent and not conserved evolutionarily across the members of the steroid/thyroid/retinoid superfamily. These regions do, however, play important roles in mediating the functions of the retinoid receptors and other superfamily members. Modifications to these nonconserved regions can have a significant impact on the transcription-modulating activity of the retinoid receptor.

shares certain common structural features, including a ligand-binding domain and a DNA-binding domain (**Fig. 2**). Each member of this superfamily recognizes a specific natural ligand (or ligands), and in response to the availability of this ligand acts to influence either positively or negatively the expression of genes that are responsive to the ligand. Over 50 members of the steroid/thyroid/retinoid superfamily have been identified. These include nuclear receptors for estrogens, androgens, glucocorticoids, mineralocorticoids, thyroid hormone (T_3), and vitamin D. Thus, all of these hormones or vitamins act similarly to regulate gene activity through their own specific nuclear receptor. Some representative members of the superfamily and their cognate ligands are listed in **Table 1**. For some members, natural activating ligands have not yet been identified, and consequently these receptors are collectively referred to as orphan receptors.

Upon binding to a cognate response element, the retinoid receptors must interact with other nuclear proteins that act as either coactivators or corepressors of transcription. These interactions are essential for regulating transcriptional rates of retinoid responsive genes. These interactions influence directly the acetylation or deacetylation of histone proteins. The binding of retinoids to their receptors enables the receptors to interact with coactivators that have histone acetyltransferase activity (**Fig. 3**). This results in the local acetylation of histone proteins and their dissociation from the gene and an opening up of chromatin structure. This allows the transcriptional machinery access to the DNA and results in enhanced rates of gene expression. In the absence of retinoid, unliganded retinoid receptors can interact with corepressors that have histone deacetylase activity (**Fig. 4**). This results in the local deacetylation of histones and a closure of chromatin structure, thus lessening the activity of the transcriptional machinery for that gene. Consequently, corepressor binding by retinoid receptors serves to diminish transcription rates.

Although some members of the steroid/thyroid/retinoid superfamily (for instance, the steroid

TABLE 1. Some members of the steroid/thyroid/retinoid superfamily of nuclear receptors and their activating ligands

Nuclear receptor	Activating ligand
Retinoid X receptor (RXR)	9-*cis*-retinoic acid
Retinoic acid receptor (RAR)	All-*trans*-retinoic acid
Vitamin D receptor (VDR)	1,25-dihydroxy vitamin D
Thyroid hormone receptor (TR)	Triiodothyronine (T_3)
Estrogen receptor (ER)	Estrogen
Progesterone receptor (PR)	Progesterone
Glucocorticoid receptor (GR)	Cortisol
Mineralocorticoid receptor (MR)	Aldosterone
Androgen receptor (AR)	Testosterone
Lipid X receptor (LXR)	Cholesterol metabolite(s)
Peroxisome proliferated-activated receptor (PPAR)	Fatty acids or metabolite(s)

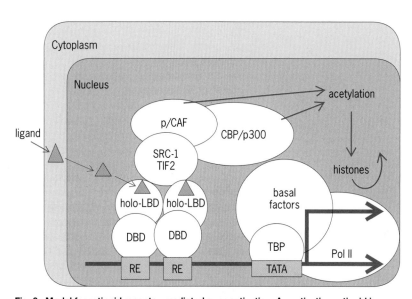

Fig. 3. Model for retinoid receptor–mediated gene activation. An activating retinoid is available to bind the LBD, yielding a holo-LBD form. This displaces corepressors that may have bound the receptor in the absence of ligand, and facilitates binding of coactivators (p/CAF, SRC-1, TIF2, and CBP/p300 in this example). These bound coactivators catalyze the local acetylation of histones associated with the DNA composing the gene. This has the effect of releasing the histones from the DNA and making the gene more accessible to the basal transcriptional machinery (the basal factors, TBP and Pol II). This has the overall effect of enhancing the actions of this gene product within the cell.

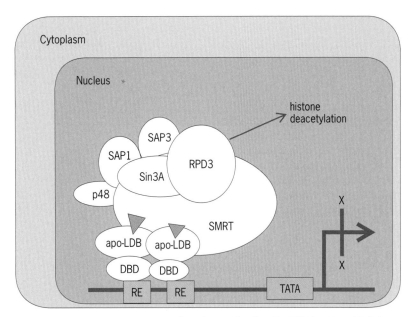

Fig. 4. Model for retinoid receptor-mediated gene silencing. The DBD is able to bind the retinoid response element (RE). In the absence of retinoid, the LBD is unoccupied or in the apo-LBD form. This allows corepressors (p48, Sin3A, SAP1, SAP3, RPD3, and SMRT) to bind to the retinoid receptor and to catalyze local deacetylation of histones. As a result, the histones bind the DNA more tightly, thus preventing the basal transcription machinery from efficiently transcribing the gene. This has the overall effect of lessening the actions of this gene product within the cell.

regulators of several hormone response systems in the body. These RXR actions upon heterodimerization to other members of the steroid/thyroid/retinoid superfamily involve mechanisms analogous to those described above for the activation of retinoid-responsive genes.

Physiologic effects. The actions that retinoids and their receptors have in regulating gene activity touch on many essential physiologic processes within the body. Living organisms require retinoids to maintain normal growth and differentiation of the embryo, to maintain epithelial linings throughout the body, to assure normal blood cell development, to maintain reproduction in both the male and the female, to maintain an immune response, to maintain healthy skin, to maintain normal brain function, and to assure proper bone growth and remodeling. At the cellular level, retinoids are needed to maintain a balance between cellular differentiation and proliferation and cell death.

A review of even a partial listing of the 500 genes reported to be responsive to retinoids provides insight into the diversity of responses regulated by retinoid receptors and their importance in maintaining the health of living organisms (**Table 2**). This importance is further underscored by targeted gene disruptions (gene knockouts) in mice where one or more of the genes encoding RARs or RXRs have been deleted. Single RAR or RXR gene knockouts have resulted in mice that show severe developmental or functional impairments in organ systems such as the cardiovascular system, the central nervous system, the reproductive system, and the gastrointestinal system. Disruption of combinations of two RAR and/or RXR genes results in embryonic lethality. Presumably, the absence of marked embryonic lethality for single-gene knockouts arises from a functional redundancy in the actions of the RARs and/or RXRs.

Clinical applications. Because of the large number of diverse genes that are regulated by retinoids through retinoid receptors, there has been considerable interest in the use, or potential use, of retinoids for treating disease. At present, both naturally occurring and synthetic retinoids are used clinically to treat skin disease. Among the retinoids commonly used in clinical dermatology are 13-*cis*-retinoic acid and all-*trans*-retinoic acid. Natural and synthetic

receptors) act solely as homodimers, the RARs and RXRs modulate retinoic acid–responsive transcription either as homodimers or as heterodimers. It is established that the RXR subtypes of retinoid receptors play important roles in helping some other members of the superfamily to regulate gene activity. The RXRs must serve as a partner in such a receptor pair if the pair is to recognize a specific response element within a gene. The RARs, the two thyroid hormone receptors (TR), the vitamin D receptor (VDR), and the three peroxisome proliferated-activator receptors (PPAR) heterodimerize with an RXR molecule in order to regulate their respective responsive genes. This implies that retinoids, through the actions of the RXR nuclear receptors, are needed to assure activation of genes by thyroid hormone, vitamin D, and some essential fatty acid metabolites for these hormones or substances to be active within the body. Thus, retinoids and the RXRs can act as master

TABLE 2. Examples of genes regulated by retinoid receptors	
Substance encoded in gene	Action
Oxytocin	Needed for reproduction
Growth hormone	Needed to maintain growth
Phosphoenol pyruvate carboxykinase	Involved in glucose metabolism
Class I alcohol dehydrogenase	Needed for alcohol oxidation
Tissue transglutaminase	Needed to regulate cell growth and death
Laminin B1	Needed to facilitate cell-cell interactions
Matrix gla-protein	Needed for bone formation
Several keratin genes	Needed for healthy skin formation
Cellular retinol-binding protein, type I	Active in retinoid metabolism
RAR-β	Affects retinoid action
Cytochrome 26	Catalyzes catabolism of all-*trans*-retinoic acid
Several hox genes	Needed to control embryologic development
Dopamine D2 receptor	Needed in central nervous system function

retinoids have been used effectively in the treatment of skin disorders such as severe cystic acne, psoriasis, many cutaneous disorders of keratinization, several dermatoses, acne vulgaris, and the damaging effects of sun exposure. Although retinoids are highly effective for use in clinical dermatology, their systemic use is associated with a high incidence of birth defects (teratogenic effects), and they must never be taken by pregnant women or women considering pregnancy.

Since retinoids have an important role in regulating cellular proliferation and differentiation, there has long been interest in their use as cancer chemopreventive and chemotherapeutic agents. The most successful use of retinoids as cancer therapeutic agents has been for treatment of some forms of acute promyelocytic leukemia (APL). Based on the observation that all-*trans*-retinoic acid treatment can induce the differentiation of cultured malignant cells harboring this chromosomal translocation to a normal phenotype, all-*trans*-retinoic acid differentiation trials in APL patients were initiated. Complete remission rates of about 90% were observed when patients were treated daily with all-*trans*-retinoic acid. Unfortunately, within a few months after achievement of complete remission almost all patients developed a resistance to all-*trans*-retinoic acid and relapsed. This resistance arises through the induction of the gene for cytochrome 26 (and possibly other genes), an enzyme that rapidly catabolizes the large doses of all-*trans*-retinoic acid being administered to the patients. These findings prompted clinicians to administer to APL patients treatments combining all-*trans*-retinoic acid and intensive chemotherapy.

Retinoids, administered either as all-*trans*- or 13-*cis*-retinoic acid, have been reported to have clinical activity in cancer therapy and prevention in several other settings. These include treatment of the premalignant lesions of oral leukoplakia, actinic keratosis, cervical dysplasia, and xeroderma pigmentosum. Reductions in second hepatocellular and aerodigestive tract cancers are reported to follow retinoid treatments. Some established malignancies, including juvenile chronic myelogenous leukemia and mycosis fungoides, are reported to respond to retinoid-based therapy. These studies suggest that in the future retinoids may acquire relatively widespread usage for preventing and treating cancer. This, however, will likely require the development of new and effective synthetic retinoids that lack the toxicity of natural retinoids and do not induce their own catabolism.

Administration of all-*trans*-retinoic acid to a rat model used to study emphysema experimentally was found to induce differentiation of new lung tissue, thus lessening the severity of the disease in this animal. Based on these data, a clinical trial of all-*trans*- and 13-*cis*-retinoic acid as possible therapeutic agents in human emphysema patients has been initiated.

Much recent work in animal models suggests that the pharmacological use of retinoids to target retinoid receptor action will be of benefit to patients with cardiovascular disease and with type II diabetes. Much of this interest is focused on the potential use of RXR-specific agonists and antagonists (rexinoids) to influence the actions of RXRs but not RARs. This is because RXRs must heterodimerize with PPARs if PPAR-responsive genes are to be activated within cells. It is well established that PPAR-responsive genes are important in the causation and prevention of both cardiovascular disease and type II diabetes. Many drugs that are presently used to treat cardiovascular disease and type II diabetes are ligands for the PPARs and act by influencing genes regulated through PPAR response elements. Since PPAR-responsive genes must be regulated by PPAR-RXR heterodimers, it is hypothesized that retinoids which influence RXR actions also might have an effect on PPAR-responsive gene activity. Several studies exploring this hypothesis have been carried out in rat and mouse models for human disease. Data from these studies indicate that rexinoids have efficacy for lowering plasma lipid levels, for preventing restenosis injury, and for increasing insulin responsiveness. Clinical trials to demonstrate rexinoid efficacy in human patients are being initiated.

The interest in retinoids as potential drugs for use in treatment of disease and in the retinoid receptors as targets for these drugs is based on the understanding that retinoids are potent regulators of cell proliferation and differentiation and cell death. However, because retinoids are such potent modulators of cellular actions and because they influence many diverse physiologic actions, retinoids can have many undesirable toxic effects when administered in pharmacologically effective doses. The toxicity of retinoids is the major factor that impedes their widespread use for treating disease. Nevertheless, there is a large body of clinical research focused on using retinoids pharmacologically to treat or to prevent chronic disease.

William S. Blaner

Bibliography. P. Chambon, A decade of molecular biology of retinoic acid receptors, *FASEB J.*, 10:940–954, 1996; IARC Working Group on the Evaluation of Cancer Preventive Agents, *IARC Handbooks of Cancer Prevention*, vol. 3: *Vitamin A*, World Health Organization, International Agency for Research on Cancer, Lyon, 1998; IARC Working Group on the Evaluation of Cancer Preventive Agents, *IARC Handbooks of Cancer Prevention*, vol. 4: *Retinoic Acid*, World Health Organization, International Agency for Research on Cancer, Lyon, 2000; M. A. Livrea (ed.), *Vitamin A and Retinoids: An Update of Biological Aspects and Clinical Applications*, Birkhauser Publishing, Basel, 1999; H. Nau and W. S. Blaner (eds.), *Retinoids: The Biochemical and Molecular Basis of Vitamin A and Retinoid Action*, vol. 139 of *Handbook of Experimental Pharmacology*, Springer-Verlag, Heidelberg, 1999; M. B. Sporn, A. B. Roberts, and D. S. Goodman (eds.), *The Retinoids*, Academic Press, New York, 1984; M. B. Sporn, A. B. Roberts, and D. S. Goodman (eds.), *The Retinoids: Biology, Chemistry, and Medicine*, 2d ed., Raven Press, New York, 1994.

Rotifera

The phylum Rotifera consists largely of free-living animals less than 1 mm long that are common in aquatic ecosystems throughout the world. Although most rotifers have fewer than 1000 cells, they have muscles, ganglia, sensory organs, structures for feeding and swimming, and digestive and secretory organs. Rotifers are characterized by a ciliated head structure, the corona, used for locomotion and food gathering, and a muscular pharynx, the mastax, used to process food. All rotifers have a syncytial epidermis with an intracytoplasmic lamina that is unique among metazoans. Rotifers employ a wide variety of reproductive modes: some species reproduce only sexually, some alternate between sexual and asexual reproduction, and others reproduce only asexually. The successful evolution of asexuality in rotifers renders them important model organisms in studying why sexual reproduction is the dominant form of reproduction in animals.

Phylogeny. The phylogeny of the Rotifera and their closest relatives has recently undergone significant revision. The phylum has historically been composed of three classes: Seisonidea, Monogononta, and Bdelloidea. Evidence from recent molecular phylogenetic and ultrastructural studies strongly suggests that a fourth group, Acanthocephala—long regarded as a separate but closely related phylum—is more related to some rotifer classes than to others and is therefore best regarded as a superclass within the phylum Rotifera.

The superclass Acanthocephala is composed of more than 1000 described species divided into three or four classes and more than 25 families. All acanthocephalans are obligate endoparasites of vertebrates and arthropods and lack the corona and mastax typical of other rotifers. There are only two described species in Class Seisonidea; both are found on the legs and carapace of individuals of the crustacean genus *Nebalia*, and have a simple corona with minimal ciliation and a specialized mastax. Class Monogononta is the largest rotifer group, with considerable morphological variation among the more than

1600 described species classified into three orders and more than 20 families. There are more than 360 described species in Class Bdelloidea, classified in three orders and four families. Bdelloidea and Monogononta together form the superclass Eurotatoria; these rotifers possess the corona, mastax, and overall form typical of rotifers.

Morphological evidence suggests that the closest relatives of rotifers are the phylum Gnathostomulida and a newly discovered group, the Micrognathozoa (currently defined by only one described species, *Limnognathia maerski*). Some biologists believe that gnathostomulids, the micrognathozoan, and the seisonid, acanthocephalan, and eurotatorian rotifers should be considered a single phylum, termed Gnathifera, united by the specialized jaws common to all these groups. These jaw-bearing taxa may be loosely allied with the phyla Gastrotricha and Platyhelminthes, forming a superphylum assemblage that has been termed Platyzoa. Platyzoa, in turn, may be a sistergroup to a collection of phyla called the Lophotrochozoa, or it may be the most basal branch in the evolution of triploblast metazoans (**Fig. 1**). Molecular phylogenetic studies have only begun to focus on the relationships between the platyzoan phyla and their relationship to other metazoans. Rotifers, as the best studied and easiest to examine of this group, will play an important role in determining the validity of Platyzoa and its placement within the Metazoa.

Reproduction and development. Species of seisonid and acanthocephalan rotifers are gonochoristic; that is, offspring are produced by the union (syngamy) of haploid eggs and haploid sperm from diploid females and males (**Fig. 2a**). Females and males are of similar size. Species of monogonont rotifers usually reproduce by parthenogenesis, with diploid females producing diploid female offspring by mitotic division of oocytes (Fig. 2b). Oogenesis occurs through a single mitotic division of the oocyte mother cell, giving rise to the egg and a polar body. Under certain environmental conditions, parthenogenetic females produce female offspring capable of producing haploid eggs by meiosis. If unfertilized, these haploid eggs develop into haploid males. In most monogonont species, males are much smaller than females, and may be so highly reduced as to lack a digestive system. If a male fertilizes a haploid egg, the egg develops into a diploid parthenogenetic female. This type of sexual reproduction is called arrhenotoky, and the reproductive cycle is generally termed cyclical parthenogenesis. Parthenogenesis appears to be the only means of reproduction used by any species in the class Bdelloidea; no sexual cycle has ever been observed in bdelloids (Fig. 2c). Oogenesis occurs through two mitotic divisions of the oocyte mother cell, giving rise to the egg and two polar bodies. There is no pairing of chromosomes or reduction of chromosome number.

Development in rotifers is eutelic (cells divide in a predetermined sequence during embryogenesis without further cell division during the life of the

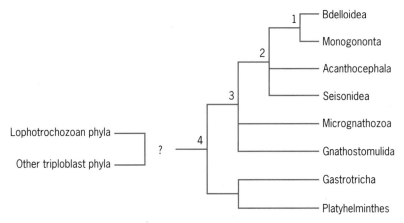

Fig. 1. Phylogenetic relationships of rotifer groups and their relatives. 1. Eurotatoria, 2. Rotifera, 3. Gnathifera, 4. Platyzoa.

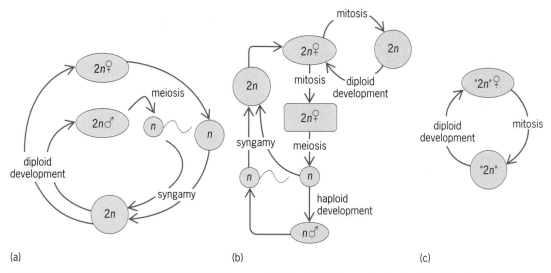

Fig. 2. Reproductive modes used by different rotifer groups. (*a*) Gonochoristic reproduction in seisonid and acanthocephalan rotifers. (*b*) Parthenogenesis alternating with sexual reproduction in monogononts. (*c*) Obligate parthenogenesis in bdelloids. 2*n* represents diploidy, *n*, haploidy; bdelloids are designated "2*n*" because the concept of diploidy loses meaning in the absence of meiosis.

rotifer) and, in most rotifers, direct (newly hatched rotifers grow in size to adults without significant morphological change). All acanthocephalans and some monogononts have a juvenile stage that is morphologically distinct from the adult form. The eggs of acanthocephalans remain viable for many months, even in hot dry soil or in ice. The fertilized diploid eggs of monogononts can survive desiccation and hatch many years later. Many species of bdelloid rotifers are capable of surviving desiccation as adults, a process called anhydrobiosis. As the medium around them evaporates, bdelloids contract and exude nearly all water from their bodies. In the anhydrobiotic state, they are metabolically inactive and are able to withstand vacuum and extreme temperatures. When rehydrated, they revive and resume their normal lifespan (although in some species there can be considerable mortality). The ability to undergo anhydrobiosis allows bdelloid rotifers to survive in ephemerally aquatic and other marginally habitable environments, such as mosses or temporary pools.

Ancient asexuality of bdelloid rotifers. Although asexual reproduction arises relatively frequently in some animal and plant groups, fewer than 1 in 1000 described species is obligately parthenogenetic. Most such species have arisen only recently and rarely constitute entire genera or families. This has led to a general observation that parthenogenesis is a "dead end" in evolution. Class Bdelloidea, the largest animal taxon in which males and sexual reproduction are unknown, appears to be a drastic exception to this rule and has been called an "evolutionary scandal" by the evolutionary biologist John Maynard Smith.

Bdelloid rotifers were observed by Antony van Leeuwenhoek in the late 1600s and have been favorite subjects of amateur and professional naturalists ever since. The absence of male bdelloids was remarked upon by Charles Hudson in his 1886 treatise on rotifers. Since that time, the lack of males, meiosis, or hermaphroditic structures in bdelloids has become widely accepted, but this evidence for the lack of sexual recombination is circumstantial, particularly as sexual cycles have been found in other putatively asexual taxa. Recently, positive evidence for the parthenogenetic evolution of bdelloid rotifers has come from comparative molecular studies of the genomes of bdelloids and rotifers known to reproduce sexually. The expectation was that an individual of a species that evolved without meiosis from a diploid sexual ancestor will have a different pattern of nucleotide divergence in its genome than an individual of a species that evolved with sexual reproduction.

In all species, nucleotide differences between alleles arise at a relatively constant rate from mutation. Most of these mutations will be deleterious and will be eliminated from the population by natural selection. Some will have no effect on fitness; alleles can therefore continuously accrue these neutral mutations. However, in sexual populations alleles segregate in meiosis and, due to random genetic drift, are continuously lost. This minimizes the amount of nucleotide divergence that can accumulate between the alleles found in any individual genome. In all sexual species that have been studied, including nonbdelloid rotifers, heterozygosity at the nucleotide level is low, approximately 1%. However, if a species abandons sexual reproduction and evolves parthenogenetically, neutral mutations will accumulate because segregation and allelic drift no longer occur (if the species also lacks processes that can result in homozygosis in the germline, such as gene conversion or mitotic crossing-over). As parthenogenetic evolution continues, neutral sequence divergence between formerly allelic sequences will accumulate proportionally to the amount of time since sex was abandoned.

In keeping with this prediction, the genomes of individual bdelloid rotifers do not contain gene copies as similar to each other as the alleles in rotifers known to reproduce sexually. Instead, for every gene that has been examined, bdelloid genomes contain at least two copies that are highly divergent. In addition, the pattern of nucleotide differences between the copies indicates that different bdelloid species inherited pairs of formerly allelic sequences, as would be expected if the species descended from a common parthenogenetic ancestor. Comparison of the amount of accumulated nucleotide divergence between these formerly allelic sequences and sequences of the same gene in species of vertebrates or arthropods for which there is a fossil record indicates that sexual reproduction may have been abandoned in the evolution of Bdelloidea approximately 100 million years ago.

In addition to the high divergence between formerly allelic sequences and their inheritance in multiple bdelloid species, it has recently been shown that the genomes of bdelloid species, unlike all other animal taxa examined, lack retrotransposons—transposable genetic elements that multiply via an RNA intermediate. This is consistent with the theory that, because they are a major source of deleterious mutation and are transmitted only sexually, no long-term parthenogenetic lineage should possess retrotransposons.

For background information *see* BDELLOIDEA; CROSSING-OVER (GENETICS); ORGANIC EVOLUTION; RECOMBINATION (GENETICS); REPRODUCTION (ANIMAL); ROTIFERA in the McGraw-Hill Encyclopedia of Science & Technology.   David B. Mark Welch

Bibliography. G. Bell, *The Masterpiece of Nature: The Evolution and Genetics of Sexuality*, University of California Press, Berkeley, 1982; D. B. Mark Welch and M. Meselson, Evidence for the evolution of bdelloid rotifers without sexual reproduction or genetic exchange, *Science*, 288:1211–1215, 2000; J. Maynard Smith, Contemplating life without sex, *Nature*, 324:300–301, 1986; E. Mayr, *Animal Species and Evolution*, Harvard University Press, Cambridge, MA, 1963; R. L. Wallace and T. W. Snell, Rotifera, in J. H. Thorp and A. P. Covich (eds.), *Ecology and Classification of North American Freshwater Invertebrates*, Academic Press, San Diego, 1991; M. J. D. White, *Modes of Speciation*, Freeman Press, San Francisco, 1978.

Smart structures

A high-performance fighter aircraft is a self-contained unit with its own energy source in the form of the fuel. However, a pilot is normally needed to take off, perform maneuvers, and land the aircraft. A pilotless vehicle that could automatically perform these tasks would be a smart aircraft. For an example of a smart structure, consider that the fighter aircraft is capable of flying and performing maneuvers at subsonic, transonic, and supersonic speeds, and different airfoil cross sections or shapes of the wing are more efficient at different speed regimes. The wing structure is a smart structure if any of the following can be achieved:

1. The shape of the airfoil can be changed while the aircraft is flying to optimize for different speeds.

2. If the aircraft encounters a gust, the shape of the wing can be changed and active damping can be added to alleviate the effect of the gust.

3. If the wing suffers damage or develops a crack, it can automatically detect the damage and repair itself.

4. The aircraft can be made more agile in performing selected maneuvers by changing the wing shape as needed, and damping can be reduced as needed.

5. The active or real-time shape change can be used to enhance the flight envelope by delaying aeroelastic instabilities and alleviating buffet-induced vibrations.

In other words, a smart structure should respond automatically to a changing external environment (like the gust), a need (like an agile maneuver), or a changing internal structural configuration (like damage or a failure). The automatic response (in the form of shape changes, active damping, or damage repairs) is to be achieved by the action of built-in actuators that receive their commands from controllers or microprocessors in response to the needs detected by sensors. The controllers reside in computers or microprocessors. However, the actual forces and displacements, which are needed for shape changes, vibration control, noise control, and flap deflections, are delivered by actuators. At present, some usable sensors and controllers are available.

Thus, the most important component of smart structures that needs further development is the actuator. The actuator should be an integral part of the structure and must be capable of blending into the structure in a seamless fashion. The actuator should respond quickly, should not be heavy, and should consume very little energy. These requirements eliminate conventional actuators such as bulky hydraulic actuators and electric motors. A need for a new class of actuators has stimulated a significant amount of research that seeks to identify actuator materials, and to develop and design procedures for fabricating actuator assemblies. Currently, the smart actuator materials that are being considered include piezoelectric materials, magnetostrictive materials, and shape memory alloys. Other candidate materials include electrostrictive materials and electrorheological fluids.

Smart materials. The direct piezoelectric effect is the creation of an electric charge by an applied stress to a solid; it was discovered in 1880. The converse piezoelectric effect, the production of a displacement (or strain) by an applied electric field, is important for actuators. The piezoelectric effect was observed only in single crystals until 1940, when it was discovered in poled polycrystalline ceramics. The term "poling" means application of a strong electric field to the prepared ceramic material, such as lead zirconate titanate. At present, piezoceramic materials are very desirable candidate actuator materials. A piezoceramic actuator was first used, in 1978, to control the vibrations in an airborne laser system. Researchers have recently found that piezoceramic actuators can deliver large blocked forces (that is, large forces are created when they are prevented or blocked from expanding), and have a high frequency response but small displacements. Yet, these actuators have been successfully used in many applications. The current challenge is to convert the work done by these large forces at high frequencies in order to deliver usable energy (control authority) at low frequencies, large displacements, and moderate forces.

Shape memory alloys are materials that have the ability to return to a previously specified shape and size when exposed to a thermal field. The shape memory effect is observed in metallic materials like the nickel-titanium alloys and some copper-base alloys. When such a material is first plastically deformed at a low temperature and then exposed to a high temperature, the material will return to its shape prior to the deformation. This is known as one-way memory. Some materials can exhibit shape changes during cooling, and are known as two-way memory materials. To produce an actuator, the material is constrained in the desired shape and heated to approximately 500°C (900°F) for a specified period of time, typically 2 to 5 minutes. Upon cooling, the material returns to an undeformed shape. When reheated to a much lower temperature, the material returns to the specified shape. These materials are strong and can actuate large strains or displacements. However, the frequency or time response is very slow. This makes it difficult to use them in vibration and noise control.

Some materials, such as terfenol-d, display magnetostriction. Such materials experience a change of shape when subjected to a magnetic field. In practical applications, the magnetic field is produced by an electric field. These materials cannot deliver the high frequencies offered by piezoceramics, but offer a larger strain than piezoceramics.

In some applications a magnetic field may not be desirable. Electrostrictive materials, which are known as relaxor ferroelectrics, produce strains (comparable to piezoceramics) that vary nonlinearly with the electric field. The properties, however, are very sensitive to the temperature. Electrorheological fluids consist of suspensions of selected fine particles in an insulating fluid. The application of an electric field changes the viscous properties of the fluid. Their applications in aerospace structures are still being explored.

Applications. Many applications of smart structures are being pursued. Possible applications in the aerospace field include vibration control, noise control, structural health monitoring, control of aeroelastic instabilities, and the development of an active flexible wing to increase the flight envelope, to improve the maneuvering capability, and to control the

Fig. 1. Buffet in an aircraft.

aerospace vehicle. Some of the sought applications involve helicopters. One of the applications will be summarized to explain the procedure for the development of a smart structure.

Fighter aircraft often execute maneuvers at high angles of attack (the angle between the wing and the flight path). Then vortices are created at the leading edges of the wings. Because of the geometry of the wing, the fuselage, and the forward speed, these vortices can produce oscillatory loads on the vertical tails. This behavior, called buffet, has been known since 1930 (**Fig. 1**). It can cause undesirable oscillations, and fatigue damage, and can limit the flight envelope.

Smart structures can be designed to suppress these buffet-induced vibrations. First, an actuator with sufficient control authority was needed. Such a device was developed in the form of an offset piezoceramic stack actuator (**Fig. 2**). The stack is attached to the structure that is being controlled and generates forces when it is partially blocked from expanding. It is mounted at a selected offset distance from the structure in order to produce from the generated forces the bending and torsional moments on the structure that are needed to control the structure effectively. The next steps were to integrate this actuator into the vertical tail structure, select a

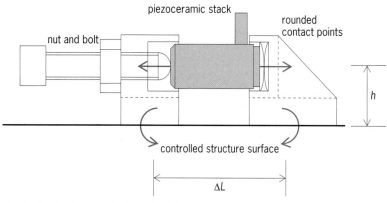

Fig. 2. Structural assembly of an offset piezoceramic stack actuator. The nut and bolt provides mechanical decompression. Contact points are rounded to avoid tensile loads. h = offset, ΔL = elongation.

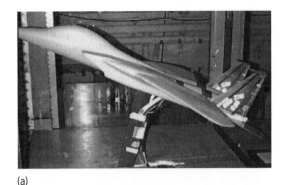

(a)

(b)

Fig. 3. Wind tunnel model built to demonstrate the effectiveness of a smart structure in reducing buffet-induced vibrations. (*a*) Side view. (*b*) Tail with smart structural elements.

sensor design, and implement a controller that resided in a computer. An accelerometer was selected as a sensor, and controllers were designed, based on the principles of acceleration feedback controllers. To demonstrate the effectiveness of the de-

signed smart structure, a scale model of the selected aircraft was built with flexible tail surfaces. The actuator, sensor, and controller were integrated into the vertical tail structure of the model (**Fig. 3**). The controllers were implemented by using digital signal processors and by writing the needed software. The closed-loop model, with smart structural elements, achieved a significant reduction in buffet-induced vibrations (**Fig. 4**).

Many applications of smart structures are also being pursued in fields other than aerospace. These include automotive systems, civil structures, machine tools, and bioengineering structures. Another interesting application is the use of the smart-structure concept to improve the acoustical qualities of musical instruments like the guitar and the violin.

For background information *see* AEROELASTICITY; ELECTROSTRICTION; FLIGHT CHARACTERISTICS; FLIGHT CONTROLS; MAGNETOSTRICTION; PIEZOELECTRICITY; SHAPE MEMORY ALLOYS in the McGraw-Hill Encyclopedia of Science & Technology.

Sathyanaraya Hanagud

Bibliography. I. Chopra, Review of current status of smart structures and integrated systems, *Proceedings of the SPIE Conference on Smart Structures*, SPIE vol. 2717, pp. 20–62, 1996; S. Hanagud et al., Model based simulation of buffet-induced vibration control of a F/A 18 vertical stabilzer, *Proceedings of the AIAA SDM Conference*, AIAA Pap. 2001-1352, pages 1352-01 to 1352-10, 2001; S. Hanagud et al., Tail buffet alleviation of high performance twin tail aircraft using piezo-stack actuators, *Collected Technical Papers of the AIAA Structural Dynamics and Materials Conference*, AIAA-99-1320, 1999; S. Hanagud, M. B. Obal, and M. Meyyappa, Electronic damping techniques and active vibration control, *Proceedings of the AIAA SDM Conference*, AIAA Pap. 85-0752, pp. 443–453, 1985.

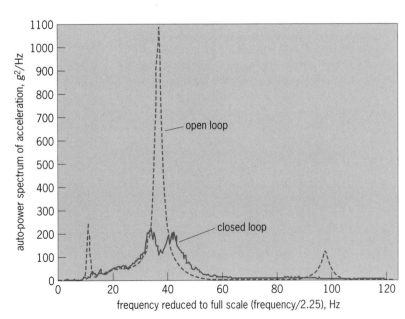

Fig. 4. Comparison of vibration levels of open-loop (without smart structure) and closed-loop (with smart structure) wind tunnel models at $20°$ angle of attack.

Soft computing

The family of emergent methods that imitate human intelligence have formed a new subject area called soft computing. Since the development of artificial intelligence, there have been many attempts to create devices that mimic intelligent behavior. The goal is to make tools provided with some humanlike capabilities (such as learning, reasoning, and decision making) to solve complex problems, not to build machines with the capacity to compete with humans. Soft computing replaces the traditional time-consuming and complex techniques of programming with more intelligent processing techniques. Soft computing uses the human mind as a model and aims to formalize the cognitive processes used to accomplish daily tasks.

Emerging methodologies. The three main elements of soft computing are fuzzy logic (FL), neural networks (NNs), and probabilistic reasoning (PR). Fuzzy logic nearly reflects the way that humans reason, using approximate instead of precise rules, and perception instead of measurement. Neural networks extract their principles from brain sciences to model recognition, learning, and process planning. Probabilistic reasoning includes genetic algorithms (GAs), belief networks, parts of learning theory, and chaotic systems. Genetic algorithms take the knowledge from research in genetics, modeling Darwin's natural selection principles. In short, fuzzy logic is related to imprecision, neural network to learning, and probabilistic reasoning to uncertainty. As such, these methodologies are complementary instead of competitive. Intelligent systems are designed to deal with highly variant, complex, noisy, and vague problems in areas such as management, business, manufacture, image recognition, adaptive instrumentation, control processes, robotics, diagnostics, and intelligent databases. Their application allows a variety of smart products such as washing machines, video cameras, microwave ovens, and elevators. For industrial applications these systems are used for automation and quality control.

Fuzzy logic. This was developed by Lotfi Zadeh in 1965. Fuzzy logic belongs to the class of multivalued logics, so named because they allow more than the values of true and false of the older classical logics. Fuzzy logic is a method that allows the easy coding of human knowledge in such a way that it can be understood and used by computers. In fuzzy systems, knowledge is put in the form of conditional rules, such as IF (vehicle ahead is close) AND (speed is fast) THEN (press brake hard). In other words, fuzzy logic allows the phrasing of rules by means of everyday language. However, no logic is involved in fuzzy logic applications; instead they are supported by mathematics, function manipulation, and evaluation.

The first controller based on fuzzy logic was developed by Mamdani and Assilian in 1975. They built it following Zadeh's ideas, and it was applied to control the operation of a simple steam engine. Industrial applications began in the early 1980s with the use of fuzzy control for cement plants in Denmark and the subway system of Sendai, Japan. The first commercial product using fuzzy logic was an electronically controlled shower produced in 1987. This was followed by the first fuzzy washing machine in 1989. The use of hybrid techniques began shortly after, combining neural networks with fuzzy logic in a great variety of consumer products. These products were provided with the capacity to adapt and learn from experience. By 1990, the number of consumer smart products began to grow considerably. By July 1991, the Omron company had obtained 700 patents for devices with fuzzy logic. In 1993, the Japanese reported more than 1500 industrial and commercial applications using fuzzy logic. Numerous intelligent commercial products that use fuzzy technology are available in Japan, and many of them are available around the world. Fuzzy logic was "invented" in the United States; however, the general reaction of the American scientific and business communities to the success of applications of the new methodology was initially negative. Recently, motivated by the economic success of the Japanese fuzzy technology products, American companies have been accepting and even publicly disclosing the inclusion of fuzzy components in their products. Also, the scientific community has recently accepted the validity and utility of fuzzy logic, as a complementary alternative to existing technologies.

Neural networks. Artificial neural networks are systems composed of many simple parallel processing elements whose function is to acquire, store, and utilize experiential knowledge. Their process is similar to the brain's in two ways: knowledge is acquired by the network through a learning process; and it is stored by interneuron connection strengths known as synaptic weights. Neural networks are characterized by a great number of very simple processing elements (similar to neurons) and weighted connections among the elements (that codify the knowledge of the network). They are highly distributed, controlled in parallel, and capable of learning internal representations automatically (by the application of a selected training method).

The first research in neural network was presented by McCulloch and Pitts in 1943. In 1957, Frank Rosenblatt began developing a neural network model called the Perceptron. In 1969, Marvin Minsky and Seymour Papert wrote the book *Perceptrons,* in which they showed that Perceptron development was completely useless. Due to the negative criticisms, during the 1970s and up until 1982, research in neural network practically disappeared. Despite the book, some researchers kept working on neural networks. In 1982, several events allowed the reemergence of interest in neural networks. These included the publication of an important work by John Hopfield; convening of the United States-Japan Conference on Cooperative/Competitive Neural Networks; and the development of thinking computers for robotics applications. Further developments gave birth to new architectures, which were given the more generic name of connectionist architectures.

Linear neural networks are applied in telecommunications, mainly in modems to control sound and vibration. They are also applied for the control and acceleration of atomic particles. Nonlinear neural networks are used for the following applications:

Pattern classification. This is used for the detection of credit card fraud, the recognition of machine-printed characters and hand writing, quality control, exploration for new oil deposits, the war against drugs, and many medical applications.

Prediction and financial analysis. Neural networks are used for financial prognoses, portfolio management, loan approvals, real estate analysis, marketing analysis, and seat allocations in airlines.

Control and optimization. These include various applications designed to control semiconductor processes, chemical processes, oil refinement, and steel production. Optimization methods are used in chemical food formulations.

There is a very close relationship between neural networks and fuzzy systems, because both work with some degree of imprecise information, in a non-well-defined space without clear and deterministic limits. In fact, a bidirectional relationship can be established between fuzzy logic and neural networks, since it is possible to use neural networks as membership functions, or to optimize certain parameters in fuzzy systems. However, fuzzy logic systems can be used to model neurons specialized in the processing of vague or ambiguous information.

Probabilistic reasoning. This section discusses only one aspect of probabilistic reasoning: genetic algorithms. These are searching techniques based on biological principles supported by the theory of evolution (that is, the survival of the fittest). In the 1970s, John Holland established the basis of genetic algorithms as a process able to emulate nature. The main objectives were to design artificial systems software with the capacity to incorporate the important mechanisms of biological systems and to explain the adaptive processes of natural systems.

Genetic algorithms turn a population of individuals, according to a fitness function (indicating how strong an individual is with respect to others), into a new population. Chromosomes are represented in computers by chains of bits. Each individual (or chromosome) in a population represents a possible solution to a problem. To solve a problem, a solutions space search is needed, using information that leads to the desired objective. The most apt individual in a population is the one with the greatest ability to find the solution to a given problem. A practical genetic algorithm is based on three operators. Reproduction is a process where individual strings are copied, based on how good they are with respect to the others. Crossover allows the creation of new individuals from the mating of two parents. Mutation allows the introduction of information not present in the population. The parameters to control genetic algorithms are the size of the population and the maximum number of generations (called a run). Every genetic algorithm run requires the specification of a completion criterion and a designation method to determine when to stop it. Each new generation is an approach toward the solution of the problem at hand.

Genetic algorithms are used in areas of optimization, machine learning, automatic programming, economy, ecology, and social systems. In optimization they are used for function optimization, the traveling agent problem, image processing, job shop scheduling, and control systems. In machine learning they are used to syntactically learn a set of simple IF-THEN rules, as well as to generate weights for neural networks, rules for classification systems, symbolic production systems, and sensors for robots.

Smart future. Smart devices in the future will have high machine intellectual quotients (MIQ) and will be completely different from present intelligent machines. There will be small and fast computers, and the huge systems and networks also will become smart. Machines will be reduced in size, have finer sensors and signal processors, and be able to generate their own fuzzy rules, based on their own experience.

For background information *see* ARTIFICIAL INTELLIGENCE; AUTOMATION; FUZZY SETS AND SYSTEMS; GENETIC ALGORITHMS; HUMAN-MACHINE SYSTEMS; INTELLIGENT MACHINE; NEURAL NETWORK; ROBOTICS in the McGraw-Hill Encyclopedia of Science & Technology.   Carlos Alberto Reyes-Garcia

Bibliography. D. Goldberg, *Genetic Algorithms in Search, Optimization, and Machine Learning*, Addison-Wesley, 1989; R. Jang, C. Sun, and E. Mitsumi, *Neuro-Fuzzy and Soft Computing: A Computational Approach to Learning and Machine Intelligence*, Prentice Hall, 1997; B. Kosko, *Fuzzy Thinking*, Hyperion, 1993; J. R. Koza, *Genetic Programming: On the Programming of Computers by Means of Natural Selection*, MIT Press, 1992; C.-T. Lin and G. Lee, *Neural Fuzzy Systems: A Neuro-Fuzzy Synergism to Intelligent Systems*, Prentice Hall, 1996; T. Ross, *Fuzzy Logic with Engineering Applications*, McGraw-Hill, 1995; A. Sangalli, *The Importance of Being Fuzzy*, Princeton University Press, 1998.

Soil erosion reduction

The human population may double by midcentury. Coupled with improved living standards in underdeveloped nations, this growth will demand unprecedented increases in agricultural production. The great threat to meeting these needs is decreased crop productivity caused by soil erosion. Agricultural productivity and production value are highest in irrigated arid areas, which tend to have shallow, highly erodible soils. Thus, the agricultural systems most capable of meeting future needs are also the most threatened by erosion. Developing effective erosion control methods to protect the sustainability of the Earth's soil is of utmost importance.

Erosion processes. Erosion results from the interaction of fluids (air and water) and soil. Erosion process mechanics involve the transfer of fluid kinetic energy to soil and the resulting detachment, transport, and deposition of soil particles. Wind causes about one-third and water two-thirds of all erosion. Which type dominates depends on the climate, landscape, and land ecology and management. Wind and water erosion control techniques share many soil and land management principles, especially maintaining surface roughness and anchoring or protecting soil with vegetation, organic residues, or other amendments. Catastrophic erosion occurs more commonly with water than wind.

Eroding water originates as precipitation (including snow-melt) or irrigation. Erosion cannot initiate detachment without kinetic energy from water-drop impact or shearing runoff flow. The ease of detachment is influenced by soil and water chemical and physical properties, which affect soil particle cohesion and kinetic energy transfer. Transport cannot occur without runoff. Consequently, nearly all erosion prevention methods depend on soil structure stabilization, armoring, and prevention or management of the runoff volume or velocity. Deposition occurs when the runoff energy decreases, either because of reduced flow velocity or because of reduced flow volume due to infiltration or decreased inflow.

Control methods: rain-fed versus irrigation agriculture. Erosion control in rain-fed agriculture includes tillage to increase surface roughness or infiltration; strip tillage; reduced tillage and no-tillage to preserve surface vegetative residue; slope and slope-length reduction via terracing, contour cropping, intercropping, and narrow rows to provide canopy coverage; grassed waterways; tile drainage; and preserving soil organic matter to promote earthworms and aggregation, which create macropores and reduce runoff. Many of these erosion control approaches can be used in irrigated agriculture as well. However, although erosion processes are identical for water from precipitation or irrigation, the process dynamics and interaction of the process components differ between precipitation- and irrigation-induced erosion. These differences stem from soil and water chemistry interactions, wetting rates, splash rates, applications, and infiltration patterns. The effects of these differences vary greatly among irrigation types.

Some important differences between precipitation- and irrigation-induced erosion are easily identified. Water with high electrical conductivity or a high ratio of dissolved calcium to sodium salts aids interparticle attractions, cohesion, and flocculation, reducing particle detachment and dispersion. Conversely, the absence of electrolytes or the dominance of sodium ions (lower charge and larger hydrated radius than calcium ion) disperses particles, facilitating detachment and increasing erosion. Nonsprinkler irrigation has no splash energy to detach soil. In precipitation-induced erosion, the soil is wet before runoff begins and the runoff volume accumulates downslope. In furrow irrigation, hot dry soil is instantly hydrated by inflow and infiltration decreases the runoff volume downslope.

Various means of reducing erosion have been developed that are unique to irrigation, such as managing the physical application of water to the soil (for example, by using small close-spaced furrow streams; sprinkling with smaller drops, reducing runoff; and using precise scheduling and volume application control to prevent overirrigation). The irrigation water's electrolyte chemistry can be adjusted to overcome salt effects or to adjust the ratio of specific cations. This is done by adding calcium salts (such as gypsum) or by blending water from multiple sources (conjunctive water use).

Polyacrylamides. Among the newest and most successful erosion control technologies for irrigation is the use of natural and synthetic polymers in irrigation water. Acid cheese whey, chitosan (shellfish by-product) compounds, emulsified cellulose microfibrils, and ultrahigh-molecular-weight synthetic polymers, particularly water-soluble polyacrylamide (PAM), are effective. PAM, currently the most economical of these polymers, reduces erosion 94% in furrow irrigation and 75% in sprinkler irrigation. High efficacy, low cost, and easy application led to successful PAM use on a million United States acres in 1999. Use in the United States and elsewhere is growing rapidly. PAM formulations vary greatly, depending on molecular weight, conformation, functional group substitutions, charge type, and charge density. The PAMs used for erosion control encompass only water-soluble, noncrosslinked (linear) PAMs, and not gel-forming superwater-absorbent, crosslinked PAMs. Application cost plus environmental and human hygiene considerations have limited erosion-control PAMs to anionic formulations of 4–15 mg per mole. Research suggests that 12–15 mg per mole compounds are the most effective for erosion control.

Anionic PAMs bind anionic soil particles by bridging with cations of small hydrated radius, particularly divalent calcium cation. PAM improves soil aggregate cohesion, resisting detachment and dispersion, preserving structure and soil surface roughness. This also maintains pore continuity to the soil surface by preventing dispersion and redeposition of particles that block pores.

Application. The modes of PAM application for erosion control in irrigation are covered in the 2000 U.S. Department of Agriculture–Natural Resource Conservation Service (USDA-NRCS) PAM conservation practice standard. For furrow irrigation, PAM can be dissolved in the water before reaching the furrow, or placed on the soil in the first 1–2 m of the furrow below the water inlet. When PAM is dissolved in the water supply, the best results are obtained at a concentration of 10 ppm PAM as the water first crosses the field, then ceasing PAM application when runoff begins. Placing a dry powder "patch" directly on the soil is the easiest and probably most widely used method. The application rate for the initial patch treatment is an area equivalent rate (based on furrow spacing and length) of 1–2 lb/acre. A variation on the patch is use of PAM tablets in the furrow. However, the tablets sometimes dissolve erratically. On nearly level fields (slopes <0.2%), continuous application of 1 ppm PAM works well. For all methods, if the soil remains undisturbed after initial PAM treatment, subsequent irrigation usually requires half (or less) the initial PAM concentration applied. Season-long erosion control for irrigation furrows can usually be achieved using 3–5 lb of PAM per acre. Sprinkler irrigation wets a larger soil area than furrow irrigation and involves splash. PAM in an equivalent water application depth of 18–20 mm (3/4 in.) at a rate of 2–4 kg/hectare (2–4 lb/acre) controls sprinkler-induced erosion. Season-long control with sprinklers requires continued PAM application until canopy coverage;

however, application rates of subsequent irrigations can be greatly reduced.

Because soil surface structure is kept porous when using PAM, infiltration rates are generally higher with PAM-treated water on fine to medium-textured soils (clays and loams). Farmers can use PAM to improve infiltration precision and uniformity in surface and sprinkler irrigation systems and reduce runoff and runon problems. In many settings, the infiltration management potential is a greater incentive than erosion control to adopt PAM use.

Environmental restrictions. Soil amendment registrations, environmental regulations, and USDA-NRCS guidelines restrict erosion-control PAMs to anionic forms containing <0.05% unreacted acrylamide monomer (AMD). Neutral or cationic PAMs can harm certain microorganisms or aquatic species. Cationic PAMs adhere to hemoglobin-bearing fish gills, causing suffocation. Anionics are safe at (and well beyond) the prescribed erosion-controlling concentration. Acrylamide monomer, a neurotoxin, poses no health or environmental risk at the rates and concentrations specified, and it is removed rapidly from the environment (hours to days) by microorganisms. High-purity anionic PAMs are used in municipal water treatment, food processing and packaging, pharmaceutics, and animal feeds. Erosion control represents 1–2% of PAMs or related polymers used annually in the United States for paper manufacture, mining, and sewage treatment.

Environmental benefits. The environmental benefits of PAM-based erosion prevention are well documented. PAM reduces nitrogen and phosphorus (eutrophying nutrients), biological oxygen demand (BOD), and several herbicides and pesticides by 60–80% in runoff water. Since 1998, 10–20 million tons of sediment, thousands of tons of nutrients, and hundreds of tons of herbicides and pesticides per year have been prevented from entering riparian (wildlife-supporting) waters. Recent research has shown large reductions in weed seed and microorganism loads in PAM-treated runoff. Sequestration of weed seed and microbes reduces the spread of weeds and crop diseases within and among fields, reducing their impact on crop production and lowering the need for herbicides and pesticides. Because fecal coliforms and other human hygiene-impacting microorganisms enter surface water from manure-treated fields, microbe sequestration via PAM also reduces organism-related human health threats.

Extension of PAM technology to rain-fed agriculture is difficult and may not prove economical or effective on a wide scale for various physical, chemical, and logistical reasons. However, PAM use for construction site, road cut, and mine site erosion control and for accelerating water clarification in runoff retention ponds has increased rapidly. PAM and other organic and synthetic polymer–based erosion-control and water quality–protection technologies will likely continue to improve and be implemented in the future.

For background information *see* EROSION ; IRRIGATION (AGRICULTURE); POLYACRYLONITRILE RESINS; SOIL CONSERVATION; WATER POLLUTION in the McGraw-Hill Encyclopedia of Science & Technology.
R. E. Sojka; D. L. Bjorneberg

Bibliography. F. W. Barvenik, Polyacrylamide characteristics related to soil applications, *Soil Sci.*, 158:235–243, 1994; D. L. Bjorneberg et al., Unique aspects of modeling irrigation-induced soil erosion, *Int. J. Sediment. Res.*, 15:245–242, 2000; W. J. Orts, R. E. Sojka, and G. M. Glenn, Biopolymer additives to reduce soil erosion-induced soil losses during irrigation, *Ind. Crops Prod.*, 11:19–29, 2000; F. J. Pierce and W. W. Frye (eds.), *Advances in Soil and Water Conservation*, Ann Arbor Press, Chelsea, MI, 1998; R. E. Sojka et al., Irrigating with polyacrylamide (PAM): Nine years and a million acres of experience, in R. G. Evans et al. (eds.), *National Irrigation Symposium: Proceedings of the 4th Decennial Symposium*, American Society of Agricultural Engineers, St. Joseph, MI, 2000; U.S. Department of Agriculture, *Natural Resources Conservation Service Conservation Practice Standard: Anionic Polyacrylamide (PAM) Erosion Control, Code 450*, Washington, DC, July 2000.

Soil quality

Soil—the thin, unconsolidated, vertically differentiated portion of the Earth's surface—is ubiquitous and often ignored despite its many important environmental and life-sustaining functions. Soil is necessary for the production of food, feed, and fiber products, and supports buildings, roads, and playing fields. Soil helps to safely dispose of and process biological and industrial wastes, and it purifies and filters water that may enter drinking water supplies. Usually, soil performs more than one of these roles simultaneously.

Soil is in large but finite supply. It varies greatly in chemical and physical properties both in short distances and regionally. Some soil components cannot be easily renewed within a human time frame; thus the condition of soil in agriculture and the environment is an issue of global concern. For these reasons, an effort has been made to distinguish among the many kinds of soils and identify those best suited for specific uses. The concept of soil quality stems from the desire to evaluate soils, match appropriate management and uses for each soil, and measure changes in soil properties.

Concept. The concept of soil quality has been controversial among soil scientists because it is subjective, as well as being management- and climate-dependent. The concept has not been thoroughly tested by the scientific community, but it has been institutionalized by some government agencies despite the scientific discord surrounding it. In contrast, concepts of air and water quality are well accepted. It may seem reasonable to include soil quality as a basic natural resource. However, air and water quality are based on standard pure states against which all qualities can be measured. No ideal or "pure" soil state exists or can be measured for all possible uses, or for the many different combinations of soil types, climates, and management strategies.

In the United States, soil quality includes soil fertility, potential productivity, resource sustainability, and environmental quality. Most assessments attempted to date have been linked mainly to microbial diversity or crop yield. In Canada and Europe, contaminant levels and their effects are the primary factors determining soil quality. Most farmers and agricultural scientists who have embraced the concept of soil quality have associated it primarily with crop productivity. Some expand this to include specific indicators, such as soil surface condition, organic matter content, or microbial respiration. Critics of the concept note that, despite the best of intentions, such paradigms fail to resolve the contradiction that some soil properties associated positively with productivity have negative impacts on environmental quality.

The Soil Science Society of America Ad Hoc Committee on Soil Quality proposed that soil quality is "the capacity of a specific kind of soil to function, within natural or managed ecosystem boundaries, to sustain plant and animal productivity, maintain or enhance water and air quality, and support human health and habitation." This definition requires that the following soil functions be evaluated simultaneously to describe soil quality: (1) sustaining biological activity, diversity, and productivity; (2) regulating and partitioning water and solute flow; (3) filtering, buffering, degrading, immobilizing, and detoxifying organic and inorganic materials, including industrial and municipal by-products and atmospheric deposition; (4) storing and cycling nutrients and other elements within the Earth's biosphere; and (5) providing support of socioeconomic structures and protection for archeological treasures associated with human habitation.

Quantification of the Ad Hoc Committee's general definition is difficult because no single conserving or degrading process or property determines soil quality. Soil has both dynamic and static properties that vary spatially. In addition, soils perform various functions, often simultaneously, for which quantitative relationships between measured values and predicted responses are lacking.

M. J. Singer and S. A. Ewing have reviewed other definitions of soil quality. The existence of multiple definitions suggests that the soil quality concept is evolving. All soil quality definitions have some similarities: high-quality soils are productive and biologically active; they support plant productivity and human and animal health; and they serve in various capacities in unmanaged ecosystems. In particular, high-quality soils are those that adequately regulate water, nutrient, and energy flow through the environment, while providing buffers against undesir-able environmental changes. An integrated vision of soil quality has been offered in definitions, but its characterization and indexing have focused on limited individual aspects of the definition without attempting to integrate the conflicting functions.

Measurement. To proceed from definition to quantitative measure, a minimum data set of characteristics representing soil quality must be selected and quantified. A minimum data set may include the presence of specific biological, chemical, or physical soil characteristics; optimum levels of specific characteristics that benefit soil productivity or other important soil functions; or the absence of a property that is detrimental to these functions.

For agriculture, the goal is to measure properties that lead to a relatively simple and accurate soil ranking based on potential plant production without soil degradation. Presumably, optimal ranges of soil properties exist for meeting the quality criteria of productivity with acceptable levels of soil and environmental degradation.

Examples of dynamic soil characteristics are the size, membership, distribution, and activity of a soil's microbiological community; the soil composition, pH, and nutrient ion concentrations; and the exchangeable cation population. Soils respond quickly to changes in conditions such as water content. As a result, the optimal timing, frequency, and distribution of soil measurements vary with the property being measured.

Soil properties that change quickly present a problem because many measurements are needed to know the average value and to determine if changes in the average indicate improvement or degradation of soil quality. Unfortunately, the average quantified value of a rapidly changing property may not accurately represent the soil condition at any given time. Conversely, properties that change very slowly are insensitive measures of short-term changes in soil quality.

Some important soil characteristics are slowly renewable. Organic matter, most nutrients, and some physical properties may be renewed through careful long-term management. Certain chemical properties (pH, salinity, and nitrogen, phosphorus, and potassium content) may be altered to a more satisfactory range for agriculture within a growing season or two, while removal of unwanted chemicals may take much longer.

Physical and chemical factors. Physical factors, such as effective rooting depth, porosity or pore size distribution, bulk density, hydraulic conductivity, soil strength, and particle size distribution, are potential soil quality indicators. Other physical properties, such as structure, texture, and profile characteristics, affect management practices in agriculture but are only indirectly related to plant productivity and require large efforts to specifically correlate with crop performance. Water potential, oxygen diffusion rate, temperature, and mechanical resistance directly affect plant growth, and may be better indicators of the physical quality of a soil for production, but are difficult to measure. Nutrient availability depends on soil physical and chemical processes, and chemical characteristics. At low and high pH, for example, some nutrients are unavailable to plants and some toxic elements become more available.

Biological factors. The focus of many soil quality definitions is soil biology. Soil supports a diverse population of organisms, ranging in size from viruses to large mammals. Members of these populations usually interact positively with plants and other system components. However, some soil organisms, such as nematodes and bacterial and fungal pathogens, reduce plant productivity. Many proposed soil quality definitions focus on presence of beneficial rather than absence of detrimental organisms, although both are critically important. Various measures of microbial community viability have been suggested as measures or indices of soil quality. Community-level studies consider species diversity and frequency of occurrence of species. Scientists emphasizing the use of biological factors as indicators of soil quality often equate soil quality with relatively dynamic properties, such as microbial biomass, microbial respiration, organic matter mineralization, and organic matter content. They suggest that keystone species, taxonomic diversity at the group level, and species richness of several dominant groups of invertebrates can be used as part of a soil quality definition.

Future. Complex and evolving, the concept of soil quality brings together soil science, philosophy, politics, and policy. No single soil property represents soil quality. Soil scientists have not agreed on a single set of soil properties that can universally assess soil quality. Disagreement continues as to whether the concept is scientifically valid. It is unlikely that scientific agreement will be found to satisfy all points of view. Regardless of acceptance of the concept, or its definition, or the suite of soil variables chosen to define and quantify soil quality, soil scientists agree that it is critical to human sustainability that soils be carefully managed to provide for human health and welfare, while minimizing soil and environmental degradation.

For background information *see* AGRICULTURE; CONSERVATION OF RESOURCES; SOIL; SOIL CONSERVATION; SOIL ECOLOGY; SOIL FERTILITY in the McGraw-Hill Encyclopedia of Science & Technology.

M. J. Singer; R. E. Sojka

Bibliography. J. W. Doran et al. (eds.), *Defining Soil Quality for a Sustainable Environment. SSSA Spec. Publ.*, no. 35, Soil Science Society of America, Madison, WI, 1994; D. L. Karlen et al., Soil quality: A concept, definition, and framework for evaluation, *Soil Sci. Soc. Amer. J.*, 61:4–10, 1997; M. J. Singer and S. A. Ewing, Soil quality, in M. E. Sumner (ed.), *Handbook of Soil Science*, CRC Press, Boca Raton, FL, 2000; R. E. Sojka and D. R. Upchurch, Reservations regarding the soil quality concept, *Soil Sci. Soc. Amer. J.*, 63:1039–1054, 1999.

Sol-gel sensors

The use of sol-gel chemistry as a means to prepare inorganic and organic-inorganic composite materials has blossomed in recent years. The fundamentally intriguing chemistry, coupled with the ease with which these materials can be made, modified, and processed, has attracted the interest of a variety of scientists.

Sol-gel processing. Sol-gel processing involves the preparation of glasslike materials through the hydrolysis and condensation of metal alkoxides. Two of the most studied alkoxysilanes used for fabricating silicate materials are tetramethoxysilane (TMOS) and tetraethoxysilane (TEOS). In a typical procedure, TMOS is mixed with water in the presence of a mutual solvent such as methanol and a catalyst such as an acid (for example, hydrochloric acid), a base (such as ammonium hydroxide), or a nucleophile (such as fluoride). During sol-gel transformation, the viscosity of the solution increases as the sol (colloidal suspension of small particles) becomes interconnected to form a rigid porous network—the gel. Gelation can take place on the order of seconds to months, depending on the sol-gel processing conditions (type and concentration of catalyst, alkoxide precursors, silicon-to-water ratio, temperature). A simplified scheme for the hydrolysis [reaction (1)] and condensation [reaction (2) and/or (3)] of TMOS is shown.

$$Si(OCH_3)_4 + nH_2O \longrightarrow Si(OCH_3)_{4-n}(OH)_n + nCH_3OH \quad (1)$$

$$-\overset{|}{\underset{|}{Si}}-OH + HO-\overset{|}{\underset{|}{Si}} \longrightarrow -\overset{|}{\underset{|}{Si}}-O-\overset{|}{\underset{|}{Si}}- + H_2O \quad (2)$$

$$-\overset{|}{\underset{|}{Si}}-OCH_3 + HO-\overset{|}{\underset{|}{Si}} \longrightarrow$$

$$-\overset{|}{\underset{|}{Si}}-O-\overset{|}{\underset{|}{Si}}- + CH_3OH \quad (3)$$

Sol-gel processing provides the ability to make materials in various forms. For example, the silica sol can be spin-cast or dip-coated on a suitable substrate (such as a glass slide, electrode surface, or silicon wafer) to form a thin film. Alternatively, it can be poured into a suitable container such as a cuvette to form a block monolith. The block monolith can be crushed and sieved to form small particles that can be packed into tubes. Another unique feature of sol-gel processing is the ease with which the chemical properties of the material can be manipulated. For example, organoalkoxysilanes ($R—Si(OR')_3$, where $R = CH_3$, C_6H_5, $CH_2CH_2CH_2NH_2$, and so on) can be cohydrolyzed and condensed with TMOS to form organic-inorganic hybrid materials. Finally, because the sol-gel processing conditions are relatively mild, various molecules can be introduced into the porous silicate matrix by simply adding them to the sol prior to its gelation (see **illus.**). Relative to many organic polymer matrices, sol-gel-derived glasses

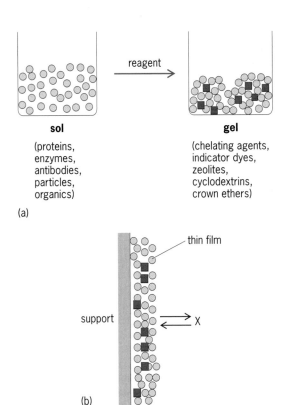

Sol-gel-derived glass with specific reagents.
(a) Encapsulation of reagents. **(b)** Use of the glass in the chemical sensing of analyte molecules X.

offer better optical transparency, chemical and mechanical stability, and permeability.

Chemical sensors. Chemical sensors are small devices that measure the concentration of a chemical species in solution or the gas phase. They generally consist of a chemically selective receptor layer bound to a suitable transducer or "signaling unit." The general methods that have been used to immobilize a receptor are impregnation, covalent attachment, and encapsulation. In impregnation, the receptor is physically adsorbed on a solid support. This method, while simple, generally suffers from loss of reagents due to leaching from the solid support. One way to eliminate this is to covalently attach the receptor to the support. This method, while effective at keeping the receptor bound to the surface, can be tedious and time-consuming and often limits the mobility of the receptor, preventing it from optimizing its reaction geometry. In encapsulation, the receptor is physically embedded in an inert matrix such as a polymer or inorganic solid. This method has received considerable attention due to its simplicity and its performance.

There has been much interest in using sol-gel-derived materials in chemical sensors because receptors of varying size and functionality can be entrapped in a stable material that is porous enough to allow an external analyte to diffuse through the network and react with the receptor (see illus.). In addition, sol-gel-derived materials can be fabricated into thin films with fast response times and recovery rates, they can be incorporated into micrometer-sized cavities for microanalytical work, or they can be used in monolithic or powder format for high-surface-area applications. Receptors that have been introduced into a silicate matrix include enzymes, proteins, chelating agents, organic dyes, crown ethers, zeolites, and cyclodextrins.

pH sensors. One of the first applications of sol-gel-derived materials in sensing applications was the development of the pH sensor. For this application, a pH-sensitive organic dye is incorporated into the porous silicate matrix by adding it to the sol prior to its gelation or by covalently attaching it to the matrix using a dye-modified alkoxysilane [dye-Si(OR)$_3$]. A change in the absorption or fluorescence spectrum of the immobilized dye with pH is the basis for sensing, and response times are seconds to minutes, depending in part on material configuration and porosity. The dynamic and pH-sensitive range is strongly dependent on the dye introduced into the matrix and the precursors used to make the sol. The sensors' pH titration curves are generally broader than those observed in solution due to the heterogeneous nature of the host material. In addition, the pK$_a$ (acid-base ionization constant) value of the dye is sometimes shifted relative to that observed for the free dye in solution. pH-sensitive sol-gel coatings have been placed on optically transparent glass, optical fibers, optical waveguides, and in micrometer-sized pipettes.

Ion sensors. Colorimetric reagents can be encapsulated in sol-gel-derived glasses and used in the quantitative determination of metal ions such as Fe^{2+}, Ni^{2+}, Al^{3+}, Cu^{2+}, and Pb^{2+}. Many of these methods are based on classical methods developed for solution analysis. The usefulness of this approach can be attributed to the fact that gel-entrapped chelating agents possess sufficient mobility to form coordination complexes with metal cations that diffuse into the matrix. For example, a colorimetric assay for iron(II) [Fe^{2+}] can be performed by first immobilizing phenanthroline in the silicate matrix. When the doped matrix is placed in a solution containing Fe^{2+}, the Fe^{2+} diffuses into the gel and forms Fe(phen)$_3^{2+}$, a colored complex, which can be detected spectrophotometrically. Similarly, nickel ions (Ni^{2+}) can be determined via the spectrophotometric detection of the colored complex Ni(DMG)$_2$ formed when Ni^{2+} reacts with gel-encapsulated dimethylglyoxamine (DMG). Other types of sol-gel ion sensors that have been developed include potentiometric sensors for the determination of alkali metals, that is, sodium and potassium ions (Na$^+$ and K$^+$), via complexation with sol-gel immobilized crown ethers; optical biosensors for small ions such as calcium, nitrate, and manganese (Ca^{2+}, NO$_3^-$, Mn^{2+}) via entrapment of proteins, enzymes, or liposomes in the silicate matrix; and electrochemiluminescence sensors for oxalate and various aliphatic tertiary amines.

Gas sensors. A variety of sensors for gas-phase analysis have been developed. Many of these are based on the immobilization of dyes, chelates, proteins, and zeolites. Among those species detected are oxygen, ammonia, water, and carbon monoxide. For example, ruthenium complexes have been immobilized into the silicate framework. In the presence of oxygen, the fluorescence from the encapsulated ruthenium complex is quenched, and the decrease in the fluorescent signal can then be related to the concentration of oxygen in the gas phase or in solution. Myoglobin (Mb) has also been immobilized within a silicate matrix. The reduced form of myoglobin reacts with dissolved oxygen to form OxyMB, and changes in the absorption spectrum can then be directly related to the oxygen content. In addition to optical investigations, electrochemical (amperometry) and surface-acoustic-wave devices have also been described. Also, zeolites have been doped into a dense silicate network and used as sensors for gas vapors.

Biosensors. Almost a decade ago, it was shown that biomolecules trapped in sol-gel glasses retained their native properties, a finding that led to development of various sensors for bioanalytical applications. The successful encapsulation of biomolecules in sol-gel glasses required more stringent control over the processing conditions compared to that required for simple doping of pH indicator dyes. High concentration of methanol and acid usually will denature most proteins and enzymes. As a result, procedures for incorporating biomolecules in a silicate matrix often have high water-to-silicon ratios and no added alcohol. In addition, buffer is typically added to the sol before protein addition to neutralize the high acidity.

Of all the enzymes and proteins that have been studied, glucose oxidase has been the most popular. Glucose oxidase (GOx) catalyzes the oxidation of glucose with oxygen to form gluconic acid and hydrogen peroxide [reaction (4)].

$$\text{B-D-glucose} + H_2O + O_2 \xrightarrow{\text{GOx}} \text{gluconic acid} + H_2O_2 \quad (4)$$

Glucose has been detected via colorimetric and electrochemical methodologies. In a typical procedure, glucose oxidase is added to the sol, which is then cast on the surface of an electrode. The electrochemical oxidation of hydrogen peroxide produced from the enzymatic reaction can be used to evaluate enzyme reactivity, which serves as the basis for chemical sensing. Also, the use of electron transfer mediators such as ferrocene to shuttle the electron between GOx and the electrode surface has been described. Both covalently attached and freely diffusing mediators have been used.

For background information *see* ANALYTICAL CHEMISTRY; BIOSENSOR; CHELATION; GEL; GLASS; MICROSENSOR; PH; SOL-GEL PROCESS; SPECTROPHOTOMETRIC ANALYSIS in the McGraw-Hill Encyclopedia of Science & Technology.   Maryanne M. Collinson

Bibliography. D. Avnir et al., Organo-Silica Sol-Gel Materials, Chap. 40 in *The Chemistry of Organic Silicon Compounds*, vol. 2, 1998; J. Brinker and G. Scherer, *Sol-Gel Science*, Academic Press, New York, 1989; M. M. Collinson, Structure, Chemistry and Applications of Sol-Gel-Derived Materials, in H. S. Nalwa (ed.), *Handbook of Advanced Electronic and Photonic Materials and Devices*, vol. 5, Academic Press, New York, 2001.

Solvated electron

The solvated electron is an excess electron localized in a cavity formed by solvent molecules. It is a transient species found everywhere in various biological and chemical systems where a charge is transferred through a medium. It plays an important role in applications such as the radiation treatment of cancer and the neutralization of toxic wastes. The structure of the solvated electron is analogous to an *F*-center (color center) found in alkali halide crystals with anion vacancy defects (**Fig. 1**). Formation of the solvated electron is usually preceded by the injection of an electron from the parent molecule into the solvent. The injection of the electron occurs via direct or indirect ionization (for example, via charge transfer to solvent) of the parent molecule, depending on the conditions under which the electron is separated from its parent. The injected electron subsequently loses its initial kinetic energy while traveling through the solvent (known as thermalization), and eventually localizes in a solvent cavity to form an equilibrated solvated electron. After the solvated electron is formed, it undergoes reactions with electron-accepting species existing in the solution that ultimately set a limit on its lifetime.

Optical properties. The solvated electron exhibits a strong absorption in the visible and near-infrared regions (**Fig. 2**). The absorption corresponds to electronic transitions from the ground to the excited electronic states. The bound excited states are localized within the solvent cavity, and the transition to these states carries a majority of the oscillator strength of the ground-state absorption. The unbound excited states form a continuum of the delocalized states,

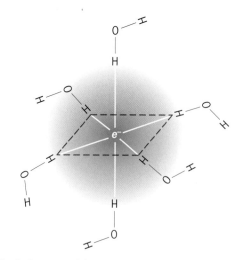

Fig. 1. Structure of the solvated electron in water—the hydrated electron. The six water molecules represent the first solvation shell immediately surrounding the electron.

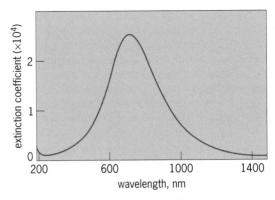

Fig. 2. Absorption spectrum of the equilibrated hydrated electron.

which are responsible for the tail in the absorption spectrum. Due to the strong absorption of the solvated electron, optical absorption spectroscopy has been used as a primary tool for its detection. The absorption wavelength of the solvated electron varies significantly, depending on the interactions between the solvated electron and the surrounding solvent molecules. Generally the absorption peak shifts to longer wavelengths as the polarity of the solvent decreases, reflecting weaker interactions between the electron and solvent molecules. The absorption spectrum of the solvated electron is also sensitive to variations of the microscopic structure of the solvent cavity, which result during the electron solvation process.

Dynamics. The mechanism of the formation of the solvated electron from an electron introduced to the solvent is a subject of intense experimental and theoretical investigations due to its importance in various chemical and biological charge transfer reactions and radiation chemistry. Earlier experimental studies on electron solvation utilized the photoionization probe method. In these experiments, an intense, ultrashort laser pulse (usually in the ultraviolet region) photoionizes the solvent molecule, injecting an excess electron into the solvent. The subsequent trapping and relaxation of the electron are monitored by measuring the time-dependent absorption of the solvated electron. Most of the studies on electron solvation have been performed in water, an important solvent in many chemical and biological systems. In water, experiments and theoretical simulations suggest that the hydrated electron is formed by either stepwise or direct relaxation from the delocalized to the equilibrated ground state on a subpicosecond time scale following fast initial thermalization. A stepwise mechanism has been suggested, involving an intermediate species that undergoes a further relaxation to form an equilibrated hydrated electron. The existence of the intermediate state is, however, uncertain.

Recently, a different type of experiment was introduced in which a preformed equilibrated solvated electron is excited with an ultrashort laser pulse in resonance with the ground-to-excited-state transition. The subsequent relaxation of the excited state

to the ground state is monitored with another ultrashort laser pulse as a function of the time delay between the two pulses. This experimental technique is denoted by a photoionization-pump probe. While the earlier photoionization probe experiments provide information on the dynamics of the localization from the delocalized quasi-free electron, pump probe experiments explore the dynamics of the excited states of the solvated electron. Pump probe studies of the solvated electron provide particularly valuable information on nonadiabatic electronic transitions in the condensed phase, which is strongly coupled to the dynamics of solvation. From the pump probe studies of the hydrated electron, an ultrafast solvation component (<100 femtoseconds), whose time scale is much shorter than that of the diffusional motion of the solvent, has been identified. This is attributed to the librational motion of the water molecules, which responds inertially to a sudden change of the charge distribution of the hydrated electron upon excitation.

Kinetics. Once the solvated electron is generated, its concentration decays continuously through reaction with the electron-accepting species existing in the solution. In pure solvents, the predominant electron-accepting species is the geminate partner of the solvated electron from which the electron is originated. For example, the hydrated electron reacts with OH radicals and H_3O^+ ions, which were fragmented from the initial geminate partner (H_2O^+) of the hydrated electron. The reaction between the solvated electron and its geminate partner is called geminate recombination. The decay kinetics of the hydrated electron, on the hundreds of picosecond time scale, is well described by geminate recombination. The kinetics of geminate recombination is dependent on the initial spatial distribution of the electron with respect to the geminate partners, since the reaction occurs through the diffusive encounter of the geminate pairs. The initial spatial distribution of the solvated electron depends on the initial kinetic energy of the injected electron, which determines the distance that the electron travels from the parent molecule before it is trapped in a solvent cavity. For this reason, the geminate recombination kinetics is strongly dependent on the conditions of the electron injection such as the ionizing photon energy and the identity of the parent molecules. The solvated electron reacts not only with its geminate partners but also with various different molecular and ionic species in solution. Thousands of reactions with the species, ranging from simple atomic ions to large organic molecules, have been reported. The kinetics of these reactions is well described by the conventional solution-phase bimolecular reaction kinetic model.

Electron transfer kinetics. The reaction of the solvated electron can be considered as an electron transfer from a solvated electron to an electron acceptor. Kinetics of various solvated electron reactions has been analyzed within the framework of the Marcus theory of electron transfer, demonstrating a direct connection to the electron transfer reaction governed by the reaction free energy (ΔG°) and reorganization energy (λ) in Eq. (1) [k is the Boltzmann

$$k_{ET} = A \cdot \exp\left(-\frac{(\Delta G^\circ + \lambda)^2}{4\lambda kT}\right) \quad (1)$$

constant, and T, the temperature].

Recently, the electron transfer reactions of precursor and excited states of the equilibrated hydrated electron have received significant attention. The reactions of these states have been exploited much less than the equilibrated hydrated electron primarily due to their short lifetimes of less than 1 picosecond. For the hydrated electron, both the localized and delocalized excited states exhibit much greater reactivity than the equilibrated ground state, when the hydrated electron is excited in the presence of electron acceptors in solution. The apparent higher reactivity of the excited states, however, has its origin in the spatially diffuse and delocalized nature of the excited-state wavefunctions, not the higher electron transfer rate between the donor-acceptor pair. Because the diffusive motion of the reactants is virtually frozen during the short lifetime of the excited state, the usual bimolecular reaction scheme is not applicable to the reactions of the excited states. The enhancement of the reactivity in the excited states is due to the increase in the number of donor-acceptor encounter pairs as the volume of space encompassed by the electron wavefunction increases upon the optical excitation. The first-order electron transfer rate within an individual donor-acceptor pair, on the contrary, decreases, as the size of the hydrated electron becomes bigger. This is consistent with the prediction from the simple nonadiabatic electron transfer rate theory, in which the electron transfer rate constant (k_{ET}) [Eq. (2), where $\hbar$ is

$$k_{ET} = \frac{2\pi}{\hbar} H_{DA}^2 \text{DWFC} \quad (2)$$

Planck's constant divided by 2π] is determined by the donor-acceptor electron matrix element (H_{DA}) and the density-of-state-weighted Franck-Condon factor (DWFC).

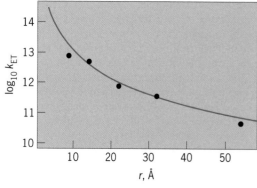

Fig. 3. Plot of the logarithm of the experimentally determined first-order electron transfer rate constant (k_{ET}) as a function of hydrated electron–electron acceptor distance (r). The solid line is the theoretically calculated k_{ET} obtained from the inverse relationship between k_{ET} and r^3.

The donor-acceptor electron matrix element, which is proportional to the donor-acceptor orbital overlap, decreases with the size of the hydrated electron. The experimentally determined electron transfer rate constants are in excellent agreement with the theoretically predicted inverse relationship between k_{ET} and volume of the hydrated electron (**Fig. 3**). A successful description of the electron transfer kinetics of the solvated electrons of various sizes with a single electron transfer model has an important implication in that it provides a unified point of view to the localized and delocalized electron transfer reactions.

For background information *see* CHEMICAL DYNAMICS; COLOR CENTERS; ELECTRON; ELECTRON-TRANSFER REACTION; FRANCK-CONDON PRINCIPLE; PHOTOIONIZATION; REACTION KINETICS; SOLUTION in the McGraw-Hill Encyclopedia of Science & Technology.

Paul F. Barbara; Dong Hee Son

Bibliography. T. W. Kee et al., A unified electron transfer model for the different precursors and excited states of the hydrated electron, *J. Phys. Chem. A*, in press; E. Keszei, T. H. Murphrey, and P. J. Rossky, Electron hydration dynamics: Simulation results compared to pump and probe experiments, *J. Phys. Chem.*, 99:22–28, 1995; X. Shi et al., Femtosecond electron solvation kinetics in water, *J. Phys. Chem.*, 100:11903–11906, 1996; D. H. Son et al., One-photon UV detrapping of the hydrated electron, *Chem. Phys. Lett.*, in press.

Sonoelastography

Physicians have long used palpation to detect hard tumors within the human body. Such palpation is limited to lesions that are close to the surface. Several ultrasound techniques have been proposed to image these tumors, including sonoelastography.

Sonoelastography is used for imaging the relative elastic properties of soft tissue. Low frequency (<1 kHz) vibrations (shear waves) are propagated through tissue while real-time Doppler techniques are used to image the resulting vibration pattern on an ultrasound scanner. The low-frequency vibration is provided by an external source, such as an audio speaker or a piston shaker, that is brought into close contact with the patient or tissue sample. An ultrasound transducer is then positioned to image the area of interest. When the propagating vibration enters the hard tumor, the amplitude of vibration decreases. Sonoelastography then images this local decrease in vibration amplitude by mapping it to a gray scale. The underlying relative elasticity is inferred from the resulting image or is estimated using advanced numerical techniques.

Wave propagation in tissue. It is well known from solid mechanics that both shear waves and longitudinal waves can propagate in an elastic medium, whereas in gas or fluid media only longitudinal waves propagate. Longitudinal waves are disturbances that change the local volume of the media. They are curl- or rotation-free and are called compression waves or irrotational waves. In contrast, shear waves are disturbances that distort the local shape of a medium but maintain the local volume. They are also sometimes referred to as equivoluminal waves. The shear modulus (ratio of applied shear stress to resulting shear strain) and tissue density determine the velocity of shear wave propagation, while bulk modulus (ratio of applied volume stress to resulting volume strain), shear modulus, and tissue density determine velocity of longitudinal wave propagation.

Shear waves are important in sonoelastography because the bulk modulus of soft tissue is about 10^5–10^6 times its shear modulus. In solid mechanics, the modulus relates the ratio of stress to strain. Stress is the ratio of the force to an area, and shear strain is the measure of the local amount of rotation of an element of the medium. Because the shear modulus of soft tissues is lower than its bulk modulus, shear waves generate larger strains.

The theory for shear wave propagation shows that when a medium is driven by an external vibration source, a small hard tumor in an otherwise homogeneous elastic medium will produce a disturbance in the vibration pattern that would not be seen if the hard tumor were not present. This can be understood intuitively by observing that a tumor with an elevated shear modulus will naturally tend to experience smaller strains, since strain is inversely proportional to the modulus.

Vibration estimation. Color flow Doppler imaging is used to create the vibration images in sonoelastography. Color flow Doppler ultrasound works by emitting a sequence of ultrasound pulses at carefully placed intervals in time. Since the target is vibrating, the transit time from the ultrasound scanner and back will vary from pulse to pulse, the target being closer or farther from the target at different times in its cycle. One simple way of estimating vibration amplitude is to track how the distance to target varies over time, subtract the minimum distance from the maximum distance, and divide by two. Such techniques are hard to implement in real time, a desirable capability in an imaging system. Fortunately, Doppler spectral variance estimation has proved to be an effective real-time method of estimating the relative vibration amplitude. Researchers at the University of Rochester have shown that the power spectrum of a vibrating target displays a linear relationship between the standard deviation of the Doppler power spectrum and the vibration amplitude. Since variance is the square of standard deviation, the resulting image will display a square-law mapping of vibration amplitude in the image when mapped to a gray scale. When the region of interest is homogeneous except for a small lesion, the resulting vibration image will reveal the underlying relative elasticity of the tissue.

Phantom and tissue imaging. Validation of this technique has been accomplished using tissue-mimicking phantoms. Phantoms are materials that have acoustical properties (speed of sound and attenuation) close to that of real tissue. The theoretical model of a small hard tumor in otherwise homoge-

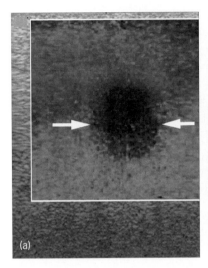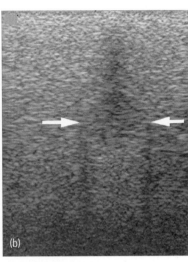

Fig. 1. Images of a 15-mm lesion in a phantom: (*a*) sonoelastography; (*b*) B-mode. The lesion is visible as a dark region indicating low vibration amplitude.

neous tissue can be tested by making the phantom homogeneous in its shear elastic properties except for small inclusions where the shear modulus is elevated. **Figure 1***a* shows a sonoelastography image of a 15-mm-diameter hard inclusion in such a phantom. The lesion is visible as a dark region indicating decreased vibration. The shear modulus of the lesion is seven times that of the background material. In the B-mode (conventional ultrasonic method) image (Fig. 1*b*) the lesion is not visible except for acoustic shadows off its boundary. The arrows indicate the same spatial locations in both images. Four pure-tone frequencies were used at the source vibration in order to reduce artifacts.

Sonoelastography has also been applied to tissue imaging. **Figure 2** shows a lesion induced in a calf liver by means of a radio-frequency (RF) ablation process. The lesion measures 27 mm in its long axis indicated by two arrows. RF ablation, a technique for destroying lesions in vivo, produces a palpably hard lesion. The long axis of the lesion after dissection measured 29 mm.

Tissue viscoelasticity. Many of the soft glandular tissues have been reported to exhibit relaxation responses when subjected to stress and hence cannot be completely characterized by the elastic constitutive equations. In the cases where a viscoelastic model more accurately describes the response of the tissue to stress, a concept from viscoelastic theory called the correspondence principle can be invoked. This principle states that the Laplace transform of the viscoelastic time domain constitutive equations correspond in form to the elastic constitutive equations. For the practical cases where the region of convergence of the Laplace transform includes the complex frequency axis, this has the effect of making the shear modulus complex and frequency-dependent. As a consequence, the stress and strain are out of phase, but as sonoelastography is interested in peak vibration amplitude, an accurate estimate of phase is not essential for image formation. Tumor imaging will still be possible as long as the

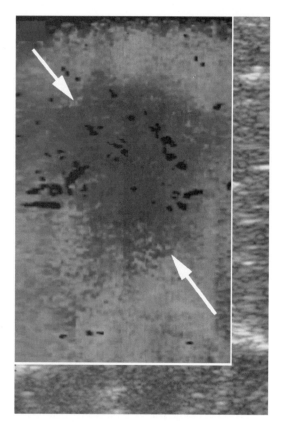

Fig. 2. Sonoelastography image of a 29-mm palpable lesion induced in an excised calf liver using a radio-frequency ablation process. The lesion is visible as an ellipse darker than the surrounding tissue.

frequency-dependent shear modulus of the lesion is higher than the frequency-dependent shear modulus of the background.

Clinical implications. Although sonoelastography has not become a routine clinical tool, numerous applications have been proposed, such as its use for diagnosing prostate cancer. The conventional ap-

proach to detecting this disease is a blood test for prostate specific antigen (PSA). However, PSA levels are not definitive in establishing the size and location of lesions. Because of the concern about metastasis, it is important to know if the cancer cells have penetrated outside the prostate. Application of three-dimensional (3D) ultrasound image acquisition to sonoelastography would provide 3D images of tumor shape, size, and location within the glandular capsule. Such images would make possible the estimation of tumor size, which some physicians believe would provide valuable information to determine the stage of the disease. **Figure 3** is an example of 3D sonoelastography of a prostate cancer; the boundary of the gland has been represented by an ellipsoid with axes equal to the measured dimensions of the gland. 3D images might also prove useful in biopsy planning.

The real-time capability of sonoelastography should also make it useful in guided biopsy. Conventional ultrasound is widely used for guided biopsy, but 20–30% of prostate cancers are isoechoic (not visible) in conventional B-mode (two-dimensional) imaging. As Fig. 1 shows, tumor detection in sonoelastography is not highly dependent on differences in echogenicity (echo signature), so guided biopsy of isoechoic lesions is another area where a contribution could be made.

For background information *see* ACOUSTICS; BIORHEOLOGY; DOPPLER EFFECT; LAPLACE'S TRANSFORM; MEDICAL IMAGING; MEDICAL ULTRASONIC TOMOGRAPHY; RHEOLOGY; SHEAR; TRANSDUCER; ULTRASONICS in the McGraw-Hill Encyclopedia of Science & Technology.   Lawrence S. Taylor; Kevin J. Parker

Bibliography. L. Gao et al., Imaging of the elastic properties of tissue—A review, *Ultrasound Med. Biol.*, 22:959–977, 1996; S. R. Huang, R. M. Lerner, and K. J. Parker, On estimating the amplitude of harmonic vibrations from the Doppler spectrum of reflected signals, *J. Acous. Soc. Amer.*, 88:310–317, 1990; K. J. Parker et al., Vibration sonoelastography and the detectability of lesions, *Ultrasound Med. Biol.*, 24(9):1437–1447, 1998; A. P. Sarvayan et al., Biophysical bases of elasticity imaging, *Acous. Imag.*, 21:223–240, 1995; L. S. Taylor et al., Three-dimensional sonoelastography: Principles and practices, *Phys. Med. Biol.*, 45(6):1477–1494, 2000.

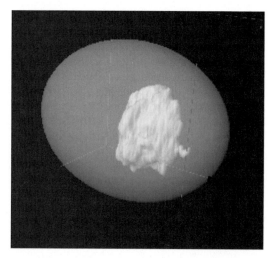

Fig. 3. Three-dimensional sonoelastography image of a cancerous lesion in a prostate excised during a radical prostatectomy. The lesion is the light-gray solid mass rendered within the gland boundary (transparent ellipsoid).

Sound propagation

In the last few years, there has been significant progress toward designing new types of transmitter arrays for acoustic and electromagnetic radiation. The radiation from these arrays would propagate without spreading and without diminishing in intensity for greater distances than previously was possible. A promising approach to the design of such arrays is through a new class of three-dimensional solutions of the relevant wave equations, which have been named acoustic Bessel bullets. These arrays could have wide-ranging applications in such fields

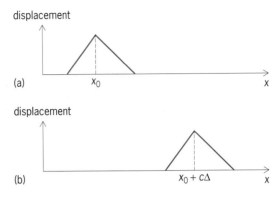

displacement

(a) x_0 x

displacement

(b) $x_0 + c\Delta$ x

Fig. 1. Displacement of a string versus position at two different times. (a) Time $= t_0$. The wave peak is located at $x = x_0$. (b) Time $= t_0 + \Delta$. The wave peak has moved to $x = x_0 + c\Delta$, where $c =$ wave speed.

as radar, sonar, biomedical ultrasonics, and nondestructive testing.

D'Alembert's solution. In 1750 Jean d'Alembert developed a general wave solution to the linearized small-amplitude one-dimensional wave equation. The well-known plane-wave solution consists of two traveling waves of arbitrary shape which travel in opposite directions with a constant wave speed. An important characteristic of the wave or pulse solution is that a pulse of energy exhibits no change in shape or amplitude as it propagates at the speed that is characteristic of small-amplitude waves in the medium. A simple example of such wave motion is a transverse wave traveling on a string under tension. **Figure 1** illustrates "snapshots" of the displacement of a string versus position at two different times. It is apparent that the wave is traveling in the $+x$ direction and exhibits no change in shape or amplitude.

Spatial spreading and directed energy. In contrast to wave propagation in one dimension, waves or pulses propagating in two and three spatial dimensions exhibit spatial spreading and thus a change in pulse shape and amplitude. Surface water waves associated with dropping a pebble in a pond or a periodic wave generator–plunger are simple examples of a two-dimensional wave field. The sound field emanating from a loudspeaker is a simple example of a three-dimensional wave field. As the distance from the source (pebble or loudspeaker) increases, the peak wave amplitude decreases since the energy is spread out over a larger area and the wave shape may also change. **Figure 2** illustrates the spatial spreading or divergence of sound from a harmonically excited horn loudspeaker where the acoustic field in a longitudinal plane in front of the loudspeaker was scanned with a microphone. A neon bulb, attached to the scanning microphone, was excited by the combined output of the microphone and a harmonic reference signal. The photographic long-term exposure of the variations in the brightness of the microphone–lamp combination provides a picture of the propagating sound field in the scanned area.

In contrast to the pebble in the pond, where the energy spreads out uniformly in all directions from the source, in many applications of interest it is important to direct energy from a source in a specific direction. A simple example is a communication system between a source and a receiver. The ability to transmit energy in a specific direction with minimal spatial spreading can provide a desired covert communication link or a means of minimizing cross interference among receivers at different spatial locations. In many other applications, it is also of interest to focus energy from a source to a target in order to affect the target. Common examples include the focusing of ultrasonic energy to destroy tissue, for example, lithotripsy, and the focusing of electromagnetic energy for various defense-related purposes.

Pulse-echo systems. Pulse-echo systems are used in widely different contexts for locating, tracking, or identifying targets via the transmission and detection of pulsed waves, which may be either electromagnetic, acoustic, or elastic in nature. Radar, sonar, biomedical ultrasonics, and nondestructive testing are a few of the areas where pulse-echo systems are utilized extensively. In all of these systems, energy is transmitted from a source to a target of interest. A generic pulse-echo system consists of a transmitter and a receiver. The purpose of the transmitter is to generate a concentrated directional wave field in the direction of interest, whereas the purpose of the receiver is to detect the echo from the target and determine distance to the target in addition to other characteristics of the target, such as velocity, shape, and composition.

The transmitted wave field generally exhibits both a near-field region and a far-field region. The far-field region exhibits the spreading characteristics of three-dimensional wave fields noted earlier. For the case of a uniformly excited planar array transmitting a sinusoidal signal, the radiated energy in the near field is confined to a cylindrical region (**Fig. 3**) when the dimensions of the array are large relative to the wavelength. The wave field at each point in this cylindrical near-field region is the result of a superposition of waves from each point in the planar array. The net result of the superposition is a coherent addition

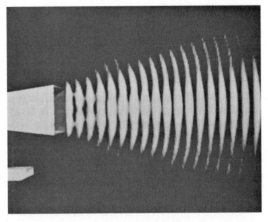

Fig. 2. Diverging sound field from a loudspeaker. (*From W. E. Kock, Seeing Sound, John Wiley, 1971*)

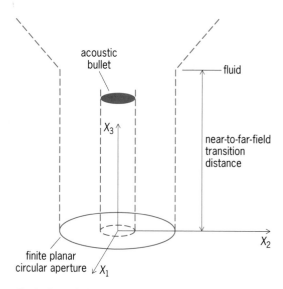

Fig. 3. Acoustic bullet launched from a finite planar circular aperture.

of waves in the near-field cylindrical region directly in front of the array, in contrast to significant cancellation effects in the external region. In the near-field region, the spatially averaged field amplitude in any plane transverse to the direction of propagation is approximately constant and independent of the distance from the aperture (that is, the transmitting planar array). Beyond the near-to-far-field transition distance, defined as the Rayleigh distance, the field amplitude varies inversely with the range due to spreading; that is, the field exhibits the ubiquitous law of spherical spreading. A simple estimate of the Rayleigh distance is the ratio of the area of the transmitting aperture divided by the wavelength.

Localized waves and Bessel bullets. As a result of the interest in extending the near-to-far-field transition distance beyond the Rayleigh distance for transmitting arrays of both electromagnetic and acoustic

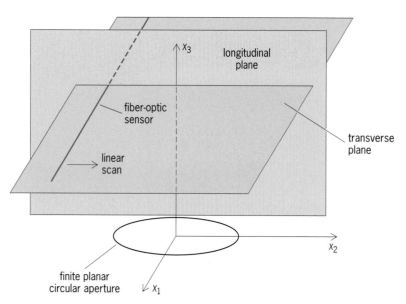

Fig. 4. Acoustooptic tomography for acoustic Bessel bullets.

radiation (including transmission in air and water, and biomedical applications), there has been considerable activity over the past few decades in developing new approaches to transmit energy. Much of this activity has been centered on searching for new solutions of the multidimensional wave equation, which describe the propagation of multidimensional wave and pulse phenomena. In contrast to the d'Alembert solution, these lesser-known localized wave solutions to the multidimensional wave equation offer some promise of eventually being able to design transmitting arrays that will generate signals that can be said to break the Rayleigh barrier. The term "localized" is used here to denote nonplanar wave fields which do not exhibit the usual spherical spreading law noted above; that is, the energy in the wave field is localized in space and time.

J. N. Brittingham discovered the first localized wave solution in 1983, which he termed the focus wave mode. A free parameter in the solution determines the overall characteristics of the field, which resembles a transverse plane wave at one extreme and a narrow spatially transverse pulse at the other extreme. R. W. Ziolkowski subsequently used the focus wave mode solution as a kernel for constructing new localized wave solutions. Most recently P. R. Stepanishen developed a new class of localized waves, which maintain their shape and amplitude as they propagate in space. These three-dimensional localized waves, which were designated as acoustic Bessel bullets, exhibit the same characteristic as the d'Alembert solution to the one-dimensional wave equation; that is, the waves exhibit no change in shape about the pulse center as they propagate.

Although the exact localized wave fields cannot be generated in all space from a finite planar aperture, the general characteristics of localized wave fields can be realized in a limited space-time region. In the near-field region the bullet propagates with no change in shape at a constant wave speed, and the lateral extent of the acoustic bullet may be smaller than the aperture. For example, from a finite circular aperture it is possible to launch an acoustic Bessel bullet with a smaller support region than the aperture itself in the near-field regions (Fig. 3). However, for this case, physical arguments imply that the field must display near- and far-field regions. In the near-field region the field must exhibit properties similar to those observed for the infinite-aperture case, and in the far-field region the field must exhibit an inverse-range dependence and spatial spreading. However, as a result of the inherently wide-band nature of the localized wave fields, the near-to-far-field transition range is not as clearly defined as the Rayleigh distance associated with the narrow-band harmonic fields discussed above.

Analytical and experimental efforts are under way to more completely investigate the near-to-far-field transition properties of localized wave fields. The analytical efforts are based on the use of space-time impulse response methods of analysis, which provide a direct means of numerically evaluating the evolution of the space-time localized wave fields that are transmitted from finite planar apertures. Acousto-optic experiments are being conducted to verify the analytical field predictions. In these experiments, a thin fiber-optic pressure sensor is used to measure a line integral of the acoustic pressure field in a transverse plane at a fixed distance from an aperture (**Fig. 4**). A line scan of such measurements can then be used to tomographically reconstruct the full temporal and spatial pressure field in the transverse plane. In contrast to the field in the longitudinal plane, the resulting space-time pressure distribution in the transverse plane can be readily backward projected toward the source or forward projected using available projection algorithms to determine additional properties of the near- and far-field regions of localized wave fields. The purpose of these joint investigations is to resolve the outstanding issue of whether it is possible to break the Rayleigh barrier using a localized wave approach.

For background information *see* ANTENNA (ELECTROMAGNETIC RADIATION); DIRECTIVITY; ELECTROMAGNETIC RADIATION; FIBER-OPTIC SENSOR; MEDICAL ULTRASONIC TOMOGRAPHY; RADAR; SONAR; SOUND; WAVE MOTION in the McGraw-Hill Encyclopedia of Science & Technology.   Peter R. Stepanishen

Bibliography. J. N. Brittingham, Focus wave modes in homogeneous Maxwell's equations: Traverse electric mode, *J. Appl. Phys.*, 54:1179–1189, March 1983; W. E. Kock, *Seeing Sound*, John Wiley, New York, 1971; J. W. S. Rayleigh, *The Theory of Sound*, Dover, New York, 1945; P. R. Stepanishen, Acoustic Bessel bullets, *J. Sound Vib.*, 222(1):115–143, 1999; P. R. Stepanishen, Acoustic bullets/transient Bessel beams: Near to far field transition via an impulse response approach, *J. Acous. Soc. Amer.*, 103:1742–1751, April 1998; P. R. Stepanishen and J. Sun, Acoustic bullets: Transient Bessel beams generated by planar apertures, *J. Acous. Soc. Amer.*, 102:1955–1963, December 1997; R. W. Ziolkowski, Exact solutions of the wave equation with complex source locations, *J. Math. Phys.*, 26:861–863, April 1985.

Space flight

The year 2000 was marked by milestones for some major programs in the field of space flight, and the conclusion of other programs that had made outstanding contributions over the past decade. In the future, the millennial year may be remembered for the beginning of the era of permanent human habitation in space, when the International Space Station was occupied by a resident flight crew. As the International Space Station progressed toward research capability, the Russian space station *Mir*, in orbit since 1986, was visited by a crew for the last time. Also completing its mission was the *Compton Gamma-Ray Observatory*, which was deorbited in June after 9 years of groundbreaking observations. A new era began in x-ray astronomy, as the *Chandra X-ray Observatory*, orbited by the National Aeronautics and Space Administration (NASA), carried out its first full year of observations and was joined by the European Space Agency's *XMM-Newton* observatory. Together they coordinated observations of objects out to the edge of the visible universe.

Closer to home, the space observatories *Ulysses, SOHO,* and *TRACE* observed the Sun at its time of maximum activity, and NASA's *IMAGE* spacecraft and the European *Cluster II* were launched in time to observe the effect of this activity on the Earth's magnetosphere. Exploration of the solar system also progressed as the *NEAR* spacecraft became the first to orbit an asteroid and carry out extended observations, and as NASA announced a revised program for the next two decades of Mars exploration that will, among other goals, search for possible evidence of present or past life. The space probes *Galileo* and *Cassini* coordinated observations of Jupiter as *Cassini* flew by the giant planet on its way to Saturn. The year was also marked by the debut of the Minotaur, Atlas 3A, and Delta 3 launch vehicles in the United States, and the Russian Fregat upper stage.

Seven flights from the two major space-faring nations (up from four in 1999) carried 37 humans into space (15 more than in 1999), including 5 women (the same as in 1999). This brought the total number of people launched into space (counting repeat travelers) to 871 (including 89 women) or 407 individuals (36 women). Some significant space events in 2000 are listed in **Table 1**, and the launches and attempts are enumerated by country in **Table 2**.

International Space Station

A critical milestone for the International Space Station program was passed in July 2000 with the launch to the station and docking of the Russian service module Zvezda. This opened the way for the first resident crew to inhabit the station in November.

When completed, the International Space Station will have a mass of about 1,040,000 lb (470 metric tons). It will be 356 ft (108 m) across and 290 ft (108 m) long, with almost an acre (0.4 hectare) of solar panels to provide up to 110 kilowatts of power to six state-of-the-art laboratories. Led by the United States, the station draws upon the scientific and technological resources of 16 nations: Canada, Japan, Russia, the 11 nations of the European Space Agency (ESA), and Brazil.

Launches and assembly of station elements in orbit began in 1998. This second phase of the Space Station program covers station assembly up to the initiation of orbital research capability with a permanent crew. It will be followed, in 2001 and later, by phase 3, which includes further expansion and completion, adding more laboratory and habitation facilities, structural trusses, and solar power arrays. The station will require a total of 45 assembly flights, with 33 to be launched by the United States shuttle and 12 on Russian boosters. Interspersed among these

TABLE 1. Significant space events in 2000*

Mission designation	Date	Country	Event
STS 99 (*Endeavour*)	February 11	United States	Shuttle Radar Topography Mission acquired high-resolution topographic map of Earth's land areas between 60°N and 56°S latitude and tested technologies for large rigid structures.
NEAR	February 14	United States	*Near Earth Asteroid Rendezvous* probe went into orbit about asteroid 433 Eros. Renamed *NEAR-Shoemaker*, it began a year of observations.
Galileo	February 22	United States	The Jupiter orbiter encountered the satellite Io at a distance of 123 mi (198 km), the closest so far.
IMAGE	March 25	United States	NASA's *Imager for Magnetopause-to-Aurora Global Exploration* satellite uses neutral-atom, ultraviolet, and radio imaging techniques to observe dynamics of the magnetosphere and auroras.
Soyuz TM-30/Mir 28	April 4	Russia	The *Mir 28* crew was launched—the last to occupy *Mir*, April 6–June 15.
STS 101 (*Atlantis*)	May 19	United States	This second crewed logistics-supply visit to the International Space Station was intended to prepare it for docking with the Zvezda Service Module.
ISS-1R (Zvezda)	July 12	Russia	The Zvezda Service Module, with the first human living quarters for the International Space Station, was launched. It docked with the ISS under remote control on July 25.
Samba and *Salsa* (*Cluster 2*)	July 16	Russia/Europe	The first two satellites of the European Space Agency's *Cluster 2* quartet were launched on a Russian Soyuz-Fregat rocket to study time and spatial variations of particles, fields, and waves in the magnetosphere.
Rumba and *Tango* (*Cluster 2*)	August 9	Russia/Europe	Third and fourth *Cluster 2* satellites.
STS 106 (*Atlantis*)	September 8	United States	Third logistics-outfitting flight to the International Space Station.
HETE-2	October 9	United States	The *High-Energy Transient Explorer-2* satellite was launched to detect and localize gamma-ray bursts.
ISS-3A/STS 92 (*Discovery*)	October 11	United States	Shuttle *Discovery* carried the truss structure Z1 and the third pressurized mating (docking) adapter, PMA-3, which were installed at the International Space Station.
ISS-2R/*Soyuz TM-31*	October 31	Russia	*Soyuz TM-31* was launched, carrying the first resident crew (Expedition One) to the International Space Station. William Shepherd, Yuri Gidzenko, and Sergei Krikalev inhabited the station for 138 days, from November 2, 2000, to March 20, 2001.
ISS-4A/STS 97 (*Endeavour*)	November 30	United States	This shuttle mission deployed solar array wings spanning 240 ft (73 m) at the International Space Station.
Cassini	December 30	United States	The spacecraft made its closest approach to Jupiter, on its flight to Saturn in 2004. It coordinated observations of Jupiter with *Galileo* and the Hubble Space Telescope.

* Information on crewed missions provided by Jesco von Puttkamer.

missions will be logistics missions by the shuttle (for a total of 37 shuttle flights), Russian *Soyuz* flights to launch crews, and multiple *Progress* tanker flights for refueling and supplying the growing structure in orbit.

STS 101. The May 19–29 flight of the shuttle *Atlantis* involved the second crewed logistics-supply visit to the Space Station. The crew of seven included veteran Russian cosmonaut Yuri V. Usachev and the future second long-stay Space Station crew (Susan J. Helms, James S. Voss, and Usachev). The shuttle was docked to the station May 21–26. The complex then included the power and control mod-

ule FGB/Zarya and a connecting module and passageway, Node 1/Unity, both of which had been deployed late in 1998. The primary objectives of the mission were to remove and replace Zarya hardware that had failed, such as battery systems, and to reboost the station to an altitude consistent with the forthcoming Zvezda launch and docking. During a space walk on May 21–22, mission specialists Jeffrey N. Williams and Voss reinstalled a loose United States crane, assembled and installed a Russian cargo crane, replaced a communications antenna outside Unity, and installed handrails on its outside walls. On May 22 the crew entered the Space Station and

Country	Number of launches†	Number of attempts
Russia	34	(36)
United States (NASA, Department of Defense, commercial)	30	(31)
Europe (European Space Agency, Arianespace)	12	(12)
People's Republic of China	5	(5)
Japan	—	(1)
Total	81	(85)

*Data provided by Jesco von Puttkamer.
†Successful launches to Earth orbit and beyond

Zvezda is also equipped with the propulsion system that will maintain the attitude and orbit of the Space Station. The station now spanned 119 ft (36 m) in length, and its mass had grown to almost 60 tons (54 metric tons). On August 8 the automated *Progress M1-3* docked with the station, carrying fuel for Zvezda's propellant tanks, as well as equipment and supplies.

STS 106. The September 8–20 flight of *Atlantis* was also a logistics-outfitting mission to the Space Station. The crew of seven included two Russian cosmonauts. *Atlantis* was docked to the station September 9–17. During a space walk on September 10, mission specialists Edward T. Lu and Richard A. Mastracchio prepared the station exterior for future operations. There followed a week of docked operations, during which the crew worked as movers, cleaners, plumbers, electricians, and cable installers, stowing more than 6600 lb (3 metric tons) of supplies from *Atlantis* and *Progress M1-3*, and filling the *Progress* craft with trash and discarded materials, prior to its jettisoning and incinerating atmospheric reentry.

STS 92. The October 11–24 flight of *Discovery* was an assembly mission. It brought the first truss structure segment, called Z1 (for its zenith port location on Unity), and the third docking adapter, PMA-3 (pressurized mating adapter), to the Space Station. On October 14, the Z1 truss was installed on Unity,

conducted repair and replacement activities over the next 4 days, increasing the station's altitude in three separate maneuvers, transferring gear to the station, and returning other cargo to *Atlantis*.

Zvezda. The service module, named Zvezda (Star), the first Russian-funded and -owned element of the station, was launched from Baikonur Cosmodrome in Kazakhstan on July 12, and linked up with the station on July 25. The docking, executed very gently at 0.6 ft/s (0.2 m/s) by Zarya as the active partner, marked the arrival of the first human living quarters and an initial command post for the station.

Fig. 1. International Space Station with its solar arrays deployed, photographed from the space shuttle *Endeavour* following undocking on December 9, 2000. (*NASA*)

and electrical connections were hooked up and other deployment activities were accomplished during a space walk the next day. On October 16, PMA-3 was installed on Unity's nadir port during a second space walk. Two additional space walks on October 17 and 18 completed the assembly tasks for the mission.

Soyuz TM-31. On October 31, a Soyuz-U carrier lifted off from Baikonur and placed in orbit *Soyuz TM-31*, carrying the first resident crew for the Space Station. The Expedition One crew consisted of American commander William Shepherd and veteran Russian cosmonauts Yuri Gidzenko and Sergei Krikalev. On November 2, the *Soyuz* craft docked with the station and the crew entered, to begin a new era of permanent human habitation in space. The Expedition One crew would remain on the station for 138 days, until March 20, 2001.

On November 17, the automated *Progress M1-4* arrived, carrying fuel, oxygen, equipment, and supplies for the station. The carrier's automatic docking system failed, forcing Gidzenko to perform the docking under manual control.

STS 97. The November 30–December 11 flight of *Endeavour* was another assembly mission, on which were installed two solar array wings, jointly spanning 240 ft (73 m), at the Space Station (**Fig. 1**). After docking on December 2, Canadian astronaut Marc Garneau used the shuttle's remote manipulator system to park the truss segment P6 above the shuttle's cargo bay to allow for its temperature adjustment. On December 3, Garneau, assisted by Carlos I. Noriega and Joseph R. Tanner on their first of three space walks, attached P6 to the truss segment Z1, and Commander Brent W. Jett deployed the starboard solar array wing. Deployment of the other wing was delayed until the next day when the tensioning of the two array blankets on the first wing malfunctioned. Two additional space walks were conducted on December 5 and 7, the latter to correct the starboard wing's tensioning and to install an electrical potential sensor, bringing the total number of space walks in support of the station to 13. *Endeavour* undocked from the Space Station on December 9. The P6 is now providing the station with up to 62 kW of power.

United States Space Activities

The four reusable shuttle vehicles of the U.S. Space Transportation System (STS) continued carrying people and payloads to and from Earth orbit. All but one of their missions were for the supply, logistics, outfitting, and orbital assembly of the International Space Station. Numerous satellites and space probes continued to pursue scientific, defense, and commercial activities.

Space shuttle. During 2000, NASA successfully completed five space shuttle missions, two more than in 1999, bringing the total number of shuttle launches since program inception to 101.

STS 99, the February 11–22 flight of *Endeavour*, was the only shuttle mission in 2000 that did not involve the International Space Station. The primary objective of the Shuttle Radar Topography Mission (SRTM) was to acquire a high-resolution topographic map of the Earth's landmass and to test new technologies for large rigid structures and measurement of their distortions to extremely high precision. A 200-ft (61-m) radar mast was deployed from the *Endeavour*'s cargo bay and, divided into two shifts, the crew of six (including astronauts from Europe and Japan) worked on a 24-hour-a-day schedule to complete the radar mapping of about 80% of the Earth's surface. SRTM was a breakthrough in remote sensing which will produce topographic maps of the Earth 30 times as precise as the best global maps now in use.

Advanced transportation systems activities. NASA announced on March 1, 2001, that the X-33 and X-34 projects would not receive any additional funding, effectively ending its first cooperative efforts with private industry to demonstrate in suborbital flight the technologies needed for a reusable space launch vehicle. The X-33 suffered a major setback in 1999 when one of its composite (material) liquid hydrogen fuel tanks failed due to microcracking, delaying the project by perhaps several years. After review, NASA determined that the benefits of continuing both the X-33 and X-34 programs did not justify their expected cost.

NASA at the same time announced a new 5-year project, the Space Launch Initiative, also called the second-generation reusable launch vehicle program. On May 17, 2001, NASA awarded 22 contracts for developing the technologies that will be used to build an operational reusable space launch vehicle before 2015.

The second-generation reusable launch vehicles are expected to fly a hundred times a year, with a 10-member launch crew. For comparison, the space shuttle now flies 10 times or less a year and requires a 170-member launch crew. The new vehicles also are expected to cost 10 times less to operate, compared to the space shuttle.

NASA's Hyper-X program suffered a setback in 2001 when the first of three planned X-43 hypersonic test flights ended in the intentional destruction of the research vehicle after it lost control shortly after launch over the Pacific Ocean. Its hydrogen-powered, air-breathing propulsion system, scheduled for testing at Mach 7 (seven times the speed of sound), had not yet been activated. NASA immediately began an investigation. *See* HYPERSONIC FLIGHT.

In 2000, NASA continued flight-testing of the X-38 crew return vehicle. This vehicle is expected to take over the escape function for the Space Station (which currently uses a three-seat Soyuz capsule) and carry as many as seven crew members in the event of an accident.

On November 2, the latest X-38 prototype—the 131-R, an 80% completed version of the crew return vehicle—was air-dropped from a B52 from 36,000 ft (11,000 m) and successfully made a 9-minute flight, landing in California's Mojave Desert. In 2002, NASA planned to release a remotely operated X-38 from the space shuttle that would fly back to and land on Earth.

Space sciences and astronomy. Since 1995, NASA has markedly increased the number of civilian satellites it develops and launches, following the switch in development strategy, shortly after the loss in August 1993 of the *Mars Observer*, from large expensive satellites to smaller and more numerous ones.

Hubble Space Telescope. In January 2000 the Hubble Space Telescope resumed observations, following its successful servicing mission the previous month. April 24 was the tenth anniversary of the launch of this first large telescope to make observations above the distortions of the Earth's atmosphere. Hubble has continued to make discoveries at a rate that is unprecedented for a single observatory, and its contributions to astronomy are wide-ranging. Its notable achievements include measuring the size and age of the universe, observations of young galaxies from more than 10 billion years ago, and observations of disks around young stars. In its first 10 years, Hubble studied 13,670 objects, made 271,000 individual observations, and returned 3.5 terabytes of data. Its achievements have resulted in over 2651 scientific papers.

Compton. On June 4 the *Compton Gamma-Ray Observatory* was deorbited and broke up in the atmosphere. *Compton*, which had revolutionized the field of gamma-ray astronomy since its launch in 1991, had lost one of its three onboard gyroscopes. NASA feared that if a second gyroscope failed it could lose control of the craft, leading to an uncontrolled fall like that of *Skylab* in 1979. It therefore brought down *Compton* in a series of four controlled burns, so that it fell harmlessly into the South Pacific. *See* GAMMA-RAY ASTRONOMY.

HETE-2. The *High-Energy Transient Explorer-2* (*HETE-2*), launched on October 9, is a small scientific satellite designed to rapidly detect and localize gamma-ray bursts. The coordinates of bursts detected by *HETE-2* are distributed to ground-based observers within seconds of burst detection, thereby allowing detailed observations of their initial phases. The science payload consists of one gamma-ray and two x-ray detectors. *HETE-2* was launched into Earth orbit aboard a Pegasus XL winged rocket that had been dropped from an L-1011 jet aircraft near Kwajalein Atoll in the Marshall Islands, although the launch was controlled from Cape Canaveral Air Force Station in Florida. It replaced a similar satellite lost during launch in 1996. *See* GAMMA-RAY BURSTS.

Chandra. The *Chandra X-ray Observatory*, NASA's third "Great Observatory" (after Hubble and *Compton*), was launched on July 23, 1999, and began science observations that October. Its targets include early galaxies, gamma-ray bursters, the intergalactic medium, black holes at the centers of galaxies, supernova remnants, pulsars, x-ray binaries, clusters of young stars, stellar atmospheres, comets, and planets. For further discussion of the mission *see* X-RAY ASTRONOMY.

Galileo. *Galileo* has been orbiting Jupiter and its satellites since December 7, 1995. It completed its primary mission in December 1997, and a 2-year extended mission, the Galileo Europa Mission, with a flyby of Europa on January 3, 2000. NASA Headquarters then agreed to a second extension, the Galileo Millennium Mission. On February 22 the spacecraft had its closest encounter so far with Io, flying only 123 mi (198 km) above the satellite. This flyby concentrated on studies of volcanoes and the plasma torus generated along the satellite's orbit by volcanic ejecta. Encounters with Ganymede followed on May 20 and December 28. Between them, the orbiter flew its farthest from Jupiter, at about 12×10^6 mi (20×10^6 km) from the planet, on September 8, passing outside of the Jovian magnetosphere for the first time. In December, *Galileo* made joint observations with the *Cassini* probe, which was flying by Jupiter on its way to Saturn. *Galileo* made a close pass of Jupiter on December 29, while *Cassini* passed its closest to the planet on December 30. *Galileo* was inside Jupiter's magnetosphere while *Cassini* was well outside, and the pair studied how changes in the solar wind affect conditions in near-Jovian space.

In August, NASA researchers announced new evidence for an ocean of water beneath Europa's surface based on *Galileo* magnetometer measurements during flybys of the satellite. Europa's magnetic field reverses direction in response to reversals of Jupiter's magnetic field at Europa, as would be expected if there is a conducting liquid, such as salt water, beneath the surface.

Cassini. At its closest on December 30, *Cassini* was 6.1×10^6 mi (9.8×10^6 km) from Jupiter. From October 2000 through March 2001, the craft carried out a program of observations of the Jupiter system, including atmospheric dynamics and composition, Jupiter's magnetic environment, and the interactions between Jupiter and its satellites. Joint observations were conducted with the Hubble Space Telescope and *Galileo*. The spacecraft will reach Saturn in July 2004.

Mars exploration. In March, a panel studying the failure of the *Mars Polar Lander* on December 3, 1999, concluded that the touchdown sensor may have been triggered prematurely by the unfolding of the craft's landing legs, and the craft, sensing it had already landed, may have failed to fire its braking rockets. More broadly, the panel concluded that the failure was caused by significant underfunding and understaffing.

Following the failures of both the *Mars Polar Lander* and the *Mars Global Surveyor* in late 1999, NASA decided to postpone its next landing mission, and to send a single orbiter to Mars in 2001. The *Mars Odyssey* spacecraft was prepared for launch on April 7, 2001. In October 2000, NASA announced a revamped Mars program for the next two decades which is broader and slower-paced than the one it replaces. It alternates launches of orbiters and landers, allowing additional time for recovery should one of the vehicles fail.

Mars Global Surveyor continued to produce high-resolution images of Mars. The Mars Orbiter Laser Altimeter counted numerous scientific achievements. Among them was the production of the highest-integrity global topographic model for any planet, including Earth. It reveals the 19-mi (30-km) dynamic range of topography, the pole-to-pole slope that controlled the transport of water in early Martian history, and the flat northern hemisphere that may represent the location of a large ancient ocean. In June, scientists announced that gullies seen on Martian cliffs and crater walls in a small number of images from the Mars Orbiter Camera suggest that liquid water has seeped onto the surface in the geologically recent past.

NEAR-Shoemaker. On February 14, the *Near Earth Asteroid Rendezvous* (*NEAR*) spacecraft was placed in orbit about the asteroid 433 Eros. It was renamed *Near-Shoemaker* and began a year of observations. For further discussion of the mission *see* ASTEROID.

Deep Space 1. NASA's *Deep Space 1* was launched on October 24, 1998, into an elliptical solar orbit to test several advanced technologies for future interplanetary science missions, particularly the ion engine, which provides high-efficiency, low-thrust power. In November 1999, after encountering the asteroid 9969 Braille, the craft lost the use of its star tracker, throwing a planned encounter with Comet Borrelly in September 2001 into doubt. Without the star tracker, the craft could not tell which way it was pointing, and navigation seemed impossible. However, mission engineers were able to reprogram the spacecraft to use its science camera, the Miniature Integrated Camera Spectrometer, to substitute for the star tracker, even though it had not been designed to work with the orientation system and its field of view was only about one-hundredth the area of the tracker's. *Deep Space 1* was then able to resume using the ion engine on June 21, 2000, gradually modifying its orbit to encounter the comet. *See* ELECTRIC PROPULSION.

Stardust. The *Stardust* probe was launched on February 7, 1999. Between January 18 and 21, 2000, the spacecraft fired its engine three times so that it could swing past Earth in January 2001 on its way to meet Comet Wild-2 in 2004. On February 22, the spacecraft's aerogel collector was deployed. It is collecting interstellar dust, including recently discovered dust streaming into the solar system from the direction of Sagittarius. It is to collect dust and carbon-based samples during its encounter with the comet and to return all its samples to Earth in 2006.

SOHO. The European–United States *Solar and Heliospheric Observatory* (*SOHO*) continued to send back spectacular images of solar flares as the Sun reached its maximum activity, observing a series of large eruptions in late November that bathed Earth in high-energy particles. In December, *SOHO* began

a series of stereo solar wind studies with the *Ulysses* probe, then flying over the solar south pole.

Using the technique of helioseismology, which uses ripples on the Sun's surface to probe its interior, *SOHO* scientists for the first time imaged storms on the side of the Sun facing away from the Earth. This technique may allow a week's advance warning before these storms rotate to the Sun's near side, bombarding the Earth's magnetosphere with radiation and particles that are capable of disrupting satellites, telephone and radio communications, and power systems. In June, scientists announced a model, based on observations from *SOHO* and the *WIND* spacecraft, that reliably predicts the time required for electrified gas clouds from coronal mass ejections to traverse the space between Sun and Earth, allowing more accurate prediction of ensuing space storms.

Ulysses. From its orbit outside the ecliptic plane the European–United States *Ulysses* spacecraft continues to study the solar wind and solar magnetic field, as well as the intensity of comic rays, and is used to help triangulate positions of gamma-ray bursts. From September 8, 2000, through January 16, 2001, the probe was within $7°$ of the direction of the solar south pole (though several hundred million miles away from it). This pass, and the north polar pass from September 3 to December 12, 2001, come at solar maximum, the time of greatest solar activity, and complement the observations from the *Ulysses* polar passes at solar minimum in 1994–1996. In April, scientists announced that data from the solar-wind ion-composition spectrometer and the magnetometer indicated that the craft had flown through the tail of Comet Hyakutake on May 1, 1996, when the comet was close to the Sun. *Ulysses* was then about 300 million miles (500 million kilometers) from the nucleus of the comet, far beyond the visible tail, indicating that comet tails may be much longer than previously believed. In June, *Ulysses*' mission was extended to its return to the vicinity of Jupiter in late 2004.

TRACE. The *Transition Region and Coronal Explorer* (*TRACE*) spacecraft images the solar atmosphere from the photosphere to the corona, with high temporal and spatial resolution. *TRACE* has sent back spectacular images of hot, electrified coils of gas known as coronal loops (**Fig. 2**). In September, scientists announced that these observations have revealed an important clue to a long-standing mystery, the location of the heating mechanism that makes the corona 300 times hotter than the Sun's visible surface. They show that most of the heating occurs low in the corona, within 10,000 mi (16,000 km) of the Sun's visible surface, near the bases of the loops where they emerge from and return to the Sun's surface, rather than uniformly throughout the loops, as had been previously assumed.

FUSE. In January 2000 it was reported that NASA's *Far Ultraviolet Explorer* (*FUSE*), launched on July

Fig. 2. Coronal loop, imaged by the *Transition Region and Coronal Explorer* (*TRACE*) spacecraft. An image of the Earth is superimposed to indicate the scale. (*NASA*)

24, 1999, had observed large amounts of O VI (oxygen with five of its eight electrons stripped away) in the halo of hot gas that surrounds the Milky Way Galaxy. Scientists reasoned that the only way to make this amount of O VI was through collision with blast waves from supernovae, indicating that supernovae generated the halo as the Milky Way Galaxy evolved.

Pioneer 10. Launched in 1972, *Pioneer 10*, the longest-lived interplanetary explorer, continues its voyage, and data continue to be received from the spacecraft via NASA's Deep Space Network. In June, 2001, *Pioneer 10* was 7.3×10^9 mi (11.8×10^9 km) from Earth. Cosmic-ray data from the spacecraft indicated that, at a distance of nearly 78 astronomical units from the Sun (nearly twice Pluto's mean distance from the Sun), it was still under the influence of solar activity and had not yet reached the cosmic-ray modulation boundary of the heliosphere. Signals transmitted by the spacecraft need nearly 11 hours to reach Earth.

Earth science. The large *Landsat 7* and *Terra* satellites launched in 1999 continued to provide valuable data.

IMAGE. NASA launched the *Imager for Magnetopause-to-Aurora Global Exploration* (*IMAGE*) satellite from Vandenberg Air Force Base in California on March 25, 2000. Its highly elliptical orbit, ranging from 621 mi (1000 km) to 28,503 mi (45,781 km) at apogee over the North Pole, takes the satellite through the most intense parts of the magnetosphere's radiation belts. Previous spacecraft explored the magnetosphere by detecting particles and fields along their paths, but *IMAGE* is the first spacecraft to view the entire magnetosphere through "plasma-colored glasses" in the words of one scientist. Of particular interest is its behavior during magnetic storms. A suite of three neutral-atom imaging instruments records the glow of fast atoms coming from throughout the Earth's magnetic field, revealing the shape and motion of plasma clouds that make up such storms. There are two ultravio-

let cameras, one for imaging auroras from space and the other for capturing global images of the plasmasphere. Finally, a radio plasma imager provides a three-dimensional view of the plasmasphere by sounding it with radio pulses. It uses four wire antennas over 800 ft (250 m) long.

NOAA 16. On September 21, 2000, an Air Force Titan 2 missile launched the National Oceanic and Atmospheric Administration's *NOAA 16* satellite into a Sun-synchronous polar orbit. It replaced *NOAA 14*, which had drifted out of its proper orbit. The second of five new Polar Operating Environmental Satellites, it images clouds; monitors polar ice, soil moisture, and air pollution; and has international search and rescue capability.

Department of Defense activities. United States military space organizations continued their efforts to make space a routine part of their operations. Their were 10 military launches from Cape Canaveral, Florida, and Vandenberg Air Force Base, California. Three of them were satellites for the Global Positioning System.

Commercial space activities. A strong ongoing demand for satellites, launch vehicles, and ground equipment driven by the expansion of the telecommunications industry continued to support the infrastructure segment of the space industry. The development of LEO satellite constellations, which require a large number of small satellites in orbit, had begun to dramatically expand the requirements for satellite launches. Two early satellite mobile-phone projects, Iridium LLC and ICO Global Communications Ltd., filed for bankruptcy protection late in 1999. Projects such as Globalstar and Orbcomm have found more success, but as of mid-2001, no such project had demonstrated a profitable combination of user demand at an attractive price.

The Sea Launch system suffered a failure on March 12, 2000, after two successes in 1999. The second stage of its Zenit-3SL rocket went off course, resulting in the loss of an ICO communications satellite. However, two subsequent Sea Launch flights were successful. Sea Launch Co. was formed in 1995 by companies in the United States, Russia, Norway, and Ukraine. The Sea Launch system consists of a floating mission control center and rocket assembly factory (Commander) in the South Pacific, a self-propelled launch platform (Odyssey) nearby, and a Russian-Ukrainian Zenit-3SL rocket.

Three new launch vehicles made their first successful flights in 2000. On January 26, the Air Force Minotaur rocket was launched from Vandenberg carrying nine satellites. Assembled by Orbital Sciences Corporation, it combines two stages of a retired Minuteman 2 intercontinental ballistic missile (ICBM) and two stages of Orbital's own Pegasus XL launch vehicle. A second Minotaur launch in July was also successful. On May 24, Lockheed Martin's Atlas 3A rocket was launched from Cape Canaveral, carrying the *Eutelsat W4* communications satellite. The first stage of the Atlas 3A, which is a modern version of the first United States ICBM, was propelled by the powerful Russian RD 180 engine, which had been used in Soviet missiles. On August 23, Boeing's Delta 3 placed the dummy *DM-F3* satellite in orbit, following launch failures in 1998 and 1999. The Delta 3 is considered an evolutionary step from the Delta 2, which has been a standard space-launch vehicle for many years, to the Delta 4 family, which is being designed primarily for the Air Force's Evolved Expendable Launch Vehicle program.

Of the 31 launch attempts by the United States in 2000 (versus 33 in 1999), 23 were on commercial expendable launchers (plus 5 space shuttles and 3 launches of the military Titan 2 and 4), with only one failure. These 23 launches comprised 7 Atlas 2A, 6 Delta-2 vehicles, 3 sea-launched Zenits, 2 Pegasus XL, 2 Minotaurs, 1 Atlas 3A, 1 Delta 3, and 1 Taurus.

Russian Space Activities

Despite the continuing slump in the national economy and political uncertainties, Russia in 2000 showed no slack in its space operations. Its 34 successful launches (out of 36 attempts) was a significant increase from the previous year's 26 (out of 28 attempts): 14 Protons, 13 Soyuz-U (2 crewed, including 4 with the Fregat upper stage), 2 Zenit-2, 2 Kosmos-3M (plus one failure), 1 Dnepr, 1 Rokot, and 1 Start-1. On December 27, a Tsiklon-3 booster carrying six communications satellites failed.

In its partnership with the United States in the development of the International Space Station, Russia launched the Zvezda Service Module, the long-awaited third building-block of the station, as well as *Soyuz TM-31*, which carried the first resident crew to the station, and two *Progress-M1* automated resupply ships.

Space station Mir. By the end of 2000, Russia's seventh space station, in operation since February 20, 1986, had circled the Earth approximately 85,000 times. However, the station remained unoccupied except for the 2-month visit of the *Soyuz TM-30* crew. In early spring, RKK-Energia leased *Mir* to the Netherlands-based firm MirCorp. The firm sought commercial customers, particularly tourists, but these plans did not materialize.

Soyuz TM-30 was launched on April 4 with the twenty-eighth *Mir* crew, Sergei Zalyotin and Alexander Kaleri. It docked to *Mir* on April 6, after the station had orbited for 233 days without occupants. The crew had the difficult task of returning the aging station to safe operating condition. After 2 weeks of laborious tests, they succeeded in sealing an air leak in a hatch between the core module's transfer compartment and the airless *Spektr* module. The crew departed in *Soyuz TM-30* on June 15 and landed the next day. The station remained unoccupied until it was deorbited on March 22, 2001.

Russian commercial activities. The Russian space program's efforts to enter the commercial market continued in 2000. On February 12, flights of the Proton launcher resumed, following suspension of

the rocket after the second Proton crash in 1999, on October 27. Between 1995 and 2000, 151 Proton and 371 Soyuz rockets were launched, with nine failures of the Proton and nine of the Soyuz. Of the 14 Protons launched in 2000, seven were for commercial customers.

The Fregat upper stage was first launched on a Soyuz rocket on February 8, 2000. It will enable Soyuz and other rockets to lift larger payloads into space. It flew again on March 20 carrying dummies of the European *Cluster II* space probes, before launching these probes in July and August.

Of the five launches of the Russian-Ukrainian Zenit rocket, three were conducted from the ocean-based Sea Launch facility Odyssey. Only one of the sea launches failed. The satellite navigation system GLONASS, Russia's equivalent of the Global Positioning System, requires 24 satellites, but there were only 18 in the constellation by late 2000.

European Space Activities

Arianespace operated a successful series of eight Ariane 4 launches, one less than in 1999, carrying nine commercial satellites for customers such as Japan, Brazil, Egypt, France, and Canada. By the end of 2000, the Ariane 4 had flown 130 times, with seven failures, from its Kourou, French Guyana, spaceport.

The first commercial launch of the Ariane 5 heavy-lift booster on December 10, 1999, carried the *X-ray Multimirror* (*XMM*) *Observatory* of the European Space Agency, later renamed *XMM-Newton*. It was followed by four successful Ariane 5 launches in 2000, carrying 11 payloads for customers such as India, the United Kingdom, and Japan.

XMM-Newton, Europe's equivalent of NASA's *Chandra X-ray Observatory*, began routine observations in June. In its first months of observations, it solved the greater part of the mystery of the x-ray background emission, showing that 80–90% of it originated from a very large number of x-ray point sources, which the telescope detected. Other observations included detailed studies of the hot x-ray-emitting gas between the members of galaxy clusters, studies of the composition of supernova remnants, and analyses of x-ray-emitting matter on the verge of disappearing into black holes.

The four identical satellites of the European Space Agency's *Cluster II* fleet were launched in pairs on Russian Soyuz-Fregat rockets on July 16 and August 9, 2000. The satellites fly in a tetrahedral formation, with separations varying from a few hundred to tens of thousands of miles, on elongated, highly inclined orbits that take them in and out of the Earth's magnetosphere. They study time and spatial variations of particles and electromagnetic fields and waves in the magnetosphere, investigating the ways in which the Sun and Earth interact by monitoring the magnetospheric bow shock, the cusp region over each magnetic pole, and the magnetotail. On November 16 the deployment of four 150-ft (45-m) wire antennas on each of the four satellites was completed.

Cluster II replaces four satellites that were lost in the failure of the first flight of the Ariane 5 rocket in 1996.

Asian Space Activities

India, China, and Japan have space programs capable of launch and satellite development activities, although India attempted no launches in 2000.

Japan. Japan progressed with its satellite production, acquisition, and launch programs. Its space plans encompass missions to the Moon and to Mars. The Mars probe *Nozomi* was launched in 1998 but, due to a thruster malfunction, its arrival date at Mars has slipped from October 1999 to December 2003 or January 2004. Japan is also participating in the International Space Station Program, with the development of the Japanese Experiment Module (JEM), now called Kibo (Hope), along with its ancillary remote manipulator system and porchlike exposed facility.

Japan's only launch in 2000 failed on February 10. A solid-fueled M-V-4 vehicle carrying the x-ray satellite *Astro E*, launched from the Kagoshima facility, suffered a malfunction in its first-stage motor, resulting in the loss of the satellite. *Astro E* was a large observatory, intended to complement *Chandra* and *XMM-Newton*.

In July the largest solar flare in a decade initiated a series of events that caused the premature demise of another Japanese x-ray satellite, the *Advanced Satellite for Cosmology and Astrophysics* (ASCA). The resulting ballooning of the Earth's uppermost atmosphere produced drag forces on the spacecraft that overwhelmed its attitude control system, leading to a series of other problems. Efforts to save the spacecraft were unsuccessful, and it entered the Earth's atmosphere on March 2, 2001, after 8 years in orbit.

China. The space program of the People's Republic continued to progress in 2000. There were five successful missions of the Long March (Chang Zheng, CZ) rocket, one more than in 1999. At the end of 2000 the Long March had achieved 22 consecutive successful launches since October 1996, following a string of failures in 1995 and 1996. The launches comprised a CZ-3A carrying the *Zhongxing-22* communications satellite, a CZ-3 carrying a *Fengyun-2* meteorological satellite, a CZ-4B carrying a *Zi Yuan(ZY)-2* remote sensing satellite, and two CZ-3A, each carrying a *Beidou* navigation satellite. The two *Beidou* satellites form China's first satellite navigation positioning system, designed to lessen the country's dependence on foreign technology.

For background information *see* ASTEROID; COMET; COMMUNICATIONS SATELLITE; GAMMA-RAY ASTRONOMY; JUPITER; MAGNETOSPHERE; MARS; MILITARY SATELLITES; SATELLITE ASTRONOMY; SOLAR WIND; SPACE FLIGHT; SPACE PROBE; SPACE SHUTTLE; SPACE STATION; SPACE TECHNOLOGY; SUN; ULTRAVIOLET ASTRONOMY in the McGraw-Hill Encyclopedia of Science & Technology.   Jonathan Weil; David Blumel

Bibliography. European Space Agency, *Press Releases 2000*; *Jane's Space Directory, 2000-2001*; J. McDowell, Mission Update, monthly column in *Sky & Telescope*; NASA Public Affairs Office, *News Releases 2000*; *State of the Space Industry—2000*, Space Publications LLC; *United States Space Directory—2000*, Space Publications LLC.

Spectroscopy (CRLAS)

Direct absorption spectroscopy is the ideal method for measuring concentration and chemical composition in a variety of environments. However, it generally requires a very stable light source, such that continuous-wave lasers must be used to achieve high sensitivity. This is unfortunate because pulsed lasers cover a larger portion of the spectrum and are considerably easier to use. The development of cavity ringdown laser absorption spectroscopy (CRLAS) has allowed researchers to use simple pulsed laser technology in direct absorption experiments, achieving sensitivities rivaling that of more complicated continuous-wave techniques.

Basic principles. CRLAS measures the rate of decay (ringdown) of light trapped in a cavity formed by two highly reflective mirrors (see **illus.**). The intensity of the light exiting the cavity decays exponentially with a decay constant, which depends on the mirror reflectivity, cavity length, and absorption of light by the sample in the cavity. Given a reflectivity R, cavity length L, and per pass absorption A, the intensity of light exiting the cavity is given by the equation $I = I_0 e^{-tc(1-R+A)/L}$, where c is the speed of light and I_0 is the initial intensity. A suitable detector placed behind one of the mirrors monitors the light exiting the cavity. The detector output is amplified, digitized, and sent to a computer for analysis.

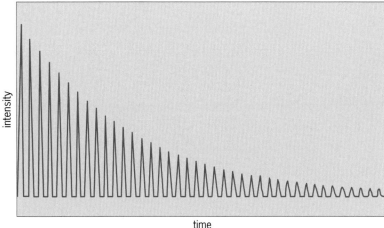

Typical cavity ringdown laser absorption spectroscopy setup and detector output. The laser pulse is injected into the cavity through one of the mirrors, and the intensity decay is monitored with a detector. By fitting the detector output to a single exponential function, per pass losses can be determined to within 1 part per million.

By determining the decay constant, it is possible to calculate the amount of light absorbed by the sample. Because the decay rate is independent of the initial light intensity I_0, CRLAS is insensitive to intensity fluctuations in the light source. As a result, pulsed lasers can be used in CRLAS experiments to achieve ultrahigh sensitivity.

The sensitivity of CRLAS depends on how accurately the decay constant can be determined. In experiments using pulsed lasers, the decay constant can be easily determined to within 1%. Given a reflectivity of 99.99%, which translates to a per pass loss of 100 photons in 10^6 for an empty cavity, the sensitivity is 1 part per million (ppm). This is a two orders of magnitude improvement over Fourier transform infrared (FTIR) spectrometers. The availability of high-reflectivity mirrors is crucial for attaining this sensitivity. With 99.95% reflectivity mirrors, the per pass losses rise to 500 ppm and the sensitivity is decreased to 5 ppm. Techniques using high-resolution continuous-wave lasers and active stabilization of the ringdown cavity can achieve sensitivities of up to 1 part per billion, albeit with much added experimental complexity.

Applications. CRLAS has been extensively used to study trace species produced in supersonic molecular beams. Metal clusters and metal-containing molecules have been studied in the ultraviolet and visible region of the spectrum. New features in the spectrum of copper and aluminum dimers were observed using CRLAS; they were invisible to "action" techniques, such as laser-induced fluorescence and resonant multiphoton ionization, due to the rapid predissociation of the excited state. With the extension of CRLAS to the mid-infrared, rotation-vibration spectra of hydrogen-bonded complexes, such as water and alcohol clusters, were observed in the bonded OH stretching region, yielding important new information on hydrogen bond cooperativity. Molecules of astrophysical interest such as carbon clusters, polycyclic aromatic hydrocarbons, and C_nH_2 chains have been studied with cavity ringdown in the infrared, visible, and ultraviolet regions of the spectrum. Studies of small amino acids have also been carried out showing that the zwitterionic (dipolar ionic) form is not present in the gas phase.

The capability of CRLAS for providing time-resolved concentration information in a variety of chemical environments makes it ideally suited for kinetics studies. As long as the concentrations of reactants and products do not change significantly over the time scale of a ringdown event (30 microseconds), the light exiting the ringdown cavity displays a single exponential decay. By varying the timing between the probe pulse and the reaction-initiating event, concentration as a function of reaction time can be determined. Using this approach, researchers have extensively studied a variety of first- and second-order reactions.

If the concentration of products or reactants changes significantly over the ringdown time scale, the light decay is no longer a simple first-order exponential. In this case, the shape of the light decay depends on how the concentration of a given species is changing while the light is trapped in the ringdown cavity. This has allowed researchers to study nitrogen trioxide (NO_3) radical reactions by using a "single shot" approach. Unlike conventional kinetics studies, wherein the concentration as a function of time is determined by varying the time delay between a reaction-initiating event and a probe pulse, a single ringdown event carries information on how a species' concentration is changing during the ringdown time scale, thus allowing reaction rates to be calculated from a single laser shot. This approach provides significant advantages over other kinetics techniques because, being considerably faster, it is not necessary to maintain the experimental conditions over a long period of time.

CRLAS has also been used extensively in trace gas analysis. Researchers have used CRLAS to observe pollutants, such as nitrogen dioxide (NO_2) and mercury (Hg), in 1 ppb and 1 ppt concentrations. However, because many current ringdown experiments use large lasers, such as neodymium:YAG pumped dye lasers, they are restricted to the laboratory. The development of compact ringdown spectrometers for use in the field is an active area of research. Compact continuous-wave diode lasers are promising candidates in this area. However, because continuous-wave lasers have coherence lengths that are much longer than the cavity length, the light interferes with itself as it reflects back and forth in the ringdown cavity. As a result, the transmission of light no longer depends only on the mirror reflectivity. In order to conduct ringdown experiments using continuous-wave sources, researchers have had to actively stabilize the rindown cavity, ensuring that the cavity length is a half-integer multiple of the wavelength. The active cavity stabilization not only greatly complicates the experiment but also precludes the use of this technique in the field, as small vibrations and air currents are enough to destabilize the cavity. Recently it was shown that this problem can be circumvented by aligning the cavity in such a way that the light undergoes hundreds of reflections before it overlaps with itself. This technique shows promise for the study of atmospheric chemistry because small continuous-wave-based ringdown spectrometers can now be placed on airplanes and balloons.

In the past decade, CRLAS has emerged as a powerful and versatile direct absorption technique. Pulsed CRLAS experiments have allowed researchers to scan large regions of the spectrum quickly and with high sensitivity, making the technique ideal for studies of short-lived species in molecular beams. CRLAS is also being developed for trace gas analysis in the field. These spectrometers differ from traditional pulsed cavity ringdown laser absorption spectrometers in that they use small continuous-wavelength lasers with limited spectral coverage and ultrahigh resolution. These small spectrometers will be invaluable for atmospheric studies, as they will enable realtime concentration measurements of species from balloons and airplanes, while continued technological development of CRLAS might even lead to the replacement of FTIR spectrometers.

For background information *see* ATOM CLUSTER; INFRARED SPECTROSCOPY; KINETIC METHODS OF ANALYSIS; LASER SPECTROSCOPY; METAL CLUSTER COMPOUND; MOLECULAR BEAMS; SPECTROSCOPY in the McGraw-Hill Encyclopedia of Science & Technology.   Raphael N. Casaes; Richard J. Saykally

Bibliography. A. O'Keefe and D. A. G. Deacon, *Rev. Sci. Instrum.*, 59:2544, 1988; W. Kenneth and M. A. Busch (eds.), Cavity-Ringdown Spectroscopy: An Ultratrace-Absorption Measurement Technique, *ACS Symp. Ser.*, no. 720, 1997; J. B. Paul and R. J. Saykally, *Anal. Chem.*, 69:287A, 1997; R. A. Provencal et al., *Spectroscopy*, 14:24, 1999.

Speech perception

Speech recognition relies on a complex interaction between the sensory information supplied by the ear and the pattern recognition by the brain. Recent advances in defining the role of sensory information in speech understanding have come from work with the cochlear implant, a sensory prosthesis that restores hearing to the profoundly deaf. Artificial electrical stimulation of the auditory nerve provides an opportunity for evaluating the interaction between the quality of sensory information and auditory pattern recognition. The results have been somewhat surprising.

Cochlear implants. Cochlear implants restore some hearing to deaf people by electrically stimulating the auditory nerve. Early models of cochlear implants used only a few electrodes and provided rudimentary sound information. However, modern multichannel cochlear implants, which have now been applied to more than 30,000 deaf people worldwide, are so successful that most cochlear implant patients can talk on the telephone with minimal difficulty. The level of speech recognition achieved with cochlear implants is astonishing and challenges current understanding of the sensory processing underlying the recognition of speech.

In the normal cochlea, frequency information is mapped onto position in the cochlea, with high frequencies processed at one end (the base) and low frequencies at the other end (the apex). Most types of deafness are caused by the loss of the hair cells in the cochlea, which are the cells that convert acoustic mechanical energy into neural activation. The cochlear implant bypasses the damaged cochlea and directly stimulates the auditory nerve fibers with electrical pulses. Modern cochlear implants contain multiple electrodes to stimulate neurons that would normally process different acoustic frequencies (**Fig. 1**).

Signal processors for cochlear implants attempt to recreate the normal acoustic frequency–place distribution of information in the cochlea (**Fig. 2**). The entire frequency range of speech is filtered into mul-

tiple, contiguous frequency bands. The envelope of the signal is extracted from each band by half-wave rectification and low-pass filtering. The extracted envelope is then amplitude-compressed and used to modulate a pulse train of biphasic electrical pulses. The pulses from adjacent electrodes are interleaved in time to avoid electrical field interactions. In contrast to the detailed spectro-temporal pattern of acoustic information in a normal cochlea, the cochlear implant presents only low-frequency envelope information from a few broad frequency bands.

In spite of the enormous differences between normal acoustic stimulation and electrical stimulation in the pattern of activity produced in the auditory nerve, implant patients can understand speech at a high level, much better than anyone predicted. The success of cochlear implant listeners in understanding speech presents a challenge for speech science, because many of the cues thought to be important for speech recognition are not available to cochlear implant patients.

Ironically, studies to understand the cues used by cochlear implant listeners have resulted in an improved understanding of the cues used by normal-hearing listeners to recognize speech. These may be classified as amplitude, temporal, and spectral cues. Speech processing for cochlear implants can be simulated in normal-hearing listeners using a noise-band vocoder. This processing is similar to that shown in Fig. 2 for implants, except that the envelope from each frequency analysis band is used to modulate a band of noise rather than an electrical pulse train. The comparison of normal hearing and prosthetic

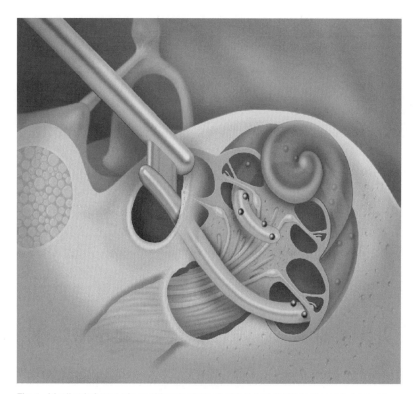

Fig. 1. Idealized picture of a cochlear implant electrode inserted into the scala tympani.

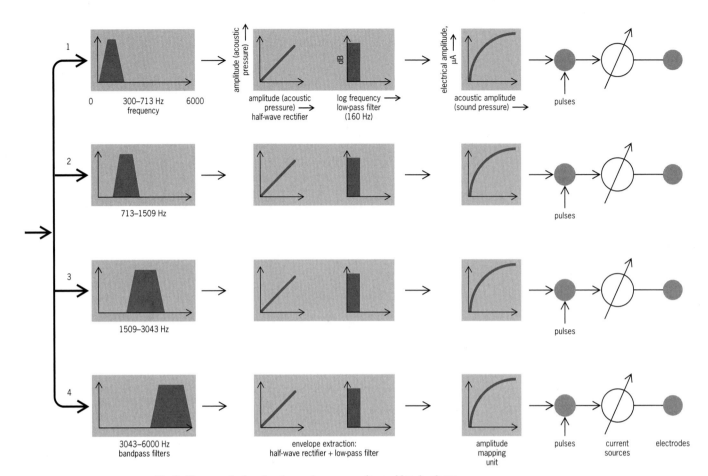

Fig. 2. Diagram of a four-band speech processor for cochlear implants.

hearing produced by electrical stimulation of the cochlea by a cochlear implant has led to a better understanding of the interplay between the brain and the ear in auditory pattern recognition.

Amplitude cues. The sound amplitude waxes and wanes as speakers talk. Vowels have more energy than consonants, for example. Some consonants have much higher amplitude than others. A transition between two speech sounds may be signaled by a brief silent period or a period with lower amplitude. These amplitude changes appear to be important cues for signaling different speech sounds. However, it has long been known that speech is highly robust to distortions in amplitude. Excellent speech recognition is possible even when all the amplitude cues are removed from the wide-band speech signal by clipping. In this case, listeners still have access to the full-resolution spectral signal, which contains a rich enough source of speech information that amplitude cues may not be of such importance. Cochlear implants have poor spectral resolution because they have only 8–20 electrodes representing the entire frequency range of speech information. It is possible that amplitude cues might assume more importance when only coarse spectral information is available, as in cochlear implants. In addition to the limited spectral resolution, cochlear implants must compress the normal acoustic dy-

namic range of 120 dB into the much smaller electrical dynamic range of 10–20 dB. In spite of these differences, experiments have shown that amplitude distortion in cochlear implants has only a minor effect on speech recognition. Experiments in normal-hearing listeners with simulated implant processing also confirm that speech can tolerate high degrees of distortion in amplitude cues, even when the listener has access to only coarse spectral resolution. In sum, recent experiments confirm that amplitude cues are of relatively minor importance in speech recognition.

Temporal cues. Temporal cues in speech can be partitioned into three classes: envelope cues in the frequency range 1–50 Hz, periodicity cues at 50–500 Hz, and fine-structure cues above 500 Hz. The envelope and periodicity categories can be detected by both normal-hearing and cochlear-implant listeners even in the absence of spectral cues. Temporal fluctuations can be detected up to 500 Hz, and produce a pitch sensation that can be used to carry musical melody even when no spectral cues are present. Temporal fluctuations faster than 500 Hz can be detected and processed only in the spectral domain. If temporal fluctuations slower than 500 Hz can be perceived, are they useful in speech? To answer this question, speech processors were designed to remove all temporal information above a selected frequency. Speech recognition was measured as a function of the low-pass-filter frequency on the envelope filter in speech processors diagrammed in Fig. 2, and in cochlear implant processors. In general, these studies have found that speech recognition is not degraded until temporal information is reduced to less than 20 Hz. Although temporal fluctuations more rapid than 20 Hz are detectable and contribute to the perceived quality of speech, they do not appear to contribute to intelligibility.

Another indication of the influence of temporal cues in speech comes from studies of cross-spectral asynchrony. Transients in speech, like a "p" consonant or a voicing glottal pulse, produce a short burst of energy across the entire frequency region of speech. It has been suggested that these cross-spectral bursts serve to synchronize the processing of speech, acting like stroboscopic clock pulses. Recent studies introduced time delays between adjacent frequency bands in speech to disrupt the synchrony of temporal events across frequency. The surprising result was that speech was tolerant of cross-spectral asynchrony up to more than 200 milliseconds.

An even more surprising result was presented by K. Saberi and D. R. Perrott, who reversed speech segments in time. They measured speech intelligibility in sentences as a function of the length of the time segments that were reversed. They observed a significant reduction in speech intelligibility only when entire 100-ms segments were reversed. When each 50-ms segment was reversed in time, speech recognition was nearly perfect.

All of these results suggest that the effective analysis time window for speech is about 50 ms in duration. Any temporal manipulation within a 50-ms window (smearing, jittering, and even reversal) produces only minimal impact on speech recognition.

Spectral cues. Differences in some speech sounds are primarily differences in spectral shape; for example, vowels may differ primarily in the frequency of the second formant (or spectral peak). Considerable research has investigated the effects of subtle changes in formant peaks on vowel recognition, and the effects of subtle formant transitions on consonant recognition. However, spectral resolution can become poor with hearing loss and with prosthetic hearing; cochlear implants represent the entire spectrum with 8–20 electrodes spaced along the cochlea. How well can speech be recognized with such a quantized representation of spectral information? Recent studies have shown the surprising result that excellent speech recognition can be achieved when spectral resolution is reduced to only four channels. Vowel, consonant, and sentence recognition scores are better than 80% correct even when the spectral detail in speech is represented by only four bands of noise, each slowly modulated in time (**Fig. 3**). This type of processing obscures much of the spectral detail in speech, including the spectral shape of vowels and formant peak frequencies. Yet speech recogni-

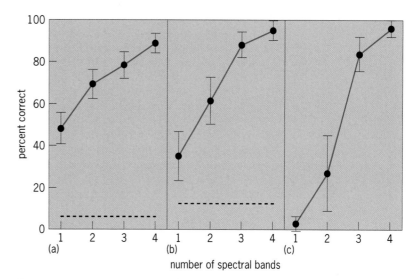

Fig. 3. Recognition of (a) consonants, (b) vowels, and (c) sentences as a function of the number of spectral channels for eight normal-hearing listeners with noise-band processors. The processing was the same as for cochlear implants shown in Fig. 2, but noise was modulated rather than electrical pulses.

tion is achieved in the absence of spectral information thought to be important.

While speech recognition is relatively robust to changes in spectral resolution, it is more sensitive to distortions in the spectral domain. Recent results demonstrate how sensitive speech recognition is to alterations in the frequency–place mapping in the cochlea. Cochlear implants and hearing aids can deliver sound frequency information to a place in the cochlea that does not normally receive information from that frequency, resulting in a shift in the frequency–place mapping. Speech recognition is strongly reduced under such conditions. It has been demonstrated that a shift in the frequency–place map of only 3 mm (about two-thirds octave) can dramatically reduce intelligibility, and that distortion in this map can result in the loss of all spectral information. Recent results have shown that amplification by a hearing aid can actually reduce intelligibility in cases where the amplification results in a local warping in the frequency–place mapping in the cochlea.

These results imply that speech pattern recognition in the brain relies on the correct spectral information coming from the normal acoustic tonotopic location. Alterations or distortions in the normal frequency–place mapping, such as those that can occur with prosthetic devices, can produce a significant reduction in speech recognition.

For background information *see* HEARING (HUMAN); HEARING IMPAIRMENT; PHYSIOLOGICAL ACOUSTICS; SPEECH; SPEECH PERCEPTION in the McGraw-Hill Encyclopedia of Science & Technology.

Robert V. Shannon

Bibliography. Q.-J. Fu and R. V. Shannon, Effect of stimulation rate on phoneme recognition in cochlear implants, *J. Acous. Soc. Amer.*, 107(1):589–597, 2000; Q.-J. Fu and R. V. Shannon, Effects of amplitude nonlinearity on phoneme recognition by cochlear implant users and normal-hearing listeners, *J. Acous.*

Soc. Amer., 104(5):2570–2577, 1998; Q.-J. Fu and R. V. Shannon, Recognition of spectrally degraded and frequency-shifted vowels in acoustic and electric hearing, *J. Acous. Soc. Amer.*, 105(3):1889–1900, 1999; S. Greenberg and T. Arai, Speech intelligibility is highly tolerant of cross-channel spectral asynchrony, *Proc. Int. Cong. Acous.*, Seattle, pp. 2677–2678, 1998; C. A. Hogan and C. W. Turner, High-frequency audibility: Benefits for hearing-impaired listeners, *J. Acous. Soc. Amer.*, 104:432–441, 1998; K. Saberi and D. R. Perrott, Cognitive restoration of reversed speech, *Nature*, 398:760, 1999; R. V. Shannon et al., Speech recognition with primarily temporal cues, *Science*, 270:303–304, 1995; R. V. Shannon, F.-G. Zeng, and J. Wygonski, Speech recognition with altered spectral distribution of envelope cues, *J. Acous. Soc. Amer.*, 104(4):2467–2476, 1998.

Splashing drops

When a drop of rain impacts a pool of water, a number of events occur within a fraction of a second. Capillary waves propagate outward from the point of impact, and a crater forms and collapses. One or more of the following events may also occur: a vortex ring may be propagated downward into the water; one or more air bubbles may be entrapped as the crater collapses; a jet of fine high-speed waterdrops may be ejected; or a thick jet that breaks into one or more large splash drops may be seen. The formation of small air bubbles during drop impact gives rise to the characteristic sound of rain and promotes the transfer of carbon dioxide (CO_2) from atmosphere to ocean. The small, high-speed droplets that form as a result of primary drop impact enhance the transfer of salt and microorganisms from the ocean to the atmosphere. Drop impact in shallow pools and puddles leads to erosion of the pool bottom, and the splash can distribute soil particles and plant viruses over a wide area. In metal smelting, splash is increasingly used to enhance heat transfer in the development of technologies to replace the blast furnace. In the galvanization of steel and spray coating with liquid metals, the effect of drop impact can influence the quality of the final product.

A. M. Worthington carried out pioneering photographic studies of splashing drops over a hundred years ago. However, it is only since around 1980 that advances in photographic and computational methods have enabled significant progress in understanding the physics of drop impact. When a drop of newtonian liquid impacts a deep pool (that is, deeper than approximately ten times the drop diameter), the different phenomena that occur are characterized by three dimensionless numbers: the Reynolds number ($Re = \rho u d / \mu$); the Froude number ($Fr = u^2/gd$); and the Weber number ($We = \rho u^2 d / \sigma$). Here d, u, ρ, and μ are the drop diameter, impact velocity, density, and dynamic viscosity. The surface tension coefficient between liquid and atmosphere is σ, and g is the acceleration due to gravity.

Low-viscosity impacts. For low-viscosity fluids, the Reynolds numbers attained by most splashing drops of interest are sufficiently high that energy losses are small and viscous effects play a secondary role in drop impact. For such impacts, the different phenomena can be summarized by plotting the drop's Weber number versus Froude number (**Fig. 1**).

Floating and bouncing drops. When the Froude and Weber numbers are small (in practice when one or more of the following occur: the drop size is small, impact velocity is low, or surface tension is high), a drop may float on, bounce off, or coalesce with the pool without significant disruption to the surface or the creation of splash. Floating and bouncing drops have been observed only for Weber numbers less than approximately 8. Floating occurs for low-energy impacts when the kinetic energy in the drop is initially unable to overcome the surface tension forces required for coalescence. Bouncing drops have been observed for drops impacting at an angle to the pool surface. For bouncing drops that impact normally, the whole drop does not bounce. Instead, when the drop begins to coalesce with the pool, it stretches and pinches off a smaller drop that is ejected, often landing some distance from the initial impact. This process may repeat itself until the drop is too small to stretch and break up further. Because the impact energy of floating and bouncing drops is small, only a small cavity is formed from the impact, although capillary waves are propagated outward.

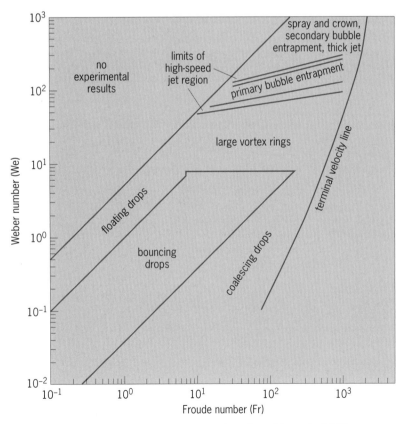

Fig. 1. Impact regimes for drops of low-viscosity (high-Reynolds-number) fluids, described on a plot in the Froude number–Weber number plane.

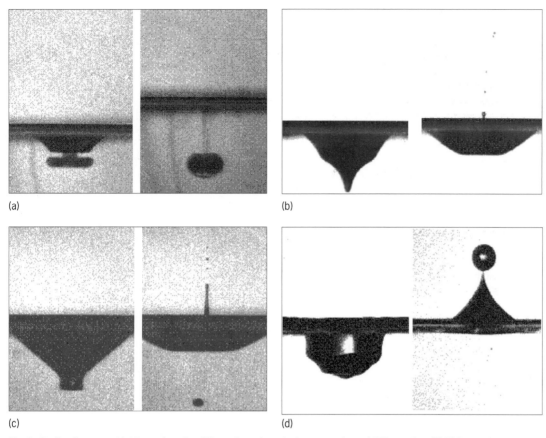

(a)　(b)

(c)　(d)

Fig. 2. Cavity shapes and jet formations for different low-viscosity impact regimes. (*a*) Vortex ring. (*b*) High-speed jet. (*c*) Bubble entrapment and high-speed jet. (*d*) Thick jet. In each regime the two images show successive stages of development of the impact process.

Coalescing drops. For coalescing drops, a significant fraction of the drop fluid is entrained into a single large vortex ring that propagates downward, sometimes becoming unstable and splitting into a number of coherent vortex structures (**Fig. 2***a*). The penetration of the vortex ring promotes mixing in the upper layers of the ocean and also enhances soil erosion at the base of shallow pools. The impact craters formed from coalescing drops are usually one to two drop diameters deep and collapse gently, with the crater center rising to fill the crater. The peak height of the rebound is only slightly higher than the original water level. This regime is sometimes called the vortex-ring regime, although it is now known that vortex rings are not confined just to coalescing drops. They may be formed over a wide range of higher-energy drop impacts, although the rings are smaller in size.

High-speed jets. For higher-energy impacts, or those with lower surface tension, the vortex ring regime makes a sudden transition to one in which fine, high-speed jets are formed on the axis of impact (Fig. 2*b*). These jets break up to form one or more small secondary drops that are ejected vertically upward with a velocity up to ten times the impact velocity. The high-speed jets are caused by the way in which the impact crater collapses, with capillary waves that travel down the crater walls being focused, resulting in a singular point on

the axis at the crater base. At this point a high-pressure region is formed that accelerates fluid to form a high-speed jet. The region of the Fr-We plane (Fig. 1) occupied by this type of high speed jet is small.

When the impact energy is slightly higher and crater slightly larger, the phase and amplitude of the capillary waves are such that the wave collapses onto the axis before it reaches the crater base and a bubble is formed (Fig. 2*c*). When the capillary wave collapses onto the axis, a high-pressure zone is formed. The entrapped bubble is accelerated downward into the water, and a high-speed jet moves upward with a lower velocity than when no bubbles are entrapped (Fig. 2*c*). The wavelength of the capillary wave determines the bubble size, which ranges from 80 to 400 μm. After formation, the bubbles oscillate to give the familiar sound of drop impact. Only a small range of raindrop sizes fall within the bubble entrapment region in the Fr-We plane (approximately 0.8–1.1-mm drops), and the bubble sizes resulting from their impact have a characteristic frequency of approximately 14 kHz, which is the observed peak in the raindrop noise spectrum. Vortex ring formation is suppressed in the bubble entrapment regime because the vorticity generated by the downward motion of the bubble tends to interact with and cancel the vorticity generated by the bubble rising back to the surface. These two sources of vorticity interfere and overwhelm the vorticity that was initially generated at impact.

For a small region of the Fr-We plane, increasing the impact energy further results in drops with high-speed jets but no bubble formation. This regime is similar to the one observed just before the onset of bubble entrapment. For the entire range of impacts in which high-speed jets are formed, a wave swell is formed as a result of the initial impact, and this wave travels radially outward from the impact site.

Sprays, crowns, and thick jets. For impact energies still higher, a thin sheet forms in the first millisecond of impact and rapidly breaks down into a horizontal spray of fine droplets. A thicker wave swell that follows breaks up to form a structure known as the crown, after its shape. The craters formed from higher-energy impact are approximately hemispherical in shape and collapse in such a way that the crater base rises before capillary waves reach the base. The end result is a thick, low-speed jet (usually called a Rayleigh jet) that rises two to five drop diameters above the free surface and breaks off one or more tip drops that have a size similar to the initial impacting drop (Fig. 2*d*). The first tip drop to form consists almost entirely of the fluid that was in the original impacting drop. This may be observed by letting a drop of milk fall into black coffee from a height of 30–40 cm (12–16 in). Vortex rings are also formed but are much smaller and less easily observed.

For very high energy impacts, a large volume of fluid is ejected almost vertically upward from the crater's edge to form a cylindrical crown. If the impact energy is high enough, the crown closes over, as a result of surface tension, and collapses on the axis, resulting in a small upward spray and a jet of fluid moving down the axis. The downward jet impacts the rising Rayleigh jet and retards its upward motion. The closed crown may entrap a large bubble that floats for a short time before breaking up.

High-viscosity impacts. When the fluid viscosity is high, the crater formed is reduced in size. For high-energy impacts, part of the drop impact energy is dissipated in the formation of a crown. A thick central jet forms which may break off a splash drop before the jet collapses back into the liquid. The collapse of this first jet results in a second cavity being formed. The collapse of the second cavity results in the formation of a thinner second jet that breaks up into a series of small splash drops.

Shallow pool impacts. A pool is shallow when the crater formed during impact is of the same dimension as the depth of the pool. The presence of the pool base causes the fluid to collapse from the sides of the crater, and the base cannot move upward as there is no water below. The crater sides converge toward the center, and a thin jet is formed that breaks up into a stream of fine droplets. For a specified value of the drop Froude number, a critical depth occurs for which the jet height is a maximum. This critical depth appears to be confined to a narrow range of pool depth regardless of fluid, and has been observed for water, glycerol, and xanthan gum solu-

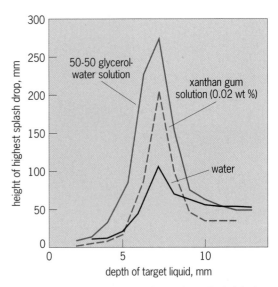

Fig. 3. Effect of water depth on the maximum jet height for different fluids. 1 mm = 0.04 in. The results are for a 4.1-mm (0.16-in.) drop falling from 300 mm (11.8 in.); that is, the Froude number is fixed and the Weber number varies with fluid.

tions (**Fig. 3**). The reasons for this behavior are not yet understood.

Numerical simulation. Numerical simulation of splashing drops is extremely difficult because the problem involves accurate tracking of a moving interface that may coalesce or fragment, accurate application of surface tension forces that occur over a length scale far smaller than that resolved by numerical methods, and density ratios between the liquid and atmosphere that are typically 1000:1. Boundary integral methods have been used successfully to model splash from just after the point of drop and pool coalescence to just before fragmentation or bubble entrapment. After fragmentation of the domain, special treatments have to be used to continue simulations, making the method unsuitable for general applications. Arguably the most robust methods for numerical simulation of splashing drops are volume-of-fluid (VOF) methods, in which a scalar function is used to represent the distribution of fluid on a regular finite volume mesh. When these methods are coupled to robust fluid solvers and recently improved methods for estimating surface tension forces, accurate numerical results can be obtained.

For background information *See* ATOMIZATION; COMPUTATIONAL FLUID DYNAMICS; FLUID FLOW; JET FLOW; PRECIPITATION (METEOROLOGY); SURFACE TENSION; VISCOSITY; VORTEX in the McGraw-Hill Encyclopedia of Science & Technology.

Murray Rudman; Jong-Leng Liow

Bibliography. A. Frohn and N. Roth, *Dynamics of Droplets*, Springer-Verlag, New York, 2000; A. Prosperetti and H. N. Oguz, The impact of drops on liquid surfaces and the underwater noise of rain, *Annu. Rev. Fluid Mech.*, 25:577–602, 1993; J. Shin and T. A. McMahon, The tuning of a splash, *Phys. Fluids A*, 2(8):1312–1317, 1990; A. M. Worthington, *A Study of Splashes*, Longmans, Green, 1908.

Star formation

Stars form from clouds of molecular gas in interstellar space. Molecules are protected from radiative dissociation by the surface layers of the cloud, in which light is absorbed by particles of dust mixed with the gas. The absence of starlight in the center of a molecular cloud allows the gas to become very cold, which in turn lowers the pressure. Lacking the support of pressure against gravity, the densest regions at the cloud's center collapse and fragment to form stars. When stars finally form, their irradiance heats the cloud from within, causing it to chemically evolve and eventually to disperse. The entire process, from giant molecular cloud to star, is estimated to take around 1–10 million years. Although the overall picture of how stars form from molecular clouds is known at this simple level, the details of the process are just beginning to be understood.

The same dust that protects the molecular cloud from the destructive effects of starlight also blocks our view of the cloud's interior. In optical telescopes, molecular clouds appear as dark patches against the background of stars. The process of star birth, which takes place mostly inside molecular clouds, is hidden from optical view. Astronomers can view the earliest stages of star birth by observing molecular clouds via the radiation emitted by molecules. The characteristic spectral lines of molecules are found in the radio portion of the spectrum, generally at millimeter and submillimeter wavelengths, which are unaffected by dust absorption. For the later stages of star birth, when the newly formed stars first illuminate the clouds from within, astronomers use radio waves from the surrounding hot, starlit gas to locate the stars. By studying many molecular clouds, at different stages of evolution, astronomers have identified several distinct stages in the star-formation process.

Cold gas clouds. The starting point for star formation is a large, diffuse molecular cloud known as a giant molecular cloud. Giant molecular clouds are among the largest objects in the Milky Way Galaxy, at 50–200 light-years in size. They are very cold, with temperatures of only 5–10 K. Giant molecular clouds are generally studied with millimeter-wave telescopes, which detect spectral lines corresponding to the rotational energy states of molecules. Because molecular hydrogen (H_2) lacks a dipole moment and has high excitation energy, even though it is the most abundant molecule, it cannot easily be seen in cold clouds. Instead proxies are used, from which the total mass of H_2 gas is inferred. The most commonly used tracer of H_2 is carbon monoxide (CO), which is also the most abundant tracer, with an abundance relative to H_2 of 10^{-4}. Carbon monoxide is easily excited in cold clouds, since its lowest energy levels correspond to effective temperatures (E/k, where E is the energy and k is Boltzmann's constant) of $\sim$5.5 K and 17 K. Less abundant molecules, such as ammonia (NH_3), formaldehyde (H_2CO), methanol (CH_3OH), carbon monosulfide (CS), and the CCS radical, can be used to trace warmer or denser clouds than carbon monoxide, depending on the physical properties of the molecule. Giant molecular clouds are large enough to be easily resolved by single millimeter-wave telescopes (as opposed to arrays).

A distinctive property of giant molecular clouds is that they appear to be gravitationally bound, in contrast to diffuse atomic hydrogen clouds of similar size, which are in pressure equilibrium with the general interstellar medium. Simple gravitational balance considerations dictate that these clouds should be collapsing and that stars should thereby form in about a gravitational free-fall time. The free-fall time for a typical giant molecular cloud is a few million years. If all molecular clouds were collapsing at this rate, the star formation rate in the Milky Way Galaxy would be 200 times larger than observed. Turbulence within clouds probably prevents them from collapsing this rapidly. Sources of turbulence are the winds from nearby young stars and the explosive gas motions caused by supernovae.

Quiescent, starless cores. Starless cores are the next stage in the life of a molecular cloud. Radio astronomers are particularly interested in starless cores because they appear to be the missing link between static giant molecular clouds and active cloud collapse. Starless cores are small, dense clumps found within a giant molecular cloud. The high opacity of the carbon monoxide line makes it difficult to observe cloud cores, since they reside within larger molecular clouds. Cores are more typically mapped out with spectral lines of molecules that are less easily collisionally excited than carbon monoxide; these molecules, so-called high-density tracers, show up only in the densest molecular regions. Carbon monosulfide and NH_2^+ are commonly used to trace starless cores. Cores tend to be small enough that resolutions of a few arcseconds are required, dictating the use of telescope arrays and aperture synthesis techniques to observe them.

Starless cores are typically a few tenths of a light-year (1 light-year $= 9.5 \times 10^{12}$ km $= 5.9 \times 10^{12}$ mi) in size (4000–10,000 astronomical units, where 1 AU $= 1.5 \times 10^8$ km $= 9.3 \times 10^7$ mi), with densities of 10^5 molecules per cubic centimeter, and masses ranging from a few hundredths to several hundred solar masses. The large range of sizes and masses reflects the substructure and clumping that is characteristic of molecular clouds on all size scales. The smallest cores, of less than 0.2 solar mass, appear not to be gravitationally bound. This finding may suggest a reason for the relative paucity of the smallest starlike bodies, brown dwarfs.

Recent studies of the masses of cloud cores reveal that they have a power-law distribution of masses, with lower-mass cores more common than high-mass cores. This mass distribution is also characteristic of the initial mass function of young stars. Thus, the mass of a molecular core may well determine the eventual mass of the star that is formed.

Turbulence appears to play a role in defining starless cores. Abundant evidence suggests that larger clouds, including giant molecular clouds and larger clumps within them, are turbulently supported: the widths of molecular lines are significantly larger than thermal linewidths, spatial structure occurs on all size scales, and core masses obey the power-law distribution noted above. Starless cores may be the end of the turbulent cascade that appears to dominate the existence of molecular clouds. It is estimated that the lifetime of this stage of the molecular cloud's life, before the formation of a central luminosity source, is less than 10^5 years.

Hot molecular cores and outflows. A sign that star formation is under way is when the temperature of the molecular cloud begins to rise. This is the stage of star formation that is associated with active accretion of mass and the formation of a protostar. Protostars are deeply embedded in dense molecular envelopes, and optically invisible. These objects are also invisible in the infrared and, strangely enough, in the abundant carbon monoxide molecule, which "freezes out" onto the surfaces of dust grains. Instead, emission lines of carbon monosulfide and formaldehyde are used to study the dense molecular envelopes of protostars. Like starless cores, hot cores must be studied with high-resolution images, requiring the use of arrays of millimeter-wave telescopes.

The high densities of hot molecular cores and the temperature gradients within them set up a complex, time-dependent cloud chemistry. Large molecules, including organic molecules, are common within these clouds. Hot cores can be as large as 10^4–10^5 AU in size for more massive young stars, of up to 10^2–10^3 solar masses. Temperatures range from 20 K in the outer parts of the envelope to 100 K in the inner parts at the distances that planets might form. These conditions are ripe for the formation of molecular masers, and hydroxyl (OH), water (H_2O), and methanol masers are often signposts of this stage.

Paradoxically, while the young stellar object is actively accreting mass from the molecular core, it is also observed to be vigorously expelling gas. A protostar can simultaneously accrete and expel gas if the accretion and expulsion take place in different directions. Carbon monoxide images reveal bipolar outflows toward young stellar objects. The central carbon monoxide line wavelength is Doppler-redshifted on one lobe of the outflow, indicating motion away from the observer, and blueshifted (motion toward the observer) on the other. Molecular outflows are often associated, as well, with optical and infrared nebulosity of high radial velocity known as Herbig-Haro objects. Outflows can be large in scale, several light-years across (10^6–10^7 AU). They can also carry away significant amount of mass. The outflow in the young stellar object G192.16-3.82 has a mass of 95 solar masses, and carries an estimated 6×10^{-4} solar mass per year of gas out of the system (see **illus.**). At the same time, a similar amount of mass is

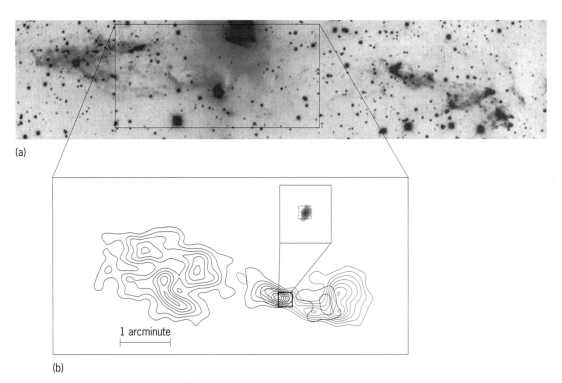

(a)

(b)

Massive young stellar object G192.16-3.82. (a) Optical picture of the outflow region showing, to the left and right, emission from ionized gas associated with shock waves (Herbig-Haro objects). The dark regions in the center are unrelated to the outflow (*from D. Devine et al., A giant Herbig-Haro flow from a massive young star in G192.16-3.82, Astron. J., 117:2919-2903, 1999*). (b) Carbon monoxide (CO) outflow. Color and black contours indicate intensity of redshifted and blueshifted CO emission, respectively. Inset shows the rotating circumstellar disk (*from D. S. Shepherd and S. E. Kurtz, A 1000 AU rotating disk around the massive young stellar object G192.16-3.82, Astrophys. J., 523:690–700, University of Chicago Press, 1999*).

presumably being accreted onto the protostar. In fact, outflows appear to be caused by accretion. The development of an outflow may be the determining factor in the final mass of the star.

When the region in between the red- and blueshifted lobes in G192.16-3.82 is examined more closely, denser gas can be seen at the intersection of the lobes (see illus.). At high resolution, this source is seen to be elongated perpendicular to the axis of the bipolar flow. Such elongated sources are often seen in outflows. The shape and motion of these central sources suggest that they are rotating disks of molecular gas. Disks are far smaller than the bipolar lobes, a few hundred astronomical units in size. Slow accretion of gas through the rotating disk is believed to be responsible for the bipolar outflow.

Compact H II regions. The formation of massive young stars is the last stage in the life of a molecular cloud. Young stars rapidly destroy and disperse any nearby molecular gas. Massive young stars, those with initial masses above 10 solar masses, are particularly destructive, producing copious amounts of ultraviolet radiation capable of ionizing hydrogen. These stars not only heat the surrounding molecular clouds but also ionize them. Ionized clouds of gas are called H II regions. The Orion Nebula is an example of an H II region, although a fairly old and visible one. The youngest H II regions are dense (10^5 molecules per cubic centimeter) gas clouds, still embedded in their natal molecular clouds and invisible to optical view. H II regions are called compact H II regions if they are a few tenths of a parsec in size (10^4 AU), or ultracompact H II regions if they are an order of magnitude smaller. Compact H II regions are overpressured with respect to their surrounding interstellar medium and will expand into the surrounding clouds. They are not expected to remain compact for more than a few hundred thousand years, at most.

The compact H II region phase is the only phase of early star formation that can be easily studied in other galaxies. Radio emission from H II regions is detectable at centimeter wavelengths with the powerful Very Large Array. Radio emission by free electrons within an H II region allows astronomers to measure the rate at which ultraviolet photons maintain the ionization of the nebulae, thus providing a good measure of the number of young stars. Such measurements have recently revealed the birth of an extraordinarily large star cluster in the nearby dwarf galaxy NGC 5253. This cluster, invisible at optical wavelengths and first detected by the Very Large Array, has on the order of 10^6 stars, packed into a region only 5 light-years across. This cluster could well be a forming globular cluster that is less than 200,000 years old. Although this cluster is quite young, still hidden from optical view by dust within the nebula, it already appears to have ionized and photodissociated its molecular cloud. Gas clouds will not again be found near this cluster until molecules form again in the winds of dying stars. *See* STARBURST GALAXY.

For background information *see* HERBIG-HARO OBJECTS; INTERSTELLAR MATTER; MOLECULAR CLOUD; ORION NEBULA; PROTOSTAR; RADIO ASTRONOMY; STAR CLUSTER; STARBURST GALAXY in the McGraw-Hill Encyclopedia of Science & Technology.

Jean Turner

Bibliography. S. C. Beck, Dwarf galaxies and starbursts, *Sci. Amer.*, 282(6):66–71, June 2000; N. J. Evans II, Physical conditions in regions of star formation, *Annu. Rev. Astron. Astrophys.*, 37:311–362, 1999; W. B. Latter et al. (eds.), *CO: Twenty-Five Years of Millimeter-Wave Spectroscopy*, Kluwer, Dordrecht, 1997; V. Manninas et al. (eds.), *Protostars and Planets IV*, University of Arizona Press, 2000; L. Spitzer, Jr., *Physical Processes in the Interstellar Medium*, John Wiley, New York 1978, reprint 1998; E. F. van Dishoeck and G. A. Blake, Chemical evolution of star-forming regions, *Annu. Rev. Astron. Astrophys.*, 36:317–368, 1998.

Starburst galaxy

A normal galaxy like the Milky Way produces a few new stars each year. Starburst galaxies create new stars much more frequently, especially in the central regions, at rates tens or even hundreds of times higher. This phenomenon is still a mystery; a large supply of dense gas collected around the nucleus of the galaxy can supply the raw material for star formation, but it is not clear why or how the gas piles up in this way. Some likely causes include the formation of a galactic bar, with stars and gas following elliptical orbits that pass near the nucleus, or an interaction with another passing galaxy that perturbs the orbital motions. A third influence may be a magnetic field; since interstellar gas is partially ionized and ionic particles tend to move with magnetic field lines, this can ease the flow of gas toward the nucleus. In the last few years, astronomical techniques have been developed that allow the magnetic field to be imaged in starburst galaxies.

Submillimeter techniques. The submillimeter region of the spectrum of light covers wavelengths of about 0.3–1 mm, with longer wavelengths continuing into the radio regime and shorter ones into the far-infrared. The Earth's atmosphere blocks a large fraction of submillimeter light coming from space, primarily by absorption in oxygen and water molecules; but by placing a telescope on a high mountain site (**Fig. 1**), it is possible to detect much of the astronomical radiation. Other technical difficulties include the need to build very large telescopes (because at long wavelengths the image details tend to be crude, due to diffraction), and the complexity of building high-frequency electronics and detection systems. It is only since about 1996 that submillimeter observing has been fully developed as an astronomical tool.

Fig. 1. James Clerk Maxwell Telescope for submillimeter observations. The main mirror is 15 m (50 ft) in diameter, and the telescope is inside a carousel, which rotates to follow sources across the sky. In normal operation a large blind is lowered across the front of the carousel as protection from sun and wind. The telescope is located just below the summit of Mauna Kea, a dormant volcano on the Big Island of Hawaii, at an altitude of about 4000 m (13,000 ft). It is operated jointly by the United Kingdom, Canada, and the Netherlands. (*Robin Phillips*)

A very new technique is submillimeter polarimetry. Normally the two perpendicular waves that make up a ray of light have equal intensity, so the light is unpolarized. However, small particles can affect light in two ways that make these waves unequal. One way is by scattering in a particular direction; this effect is utilized in polarizing sunglasses. The other way is when elongated particles emit more along their long axes. Normally in an assembly of particles the directions of these axes will be random. However, in interstellar space, if there is a magnetic field, iron-bearing grains will tend to line up across the field lines like little bar magnets. The end result is that one of the two planes of light is brighter, so the submillimeter emission from the grains depends on the orientation of the light, or is polarized (**Fig. 2**).

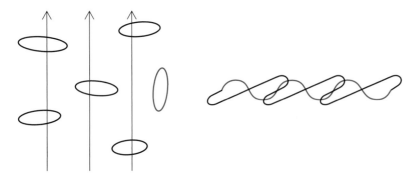

Fig. 2. Polarized light waves. Elongated dust grains (ovals) tend to spin with their long axes across the magnetic field lines (arrows), as this is the lowest energy state. The submillimeter light waves therefore have most emission in the horizontal direction in this picture (black wave at right), and less in the vertical direction (color wave at right) from the less common vertically oriented grains (represented by a single color oval). For clarity the contrast between the horizontal and vertical states has been exaggerated, and in nature polarization percentages tend to be small.

To detect the polarized light, a birefringent quartz plate is used. This flat plate retards one plane of polarization by half a wavelength with respect to the other as the light passes through the plate. The light then passes through a wire grid, which absorbs light waves that are parallel to the wires. By spinning the plate, the source plane of polarization is rotated, and the fixed grid then selectively removes a fixed plane. The result is a varying signal that is detected by a submillimeter camera, using very cold semiconducting detectors. The signals from each detector are put together to make up an image of the (polarized) sky.

Observing starburst galaxies. Submillimeter polarimetry has already been used to map interstellar clouds in the Milky Way Galaxy, and to study the magnetic fields around very young stars. Starburst galaxies are much more difficult targets, mainly because they are typically distant (so far away that light takes many millions of years to travel from them to the Earth, even at a speed of 3×10^8 meters per second). The brightest relatively nearby starburst is M82, a galaxy at 10^7 light-years, first cataloged by the eighteenth-century astronomer Charles Messier as a nebulous object. This nearby starburst was the first target for extragalactic imaging using submillimeter polarimetry.

The center of this galaxy has a bright torus (ring-shaped region) of gas and dust particles that extends over a diameter of about 1000 light-years. Most of the submillimeter emission in M82 comes from this torus, with a small fraction coming from dust grains expanding out into a halo. Although much more distant, the galaxy is comparable in average submillimeter brightness (as seen from the Earth) to typical galactic clouds such as the star-formation region in Orion's sword.

In M82, star formation is much more energetic than in the Milky Way Galaxy, and there are at least 100 stellar superclusters inside the ring. These drive high-speed energetic winds that emerge from the galaxy, creating submillimeter to x-ray emission 3000 light-years or more above and below the galactic plane. The starburst is estimated to have been proceeding for 5×10^7 years, and was probably triggered by a close encounter with the neighboring spiral galaxy M81. Some very long streamers of hydrogen gas have been observed still connecting the two.

Magnetic field in M82. The magnetic field in M82 was mapped using the James Clerk Maxwell Telescope in Hawaii (Fig. 1) and its United Kingdom–Japan Polarimeter, together with the submillimeter camera, SCUBA. The wavelengths of the observations were 0.85 mm for the polarimetry and 0.45 mm for the image. The results are shown as bars pointing in the magnetic field direction superimposed on a false-color image of the intensity in which white represents the brightest emission (**Fig. 3**). The length of the bars indicates the percentage of polarization observed. The brightest emission comes from the torus, which is seen almost edge-on so that it appears as a flat oval.

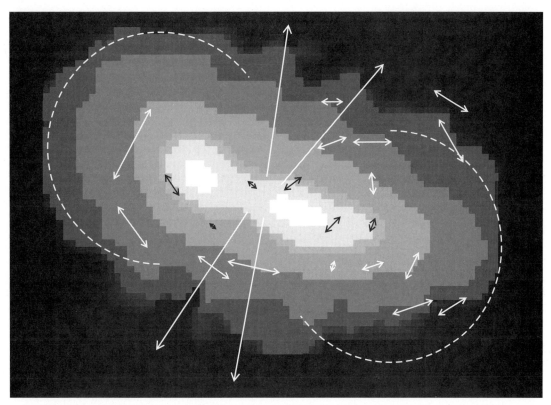

Fig. 3. M82 polarization. The image shows the brightness of the galaxy at a wavelength of 0.45 mm, from black (minimum) to white (maximum). The double-ended arrows (bars) show the directions of the magnetic field, deduced from 0.85-mm polarimetry. The length of these bars is proportional to the percentage of polarization, which lies between 0.7 and 9.1%. The long white arrows depict the winds emerging from the center of the galaxy, and the broken white lines outline the hypothesized magnetic field bubble.

At first sight the magnetic field bars point in very varied directions, but there is an underlying order. In the brightest regions, the magnetic field points in toward the nucleus. The nucleus, or the galaxy's center about which it rotates, lies between the two bright peaks in the image which correspond to two opposite sides of the torus. The directions in the magnetic field are as required to help gas flow into the reservoir that powers the starburst. This was hypothesized before the data were taken, and is an important clue that magnetic effects may indeed be important in the inward flow of gas needed for a starburst to take place.

In contrast, in the outer regions the magnetic field appears to wrap around the torus. This was a surprising result as it has not been seen in other galaxies, which mostly have flat magnetic fields lying in the galactic plane. The most likely explanation is that the powerful winds from the stellar superclusters have blown an originally flat field up into a kind of giant bubble. This is represented by the broken lines in Fig. 3. The scale of the bubble phenomenon is very large, with a diameter of at least 3000 light-years.

The percentage of the polarization is also interesting; it is not a measure of the magnetic field strength, which must be measured by other techniques, but it does reflect how tangled the magnetic field lines are within the telescope beam. If the field lines lie in many different directions, the grain emission is also oriented in many directions, so the polarization tends toward zero. For observations of M82 at 0.85 mm, the telescope beam covers a large region approximately 800 light-years across, or $0.004°$ on the sky. (For scale, the magnetic bars are plotted slightly closer than $0.002°$ apart.) The polarization percentages are observed to be smaller nearer the galactic nucleus, which suggests that the magnetic field structure is more complex than it appears; this is very likely, as the magnetic field must be dragged around by rotation of the torus, as well as under pressure from the galactic wind.

Prospects. This is the first time that the magnetic field has been imaged in the actual regions where the starburst phenomenon occurs. Polarimetry can also be done at other wavelengths, such as the near-infrared and the radio. However, near-infrared radiation traces absorption of galactic light by dust grains, and thus the polarization traces all the foreground grains in the whole galaxy, not just the torus that is seen in submillimeter emission. Radio waves trace only diffuse hydrogen gas that is also not directly associated with the starburst region. Thus submillimeter polarimetry has opened up a new window for examining magnetic fields in other galaxies. When polarimetry becomes available at submillimeter arrays of telescopes, offering resolution 10 times higher or more, more starburst galaxies will be observed, and in even greater detail.

For background information *see* GALAXY, EXTERNAL; POLARIMETRY; POLARIZED LIGHT; STARBURST GALAXY; SUBMILLIMETER ASTRONOMY in the McGraw-Hill Encyclopedia of Science & Technology.

Jane Greaves

Bibliography. R. Beck et al., Galactic magnetism: Recent developments and perspectives, *Annu. Rev. Astron. Astrophys.*, 34:155–206, 1996; J. S. Greaves et al., Magnetic field surrounding the starburst nucleus of the galaxy M82 from polarized dust emission, *Nature*, 404:732–733, 2000; R. C. Kennicutt, Jr., Star formation in galaxies along the Hubble sequence, *Annu. Rev. Astron. Astrophys.*, 36, 189–231, 1998; D. W. Weedman, Making sense of active galaxies, *Science*, 282:423–424, 1998.

Stripe phases

Stripe phases are a recently discovered type of electronic crystal (similar to Wigner crystals). The difference between electronic and conventional crystals is that the lattice of an electronic crystal is made of electrons which, being much lighter than atomic nuclei, are subject to much stronger quantum and thermal fluctuations. Therefore electronic crystals are likely to melt at much lower temperatures than conventional ones.

Stripe phases constitute a special type of electronic crystal; they are electronic liquid crystals, so they exhibit properties of both a crystal (an orientational order and, related to it, an anisotropy) and a liquid (the absence of a space periodicity). To stabilize an electronic crystal, the interactions have to quench the kinetic energy of electrons. Therefore stripes appear only in systems where interactions play a dominant role (strongly correlated systems).

Stripes in quantum Hall systems. Experimentally the stripe phases have been observed in two- or quasi-two-dimensional systems. For example, stripes have been observed in quantum Hall systems. Such systems are realized when a clean metallic film is placed in a strong magnetic field. This magnetic field confines the electrons to widely separated discrete energy levels (the Landau quantization), thus quenching their kinetic energy. Each Landau level can be occupied by a macroscopic number of particles, its total capacity being proportional to the magnetic flux through the system. The stronger the magnetic field, the fewer Landau levels are occupied, since each is endowed with an increased capacity. The ratio of the number of electrons to the capacity of a Landau level is called the filling factor, and specifies the number of Landau levels that are filled with electrons. Since the kinetic energy of electrons on a given Landau level is quenched by the magnetic field, their behavior is determined solely by interactions. This situation is very favorable for the formation of various exotic states of matter, electronic liquid crystals included. Stripe phases appear at high half-integer values of the filling factor (above about 10).

The evidence for stripe order in quantum Hall systems comes from the observed anisotropy of the resistivity tensor. This anisotropy is rather significant (the ratio of the longitudinal to the transverse conductivities is probably of the order of five).

Theoretically at absolute temperature $T = 0$, stripes in quantum Hall systems can be described as unidirectional charge-density waves, and thus share the symmetries of two-dimensional smectic liquid crystal (that is, a crystal where the translational invariance is broken in just one direction). According to general principles, smectic crystalline order is unstable at finite temperatures. Therefore at nonzero temperature the translational invariance is restored, but the rotational invariance remains spontaneously broken up to a higher temperature T_{BKT}. Below T_{BKT}, intervals between stripes widely fluctuate, but there is still a preferential direction for their orientation. Such a state is called a nematic liquid crystal. When the temperature increases, the number of dislocations in such a crystal increases, and this increase eventually leads to a phase transition (called the Berezinskii-Kosterlitz-Thouless, or BKT, transition) into a uniform state. The stripes appear at magnetic fields of intermediate strength when there

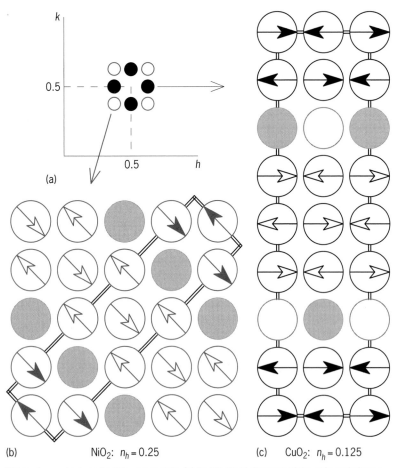

(a)

(b) NiO$_2$: $n_h = 0.25$

(c) CuO$_2$: $n_h = 0.125$

Stripe phases in doped antiferromagnets. (*a*) An idealized diagram of the spin and charge stripe pattern within a NiO$_2$ plane observed in hole-doped La$_2$NiO$_4$. (*b*) Hypothesized stripe pattern in a CuO$_2$ plane of hole-doped La$_2$CuO$_4$ with hole concentration 1/8. Arrows indicate the orientation of magnetic moments of metal atoms, which are locally antiparallel. The double line outlines the magnetic unit cell in each case.

are few occupied Landau levels. The theory of stripes on higher Landau levels has been worked out, and a transport theory has been formulated for the quantum Hall stripe phase.

Stripes in doped antiferromagnets. Experiments on the doped lanthanum nickelate and lanthanum cuprate families of materials have established that stripe phases appear as a prominent feature of doped antiferromagnets (see **illus.**). In the stripe phase, holes are concentrated in quasi-one-dimensional regions (lines) arranged periodically. These lines correspond to simultaneous discommensurations of the antiferromagnetic order. Thus the stripe phases in these materials can be understood as a system of coupled charge-density and spin-density waves. The Landau theory of such a state has been constructed. Its most distinct feature is the presence of a term describing coupling between charge and spin modulations in the Landau functional. This coupling establishes a relationship between wave vectors of the charge-density and spin-density waves. The crystalline lattice stabilizes the stripe order, pinning the stripes.

The likeliest mechanism for stripe formation in doped antiferromagnets is phase separation frustrated by long-range Coulomb forces. In an ideal antiferromagnetic state, each site is occupied by a electron in such a way that spins of electrons located at neighboring sites point in opposite directions. Such a state is insulating due to the strong Coulomb interaction which prevents double occupation of a given site. When the electron concentration is reduced by doping, the number of electrons becomes smaller than the number of sites. The unoccupied sites (holes) are usually mobile and propagate through the crystal. Since a hole can tunnel between the sites only when the spins are parallel, the antiferromagnetic arrangement makes it difficult for holes to propagate and they lose kinetic energy. To gain the energy, the holes tend to move close to each other. Without long-range Coulomb forces the holes would all get together creating a single "bubble" inside the antiferromagnetic phase (phase separation), but the Coulomb interaction prevents this. This is because the holes have electric charge and repel each other. The compromise between the tendency to create hole-rich regions and the Coulomb repulsion is achieved by the formation of an array of stripes consisting of hole-rich regions.

For background information *see* ANTIFERROMAGNETISM; CHARGE-DENSITY WAVE; CRYSTAL DEFECTS; CRYSTAL STRUCTURE; HALL EFFECT; HOLE STATES IN SOLIDS; LIQUID CRYSTALS; SEMICONDUCTOR; SPIN-DENSITY WAVE in the McGraw-Hill Encyclopedia of Science & Technology.   Alexei Tsvelik

Bibliography. S.-W. Cheong et al., Incommensurate magnetic fluctuation in $La_{2-x}Sr_xCuO_4$, *Phys. Rev. Lett.*, 67:1791–1794, 1991; M. Foegler, A. A. Koulakov, and B. I. Shklovskii, Ground state of a two-dimensional electron liquid in a weak magnetic field, *Phys. Rev. B*, 54:1853–1871, 1996; M. P. Lilly et al., Evidence for an anisotropic state of two-dimensional electrons in high Landau levels, *Phys. Rev. Lett.*, 82:

394–397, 1999; J. M. Tranquada et al., Evidence for stripe correlations of spins and holes in copper oxide superconductors, *Nature*, 375:561, 1995.

Subaru Telescope

Subaru Telescope is Japan's advanced national astronomical research facility located near the summit of the dormant volcano of Mauna Kea on the Big Island of Hawaii. The location affords a high percentage of clear nights and excellent atmospheric stability, combined with low overhead water vapor that greatly improves the quality of infrared observations. The telescope is designed to work efficiently with both visible and infrared light. Subaru is the Japanese word for the Pleiades star cluster (the Seven Sisters). It also means "get together" or "come together," appropriate for a facility that has united the Japanese scientists and is being offered for use by the international astronomy community.

Japanese astronomers began thinking about a large national optical astronomy facility in the 1970s. The National Committee for Astronomy of the Science Council of Japan presented a detailed report in 1984 recommending the Japanese National Large Telescope (JNLT) project. Critical government review began in 1986, authorizing engineering studies to test design concepts and the exchange of a Memorandum of Understanding between the Tokyo Astronomical Observatory and the University of Hawaii. The National Astronomical Observatory of Japan (NAOJ) was founded in 1988 as a national interuniversity research institute responsible for the design, construction, and operation of the JNLT. In 1991, the telescope project was renamed Subaru Telescope, and on July 6, 1992, a groundbreaking ceremony was held at the summit of Mauna Kea. Subaru Telescope gathered its first starlight on December 24, 1998, and obtained its first scientific observations one month later.

Optical configuration. Subaru provides an effective aperture of 8.2 m (26.9 ft) and uses a Ritchey-Chrétien optical design to feed light to four foci (**Fig. 1**). The Cassegrain and both Nasmyth foci operate at ~f/12.5 and offer a field of view of 6 arcminutes; the prime focus operates at f/1.9 and offers a field of view greater than 30 arcminutes.

All the mirrors of Subaru Telescope are constructed from ULE™ (titanium silicate) glass. This material has a coefficient of thermal expansion hundreds of times smaller than Pyrex™ (borosilicate) glass, giving Subaru an inherent ability to precisely maintain its optical figure during normal operation, greatly simplifying the optics control system. Most of Subaru's mirrors use aluminum as the reflective front surface coating.

Primary mirror and mirror cell. The heart of Subaru Telescope is its monolithic primary mirror. The finished mirror measures 8.3 m (27.2 ft) in diameter and 20 cm (8 in.) thick, and weighs 23 metric tons (25.4 U.S. tons). It took 3 years to produce the mirror blank, and a further 4 years to complete the

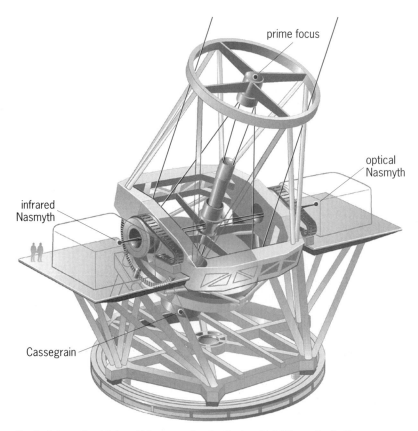

prime focus

optical
Nasmyth

infrared
Nasmyth

Cassegrain

Fig. 1. Schematic of Subaru Telescope, showing the four foci. (*Illustration by Koetsu Endo, reproduced with permission from the February 1996 issue of Nikkei Science*)

fabrication by drilling holes for the support system into its back surface and polishing the front surface to an accuracy of 14 nm.

The mirror is kept from sagging under its own weight by 261 uniformly spaced cantilever arms (actuators) plugged into the holes on the mirror's backside, with the pivot point of each actuator attached to a very rigid steel mirror cell. Every actuator contains a load sensor and computer-controlled preload spring that adjusts the force applied through the backside to correct for any small-scale variations in the shape of the mirror's front surface.

Secondary and tertiary mirrors. Subaru has three convex secondary mirrors and one flat tertiary mirror. All are constructed to be lightweight, using glass rib cores supporting front and rear glass surfaces. The choice of secondary mirror depends on which instrument is being used for observations. In particular, a silver-coated secondary is used for infrared observations at the Cassegrain focus—this mirror is equipped with a support system that can rapidly tilt the mirror through a specified angle several times per second. Such motions allow the instrument to rapidly switch between the target field and a nearby (blank) reference field to permit accurate subtraction of the very bright and rapidly changing sky background that is emitted by the Earth's atmosphere at infrared wavelengths.

Telescope mount. The telescope mount weighs 555 metric tons (612 U.S. tons), measures 22 m (72 ft) tall by 27 m (89 ft) wide (to the outside edge of

Fig. 2. Subaru Telescope is seen through the open shutters of its enclosure. A partially buried corridor connects the control building at left with the enclosure base (ESB) and an external elevator shaft. The dome of the Keck I telescope is seen in the background. (*Images copyright © 1999 Subaru Telescope, NAOJ. All rights reserved.*)

each Nasmyth platform), and uses an open-truss design to maintain precise optical alignment between the mirrors and instrumentation. The telescope is of altitude-azimuth (Alt-Az) design and uses hydrostatic oil bearings to support each axis. Precise control allows the telescope to counteract the motion due to Earth's daily rotation to an accuracy of better than 0.1 arcsecond.

The telescope can support instruments mounted at the Cassegrain focus weighing up to 2 metric tons (2.2 U.S. tons) and measuring up to 2 m (6.5 ft) on all sides. Heavier instruments can be carried on either Nasmyth platform, and instruments weighing up to 500 kg (1100 lb) can be mounted at the telescope's prime focus.

Enclosure. The telescope (**Fig. 2**) sits atop a 20-m-tall (66-ft), hollow concrete pier, surrounded by the 40-m-diameter (131-ft) concrete walls of the enclosure support building (ESB). Riding on the walls of this building is the 2000-metric-ton (2205 U.S. tons), 23-m-tall (75-ft) steel and aluminum upper structure of the enclosure that corotates with the telescope. Based on extensive physical and computer modeling, the upper structure has an elliptical horizontal cross section and vertical walls to minimize turbulence that would degrade the excellent observing properties of the site. Several robotic systems are provided within the upper enclosure to handle instrument and mirror exchanges. Two "Great Walls" flank the telescope to minimize scattered light within the enclosure and thermally isolate the telescope area

from the rest of the building. Airflow through the enclosure is controlled by a windscreen at the front and top and independently adjustable vents located along the sides and rear of the enclosure. The large annular space within the ESB contains facilities for washing and realuminizing the telescope's primary mirror. Observers and summit support facilities are housed in an adjacent control building, minimizing heat sources within the telescope enclosure. Data gathered by the telescope are transferred in near-real time via fiber-optic cables to the Hilo base facility, where the data are added to the 150-terabyte online archive.

Performance. Subaru Telescope's optics, tracking system, and enclosure work together to deliver excellent image quality. Science exposures have been made at visible and infrared wavelengths with image sizes better than 0.3 and 0.2 arcsecond, respectively. Subaru's sharpest images have been obtained using the adaptive optics system, reaching ~0.06 arcsec, close to the telescope's diffraction limit at infrared wavelengths.

Instrumentation. Subaru Telescope is very well endowed with a rich suite of seven first-phase instruments plus an adaptive optics (AO) subsystem for imaging and spectroscopy over a wavelength range of 300 nm to 20 μm.

Suprime-Cam (Subaru Prime Focus Camera). This is a mosaic of ten 2048 × 4096 charge-coupled devices (CCDs) located at the prime focus of Subaru Telescope, covering a 27 arcmin × 34 arcmin field

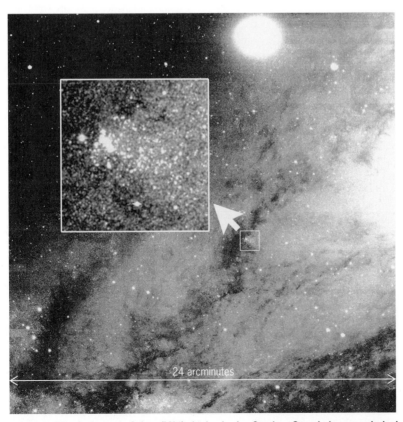

Fig. 3. Visible-light image of the Andromeda Galaxy (M31) obtained using Suprime-Cam during commissioning of the prime focus. The camera produces stellar images as sharp as 0.3 arcsecond over the whole field of view. (*Image copyright © 2000 Subaru Telescope, NAOJ. All rights reserved.*)

of view with a pixel scale of 0.20 arcsec (**Fig. 3**). This is the only prime focus camera available on any telescope larger than 4 m (13 ft).

FOCAS (Faint Object Camera and Spectrograph). This is a versatile optical imaging and spectroscopy unit used at the Cassegrain focus. Refracting optics are used for both the collimator and camera to achieve high throughput optimized for the wavelength range of 365–900 nm over a circular 6-arcmin field of view. It has four operating modes: imaging, grism spectroscopy (long slit and multislit), polarimetry, and spectropolarimetry. The multislit mode allows spectra of up to 100 objects to be taken simultaneously.

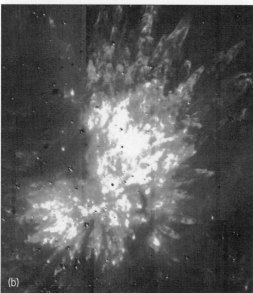

Fig. 4. Infrared images of the Orion Nebula (M42) obtained using CISCO, during the first month of operation. (*a*) The Trapezium star cluster appears near the center of the image. Toward the top is the recently discovered Kleinman-Low infrared emission region glowing in the light of warm molecular hydrogen gas. (*b*) Kleinman-Low infrared emission region. (*Images copyright* © *1999 Subaru Telescope, NAOJ. All rights reserved.*)

HDS (High Dispersion Spectrograph). This is located at the optical Nasmyth platform, providing optical spectroscopy in the range 300–1000 nm with a spectral resolution of up to 160,000. This is the highest dispersion available on any telescope larger than 4 m (13 ft).

OHS/CISCO. This is an infrared camera/spectrograph located on the infrared Nasmyth platform. CISCO (Cooled Infrared Spectrograph and Camera for OHS) is sensitive from 0.8 to 2.5 μm and provides a 2 arcmin × 2 arcmin field of view for imaging (**Fig. 4**) as well as low-resolution spectroscopy via the use of grisms. The OH suppressor (OHS) is a device inserted in the light path before CISCO to remove the OH airglow lines (contributing most of the broadband nighttime sky background) to dramatically increase the sensitivity of spectroscopic observations in the J and H atmospheric transmission windows (that is, 1.16–1.33 μm and 1.48–1.78 μm, respectively).

IRCS (Infrared Camera and Spectrograph). This is used at the Cassegrain focus. It is sensitive from 0.9 to 5.6 μm and provides up to a 1 arcmin × 1 arcmin field of view for imaging in addition to spectroscopy with grisms or a cross-dispersed echelle with a spectral resolution of up to 20,000. IRCS is also designed to deliver diffraction-limited observations with the adaptive optics system.

COMICS (Cooled Mid-Infrared Camera and Spectrometer). This provides both imaging and spectroscopic capabilities from 8 to 20 μm. The camera provides a field of view of 42 arcsec × 32 arcsec, while the spectrometer provides low- and intermediate-resolution spectra (up to $R \sim 2000$).

CIAO (Coronagraphic Imager with Adaptive Optics). This is primarily designed for the detection of faint objects near much brighter ones via the use of occulting masks. It can also be used as a general-purpose high-resolution near-infrared imager and low-dispersion long-slit spectrograph.

For background information *see* ADAPTIVE OPTICS; ASTRONOMICAL OBSERVATORY; ASTRONOMICAL SPECTROSCOPY; INFRARED ASTRONOMY; OPTICAL TELESCOPE in the McGraw-Hill Encyclopedia of Science & Technology.   Ian Shelton

Bibliography. T. Koguge and A. T. Tokunaga (eds.), *Japanese National Large Telescope and Related Engineering Developments, Proceedings of the International Symposium on Large Telescopes*, Tokyo, November 29 – December 2, 1988 Kluwer Academic, 1989; M. -H. Ulrich, (ed.), Progress in Telescope and Instrumentation Technologies, *Proceedings of the ESO Conference*, Garching, April 27–30, 1992; *Subaru News Letter*, NAOJ, vol. 1, no. 1, 1993, vol. 1, no. 2, 1995, vol. 1, no. 3, 1996; *Proceedings of the SPIE "Astronomical Telescopes and Instrumentation" Symposium*, Kona, Hawaii, March 1998 (SPIE Proceedings, vols. 3352 and 3356); *Proceedings of the SPIE "Astronomical Telescopes and Instrumentation 2000" Symposium*, 2000 (SPIE Proceedings, vol. 4008); *Publ. Astron. Soc. Japan*, 52:1–98, 2000.

Suboceanic landslides

Although normally thought of as a feature of mountainous regions, landslides can happen almost any place where the ground surface slopes. In fact, some of the largest landslides on Earth occur under water. Suboceanic or submarine landslides can involve the movement of rocks and sediments entirely beneath the sea, or they can begin as partly above-water landslides that later enter the ocean. Like open-air landslides, submarine landslides often strike steep inclines ($\sim$10°). Unlike open-air slides, submarine landslides also occur in very slightly dipping terrain (<1°). Many historical and prehistorical landslides have raked the slopes of deep ocean trenches and continental margins where strong earthquakes recur periodically. Seismic shaking probably triggered these slides. Other landslides, however, have been located on seismically quiet continental margins, such as the east and west edges of the Atlantic Ocean, and on the flanks of oceanic island volcanoes in Hawaii and the Canary Islands. Best evidence suggests that the potential for suboceanic landslides exists globally, whether in tectonically active or tectonically inactive regions. A primary hazard of submarine landslides, like their land-bound relatives, is the wasting of human-made structures along their path. The newest research, however, perceives that undersea slope failures present an additional threat—landslide-generated tsunami waves.

Landslide-generated tsunamis. The first recognized submarine landslide dates to 1929 when 300–700 km³ (72–168 mi³) of sediment slid off the top of the continental slope south of Newfoundland in a thin but broad flow that passed just west of the wreck of the *Titanic*. The mass of fluidized sediment plunged into the depths of the Atlantic at speeds near 80 km/h (50 mi/h). During its course, the landslide mass turned into a giant flow of turbulent, sediment-laden water that successively broke several transatlantic telegraph cables connecting America and Europe. The timing of the cable breaks established the speed of the landslide. Although it was not recognized until many years later, a tsunami generated by this event struck the sparsely populated coasts of Newfoundland and Nova Scotia, with waves 10 m (33 ft) and higher killing approximately 30 people.

In 1998, another landslide (this one definitely initiated by an earthquake) swept the submarine slopes north of Papua New Guinea. In contrast to the 1929 event, the landslide mass held together as a thick slab that was later found at the bottom of the slope. The New Guinea slide also raised a tsunami. On this occasion, large villages on the adjacent coastline stood in harm's way, and more than 2000 people died in the wave. Because the 1998 wave seemed to be too big to have been generated by the earthquake alone, researchers suspected and later verified that the tsunami had a "landslide assist." Subsequently, scientists implicated submarine landslides in other historical tsunamis that looked too large to have been

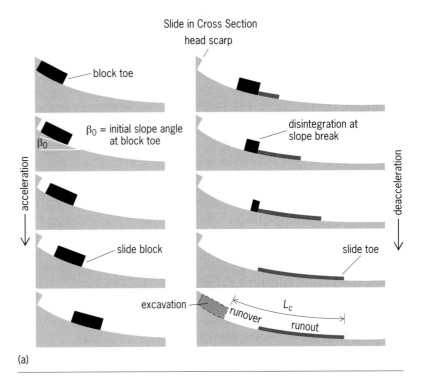

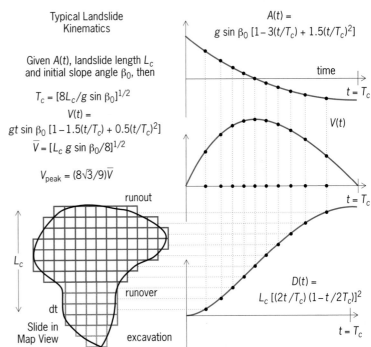

Fig. 1. Typical suboceanic landslide. (*a*) Cross-section view. (*b*) Map view. Curves to the right describe a mathematical model of the acceleration *A*(*t*), velocity *V*(*t*), and distance *D*(*t*) of the slide front versus time.

initiated by the earthquakes originally blamed for them. Most notably, studies revealed that the source region of the April 1, 1946, Unamak Island (Aleutian Islands) tsunami contained a prominent fresh landslide similar to that discovered off Papua New Guinea.

The 1946 tsunami, one of the biggest in history, caused major damage and death as far away as Hawaii

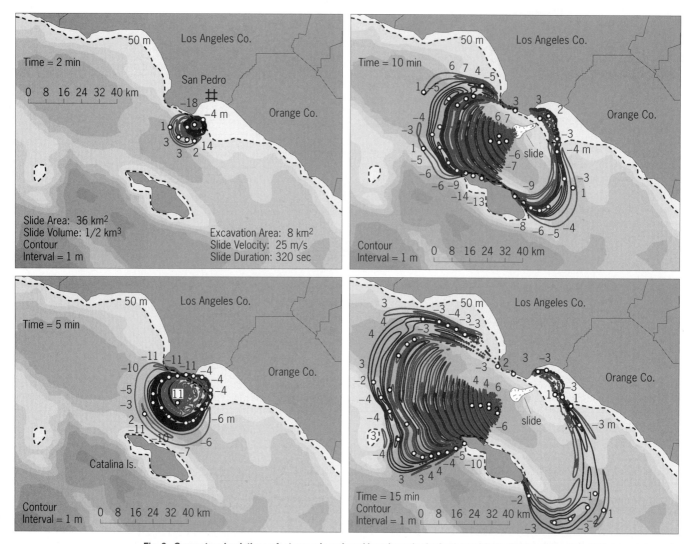

Fig. 2. Computer simulations of a tsunami produced by a hypothetical suboceanic landslide off the United States west coast near Los Angeles. Panels show the waves at 2, 5, 10, and 15 minutes after the start of the event. Contour interval is 1 meter. During travel, tsunamis form long trains of both upward and downward waves. The color and black markings distinguish elevated and depressed ocean surfaces, respectively. The numbers sample wave heights (positive above initial sea level; negative below initial sea level) in meters.

and the Marquesas (French Polynesia). Even this event, however, is likely to have been dwarfed by tsunamis produced by prehistoric submarine landslides. The Storegga landslides, the largest continental slope failures documented by geological observation, struck off the coast of central Norway between 5000 and 50,000 years ago. The largest of these had a volume 10 times greater than the 1929 Newfoundland landslide and transported material 500 km (310 mi), halfway to Greenland. Curiously, the existence of the Storegga landslide came to light partly because of the identification of layers of tsunami-deposited sediment as far away as Scotland and Holland. Submarine landslide potential near Storegga continues to receive a great deal of scientific study due to the presence there of Norway's richest reserves of petroleum and the country's huge offshore infrastructure in drilling platforms and seabed pipelines. A Storegga slide today would destroy many millions of dollars worth of infrastructure.

Computer experiments show that the efficiency of tsunami generation increases with the speed and volume of the landslide. It is likely that landslides even bigger and faster than Storegga have produced larger tsunamis. The "kings" of suboceanic landslides are giant slope failures on oceanic islands. (The existence of these landslides is known from their residual debris.) Starting far above sea level, "flank collapse landslides" as much as 3 km (2 mi) thick, sliding down into the water at speeds nearing 360 km/h (224 mi/h), could induce waves with initial heights of several hundred meters. Some scientists believe that coral rubble beds discovered 100–200 m (330–660 ft) up the side of the Hawaiian island of Lanai were actually laid down by one such tsunami. Ongoing and recent movements of the flanks of a number of oceanic island volcanoes, including Kilauea in Hawaii and the Cumbre Vieja on La Palma in the Canary Islands, hint that one of these may break down during an eruption in the not-too-distant future. Fortunately, giant flank collapses are rare; however, volcanic landslides of a few cubic kilometers' volume have punctuated the historical past. Those at Japan's Oshima-Oshima volcano in 1741, and Papua New Guinea's Ritter Island volcano in 1888, rolled 4-m (13-ft) tsunamis onto coastlines 600-1000 km (360-600 mi) distant.

Landslide physics. To understand suboceanic landslides scientifically, their kinematics must be quantified in an elementary but realistic manner. Landslide kinematics specify the position, velocity, and thickness of the material as it moves downslope. **Figure 1a** shows a typical suboceanic landslide in cross section. Usually, landslides begin when a slide block pulls away from a slope, leaving an excavation with its signature head scarp, a cliff tens to hundreds of meters high. The block accelerates downhill and crosses, more or less as a coherent mass, a region of the slide scar termed the runover. At the slope break where the incline begins to level out, the slide mass often disintegrates, if it has not already, into a runout sheet. The runout is much thinner than the block but covers a much larger area. Material excavated at the head of the slide carries across the runover and deposits in the runout. The runout represents the deceleration phase of the slide that terminates at the slide toe.

Figure 1b illustrates a typical suboceanic landslide in a map view. Submarine landslides often have an upslope neck containing the excavation and the runover, and a broad downslope fan containing the runout. The curves to the right of the slide map illustrate a mathematical model of the history of slide motion. In particular, the middle velocity curve $V(t)$ shows the acceleration and deceleration phases mentioned above. These formulas for slide motion, together with the slide shape and thickness, constitute the kinematics needed to compute slide-generated tsunami waves. Note that in this description, landslide duration T_c and mean slide velocity $\overline{V}$ depend on the length L_c of the slide from the block toe to the slide toe and the initial slope angle β_0 just below the slide block, as shown in Eqs. (1) and (2). The g

$$T_c = \sqrt{8L_c/g \sin \beta_0} \qquad (1)$$

$$\overline{V} = \sqrt{gL_c \sin \beta_0/8} \qquad (2)$$

represents the acceleration of gravity ($g = 9.8$ m/s²; 32 ft/s²). Most submarine slides span dimensions of $L_c = 5$-500 km (3-310 mi) on slopes β_0 of 1-10°. Formula (1) predicts slide durations of a couple of minutes to over an hour. Formula (2) predicts average slide velocities of 36-540 km/h (22-336 mi/h). While these velocities seem high, be aware that landslide speeds usually lag the $\sqrt{gh}$ speed of the tsunami in an overlying ocean of depth h.

Figures 2 and **3** picture tsunamis generated from two submarine landslides—one small, and one huge. Imagine that a small slide occurred on the continental slope just off San Pedro, the port facility for Los Angeles, California, and that the slide had a 0.5 km³ vol-

ume, $L_c = 8$ km, $T_c = 320$ s, $\overline{V} = 25$ m/s, and a mean excavation thickness of 60 m. Although this slide is hypothetical, detailed sonar images of nearby sea floor reveal several landslide scars of similar scale— a fact not overlooked by the San Pedro port directors. Above the slide, tsunami waves from this event reach 10-15 m (30-45 ft) height, but because the shoreline is so close (just 5-10 minutes' travel time), the waves cannot spread out and disperse much before they beach. Worldwide, 0.5-km³ (0.12-mi³) landslides might befall some continental margin every few decades, so their hazard is palpable. Arguably, nearshore suboceanic landslides could drop 3-4 m (10-13 ft) waves with little warning on just about any coast that has good exposure to the sea.

Figure 3 shows contours of the tsunami expected from a lateral collapse of the Cumbre Vieja volcano in the Canary Islands. Recall that flank collapses represent the worst-case submarine landslide. This La Palma slide involves 500 km³ (120 mi³) of material running out 60 km (36 mi) at a mean speed of 100 m/s (328 ft/s). Although this incident is also hypothetical, the Canary Island chain has witnessed 10 comparable landslides in the past million years. Considering all of the oceanic volcanoes in the world, a La Palma-scale collapse might occur somewhere once every 10,000 years or so. Some computer models predict

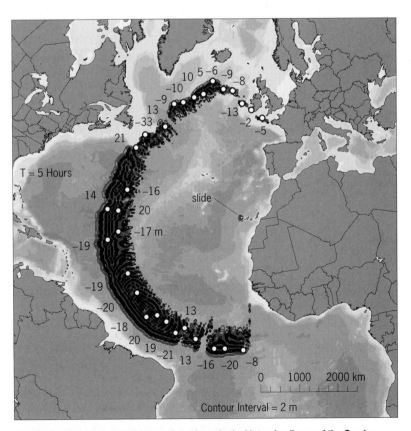

Fig. 3. Computed tsunami 5 hours after a hypothetical lateral collapse of the Cumbre Vieja Volcano on La Palma, Canary Islands. Landslides of this volume (500 km³) and velocity (100–150 m/s) have sent damaging waves across entire ocean basins. The color and black markings distinguish elevated and depressed ocean surfaces, respectively. The numbers sample wave heights (positive above initial sea level; negative below initial sea level) in meters.

that waves generated from collapse-scale submarine landslides could traverse entire ocean basins and retain heights of 20 m (66 ft).

Outlook. Evidence argues that suboceanic landslides are as common as landslides on land, and that the behaviors of the two groups share many elements. Like those on land, suboceanic landslides span several orders of magnitude in size and volume. Unlike landslides on land, submarine landslides lie hidden largely under the sea and so conceal certain mysteries, not the least of which is a potential to parent tsunamis. Vastly improved technologies in multibeam sonar can now quickly map large swaths of the sea floor to 1-m (3.3-ft) resolution, easily spotting slumps and slope failures. With the expanded use of these instruments, perhaps within a decade scientists will have a much better census of submarine landslides and a clearer picture of their hazards.

For background information *see* EARTHQUAKE; GEOPHYSICS; KINEMATICS; LANDSLIDE; SEISMIC RISK; TSUNAMI; VOLCANO in the McGraw-Hill Encyclopedia of Science & Technology.   Steven N. Ward; Simon Day

Bibliography. S. J. Day et al., Recent structural evolution of the Cumbre Vieja volcano, La Palma, Canary Islands, *J. Volcan. Geotherm. Res.*, 94:135–167, 1999; S. N. Ward, Landslide Tsunami, *J. Geophys. Res.*, 106:11,201–11,215, 2001; S. N. Ward, Tsunamis, in R. A. Meyers (ed.), *The Encyclopedia of Physical Science and Technology*, Academic Press, 2001.

Sun-climate connections

Measuring 1.4 million kilometers (870,000 mi) in diameter, estimated to be 5 billion years old, and having a core temperature of 15 million degrees Kelvin (27 million degrees Fahrenheit), the Sun is the primary energy source that creates and drives the Earth's climate. Major features of changes in climate are influenced over long periods by characteristics of the Earth's orbit around the Sun. These characteristics determine the amount of energy that the Earth receives and where that energy is concentrated. For example, the eccentricity of the Earth's orbit (that is, its deviance from a perfectly circular orbit) has a period of 90,000–100,000 years. Seasonal contrasts in temperature should be greater when the eccentricity of orbit is greatest. One cycle of change in the inclination, or tilt, of the Earth's axis relative to the orbital plane occurs on average every 41,000 years. All else being equal, when the tilt is at its maximum, seasonal temperatures should vary most. It takes approximately 21,000 years to complete one cycle of the precession of the equinoxes (that is, the time of year in which the Earth is closest to the Sun marches through the year in a 21,000-year period.) The Sun at present is closest to the Earth during the Northern Hemisphere winter. Ten thousand years from now, the Sun will be closest to the Earth in the Northern Hemisphere summer. Glacial and interglacial periods are delineated fairly well by these astronomical parameters, but these are not the sole factors behind climate variability and change.

Solar variability. Research that seeks to identify and explain solar-induced effects on the Earth's climate often looks first at the variability of the Sun itself. There are several apparent cycles of solar activity that have attracted most attention. The most popular are the nominal solar 27-day period of rotation, the 11- and 22-year sunspot cycle reflective of magnetic change, and the 80-year Gleissberg cycle in the amplitude of the 11-year sunspot cycle. (Sunspots are dark, cooler areas of the Sun's surface, produced by intense magnetic fields, typically lasting several days.) Consecutive 11-year cycles show a reversal in magnetic characteristics, so the full period of the magnetic cycle is 22 years. The difference between maximum and minimum periods of the solar cycles in terms of radiative flux (rate of energy emission) is about $+/- 0.1\%$, which translates to about 0.24 W/m^2 contribution to climate forcing (a disturbance in the Earth's radiative balance with space, having the potential to change the climate); this converts to just over a $0.1°$C change in global surface temperature. By contrast, estimates of climate forcing by the introduction of radiatively important trace gases (greenhouse gases, which reduce Earth's heat radiation to space) since the dawn of the industrial age is about 2.45 W/m^2. However, it is believed that during the Maunder Minimum (of sunspot activity) between 1645 and 1715, which occurred during the coldest part of the Little Ice Age, solar radiative flux may have been considerably less, perhaps a decrease of 0.2% or more (perhaps a reduction of 0.5 W/m^2). The Maunder Minimum notwithstanding, if one assumes that first-order radiative forcing of this magnitude is the only solar factor affecting climate change, it is not large enough to explain the observed rise in global mean temperature over the past century.

Solar influence on temperature. The Sun and many other factors have influenced the mean near-surface temperature of the Earth's atmosphere in the last four centuries (see **illus.**). The total solar irradiance reconstructed from the orbital and solar-cycle arguments, along with recent measurements, tracks the surface temperature quite well, with two important exceptions. It is believed that a period of intense volcanic activity threw aerosols and dust high into the atmosphere, significantly blocking the Sun's energy from the Earth, and led to the observed decrease in surface temperature in the mid-1800s. The second exception began around 1980 and continues to this day. The rapid rise in surface temperature is attributed to the introduction of the greenhouse gases that absorb the long-wave radiation emitted by the Earth.

While the global mean surface temperature (see illus.) clearly shows a relationship with solar irradiance, other studies have shown statistical relationships between solar activity and regional patterns of temperature as well. Negative correlations have been reported between sunspots and annual mean

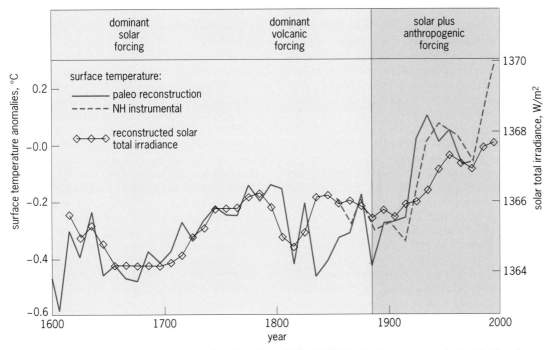

Global mean surface-air temperature and solar irradiance from 1600 to 2000. The Earth's temperature is depicted from two sources: the period from the midnineteenth century to 2000, recorded with meteorological instruments; and the prior period, where only proxy temperatures are available, reconstructed from ice cores, lake sediments, tree rings, and other sources in the paleoclimate record. (*After J. Lean, The Sun's variable radiation and its relevance for Earth, Annu. Rev. Astron. Astrophys., 35:33–67, 1997*)

temperatures in equatorial and temperate zones, often explained as the effect of dark sunspots in blocking radiation from the Sun, but the arid subtropics are positively correlated. Studies of temperature variability in northeastern North America have been correlated with high sunspot activity in the 11-year cycle. The lengths of the growing season in England and the persistence of ice along the shores of Iceland have also been correlated with the 11-year cycle.

Solar influence on rainfall. As early as 1923, the global distribution of annual rainfall was related in statistical analyses to periods of high and low sunspot number. About the same time, other analyses depicted relationships between sunspot cycle and the flow of the Nile and Parana rivers and the water level of Lake Victoria. More rainfall appeared to occur at high latitudes and near the Equator during periods of high solar activity indicated by the number of sunspots. Drought in the United States has frequently been correlated to the 22-year sunspot cycle.

The frequency of thunderstorms in the Mediterranean Sea and in North America has been correlated with solar flares (explosions on the Sun's surface, usually occurring near sunspots, which release massive amounts of energy). And the positions of storm tracks, according to other investigations, have been statistically linked to the sunspot cycle.

Solar influence on atmospheric pressure. There have been frequent studies of atmospheric pressure patterns and possible connections to solar activity. Results have varied. For example, some studies have shown that when solar flare and related geomagnetic activity are high, surface air pressure in Greenland and Iceland rises while European surface pressure drops. Other studies have not been able to confirm such strong correlations between solar activity and surface pressure. However, subsequent analyses of 500-millibar-level (midway through the vertical mass of the atmosphere) pressure changes have tended to bolster the argument. Active periods of solar flares and geomagnetic disturbances in western Germany, eastern Siberia, and the Gulf of Alaska have been correlated positively with pressure increases, while the opposite effect is seen in regions of the Kara Sea, southern Japan, and south of Iceland. Atmospheric low-pressure troughs and areas of atmospheric vorticity have also been linked statistically with solar-induced geomagnetic disturbances. Some studies have correlated high sunspot numbers with high atmospheric pressure over the world's continents in winter, and high pressure over the oceans in summer. Other research has connected solar activity with tropospheric winds, the position of high-pressure ridges, and the location of the Icelandic Low. Of course, if atmospheric pressure is influenced by the Sun, the effects are very likely to be reflected in temperature, precipitation, and other meteorological variables.

These numerous relationships are largely statistical in nature. Without an understanding of the physical causes of correlation, they will remain suspect, perhaps the result of chance or of biases in the selection of variables or analytical methods. Current research is focusing on the causal mechanisms for solar influences on weather and climate. This new thrust is made possible by long-running satellite-based observational programs and the payoff from two decades of intensive international climate research.

Contemporary directions of study. Global atmosphere-ocean circulation model experiments that include atmospheric chemistry show that an increase in stratospheric ozone occurs when increases in solar ultraviolet (UV) radiation which are typical of that measured from solar minimum to solar maximum are introduced to the model. Such model experiments have produced 5% increases in ozone, and observational evidence shows similar results in the real atmosphere. Consequences of increased ozone in these experiments include a 0.5–1.0°C increase in middle and upper stratosphere temperature and up to a 5.0°C increase in the lower stratosphere. The ozone also increases heating rates in the Northern Hemisphere by more than 0.25°C per day from a period of solar minimum to solar maximum. In turn, tropospheric winds in the model increase and, again with model results backed by observational evidence, winter temperatures can subsequently increase by some 5°C in the Northern Hemisphere.

Study of stars similar to the Sun has shown that long-term solar irradiance may actually vary up to 0.7%. This is significantly more than the aforementioned 0.1% difference currently observed between periods of solar maximum and solar minimum activity.

One theory suggests that cosmic rays (electrons and nuclei of atoms, largely hydrogen) enter the atmosphere, creating cloud condensation nuclei. This may be the mechanism responsible for observed changes in low-latitude cloud cover that have been correlated with the solar cycle.

During periods when solar particles enter the atmosphere in great number, oxides of nitrogen and hydrogen are produced in the upper atmosphere at polar latitudes. These solar particle events usually occur near the time of maximum solar activity in the solar cycle. Odd nitrogen (NO_x) that is created diffuses into the lower stratosphere and becomes involved in ozone chemistry; this may reduce stratospheric ozone by some 20%. Such reductions may then lead to changes in upper-atmosphere wind patterns and climate on time scales coinciding with the solar cycle. Similar effects involving odd nitrogen occur during periods of extreme energetic electron precipitation into the atmosphere. This can occur as the time of solar minimum nears, when the solar wind increases. Significant energetic electron precipitation events causing increases in odd nitrogen have been observed.

Analyses of the global oceans show that average upper-ocean temperature changes of 0.1°C are in phase with changes in solar irradiance on decade-to-century time scales. These studies indicate that solar irradiance varying over decadal time scales can amplify the global average upper-ocean temperature response to solar forcing two to three times that anticipated from well-known laws of radiation. Over a century or so, this could lead to changes in the effective radiative forcing of the Sun that rival or exceed the estimated increase in net radiative forcing due to the past century's increase in greenhouse gases and aerosols.

Other studies are detailing the statistical relationships between the 11-year solar cycle and the structure and temperature of the Earth's atmosphere. During periods of solar maximum, upper-air temperatures, for example, are higher than when solar activity is low. Near the Equator the difference in temperature between solar minimum and solar maximum is almost 2°C averaged around the globe. The effect appears greatest in the tropical and subtropical summers. These findings are consistent with the modeling studies mentioned previously that suggest atmospheric ozone chemistry may be a strong factor.

For background information *see* CLIMATE HISTORY; HEAT BALANCE, TERRESTRIAL ATMOSPHERIC; SOLAR ENERGY; SOLAR MAGNETIC FIELD; SOLAR RADIATION; SOLAR WIND; SUN in the McGraw-Hill Encyclopedia of Science & Technology.　　　　William A. Sprigg

Bibliography. J. Lean, The Sun's variable radiation and its relevance for Earth, *Annu. Rev. Astron. Astrophys.*, 35:33–67, 1997; National Academy of Sciences, *Solar Influences on Global Change*, 1994; National Academy of Sciences, *Solar Variability, Weather, and Climate*, Studies in Geophysics, 1982; J. Pap et al. (eds.), *The Sun as a Variable Star: Solar and Stellar Irradiance Variations I*, Cambridge University Press, 1994; W. Schroder (ed.), *Long and Short Term Variability in Sun's History and Global Change*, Science Edition, 2000.

Sunflower

Sunflower has become an important world crop for edible oilseed and food production. Because it has such a wide range of adaptability, it has had a major impact on agricultural systems worldwide. It continues to increase in popularity as new and expanded uses for the crop are developed.

History. Sunflower is native to North America, and were widely used by Native Americans as food, dye, and medicine, and sometimes in religious ceremonies. Native Americans were the first plant breeders; they selected and propagated different plant types for specific uses. Sunflower as a crop was probably first grown in New Mexico and Arizona about 3000 BCE, and may have predated corn as a food crop. Sunflower was cultivated or harvested wild by Native Americans from the Arctic Circle to the tropics and from the Atlantic to the Pacific Ocean. Sunflower seeds were first taken to Europe by Spanish explorers, with the first published description of sunflower in Europe reported in Spain in 1568. The use of sunflower primarily as an ornamental with secondary use as a food crop quickly spread throughout Europe. Peter the Great is credited with introducing the crop to Russia where, over time, plant selections were made and the uses of sunflower for commercial agricultural production developed primarily into oilseed and nonoilseed types. Sunflower was reintroduced into North America, especially Canada and the northern Great Plains, by Russian immigrants. Because of its origin, sunflower was sometimes referred to as "Russian peanuts." In addition to the traditional oil and food uses, sunflower was sometimes grown for livestock feed as silage or as fireplace fuel.

Today the sunflower species (*Helianthus annuus* L.) is grown worldwide, with major areas of production in the United States, Europe, Argentina, China, Russia, and India. Sunflower produces a high-quality oil that can be used for cooking, while both the whole seed and dehulled seeds of nonoilseed sunflower are used as human snack food and birdseed. Sunflower plants are also widely grown in gardens as ornamentals, and in the United States they are commonly used in floral arrangements.

Characteristics. The genus *Helianthus* consists of about 50 species, all of which are found in North America. Some species, such as *annuus*, are widely distributed, while others are rare and occur only under certain environmental conditions. There exists among and within each species a wide diversity of flower colors, plant types, and other characteristics.

A unique characteristic of *Helianthus* (from Greek *helios*, sun, and *anthus*, flower) is that, prior to flowering, the head and leaves follow the Sun. This characteristic is known as heliotropism. Sunflowers face east in the morning and follow the Sun, usually lagging about 12° behind. By evening the head and leaves face west, turning again toward the east between 3:00 and 6:00 a.m. The mechanism that causes heliotropism is thought to be related to differential auxin accumulation in different parts of the plant. When the plants reach anthesis (the flowering period), heads generally tilt toward the northeast in the Northern Hemisphere and toward the southeast in the Southern Hemisphere.

The floral head of sunflower consists of two types of flowers—ray flowers and disk flowers, serving completely different purposes. The ray flowers consist of the showy petals that adorn the outside ring of the head (see **illus.**). They are sterile and do not produce seed. Their main function appears to be to attract pollinators such as bees, other insects, and birds to the plant. Ray-flower colors differ among and within species, with the most common colors being yellow, lemon yellow, orange-yellow, or reddish. The disk flowers are the small rings of flowers that constitute most of the head (see illus.). Each flower consists of male and female parts, and seed is produced after fertilization. Anthesis, or flowering, of the disk flowers begins at the outer ring of the head and progresses in a circular manner toward the center. Usually this process will occur over a period of seven to ten days. As soon as all the disk flowers have completed anthesis, the ray flowers immediately wilt and begin to fall off.

Production. Sunflower is an extremely adaptable plant that can grow under a wide range of environmental conditions. As a result, cultural practices vary among and within countries. Sunflower response to varying environmental conditions is usually characterized by changes in plant height, head size, number of seeds per head, and seed size or weight. Both oil concentration and composition can be influenced by environmental conditions. Oil concentration is an important economic factor, because producers are sometimes paid a bonus for the production of seed with high oil content.

Sunflower is grown mostly under rain-fed conditions. A characteristic that makes sunflower so productive and adaptable is its extensive root system. Sunflower reaches its maximum rooting depth near anthesis. Rooting depths of 6 ft (1.8 m) are common, with depths of 9 ft (2.7 m) reported. The ability to extract water from these depths has given sunflower the reputation of being drought-tolerant. The deep root system has also made sunflower extremely adaptable to areas where water is a limiting factor. In much of the northern Great Plains, sunflower has provided producers an alternative to continuous small-grain production. This is important because it provides the producers an opportunity to rotate crops and diversify their sources of income. It has also provided producers with the opportunity to spread out planting and harvesting operations, allowing each crop to be sown and harvested at specific dates to maximize yield and labor efficiency. Sunflowers are sown and harvested using the same type of equipment used for small grains and many other crops.

Plant breeding goals. Almost all sunflower is grown for commercial production are hybrids that typically outyield older open-pollinated types by about 20%. The production of hybrids was made possible by the discovery of cytoplasmic male sterility. Cytoplasmic sterile sunflower genotypes (lines) do not produce pollen and are used as female parents. These lines are sown in alternating strips with pollen-producing male lines. The resulting pollination of the female lines results in hybrid seed production. Seed-production fields should be isolated by at least a mile (0.6 km) from other sunflower fields; from volunteer sunflower from adjacent fields; and from wild sunflower growing in roadsides, ditches, and pastures.

The floral head of sunflower consists of small rings of fertile disk flowers that produce seed, and showy petals of sterile ray flowers that adorn the outside ring and attract pollinators.

Unwanted contamination by pollen other than from the desired male lines will result in contaminated hybrid seed. Both public and private plant-breeding programs have made significant progress by developing hybrids with desirable agronomic and quality characteristics. The goals of almost all plant-breeding programs are increased yield and pest resistance. Plant-breeding efforts to increase yield have been directed primarily toward incorporation of disease resistance, and to a lesser extent insect resistance, into commercial hybrids.

Because sunflower is native to North America, producers in both the United States and Canada have been faced with a wide range of problems involving insects and diseases that have developed over hundreds of years on wild sunflower. The most serious diseases are leaf rust, white mold, downy mildew, verticillium wilt, and phomopsis. White mold, also called Sclerotinia, is of special concern because it attacks most nongrain crops, reducing crop rotation options. Although a number of insects can cause serious problems, they have received less attention; infestations are more sporadic, and can be controlled to a certain extent by the use of insecticides. Other factors that are the focus of plant-breeding efforts are maturity, stalk strength, self-comparability (the ability of a hybrid to self-pollinate), protein content, height and head type, and inclination. Many of the native species of sunflower have been used as a source of genetic resistance for insect and disease problems in commercially grown hybrids.

Additional goals of plant breeders of oilseed sunflower are to increase oil content and sometimes to modify oil quality for different markets. For nonoilseed hybrids, achene size, shape, and color are common goals of breeding programs. In the nonoilseed types, cultural preferences exist for achene color, size, and shape. For example, the preferred seed color and shape for markets in Turkey are different from those in Italy. As a result, different hybrids are developed for different markets.

Alternative uses. In addition to the typical oil and food uses, alternative uses of the crop and its byproducts are continually being developed. For example, whole sunflower seeds, the meal left after oil extraction from the seed, sunflower screenings, silage, and sunflower hulls are used for animal feed. In addition, sunflower hulls and other sunflower biomass have been used as a source of fuel. The use of sunflower and other vegetable oils as a replacement for or as a supplement to petroleum fuel products in diesel engines has been studied extensively. A common problem with using vegetable oils as fuel has been the buildup of carbon in engine parts, so manufacturers are reluctant to provide extended warranty on engines that use vegetable oil fuels. Other potential uses for sunflower include food coloring from the hulls and pectin from harvested heads.

For background information *see* AGRICULTURAL SCIENCE (PLANT); BIOMASS; BREEDING (PLANT); PLANT PHYSIOLOGY; SUNFLOWER in the McGraw-Hill Encyclopedia of Science & Technology.   Albert A. Schneiter

Bibliography. C. B. Heiser, Jr., *The Sunflower*, University of Oklahoma Press, Norman, 1976; A. A. Schneiter (ed.), Sunflower Technology and Production, *Amer. Soc. Agron. Monog.*, no. 35, 1997.

Superconducting balls

Superconductors strongly repel a magnetic field. This phenomenon, known as the Meissner effect, is the scientific basis for the magnetically levitated (maglev) superconducting trains, which can move at a speed of 500 km/h (310 mi/h). According to electromagnetic theory, electric and magnetic fields are intricately related. Thus, the interaction between superconductors and an electric field is of great interest. Fritz London first studied the issue in 1935 and proposed that superconductors and conventional conductors might react to electric fields differently. However, due to the difficulties associated with these types of experiments, this issue remains controversial and unresolved. The recent discovery of the formation of superconducting balls in a strong electric field illustrates the rich physics of this area.

Experiment. Application of a uniform electric field defines a preferred direction in space. In this situation, metallic particles bounce between the two electrodes that generate the electric field, or tend to line up in chains in the field direction. It is against common sense that an electric field could drive high-temperature superconducting particles together to form a round ball. Yet, this occurred in a recent experimental finding that revealed a new property of superconductivity.

The particles used in this experiment were $YBa_2Cu_3O_{7-x}$ (99.99% purity), $NdBa_2Cu_3O_x$ (99.9% purity), $Bi_2Sr_2CaCu_2O_{8+x}$ (99.9% purity), and $YBa_2Cu_3O_x$ (99.9% purity) with transition or critical temperatures (T_c) around 92, 94, 84, and 87 K (-294, -290, -308, and $-303°F$) respectively. Particle size was in the micrometer range (**Fig. 1**).

The experimental setup involved a horizontal Teflon microscope slide on which was mounted two parallel brass electrodes 10 mm wide and 4 mm apart. Teflon spacers were used between the electrodes to form a cell in a region of uniform electric field. Only

Fig. 1. Scanning electronic micrograph of $NdBa_2Cu_3O_x$ particles.

a small amount of a slurry containing liquid nitrogen and high-temperature superconducting particles was needed to fill the cell, measuring 4 mm × 5 mm horizontally and 5 mm in vertical depth. The particle volume fraction was about 10%. The whole cell was submerged in liquid nitrogen, and the top surface of the cell was open to allow liquid nitrogen to flow in so that a constant temperature was maintained at 77 K (−321°F). The experiment was recorded by a high-speed camera that could take up to 1000 frames per second, enabling the dynamic process to be examined in detail.

The high-temperature superconducting particles were well dispersed before the electric field was applied. When a strong dc electric field was turned on, the particles first bounced between the two electrodes as a cloud. During the collision, they aggregated into one big ball in milliseconds. The ball then bounced between the two electrodes at a high speed as one metallic particle. A typical ball had a radius of about 0.25 mm and was closely packed (**Fig. 2**). As a particle's average size was only 1–2 micrometers, this ball consisted of several million particles.

The ball's speed was estimated to be about 20 cm/s (8 in./s) from the video recording. As soon as the ball touched the solid-brass electrode, it quickly discharged and acquired an opposite charge, then bounced back to the other electrode. The whole process for the ball to strike and bounce back from one electrode took only a couple of milliseconds. Therefore, the average acceleration in this collision was at least 10 times greater than gravity (9.8 m/s²; 32 ft/s²), equivalent to the impact force on an automobile when it hits a soft bump at a speed of 24 km/h (15 mi/h). Under such an enormous impact force, several million particles are still able to

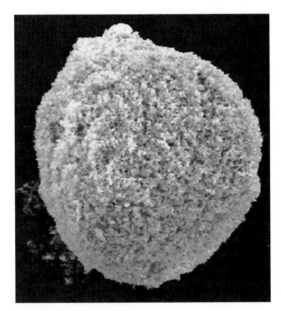

Fig. 2. Scanning electronic micrograph of a ball of $Bi_2Sr_2CaCu2O_{8+x}$ particles.

hold together, indicating that the cohesive force is very strong. However, the ball was fragile after it was removed from the liquid nitrogen, although the granular friction may sustain the ball. If the particle concentration was high, two or more balls could be formed under a strong electric field. During the whole process, the electric current density was very low, about several microamperes per square centimeter.

Ball formation has also been found in a strong ac field of low frequency. The phenomenon is quite similar, except that the high-temperature superconducting particles first form chains in an ac field. As the electric field is increased above a critical value, the chains break and the particles assemble together to form balls.

In order to verify that the ball formation is a result of superconductivity, several experiments were conducted. First, the same cell with the same particles was placed in silicon oil. At room temperature, under the same field, these particles formed only chains. This behavior is easy to understand. These particles are ceramics at room temperature, and the ceramic-oil suspensions are electrorheological fluids. Under a strong electric field, the dielectric particles are polarized so that they form chains along the field direction. Tests also confirmed that normal metallic powders, such as copper, iron, and aluminum, do not form balls in liquid nitrogen under an electric field. Instead, they bounce between the two electrodes to transport charges and move separately in the same way as at room temperature.

Liquid argon was used to raise the temperature from 77 K (−321°F) through the transition temperature of BiSrCaCuO. The boiling point of argon is 87.3 K (−303°F), just above the transition temperature of BiSrCaCuO. When the temperature was below the transition temperature, a big ball formed and bounced between the two electrodes. The temperature was then raised above the transition temperature, and the high-speed camera was used to examine the situation. Two frames were captured in sequence. The first frame showed that the big ball was going to strike an electrode (**Fig. 3***a*). The next frame showed that in less than 1 millisecond the ball disappeared; it broke into pieces in the collision (Fig. 3*b*). The conclusion from these observations is that once the particles are no longer in the superconducting state the strong force binding these particles disappears.

Theoretical explanation. To understand the phenomenon, several relevant facts need to be summarized. Ball formation requires a strong applied electric field. Without such a field, the high-temperature superconducting particles are just dispersed in liquid nitrogen. If the applied electric field is not strong enough, the high-temperature superconducting particles do not form balls, and just bounce between the two electrodes in the dc field or form chains in the ac field. Therefore, the above phenomenon is deeply related to the interaction between superconductors and a strong electric field. The shape of these

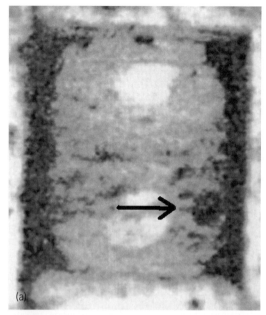

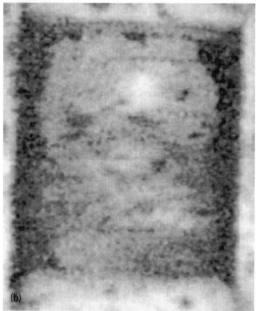

Fig. 3. Destruction of a superconducting ball as the temperature of liquid argon is raised above the ball's transition temperature. (*a*) Ball moving toward an electrode; arrow indicates the ball and its direction of motion. (The upper ball is not moving.) (*b*) After collision with the electrode, the ball has disappeared, being broken into pieces in the collision.

balls is round, different from the structures formed under electric polarization, which are chains or thick columns along the field direction.

A sphere has the least surface area in comparison with all other geometric objects of the same volume. The round shape is usually associated with a positive surface energy. For example, water droplets are spherical, as water surface tension requires minimization of the surface area. Hence, the formation of superconducting balls is also due to a positive surface energy that is induced by a strong electric field.

When the high-temperature superconducting particles touch an electrode, they pick up charges. Static charges stay at the particle surface. In a classical physics course, this surface layer is treated as being two-dimensional. In reality, this charged layer has a thickness, which is also the penetration depth of the electric field into the superconductor. High-temperature superconducting materials can be regarded as two-dimensional superconducting layers weakly coupled by Josephson tunneling along the *c* axis of the crystal structure. Josephson tunneling is a phenomenon that characterizes the quantum correlation of two separated superconductors. In superconductivity, an important quantity, coherence length, provides a range within which the superconducting electron density cannot change drastically. Along the *c* axis, the coherence length is very short. When the electric field penetration depth is comparable or longer than the coherence length along the *c* axis, the electric field produced by the surface charge layer affects the lower layers and leads to a voltage bias. If this biased voltage is stronger than a threshold, it turns off the Josephson coupling and produces a positive energy. This positive energy is equal to the loss of Josephson coupling energy and is associated with the surface charge; therefore, it is a positive surface energy. The formation of superconducting balls is the result of minimization of this surface energy. Once these particles aggregate together, the free charges move to the outer surface and there is no electric field inside the aggregated ball. As the surface energy is associated only with the free charges, the ball formation greatly reduces the surface energy. The above theory can help to explain the phenomenon and make useful predictions.

The discovery of superconducting granular balls in a strong electric field indicates that the field of interaction between superconductors and a strong electric field has rich physics and requires much more investigation. Currently, extensive theoretical and experimental research in this area is going on. Increased understanding of this area may also lead to many important scientific and industrial applications, such as superconducting film growth and maglev trains.

For background information *see* ELECTRIC FIELD; JOSEPHSON EFFECT; MEISSNER EFFECT; SUPERCONDUCTING DEVICES; SUPERCONDUCTIVITY in the McGraw-Hill Encyclopedia of Science & Technology.

Rongjia Tao

Bibliography. P. W. Anderson, C-axis electrodynamics as evidence for the interlayer theory of high-temperature superconductivity, *Science*, 279(5354): 1196–1198, February 20, 1998; A. Barone and G. Paterno, *Physics and Applications of the Josephson Effect*, John Wiley, New York, 1982; C. Burns, *High-Temperature Superconductivity: An Introduction*, Academic, 1991; R. Tao et al., Formation of high temperature superconducting balls, *Phys. Rev. Lett.*, 83:5575–78, December 27, 1999.

Supercritical fluid technology

Most chemical processes (laboratory and industrial) occur in liquid phases used as reaction or separation media. Processes in the liquid state benefit from a comparatively fast mass and heat transfer, low viscosity, and high particle density. Dense states, however, can also be generated in compressed gases. At low temperature, if a gas is compressed to high density, it will condense into a liquid. Above a certain temperature, known as the critical temperature, any gas can be compressed gradually to a liquidlike density without undergoing such a transition. The resulting dense phase is called a supercritical fluid.

Many solvent properties are related to density. For a process, the solvent properties of a supercritical fluid may be tuned over a wide range by adjusting the pressure. Separation and processing by means of supercritical fluids are now carried out in various industrial operations, such as the food, pharmaceutical, and cosmetic industries. Supercritical fluid chromatography is a well-known analytical tool. New techniques use supercritical fluids in materials processing such as particle formation, the creation of porous materials, and the production of polymers with interesting morphologies. Supercritical water may be used to destroy hazardous waste.

Properties of supercritical solvents. Understanding the properties of supercritical fluids requires understanding the concept of the critical point. Whether a substance is solid, liquid, or gaseous depends on temperature and pressure (see **illus.**). The gaseous and liquid phases are separated by the liquid-vapor equilibrium line, known as the vapor pressure curve. When increasing the temperature along this line, the liquid and gaseous phases become more and more similar. Both phases become identical at the critical point, above which it is impossible to produce a liquid meniscus. The solid-liquid and solid-gas equilibrium lines do not terminate in such a point. At conditions above the critical temperature (T_c) and critical pressure (P_c), the fluid states are known as supercritical (see **illus.**). By adjusting the pressure, a wide range of densities from vaporlike to liquidlike can be attained.

The critical points of some substances used in supercritical technologies are shown in the **table**. Most processes rely on carbon dioxide (CO_2), for which both the critical temperature and critical pressure are conveniently low. Carbon dioxide is nontoxic, non-

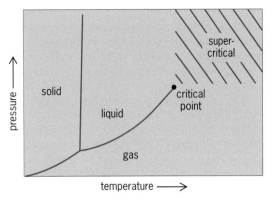

Typical pressure-temperature (*P-T*) phase diagram showing the critical-point discontinuity.

carcinogenic, nonflammable, and cheap. Few conventional liquid solvents fit this description.

Nonpolar carbon dioxide is a particularly good solvent for substances of low polarity. For more polar substances, it is of limited use, but its solvent properties can often be substantially modified by adding trace amounts of a polar solvent such as an alcohol. Other fluids shown in the table may be used as solvents for more polar substances. Some halocarbons such as trifluoromethane (CHF_3) may form excellent supercritical solvents for special purposes, but most of them are not used because they fall under the regulation of the Montreal Protocol for substances that deplete the ozone layer. The use of water as a supercritical fluid would be most desirable, but its high critical temperature and pressure (see table) and high corrosivity under such conditions exclude its routine applications. Nonetheless, supercritical water has been considered in the last two decades as a potential medium for the destruction of hazardous organic wastes.

All supercritical fluids are compressed gases with high potential energy, and are operated at high pressures and often also at elevated temperatures. This requires considerable safety precautions and special equipment with both industrial and laboratory reactors. Moreover, large-scale operations lead to considerable compression costs. As a result, processes using conventional liquids may be lower-cost than those using supercritical solvents. Still, solvents such as carbon dioxide are important for developing "clean" technologies and "green chemistry." An increasing number of solvents are expected to be replaced in the future because of environmental and

Some fluids used in supercritical technologies		
Fluid	Critical temperature	Critical pressure
Carbon dioxide (CO_2)	31°C (88°F)	7.4 MPa (73 atm)
Ethane (C_2H_6)	32°C (90°F)	4.9 MPa (48 atm)
Nitrous oxide (N_2O)	36°C (97°F)	7.3 MPa (72 atm)
Propane (C_3H_8)	97°C (207°F)	4.2 MPa (41 atm)
n-Pentane (C_5H_{12})	197°C (387°F)	3.4 MPa (34 atm)
Water (H_2O)	374°C (705°F)	22.1 MPa (218 atm)

health concerns as well as legislative restrictions. For these reasons, major applications of supercritical fluids are presently in extractions in the food (for example, coffee, tea, hops, or spices), pharmaceutical, and cosmetic (for example, fragrances) industries, where clean technologies are mandatory. The same is true for the production of high-value chemicals such as vitamins. *See also* GREEN CHEMISTRY.

Supercritical fluid extraction. Supercritical fluid extraction is a separation process that uses a supercritical fluid for extracting substances from solids (leaching). In principle, countercurrent multistage extractions can also be used for the fractionation of liquids. Industrial supercritical fluid extractions use mainly carbon dioxide as a solvent.

In the decaffeination of coffee, carbon dioxide is pumped through a cell containing the ground coffee at a typical pressure of 200 atm (20 MPa) and a temperature of 50°C (122°F)—conditions at which carbon dioxide dissolves caffeine, but not the compounds that give the coffee its flavor. After extraction of the caffeine the pressure is reduced below the critical pressure. Caffeine is not soluble in carbon dioxide at low pressures and thus precipitates. The carbon dioxide is recycled by compression.

Supercritical fluid chromatography. In chromatography, a mixture of substances is transported by a mobile carrier over a stationary surface. Mass transfer processes between the mobile and immobile phases lead to different velocities of the different solutes along the surface, thus leading to separation. In supercritical fluid chromatography, supercritical fluids such as carbon dioxide are used as mobile phases. Compared to conventional gas chromatography at ambient pressure and to liquid chromatography, the pressure can be used as a means for changing the solvent power of the mobile phase. Apart from analytical applications, preparative separations of high-value chemicals are possible.

Materials processing. By exploiting the peculiar solvent properties of supercritical fluids, the size of particles and their morphology can be tailored. For example, finely dispersed microparticles with a narrow size distribution can be precipitated under mild operating conditions by rapid expansion from supercritical solutions (RESS). In a related method, solids are precipitated from a liquid solution by contacting it with a supercritical fluid in which the solid is sparsely soluble [called supercritical antisolvent fractionation (SAS) of gas antisolvent precipitation (GAS)]. This tuning of the particle size is possible because nucleation depends on the supersaturation of the solutions, which in supercritical systems can be manipulated via the pressure. Potential applications include the controlled production of stable particles of labile pharmaceuticals, the formation of polymer-drug composites for the controlled release of pharmaceutical drugs, and the processing of polymers into useful morphologies.

Another important process is tobacco expansion, whereby tobacco leaves are impregnated with a high-

pressure gas. After a sudden pressure release, the leaves are torn apart into very voluminous pieces, decreasing not just the amount of tobacco but also nicotine and tar in a cigarette.

Yet another promising technique is fiber dyeing in a supercritical solvent such as carbon dioxide, which eliminates harmful solvent waste and allows the recovery of the remaining dyes. Efforts are also being made toward spray-painting cars.

Chemical reactions and supercritical water oxidation. The possibility to manipulate solvent power and phase behavior by changing the pressure, combined with the rapid heat and mass transfer of supercritical fluids, opens new pathways for chemical reactions and for finding optimum reaction conditions. A classical large-scale process is the production of low-density polyethylene, where highly compressed ethylene is the solvent and reactant.

New reaction pathways have been observed in supercritical aqueous media. In contrast to water at ambient conditions, supercritical water can homogenize large amounts of nonpolar organic substances, as well as inorganic substances such as oxygen, making them available for chemical reaction, and specifically for oxidation processes. For example, both flameless oxidation and flaming combustion can take place if organic substances are transferred into supercritical water that contains some dissolved oxygen. These properties are used in the supercritical water oxidation process (SCWO) for the destruction of hazardous chemical wastes. There are several pilot SCWO plants, and the first commercial plant, Huntsman Chemical, opened recently in Houston, Texas.

For background information *see* COFFEE; CRITICAL PHENOMENA; EXTRACTION; FOOD ENGINEERING; PHASE EQUILIBRIUM; POLYMERIZATION; SUPERCRITICAL CHROMATOGRAPHY in the McGraw-Hill Encyclopedia of Science & Technology.   Hermann Weingärtner

Bibliography. G. Brunner, *Gas Extraction*, Springer, New York, 1994; P. G. Jessop and W. Leitner (eds.), *Chemical Synthesis Using Supercritical Fluids*, Wiley-VCH, Weinheim, 1999; E. Kiran, P. G. Debenedetti, and C. J. Peters (eds.), *Supercritical Fluids: Fundamentals and Applications*, NATO Science Series E, vol. 366, Kluwer Academic, Dordrecht, 2000; M. Perrut, Supercritical fluid applications: Industrial developments and economic issues, *Ind. Eng. Chem. Res.*, 39:4531, 2000; G. M. Schneider, High-pressure investigations on fluid systems; A challenge to experiment, theory and application, *J. Chem. Thermodyn.*, 23:301, 1991; R. W. Shaw et al., Supercritical water, *Chem. Eng. News*, 69:26, 1991.

Supply chain management

Beginning with the work of Ford W. Harris in 1915 on the economic order quantity (EOQ) model, many researchers developed a variety of mathematical models for minimizing the costs associated with holding inventories (raw materials, components, subassemblies, work in process, and finished goods) in industries and businesses. The subject dealing with these problems was initially called inventory control. These models were essentially single-decision-maker models involving one item. In those days, several reasons for holding sizable inventories were given, including economies of scale; uncertainties in demand, supply, delivery lead times, and prices; and a desire to hold buffer stocks as a cushion against unexpected swings in demand and to assure smooth production flow.

Starting in the 1950s, Japanese manufacturers (in particular Toyota) initiated the just-in-time (JIT) philosophy to reduce work-in-process inventories to a minimum, and implemented it using a simple *Kanban* (which means card or ticket in Japanese) system to track the flow of in-process materials through the various operations. In the late 1960s, Toyota extended the JIT philosophy to reduce all inventories to a minimum by developing collaborative working relationships with its component suppliers and distributors with the aim of encouraging them to make and accept small and frequent JIT deliveries; providing for careful monitoring of quality and workflow; and ensuring that products were produced or received only as they were needed.

In today's world of rapid technological developments, frequent design changes, and shorter product cycles, carrying as little stock as possible is crucial. The more one relies on stock, the more difficult it will be to accomodate design changes. That's why the JIT philosophy is now being integrated into the overall business strategy worldwide, changing the nature of manufacturing and business dramatically. It has expanded beyond the walls of a factory or shop to include the capabilities, skills, and cooperation of its suppliers and the insights of its customers. This new expanded system is now referred to as the supply chain. Supply chain management comprises planning and processing orders; handling, transporting, and storing all materials purchased, processed, or distributed; and managing inventories in a harmonious, coordinated, and synchronized manner among all the players on the chain to build to order (to fulfill customer orders as they arise) rather than build to stock (to build up stock level to fulfill anticipated future demand).

Strategic partnering. As part of the collaboration with their suppliers, many companies are adopting the practice of vendor managed inventories in which the company provides warehouse space to its suppliers for storing their components, to be delivered to the company as demand arises (demand pull basis). With such an arrangement, the process of ordering components usually takes the following form:

1. Before each quarter the company informs the supplier of the aggregate quantity of the component that they expect to order during that quarter. This is to let the supplier know how much of their production capacity to dedicate to the manufacture of the buyer's components.

2. Before the beginning of each week the company provides the supplier with a revised estimate of the quantity of the component to be ordered that week. This is to help the supplier plan shipments of the component from its manufacturing facilities, which may be far away, to the warehouse space in the company, and be ready to deliver according to orders placed each day.

3. Each workday morning the company puts in an order for the quantity of the component to be delivered that day. This quantity is usually delivered within approximately 4 hours.

Instead of ordering once daily, some companies order once in every planning period (maybe a shift or half-a-shift). For the purpose of this discussion, a day will be used as the planning period.

In this mode of operation, it is critical to maintain good databases on demand, production, quality, and inventory levels, and to develop Web-based interfaces containing relevent information to which all players on the supply chain have access.

Demand distribution. The key to making this whole process run smoothly is accurate forecasting of the component demand each day (or whatever planning period is used).

The actual demand is usually a random variable with a probability distribution that can be estimated from past data. The range of variation of daily demand is divided into a convenient number of demand intervals (in practice about 10–25) of equal length, and the relative frequency of each interval is defined to be the proportion of days during which the observed demand lies in that interval. The chart, obtained by marking the demand intervals on the horizontal axis and erecting a rectangle on each interval with its height along the vertical axis equal to the relative frequency, is known as the relative frequency histogram of daily demand or its empirical distribution. The relative frequency in each demand interval I_i is an estimate of the probability p_i that the daily demand lies in that interval (**Fig. 1**).

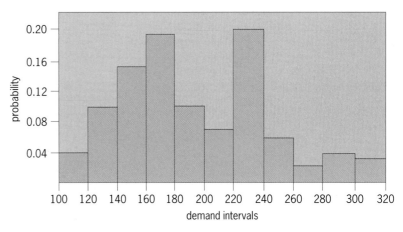

Fig. 1. Relative frequency histogram for daily demand for a major component at a plant.

Let $I_1, \ldots, I_n$ be the demand intervals with $u_1, \ldots, u_n$ as their midpoints; and $p = (p_1, \ldots, p_n)^T$, the probability vector in the empirical distribution of daily demand. Let Eqs. (1) apply.

$$\mu_D = \sum_{i=1}^{n} u_i p_i$$

$$\sigma_D = \sqrt{\sum_{i=1}^{n} p_i (u_i - \mu_D)^2} \qquad (1)$$

Then μ_D is an estimate of the expected (or average or mean) daily demand, and σ_D is an estimate of the standard deviation of daily demand (which is a measure of the variability) of the component.

In inventory control, the demand distribution is usually approximated by the normal distribution, a continuous distribution that is symmetric around its mean. The normal distribution (**Fig. 2**) is completely specified by two parameters—the mean μ and the standard deviation σ.

One of the theoretical advantages that the normality assumption confers is that when the demand distribution changes (as might arise after a product promotion), one has to change only the values of the mean and the standard deviation in the models. In practice, almost always it is only the value of the mean that is changed; the standard deviation is usually assumed to remain unchanged.

Determining the daily order quantity. The distribution of daily demand is used to determine the order quantity Q for a day, to balance the expected cost of overage (ordering too much) and underage (ordering too little). The quantity q left over at the end of this day has to support production the next morning until the delivery ordered on that day arrives. If q is too small, the company may be forced to shut down the production line until delivery of the order occurs. So, the company sets a safety level D_1

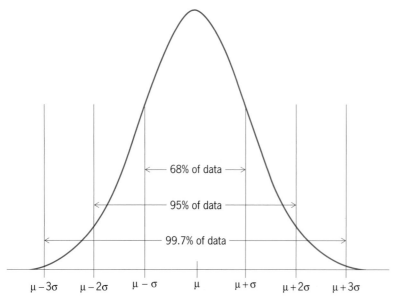

Fig. 2. Normal distribution with mean μ and standard deviation σ. The interval of $\mu \pm 3\sigma$ is associated with a probability of 0.997 in the normal distribution.

for q and a penalty of $\$c_s$ representing the shortage penalty incurred if $q \leq D_1$.

If q is too large, it may exceed the amount of convenient storage space near the production line that is allocated for this component. So, the company sets up a desired upper limit D_2 for q, and an excess stock penalty of $\$c_e$ per unit that q is over D_2. The total expected penalty as a function of the order quantity Q is expressed in Eq. (2), which can be computed

$$g(Q) = c_s[\text{probability } (q \leq D_1)]$$
$$+ c_e[\text{expected value of excess of } q \text{ over } D_2] \quad (2)$$

very easily for any given value of Q using the distribution of daily demand. To determine the optimal Q, the values of $g(Q)$ for various values of Q in a range around the mean demand are computed, and the optimum value of Q is considered to be the one that gives the smallest value for $g(Q)$. This is the most commonly used single period model for determining order quantities.

The weekly and the quarterly order amounts are determined by a similar procedure using the distributions of weekly and quarterly demands. Since there may not be enough past data to estimate weekly and quarterly demand distributions directly from data, these distributions are usually derived from the distribution of daily demand using convolutions or simulation.

Forecasting demand. Successful inventory management systems depend heavily on good demand forecasts. History shows many examples of firms benefiting from accurate forecasts and paying the price for poor forecasting.

The purpose of forecasting is commonly misunderstood to be that of generating a single number. This misunderstanding is created because all existing demand forecasting methods output only an estimate of the expectation of future demand. So, these methods are useful only when changes in the probability distribution of demand can be captured by the value of a single parameter, the expectation. All these methods that forecast only the expected value seem very inadequate to capture all the dynamic changes occurring in the shapes of probability distributions of demand.

Another important factor today is the highly competitive environment and the rapid rate of technological change that is shortening product life cycles. When a new product is introduced in the market, it enjoys growing demand due to gradual market penetration for some time, followed by a short stable period, and finally a period of declining demand. Because of this constant change, multiperiod stochastic (probabilistic) inventory models based on a stable demand distribution do not seem to be appropriate for application. Single period models of the type discussed above, combined with frequent updating of the demand distribution based on the most recent data, offer the most practical value.

Considering these arguments, it has been suggested that a better strategy is to approximate the probability distribution of demand by its empirical distribution obtained from past data, which in this context is called the discretized distribution of demand. This is the initial distribution at the time it is computed. This distribution is periodically updated using recent data.

Updating the demand distribution. Let $I_1, \ldots, I_n$ be the demand intervals and $p = (p_1, \ldots, p_n)$ the vector of probabilities associated with them in the present discretized demand distribution. In updating, changing the values of all the p_i makes it possible to capture any change in the shape of the distribution subject to the discretization used (that is, the division of the range of variation into the demand intervals $I_1, \ldots, I_n$).

At the time of updating, the data used are based on the demand over the most recent k planning periods (if the planning period is a day, for example, k could be about 50). Let f_i represent the number of planning periods among the most recent k for which the demand was in the interval I_i, and $r_i = f_i/k$, $i = 1$ to n. Then $r = (r_1, \ldots, r_n)^T$ is the relative frequency vector corresponding to the most recent k planning periods. It represents the estimate of the probability vector corresponding to the most recent demand distribution, but it is based on too few (only k) observations.

Let $x = (x_1, \ldots, x_n)^T$ denote the updated probability vector corresponding to the updated demand distribution. An estimate can be taken of x to be the optimum solution of the quadratic program (QP) in notation (3), where α is a weight between 0 and 1.

$$\text{Minimize} \quad \alpha \sum_{i=1}^{n} (p_i - x_i)^2 + (1 - \alpha) \sum_{i=1}^{n} (r_i - x_i)^2$$

$$\text{subject to} \quad \sum_{i=1}^{n} x_i = 1 \tag{3}$$

$$x_1 \geq 0 \quad \text{for all } i$$

Typically $\alpha = 0.9$ works well. The reason for choosing the weight for the second term in the objective function to be small is that the relative frequency vector, r, is based on only a small number of observations. Since this quadratic model minimizes the weighted sum of squared forecast errors over all demand intervals, when used periodically after every few planning periods, it has the effect of tracking gradual changes in the demand distribution.

If the mean demand is increasing (decreasing) since the time that the original demand intervals were set up, new demand intervals open up to the left (right) and those at the other end will tend to have their associated probabilities become close to 0.

The above problem is a convex quadratic program which has a unique optimum solution that can be computed very efficiently using any commercially available mathematical programming software.

This QP model for updating the demand distribution is backward-looking in that it can only pick up changes showing up in recent data. In some dynamic industries where rapid changes are common, it is

appropriate to also use forward-looking strategies, which incorporate the vision of experienced people and expert judgments for updating the demand distribution.

For background information *see* INVENTORY CONTROL; MATERIAL RESOURCE PLANNING; OPERATIONS RESEARCH; PRODUCTION PLANNING; STOCHASTIC PROCESS in the McGraw-Hill Encyclopedia of Science & Technology. Katta G. Murty

Bibliography. K. G. Murty, *Supply Chain Management in the Computer Industry*, Technical Report, Dept. IOE, University of Michigan, Ann Arbor, September 1998; D. Simchi-Levi, P. Kaminsky, and E. Simchi-Levi, *Designing and Managing the Supply Chain*, Irwin McGraw-Hill, Boston, 2000.

Sustainable forest management

Sustainable forest management once meant sustained yield, which, according to the Society of American Foresters' *Dictionary of Forestry*, is "the amount of wood a forest can continuously produce at a given intensity of management." During the 1990s the scope of sustainable forest management broadened. Instead of viewing the forest as the source of any one economic product (for example, timber, paper, or mushrooms) or service (for example, recreation), sustainable forest management now recognizes the full range of environmental, social, and economic values of the forest. Its mission is to integrate the management of these values so that none are neglected, ensuring that the forest remains both useful and healthy into the future.

Criteria and indicators. As interpretations of sustainable forest management evolved, criteria and indicators were developed to provide a common language for discussing and evaluating progress toward implementing sustainability. According to the Canadian Forest Service in its Criteria and Indicators for the Conservation and Sustainable Management of Temperate and Boreal Forests: The Montreal Process (1995), a criterion is "a category of forest-related conditions or processes by which sustainable forest management may be assessed." The criteria reflect a series of broadly held values related to the environmental, economic, and social functions of forests. For example, one often-used criterion calls for the conservation of biological diversity; another calls for promoting sustainable human settlements in the developing world. Each criterion is accompanied by a set of related indicators.

An indicator is defined as "a measure (measurement) of an aspect of the criterion." Defined at the national, regional, and state levels, indicators generally are not directly applicable at the working level. Some indicators are quantitative, such as the number of forest-dependent species or the area and percent of forestland with significant soil erosion. Others are qualitative or descriptive, such as those related to tax policies, investments, and legal frameworks. All indicators provide information about present forest conditions and can demonstrate trends when observed over time. There are no absolute standards established for any indicator. Their relative value is inferred by examination of trends and subsequently established policies based on the trends.

Criteria and indicators serve as a commonly agreed-to set of measurements used to assess and monitor trends, providing the necessary information to formulate national policies to promote sustainable forest management. On an international scale, criteria and indicators are a means for improving dialogue about forest conditions and for strengthening the technical understanding of changes resulting from various management options.

International programs. World leaders at the 1992 United Nations Conference on Environment and Development in Rio de Janeiro adopted the Statement of Forest Principles and Agenda 21, pioneering the international discussion of sustainable forest management. The Forest Principles acknowledged the international significance of forests' multiple values and recognized forests as one of the keys to sustainable development worldwide. Agenda 21 was an action plan for achieving sustainable development worldwide by 2000. To work toward this plan, countries needed a common framework for measurement and discussion. The International Tropical Timber Organization's Criteria for the Measurement of Sustainable Tropical Forest Management, presented at the same conference, provided one framework through a set of criteria and indicators for sustainable tropical forest management. In addition to six working-level criteria, the document contained five national criteria: (1) forest resource base, (2) continuity of flow, (3) level of environmental control, (4) socioeconomic effects, (5) institutional framework. The Conference on Environment and Development also established the United Nations Commission on Sustainable Development, and called for participating countries to report their progress at a 1995 meeting.

Subsequently, Canada sponsored an international seminar on the sustainable development of temperate and boreal forests as part of the September 1993 Conference on Security and Cooperation in Europe. At this meeting, held in Montreal, international organizations, both governmental and nongovernmental, began to define sustainable management criteria and indicators for temperate and boreal forests. The five criteria on environmental conservation developed were the basis for the Helsinki Process and the Montreal Process formed later.

Helsinki Process. At the Ministerial Conference on Forests in December 1993, the Pan-European countries developed the Helsinki Process, also known as the Pan-European Process on Criteria and Indicators for Sustainable Forest Management. It embraced the five environmental criteria developed in September at the Conference on Security and Cooperation and included a sixth criterion that addresses socioeconomic issues of sustainable forest management. This process focuses on temperate, boreal, and mediterranean forests.

Montreal Process. Twelve countries that did not participate in the Helsinki process—including Argentina, Australia, Canada, Chile, China, Japan, Mexico, New Zealand, the Republic of Korea, Russian Federation, the United States, and Uruguay—developed the Montreal Process on Criteria and Indicators for the Conservation and Sustainable Management of Temperate and Boreal Forests. The Montreal Process, held in June 1994, is more specific than the Helsinki Process and contains seven criteria and 67 indicators. The first six criteria are comparable to the Helsinki Process. The seventh looks at legal, institutional, and economic factors.

1. Conservation of biological diversity
2. Maintenance of productive capacity of forest ecosystems
3. Maintenance of forest ecosystem health and vitality
4. Conservation and maintenance of soil and water resources
5. Maintenance of forest contribution to global carbon cycles
6. Maintenance and enhancement of long-term multiple socioeconomic benefits to meet the needs of societies
7. Legal, institutional, and economic framework for forest conservation and sustainable management

The Montreal Process is applicable at the national level across all ownership types. In 1995, the United States committed to using the Montreal Process Criteria and Indicators as the principal means for reporting forest conditions nationally. To encourage support for the criteria and indicators, a broad scope of stakeholders—including state and federal government agencies, nongovernment organizations, private landowners, professional associations, industry, indigenous people, and academic institutions—have been involved in the criteria and indicator implementation process and the further discussion of sustainable forest management. At the international level, countries that have endorsed the Montreal Process have committed to sustainable forest management of their forests.

Other international programs. The Tarapoto Proposal for Criteria and Indicators for Sustainability of the Amazon Forests was adopted in February 1995. The Proposal's framework has one criterion and seven indicators at the global level, seven criteria and 47 indicators at the national level, and four criteria and 22 indicators at the forest management level. The Dry-Zone Africa Process (1995) has seven criteria and 47 indicators for sub-Sahara Africa. The Near East Process (1996) for the Middle East consists of seven national-level criteria and 65 indicators. In addition, the Lepaterique Process of Central America (1997) has eight criteria and 52 indicators for the national level and four criteria and 40 indicators at the regional level. Combined efforts around the criteria and indicators currently involve more than 100 countries.

The future. Once criteria and indicators were developed and endorsed, countries began reporting available data. At the 11th World Forest Congress in October 1997, the Montreal Process countries presented First Approximation Reports. The reports highlighted data gaps and provided baseline information for the criteria and indicators. For instance, in the United States, federal, state, and local governments own 42% of the forest land, and industry and nonindustrial private landowners own the remaining 58%. Much biological and physical data are available for public land, while limited data are available for private land. Consequently, the data for the first five criteria are limited in scope and present information on less than 50% of the forest land in the United States.

The reports also revealed the need to improve data collection systems and incorporate nongovernment data, such as the Nature Conservancy Classification and Stewardship databases, into future national reports as well as in the First Montreal Process Forest Report to be published in 2003. As countries move to collect data and monitor and report on indicators more comprehensively, active participation of all stakeholders is critical.

Another challenge as criteria and indicator data become available is interpreting trends and making required changes to foster sustainable forest management. It is inadvisable at this point to draw conclusions on national trends or revise policy based on the data acquired so far, which have been measured using inconsistent definitions or methodologies. Even as such trends are interpreted, moreover, changes will be made not only to policy but also to the criteria and indicators themselves. The criteria and indicators are dynamic and meant to change as the understanding of sustainable forest management evolves. Based on the evolving process of regional criteria and indicator development, it is understood that measurement and analysis of collected and reported data will require continuous interpretation and modification to understand and appreciate movement toward sustainable forests.

Long-term commitment by governments and stakeholders is necessary to achieve meaningful results. Forests are complex and dynamic ecosystems; as our understanding improves with experience and knowledge, it leads to more effective approaches to forest assessment and management. Work on the criteria and indicators involves continuous adaptation to changing societal needs, new information, experience, and greater capabilities. As a potential leading innovation in forest management, the criteria and indicators will make a significant contribution only if countries use them to establish national policy changes in response to indicator trends. The citizens and decision-makers in each country will therefore determine the ultimate contribution made by the use of criteria and indicators toward the protection and management of their forests and in the composite of the world's forests.

For background information *see* CONSERVATION OF RESOURCES; FOREST AND FORESTRY; FOREST MAN-AGEMENT; FOREST TIMBER RESOURCES in the McGraw-Hill Encyclopedia of Science & Technology.

Laurie Schoonhoven; Michael P. Washburn

Bibliography. M. D. Brown, The Montreal Process: Criteria and Indicators for the Conservation and Sustainable Management of Temperate and Boreal Forests, in *Proceedings of the Penn State School of Forest Resources Issues Conference, Forest Sustainability: What's It All About*, pp. 54–57, Pennsylvania State University, 1997; *The Central American Process of Lepaterique*, Central American Council of Forestry and Protected Areas, Tegucigalpa, Honduras, 1997; *Criteria and Indicators for the Conservation and Sustainable Management of Temperate and Boreal Forests: The Montreal Process*, pp. 3–6, Hull, Quebec, Canada, Canadian Forest Service, 1995; *Criteria and Indicators for the Sustainable Forest Management in Dry-Zone Africa*, UNEP/FAO, Dry-Zone Africa Process, Nairobi, Kenya, 1995; *Criteria and Indicators for the Sustainable Forest Management in the Near East*, FAO Regional Office for the Near East, Cairo, Egypt, 1996; *European Criteria and Most Suitable Quantitative Indicators for Sustainable Forest Management*, adopted by the first expert level follow-up meeting of the Helsinki Conference, Geneva, Switzerland, 1994; *Forests for the Future Montreal Process and Criteria and Indicators*, Ottawa, Canada, Montreal Process Working Group, 1998; J. A. Helms, *The Dictionary of Forestry*, Society of American Foresters, Bethesda, MD, 1998; International Tropical Timber Organization (ITTO), *Criteria for the Measurement of Sustainable Tropical Forest Management*, ITTO, Yokohoma, Japan, 1992; *A New View of Our Forests: Questions and Answers Related to the Criteria and Indicators for the Conservation and Sustainable Management of Forests*, pp. 1–3, Washington, DC, Roundtable on Sustainable Forests, 2000; *The Tarapoto Proposal: Criteria and Indicators for the Sustainability of the Amazonian Forest*, Amazon Cooperation Treaty, Tarapoto, Peru, 1995; *UNCSD 95 Guidelines for National Information, Agenda 21/Chapter 11: Combating Deforestation and "Non-Legally Binding Forest Principles,"* 1995; United Nations, *Review of Sectoral Clusters, Second Phase: Land, Desertification, Forests and Biodiversity, "Combating Deforestation" and the Non-Legally Binding Authoritative Statement of Principles for a Global Consensus on the Management, Conservation and Sustainable Development of All Types of Forests*, Report of the Secretary-General, E/CN.17/1995/3, 1995.

Sustainable nuclear energy

Sustainable development that meets the energy needs of the present without compromising the ability of future generations to meet their needs is a major consideration of governments, nongovernmental groups, and intergovernmental agencies. Projected energy demand growth throughout the twenty-first century will be driven not only by population increase but also by a dramatic rise in world-average energy use per capita as the economies and associated living standards rise in developing countries. Recent energy demand projections combine numerous regional scenarios to attempt to encompass plausible global energy use growth rates over the coming decades. Realistic scenario assumptions concerning resource mix (such as fossil, solar, wind, nuclear) and ecological constraints are used. Based on those studies, the projection is for a global twofold increase in energy requirements by 2050 and a fourfold increase by 2100. The energy supplied by nuclear reactor technology (compared to fossil fuels, which have a 50-year to several-century supply potential) is sufficient for more than a thousand years, even after accounting for worldwide equity in energy use and population growth.

Reactor technology. Advanced fast reactors utilizing high-energy neutrons to sustain the fission chain reaction are capable of burning the abundant isotopes of uranium and thorium using a two-step process. They first convert the uranium-238 to plutonium-239 (or thorium-232 to uranium-233) by neutron capture; then they fission the ^{239}Pu or ^{233}U in situ. This approach requires that the fuel be subjected to multiple exposures in the reactor via recycling through a chemical partitioning step which extracts as a waste stream the relatively light-element fission products from the partially consumed fuel, returns the nonfissioned fissile elements to the reactor, and tops off the fuel with new uranium feedstock.

If fully exploited in advanced reactors, the recoverable uranium and thorium ores constitute a sustainable fuel resource base for meeting all of the world's energy needs for millennia. Indeed, the uranium which has been mined and refined already and is now in storage would by itself meet the world's energy needs for half a millennium. Over and above the ore in the Earth's crust, uranium at dilute levels in the ocean's waters extends the nuclear energy supply potential to ten millennia. Though this resource is not needed for centuries, a search for a cost-effective means to extract it from the ocean is being undertaken.

Reactors and fuel cycles capable of burning the abundant isotopes of uranium and thorium via the two-step process were recognized from the beginning of the nuclear age as the best long-term configuration for nuclear energy supply; their development has progressed steadily and has reached a level of maturity and prototype deployment appropriate for initiating commercial deployment. But their cost is not yet competitive with that of the current generation of nuclear power plants and fuel cycle.

Nuclear ecology. Nuclear energy produced in the current generation of power plants exploits less than 1% of the energy potential of each ton of uranium mined because it uses as fuel only the rare isotope, ^{235}U, of the uranium ore; the prevalent isotope, ^{238}U, is not directly fissionable in current reactor types, which sustain a fission chain reaction based on low-energy neutrons. The ^{238}U remains unexploited as a dominant (96 weight percent) component in the spent fuel discharged. If it continues to be used only at the low efficiency achievable in current reactor types, the world's recoverable uranium resources (~18 million tons or 16.3 million metric tons) would be consumed in less than a century.

In current-generation nuclear power plants, since the fuel is burned incompletely not only the fission products but also unfissioned radioactive heavy elements remain in the discharged fuel that is sent to waste. The radioactivity of this incompletely fissioned fuel persists for tens of thousands of years. It is these extremely long-lived radiotoxic heavy transuranic atoms rather than the fission products (which decay away in several centuries) that raise public concern regarding nuclear waste disposal sites.

However, total fission consumption of the uranium or thorium feedstock by means of multiple recycle into advanced reactors to extract the full fuel energy benefit would, at the same time, achieve a long-term ecological balance. In that case, all the transuranics would be gone, and the long-term radioactive toxicity contained in the fission product waste would just balance the natural radioactive toxicity originally contained in the ore. The toxicity balance would take several centuries to achieve because the fission products are initially highly radioactive, however, all but a few decay to stable, nonradioactive isotopes within about 300 years. After that period the total combined long-lived radioactivity is no larger than the long-lived radioactivity of the original ore. The small volumes of fission product waste can be immobilized in durable waste forms and sequestered for several centuries to allow time for them to decay to background levels; a several-century sequestration of small volumes of fission product wastes is recognized to lie within technological and institutional capability, and it stands in contrast to current expectation to sequester partially burned nuclear fuel for thousands of years.

Hydrogen as an energy carrier. A hydrogen-based, nuclear-driven global energy supply infrastructure was proposed in the 1970s as the sustainable solution for global growth. The visionary energy supply infrastructure would rely on nuclear fission for the essentially infinite resource in the energy supply chain. Heat from fission would be used to manufacture hydrogen by cracking water into its hydrogen and oxygen constituents. Hydrogen would be used as a synthetic chemical energy carrier in combination with electricity. Together they would service both the electric and nonelectric energy sectors, coupling to high-conversion-efficiency combustion turbines and fuel cells for the electrical fraction of the energy market.

The entire energy supply chain would become ecologically neutral because fission energy is stored in the electrical or the hydrogen energy carrier without production of greenhouse gases. The energy delivery and utilization links in the supply chain would be ecologically neutral because the hydrogen manufactured by cracking water, when burned in fuel cells or gas turbines to harvest the energy, would produce only water as the combustion product. The water would recycle to the hydrogen production plants via natural ecological processes. By using fast reactors and multiple recycle for complete fission consumption of the uranium fuel resource, only fission products would be produced as waste, which after three centuries of sequestration, would have no more radiotoxicity left in it than that of the original uranium ore. The resource is sufficient to last for millennia. All links of the energy cycles are ecologically neutral.

In pursuit of this vision, research was intensified in the late 1960s and early 1970s on means to crack water using electricity or using high temperature. The direct heat methods rested on thermochemical cycles where heat ($1300-1800°F$; $700-1000°C$) drives a series of chemical reactions producing hydrogen and oxygen while the reagents recycle in a closed cycle and a waste stream is not created. The research activity peaked in the 1980s and has continued at a lower level. Thermochemical processes remain at a preprototype level of maturity. High-efficiency water electrolysis cells that use electricity to crack water have been commercialized.

Infrastructure. The current nuclear architecture is based on economy of scale: large plants of substantial initial cost. Future nuclear energy, however, must include deployment in developing countries where the indigenous infrastructures may initially be sparse and competition for development financing may be severe. New nuclear energy must include offerings for small increments in deployment. Plants should be available in modular size as well as at the traditional large-scale gigawatt-electric size. Strategies for mass-producing modular fast neutron reactor plants of simplified design economically are under development.

Recycling. The current nuclear architecture is based on fission of ^{235}U with incomplete consumption of the uranium ore's energy content and disposal as waste of incompletely fissioned material that have very long radioactive half-life. New nuclear energy technology is based on multiple recycle to total fission consumption, and will achieve an ecologically neutral fuel cycle with relatively short-lived fission products as the only waste stream. Advanced recycle technology is being designed that would cleanly separate fission products destined for the waste stream from a commixed recycle product of all unfissioned heavy elements and uranium, which would be returned to the reactor. All material handling must be done remotely; simple, few-step processes are used. Trace fission product carryover to the recycle fuel and the commixing of all transuranic elements makes the recycle fuel highly radioactive and as unattractive for military use as that which is discharged from current-generation reactors. Advanced recycle technologies having these characteristics are under development in many countries having a nuclear component in their current energy supply.

As nuclear energy deployment extends into more countries, centralized regional fuel cycle support facilities may be deployed. Such facilities, operated under international oversight, would service hundreds of power plants in a large surrounding region. The expense of placing an entire fuel cycle infrastructure (fuel manufacture, recycle, waste management) could be shared (or the service could be purchased) by those nations having no capacity or no desire to deploy the entire infrastructure on their own. All bulk handling of fissile materials during fuel fabrication, recycle, and final waste management would be conducted at only a few sites worldwide and under international oversight. These processing sites would employ recycle technologies which minimize weapons proliferation vulnerabilities by coextraction of all fissionable elements and incomplete fission product separation for the fabrication of reload fuel for shipment to the reactor sites. Because all fuel shipments—even fresh fuel—would be of highly radioactive commixed compositions, the current level of proliferation resistance would be preserved even as the scope of power plant deployment expanded. Such an institutional approach could make the benefits of sustainable nuclear energy available to a larger fraction of the world's population.

For background information *see* ENERGY; NUCLEAR ENGINEERING; NUCLEAR FISSION; NUCLEAR FUEL CYCLE; NUCLEAR FUELS REPROCESSING; NUCLEAR POWER; NUCLEAR REACTOR; RADIOACTIVITY in the McGraw-Hill Encyclopedia of Science & Technology.

David C. Wade

Bibliography. International Atomic Energy Agency, *Second Scientific Forum: Sustainable Development—A Role for Nuclear Power?*, September 1999; International Nuclear Societies Council, *A Vision for the Second Fifty Years of Nuclear Energy: Vision and Strategies*, published by American Nuclear Society, La Grange, IL, 1996; J. Laidler et al., Chemical partitioning technologies for an ATW system, *Prog. Nucl. Energy*, 38(1/2):65-80, 2001; C. Marchetti, On hydrogen and energy systems, *Int. J. Hydrogen Energy*, 2:3-10, 1976; N. Nakicenovic et al. (eds.), *Global Energy Perspectives*, Cambridge University Press, 1998; H. Sekimoto, Physics of future equilibrium state of nuclear energy utilization, *Proceedings of the International Conference on Reactor Physics and Reactor Computations* (p. 515), Tel Aviv, Israel, January, 23-26, 1994.

Systems architecture

Application of the systems sciences has facilitated remarkable technological achievements, ranging from the International Space Station to our household computers. Both are tremendously complex and involve the orchestration of engineers, scientists, and technicians from a variety of disciplines. The relatively new discipline of systems engineering focuses on pulling together, or integration, of the efforts of mechanical engineers, electrical engineers, software engineers, scientists, and technicians for team achievements. The master plan or "blueprint" that guides this common work effort is a systems architecture.

Systems architecting, according to Eberhardt Rechtin, is the process of creating complex, unprecedented systems. Small systems, or systems that have been built before, may not require a formal architecture. The systems architect, according to Alexander H. Levis, is the master builder, responsible for designing and conceptualizing the system; is knowledgable about each of the building components; and has a good understanding of the relationships among those components. The architectural process works top-down from the abstract to the concrete, resolving ambiguities and inconsistencies in an iterative fashion.

A system can be defined to be a set of interacting components in which the behavior of each component contributes to the behavior of the whole set. An architecture can be defined as the structure of components of a system, their relationships, and the principles and guidelines governing their design and evolution over time. A framework is an enclosing structure used to support an object under construction; an external work platform; or a basic arrangement, form, or system. A three-stage process of systems architecting, as described by Levis and Lee W. Wagenhals, is typical: analysis phase, synthesis phase, and evaluation phase (see **illus.**). The analysis phase examines the mission of the system to be built, formulating an abstract concept of operations. The functional requirements of the system are derived from the mission and operational concept. The operational concept is decomposed into a functional architecture, incorporating models of the system processes, data, and rules. The analysis continues with the development of a physical architecture, an organizational model, and a dynamics model. The synthesis phase involves the mapping of these constructs onto an executable model. The executable model is used in the evaluation phase to compute measures of performance and measures of effectiveness.

Architecting as art and discipline. Rechtin and Mark W. Maier consider the art of architecting to be nonanalytic, inductive, difficult to certify, and less understood than most other engineering activities. According to them, the greatest architectures are the product of a single mind—or of a small, carefully structured team—along with a responsible and patient client, a dedicated builder, and talented designers and engineers. They point out four competing factors involved in systems architecting: risk, schedule, performance, and cost. Levis discusses the use of architectures as a way of managing uncertainty. The academic community addresses systems architecting within several engineering disciplines, especially systems engineering. There are a number of contemporary efforts to develop useful frameworks and procedures to support successful identification of trustworthy system architectures.

Architectural frameworks. J. A. Zachman addresses a spectrum of architectural models, from the perspective of the enterprise that will be using the system, in a two-dimensional landscape. One dimension addresses the view, which is specified by the model developers: planner, owner, designer, builder, and subcontractor. The second dimension answers who, what, when, where, and how. Each cell is a unique architectural view of the system, which is comparable to the various architectural views used in the design and construction of today's buildings, such as structural, mechanical, and electrical views. Each of these views communicates specifications about the product from the perspective of a specific party.

Enterprise architectures. The Open Group Architecture Framework (TOGAF), produced by an industry-wide organization of suppliers and buyers, presents generic services and functions that provide a foundation on which specific architectures and architectural building blocks can be built. This framework includes the TOGAF Standards Information Base (SIB), which is a collection of open industry standards for use in defining the particular services and other components of an organization-specific architecture. Whether or not an enterprise draws from TOGAF for its architectural resources, it should expect to formulate an enterprise technical reference model, a set of enterprise mandatory standards, and an architectural framework. These resources will provide a basis on which individual systems architectures may be produced so that they participate in a desirable enterprise systems ecology.

Component technology. Component-based framework solutions are partial software implementations, specifying the nature of the framework and the way to extend the framework with pluggable components. "Pluggable" means that the component can be assimilated into the system without modification. This requires that the sockets into which the components are to be inserted match the component's construction. Jon Hopkins defines a software component to be a physical packaging of executable software with a well-defined and published interface. Software reuse and software maintenance over the system's life are the motivation for component-based systems development.

EAI and COEs. Enterprise application integration (EAI) has achieved considerable interest and following in the practicing community, in the standards communities, and among commercial software

Analysis	Synthesis	Evaluation
mission operational concept functional architecture physical architecture organizational model dynamics model	executable model	measures of performance measures of effectiveness

Phases of system architecting.

suppliers. One view of an enterprise application integration environment is that the enterprise application consists of one all-encompassing package, which is closely related to the enterprise foundation programs implementing an enterprise-wide common operating environment (COE). The U.S. Department of Defense has implemented a Defense Information Infrastructure (DII) COE, which serves as the cornerstone of its command-and-control and combat-support computer software. The DII COE consists of a collection of segments, or components, that comply with preestablished integration and run-time specifications, or systems ecology rules, which permit them to plug into a system and play without further modification.

The benefits of this DII COE include facilitated software integration, reuse of common support applications, adherence to the enterprise architecture, a common end-user look and feel, common system management interface, and the integration of dissimilar platforms into a supported information infrastructure through a certification process. One community of interest built on the DII COE foundation program is the Global Command and Control System (GCCS), which incorporates the force planning and readiness assessment applications required by battlefield commanders to effectively plan and execute military operations. The GCCS Common Operational Picture correlates and fuses data from multiple sources to provide war-fighters the situational awareness needed to act and react decisively.

Electronic commerce and the Internet. An enterprise is not just interested in the proper functioning of its payroll system. It needs the payroll system to interoperate with other corporate systems, such as personnel and finance, with which the payroll system shares common data. When the enterprise is involved in electronic commerce, it is vital that the enterprise systems which control inventory, shipping, ordering, billing, and the Web presence to customers, communicate with one another. If an enterprise with a Web presence fails to integrate its systems with its commercial suppliers, shippers, and financial institutions, it will be unable to keep up with real-time demands of the customer base, and will fail. Architectures and frameworks for electronic commerce are valuable commodities in contemporary trade and professional journals. Standards supporting electronic commerce include Electronic Data Interchange (EDI) and Extensible Markup Language (XML). *See* XML (EXTENSIBLE MARKUP LANGUAGE).

Issues and trends. An organization or enterprise needs to ensure that the multiple information systems that support the organization contribute desirable characteristics to the ecology of their information infrastructure. That is just what the systems architectures produced in accordance with the enterprise technical reference model, the adopted standards, and the prescribed architectural framework are intended to accomplish.

Interoperability. A major issue addressed by architectures is interoperability among systems. When the enterprise is the Department of Defense, interoperability among the command-and-control systems and the combat-support systems is vital to mission support and the nation's defense. In a large enterprise, objective acquisitions executed without consideration for system ecology may result in dissimilar or nonhomogeneous computer platforms which fail to interoperate. Major issues have been associated with interoperability among heterogeneous platforms, and with application software portability.

Use of commercial software. Because of the large investments required to develop application software, many organizations seek to maximize the use of commercially available application software. Often one commercial product does not satisfy the requirement completely, but the integration of multiple commercial products into one application solution is preferred. In a sense, this amounts to sharing the development costs of a software product among the many customers of the product. However, the burden of integration of multiple products is on the customer. There can be a variety of complications with this approach over the life of a system. Some products may be discontinued and others may have major upgrades. The products may no longer solve the original problem or talk together.

As the products grow apart in terms of functionality, and with regard to their points of integration, the burden of reintegrating the new product rests solely on the customer. To resolve this issue, some commercial software integration organizations have mandated, as a prerequisite for doing business, a process of negotiation regarding these issues among the multiple commercial application software suppliers. Such an approach is used for the support of products within a common operating environment, which is defined in advance by the integrator. The integrator then markets the preintegrated package to customers, a process which is viable only in the case when the integrator represents a sufficiently large customer opportunity to be judged worthwhile by the software publishers.

Security and privacy. Perhaps security and privacy are the most serious issues facing all enterprises concerned with using the Internet for commerce. The need for global access by employees for internal operations, as well as by partners and customers for commercial transactions, introduces these vulnerabilities. Encryption and public-key infrastructure (PKI) authentication technology are in use today. Smart cards are being implemented by some organizations. Biometrics is expected to play a role in this area in the near future.

For background information *see* COMPUTER SECURITY; INFORMATION SYSTEMS ENGINEERING; SOFTWARE; SOFTWARE ENGINEERING; SYSTEMS ARCHITECTURE; SYSTEMS ENGINEERING; SYSTEMS INTEGRATION in the McGraw-Hill Encyclopedia of Science & Technology.   Bernard Sharum

Bibliography. B. H. Boar, A blueprint for solving problems in your IT architecture, *IEEE IT Pro*, 1(6):23–29, November/December 1999; B. H. Boar,

Constructing Blueprints for Enterprise IT Architectures, Wiley, New York, 1999; M. Boster, S. Liu, and R. Thomas, Getting the most from your enterprise architecture, *IEEE IT Pro*, 2(4):43–50, July/August 2000; J. Hopkins, Component primer, *Commun. ACM*, 43(10):27–30, October 2000; S. Koushik and P. Joodi, E-business architecture design issues, *IEEE IT Pro*, 2(3):38–43, May/June 2000; A. H. Levis, System architectures, in A. P. Sage and W. B. Rouse (eds.), *Handbook of Systems Engineering and Management*, Wiley, New York, 1999; A. H. Levis and L. W. Wagenhals, C4ISR architectures: I. Developing a process for C4ISR architecture design, *Sys. Eng.*, 3(4):225–247, 2000; M. W. Maier and E. Rechtin, *The Art of Systems Architecting*, 2d ed., CRC Press, Boca Raton, FL, 1997; E. Rechtin, *Systems Architecting: Creating & Building Complex Systems*, Prentice Hall, Upper Saddle River, NJ, 1991; J. A. Zachman, Extending and formalizing the framework for information systems architecture, *IBM Sys. J.*, 31(3):590–616, 1992; J. A. Zachman, A framework for information systems architectures, *IBM Sys. J.*, 26(3):276–292, 1987.

Tallgrass prairie

Tallgrass prairie is one of the most diverse and productive types of grassland in North America. Historically, tallgrass prairie stretched from Canada to Texas and from Kansas to Ohio. West of the tallgrass prairie lie drier, less productive grasslands, and to the east are deciduous forests. Throughout the presettlement range of this grassland, tallgrass prairie was characterized by three interacting forces: a variable climate that on average provided sufficient precipitation to support forest but was prone to drought; frequent fire that promoted grass growth but suppressed woody species; and grazing by large ungulate herbivores (bison). Today, tallgrass prairie is termed America's most endangered ecosystem because more than 95% of the original prairie has been lost to row crop agriculture, and what remains has been highly fragmented.

Species diversity. Tallgrass prairie is defined throughout much of its range by the dominance of warm-season (C_4) perennial tall grasses. These plants, which include big bluestem, *Andropogon gerardii*, and Indian grass, *Sorghastrum nutans*, may have flowering culms (stems) that rise to more than 2 m (6 ft) above the ground in late summer. These and other grasses process most of the energy that flows through the ecosystem. Indeed, the grasses may constitute 70–90% of the total aboveground plant biomass in many prairies. However, the number of grass species is low compared to the C_3 perennial forbs (broad-leaved herbaceous dicots), of which there may be several hundred species. The forbs contribute to the high plant diversity in tallgrass prairie. This diversity in flora, in turn, allows for greater diversity in prairie fauna, both as consumers of plants and as pollinators (most grasses are wind-pollinated, while most forbs are insect-pollinated).

Despite the diverse assemblage of plant and animal species that define the tallgrass prairie, few are endemic to this grassland. Instead, the flora and fauna are "borrowed" from adjacent biomes. The lack of endemic species reflects the youth of this grassland, which is generally regarded as originating after the retreat of the last glaciers at the end of the Pleistocene 10,000 years ago. This mix of plant and animal populations is also temporally very dynamic, a product of the variable climate, the uncertainty of fire, and grazing bison.

Climate. Climatically, tallgrass prairie is a mesic (moderately wet) grassland. Throughout most of its range, precipitation is strongly seasonal with spring/summer rainfall more abundant than winter snow. Annual precipitation amounts may vary from 30 to 40 in. (75–100 cm). However, less of this moisture is available to the biota in grasslands than in other ecosystems with comparable precipitation amounts, because of the openness of the landscape, which provides little protection against winds and solar radiation. As a result, evaporative demand is high, and this factor, when combined with the large amount of transpiring leaf area, leads to frequent periods of water stress in the summer. Moreover, variability in rainfall amounts is quite high: years with extremely high rainfall are common as are single-year to multi-year droughts. Air temperatures in the tallgrass prairie can also be described as extreme with very warm summers and cold winters.

Fire. Tallgrass prairie has been described as a fire-maintained grassland. Frequent fire is certainly an inherent feature of the prairie with abundant historical evidence of fires ignited both by lightning and by the native peoples. The abundance of fine fuels and ignition sources, the nearly unbroken terrain of the Great Plains, and high wind speeds provide all the elements necessary for frequent and extensive conflagrations. Studies have shown that fires may occur at any time of year, even when the grassland appears green in midsummer. However, it is likely that dormant-season fires, when the fine grassland fuels are driest, would have been most common historically.

Recent research in the remaining tracts of tallgrass prairie provides ample evidence of the importance of fire for this ecosystem. Fire is important for keeping trees and adjacent forests from encroaching into tallgrass prairie. Many tree species are killed by fire, or if they are not killed, they are damaged severely because their active growing points are aboveground. Grassland plants survive and even thrive after fire because their buds are belowground, where they are protected from lethal temperatures. In tallgrass prairie, fire results in an increase in growth of the grasses and a greater production of plant biomass. This occurs because the buildup of dead biomass (mulch) from previous years inhibits growth and fire removes this mulch layer. In addition, many successive annual fires may reduce plant species diversity, allowing the grasses to become even more dominant. Maximum dominance by the grasses in annually burned prairie is driven by an environment of high light but low water and nitrogen availability, compared to sites without fire. Fire at intermediate frequencies (2–4-year intervals) may allow for maximum species diversity of plants and animals.

Most grassland animals are not harmed by fire, particularly if the fires occur during the dormant season. Those animals living belowground are well protected, and most grassland birds and mammals are mobile enough to avoid direct contact with fire. Insects that live in and on the stems and leaves of the plants are the most affected by fire. But these animals have short generation times and populations recover quickly.

Grazing. Grazing by herbivores is an important phenomenon both above- and belowground in tallgrass prairie. Important herbivores include a myriad of invertebrates along with, historically, herds of bison, elk, and antelope (see **illus.**). Of these, bison are thought to be the most important, with numbers estimated at between 30 and 60 million in the Great Plains. Domestic cattle have replaced bison in much of the remaining grasslands today. Nonetheless, where bison still exist, they are considered keystone species in tallgrass prairie based on the disproportionately large effects they have on this ecosystem. The bison diet is primarily grasses. Because burned prairie has the highest proportion of grasses, bison are attracted to burned sites. The targeted grazing of grasses by bison allows more space and resources to be available for the normally less

abundant species (forbs). As a result, plant species diversity may be higher in grazed grasslands compared to those without bison. Bison also remove fuel aboveground and may lessen the frequency and intensity of fires. This too may increase plant species diversity. Finally, bison accelerate the conversion of plant nutrients from forms that are unavailable for plant uptake to forms that can be readily used. Essential plant nutrients, such as nitrogen, are bound for long periods of time in unavailable (organic) forms in plant foliage. Microbes slowly decompose these plant parts, and the nutrients they contain are only gradually released in available (inorganic) forms. This decomposition process may take more than a year or two. Bison consume these plant parts and excrete a portion of the nutrients in plant-available forms. This happens very quickly compared to the slow decomposition process, and nutrients are excreted in high concentrations in small patches. Thus, grazers may increase the availability of potentially limiting nutrients to plants as well as alter the spatial distribution of these resources. Many forbs require higher nutrient levels than the grasses, so this too is a mechanism by which bison grazing may increase plant species diversity.

Soils. In addition to fire, grazing, and climatic extremes, tallgrass prairie is noted for its deep, fertile soils. Indeed, the conversion of the central United States from an unbroken grassland dominated by perennial plants to agricultural ecosystems dominated by annual plants is the primary reason for the dramatic loss of tallgrass prairie. This conversion was driven by the ease of exploitation of the fertile soils underlying tallgrass prairie.

Bison grazing on the tallgrass prairie. (*Konza Prairie Biological Station, Kansas State University*)

Tallgrass prairie soils are fertile because of the large stores of organic carbon and nitrogen they contain. And these stores are a direct result of plant responses to the unique combination of fire, grazing, and climatic extremes that shaped and maintained this grassland. The perennial plants that dominate the prairie allocate much of the carbon they gain from photosynthesis belowground (up to twice as much as aboveground). This allocation pattern differs dramatically from forests, where most carbon is stored aboveground. Belowground these reserves are protected from loss via fire or grazers. Moreover, extensive root systems in prairie plants are necessary for growth during periods of water stress and for survival during extended droughts. Nitrogen, which is needed in leaves for photosynthesis, is also stored belowground during the dormant season to protect it from loss via fire and grazing. Thus, tallgrass prairie plants allocate tremendous amounts of biomass containing carbon and nitrogen belowground. As belowground parts die, they decompose slowly; and it is this organic matter, rich in carbon and nitrogen, that imbues prairie soils with their characteristic dark brown or black appearance.

Due to the endangered status of tallgrass prairie today, as well as the recognized value of many of its attributes, tallgrass prairie conservation and restoration efforts are under way throughout much of this grassland's original range. Although certain characteristics of the tallgrass prairie can be restored, those features dependent on large-scale phenomenon, such as large migrating bison herds or extensive wildfires, will likely never be recognized.

For background information *see* BIOME; BISON; BLUESTEM GRASS; DROUGHT; GRASSLAND ECOSYSTEM; POSTGLACIAL VEGETATION AND CLIMATE; SOIL in the McGraw-Hill Encyclopedia of Science & Technology.

Alan K. Knapp

Bibliography. D. I. Axelrod, Rise of the grassland biome, central North America, *Bot. Rev.*, 51:163–201, 1985; A. K. Knapp et al., *Grassland Dynamics: Long-Term Ecological Research in Tallgrass Prairie*, Oxford University Press, New York, 1998; A. K. Knapp et al., The keystone role of bison in North American tallgrass prairie, *BioScience*, 49:39–50, 1999; P. G. Risser et al., The True Prairie Ecosystem, *US/IBP Synthesis Ser.*, no. 16, Hutchinson Ross, Stroudsburg, PA, 1981; F. Samson and F. Knopf, Prairie conservation in North America, *BioScience*, 44:418–421, 1994; J. H. Shaw and M. Lee, Relative abundance of bison, elk, and pronghorn on the southern plains, 1806–1857, *Plains Anthropol.*, 42:163–172, 1997.

Terahertz technology

Over the past century, the electromagnetic spectrum (**Fig. 1**) has gradually been filled with radiation generated by various technologies. Communications is one of the main applications; radio frequencies (10^5–10^9 Hz) have been used for radio and television,

while gigahertz (10^9 Hz) frequencies are commonly used in mobile telephony, and even higher-frequency (10^{14} Hz) visible light is used to send Internet communications through optical fibers. Imaging technology, which uses different regions of the light spectrum, is also increasingly prevalent, from x-ray scanners at airports to magnetic resonance imaging (MRI) machines that display the organs inside the body with immense detail. In spite of such progress, new communications and imaging modalities which exploit different physical principles may be required to attack many issues such as lack of bandwidth for high-speed Internet links, or lack of sensitivity, high equipment cost, and safety concerns in imaging. There is one area of the spectrum, the so-called terahertz (10^{12} Hz) gap, that remains largely unexplored. The gap has traditionally referred to a lack of bright sources or coherent, sensitive detectors in the frequency range 0.05–20 THz (Fig. 1). This region lies at the boundary between where electronic devices such as high-speed transistors operate in the microwave region, and photonic devices such as lasers operate in the infrared and visible regions of the electromagnetic spectrum. *See* PHOTONICS.

Generation of terahertz radiation. The solution is provided by semiconductor physics; irradiating specially designed semiconductor structures with visible light from a laser causes the semiconductors to reradiate the light in the terahertz frequency range. Two effects are now commonly used to down-shift the frequency of such visible light into the terahertz region: photocurrent techniques and visible pulse rectification. Both techniques rely on the irradiation of a semiconductor with visible laser pulses of sub-picosecond (1 ps = 10^{-12} s) duration.

The first breakthrough occurred in the 1980s with the demonstration that photoconductive emitters—so-called Auston switches—could be used to generate coherent terahertz pulses of subpicosecond duration and high spectral brightness. In a photoconductive emitter, terahertz radiation is generated when the electron-hole pairs in a semi-conductor, created by the visible pulse of light, are accelerated by a bias applied between two surface electrodes. The resulting transient photocurrent, proportional to the charge acceleration, radiates at terahertz frequencies (**Fig. 2**), with the terahertz electric field E_{THz} proportional to the rate of change of current.

Alternative means of converting subpicosecond pulses to terahertz pulses are based on rectification of visible pulses arising from the second-order susceptibility $\chi^{(2)}$ of a semiconductor crystal, which governs the way in which excited electrons vibrate in the crystal. Visible pulses can have a frequency bandwidth in excess of 10 THz. Thus, the pulses contain different visible frequencies, ω_1 and ω_2, which beat against each other to create a time-dependent polarization P of electrons in a semiconductor crystal having a large second-order susceptibility (such as zinc telluride or gallium arsenide), with components at the terahertz difference frequency $\omega_{THz} = |\omega_1 - \omega_2|$. The excited electrons reradiate a terahertz

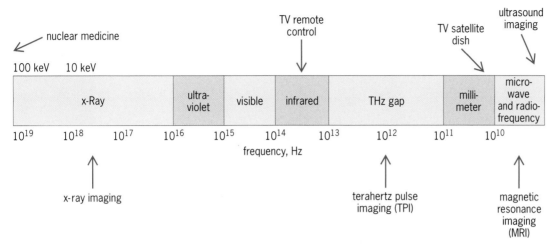

Fig. 1. Position of the terahertz gap in the electromagnetic spectrum. Frequency ranges in which conventional communications and imaging technologies operate are also shown.

pulse at a series of frequencies ω_{THz} corresponding to the different frequency components in the incident visible pulse. The terahertz electric field is proportional to the visible power: $E_{THz} \propto \chi^{(2)} E(\omega_1) E(\omega_2)$, where $E(\omega_1)$ and $E(\omega_2)$ are the electric fields associated with the components of the visible pulse at frequencies ω_1 and ω_2.

Terahertz pulse detection. These terahertz pulses have two main advantages that make a myriad of applications possible. The first advantage is that although the time-averaged power is low (less than 10^{-3} watt), most of the energy is concentrated in the pulse itself, which is spectrally bright as a result. Detection is accomplished using a time-gated technique that allows the pulse to be measured during its duration but turns off the detector at all other times, eliminating noise and background sources of interference. This gating is provided by a portion of the same laser beam used in generation (Fig. 2), and is a natural extension of the detection technique used in pulsed radar systems operating at lower (megahertz and gigahertz) frequencies. Such detectors operate at room temperature and are capable of detecting noise-equivalent powers at the 10^{-18}-watt level. This is orders of magnitude less than conventional bolometric detectors in this frequency range, which are bulky, expensive, and rely on cryogenic cooling.

Detection of terahertz pulses can be thought of as the inverse of the generation process in many respects. Techniques include photoconductive detection, which is the inverse of photoconductive generation; incoming terahertz pulses induce a current between electrodes on a semiconductor in which electron-hole pairs have been created by a visible gating pulse. A second detection technique, free-space electrooptic sampling (EOS), has similarities to generation using visible rectification. A terahertz pulse incident on a suitable semiconductor (for example, zinc telluride) induces a change in the refractive index (speed of light) in the semiconductor that is proportional to the amplitude of the terahertz elec-

tric field E_{THz} as well as the same second-order susceptibility $\chi^{(2)}$ that governs generation. This change in refractive index is probed with readily available optical components and photodetectors that allow a comparison of how the different polarization components of the visible gating beam travel through the detection crystal, which yields the change in index. Due to the instantaneous response of the semiconductor detector, electrooptic sampling in principle has an extremely high terahertz bandwidth, whereas photoconductive technology is an older technique that is used primarily at low frequencies (0.05–3 THz) due to its limited bandwidth.

A second crucial advantage of terahertz pulses is that both the generated pulse and the detector are coherent, meaning that both the phase and the amplitude of the terahertz pulse can be detected. This

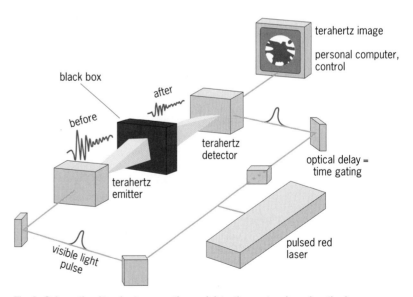

Fig. 2. Schematic of terahertz generation and detection system based on the frequency down-conversion (terahertz generation) and up-conversion (terahertz detection) of visible light. Time-gated detection techniques are used to measure terahertz pulses which pass through a black plastic box containing a spider, and the terahertz image is displayed on the computer screen.

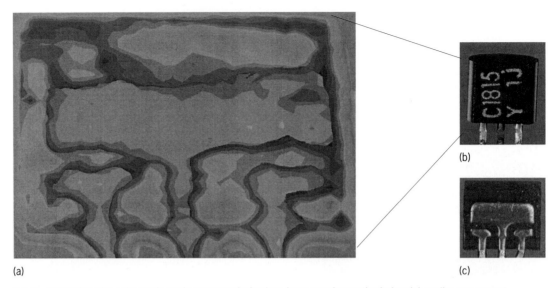

(a)

(b)

(c)

Fig. 3. Application of panchromatic terahertz transmission imaging to nondestructive industrial-quality assessment. (*a*) Terahertz image of a packaged transistor, allowing the internal leads to be examined. The image represents the peak value of the terahertz electric field at each pixel. (*b*) Visible image of the device before terahertz imaging. (*c*) Visible image of an identical device that was opened (cut-away view) to verify the accuracy of the terahertz image.

capability allows both amplitude (absorption) and phase (time-of-flight) information to be exploited. Time-of-flight information originates from the fact that when the pulse travels through a medium with refractive index $n > 1$, it takes longer [with a time delay of $d(n-1)/c$] than it would if it were traveling only through air ($n = 1$) at the speed of light c over the same distance d. The time of flight of the terahertz pulse yields information on the thickness and refractive index of the medium. In addition, the time-domain pulse can be transformed using standard Fourier mathematical techniques to give the transmission (absorption) in the frequency domain. This frequency information can be used to collect spectroscopic data on the medium under investigation. Since the mid-1980s, terahertz pulses have been used in spectroscopic studies of semiconductors, gases, polar and nonpolar liquids, and deoxyribonucleic acid (DNA). Similarly, the amplitude or phase of the terahertz pulse can be modulated and manipulated to carry information through free space for communications purposes.

Imaging. Another major application of terahertz pulses is imaging. Historically terahertz pulses have much in common with radar techniques that typically operate in the gigahertz frequency range; indeed, terahertz pulse techniques can be regarded as high-frequency radar, operating typically at frequencies 100 times higher. A sequence of pulses are injected into a point on an object, and the transmitted or reflected portion of the pulses is then detected after a variable time delay introduced by the gating pulse. The object to be imaged may be translated through the beam, or alternatively the beam may be translated across the object to build up an image at various points (Fig. 2). Alternative methods using charge-coupled-device (CCD) cameras can be employed to eliminate the need for mechanical

motion of the sample and source, and other coupling techniques will allow for in vivo imaging and diagnosis of medical and nonmedical subjects. Time-domain data are displayed on a computer screen, or frequency-domain data may be examined at each pixel in the image by applying a mathematical transform to the time-domain data. Because the technique makes use of impulsive transmitted signals whose spectrum covers a wide bandwidth, frequency-domain as well as time-domain images may be obtained.

One of the most attractive features about terahertz imaging is that a variety of common materials are transparent or semitransparent in this frequency range. These include certain thicknesses of plastics (**Fig. 3**), paper, cardboard, semiconductors, and animal and human tissue. Indeed, completing the list of such materials is one of the pressing issues for the commercialization of terahertz imaging. Traditional techniques such as x-rays, magnetic resonance imaging (MRI), and ultrasound are capable of penetrating deep into the human body as well as many inanimate objects, and provide detailed images which represent many years of effort by the imaging community. As competition with such accomplishments would be difficult at this early stage of development, applications of terahertz imaging are focused on specific areas where terahertz light might have capabilities not afforded by the conventional imaging techniques. In this regard, one of the other major advantages of terahertz pulse imaging (TPI) is its diagnostic capabilities. Its ability to supply spectral data may, for example, enable diseased and healthy tissue to be imaged and diagnosed; **Fig. 4***a* shows an example of transmission images of demineralization in a human tooth.

A further advantage of terahertz imaging is the multiplicity of contrast mechanisms available that are not

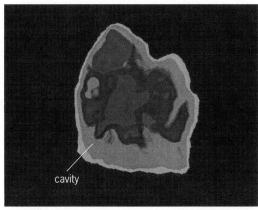

(a)

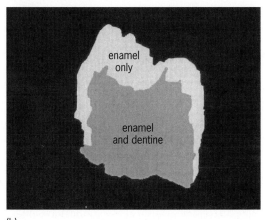

(b)

Fig. 4. Terahertz images of an extracted human tooth. The data can be manipulated to provide different terahertz images, each containing different diagnostic information. (*a*) Plot of the panchromatic terahertz transmission data, showing a different value of the absorption in the cavity region. The terahertz image represents the peak value of the terahertz pulse electric field at each point on the tooth. (*b*) Plot of time-of-flight of the terahertz pulse through the tooth, revealing the different refractive indices and thicknesses of enamel and dentine layers.

typically found in other techniques. At each pixel in an image, it is possible to plot the transmission (or equivalently absorption) over the entire frequency bandwidth of the terahertz pulse as in Fig. 4*a*, or the transmission at a single frequency band in the pulse. Alternatively, the time of flight (thickness or refractive index) of the pulse through the object can be extracted mathematically from the same data set and plotted at each pixel. Figure 4*b* shows a time-of-flight image of a human tooth, where the different refractive indices of the enamel and dentine are clearly seen, allowing enamel decay to be monitored.

Penetration depth. Two factors, however, may play the deciding role in determining the applications of terahertz light in both communications and imaging. The penetration depth of the radiation through various materials will dictate in part what items can be imaged or diagnosed. Typically, millimeters to centimeters of dry, nonmetallic materials can be probed, but the penetration of terahertz radiation in materials with high water content is usually limited

to a couple of millimeters or less beneath the surface. Medical applications are currently limited to dental and dermatological imaging, with possibilities for endoscopic imaging inside the body. For nonmedical applications, spectroscopic examination of biofluids along with imaging of semiconductors, ceramics, food processing ingredients, and a variety of other materials for quality assessment and control are areas where penetration depth does not present a limitation.

Cost. The second important point for commercial focus is cost. The development of alternative, inexpensive, and compact systems of coherent terahertz radiation, both in terms of bright sources and of sensitive detectors, would dramatically increase the number of markets for the technology. The use of the generation techniques described above with continuous-wave (unpulsed) laser diodes is one avenue. Others include the extension of high-speed microwave devices to terahertz frequencies and the construction of terahertz lasers from semiconductor structures. At stake is the last unconquered region of the light spectrum.

For background information *see* CHARGE-COUPLED DEVICES; COHERENCE; ELECTROMAGNETIC RADIATION; ELECTROOPTICS; INFRARED SPECTROSCOPY; MEDICAL IMAGING; NONLINEAR OPTICS; OPTICAL PULSES; PHOTOCONDUCTIVITY; RADAR; REFRACTION OF WAVES; SUBMILLIMETER-WAVE TECHNOLOGY in the McGraw-Hill Encyclopedia of Science & Technology.

D. D. Arnone

Bibliography. D. D. Arnone, C. M. Ciesla, and M. Pepper, Terahertz imaging comes into view, *Phys. World*, 13(4):35–40, April 2000; D. H. Auston, K. P. Cheung, and P. R. Smith, Picosecond photoconducting hertzian dipoles, *Appl. Phys. Lett.*, 45:284–286, 1984; D. R. Grischkowsky, An ultrafast optoelectronic THz beam system: Applications to time-domain spectroscopy, *Opt. Photonics News*, pp. 21–28, May 1992; B. B. Hu and M. C. Nuss, Imaging with terahertz waves, *Optics Lett.*, 20:1716–1718, 1995; D. M. Mittleman, R. H. Jacobsen, and M. C. Nuss, T-ray imaging, *J. Sel. Top. Quantum Electr.*, 2:679, 1996; P. Uhd Jepsen et al., Detection of THz pulses by phase retardation in lithium tantalate, *Phys. Rev. E: Rapid Commun.*, 53:R3052–R3054, 1996.

Text-to-speech systems

Text-to-speech (TTS) systems are is used to transform written text in a given language into its spoken equivalent. That is, a TTS system converts an input sentence, which is usually a sequence of letters (as in English) or symbols (as in Chinese), into machine-generated speech. The input sentence may include details of linguistic representation such as phrasing, intonation, and stress. Unlike limited vocabulary systems, which produce only a small predefined set of possible utterances, open-domain TTS systems must be able to intelligently handle any input text with an unlimited set of vocabularies.

One important application of TTS is human-machine communication. For decades, science fiction writers have provided images of people and computers communicating via human language. In the movie *Star Trek*, for example, the captain simply asks the computer in spoken English to check the status of his spaceship. After checking the status, the computer reports it, again in spoken English. In this simple human-machine interaction, there are three basic technical components involved. First, the machine must understand a human language by means of automatic speech recognition (ASR). Second, it must find the answer and prepare the report in that language. Finally, the computer must speak or read the prepared answer. TTS is an essential technology for this interaction.

Other TTS applications can be found in situations that are more realistic. The explosive expansion of the Internet has enabled an immense number of computers to interact across data networks. E-mail messaging has become very popular in data network environments. In telecommunications, the number of wireless communication devices has increased exponentially. As computer and telephone networks converge, spoken language technologies such as ASR and TTS are playing a major role in their integration. A TTS system can read an e-mail or fax message to its recipient over the telephone, and Web pages can be read to a user over a wireless phone. In offices, TTS systems can read typed or dictated documents aloud for proofreading. A TTS voice response system can generate spoken prompts, and is more practical than using prerecorded human voices. In the case of small, wearable computers, there is no room for an output device such as display or printer. TTS systems can be used as an output method for such diminutive computers. For educational purposes, TTS systems are useful for reading and foreign language instruction. TTS systems can also be used to read the content on a computer screen to a visually impaired user. TTS is ideally suited to the needs of users whose eyes or hands are occupied with other tasks. In cars, a TTS system can read the travel directions generated by a global positioning system and map computer so that the driver can remain focused on the road.

Building a TTS system requires multidisciplinary research and development efforts. The contributing fields include, but are not limited to, linguistics, phonetics, acoustics, statistics, digital signal processing, electrical engineering, and computer science. The process of converting digital written text to speech can be broken down into three tasks: (1) text analysis, (2) prosodic modeling, and (3) speech synthesis.

Text analysis. The text analysis module converts input text into a sequence of phonemes. To understand the functions of the text analysis module, consider the sentence: "*Dr. Smith lives at 1595 Smith Dr., St. Louis, MO.*" The text analysis module must determine the following. The first *Dr.* should be read as *doctor*, while the second must be pronounced *drive*. The verb *lives* rhymes with *gives*, as opposed to the plural noun *lives*. The number *1595* has to be pronounced either *one five nine five* or *fifteen*

ninety-five, but not *one thousand five hundred and ninety five*. *St.* represents *Saint* rather than *street*. Finally, *MO* has to be expanded to *Missouri*, not *emoh*. This process is called text normalization. The next step is a process called text-to-phoneme conversion; that is, each normalized word must be converted into a sequence of phonemic symbols. Using the International Phonetic Alphabet, the example above can be represented as the phoneme sequence [dɑktɔr smiθ livz...]. Rule-based or dictionary lookup methods are extensively used for text-to-phoneme conversion.

Prosody generation. The task of the prosody module is to determine the pitch, intonation, and duration of each phoneme in the phoneme sequence that is obtained from text analysis. This is not a simple task because many factors affect prosody. Moreover, the effects of these factors are complex and not yet fully understood. Rule-based or statistically based methods are used for prosody generation. One example of rule-based intonation models is the superpositional intonation model. It computes intonation by adding three types of time-dependent curves: a phrase curve, an accent curve, and a perturbation curve. The phrase curve depends on the type of phrase, such as declarative or interrogative. The accent curves represent each individual accent group, and the perturbation curve captures the effects of obstruents (consonants produced with an obstruction of the air flow above the larynx) on pitch in the postconsonantal vowel. For the duration generation module, statistically motivated methods such as sum-of-products approaches are widely accepted.

Speech synthesis. The synthesis module generates acoustic speech segments based on output from the text analysis and prosody generation modules. Any TTS system uses one of three main strategies: rule-based, concatenative, or hybrid. Rule-based approaches apply algorithmic rules that simulate coarticulation effects to generate speech production model parameters. They require very little disk space, and the speech parameters can easily be modified to alter the voice. Varying pitch level, speaking rate, timbre, and breathiness are simple. Rule-based approaches require extensive knowledge of sound patterns and the speech production model, however, and presently our understanding is not sophisticated enough to produce high-quality synthetic sound. Although the voice quality is not very natural, these voices can be embedded in small devices with limited system resources.

In concatenative systems, speech segments are cut from natural speech. During synthesis, the units are concatenated (linked together) to produce a target sentence, producing a natural-sounding segmental voice quality. The systems store the actual speech waveform segments, while minimizing the signal processing for prosody modification of the original recording. This may result in unnatural prosody. Generating a variety of voices can be difficult because an entirely new set of recordings may be required for a new voice. Many concatenative systems use multiple segments, recorded with different prosody, for each synthesis unit. This can improve the voice quality and prosody, but increases the inventory size dramatically, and the unit selection algorithm becomes computationally expensive. Therefore, such systems are limited to server applications.

Hybrid approaches involve concatenating segments of natural speech as in concatenative approaches. However, the segments are represented in terms of speech production model parameters, such as linear predictive coefficients, rather than the waveform itself. These approaches share the advantages of rule-based as well as concatenative approaches. Parametric representation of speech, as in rule-based approaches, enables more flexible control over voice characteristics and prosody. One can obtain a more natural prosody than in pure concatenative approaches. On the other hand, since the parameters are extracted from segmented real speech, they produce higher voice quality than rule-based approaches do. However, the segmental voice quality is not as good as that generated by concatenative synthesis since most signal processing algorithms for prosody modification cause slight degradation of output speech quality.

Further issues. Although it is generally agreed that state-of-the-art TTS systems produce very natural and intelligible sound, any human listener can determine immediately whether a speech sample is synthetic or real. Machine-generated prosody and voice quality are not as natural as human speech. In particular, TTS systems cannot generate the prosody to emphasize and deemphasize the appropriate words. For example, the sentence "*My name is Minkyu*" can have an emphasis on either *my*, *name*, *is*, or *Minkyu*, depending on the context. Most current text-to-speech systems are not sophisticated enough to automatically make these kinds of contextual distinctions. Additional work on discourse analysis must be done. TTS systems cannot synthesize speech sounds with emotion, so distinctions such as depressed, cheerful, and excited voices are currently unavailable. The process of designing a TTS system for more than one voice has not yet been fully automated. Consequently, intensive human intervention is still required.

For background information *see* CHARACTER RECOGNITION; ELECTRONIC MAIL; PHONETICS; SPEECH PERCEPTION; VOICE RESPONSE in the McGraw-Hill Encyclopedia of Science & Technology.　　Minkyu Lee

Bibliography. T. Dutoit, *An Introduction to Text-to-Speech Synthesis*, Kluwer, Dordrecht, 1997; D. Klatt, Review of text-to-speech conversion for English, *J. Acous. Soc. Amer.*, 82:737–793, 1987; R. Sproat (ed.), *Multilingual Text-to-Speech Synthesis: The Bell Labs Approach*, Kluwer, Dordrecht, 1998.

Three-body Coulomb problem

Although the analytic solution of the wave function for the isolated hydrogen atom played a pivotal role in establishing the new quantum theory during the early part of the twentieth century, no corresponding solutions exist for systems with three or more charged particles. A complete solution of the three-body Coulomb problem requires a solution of Schrödinger's wave equation for a function that gives the probability that three charged particles are moving away from each other with precisely specified energies and directions. The three-body Coulomb problem plays a central role in understanding the physics of collisional ionization, a basic process that underlies many physical phenomena, from the glow of fluorescent lights to the chemistry of interstellar gases. Recently, a computational framework for solving ionization problems has been provided, and supercomputers have been used to produce numerical solutions of the ionization of a hydrogen atom by electron impact that reveal the full dynamics of this problem.

Quantum mechanics of the problem. The nonrelativistic quantum mechanics of two-electron atoms has a long history, beginning with the pioneering work of Egil A. Hylleraas on bound states in the 1930s that led to Chaim L. Pekeris's accurate determination of the bound states of the helium atom in the late 1950s. In contrast to such bound-state problems, where the electrons are spatially confined to a finite volume about the nucleus in so-called stationary states, electron-atom collisions refer to processes in which an initially free electron impinges on an isolated atom. Such problems, which must be solved using the tools of quantum scattering theory, are intrinsically more difficult to solve than bound-state problems, because the associated wave functions are not confined, but extend over all space. It was not until 1961 that Charles Schwartz obtained an accurate numerical solution to the simplest collision problem in a two-electron system, scattering of an electron by a hydrogen atom without energy exchange. Since then, the effort has been extended to so-called inelastic collisions in which a portion of the incident electron energy is transferred into internal excitation of the atom, and a number of benchmark calculations of excitation probabilities and associated scattered electron angular distributions have been carried out. For such problems, the traditional approach has been to expand the unknown solution of the Schrödinger equation in terms of the known wave functions of the target atom—the so-called close-coupling method. Such calculations have have reached a level of sophistication where it is possible, for the electron-hydrogen atom system, to compute these excitation probabilities with a greater accuracy than can be achieved in most experiments, even at collision energies above that required to ionize the target atom.

Three-body breakup. When an electron with more than 13.6 eV kinetic energy collides with a hydrogen atom, it can tear the bound atomic electron from its stationary orbit, producing a final state in which three charged particles emerge from the interaction region. Such breakup collisions are considerably more difficult to describe quantum-mechanically than elastic or excitation collisions, because in breakup events energy can be shared continuously

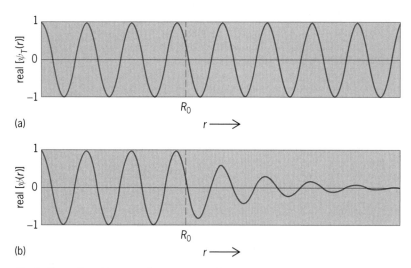

(a)

(b)

Fig. 1. Illustration of how exterior complex scaling transforms outgoing waves into exponentially decaying functions. (a) Real part of the function $f(r) = e^{ikr}$, representing an outgoing wave. (b) Real part of the transformed wave function $f[R(r)]$, under the mapping given in the text.

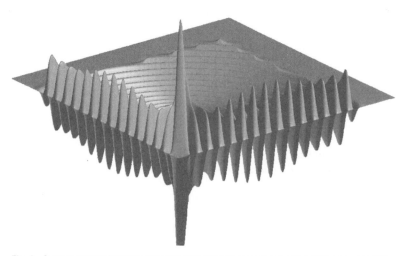

Fig. 2. Component of the wave function that describes a subset of the outcomes possible when an electron hits a hydrogen atom. The horizontal axes represent the distances of each of the two electrons from the proton. The spherical wavefronts give the probability for ionization events, while the sharper features near the axes correspond to two-body (elastic and excitation) collisions.

among the fragments. Thus the number of final outcomes, or channels as they are referred to in scattering theory, is continuously infinite in a breakup collision. When the collision fragments are charged, as they are in an ionizing electron-hydrogen atom collision, then the problem is aggravated by the long-range nature of the Coulomb forces between the particles, which can lead to an exchange of energy between the emerging electrons at distances far from the initial point of collision. The formal theory of three-body breakup in Coulomb systems was formulated in the early 1960s by Raymond K. Peterkop and by M. R. H. Rudge and M. J. Seaton. While aspects of that formal theory have been successfully incorporated into a number of approximate methods that have produced useful results for intermediate- to

high-energy collisions (100 eV and above), it has not yet provided a viable computational path for any accurate, first-principles approach to ionization. The form of the wave function in which all three particles are widely separated places a boundary condition on the wave function that is so intractable that no known numerical approach to solving the Schrödinger equation has successfully incorporated it explicitly.

Extended close-coupling methods. Recent first-principles approaches to this problem generally fall into one of two broad categories. One approach has been to extend traditional close-coupling methods to the ionization problem by replacing the true continuum of ionized hydrogenic target states with a finite number of discrete, normalized, positive-energy pseudostates, while treating the incident free electron with conventional two-body scattering boundary conditions. While such approaches have been very successful in predicting total ionization probabilities, attempts to predict various "differential" quantities, such as the way that energy is shared between the exiting electrons or the angles at which they have been ejected, have been more problematic, especially at low collision energies. This area continues to be one of active research, and there are still many puzzles and unanswered questions that need to be sorted out, but there are strong indications that the problems have to do with the inappropriate use of two-body boundary conditions to treat a true three-body problem.

Exterior complex scaling. Because of the intractability of the boundary conditions dictated by the formal theory, there has also been considerable interest in the development of computational approaches that do not rely on the explicit enforcement of asymptotic boundary conditions. One approach that has been particularly successful relies on a mathematical device called exterior complex scaling to greatly simplify the boundary conditions on the wave function and to allow one to solve the time-independent Schrödinger equation just as if one were solving for an ordinary bound state. To accomplish this, one makes use of the fact that the portion of the true wave function that describes all scattering events has the form of an outgoing wave at large distances from the nucleus. This wave can, in turn, be transformed into an exponentially decaying function by allowing the coordinates associated with the two electrons to take on complex values. Exterior complex scaling is the name of the mathematical transformation that accomplishes this task.

Figure 1 gives an example of how the transformation works in a simple one-dimensional case, in which the outgoing wave is $f(r) = e^{ikr}$, where $i = \sqrt{-1}$, r is the electron's distance from the nucleus, and k is a real constant (the wave number, proportional to the electron's momentum). The real part of the function $f(r)$ is shown in Fig. 1. When the distance r is a real number, the real part of $f(r)$, shown in Fig. 1a, is a simple sinusoidal function that oscillates with a constant amplitude. When the distance

r is mapped into the complex plane under the transformation

$$r \rightarrow R(r) = \begin{cases} r, & r \leq R_0 \\ R_0 + (r - R_0)e^{i\theta}, & r > R_0 \end{cases}$$

the mapped function, $f[R(r)]$, decays exponentially when r is greater than R_0, as shown in Fig. 1b.

By applying this transformation to the Schrödinger equation, and then introducing a numerical grid, the problem can be reduced to the solution of a system of complex, linear equations. Because one is dealing with a Coulomb system, the region of space over which the physical wave function must be known—that is, the region where the electron coordinates are real—is quite large, so the complex scaling must be performed at distances far from the nucleus. This leads to very large systems of linear equations (several million by several million) whose solutions require carefully designed algorithms that can be solved only on large supercomputers. This approach has produced the first essentially exact solution to the electron-hydrogen atom ionization problem, with detailed differential ionization probabilities in exquisite agreement with experiment.

Figure 2 depicts a portion of the wave function for the electron-hydrogen system that gives a striking visualization of the ionization process. The quantity shown is the probability amplitude for the two electrons, each in a specific angular momentum state, as a function of their radial distance from the nucleus. Many such components, corresponding to different values of the individual angular momenta, are needed to describe the full wave function for the electron-hydrogen system. In a collision that results in ionization, both electrons escape and their radial distances from the nucleus become large. This is depicted by the spherical wave fronts that characterize the wave function in the region between the axes. Near the axes, one of the electrons is close to the nucleus and the wave function displays different structure that corresponds to collisions where only one electron escapes and the other remains bound to the nucleus.

Time-dependent methods. Another promising approach is based on solving the time-dependent Schrödinger equation. In this method, one starts with a localized wave packet that represents an electron incident on the hydrogen atom and propagates it on a finite grid for a sufficiently long time until the collision is essentially over. Ionization information is obtained by analyzing the final propagated wave packet. Because electrons are light particles, the wave packet rapidly spreads as it is propagated, so the grids that must be used can become quite large. However, since this is a propagation method, and in contrast to exterior complex scaling does not involve solving linear equations, much larger grids can be used than would be practical with methods that require solution of linear equations. Methods for calculating detailed differential ionization probabilities have yet to reach the same stage of development as with the time-independent methods, but this is also an active area of research and rapid progress can be expected.

Extension to more complex systems. Although the exterior complex scaling and time-dependent methods represent completely independent approaches to the problem of electron-impact ionization, they share one critically important feature: Neither method relies on detailed specification of scattering boundary conditions in producing the quantities from which collision information is ultimately extracted. This has been the key to a computational solution of the three-body Coulomb problem. Both approaches are computationally intensive, however, and cannot be readily extended to systems with more that two (or perhaps three) electrons. The challenge for studying ionization in more complicated systems will be to develop hybrid techniques that combine the essential features of these approaches with more traditional methods for tackling many-body problems.

For background information *see* ATOMIC STRUCTURE AND SPECTRA; NONRELATIVISITIC QUANTUM THEORY; QUANTUM MECHANICS; SCATTERING EXPERIMENTS (ATOMS AND MOLECULES); SCHRÖDINGER'S WAVE EQUATION in the McGraw-Hill Encyclopedia of Science & Technology.   Thomas N. Rescigno

Bibliography. R. G. Newton, *Scattering Theory of Waves and Particles*, 2d ed., Springer, New York, 1982; R. K. Peterkop, *Theory of Ionization of Atoms by Electron Impact*, Associated University Press, Boulder, 1977; J. Taylor, *Scattering Theory*, John Wiley, New York, 1972.

Tool evolution

Until 40 years ago, when Jane Goodall began her field observations of wild chimpanzees in Gombe National Park in Tanzania (then Tanganyika), human beings were thought to be the only species to have technology, that is, to make and use tools as part of a material culture. The four living forms of great apes (bonobo, *Pan paniscus*; chimpanzee, *Pan troglodytes*; gorilla, *Gorilla gorilla*; orangutan, *Pongo pygmaeus*) were known to use tools in captivity, but this was deemed unnatural. Research published by the German psychologist Wolfgang Köhler in 1917 showed that laboratory chimpanzees spontaneously made and used tools to solve problems, if given the right raw materials and setting. Today, we know that some apes in nature have a rich repertoire of tools that are put to daily use.

Background. If a tool is defined as an external, portable object used to achieve a goal more efficiently, then many species of animals are tool users, including parasitoid wasps, California sea otters, and Galápagos finches. However, few of these species make tools, and almost all employ one type of specialized tool use, but no others. The best performance by a non-ape is probably that of the Caledonian crow, which makes two kinds of tools to extract insect larvae from rotten wood.

Diversity. Great apes have both tool kits and tool sets. A tool kit is a repertoire of tools of different sizes, shapes, materials, and functions. Each population of chimpanzees has a different collective tool kit, so that we can compare their elementary technologies across different parts of their native Africa. A tool set is two or more different tools used in a combined obligatory sequence to solve a problem. For example, a chimpanzee may first use a digging stick, then a flexible probe of vine to obtain underground insects.

The hallmark of ape technology is plasticity. A single raw material, such as a leaf, can function as a variety of tools, such as a napkin, drinking sponge, grooming stimulator, courtship signal, or probe to get insects for food. Conversely, a tool can be made from a variety of materials. For example, an insect probe can be made from bark, vine, grass stem, or twig, as well as a leaf. Tools are used not only in subsistence but also in hygiene, self-maintenance, and social and sexual life. Flexibility means that diversity occurs across individuals, families and lineages, communities, populations, subspecies, or species. Comparisons at these levels sometimes yield unexpected results.

Of the four types of great ape, only the chimpanzee shows ubiquitous, elementary technology. Among these African apes, found in more than 40 populations, tool use and tool making are universal and, hence, species-typical. Surprisingly, the closely related bonobo (or pygmy chimpanzee) shows almost no customary tool use, and not a single studied population shows tool use in foraging. (This is puzzling, because bonobos in captivity appear to be just as adept at tool use as chimpanzees.) The other great ape of Africa, the gorilla, shows no customary technology at all, although many populations live in the same equatorial forests as do tool-using chimpanzees. Until recently it could be claimed that not enough field study had been done to draw such negative conclusions about bonobos and gorillas, but this is no longer true. In any case, even if behavioral observations are sparse (because these apes are shy of human contact), the tools would be found as artifacts if present.

The orangutan, the only living Asian great ape, presents a different picture. Some populations that have been well studied for decades show virtually no tool use, while others show it as often as chimpanzees. Since all of the species of great apes appear to be equally intelligent, at least as determined by standardized testing in the laboratory, the explanation for their different performances in nature seems likely to lie in socioecology. This is a research challenge awaiting elucidation. Primatologist Carel van Schaik hypothesized that the key is sociality, as this allows greater opportunity for learning skills from others.

Distribution. The chimpanzee is widespread across tropical Africa, and elementary technology is found in all types of habitat, from rainforest (Gabon) to savanna (Senegal). All three subspecies of chimpanzees—eastern, central, and western—are tool users, but subspecific differences exist. Only the far western race, from Ivory Coast westward, uses hammers and anvils to crack nuts, although the same species of nut-bearing trees and appropriate raw materials are found in central Africa, where the apes ignore them.

The bonobo is found only in Congo (formerly Zaire), where it has been studied at a handful of sites. The habitats are suitable for tool use; perhaps it is absent because bonobos have no ape competitors, unlike chimpanzees and gorillas which coexist and compete over much of their species' ranges.

None of the three subspecies of gorilla—mountain, eastern lowland, and western lowland—uses tools. Their greater size allows them to solve some problems by brawn rather than brains. For example, they use their teeth to crunch nuts that chimpanzees crack open with tools.

The formerly wide species range of the orangutan is now limited to only two islands in the Malay Archipelago, Borneo and Sumatra. The subspecies in Borneo uses no customary tools, while several populations in Sumatra use tools of vegetation to process fruit and to extract insects.

Adaptive importance. How important is technology to the apes? Consider the oil palm (*Elaesis guineensis*), a widespread species in Africa that produces a high-energy fruit. Some populations of chimpanzees ignore it altogether, thus missing out on a high-quality foodstuff. Others eat the oily husk but discard its hard-shelled nut. Others use a hammer and anvil of wood or stone to crack open the nut to get the proteinaceous kernel (see **illus.**).

A female chimpanzee in Liberia positions an oil palm nut on a log anvil before cracking it open with a stone hammer. (*Copyright by William McGrew*)

Throughout their species range, chimpanzees use tools to exploit the Macrotermitine termites, which live in large earthen mounds. These termites are secure against primate predators, except for those chimpanzee populations that dig them out with sticks or fish them out with probes. The latter technique, first described by Goodall, requires much sensory-motor skill, as the subterranean prey must be extracted from tiny holes, using tactile feedback. Thus, a prey that would be otherwise unavailable is exploited by elementary technology. Other chimpanzees use similar probes to extract wood-boring ants from arboreal nests.

When surface water is not available, chimpanzees fashion sponges of wadded-up leaves to soak up runoff water in tree holes. In aggression directed against competitors or predators, chimpanzees use stones or sticks as missile, club, spear, or flail.

Not all tools are so directly functional: some apes noisily clip leaves with their teeth and lips, not to consume the foliage but to attract the attention of a potential mate. Such use in courtship may be symbolic, for the leaf per se means nothing; instead the leaf arbitrarily serves as a reference to sexual arousal and mating intention.

Chimpanzees also ingest vegetation for medicinal, not nutritive, purposes; specific parts of certain species of plants are used to treat symptoms of illness and parasitic infection.

Clearly tool-using apes benefit from their technology by virtue of their enriched diets and defense against predators and disease. However, tool use is not always adaptive. Chimpanzees in Guinea invented pestle pounding, in which the heart of an oil palm is pounded to a sappy pulp, then consumed. The result in the short term is a tasty meal; in the long term, because the palm is killed, the resource base for the apes is reduced.

Material culture. Primate tool use seems to be technological—that is, it is not simply behavioral but dependent on knowledge and imbued with meaning. If so, it can be considered material culture, in the same way that anthropologists study human societies and compare them cross-culturally. Since tool use is concrete and recordable, capable of analysis even by archeological techniques when only artifacts remain, it has become the focal point of the emerging field of cultural primatology. (The word "culture" has many meanings, even in anthropology. Here is meant a phenomenon that is learned from others and becomes a normative standard characterizing a collective unit, such as a community, resulting in customs, rites, traditions, taboos, and institutions.)

Like cultural anthropology, cultural primatology has passed through stages in its development. The first, descriptive stage was one of natural history. Pioneering primatologists such as Goodall, Toshisada Nishida, and Yukimaru Sugiyama painstakingly recorded their encounters with wild chimpanzees, often starting with anecdotes. The second stage was one of ethnography, in which field workers such as Christophe Boesch and William McGrew systematically took quantitative data, often working with several populations of apes. The third stage, of ethnology, is only just beginning. Recently, nine chimpologists with more than 150 years of fieldwork among them published an analysis of 65 traits shown differentially in seven chimpanzee communities. More than half of these traits, most of them involving tool use, showed cultural variation. Finally, scientists such as Tetsuro Matsuzawa have begun experimental studies of material culture in wild apes, such as seeking to follow the spread of technology by migration.

Human evolution. The evolutionary origins of human technology are undeniably evident in the first appearance of flaked stone tools in the archeological record dated at about 2.6 million years ago. Yet the range of tool use found in living great apes suggests that our ancestors had elementary technology earlier, at least as long ago as the last common ancestor of living apes and humans, at 5-6 million years ago. What is missing from the prehistoric record are all the tools made of perishable raw materials, such as wood, bark, grass, skin, and bone. We can never reconstitute these from the Plio-Pleistocene, but much can be learned by modeling our ancestors' technology from what apes do now. For example, the characteristic wear patterns of percussion on archaic hammers and anvils may signal that not just bones but also nuts were smashed. On the other hand, knapping stone—using a tool to make another tool, such as striking a cobble with a hammer to produce a flake from its core—is a skill shown by our ancestors 2.6 million years ago but never yet seen in wild apes.

For background information *see* ANIMAL EVOLUTION; APES; EARLY MODERN HUMANS; PRIMATES in the McGraw-Hill Encyclopedia of Science & Technology.
W. C. McGrew

Bibliography. A. Berthelet and J. Chevaillon, *The Use of Tools by Human and Non-human Primates*, 1993; J. Goodall, *The Chimpanzees of Gombe: Patterns of Behavior*, 1986; W. C. McGrew, *Chimpanzee Material Culture: Implications for Human Evolution*, 1992; K. D. Schick and N. Toth, *Making Silent Stones Speak*, 1993.

Top quark

As the result of a half century of study of elementary particles, physicists today believe that all matter and energy at currently accessible energy scales consists of a small set of fundamental particles whose interactions are described by the relatively simple rules of a quantum field theory. The list of these particles and the rules for their basic interactions constitute what is now known as the standard model. At present, gravity is the only experimentally known force that does not have a place in this model.

An important prediction of the standard model was the existence of the massive top or t quark which, after 1977, was the only one of the six quarks in the model that had not been observed. Since 1994, this prediction has been verified by experiments at Fermilab (near Chicago). These experiments have identified the top quark by studying the products of

Three generations of quarks and leptons in the standard model		
Generation	Quarks (electric charge, mass)	Leptons (electric charge, mass)
1	u (+2/3, ~5 MeV) d (−1/3, ~10 MeV)	e^- [electron] (−1, 0.5 MeV) ν_e [electronneutrino] (0, very light)
2	c (+2/3, ~1250 MeV) s (−1/3, ~125 MeV)	μ^- [muon] (−1, 106 MeV) ν_μ [muon neutrino] (0, very light)
3	t (+2/3, ~175,000 MeV) b (−1/3, ~4200 MeV)	τ^- [tau] (−1, 1777 MeV) ν_τ [tau neutrino] (0, very light)

high-energy proton-antiproton collisions, and have begun to investigate its properties.

Particles in the standard model. Particles in the standard model are considered to be fundamental because they are pointlike, which means that they have no spatial extent; the less fundamental atoms and nuclei that do have a length scale are bound states of these pointlike fundamental particles, and their sizes are determined by the masses of the fundamental particles and the strength of their interactions. For example, a hydrogen atom (size about 10^{-8} cm) is a bound state of a fundamental, pointlike electron and a proton; the proton (much smaller, size about 10^{-13} cm) is a bound state of three fundamental, pointlike quarks.

The fundamental particles in the standard model are divided into sets according to the angular momentum that each carries (a half-integer multiple of Planck's constant, called the particle's spin) and the type of interactions in which they participate. There are 12 spin-1/2 particles in the standard model, which are the 6 charged quarks, 3 charged leptons, and 3 neutral leptons, called neutrinos. The standard model also implicitly includes their antiparticles, which have the opposite electric charges. Five spin-1 particles are known: the photon, the gluon, and the W^+, W^-, and Z^0 bosons. A final, crucial component of the standard model is a single spin-0 particle, the Higgs boson; interactions with the Higgs boson are speculated to be the origin of the masses of the other fundamental particles. The Higgs boson has not yet been found, presumably due to a rather large mass. The discovery of this single missing particle in the standard model is one of the most important goals of experimental high-energy physics.

The spin-1/2 particles in the standard model are sometimes referred to as matter particles and the spin-1 particles are referred to as forces, because most of the matter encountered in the everyday world is made of bound states of the spin-1/2 set, and these mainly interact through the exchange of particles in the spin-1 set. The spin-1/2 particles are separated into quarks, which interact most strongly through the interchange of particles called gluons, and leptons, which do not exchange gluons. The forces due to gluon exchange are remarkable in that they increase with distance, so that a quark can never be separated completely from other strongly interacting particles. This feature, which is known as confinement, is the reason that individual quarks cannot be isolated.

Six quarks. One puzzling feature of the standard model is a repeating pattern of sets of spin-1/2 particles, containing two quarks and two leptons, at successively higher mass scales. The three known sets, which are called generations, are listed in the **table**. The six quark labels u, d, c, s, t, and b are abbreviations for the names up, down, charm, strange, top, and bottom; these names are mainly of historical significance. For comparison, the proton mass in these units is 938 MeV; 1000 MeV is also called 1 GeV.

The generation pattern can be seen in the presence of a set of two quarks with electric charges +2/3 and −1/3, which are partnered with a pair of leptons. One might suspect that this pattern continues indefinitely, with ever higher masses of new quarks. Fortunately this can be tested by studying the decays of the heavy neutral spin-1 Z_0 boson, which decays to pairs of the spin-1/2 particles and their antiparticles, including neutrino + antineutrino. Since the neutrinos are very light (until recently they were thought to be completely massless), the Z_0 can decay into pairs of light neutrinos of each generation. Careful studies of the decay rate of the Z_0 have shown that it decays into just three types of light neutrino-antineutrino pairs. Thus, assuming that all fermions occur in these generations, there are just three such generations, and hence just 6 quarks. The progressively more massive c and b quarks were found in 1974 and 1977, respectively. The search for the final quark was a major activity in experimental particle physics in the 1980s and early 1990s.

Discovery of the final t quark. The search for the t quark was made more difficult by the lack of understanding of the precise origin and pattern of quark masses. It is assumed in the standard model that the quarks acquire their masses as a result of interactions with the still undiscovered Higgs boson. Unfortunately there is a free parameter for each quark in this interaction, so within the standard model the quark masses are undetermined. Experimentalists in the 1980s and early 1990s could search for evidence of the t quark only at higher and higher energies, hoping that eventually they would find evidence for what has now proven to be a surprisingly high-mass final quark.

Types of collision experiments. Techniques for discovering new quarks involve two rather distinct types of particle colliders. The cleanest in terms of simplicity of interpretation collide electrons (e^-) and antimatter electrons (e^+, known as positrons). In the collision the $e^- e^+$ pair is destroyed and a photon (particle of light) or a Z_0 boson is created, and these in turn make a quark-antiquark pair ($q + \bar{q}$). This process is quite well understood, and one need only have adequate energy in the collision to make the new quarks. This was the procedure used at Stanford Linear Accelerator Center (SLAC) to discover the charm quark, and it is currently being exploited to study bottom quarks and their antiquarks. Unfortunately the t quark proved to be very heavy (over 180 times the proton mass); making two such objects ($t + \bar{t}$) was beyond the reach of the $e^- e^+$ machines used in the earlier top quark searches.

The second search approach is to collide pairs of strongly interacting particles, such as proton + antiproton. This type of experiment is much harder to interpret, since the proton itself is a complicated bound state of at least three quarks, held together by the exchange of gluons. Hitting two such objects together was described by Richard Feynman as like hitting two watches together at high velocity to see what comes flying out. The answer is that many light quarks, antiquarks, and gluons are made in the collision, and they emerge from the collisions in what are termed jets of strongly interacting particles, which mainly contain light quarks and antiquarks from the first generation. It is only very rarely (about once in 10^7 collisions in recent experiments at Fermilab) that these collisions produce the very heavy top quark. These very rare top-producing collisions mainly involve annihilation of light quark-antiquark pairs and gluon pairs in the colliding proton and antiproton, which produce a top quark and antiquark pair ($t + \bar{t}$).

Identification of t quarks. Once a top quark is made, identification is a difficult problem. A quark decays only into other, lighter types of quarks through the relatively slow weak interaction. All the experiments which have seen top quarks to date have used proton-antiproton collisions, and the members of the $t + \bar{t}$ pair made in the collision separate, undergo weak decays, and are identified by detection of the unusual products of the weak decay of a very massive particle.

One of the puzzles of the pattern of quark generations in the table is the relative probabilities of the various weak decays from one type of quark to another; these relative strengths of quark decays are known as the Cabibbo-Kobiyashi-Maskawa (CKM) matrix. In a weak decay a quark may turn into a lighter quark and emit a heavy, charged W boson, which in turn decays into a pair of fundamental spin-1/2 particles. For some reason these weak decays prefer to remain within a generation; thus a charm quark c (second generation) decays preferentially to an s quark (also second generation). The s quark, however, must decay to the u and d quarks

(first generation), because only they are lighter in mass. This preference for weak decays within generations is much stronger within the third generation; the weak decay $t \rightarrow b W^-$ is strongly preferred in t quark decays, which are entirely within the third generation. The b quark produced in this $t \rightarrow b W^-$ decay in turn decays weakly, preferentially to c (second generation, but lighter), which decays preferentially to s (also second generation), which may then decay weakly to u (first generation, but lighter). Thus after the t quark has been made, a long cascade of weak decays through different quark types follows, which would be essentially impossible to reconstruct against the much larger background of light quarks and antiquarks produced by the strong interaction in the same collision.

The strategy for identification of t quarks that has proven successful is the detection of high-energy leptons as a secondary product of weak decays. The initial t quark decays as $t \rightarrow b W^-$, and the heavy W^- boson then has a large probability of decaying into a charged lepton and neutrino, for example, $W^- \rightarrow e^- + \nu_e$. The resulting electrons with energies around 50 GeV are quite easy to identify experimentally, and the distribution of energies of e^- (or μ^- or τ^-) observed in this process can be used to infer the mass of the t quark that was the start of the decay chain. These experiments are complicated by the fact that the accompanying neutrino is not observed, and is instead identified as missing energy in the decay.

Fermilab experiments. To date, only two experimental collaborations have observed the top quark, the D0 and CDF groups. Both operate at the high-energy Tevatron antiproton beam at Fermilab (the energy of the colliding proton and antiproton available to make new particles is 1.8 TeV, about 2000 proton masses). A series of increasingly accurate measurements of t quark (and t antiquark) decays has been carried out since the initial observation at Fermilab in 1994, mostly using the lepton identification strategy discussed above. The current best estimate for the mass of the t quark, based on a joint study by the two Fermilab groups, is $m_t = 174.3 \pm 5.1$ GeV.

Implications of t quark properties. The most interesting application of t quark studies may be what they reveal about the single as yet undiscovered particle in the standard model, the Higgs boson. Relations between particle masses predicted by the standard model make it possible to infer the mass of the Higgs boson, given other quantities such as the masses of the t quark and the charged W bosons. At present these masses are not known accurately enough to give a precise Higgs boson mass estimate, although an encouragingly low value of about 100 GeV is suggested. This would mean that the Higgs boson is within reach of experiments planned for the near future. Details of the decay branching fractions and angular distributions of t quark decay products are also being studied, since any discrepancy with theoretical predictions could imply the existence of new particles beyond those of the standard model.

For background information *see* ELEMENTARY PARTICLE; HIGGS BOSON; LEPTON; NEUTRINO; PARTICLE ACCELERATOR; QUARKS; STANDARD MODEL in the McGraw-Hill Encyclopedia of Science & Technology.

Ted Barnes

Bibliography. I. J. R. Aitchison and A. J. G. Hey, *Gauge Theories in Particle Physics: A Practical Introduction*, Hilger, 1990; Particle Data Group, 2000 Review of Particle Properties, *Euro. Phys. J.*, C15:1–878 (see especially pp. 385–390), 2000; C. Quigg, *Gauge Theories of the Strong, Weak, and Electromagnetic Interactions*, Perseus, 1983.

Trace elements (soils)

The total production of fly ash (coal combustion by-products) in the United States was 57 million tons in 1998. The current use of fly ash as a component of cement, concrete blocks, and other construction materials accounts for only about 35% of production, with the remainder being disposed of in landfills or settling ponds. Possible changes in environmental regulations and depletion of landfill capacity are likely to increase the cost of disposal.

Since fly ash contains substantial amounts of plant nutrients (such as calcium, potassium, magnesium, manganese, molybdenum, copper, and zinc) and has the capacity to neutralize soil acidity, it can be used in agriculture and land reclamation as a soil amendment and source of nutrients. However, temporary boron toxicity to plants and excessive buildup of arsenic in soil amended with fly ash containing high concentrations of these elements may cause crop yield reduction and adverse environmental effects. In addition, high levels of molybdenum and manganese, induced phosphorus deficiencies, and very low nitrogen and potassium concentrations make fly ash an unbalanced source of plant nutrients.

Likewise, land use of biosolids (municipal sewage) is problematic because of nutrient imbalances and environmental concerns. Therefore, the concept of mixing fly ash with biosolids has been suggested to take advantage of the complementary properties of both by-products. For example, the alkalinity of the fly ash can neutralize the acidity of biosolids, immobilize trace metals present, and complement the greater nitrogen and phosphorus contents of biosolids. The adsorptive capacity of the biosolids may immobilize certain trace elements (such as arsenic and boron) present in fly ash, while fly ash may retard the production and leaching of nitrates (NO_3) and phosphates (PO_4) from biosolids. Since different trace elements are usually present in elevated concentrations in fly ash (such as arsenic) and biosolids (such as cadmium and zinc), the mixing process can reduce total concentrations of trace elements through dilution to levels below U.S. Environmental Protection Agency (EPA) limits (**Table 1**). Greenhouse and field experiments show that balanced, high-organic fertilizer and soil conditioners free from negative environmental side effects can be formulated from fly ash and biosolids.

Concentrations in soils and plants. Since arsenic and boron are the two elements having the greatest potential to create problems when mixtures of ash and biosolids are land-applied and since cadmium and lead are the most common metallic contaminants, the discussion will focus on these four elements.

Soils can become contaminated with arsenic from various sources, including mining and smelting emissions, pesticides, and contaminated irrigation water. Fly ashes generally represent a low level of arsenic contamination risk. Even though fairly low concentrations of arsenic are considered toxic, arsenic does not pose a high risk of phytotoxicity for most agricultural crops. Arsenic and lead in soils behave similarly in that plants do not readily absorb these elements or transport them into the edible tissues. Even when the soil is sufficiently contaminated to cause arsenic toxicity, plants usually show little or no increase in arsenic. For example, studies conducted in the United Kingdom showed that although soils contained as high as 890 mg arsenic/kg, all vegetables grown in the soil met the British statutory limit of 1 mg arsenic/kg of fresh vegetable weight. Currently, there are no statutory guidelines for arsenic levels in food crops in the United States.

The rationale for using mixtures of fly ash and biosolids on arable land is to improve soil properties and provide plants with a slow-release source of nutrients with minimal environmental risk. Such mixed applications do not cause an excessive plant uptake of cadmium and lead from treated soil, even

TABLE 1. Concentrations of trace elements (mg/kg, dry wt) in fly ashes, biosolids, and mixtures of fly ashes and biosolids from sources in the southeastern United States[*]

	As	Cd	Cr	Cu	Hg	Mo	Ni	Pb	Se	Zn
Fly ash 1 (FA1)	58	0.4	43	53	0.3	17	24	20	8	214
Fly ash 2 (FA2)	112	0.5	58	52	0.5	18	26	59	28	102
Biosolids	3.5	2.0	42	197	nd[†]	7	50	36	1	814
FA1 + biosolids	40	0.9	43	101	nd	14	33	25	6	414
FA2 + biosolids	76	1.0	53	100	nd	14	34	51	19	339
EPA limits	75	85	3000	4300	57	75	420	840	100	7500

[*]From *Utilization of Coal Combustion By-products in Agriculture and Land Reclamation*, EPRI (Palo Alto), Southern Company, Birmingham, AL, TR-112746, 1999.
[†]nd = no data.

TABLE 2. Concentrations of trace elements (mg/kg, dry wt) in soil treated with mixtures of fly ashes and biosolids*

Treatment	Depth, cm	As	Cd	Cu	Cr	Mo	Ni	Pb	Se	Zn
Control	0–20	1.38±0.26	0.03±0.01	6.8±2.1	16.7±4.3	0.09±0.08	7.0±1.7	26.7±22.3	0.71±0.14	27.3±4.5
	21–50	1.29±0.02	0.01±0.01	9.0±1.8	20.3±2.6	0.10±0.06	8.5±1.4	12.1±1.6	0.50±0.09	24.3±3.9
FA1 +	0–20	2.95±1.11	0.11±0.04	12.8±4.3	16.9±6.1	0.57±0.30	10.7±3.6	16.3±3.8	0.86±0.27	48.8±42.7
biosolids	21–50	1.38±0.36	0.01±0.01	10.1±4.8	29.2±17.0	0.12±0.06	10.4±4.4	12.5±1.9	0.56±0.09	44.3±9.9
FA2 +	0–20	3.26±0.67	0.08±0.01	11.0±2.8	22.3±7.3	0.27±0.09	9.8±3.2	20.5±6.9	1.18±0.23	37.1±11.5
biosolids	21–50	1.68±0.13	0.01±0.01	11.3±1.4	23.7±1.0	0.10±0.03	9.7±4.2	13.7±2.4	0.53±0.10	28.5±10.0

*From *Utilization of Coal Combustion By-products in Agriculture and Land Reclamation*, EPRI (Pao Alto), Southern Company, Birmingham, AL, TR-112746, 1999.

in the case of the radish, a root crop known for its tendency to accumulate trace elements. Even though concentrations of boron and arsenic increased in some plant species tested (such as turnips), especially at higher application rates, no symptoms of phytotoxicity were observed. Mixing fly ashes with biosolids was an effective method of reducing boron and arsenic availability in soils. The reduction in element mobility was probably a result of arsenic and boron binding to the organic matter in the biosolids. Arsenic concentrations in wheat plants grown in contaminated soils depend strongly on the source of arsenic added. Arsenic levels in plants grown on soil treated with an ash-sludge mixture were substantially lower than in the plants treated with sodium-arsenite or fly ash alone, confirming that mixing fly ash with biosolids mitigates trace-element uptake by plants from by-product–amended soils. Although elevated levels of arsenic, cadmium, and lead were found in radish, turnip, and wheat plants grown on soils receiving high amounts of ash-biosolid mixes, the concentrations of these elements were still below the levels that may cause environmental or health concern. Consequently, arsenic, cadmium, and lead do not pose a significant risk to the food chain when fly ash-biosolid mixtures are applied on agricultural land.

Long-term field experiments with corn, wheat, and sorghum in which fly ash–biosolid mixtures were applied at 90–110 metric tons/hectare (dry wt) showed that the concentrations of arsenic, cadmium, copper, molybdenum, and zinc increased slightly in the plow layer (0–20 cm; 0–8 in.) but not in the subsoil (21–59 cm; 8–23 in.) as a result of ash-biosolids treatment (**Table 2**). It was concluded that no measurable movement of trace metals occurred from plow layers to deeper layers of ash-biosolids–treated soils. Consequently, risk of metal leaching into the ground water from the amended soil seems negligible.

Concentrations of arsenic, copper, molybdenum, and selenium in corn plants from the ash-biosolids treatments were somewhat higher than in the control plants, but were far below phytotoxicity thresholds and did not create a food chain contamination problem. Concentrations of trace elements in mature corn grain were even lower than those of young leaves and were similar to background levels reported for grain from productive fields of the United States (**Table 3**). Consequently, relatively high rates of fly ash-biosolids mixtures (90–110 metric tons/ha) do not affect the trace-element composition of mature corn plants and do not appear to contaminate the food chain or cause plant toxicity.

Environmental risk assessment related to arsenic. The health hazard related to arsenic (the most problematic element in ash-biosolid mixtures) in soils has been quantified in an environmental risk assessment according to the EPA path for direct ingestion of arsenic-rich soil by children. Children of ages 1 to 6 are regarded as the most highly exposed individuals (HEI). If a certain level of arsenic in soil is proven

TABLE 3. Concentrations of trace elements (mg/kg, dry wt) in corn*

Treatment	As	Cd	Cr	Cu	Mo	Ni	Pb	Se	Zn
Young leaves									
Control	0.1	0.05	0.4	10	0.1	0.9	0.4	<0.01	20
FA1 + biosolids	0.2	0.04	0.3	12	0.4	0.5	0.4	0.2	27
FA2 + biosolids	0.3	0.05	0.3	13	1.1	0.4	0.2	0.3	34
Mature grain									
Control	0.02	0.002	<0.02	1.7	0.3	0.9	<0.01	<0.02	20
FA1 + biosolids	0.02	0.002	<0.02	1.5	0.2	0.5	<0.01	<0.02	23
FA2 + biosolids	0.01	0.001	<0.02	1.5	0.3	0.4	<0.01	<0.02	26
U.S. productive fields	nd[†]	0.004	nd	1.9	0.1	0.2	0.01	0.01	25

*From *Utilization of Coal Combustion By-products in Agriculture and Land Reclamation*, EPRI (Palo Alto), Southern Company, Birmingham, AL, TR-112746, 1999.
[†]nd = no data.

to be safe for the HEI, the soil is also safe to other target organisms.

Such an environmental risk assessment shows that the allowable soil arsenic loading would be 40 kg arsenic/ha, which is far below the levels reached at application rates of 90–110 metric tons/ha of the mixtures (Table 2).

For background information *see* ARSENIC; BORON; CADMIUM; COAL; LEAD; RISK ANALYSIS; SEWAGE SOLIDS; SOIL; SOIL CHEMISTRY in the McGraw-Hill Encyclopedia of Science & Technology.

M. E. Sumner; S. Dudka

Bibliography. L. R. Chaney and J. A. Ryan, *Risk Based Standards for Arsenic, Lead, and Cadmium in Urban Soils*, Dechema, Frankfurt, Germany, 1994; Environmental Protection Agency, Sec. 503, Vol. 58, No. 32, Washington, DC, 1993; R. F. Korcak, Utilization of coal combustion by-products in agriculture and horticulture, in D. L. Karlen et al. (eds.), Agricultural Utilization of Urban and Industrial By-products, *ASA Spec. Publ.*, no. 58, ASA, CSSA, SSSA, Madison, WI, 1995; M. E. Sumner and S. Dudka, Fly ash-borne arsenic in the soil-plant system, in K. S. Sajwan et al. (eds.), *Biogeochemistry of Trace Elements in Coal and Coal Combustion By-products*, pp. 269–278, Kluwer Academic, New York, 1999; *Utilization of Coal Combustion By-products in Agriculture and Land Reclamation*, EPRI (Palo Alto), Southern Company, Birmingham, AL, TR-112746, 1999.

Trace-isotope analysis

Much can be learned from studying the concentrations of the ubiquitous long-lived radioactive isotopes. W. Libby and coworkers first demonstrated in 1949 that trace analysis of carbon-14 (^{14}C; half-life = 5730 years, isotopic abundance ~1×10^{-12}) can be used for archeological dating. Since then, two well-established methods, low-level counting and accelerator mass spectrometry, have been used to analyze many different trace isotopes and to extract valuable information encoded in the production, transport, and decay processes of these isotopes. The impact of ultrasensitive trace-isotope analysis has reached a wide range of scientific and technological fields.

A new method, atom trap trace analysis, was recently developed by Z.-T. Lu and coworkers and used to analyze two rare isotopes, krypton-81 (^{81}Kr) and krypton-85 (^{85}Kr), with isotopic abundances at about the parts-per-trillion level. This method promises to enhance the capabilities and expand the applications of ultrasensitive trace-isotope analysis.

Rare krypton isotopes. Krypton gas in the Earth's atmosphere has six stable isotopes with abundances ranging from 0.35 to 57%, and two long-lived radioactive isotopes, ^{81}Kr (half-life = 2.3×10^5 years, isotopic abundance ~6×10^{-13}) and ^{85}Kr (half-life = 10.8 years, isotopic abundance ~1×10^{-11}). There are about 20,000 ^{81}Kr atoms and 300,000 ^{85}Kr atoms in 1 liter STP (standard temperature and pressure) of air. As a result of absorbing and trapping air, 1 kilo-gram of surface water or ice contains roughly 1×10^3 ^{81}Kr atoms.

Krypton-81 is produced in the upper atmosphere by cosmic-ray-induced spallation and neutron activation of stable krypton isotopes. As a result of its long lifetime in the atmosphere, ^{81}Kr is well mixed and distributed over the Earth with a homogeneous isotopic abundance. Human activities with nuclear fission have had a negligible effect on the ^{81}Kr concentration, largely because the stable bromine-81 (^{81}Br) shields ^{81}Kr from the neutron-rich isotopes that are produced in nuclear fission. These physical and chemical properties make ^{81}Kr an ideal tracer for dating ice and ground water in the range from 100,000 to 1,000,000 years, which is too old to be dated with ^{14}C.

Krypton-85 is a fission product of uranium-235 (^{235}U) and plutonium-239 (^{239}Pu). In fact, most of the ^{85}Kr found in the environment was released by nuclear-fuel reprocessing plants. As a result of this human activity, its abundance in the atmosphere has increased by 6 orders of magnitude since the 1950s. It has been used as a general tracer to study air and ocean currents, date shallow ground water, and monitor nuclear-fuel reprocessing activities.

Existing techniques. Low-level counting (LLC), which identifies radioactive isotopes by detecting their nuclear decays, is commonly used to analyze ^{85}Kr. LLC is often carried out in a specially designed underground laboratory in order to avoid background due to cosmic-rays and the radioactivity present in common materials. This technique was also used in the first observation of atmospheric ^{81}Kr, but this is no longer possible because of the overwhelming decay activity of ^{85}Kr in the present-day environment.

Accelerator mass spectrometry (AMS) counts atoms instead of decays. Its efficiency and speed are not fundamentally limited by the long half-lives of isotopes, nor is it affected by radioactive backgrounds in the environment or in samples. W. Kutschera and coworkers recently succeeded in counting ^{81}Kr in natural samples using a high-energy cyclotron and, for the first time, realized ^{81}Kr-dating of ancient ground water.

Laser-based techniques, such as resonance ionization spectrometry (RIS) and photon burst mass spectrometry, have the potential of being simple and efficient. Using RIS, G. Hurst and coworkers counted ^{81}Kr atoms in an enriched sample with over 50% efficiency. So far, neither of these two laser-based techniques has demonstrated the isotopic selectivity required to count ^{81}Kr or ^{85}Kr in a natural sample.

Atom trap trace analysis (ATTA). ATTA is a new laser-based atom-counting method in which individual atoms of a chosen isotope are captured and detected with a laser trap. Trapping krypton atoms in the $5s[^3/_2]_2$ metastable level (lifetime approximately 40 s) is accomplished by repeatedly exciting the atoms with laser beams whose frequency is tuned to the resonance of the $5s[^3/_2]_2$–$5p[^5/_2]_3$ transition. When on resonance, a laser beam of moderate power

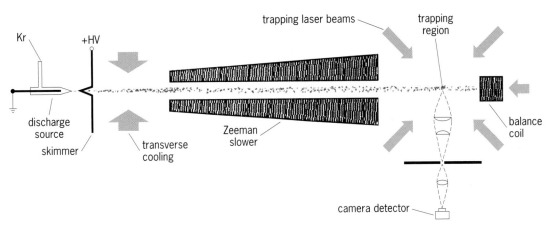

trapping laser beams

trapping region

Kr

+HV

discharge source

skimmer

transverse cooling

Zeeman slower

balance coil

camera detector

Fig. 1. Total length of the ATTA apparatus is about 2.5 m (8 ft).

($\sim$10 mW/cm^2) can exert an impressive force on the atoms. In average, a single cycle of photon absorption and emission changes the velocity of a krypton atom by 6 mm/s due to the recoil of the atom. The force, generated by fast ($\sim$10^7 s^{-1}) repetition of such recoils, can decelerate a krypton atom at $\sim$6 $\times$ 10^4 m/s^2. In the analysis, a krypton gas sample is injected into the system through a discharge region, where a fraction of the atoms are excited into the 5s[$^3/_2$]$_2$ level via electron-impact excitation. The thermal atoms are then transversely collimated, decelerated, and captured into a trap with laser beams (**Fig. 1**). A trapped atom scatters photons from the laser beams at a rate of approximately 10^7 s^{-1} and appears as a bright dot in the center of the vacuum chamber. A sensitive photon detector is used to measure the fluorescence and count the trapped atoms (**Fig. 2**).

This process is isotopically selective because atoms of different isotopes resonate at different frequencies. When the laser frequency is tuned to the resonance of a particular isotope, only atoms of this isotope are trapped and detected. Previous efforts to develop a laser-based technique have encountered serious problems as a result of contamination from

nearby abundant isotopes or isobars. ATTA is immune from this contamination for several reasons: fluorescence is collected only in a small region (diameter 0.5 mm) around the trap center so that background scattered light can be easily filtered out; a trapped atom is cooled to a speed below 1 m/s so that its laser-induced fluorescence is virtually free of spreading due to the Doppler effect; the long observation time allows the atom to be unambiguously identified; and trapping allows the temporal separation of capture and detection so that both capture efficiency and detection sensitivity can be optimized. ATTA has been used to analyze both ^{81}Kr and ^{85}Kr in an atmospheric krypton sample. The recorded atom counts of the two rare isotopes contain no contamination from other isotopes, elements, or molecules. Therefore, ATTA can tolerate impure gas samples, and does not require a special operation environment.

Potential improvements. The current ATTA setup captures approximately sixty ^{85}Kr atoms or five ^{81}Kr atoms per hour from an atmospheric krypton sample. The capture efficiencies can be calibrated with enriched samples of known isotopic abundance to correct for any isotope-dependent effects and to measure isotopic ratios in unknown samples. For example, in ^{81}Kr dating, a known amount of ^{85}Kr can be mixed into the sample, thus allowing the ^{81}Kr abundance to be extracted by measuring the ratio ^{81}Kr/^{85}Kr.

The system has achieved an overall efficiency of 1 $\times$ 10^{-7}. In other words, 10 million ^{81}Kr (or ^{85}Kr) atoms need to be injected in order to detect one ^{81}Kr (or ^{85}Kr) atom in the trap. Use of this system to measure the abundance of ^{85}Kr to within 10% would require 2 hours and a krypton sample of 3 cm^3 STP, while measurement of ^{81}Kr to within 10% would require 2 days and a sample of 60 cm^3 STP. Improvements such as using a colder and more intense source of metastable krypton atoms, and recirculating the krypton gas, are under investigation. Instead of using a discharge, the metastable level of krypton can also be populated via photon excitations. With a suitable laser or lamp, the efficiency of

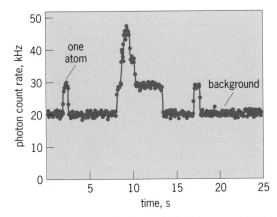

Fig. 2. The abrupt signal changes in the fluorescence level indicate the arrival or departure of individually trapped krypton atoms.

exciting krypton atoms to the metastable level could be much higher than the $\sim 10^{-4}$ currently achieved with a discharge.

Applications. Trace-isotope analysis has become an essential tool in modern science. In basic research, isotope tracers are used to detect solar neutrinos, study cosmic rays, and search for exotic particles and rare radioactive decays. In environmental science, isotope tracers are used to track atmospheric, oceanic, and ground-water currents and to help understand Earth's climate. In archeology and geology, various long-lived isotopes are used to determine ages and to help understand the causality of historical events. Isotope tracing is also widely used in biology and medicine. Furthermore, fission isotopes are monitored to assess the contamination of the environment either by the regular operation of a nuclear facility or by a nuclear accident. It is also a means to verify compliance with nuclear nonproliferation treaties.

ATTA can be applied to many different isotopes for the aforementioned applications. Laser trapping is well established on alkali, alkali earth, and noble gas elements. Trapping of other elements is generally more difficult due to both the complexity of their ground-level structures and the lack of suitable ultraviolet (UV) lasers. Some of these problems could be overcome with future advances in UV laser technology.

For background information *see* ACCELERATOR MASS SPECTROMETRY; DATING METHODS; HALF-LIFE; ISOTOPE; ISOTOPE DILUTION TECHNIQUES; ISOTOPE SEPARATION; KRYPTON; LASER COOLING; LOW-LEVEL COUNTING; MASS SPECTROMETRY; PARTICLE TRAP; RADIOACTIVE TRACER; RADIOACTIVITY; RADIOCARBON DATING; RADIOISOTOPE; RESONANCE IONIZATION SPECTROSCOPY; TRACE ANALYSIS in the McGraw-Hill Encyclopedia of Science & Technology.
Zheng Tian Lu

Bibliography. C. Y. Chen et al., Ultrasensitive isotope trace analyses with a magneto-optical trap, *Science*, 286:1139–1141, 1999; S. Chu, The manipulation of neutral particles, *Rev. Mod. Phys.*, 70:685–706, 1998; P. Collon et al., Kr-81 in the Great Artesian Basin, Australia: A new method for dating very old groundwater, *Earth Planet. Sci. Lett.*, 182: 103–113, 2000.

Trait reconstruction (evolutionary biology)

For any collection of species, given information on their attributes and a phylogeny that describes their shared hierarchy of descent, it is possible to reconstruct the characteristics of the ancestors of the species. This is an intriguing idea that offers the possibility of glimpsing the past, discovering how traits evolve, and understanding their function. Reconstructions of the ancestral states of organisms are increasingly used to investigate ancient features of life on Earth and to test ecological and evolution-

ary hypotheses. They are also being used to reconstruct proteins and genes that existed millions of years ago.

Reconstructed ancestral states complement traditional paleontological approaches to studying the past and can be used to investigate traits that do not fossilize, such as behavior or physiology. New statistical approaches to reconstructing ancestral states incorporate explicit models of trait evolution with the capability to detect historical directional trends. This makes it possible to infer ancestral features that lie outside the range of features observed in the contemporary data.

Models of trait evolution. Species' attributes can be broadly categorized into discrete traits and continuous traits. Discrete traits adopt a finite and typically small number of states and may or may not be ordered. The presence or absence of some feature is a binary discrete trait. Living solitarily, in a relationship with one other, or in a group is an example of an ordered discrete trait. Gene sequence or amino acid sequence data are additional examples of discrete traits. Other traits such as wing length, geographic range size, body size, brain volume, age at maturation, running speed, and body temperature, which constitute a continuously varying feature of the organism or its environment, are known as continuous traits.

Parsimony methods. Reconstructing the ancestral character states of organisms requires making assumptions about how traits evolve. Historically, researchers have used the method of maximum parsimony to reconstruct ancestral states. Parsimony methods choose the set of reconstructed ancestral states that minimizes the total amount of evolutionary change required to explain the distribution of traits at the tips of the phylogenetic tree. The model of trait evolution that is implicit to the parsimony method is that evolutionary change is rare.

Figure 1 shows a phylogenetic tree of three species, each of which can adopt a value of 0 or 1 for some discrete trait. Four reconstructions of the two ancestral nodes of the tree are possible, but the reconstruction that places a 1 at the root and a 1 at the common ancestor to species 1 and 2 minimizes the total amount of evolution (the number of changes) that is implied. For larger phylogenetic trees, the most parsimonious solution can be difficult to find because the number of possible solutions is vast. Therefore, a number of rules of thumb and modifications to the basic parsimony approach are used to sort through the possible reconstructions.

Continuously variable traits can also be reconstructed by parsimony. However, the criterion that is minimized is usually the square of the amount of inferred change along the branches of the phylogenetic tree, hence the term squared-change parsimony. Other criteria, such as minimizing the absolute value of change, are also possible. In each case, computer search procedures are employed that seek the

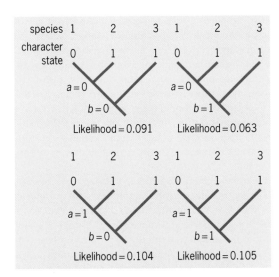

Fig. 1. All possible ancestral states on a three-species phylogenetic tree. The maximum parsimony solution corresponds to the maximum likelihood solution (tree in lower right) and requires one event of evolution.

set of values at the ancestral nodes that minimizes the chosen criterion.

Parsimony approaches work well when change is rare or when branches of the phylogenetic tree are short. (Branch lengths may be measured in units of time, or in, for example, units of numbers of gene-sequence changes.) Under these circumstances, the probability of a character changing in a branch of the phylogenetic tree is not strongly related to the branch's length. However, parsimony methods can perform poorly when rates of character evolution are high and the phylogenetic tree includes some long branches. Parsimony methods also fail to provide estimates of the uncertainty associated with a reconstructed ancestral state, although an error rate of sorts may be possible to calculate in principle. Estimates of uncertainty can be important in demonstrating which alternative ancestral states can be safely ruled out.

Markov-transition process. Statistical approaches to trait evolution provide an alternative to parsimony methods. They do not presume that change is rare or common, but estimate the rate of character evolution from the observed data. Lengths of the branches of the phylogenetic tree are taken into account when estimating rates of change. Statistical approaches also routinely provide a measure of the uncertainty of a reconstructed character.

The Markov-transition process is the statistical model most widely used to describe the evolution of discrete traits. Examples of its application can be found in studies of protein evolution, sexual selection, and habitat and diet preferences. The Markov approach estimates the rates at which a discrete character makes transitions among its possible states as it evolves. If the trait is binary, it is necessary to estimate the rate at which it makes evolutionary transitions from state 0 to state 1 and from state 1 to state

0. These are sometimes referred to as forward and backward transitions, respectively. If a trait can adopt three states, then three forward and three backward transition rates must be estimated, and so on.

The transition rates are sufficient to calculate the most probable states at ancestral nodes of the phylogenetic tree. Statistical approaches choose the most probable ancestral states by seeking maximum likelihood solutions. The maximum likelihood solution in this context is proportional to the probability of observing the inferred ancestral state or states given the phylogeny and the model of evolution presumed to characterize the trait. Applied to the data of Fig. 1, the Markov-transition-rate model returns a likelihood associated with each of the four possible assignments of ancestral states to the two nodes. Here the maximum likelihood solution also corresponds to the most parsimonious solution, although this is not always the case.

Constant-variance random-walk model. For continuously varying traits, the statistical approach to reconstructing ancestral states typically makes use of the constant-variance random-walk model (sometimes called Brownian motion). In the conventional random-walk model, traits evolve each instant of time (*dt*) with a mean change of zero and a constant variance. Time may be chronological or some other unit of divergence such as genetic distance. The expected amount of long-term evolutionary change for a species is zero because the "walk" is presumed to move randomly in either a positive or negative direction. However, random-walk models show a surprising amount of deviation from the expected value of zero, such that some random walks have diverged greatly in a positive or negative direction.

The standard random-walk model cannot detect any directional trends of trait evolution along the branches of a phylogenetic tree. Historical trends such as a phyletic increase in size (for example, Cope's law) will be masked. Consequently, this model always estimates the ancestral state at the root of the tree as falling somewhere within the range of observed values in the species data. Ancestral states obtained from squared-change parsimony are equivalent to those obtained from the constant-variance random-walk model.

Directional models. Recent directional statistical models for continuous traits can detect historical trends of trait evolution. If a directional trend exists, species that have diverged more from the root will tend also to have changed more in a given direction; for example, they will be larger or mature earlier. Under these circumstances, the directional model will reconstruct ancestral states more accurately and, importantly, it can use the trend to reconstruct the character state at the root of the tree to lie outside the range of observed values in the data.

Applications. Parsimony and statistical methods can both be applied to determine the evolution of discrete and continuous traits. However, the two methods often yield different results.

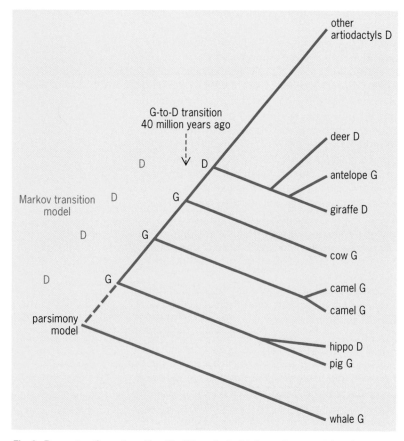

Fig. 2. Reconstructions of position 38 of the artiodactyl ribonuclease protein using parsimony and the statistical Markov-transition model (G is the amino acid glycine, D is aspartic acid). Maximum parsimony reconstructions find glycine as the most probable ancestral state, with a G-to-D transition occurring approximately 40 million years ago. However, maximum likelihood analyses using the Markov-transition model reconstruct aspartic acid as the most probable ancestral state.

Discrete traits: ribonuclease evolution. Maximum parsimony reconstructions of ancient artiodactyl (grazing mammals) ribonuclease—an enzyme involved in foregut digestion—assign glycine (G) as the ancestral state at a key amino acid position, with a transition to aspartic acid (D) about 40 million years ago (**Fig. 2**). This transition may correspond to the adoption of true ruminant digestion. However, maximum likelihood reanalyses of the same data using the Markov-transition model reveal high rates of character evolution at this site: the estimated half-life for replacement of glycine by aspartic acid is approximately 42 million years and for aspartic acid by glycine is 68 million years, over a tree length spanning 450 million years. The relatively short half-lives arise because the phylogenetic distribution of traits implies that more than one amino acid substitution has occurred among these artiodactyls. Likelihood methods reconstruct the most probable ancestral state as aspartic acid, opposite to the inference derived from parsimony, and imply at least four, as opposed to three, evolutionary transitions on the tree (Fig. 2).

The difference between the parsimony and likelihood reconstructions arises because likelihood analysis detects that at least two of the four D-to-

G replacements that its reconstructions minimally imply, occur in long branches (those leading to camels and cows). These changes are not improbable owing to the high rates of evolution. Parsimony, by definition, prefers the solution of three events to four, even though these events are reconstructed to occur in comparatively short branches (short amount of time), and require one reversal (a D-to-G) transition. Maximum likelihood analysis considers these changes relatively improbable because this method seeks the most probable explanation of the observed data (including the tree and branch lengths), not necessarily the solution with the fewest events. The Markov-transition model in a likelihood framework has an advantage in that it takes into account the length of branches in the tree and can adjust to low or high rates of change. The ribonucleases may not be an isolated case: parallel and convergent evolution (which may imply a relatively high number of trait transitions) at the molecular level may be more common than once believed, and examples of high rates of morphological trait evolution are easy to find.

Continuous traits: hominid brain-size evolution. The hominids are those species that branched perhaps 5 million years ago from the common ancestor of modern chimpanzees and humans. Their phylogenetic relationships are uncertain, but one possible phylogeny is shown in **Fig. 3**. The lengths of the branches are in units of millions of years. The species *Australopithecus afarensis* (commonly known as Lucy) is placed arbitrarily close (0.001 million years) to the root of the tree, dated about 3.1 million years ago.

The tips of the tree refer, except in the case of modern humans (*Homo sapiens*), to fossil species. In several cases a fossil species is found over a considerable proportion of the time spanned by the tree, but this tree makes use of only the latest (most recent) data point. Because all but one of the hominid species is extinct, the total path lengths from the root are different. This makes it possible to seek relationships between cranial capacity and time. As the figure shows, hominid brain size has generally increased with time. This may mean that conventional approaches to reconstructing ancestral states that do not allow for directional trends, such as squared-change parsimony or conventional random walks, would reconstruct the probable ancestral states poorly.

Figure 3 compares the measured brain sizes of fossil species with the reconstructed brain sizes of ancestral species derived from squared-change parsimony (equivalent to a neutral random walk) and from a directional random-walk model. Squared-change parsimony consistently reconstructs ancestral cranial capacities to be larger than one of the descendants. These predictions seem particularly implausible, implying at least six reversals of the trend toward increasing brain size. The directional model by comparison always predicts that the ancestral species has a smaller brain size than its descendant species.

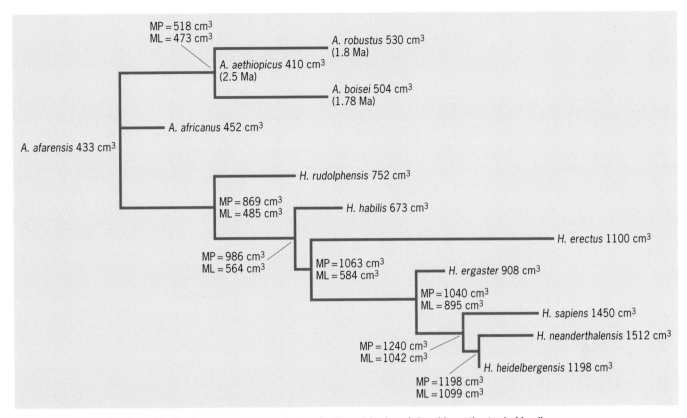

Fig. 3. A phylogeny of the hominids showing the measured brain sizes (endocranial volume is in cubic centimeters) of fossil species and the reconstructed brain sizes of ancestral species. Reconstructions labeled MP are derived from maximum parsimony; those labeled ML are derived from the maximum likelihood statistical model, which incorporates a directional trend of trait evolution.

Challenging conventional beliefs. Using these models, ancestral states of organisms can be theoretically reconstructed, giving a picture of what organisms were like millions of years ago. Parsimony methods, although perhaps still the more widely used, are gradually being replaced by statistical approaches that can incorporate sophisticated models of trait evolution. Already these methods are being used to challenge many conventional beliefs and to reconstruct startling new pictures of the past. For example, reconstructed gene sequences suggest that the common ancestor of life was not adapted to the hot conditions widely believed to prevail on the early Earth; reconstructed dates for the adaptive radiations of birds and mammals suggest that they may have begun in earnest long before the extinction of the dinosaurs; and recent reconstructions of protein sequences suggest that mammals as diverse as elephants, tree shrews, and insectivores may all share a common African ancestor.

For background information *see* ANIMAL EVOLUTION; AUSTRALOPITHECUS; EARLY MODERN HUMANS; FOSSIL; FOSSIL HUMAN; PHYLOGENY; PHYSICAL ANTHROPOLOGY; PROTEINS, EVOLUTION OF in the McGraw-Hill Encyclopedia of Science & Technology.

Mark Pagel

Bibliography. P. L. Forey et al., *Cladistics: A Practical Course in Systematics*, Oxford University Press, 1992; W. Hennig, *Phylogenetic Systematics*, University of Illinois Press, Urbana, 1996; D. R. Maddison, Phylogenetic methods for inferring the evolutonary history and process of change in discreetly valued characters, *Annu. Rev. Entomol.*, 39:267–292, 1994; W. P. Maddison, Calculating the probability distributions of ancestral states reconstructed by parsimony on phylogenetic trees, *Sys. Biol.*, 44:474–481, 1995; W. P. Maddison, A method for testing the correlated evolution of two binary characters: Are gains and losses concentrated on certain branches of a phylogenetic tree, *Evolution*, 44:539–557, 1990; W. P. Maddison, M. J. Donoghue, and D. R. Maddison, Outgroup analysis and parsimony, *Sys. Zool.*, 33:83–103, 1984; M. Pagel, Inferring the historical patterns of biological evolution, *Nature*, 401:877–884, 1999; M. Pagel, The maximum likelihood approach to reconstructing ancestral character states of discrete characters on phylogenies, *Sys. Biol.*, 48:612–622, 1999; D. Schluter et al., Likelihood of ancestor states in adaptive radiation, *Evolution*, 51:1699–1711, 1997.

Tropical ecology

Tropical forests are among the most productive, most diverse, and oldest terrestrial ecosystems and have long captured the imagination of naturalists and ecologists. Tropical forests are typically located between

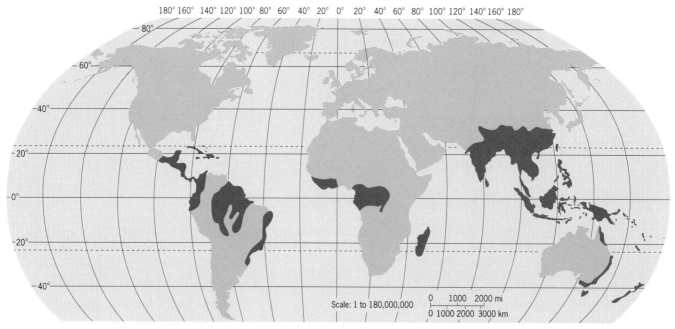

Global distribution of tropical forest ecosystems. (*Adapted from W. P. Cunningham and B. W. Saigo, Environmental Science: A Global Concern, 6th ed., McGraw-Hill, 2001*)

30°N and 30°S latitude in the continents of South America, Africa, Asia, and a bit farther south in eastern Australia and New Zealand (see **illus.**). Although tropical forests currently occupy only 7% of the Earth's land surface, they are thought to contain more than two-thirds of all higher-plant biomass and more than half of all plant and animal species.

Types of tropical forests. Because of their proximity to the Equator, most tropical forests experience little seasonal variation in day length or temperature. However, tropical forests vary in whether or not they experience distinct wet and dry seasons. They are categorized as tropical seasonal forests, which do experience wet and dry seasons, and tropical moist forests, which do not.

Tropical moist forests are characterized by abundant rainfall year-round. Cloud forests are a special type of tropical moist forest that occurs at higher elevations, with cooler temperatures and frequent fog and mist. Tropical rainforests are another type of tropical moist forest, and occur at low elevations, with warm temperatures and over 200 cm of annual rainfall spread throughout the entire year. The soil in tropical rainforests tends to be thin, acidic, and nutrient-poor. The lack of nutrients in the soil is due to the extraordinarily high productivity of the plant community. Decomposition and recycling of organic material are so rapid that the nutrients released from a fallen leaf are almost immediately incorporated back into living plants. As a result, roughly 90% of the nutrients in the system are contained in the bodies of living organisms.

In contrast to tropical moist forests and tropical rainforests, tropical seasonal forests experience distinct wet and dry seasons.

Deforestation. Burgeoning human populations in tropical regions rely increasingly on tropical forest lands for timber and food production. When forest is cleared for farms and cattle grazing, the distinction between forested and cleared land can be detected from space, and the areal extent of deforestation is monitored from *Landsat Thematic Mapper* satellites. According to the United Nations Food and Agriculture Organization (FAO), approximately 63 million hectares of tropical forest—nearly the size of Texas, and larger than France—were cleared between 1990 and 1995, corresponding to an annual deforestation rate of approximately 12.6 million hectares or 0.7% per year. If this rate of deforestation continues, more than half of all remaining tropical forests will be destroyed by 2100.

Tropical seasonal forests have been especially vulnerable to clearing by burning. In addition, these forests tend to have better soils and fewer insects and pathogens than tropical moist forests, making them more attractive for human settlement. The rate of destruction of these seasonal forests is generally higher than that of moist forests. In some areas the devastation is almost complete: less than 1% of seasonal tropical forest of the Pacific coast of Central America or the Atlantic coast of South America remain undisturbed.

The environmental impacts of deforestation can be extensive, particularly in areas with high rainfall where deforestation can result in massive soil erosion and siltation of nearby streams. If the cleared area is large, sources of seeds and other organisms are so remote that little recolonization occurs. When tropical forest is removed for agriculture, the thin soil usually cannot support sustained harvests, which means that farmers have to move on and cut new areas of forest after only a few years.

Although detecting the distinction between forested and deforested land from space is relatively easy, tropical forests are also exploited and damaged in ways that cannot be detected remotely via satellites. For example, selective logging operations reduce canopy cover and damage unharvested trees, but the damage may be difficult to detect from a satellite. As a result, even the most dramatic statistics on deforestation likely underestimate the true magnitude of degradation to forests.

Fragmentation. Critical to the future of tropical forests is not only the amount of deforestation but also the pattern in which deforestation occurs. Deforestation may leave behind either large tracts of undisturbed forest or many small patches of forest surrounded by farms, grasslands, and regenerating forest areas. The similarity between forest patches surrounded by a "sea" of disturbed habitat and true oceanic islands has led ecologists to predict that the size of patches, like that of islands, will be positively correlated with the number of species surviving in those remnant patches. According to the theory of island biogeography, small patches of forest, like small oceanic islands, will have higher extinction and lower colonization rates and are therefore predicted to lose species until a new equilibrium is attained. Many studies have examined whether forest patch area is related to the number of resident species. For birds and butterflies, small patches do tend to harbor fewer species, but for tree species the pattern is much less clear. One reason for the lack of pattern for trees may be that in many species individual trees can live for centuries. Thus, there may be a long lag between the time at which the species is doomed to extinction and the time when the last individual tree of that species disappears from a patch of forest.

In addition to having a small area, small patches have a relatively high perimeter-to-area ratio and are therefore more susceptible to what ecologists call edge effects. Forest edges that abut a disturbance generally experience increased light penetration, increased temperature, decreased moisture, and higher exposure to wind. Because many tropical forest species are adapted to the stable conditions that occur within a closed canopy forest, these species may suffer increased stress and mortality if they are close to the edge of a forest patch. In fact, increased tree mortality has been detected up to 300 m into forest patches as a result of proximity to an edge, presumably due to increased exposure to wind. Thus, the effects of deforestation can continue well after the deforestation event because the patches of remnant forest that manage to avoid deforestation often unravel and degrade gradually due to edge effects.

Other forms of disturbance. Disturbance has long been recognized as an important natural feature of tropical forests. For example, tree falls, which open up gaps in the otherwise closed forest canopy, have

been identified as one factor that helps promote the incredible diversity of tropical life. Mature tropical canopy tree species can live from one to several centuries. Although young trees may germinate beneath the canopy, they generally remain small saplings in the shady understory. Tree-fall disturbances, ranging in size from a single tree crown to many hectares, open the forest canopy and allow young saplings of sun-loving species to begin rapid growth. Once the tree-fall gap is filled, species composition of that patch can remain stable for a very long time.

Tropical ecologists have long believed that gap dynamics (the formation and subsequent filling in of tree-fall gaps) are instrumental in maintaining the high species diversity of tropical forests. This belief is founded in the intermediate disturbance hypothesis, which predicts reduced diversity if disturbance is either too high (because few species can get established) or too low (because highly competitive species will exclude weaker competitors). However, from a remarkable series of exhaustive surveys conducted on a 50-ha forest plot in Panama, biologists recently found that there was no relationship between either the size or the frequency of gap formation and the diversity of tree species. This surprising result may be in part due to the relative rarity of species that depend on light gaps for survival. In addition, most of the tree species in this community have relatively limited dispersal and therefore cannot always take advantage of new openings in the forest canopy.

Tropical ecologists, like all other ecologists, are paying increasing attention to large-scale disturbances such as global climatic oscillations, severe drought, and consequent fire disturbance. Although the predictions of global climate change models remain controversial, many agree that the frequency of climatic oscillations such as the El Niño Southern Oscillation (ENSO) will likely increase in frequency and intensity. Indeed, two recent ENSO events (1982–1983 and 1997–1998) were among the strongest ever recorded. ENSO events can cause severe droughts that trigger large forest fires. For example, during the 1997–1998 ENSO, fires swept over approximately 2 million hectares in Indonesia, destroying 100,000 ha of primary forest and causing air pollution problems as far away as Sri Lanka and northern Australia. If predictions regarding ENSO events are accurate, worse and more frequent droughts in some tropical forest regions and larger and more destructive fires can be anticipated. The significance of these major climatic events is best appreciated by looking at effects on particular forests. For example, the 1997–1998 ENSO, which caused severe drought in tropical forests worldwide, increased annual mortality of Malaysian trees sixfold (from 1% to 6%).

The 1997–1998 ENSO event also resulted in severe drought and huge fires in northwestern Borneo. The combination of drought and fire resulted in high tree mortality, but the effects went far beyond immediate mortality. For instance, biologists noted that this

ENSO drastically altered the interactions between eight species of dioecious figs (having male and female reproductive organs on different individuals) and their pollinators, with a cascade of secondary effects. Each species of fig tree is pollinated by only one species of fig wasp, and all eight of the fig species had stable pollinator populations up until January 1998. However, during the drought, the figs dropped their leaves and inflorescences (flower clusters), and failed to initiate new inflorescences. The gap in the availability of flowers lasted for more than 2 months, roughly twice the life span of the fig wasp pollinators. All pollinators of these figs went locally extinct by the end of March 1998 and still had not recolonized seven of the eight fig species by October 1998. Without fig wasps, fig trees cannot bear fruit. Fig trees are an important food source for many different mammalian and bird species during a time of year when few other trees bear fruit. Small fruit bats, once common in this area, have since disappeared from the site. Although the ultimate fate of figs in Borneo is still unknown, this example shows how dramatic climatic disturbances can resonate throughout tropical forest ecosystems in surprising ways.

Deforestation impact. Deforestation and fragmentation will likely exacerbate the impacts of global climate change and fire. First, deforestation contributes roughly 2 billion tons of carbon to the global atmosphere each year, and roughly 70% of this is due to burning of biomass in the tropics. Thus, tropical deforestation is a major source of the greenhouse gases, carbon dioxide and methane, that are contributing to an increase in global temperature. Deforestation also reduces the size of an important carbon "sink." Tropical humid forests can be so productive that they store (pull out of the atmosphere) as much as 0.34 ton of carbon per hectare per year. Thus, destruction of tropical forest will accelerate global warming problems both because carbon will be released into the atmosphere from the burning and decomposition of existing forest biomass and because deforestation eliminates a carbon sink that would have removed large quantities of carbon from the atmosphere if left intact.

Second, most climate-change models predict even worse and more frequent droughts. Because forest fragmentation results in higher surface runoff and a more open canopy, fragmentation may exacerbate drought conditions and increase the likelihood of fire. After fire moves through an area, tree mortality can remain high for several years, causing a buildup of flammable materials, an even thinner canopy, and even drier conditions. Thus, there is a positive feedback loop between fire and susceptibility to future fire. For example, tropical forests in the Brazilian Amazon that had previously burned once were almost twice as likely to burn again. In the northeastern Amazon region, fires are recurring approximately every 7-14 years, which is too frequent for forest recovery. The current fire regime, if left unchecked, could cause an irreversible transformation of forests to grasslands or scrub.

Future. Increasingly, governments of tropical nations are recognizing the importance of conserving their dwindling forest resources and, consequently, are setting aside forests in nature reserves. However, the amount of land devoted to forestry is usually vastly larger than that set aside in preserves. To bolster forest protection, conservation organizations are offering tropical nations an economic incentive for conservation in what are called debt-for-nature swaps. Conservation groups purchase debts owed to other, usually developed, nations at greatly discounted prices and then offer to forgive the debt if the tropical nation agrees to set aside land for conservation. Because opportunities for the establishment of reserves are limited, an alternative approach is the restoration of tropical forests that are already degraded. Reforestation efforts will likely be increasingly common as industries realize that this can be a cost-effective way to satisfy regulations regarding carbon emissions.

For background information see CLIMATE MODELING; EL NIÑO; FOREST ECOSYSTEM; FOREST FIRE; FOREST REFORESTATION; GREENHOUSE EFFECT; LANDSCAPE ECOLOGY; RAINFOREST; RESTORATION ECOLOGY; TROPICAL METEOROLOGY in the McGraw-Hill Encyclopedia of Science & Technology.

Michelle Marvier

Bibliography. M. A. Cochrane et al., Positive feedbacks in the fire dynamic of closed canopy tropical forests, *Science*, 284:1832-1835, 1999; Food and Agriculture Organization, *State of the World's Forests: 1999*, Rome, 1999; R. D. Harrison, Repercussions of El Niño: Drought causes extinction and the breakdown of mutualism in Borneo, *Proc. Roy. Soc. London B*, 267:911-915, 2000; S. P. Hubbell et al., Light-gap disturbances, recruitment limitation, and tree diversity in a neotropical forest, *Science*, 283:554-557, 1999; W. F. Laurance et al., Rain forest fragmentation and the dynamics of Amazonian tree communities, *Ecology*, 79:2032-2040, 1998; M. Nakagawa et al., Impact of severe drought associated with the 1997-1998 El Niño in a tropical forest in Sarawak, *J. Trop. Ecol.*, 16:355-367, 2000; O. L. Phillips et al., Changes in the carbon balance of tropical forests: Evidence from long-term plots, *Science*, 282:439-442, 1998.

Ultracold neutrons

The lifetime of a free neutron (which beta-decays into a proton, an electron, and an antineutrino) is an important parameter for understanding both the weak nuclear force and the creation of matter during the big bang. The neutron lifetime (τ_n) is defined as the time required for a population of neutrons to decay to the fraction $1/e$ of its original value, where e is the base of the natural logarithms, 2.718. Precise measurements are difficult because neutrons are electrically neutral, have minimal interaction with most materials, and survive for a relatively long time (approximately 15 minutes) in comparison with other unstable elementary particles. These characteristics make neutrons rather slippery subjects for experimentalists who wish to study and observe their decay; and although past efforts at confining these particles have been successful, the resulting lifetime measurements are now plagued with systematic errors which limit their precision. Recently, a new techniqe has been developed which employs magnetic confinement of the neutrons and offers scientists the possibility of measuring the neutron lifetime in a manner which should be free from these systematic effects.

Neutron lifetime. The neutron was discovered in 1932 by James Chadwick and has been studied extensively by physicists. It is a composite particle with a mass slightly larger than that of the proton. When bound together with protons inside the nucleus of an atom, neutrons are stable. Once removed from the nucleus—in this particular case, through the fission of uranium within a nuclear reactor—neutrons beta-decay into protons, a process which emits an electron and an electron antineutrino as byproducts.

The lifetime of the neutron provides a critical input parameter for understanding the creation of matter. During the initial stages of the universe's expansion and cooling, protons and neutrons were formed. Initially, the universe was hot enough that neutrons and protons freely interchanged. Once the universe cooled enough that the interchange "froze out," two processes—light-element formation and neutron decay—began to compete with each other for available neutrons. Light elements are formed when a proton joins with a neutron to form deuterium (^{2}H), which in turn joins with a second deuterium to form a helium (^{4}He) atom. At the same time, neutrons also decay into protons. The end result of these competing processes is that all remaining neutrons are bound in the nucleus of atoms. The value of the neutron lifetime is important for determining the freeze-out time, which in turn set the final neutron-to-proton ratio of the early universe.

The neutron lifetime is also important for understanding the weak nuclear force, which is one of the four fundamental forces in nature and is responsible for the beta decay of all radioactive nuclei. The neutron is an ideal system for scientists to use in these studies, because it has the simplest nuclear beta decay and thus provides a theoretically "clean" system in which the weak interaction can be observed. Two coupling constants, the weak vector and axial vector, determine the behavior of the weak force. A measurement of the neutron lifetime, in combination with a measurement of a second decay parameter (such as the difference between the number of electrons which are emitted in a direction parallel to the neutron spin as opposed to antiparallel), fixes both coupling constants. These parameters are important not only for understanding the weak interaction in general but also in precise testing of the Standard Model of elementary particles.

Measuring the lifetime. Previous measurements of the neutron lifetime can be categorized into one of two types of measurements, beam or bottle. In a beam measurement, a cold neutron beam passes continuously through a decay region of known volume, and each neutron decay occurring within that volume is counted. The neutron lifetime is given by the ratio of the number of decays per second to the number of neutrons in the decay region. In the most precise measurement to date using this method ($\tau_n = 889.2$ s $\pm$ 4.8 s[2]), the major uncertainty, or "margin of error," is attributable to systematic effects in the measurement of the neutron flux (number of neutrons which pass through the volume).

In a bottle-type measurement, neutrons are loaded into a containment vessel and stored for a variable length of time before being counted. The neutron lifetime can be extracted from the dependence of the detected neutron population on the storage time. Several storage techniques have been used to make such measurements, the most precise of which (yielding values of $\tau_n = 882.6$ s $\pm$ 2.7 s[3], 888.4 s $\pm$ 3.3 s[4], and 885.4 s $\pm$ 1.0 s[5]) have relied upon the total internal reflection of ultracold neutrons (neutrons cold enough that their wave properties become important) from material surfaces. Bottle-type experiments have traditionally used physical containers composed of materials such as stainless steel or beryllium as storage vessels for confining the neutrons. Significant improvements to either the beam or bottle methods are not foreseen.

Magnetic trapping. Trapping of neutral and charged particles has recently become an invaluable tool for the study of elementary particles because the trapped species can be isolated from perturbing environments. Although the neutron has no charge, its magnetic moment behaves like an extremely small bar magnet, imposing a force on the neutron which can be used to confine it in a magnetic field. The technique of magnetic trapping offers two principal advantages over previous methods used to measure the neutron lifetime. First, ultracold neutron decay events can be counted continuously (by detection of scintillations from decay electrons as they recoil through liquid helium). Second, magnetic confinement in three dimensions eliminates interaction between the neutrons and the walls of a storage vessel that previously served to limit bottle-type measurements. Ultimately, lifetime measurements which may be conducted utilizing the magnetic trapping method should boast a significantly greater certainty than those in the past.

Methodology. In order to magnetically trap a neutron, its total energy must be reduced to less than the trapping potential while it is located inside the magnetic trap. The "superthermal" technique used in producing ultracold neutrons serves as such a dissipation mechanism. Neutrons with a wavelength near 0.89 nanometer can be scattered to near-rest by emission of a single phonon in superfluid helium. The rate for the inverse process, upscattering by absorption of a phonon, scales with the temperature

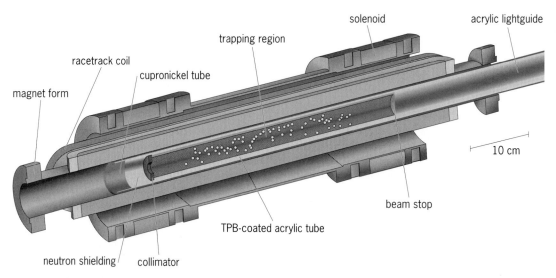

The neutron beam enters from the left, it is collimated, and unscattered neutrons are absorbed in the materials surrounding the trapping region. Neutrons scattered to near-rest and in the appropriate spin state are trapped until they decay. Upon decay, the electron recoils in the helium, producing scintillation light which is transported out of the trapping region using acrylic lightguides to detectors at room temperature. (*From P. R. Huffman et al., Nature, 403:62, 2000; reprinted by permission from Nature, http://www.nature.com, copyright 2000, Macmillan Magazines Ltd.*)

of the superfluid bath and can be highly suppressed by cooling the helium to below 100 mK. This allows neutrons with energies less than the trap depth (approximately 1 mK) to remain out of thermal equilibrium with the warmer liquid helium for times much longer than the neutron lifetime.

Once trapped, the neutron travels undisturbed in the helium until it decays. The resulting decay electron then recoils through the helium, producing ionization tracks which quickly recombine to form thousands of He_2^* molecules. These molecules decay radiatively in less than 10 nanoseconds, giving rise to a pulse of extreme ultraviolet (EUV) light (wavelength centered around 80 nm). The EUV light is converted to visible light using an organic dye (tetraphenyl butadiene, TPB) and transported out of the trap to a detector at room temperature. Each decay can be detected as it occurs, and the neutron lifetime can be extracted from the decay rate (see **illus.**).

In a proof-of-principle measurement, scientists have been successful in verifying the theoretical predictions regarding the superthermal loading process and validating the technique of magnetic trapping, with preliminary measurements yielding results consistent with the presently accepted value of the neutron lifetime.

Outlook. To move the magnetic trapping technique from proof-of-principle stages to application in a competitive measurement of the neutron lifetime, scientists must first increase the "holding capacity" of the trap. Experimentalists are building a new magnet which is expected to increase both the size and depth of the magnetic trap, allowing for containment of approximately 14,000 ultracold neutrons per loading cycle—an increase of a factor 25 in the total number of trapped ultracold neutrons demonstrated in the initial proof-of-performance tests. The

larger trap not only will provide a larger quantity of neutrons for observation but will also increase the ratio of signal to background events, providing a clearer "picture" for scientists to study. Along with other minor incremental improvements, the higher-capacity trap should allow scientists to make a substantially improved measurement of the neutron lifetime in the near future.

Magnetic trapping of ultracold neutrons has been demonstrated as an experimentally feasible and valid method of containing neutrons, and should allow scientists to significantly improve the precision of the neutron lifetime measurement by eliminating the systematic effects which have played a role in limiting previous beam and bottle measurements. Although there are certain loss mechanisms specific to this method, such as thermal upscattering and ^{3}He absorption, experimentalists are able to suppress them to a high level. With minor technical improvements to the trap and a higher flux source of cold neutrons, scientists believe that a 10^{-5} measurement of the neutron lifetime should ultimately be possible. In the near term, improvements such as increased trap size and depth should allow an improved accuracy measurement of τ_n at the 5×10^{-4} level. Magnetic trapping of neutrons provides scientists with more certain parameters in studying both the big bang and weak nuclear force concepts, and has the potential to be used for other applications such as searches for time-reversal violation and the measurement of other neutron decay parameters.

For background information *see* BETA PARTICLES; BIG BANG THEORY; ELECTRON; MAGNETIC FIELD; NEUTRINO; NEUTRON; PROTON; REACTOR PHYSICS; STANDARD MODEL; WEAK NUCLEAR INTERACTIONS in the McGraw-Hill Encyclopedia of Science & Technology.

Paul Huffman

Bibliography. S. Arzumanov et al., Neutron lifetime value measured by storing ultracold neutrons with detection of inelastically scattered neutrons, *Phys. Lett. B*, 483:15-22, 2000; J. Byrne et al., A revised value for the neutron lifetime using a Penning trap, *Europhys. Lett*, 33:187-192, 1996; P. R. Huffman et al., Magnetic trapping of neutrons, *Nature*, 403: 62-64, 2000; W. Mampe et al., Measuring the neutron lifetime by storing ultracold neutrons and detecting inelastically scattered neutrons, *JETP Lett.*, 57:82-87, 1993; V. V. Nesvizhevskii et al., Measurement of the neutron lifetime in a gravitational trap and analysis of experimental errors, *Sov. Phys. JETP*, 75:405-412, 1992.

Ultradeep xenoliths

Xenoliths are fragments of rock that occur as inclusions within volcanic rocks (formed by crystallization of a magma on the Earth's surface) or within plutonic host rocks (formed by crystallization of magma in the interior of the Earth). They are classified as cognate or exotic. Cognate xenoliths are closely related to their igneous (magma) host and are interpreted to have crystallized at relatively shallow depths in crustal magma chambers. Exotic xenoliths are unrelated to their magma host and so are believed to have originally crystallized far from their host rock, thus providing important information about the architecture of the crust and mantle traversed during movement of the magma.

Garnet-rich exotic xenoliths that contain ultrahigh-pressure minerals derived from depths of 400–660 km (240-396 mi) have recently been discovered in the Solomon Islands in the southwestern Pacific and at the Monastery Mine kimberlite pipe (type of rock-filled volcanic unit) in South Africa. These are the deepest-origin xenoliths yet recognized on the Earth. The region sampled by these xenoliths (that is, the region where they are believed to have initially crystallized) is called the mantle transition zone. It occurs between the upper and lower mantle (**Fig. 1**), represents about 11% of the mass of the mantle, and is a significant, yet poorly understood portion of the Earth. The significance of this mantle region was recognized almost 50 years ago by Frank Birch, who said that understanding the mantle transitional zone is the key to understanding the dynamics of the Earth's interior. Recent modeling of the interior of the Earth, using seismic tomography, computer simulation, and experimental data, confirmed the important role played by the transition zone, particularly its effect on the transfer of material from the lower to the upper mantle (associated with the rise of thermally buoyant mantle plumes) and on the deep subduction of high-density oceanic lithosphere into the lower mantle.

Xenolith origins. Ultramafic xenoliths entrained in mantle-derived magmas are the most important source of direct information on the composition and mineralogy of the shallowest part of the Earth's man-

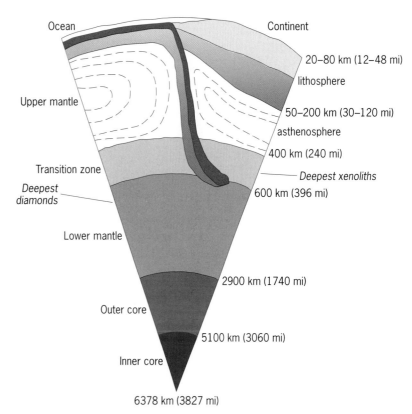

Fig. 1. Earth section showing compositional layers defined from seismic data. A slab of oceanic lithosphere on the left is subducting beneath continental lithosphere on the right. With increasing pressure, the basaltic component of the slab is first transformed into eclogite and then into garnetite.

tle, at depths less than 200 km (120 mi). These xenoliths, called peridotites, are composed of magnesium-rich silicates, such as olivine, orthopyroxene, and clinopyroxene, as well as garnet, aluminous spinel, and chromite. Peridotite xenoliths with even-grained textures are derived from the shallow, nonconvecting lithospheric mantle and record pressures of crystallization of 2.6-4.1 gigapascals (1 GPa = 10 kbar = 9.87×10^3 atm), equivalent to depths of 100-150 km (60-90 mi). Pyroxenite xenoliths composed of clinopyroxene or orthopyroxene with minor amounts of olivine and garnet are also believed to originate in the subcontinental lithospheric mantle. A less common suite of sheared peridotite xenoliths contains mineral assemblages that indicate derivation from depths of 150-200 km (90-120 mi). Such strongly deformed xenoliths are derived from the convecting asthenospheric mantle that underlies the lithosphere (Fig. 1). Some xenolith suites also contain eclogite (high-pressure assemblages of sodium-bearing clinopyroxene, Mg-Fe-rich garnet, rutile, and quartz) that is interpreted to be derived from deeply subducted oceanic basaltic crust.

Monastery xenoliths. In 1970, it was suggested that xenoliths from the Monastery Mine kimberlite pipe in South Africa may have originated from significantly greater depths—more than 300 km (180 mi). These distinctive xenoliths exhibited delicate rodlike intergrowths of Mg-rich ilmenite in a matrix of clinopyroxene (**Fig. 2**). This texture was interpreted to have

Fig. 2. Possible deep xenolith from Monastery Mine showing exsolution lamellae (precipitated layers) of Mg-ilmenite in a host of clinopyroxene.

resulted from crystallization of ilmenite and clinopyroxene from an original homogeneous high-pressure garnet, in which octahedrally coordinated Al^{3+} was replaced by $(Fe+Mg)^{2+}$ and Ti^{4+}. As high-pressure garnets of this composition had been synthesized in the laboratory at pressures greater than 10 GPa, these xenoliths were regarded at that time to be the deepest mantle rocks ever recognized at the Earth's surface.

The other source of information regarding an ultradeep origin of mineral inclusions (in diamonds and more recently of mantle xenoliths) has come from the garnet mineral majorite. Majorite $[Mg_3(Fe,Al,Si)_2-Si_3O_{12}]$ forms in the deep upper mantle in response to the gradual dissolution of pyroxene into the garnet structure due to increasing pressure. Experimental studies in both ultramafic and basaltic compositions have shown that initial dissolution of pyroxene commences at pressures equivalent to depths of 250 km (150 mi) and is completed at depths of 450 km (270 mi). As pressure increases, garnet become progressively enriched in SiO_2 and depleted in Al_2O_3 and Cr_2O_3.

In 1985, majorite garnet was first identified as microscopic inclusions in diamonds from the Monastery kimberlite pipe by R. O. Moore and J. J. Gurney, the same location as the high-pressure pyroxene-ilmenite intergrowths discussed above. Majorite, together with other ultrahigh-pressure minerals, has subsequently been recognized in diamonds from Brazil, Guinea, Russia, and Canada. These occurrences confirm the existence of diamond in the deep mantle.

Possible existence of majorite in eclogite (basaltic) and peridotite (ultramafic) xenoliths from the Jagersfontein kimberlite pipe in South Africa was indicated by garnets that exhibited delicate needlelike intergrowths of either clinopyroxene or orthopyroxene. These xenoliths were also interpreted to come from depths of 300–400 km (180–240 mi), near the top of the transition zone.

Malaita xenoliths. An extensive suite of majorite-bearing garnetite and rare garnet peridotite xenoliths was recently discovered in volcanic pipes on Malaita in the Southwestern Pacific. They provide a unique opportunity to study the Earth's transition zone. The xenoliths contain not only majorite and Mg-ilmenite/pyroxene intergrowths but also a whole range of minerals with ultrahigh-pressure chemistries, including Ca perovskite, jadeite-rich clinopyroxene, picroilmenite, zircon, Fe-Ni sulfides, a silicon dioxide phase, hollandite (potassium barium titanate), and microdiamond. In addition, the Al-Ca and Al phases and Si-Al Mg perovskite compositions that have been reported in a number of experimental studies of basaltic compositions up to 27 GPa are also present.

The Malaita majorite-bearing garnetite xenoliths range in size up to 30 cm (12 in.) in diameter (**Fig. 3**). They are interpreted as the high-pressure recrystallization products of subducted oceanic basalts. Majorite occurs predominantly as ameboid patches and interconnecting veins within sodium- and titanium-bearing pyrope garnet $[Mg_3Al_2(SO_4)_3]$. Majorite also occurs rarely in small garnet peridotite xenoliths. Malaita majorite Si and Cr + Al ranges are identical to the range of compositions recently discovered in garnetite xenoliths from the Monastery kimberlite (**Fig. 4**). A comparison of chemical compositions is given in the **table**. Compared to all other reported diamond inclusion majorite compositions, the Malaita and Monastery xenoliths include significantly more silicic compositions (Si^{4+} up to 3.82 per formula unit). They were thus derived from much greater depths, in the pressure range of 12–27 GPa.

Fig. 3. Xenoliths from Malaita. (*a*) Pyroxene Mg-ilmenite xenolith from kimberlite. Note the similarity to the Monastery xenolith in Fig. 2. (*b*) Garnetite xenolith, containing Na-Ti-rich pyrope garnet, majorite, and other high-pressure phases from the mantle transition zone.

Comparison of pyrope garnet and majorite compositions for ultradeep xenoliths from Malaita and Monastery*

	Pyrope		Majorite										
	Monastery	Malaita	Monastery	Malaita	Monastery	Malaita	Monastery	Monastery	Malaita	Monastery	Malaita	Monastery	Malaita
SiO_2	42.11	41.48	44.61	45.03	46.99	47.04	47.88	49.79	49.86	51.56	51.09	52.27	53.29
TiO_2	0.34	0.7	1.75	0.66	0.59	0.72	1.01	0.52	0.55	0.43	0.6	0.51	0.27
Al_2O_3	23.18	22.71	14.16	17.91	14.46	14.28	11.86	8.49	8.15	6.39	6.07	5.53	3.88
Cr_2O_3	0.12	0.03	0.46	0.24	0.06	0.18	0.16	0.04	0	0.08	0.01	0.09	0.01
FeO	11.45	11.2	9.26	12.41	12.09	11.41	11.42	14.07	14.19	12.12	12.97	12.43	13.55
MnO	0.31	0.3	0.27	0.36	0.27	0.33	0.42	0.47	0.37	0.41	0.42	0.32	0.46
MgO	18.27	17.83	18.59	23.29	22.83	24.81	23.49	24.48	25.38	26.55	26.52	27.05	27.29
CaO	4.34	4.71	9.96	0.42	2.14	1.16	3.47	1.96	2.01	1.80	1.86	1.76	1.92
Na_2O	0.08	0.09	0.22	0.02	0.03	0.06	0.07	0.18	0.29	0.01	0.05	0.03	0.13
K_2O	0.00	0	0.02	0	0.00	0	0.00	0.01	0	0.01	0	0.02	0.00
Total	100.20	99.04	99.30	100.3	99.46	99.99	99.78	100.02	100.8	99.37	99.59	100.00	100.79
Si^{4+}	3.014	3.008	3.264	3.200	3.373	3.345	3.437	3.592	3.574	3.701	3.680	3.731	3.797
Ti^{4+}	0.018	0.038	0.097	0.035	0.032	0.038	0.055	0.028	0.030	0.023	0.032	0.027	0.014
Al^{3+}	1.955	1.940	1.221	1.500	1.223	1.197	1.004	0.722	0.689	0.541	0.516	0.466	0.326
Cr^{3+}	0.007	0.002	0.027	0.014	0.003	0.010	0.009	0.002	0.000	0.004	0.000	0.005	0.000
Fe^{2+}	0.686	0.679	0.567	0.737	0.726	0.679	0.686	0.849	0.851	0.728	0.781	0.742	0.807
Mn^{2+}	0.019	0.018	0.016	0.022	0.017	0.020	0.025	0.029	0.023	0.025	0.025	0.019	0.028
Mg^{2+}	1.950	1.927	2.027	2.467	2.442	2.630	2.514	2.632	2.711	2.840	2.847	2.878	2.898
Ca^{2+}	0.333	0.366	0.781	0.032	0.164	0.088	0.267	0.152	0.154	0.139	0.144	0.135	0.146
Na^+	0.011	0.013	0.032	0.003	0.004	0.009	0.010	0.026	0.040	0.001	0.007	0.004	0.017
K^+	0.000	0.000	0.002	0.000	0.000	0.000	0.000	0.001	0.000	0.001	0.000	0.002	0.000
Total	7.992	7.990	8.032	8.009	7.984	8.016	8.006	8.031	8.072	8.004	8.033	8.009	8.035
Si	3.014	3.008	3.264	3.200	3.373	3.345	3.437	3.592	3.574	3.701	3.680	3.731	3.797
Al+Cr	1.962	1.942	1.248	1.514	1.227	1.207	1.012	0.724	0.689	0.545	0.516	0.471	0.326

* Structural formulas are based on 12 oxygens. Majorite data are arranged by increasing pressure from 10 to 22 GPa.

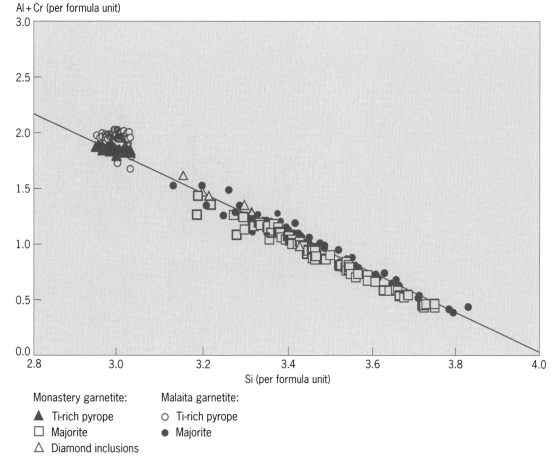

Fig. 4. Garnet and majorite chemical structural data (based on 12 oxygen atoms) for garnetites from Malaita and Monastery kimberlite.

This is significantly deeper than majorite diamond inclusions reported from the Monastery Mine [9 GPa (262 km) to 15 GPa (413 km)] and the Jagersfontein Mine [7 GPa (204 km) to 11 GPa (303 km)].

Implications. The ultradeep Malaita and Monastery xenoliths are the first confirmed rock samples yet recovered from the mantle transition zone. They provide direct information regarding the mineralogy, composition, and isotopic evolution of this region of the Earth. From the present evidence, two key issues are clear regarding the transition zone.

First, the presence of titanium and sodium-rich eclogitic pyropes in the Malaita and Monastery garnetite xenoliths, with compositions identical to eclogite pyrope inclusions in diamond, shows the metastability of pyrope in the transition zone. Metastability is the thermodynamic term that describes the persistence of a mineral into an environment where it would normally be expected to react to form one or more new minerals. In the transition zone, the metastable occurrence of pyrope probably occurs because of kinetic factors, such as the depression of temperature gradients (isotherms) in ultradeep subduction zones, as well as the relatively anhydrous (fluid-free) nature of deeply subducted slabs of oceanic crust and mantle. The region above the base of the upper mantle that is characterized by extremely rapid increases in velocities of both primary and secondary seismic waves may reflect the presence of metastable pyrope.

Second, the discovery of garnet-rich ultradeep xenoliths indicates that the transition zone may be dominated by the high-pressure components of subducted oceanic crust rather than peridotite. Thus, the transition zone is a volumetrically significant geochemical reservoir (11% of the mass of the mantle) that differs in composition from most of the upper mantle. As minerals in transition zone xenoliths, such as Mg-rich ilmenite, can incorporate substantial amounts of niobium and tantalum in their crystal structure, the transition zone probably represents the "missing geochemical reservoir" that is required to explain the apparent deficit of these elements in continental crust.

For background information *see* BASALT; CHROMITE; DIAMOND; EARTH INTERIOR; ECLOGITE; GARNET; IGNEOUS ROCKS; HIGH-PRESSURE MINERAL SYNTHESIS; ILMENITE; JADEITE; MAGMA; OLIVINE; ORTHORHOMBIC PYROXENE; PERIDOTITE; PEROVSKITE; PLATE TECTONICS; PLUTON; PYROXENE; SPINEL; VOLCANO; XENOLITH; ZIRCON in the McGraw-Hill Encyclopedia of Science & Technology.

Kenneth D. Collerson

Bibliography. F. Birch, The earth's mantle: Elasticity and constitution, *Trans. Amer. Geophys. Union*, 35:79–85, 1954; K. D. Collerson et al., Rocks from the mantle transition zone: Majorite-bearing xenoliths from Malaita, southwest Pacific, *Science*, 288 (5469):1215–1223, 2000; B. S. Kamber and K. D. Collerson, Role of "hidden" deeply subducted slabs in mantle depletion, *Chem. Geol.*, 166:241–254, 2000; R. O. Moore and J. J. Gurney, Pyroxene solid solution in garnets included in diamond, *Nature*, 318:553–555, 1985; A. E. Ringwood, The garnet-pyroxene transformation in the Earth's mantle, *Earth Planet Sci. Lett.*, 2:255–263, 1967; A. E. Ringwood and J. F. Lovering, Significance of pyroxene-ilmenite intergrowths among kimberlite xenoliths, *Earth Planet. Sci. Lett.*, 7:371–375, 1970; V. S. E. Sautter, X. X. Haggerty, and S. Field, Ultradeep (>300 kilometers) ultramafic xenoliths: Petrological evidence from the transition zone, *Science*, 252:827–830, 1991; T. Stachel, G. P. Brey, and J. W. Harris, Kankan diamonds (Guinea) I: From the lithosphere down to the transition zone, *Contrib. Mineral. Petrol.*, 140:1–15, 2000; H. Tsai et al., Mineral inclusions in diamond: Premier, Jagersfontein and Finsch kimberlites, South Africa, and Williamson Mine, Tanzania, *2d International Kimberlite Conference, American Geophysics Union*, 1:16–25, 1979.

Ultralow-velocity zones (seismology)

Scientists who analyze waves that emanate from earthquakes (seismologists) have recently discovered new methods to study the boundary between the Earth's mantle and core in unsurpassed detail. Some 2900 km (1800 mi) below the Earth's surface, the core-mantle boundary contains important clues to many unanswered questions about the Earth's formation, evolution, and present internal processes. Analysis of seismic waves has revealed that, in some places, there is a thin mushy layer right at the core-mantle boundary. Seismic waves that travel in this layer are slowed down, giving rise to the name ultralow-velocity zone (ULVZ). The correlation between the geographic locations of the core-mantle boundary mush zones and some types of volcanoes suggest that ULVZs may be the source of volcanoes in regions such as Hawaii and Iceland. The importance of ULVZs in both mantle and core processes is apparent; possible physical scenarios include melting of the very base of the Earth's mantle, a blurring of the core-mantle boundary itself, and even sedimentation processes in the outermost core on the underside of the core-mantle boundary.

Seismology as an imaging tool. The most direct approach for studying the intricacies of the Earth's interior is through seismic wave analysis. Such waves propagate through the entire interior of the Earth. Just as the path of light bends as it passes from air into water, seismic energy passing from the Earth's mantle into the core experiences slight changes in direction. Some of this energy bounces off interfaces with different bulk Earth properties (reflection), some of it travels straight through material and can bend (refraction), and some of it travels along boundaries between distinctly different materials (diffraction).

SPdKS waves. A demonstration of these properties of seismic waves comes from a particularly useful seismic wave for deep Earth study called SKS. It traverses the Earth's mantle as an S-wave (secondary or shear wave), converts to a P-wave (primary or compressional wave) at the core-mantle boundary, and travels through the core as a P-wave (represented by the symbol K in SKS), returning to the mantle for the final leg of its journey as an S-wave (**Fig. 1***a*). When SKS encounters the core-mantle boundary at a critical propagation angle, its energy completely converts to a P-wave that diffracts (d) horizontally along the core-mantle boundary, then dives into the core and follows a path similar to SKS. This wave is called SPdKS, and permits the investigation of the details of the core-mantle boundary at localized spots where SPdKS enters and exits the core. SPdKS is directly affected by any changes in seismic velocity at the core-mantle boundary, which alters the short segment of diffracting P-wave energy (Pd in Fig. 1*a*). To document peculiarities in SPdKS behavior, it is compared to SKS, which travels a very similar path through the Earth. In fact, it was through study of SKS waves that SPdKS behavior was first noted.

ScS, ScP, and PcP waves. Seismic waves reflected off the core-mantle boundary (identified by the symbol c) are also useful for detecting thin layering at the bottom of the mantle. S- and P-waves that reflect off the core-mantle boundary, as well as S-waves that convert to P-waves upon reflection, show precursory energy if any low-velocity material exists at the core-mantle boundary. Seismic energy bounces off the top of the ULVZ layer and arrives at the surface before the energy that reflects off the core-mantle boundary. Recent research has used these reflected waves, along with SPdKS, to study ULVZ structure.

Synthetic seismograms. The approach most commonly used in characterizing ULVZ structure is to compare seismic observations with predictions. Observations are the seismic waves emitted from earthquakes and recorded by seismometers around the world. Predictions are computations of what a seismogram would look like for a given model of the Earth, using princi-

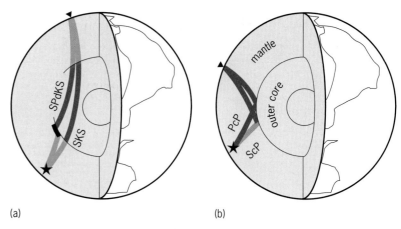

Fig. 1. Cross section of the Earth showing the mantle and core. (*a*) Seismic waves SKS and SPdKS. The hypothetical earthquake (star) generates the energy that travels through the Earth to the seismographic recorder (triangle). The short black segment of the SPdKS ray path enables study of any possible layer between the mantle and core. (*b*) Seismic waves ScP and PcP. The energy from these waves reflects off the core-mantle boundary. If any additional layering is present, seismic energy will reflect off that as well, resulting in precursory energy arriving at the seismic recorder before ScP or PcP.

ples of math and physics that describe how energy propagates through an elastic medium (the Earth). These predictions are synthesized on powerful computers, and are called synthetic seismograms. This process of comparing synthetic to observed seismograms is called synthetic modeling; if the data and predictions are similar, then the model of the Earth used in the calculations is a possibility.

Geographic distribution of ULVZ. Earthquakes predominantly occur in thin belts at the Earth's surface, which define the boundaries between 15 or so plates of the outermost shell of the Earth (the lithosphere). These earthquakes are recorded by seismometers—sensitive devices that detect ground motion—and are predominantly restricted to continents which cover about one-third of Earth's surface area. The geographical limitations of uneven earthquake and

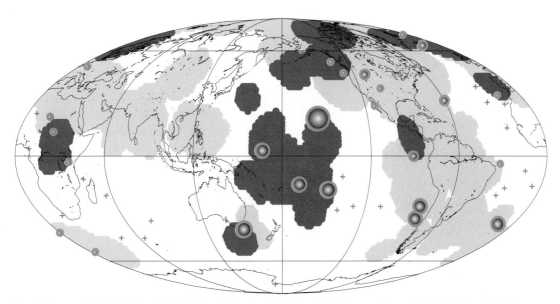

Fig. 2. Geographic distribution of ULVZ (dark color) and regions where ULVZ have not been detected (light color). These are compared to surface locations of hot-spot volcanoes (circles), which correlate well with locations of ULVZ.

seismometer distribution limit the regions of the deep interior that can be probed. Nonetheless, just under one-half of the surface of the Earth's core-mantle boundary has been probed for ULVZ structure. Roughly 25% of the sampled core-mantle boundary shows evidence for the presence of ULVZ. These areas (**Fig. 2**) are strongly correlated with regions at the surface that possess "hot-spot" volcanism—volcanoes containing magma thought to originate in the deep mantle. Regions lacking ULVZ evidence are strongly correlated with past or present subduction zones—regions where cool oceanic lithosphere plunges into the Earth's mantle.

This first-order correlation suggests a link between ULVZ occurrence and large-scale mantle convection. Relatively cold downwelling motions in the deep mantle suppress ULVZ creation, while hot material rising from the core-mantle boundary creates hot-spot volcanism and supports creation of the thin anomalous zones.

Origin of the ULVZ. The earliest ULVZ studies in the mid-1990s documented anomalous seismic data, and modeled it with low-velocity layering at the very base of the mantle. These first attempts to explain the data invoked ULVZ layers of 10–40 km (6–25 mi), showing seismic velocity reductions of 10% below "normal" mantle velocity values at those depths. Strong variations in ULVZ properties were also noted at that time. Subsequent studies have shown that much thinner layers (as thin as 1–3 km or 0.6–2 mi) can also explain some of the data, if the properties of the layer are much more extreme in comparison to the mantle. Several scenarios for the physical nature of the ULVZ have been proposed.

Partial melt of the base of the mantle. The apparent geographic correlation of ULVZ distribution with surface phenomena linked to large-scale mantle motions implies a thermal origin of the ULVZ. If the ULVZ represents a partial melt of some deep mantle constituent, then small-scale instabilities in this layer can combine to give rise to mantle plumes which feed hot-spot volcanoes. Cold downward currents in the mantle, such as those beneath subduction zones, would cool the deep mantle, preventing melting of mantle material.

Thin transition zone from mantle to core. Recently, models containing a mushy zone between the mantle and core have produced synthetic seismograms that correlate particularly well with some observations. This type of model differs from a "pure" ULVZ, in that it involves a smoother transition from the mantle to the core. In this scenario, chemical reactions between the silicate mantle and liquid iron alloy in the outer core result in a thin mixing zone (1–3 km or 0.6–2 mi thick) that "blurs" the core-mantle boundary. Thus this model is referred to as a "fuzzy" core-mantle boundary.

Thin outermost core layering. The liquid outer core of the Earth is predominantly iron, along with a minor portion of some less dense elements (a lighter component is required to satisfy the observation that the core is slightly less dense than pure iron). As the Earth cools, the solid inner core of the Earth grows; this process releases the lighter elements into the overlying liquid outer core. This process could result in the "underplating" of the lighter elements at the core-mantle boundary (a sedimentation process) since they should be significantly more buoyant than iron. It is expected that this layering would be nonuniform, since the core-mantle boundary is expected to have topography induced from convection currents in the overlying mantle, which would concentrate more sediment in "hills" on the core-mantle boundary. Seismic modeling has shown that sediment thicknesses of 1–2 km (0.6–1.2 mi) can explain the observed seismic anomalies.

The real Earth. Seismic imaging methods are always improving, and combining different methods greatly reduces uncertainties for any given model. Presently, each of the above scenarios is a viable explanation for the anomalous seismic signals. It is possible that the Earth may be a combination of all three structures (**Fig. 3**). The geographic correlation of ULVZs and surface hot spots strongly points to partial melt of the base of the mantle as an explanation. Laboratory evidence for chemical reactions between mantle and core material points to a core-mantle transition zone as a likely candidate (a fuzzy core-mantle boundary). The growth of the inner core as the Earth cools is known to release lighter elements into the outer core, opening up the possibility of sedimentation on the underside of the core-mantle boundary. The responsibility for deep-Earth scientists is to better constrain what degree each of these phenomena contributes to the anomalous waves that seismologists record at the Earth's surface. *See also* AFRICAN SUPERPLUME.

For background information *see* EARTH INTERIOR; EARTHQUAKE; LITHOSPHERE; PLATE TECTONICS; SEISMOLOGY in the McGraw-Hill Encyclopedia of Science & Technology.   Ed J. Garnero

Bibliography. B. A. Buffett, E. J. Garnero, and R. Jeanloz, Sediments at the top of the Earth's core,

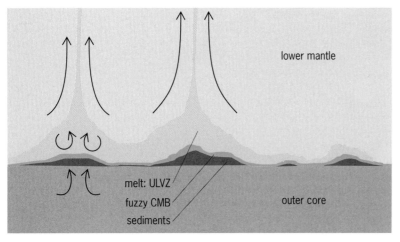

Fig. 3. A hypothetical scenario involving all three types of low-velocity layering: melt on the mantle side of the core-mantle boundary provides the genesis of mantle plumes that feed surface volcanism; sediments accrue underneath the core-mantle boundary beneath hills in the boundary; and the actual boundary itself is blurred due to chemical interaction between core and mantle material.

Science, 290:1338-1342, 2000; E. J. Garnero et al., Ultralow velocity zone at the core-mantle boundary, in *The Core-Mantle Boundary*, AGU, pp. 319-334, 1998; E. J. Garnero and R. Jeanloz, Earth's enigmatic interface, *Science*, 289:70-71, 2000; Q. Williams, J. S. Revenaugh, and E. J. Garnero, A correlation between ultra-low basal velocities in the mantle and hot spots, *Science*, 281:546-549, 1998.

Underwater sound

Sound is the most powerful remote sensing tool available for use in the sea. In contrast to electromagnetic waves, which are attenuated over very short distances underwater, sound waves at low frequencies (less than 500 Hz) can propagate over extraordinarily long ranges (thousands of miles) without significant distortion or loss of energy. The use of underwater sound as a technique for detecting submarines began during World War I and accelerated rapidly during World War II. During the Cold War, acoustic antisubmarine warfare became the principal deterrent against missile-carrying submarines roaming the high seas. It was during the latter period that ocean acoustics transcended these militarily motivated applications and emerged as a fundamental scientific discipline encompassing a broad range of topics in wave propagation physics, electrical engineering, oceanography, and geophysics. These rapidly developing scientific advances spawned the new discipline of acoustical oceanography, in which sound is used as a probe to measure oceanographic properties of the water column, the seabed, and the sea surface. Since the end of the Cold War, ocean acoustics has retained its military significance, now in the context of third-world nations potentially posing submarine threats in shallow-water areas. It is in this shallow-water context that new and exciting results have emerged in the area of tracking sound sources by measuring the phase of the low-frequency tones that they emit. These developments have general implications for the deep-water scenario as well.

Theory. In order to understand the more complicated problem of acoustic propagation in shallow water, it is useful first to consider the case of a point source radiating a single-frequency sound wave in free space. Then the acoustic pressure p consists of an outgoing spherical wave traveling with speed c and wavelength λ, and is described by Eq. (1),

$$p = \frac{\cos(\omega t - kr)}{r} = A \cos \Phi \qquad (1)$$

where t is the time, $\omega = 2\pi f$ is the angular frequency (f is the frequency), r is the distance from source to receiver, and $k = \omega/c = 2\pi/\lambda$ is the acoustic wavenumber. The signal has also been written in Eq. (1) in terms of its magnitude A and phase Φ, where $A = 1/r$ and Φ is given by Eq. (2),

$$\Phi = \omega t - kr = \omega t - \phi \qquad (2)$$

with $\phi = kr$. The magnitude reflects the decrease in intensity of the wave due to spherical spreading, while the phase describes the spherically propagating wavefront in space and time.

Measurement of source-receiver speed. These results suggest that if the frequency is known and there is a way of measuring the phase ϕ, it is possible to infer the range between source and receiver. In fact, determining the precise value of ϕ is typically impractical. A measurement of p yields only a value of the phase between $-\pi$ and π radians; and the unwrapped phase ϕ, which appears in Eq. (2), can differ from this value by any multiple of 2π, since it increases by 2π radians each time the source-receiver distance increases by one wavelength. Determining ϕ requires knowledge of an exact reference value for the phase as well as a robust phase-unwrapping algorithm, which reliably tracks the unwrapped phase as it changes through cycles of 2π radians.

But suppose that the source and receiver are moving with speed v relative to one another, so that $r = vt$ and $\phi = kvt$. Then it is possible to infer the relative source-receiver speed from measurements of the time rate of change of the phase $d\phi/dt$; Eq. (3) for $d\phi/dt$ yields Eq. (4) for v.

$$\frac{d\phi}{dt} = kv \qquad (3)$$

$$v = \frac{dr}{dt} = \frac{1}{k}\frac{d\phi}{dt} \qquad (4)$$

The quantity $kv = (v/c)\omega$ is equivalent to the Doppler shift in frequency associated with the source-receiver motion. In fact, measurement of the phase rate is a much more robust procedure than direct measurement of the phase itself. Furthermore, it is possible to infer the source-receiver range simply by integrating both sides of Eq. (4) to obtain Eq. (5),

$$r = \frac{\phi}{k} + K \qquad (5)$$

where K is an unknown integration constant corresponding to a starting range which must be independently prescribed (that is, the previously mentioned reference value for the phase).

Realistic ocean environments. A realistic ocean environment offers several significant complications compared to the simple example described above. First, the sound speed (or index of refraction) in the sea is in general not constant, but can vary with both depth and range. This characteristic immediately alters the simple spherical wave behavior associated with a point source in free space. Second, the sound waves typically interact with both the moving sea surface and the seabed, which is a complicated multilayered structure also supporting acoustic waves. All of these factors combine to create a channel, or waveguide, for the sound waves that are trapped between the surface and the bottom in shallow water or focused by the sound speed structure in deep water as they propagate outward from source to receiver.

Although these gross features of the ocean environment support the propagation of low-frequency sound to great distances with a high degree of coherence, the detailed heterogeneous character of the sea and its boundaries provides a variety of mechanisms (such as scattering) for attenuating the signals and reducing their fidelity. It is therefore not at all obvious that the phase tracking methodology described above for the point source in free space is applicable to the ocean waveguide.

The principal complication arising from the realistic oceanographic features described above is that the acoustic field now consists of a number of signals traveling along various paths through the waveguide, so that Eq. (1) is replaced by Eq. (6).

$$p = \sum_{n=1}^{N} \frac{A_n \cos(\omega t - k_n r)}{\sqrt{r}}$$

$$= A \cos \Phi = A \cos(\omega t - \phi) \quad (6)$$

Here each of the N signals has a magnitude A_n and wave number k_n corresponding to the nth acoustic normal mode of vibration excited in the waveguide. The channeling of the sound causes a less rapid geometrical attenuation of the signal, and the field now has a cylindrical spreading factor $1/\sqrt{r}$ rather than the spherical spreading factor $1/r$ in the free-space case. The total field p arises from the summation of the individual signals, and therefore both the overall magnitude A and the phase Φ depend on the individual magnitudes and phases of the component signals.

Despite this complexity, however, the following hypothesis will be adopted: the relationship between the phase rate and the source-receiver range rate described in Eq. (4) still holds true to a very good approximation. The underpinnings of this hypothesis are two fundamental physical principles, namely, the radiation condition and the paraxial approximation. The radiation condition, which is implicit in Eqs. (1) and (6), states that a source confined to a finite spatial domain produces outgoing, radiating wave fields at infinity. The paraxial approximation states that, although the source injects acoustic energy into the waveguide over a broad range of angles, only a relatively narrow band of angles (less than about 30° with respect to the horizontal) dominates the radiated field, even in shallow water. Mathematically, the paraxial approximation means that the individual wave numbers k_n in Eq. (6) can be expressed as Eq. (7),

$$k_n = k\sqrt{1 - \varepsilon_n} \approx k(1 - \varepsilon_n/2) \quad (7)$$

where ε_n is much less than one and k is a typical wave number in the water column. The application of this approximation to the sound field in Eq. (6) then leads to the result in Eq. (4).

Experiment. In March 1997, the first Modal Mapping Experiment (MOMAX I) was conducted in about 70 m (230 ft) of water off the New Jersey coast. The experimental configuration consisted of a source transmitting several pure tones in the band 50–300 Hz to a field of several freely drifting buoys, each equipped with a hydrophone, Global

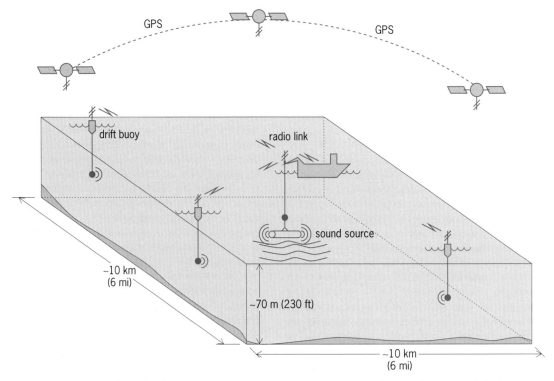

Fig. 1. Experimental configuration for the first Modal Mapping Experiment (MOMAX I). (*After G. V. Frisk, K. M. Becker, and J. A. Doutt, Modal mapping in shallow water using synthetic aperture horizontal arrays, Proceedings of the OCEANS 2000 MTS/IEEE Conference and Exhibition, Providence, RI, vol. 1, pp. 185–188, September 2000, © 2000 IEEE*)

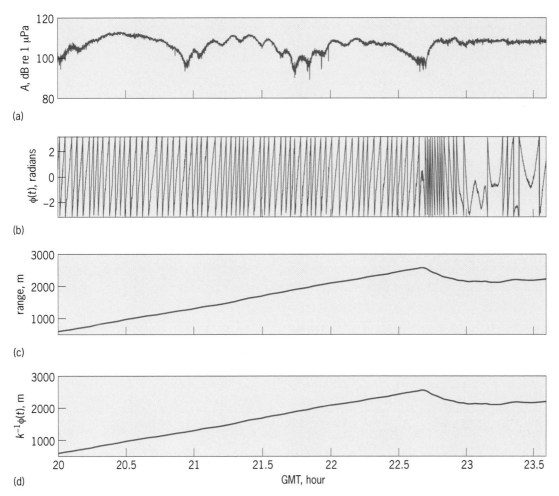

Fig. 2. MOMAX I 50-Hz data versus time. (*a*) Pressure magnitude. (*b*) Pressure phase. (*c*) GPS source-receiver range. (*d*) Product of k^{-1} and unwrapped phase. (*After G. V. Frisk, K. M. Becker, and J. A. Doutt, Modal mapping in shallow water using synthetic aperture horizontal arrays, Proceedings of the OCEANS 2000 MTS/IEEE Conference and Exhibition, Providence, RI, vol. 1, pp. 185–188, September 2000, © 2000 IEEE*)

Positioning System (GPS) navigation, and radio telemetry (**Fig. 1**). A key feature of the experiment was the establishment of a local differential GPS between the source ship and each buoy, thereby enabling the determination of the positions of the buoys relative to the ship with submeter accuracy. *See* DIFFERENTIAL GPS.

The magnitude A and phase ϕ of 50-Hz data, corresponding to a wavelength of 30 m (100 ft), are shown in **Fig. 2***a* and *b*. The magnitude displays a complex interference pattern among several modes of propagation, while a comparison of Fig. 2*b* and *c*, which shows the GPS-derived source-receiver range, suggests that the phase rate is directly correlated with the source-receiver range rate. Specifically, between 20 hours Greenwich Mean Time (GMT) and 22.7 hours, the phase wrapping (the increase of the phase through repeated cycles of 2π radians) is observed at an almost constant, positive rate, corresponding to the source and receiver opening in range at almost constant speed. At 22.7 hours, the phase rate is zero, corresponding to no source-receiver motion. Immediately following 22.7 hours, the phase rate is high and negative as the source and receiver

close in range at increased speed. This segment is followed by a period, beginning at about 23.2 hours, of slow, variable phase rate corresponding to low, variable source-receiver speeds.

In fact, when the model described in Eq. (4) is applied to these data using $c = 1490$ m/s (4889 ft/s), the excellent agreement shown in **Fig. 3** is obtained for source-receiver speeds up to 0.7 m/s (1.3 knots). Furthermore, by substituting the initial GPS range measurement at 20 hours and the 50-Hz unwrapped phase data into Eq. (5), inferred ranges in Fig. 2*d* are obtained that show close agreement with the GPS-derived ranges in Fig. 2*c*.

The phase-rate model developed in the context of these shallow-water acoustic experiments has been shown to be applicable to a broad band of frequencies (20–475 Hz), ranges up to 9200 km (4967 nautical miles), depths of several kilometers (several nautical miles), and speeds up to 3.2 m/s (6 knots). The full range of its validity, including the minimum source-receiver speed required to produce phase-rate variations that dominate fluctuations induced by oceanographic effects, has yet to be determined. The possibility of relaxing the assumption of knowing the

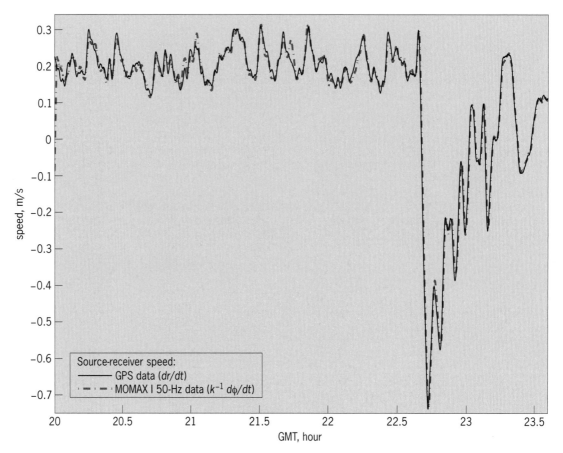

Fig. 3. Comparison of GPS-derived source-receiver speed with phase-model results obtained using MOMAX I 50-Hz data. (*After G. V. Frisk, K. M. Becker, and J. A. Doutt, Modal mapping in shallow water using synthetic aperture horizontal arrays, Proceedings of the OCEANS 2000 MTS/IEEE Conference and Exhibition, Providence, RI, vol. 1, pp. 185–188, September 2000, © 2000 IEEE*)

source frequency, perhaps by processing data from several receivers simultaneously, is also being examined.

For background information *see* ANTISUBMARINE WARFARE; DOPPLER EFFECT; SATELLITE NAVIGATION SYSTEMS; SOUND; UNDERWATER SOUND in the McGraw-Hill Encyclopedia of Science & Technology.

George V. Frisk

Bibliography. C. A. Coulson and A. Jeffrey, *Waves: A Mathematical Approach to the Common Types of Wave Motion*, Longman, London, 1977; G. V. Frisk, *Ocean and Seabed Acoustics: A Theory of Wave Propagation*, Prentice Hall, Englewood Cliffs, NJ, 1994; J. B. Keller and J. S. Papadakis (eds.), *Wave Propagation and Underwater Acoustics*, Springer-Verlag, Berlin, 1977; H. Medwin and C. S. Clay, *Fundamentals of Acoustical Oceanography*, Academic, Boston, 1998.

Wastewater reuse

Treated wastewater effluent can be used for beneficial purposes such as irrigation, industrial processes, and municipal water supply. Several developments have prompted wastewater reuse, including short-ages of fresh water, stringent quality requirements for treated wastewater effluent, and advances in treatment technology.

Traditional sanitary engineering. Traditional sanitary engineering deals with projects involving the municipal water supply and wastewater disposal (**Fig. 1**). Cities usually obtain their fresh-water supply from surface or ground-water sources. After use, about 80% of the fresh water becomes wastewater, which is collected by a sewerage system and conveyed to a treatment plant. The wastewater is treated and then discharged as effluent into receiving water, such as a river, an estuary, coastal water, or a ground-water aquifer.

Prior to 1970, most cities in the United States and other countries provided only primary wastewater treatment. Primary treatment removes suspended solids by a sedimentation process. In response to demand for pollution control and environmental protection, the U.S. Congress enacted the 1972 Federal Water Pollution Control Act Amendments (now called the Clean Water Act). The objective of the act is to restore and maintain the chemical, physical, and biological integrity of the nation's water supply. To achieve this objective, the act establishes nationwide minimum treatment requirements for all

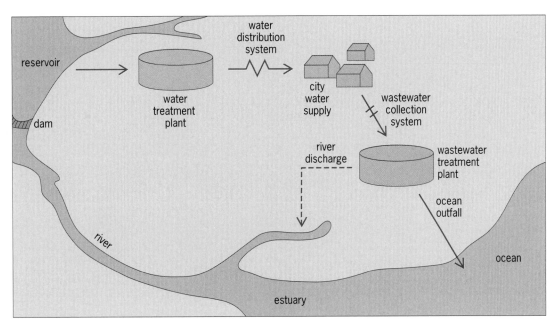

Fig. 1. Traditional sanitary engineering system.

wastewater. For municipal wastewater discharge, the minimum requirement is secondary treatment, which removes 85% of both biochemical oxygen demand (BOD) and total suspended solids. BOD is a measurement of oxygen-demanding organic wastes. Secondary treatment is achieved by biochemical decomposition of organic waste materials, followed by further sedimentation. In situations in which these minimum treatment levels are not sufficient, the act requires additional treatment.

Guidelines and treatment requirements. In 1995, about 140 billion gallons (530 billion liters) of fresh water per day were withdrawn for irrigation purposes and only about 40 billion gallons (150 billion liters) of fresh water per day for the United States municipal water supply. Irrigation is a consumptive use of water, as most of the water is lost by evaporation and transpiration and is not available for reuse. The municipal water supply is a nonconsumptive use of water, since about 80% of it eventually becomes wastewater. As the shortage of fresh water becomes more serious and the quality of wastewater effluent improves, reuse of municipal wastewater becomes one of the most attractive water resources management alternatives.

Municipal wastewater reuse is classified as nonpotable and potable. Nonpotable reuse applications include irrigation, industrial processes and cooling, and recreational impoundment (in lakes or reservoirs for public use). Potable reuse water is sent back to water supply reservoirs or directly to water treatment plants.

There are no federal regulations governing wastewater reuse in the United States. All regulations are enforced at the state level. The Hawaii Department of Health has guidelines that specify three types of reclaimed water, and the treatment and limitations of each. The highest-quality reclaimed water, designated R-1 water, is secondary-treated wastewater that has undergone filtration and intense disinfection. It is deemed acceptable for public contact and can be used for nearly all nonpotable purposes. Next in quality is secondary-treated wastewater that has undergone disinfection; it is designated R-2 water. The lowest-quality reclaimed water, designated R-3 water, is undisinfected secondary-treated effluent. The use of R-2 and R-3 water is restricted.

Fecal coliform bacteria found in the intestinal tract of warm-blooded animals are used as an indicator of microbiological purity in the United States. The amount of fecal coliform in a water sample is measured in colony-forming units (CFU). Most states set a maximum of 23 CFU/100 milliliters for unrestricted irrigation reuse or irrigation of edible crops, sports fields, and public parks; California and Arizona require 2.2 CFU/100 milliliters for higher-quality reclaimed water that can be used for recreational impoundment.

Artificial ground-water recharge of reclaimed water into potable aquifers (indirect potable recharge) can be accomplished by using injection wells or by surface spreading. The reclaimed water must meet drinking water standards if it is recharged into potable aquifers; it must meet drinking water standards after percolation through the vadose zone if it is recharged by surface spreading. The vadose zone (or unsaturated zone) occurs immediately below the land surface, where soil pores (or interconnected openings) may contain either air or water. A saturated zone in which all soil pores are full of water usually underlies the vadose zone. While underground water occurs in both vadose and saturated zones, the term "ground water" applies only to the saturated zone. Secondary treatment with filtration and

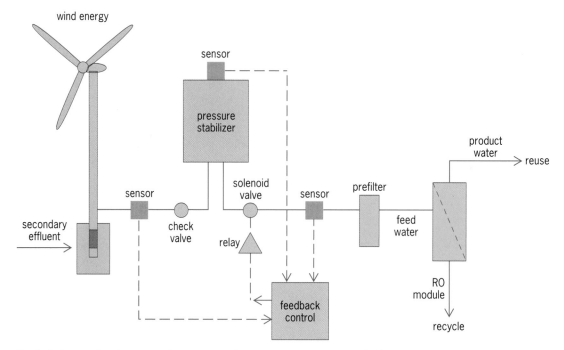

Fig. 2. Prototype of a wind-powered reverse osmosis system to process wastewater for reuse.

disinfection is required before recharging reclaimed water. Sometimes, other advanced treatment methods, such as carbon adsorption and precipitation, are applied to further remove dissolved solids or salts, nutrients, and organic and inorganic chemicals.

In the United States, water is traditionally disinfected with chlorine gas. However, there has been increasing concern about the adverse health effects caused by chlorination by-products such as trihalomethanes and chloroform. Pressure-driven membranes that remove microorganisms effectively are a desirable disinfection alternative to chlorine. Such membranes are also effective in removing dissolved solids, natural organic matters, and organic and inorganic chemicals from secondary-treated wastewater. Therefore, the advancement of pressure-driven membrane technology would significantly enhance the prospect of wastewater reuse.

Pressure-driven membrane technology. Pressure-driven membrane processes are divided into four groups based on membrane pore size: microfiltration (MF), ultrafiltration (UF), nanofiltration (NF), and reverse osmosis (RO). The required pressure for each membrane process is inversely proportional to the pore size. The membrane process to remove the largest particle (MF) requires the least pressure, and the membrane process to remove the smallest particle (RO) requires the greatest pressure. MF and UF are sieve processes whereby particles are removed because the membrane size openings are smaller than the particles. When two solutions of salts are separated by a semipermeable membrane, osmosis allows water molecules to pass through the membrane from the solution of higher salt concentration into the solution of lower salt concentration. The RO process, which reverses the direction of water, can be accomplished by applying to the higher-salt-concentration solution a pressure that is higher than the natural osmotic pressure. NF uses sieve and RO processes.

RO was developed in the 1960s with the advent of the asymmetric cellulose acetate membrane. Since then, its popularity has steadily risen because of further advances made in membrane technology with the use of different materials and membrane configurations. These advances have led to the development of ultralow-pressure RO membranes that require decreased operating pressures without sacrificing their salt rejection capabilities. Progress has also been made in overcoming membrane fouling and extending the life of the membrane.

Pressure-driven membranes are ideal for post-secondary-treatment of municipal wastewater in coastal areas. In these areas, brackish water or seawater enters into the sewer infrastructure due to infiltration and inflow, causing high salinity or total dissolved solids (TDS) in the wastewater flow into treatment plants. Because secondary treatment removes very little TDS, traditionally treated effluent would have a high salt content that would reduce the effectiveness of disinfection by chlorine. The wastewater would also be unsuitable for irrigation because a salinity can adversely affect plants and soil. Pressure-driven membranes can provide effective disinfection as well as TDS removal.

Pressure-driven membrane technology is the key for expanding wastewater reuse in the future. San Diego is developing a plan to use RO, along with a few pretreatment and posttreatment processes, to reclaim and purify its wastewater effluent before it is returned to the city's water supply reservoir. Water management authorities in the Netherlands are also studying the use of membrane technology in waste-water treatment.

The high amount of energy required to create and maintain feed water pressure is a major problem for the expanded use of pressure-driven membranes in water and wastewater treatment systems. To address this problem, new technology is being developed that integrates the use of natural energy with pressure-driven membranes. **Figure 2** shows a prototype of a wind-powered RO system that was constructed on Coconut Island off the windward coast of Oahu, Hawaii. This system consists of a 30-ft-tall (10-m) multivaned windmill, an ultralow-pressure RO membrane, a flow/pressure stabilizer, a prefilter, and a feedback control mechanism. Under a moderate wind speed of 5 m/s, this system can process a flow of 22 m^3/day and reduce the TDS content from 3100 mg/L to 50 mg/L. Another new technology is an ocean-wave-powered RO desalination system that was successfully tested in the Caribbean islands by University of Delaware researchers.

The field of water and wastewater engineering is undergoing rapid changes. It is expected that, through proper source control and treatment, the difference between water and wastewater will diminish so that they will eventually be recognized as the same kind of resources.

For background information *see* DESALINATION; FILTRATION; MEMBRANE SEPARATIONS; OSMOSIS; SEWAGE TREATMENT; WATER DESALINATION; WATER TREATMENT; WIND POWER in the McGraw-Hill Encyclopedia of Science & Technology.   Clark C. K. Liu

Bibliography. H. Bouwer, Role of groundwater recharge in treatment and storage of wastewater for reuse, *Water Sci. Technol.*, 24(9):295–302, 1991; Y. Kamiyama et al., New thin-film composite reverse osmosis membranes and spiral wound modules, *Desalination*, 51:79–92, 1984; D. Okun, Distributing reclaimed water through dual systems, *J. AWWA*, 89(11):52–64, 1997; T. Richardson and R. Trussel, Taking the plunge, *Civ. Eng.*, 67(9):42–45, 1997.

Waterjet (mining)

The resurgence of interest in hydroexcavation—the use of high-pressure waterjets for cutting and removing material—involves new areas of development that have taken it a long way from its original use. While hydroexcavation was first used for bulk material removal, a combination of mechanical and waterjetting tools followed, and recently waterjets have been used for localized fragmentation in drilling and rock slotting applications. Rock slotting is the practice of cutting a channel in rock, generally to provide a free surface that the rock can be broken into, or to isolate a mass of rock that can then be removed.

Mining applications. The use of waterjets in mining began with the low-pressure flushing of relatively soft materials in the goldfields of California. In those days, the pressures were generated by gravity, and the power to mine the soft ore was generated by using large water volumes and high flow rates. With monitor nozzles that could exceed 3 m (10 ft) in length, jets were produced that could reach several tens of meters, and break down the hillsides into the constituent grains of material. This made it easy to extract and collect the gold from the resulting slurry. The technique was then adapted for underground mining of coal and other minerals with the use of pumped water, increasing jet pressures to 10 megapascals (1450 lb/in.^{2}). In the conditions where this technique worked best, steep dipping coal seams (layers of coal that appear to cross the tunnel at an angle), the jet stream could reach up into the coal a distance of 30 m (100 ft) from the operating nozzle, washing the coal down into collection troughs, or flumes, in the access tunnels. Since no operators or equipment were needed in these mining areas, safety and productivity were significantly improved. However, the conditions required for effective use of this tool—particularly the need for dipping seams—have minimized its impact on the industry in the United States. Mines in eastern Europe and Asia have adopted the technology, and where conditions are right it is very effective.

Waterjets were first used in the Western Hemisphere in combination with conventional mechanical cutting tools. By running a high-pressure waterjet into the crushing zone (the volume of material immediately under a mechanical cutting tool), a significant gain in the performance of the machine was shown, relative to that of the cutter alone. Given that the initial successful experiments were in the hard quartzites of South Africa, it has been interesting to see that the focus of tool development has been in cutting softer rock. The definition of soft rock is relative, and the major advantage for the combination has been in cutting the harder rocks found in coal mining. The reduction in cutting hazards and the ability to use a lower-weight machine have been strong benefits to conversion. Typically, the jet pressures used have been around 70 MPa (10,000 lb/in.^{2}).

Concrete repair. Large-volume excavation generally requires higher volumes of water, which has proved to be a negative incentive. While progress in mining excavation has slowed, the peculiar advantages of waterjets have brought them an increased share of the concrete repair market. Where concrete is damaged, it is most frequently cracked. Thus by applying a pressurized stream to the cracked concrete, it is possible (if the pressure is correctly tailored) to remove the damaged concrete, while leaving the intact concrete in place.

The advantage of the tool goes beyond this. The overall low impact force reduces the vibrations imposed on the reinforcing rods located in the concrete, which are vulnerable under jackhammer attack. Further, the waterjet removes the soft phase, or cement matrix, while leaving the aggregate largely undamaged (it is removed by losing surrounding support). This means that the surface left where layers have been reduced is much rougher than that achieved conventionally. As a result, there is a much better cohesion between the repair layer and the underlying concrete. This improved bonding makes the interface less vulnerable and extends the life of the repair. The growth of this, and the industrial high-pressure cleaning market, have meant that the pumping and other support equipment, which were items of high cost and poor reliability over a decade ago, are now less expensive and equipment now reliably operates commercially at pressures above 280 MPa (41,000 lb/in.^{2}). With the growth of these industrial markets, the relatively small market that mining represents has limited development in this field.

Waterjets do, however, have several unique advantages, such as the ability to transmit energy down very small flexible conduits, and the low levels of force required to direct the stream. Thus, in specialized operations, high-pressure tools are starting to find a market.

Quarrying. The first market for specialized high-pressure tools was in the use of waterjets to cut granite (see **illus.**). Although first investigated a number of years ago, it has been only recently that waterjet quarrying of stone has become a commercial reality. As with other applications, there is a significant trade-off between pressure and flow rate choices in selecting the best tool for the operation. Waterjets can cut into typical granite, based on crack growth around the crystal interfaces, at pressures of 70–100 MPa (10,000–14,500 lb/in.^{2}). One of the earlier demonstrations that this was a practical tool came with the carving of the Missouri Stonehenge at the University of Missouri-Rolla. More recently, this has been followed by the carving of the Millennium Arch, a sculpture in which the figures of a man and a woman have been carved from the legs of a 5-m-high (16-ft) arch and the figures have then been polished and positioned some 15 m (50 ft) away.

The precision cutting and minimization of waste that this demonstrates has particular benefit in the dimension stone (cut into shaped blocks) business. Although the granite resources apparently available to quarry owners are large, at any given site this is only superficially true. The granite is not a consistently homogeneous rock. It has varying properties with direction, and frequently contains flaws, fissures, or layers of rock of different color and consistency, which reduce the volume available to market. Given that the primary quarrying method uses flame torches, which channel into the rock with a slot wider than the sole of the operators shoe, and add to this the zone of heat-weakened rock left on the side of the cut, the cutting operation is expensive. Add to this the health and safety concerns of the "old" technology—noise levels of around 140 dB and the generation of fine clouds of dust with the risk of respiratory problems—and a clear technical need becomes evident.

Over the past 4 years this need has increasingly been met by the use of an oscillating jet lance, which cuts into the granite at a pressure of around 250 MPa (38,000 lb/in.^{2}). A flow rate of about 27 liters/min (7 gal/min) is directed through a single nozzle, which oscillates over the face of the slot as it is fed up and down its length. Typically, the slots are some 3–5 m (10–16 ft) deep, and the machine has been automated, so that the lance will advance into the cut to an initial distance of some 6 m (20 ft). After this, the machine can be moved forward again, so that overall lengths of more than 30 m (100 ft) have been created as primary cuts in the quarry floor. From these, blocks can then be isolated and removed from the solid for processing into slabs for commercial use. Cutting rates vary between 1 and 1.75 m^2/h (11 and 19 ft^2/h) as a function of the granite type, which is roughly 25–50% higher than the burner production rates in the same stone. The slots are some 4–7 cm (1.6–2.8 in.) wide, but because of the lack of damage to the walls, the block surfaces can be used as the starting surface for subsequent processing.

The machines that have been installed in both the United States and Europe are fully automated so that several machines may be controlled by a single

High-pressure waterjets are used to outline granite blocks in a quarry. (*NED Corp.*)

operator who does not have to be present during the operation, but who may be summoned at the end of a cut or if a problem arises. The noise level (98 dB) is considerably less than that of the flame burner. Depending on the granite, the cost benefit will also vary, but figures of $49.62/m^2 as opposed to $73.83 m^2 for a burner have been quoted.

Drilling. Slot cutting is a narrow niche market for waterjet use. The flexibility of the tool in applying high cutting pressures at the end of a small, possibly flexible cutting head makes the tool potentially useful for drilling. The most significant work in this field is taking place in Australia, although much of the pioneering effort took place in the United States. To illustrate the benefits that can come from the use of waterjets, consider the need to drill out from vertical well bores in order to more effectively recover oil or gas from a reservoir. A flexible high-pressure hose can make a turn from vertical to horizontal in a tight radius (around 20 cm; 8 in.) and transmit power to a drill so that it might advance over half a kilometer into coal to allow recovery of methane. Premining methane has both economic and safety benefits which are now being pursued commercially after development, first at the University of Missouri-Rolla, and then at the Center for Mining Technology & Equipment in Brisbane, Australia.

For background information *see* COAL MINING; DRILLING AND BORING, GEOTECHNICAL; GRANITE; MINING; PLACER MINING in the McGraw-Hill Encyclopedia of Science & Technology.   David A. Summers

Bibliography. A. W. Momber, *Water Jet Applications in Construction Engineering*, Balkema, Rotterdam, 1998; D. A. Summers, *Waterjetting Technology*, E & FN Spon, 1995; R. A. Tikhomirov et al., *High-Pressure Jet Cutting*, transl. from the Russian by V. Berman, ASME Press, New York, 1992.

Wave-particle duality

A fundamental tenet of quantum mechanics is that every particle also has a wave nature and every wave also has a particle nature, at least in principle. For light, which in classical physics is an electromagnetic wave, the particle nature was first postulated by Albert Einstein in 1905, after Max Planck's introduction of the quantum of action in 1900. The particles of light are now called photons and have found abundant experimental confirmation.

For massive particles, the wave nature was first postulated by Louis de Broglie in 1924. For a particle of mass m and speed v, the de Broglie wavelength is $\lambda_{dB} = h/mv$, where $h = 6.6 \times 10^{-34}$ J · s is Planck's quantum of action.

The wave nature of matter has found experimental confirmation for many, very diverse particles, from electrons and neutrons to atoms and most recently even for molecules as complex as the fullerenes C_{60} and C_{70}. While diffraction of matter waves, particularly of electrons and neutrons, has become a standard tool in many areas such as nuclear physics, atomic physics, and solid-state physics, the wave-particle duality itself continues to be of fundamental philosophical significance.

Double-slit experiment. The essence of wave-particle duality emerges in discussing the famous double-slit experiment (**Fig. 1**). As a Gedanken (thought) experiment, it was already a tool in the debate between Einstein and Niels Bohr about epistemological questions in quantum mechanics. Radiation, be it light or massive particles, passes a diaphragm with two openings. Behind the diaphragm, fringes are observed which can be explained as being due to interference of waves passing the two slits. The interference is also observed when the intensity is so low that the particles are detected one by one. The observed interference fringes on the detection screen can be calculated via the appropriate wave function, whose square gives the probability to observe a particle at a given location. This naturally implies that the interference fringes will also be obtained if they are recorded individual particle by individual particle.

The philosophical conundrum arises when preconceived classical notions are applied to an analysis of this experiment. For example, if the particles are thought of as localized entities, they must pass through either slit to arrive at the observation screen. But how can an individual particle know whether the other slit is open or not? The modern interpretation, suggested by Bohr, is that one should not talk about a specific property of a quantum object

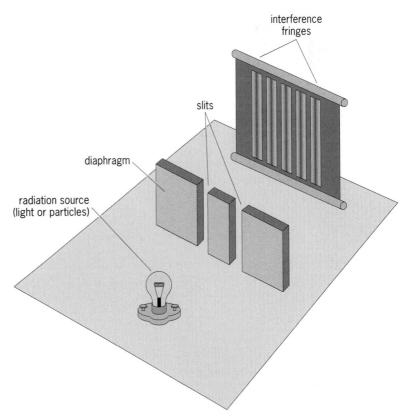

Fig. 1. Principle of the double-slit interference experiment, showing wave-particle duality.

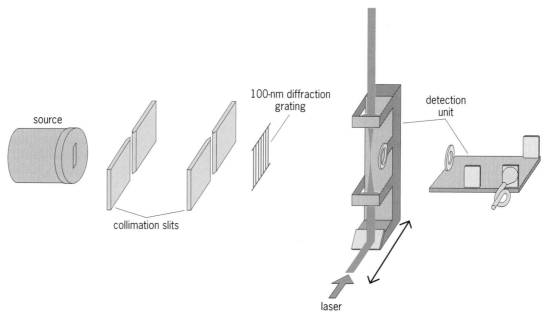

Fig. 2. Principle of the fullerene interference experiment.

without explicitly specifying the apparatus to determine that property.

The experimentalist therefore decides between the wave and the particle properties in the experiment by choosing the appropriate apparatus. Nature then gives a random answer for the single outcome within the probabilities given by the wave function. For example, with equal probability either one of the two slits will result as the location of the particle, should it be measured. Alternatively, if there is an interference pattern, the position where the particle will show up is random.

Conditions for quantum interference. On a much deeper level, wave-particle duality is just one example of complementarity in quantum mechanics, a notion also introduced by Bohr. Two concepts are complementary if complete knowledge of the one implies complete ignorance of the other and vice versa. Clearly, there are also intermediate states possible. Partial but fuzzy knowledge of which path the particle went through implies a still visible interference pattern of reduced contrast.

Quantum interference therefore arises when it is impossible to know—even in principle—which path the particle took. One might think that this requires the knowledge of an external observer. But actually, full quantum interference is seen only when there is no information present anywhere in the universe as to which path the particle took, independent of whether we care to take notice of that information or even whether we are capable of reading it or not.

Observation of quantum interference is expected to be increasingly difficult with increasingly large objects. The main reason is that quantum interference implies that a system is isolated from the environment; yet the larger an object becomes, the more degrees of freedom it has, that is, the more easily it exchanges information with the environment.

Fullerene interference experiment. The experiment with the most massive and complex objects showing de Broglie interference thus far (**Fig. 2**) was successful with the fullerene molecules C_{60} and C_{70} (although the curves shown here are for C_{60} only). The fullerenes were evaporated in an oven at a temperature of around 900 K (1160°F). They passed through fine collimation slits and transversed a silicon nitride grating whose bars were spaced at a period of 100 nanometers with 50-nm openings. The interference pattern was observed by scanning a very fine ionizing laser beam across the molecular beam. The ions were recorded as a function of the laser position.

This experiment has a number of interesting features: (1) The de Broglie wavelengths of 2–4 picometers are more than 100 times smaller than the size (1 nm) of the fullerenes, indicating that size itself is not crucial. (2) In the experiment resulting in **Fig. 3a**, the fullerenes still have their full thermal velocity distribution. Nevertheless, it is possible to clearly observe the interference pattern. (3) Because of the high temperature of the fullerenes, each individual molecule is in a quantum state different from all other molecules, as their rotational and vibrational degrees of freedom are highly excited. By that very feature alone, different fullerenes cannot interfere with each other. The experiment is truly a single-particle interference phenomenon. This is also guaranteed by the low intensity of the fullerene beam in the experiment. (4) Most importantly, again because of the high temperature of the fullerenes, they are not completely isolated from the environment. Indeed, each fullerene is expected to emit a few photons on its path from the oven to the plane of observation.

Then, why do these photons not disturb the interference pattern? The answer is obtained simply

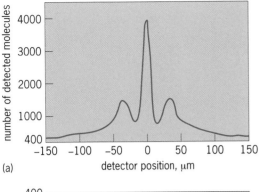

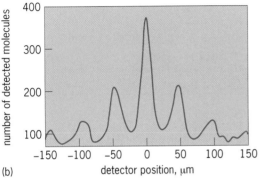

Fig. 3. Interference patterns of C_{60} molecules at a distance of 1.2 m (4 ft) after the 100-nm grating. (*a*) Interference pattern of molecules that have a full thermal velocity distribution. Besides the central peak, one interference minimum and one maximum can be seen on each side. The minimum is due to destructive interference of waves passing through neighboring slits. The first maximum on each side is due to constructive interference. (*b*) Interference pattern under similar conditions but with velocity selection (wavelength, $\lambda \sim 5$ pm, $\Delta\lambda/\lambda \sim 0.16$).

arated on a scale comparable with the wavelength of the emitted photons. The transition between quantum and quasi-classical physics could then be studied in detail. Another interesting direction of research would be to extend the techniques to the study of interference of biological macromolecules. It seems to be possible to develop methods which are scalable in mass. Thus one could investigate experimentally whether there are any fundamental obstacles for the observation of quantum interference of massive and complex mesoscopic objects. While it is not expected that any fundamental limits will be encountered, further investigations of quantum phenomena in the mesoscopic domain will certainly push forward the frontiers of current knowledge and also of coherent manipulation technologies.

For background information *see* COHERENCE; FULLERENE; INTERFERENCE OF WAVES; NONRELATIVISTIC QUANTUM THEORY; QUANTUM MECHANICS; QUANTUM THEORY OF MEASUREMENT; UNCERTAINTY PRINCIPLE in the McGraw-Hill Encyclopedia of Science & Technology.

Markus Arndt; Olaf Nairz; Anton Zeilinger

Bibliography. M. Arndt et al., Wave-particle duality of C-60 molecules, *Nature*, 401:680–682, October 1999; N. Bohr, Discussions with Einstein on epistemological problems in atomic physics, in P. A. Schilpp (ed.), *Albert Einstein: Philosopher-Scientist*, Library of Living Philosophers, Evanston, 1949; R. P. Feynman, R. B. Leighton, and M. L. Sands, *The Feynman Lectures on Physics*, vol. 3, Addison-Wesley, 1964.

by applying the criterion that interference will be observed if there is no information about the path taken by the particle. Indeed, the emitted photons are expected to have a wavelength of a few micrometers, which is much larger than the spatial separation of the interfering paths through the diffraction grating, which amounts to 100 nm. Under these circumstances no optical instrument can resolve the path separation, which means that the photons carry no useful path information and the interference pattern can still persist. Only if very many long-wavelength photons were emitted, or a few photons of much shorter wavelength, would sufficient path information be carried into the environment and result in a destruction of the interference pattern.

While Fig. 3*a* still is of limited contrast due to the large velocity spread, a narrow velocity class was selected in a more recent experiment. As expected, almost perfect interference contrast was observed (Fig. 3*b*).

Further experiments. Future experiments will have to focus on various points. It will be interesting to investigate the possibility of destroying the interference pattern through a controlled coupling with the environment. This could be done either by heating the fullerenes up a great deal or by building an interferometer where the interfering beam paths are sep-

Wind turbines

Over the centuries windmills and wind turbines have been used to grind grain, pump water, and generate electricity. The earliest windmill design on record dates back to the tenth century in West or Central Asia, where vertical-axis windmills were used for grinding corn. Windmills were primarily used for milling and pumping until 1888, when Charles F. Brush, an inventor and manufacturer of electrical equipment in the United States, designed and built the first windmill for the generation of electricity. For over 20 years his 12-kW horizontal-axis windmill was used to charge batteries for lights on his estate in Cleveland, Ohio. It took another century of developments in structural, aerodynamic, and electrical science and engineering to progress from the massive Brush windmill with its multiblade rotor to the sleek two- and three-bladed megawatt wind turbines of today. It was not until the last part of the twentieth century that a resurgence of interest in wind power sparked by the energy crisis and government tax incentives led to the development of relatively simple but efficient wind turbines and the first wind plants in California. After a bumpy start in the 1980s, the worldwide capacity of installed wind power grew rapidly to 12,455 MW at the end of 1999, with 20% (2490 MW) of this wind power residing in the United

States. This growth rate has not shown any sign of subsiding, and by the year 2007 the American Wind Energy Association (AWEA) predicts there will be nearly 48,000 MW of new capacity.

Characteristics. A large variety of turbines have been tested and operated over the years, but the wind industry appears to be slowly converging on a single configuration for utility-scale wind turbines. With the steady growth in turbine size, the horizontal-axis wind turbine (HAWT) has become the preferred configuration, with the growth-limited vertical-axis turbine (VAWT) becoming a configuration of the past. Furthermore, most newly developed horizontal-axis turbines have the rotor upwind of the support tower instead of downwind. Downwind configurations are potentially simpler and lighter because of the inherent yaw stability of the rotor (that is, the rotor yaws freely to face into the wind without active control) and the reduced danger of tower strikes by the blades. The latter characteristic allows for more flexible lightweight blades that reduce loads, turbine weight, and cost. However, the rotor must operate in the wake of the tower, and the resulting rotor-wake interaction causes noise, a loss in power, and unsteady blade loading. With wind turbines pushing toward output levels of 1 MW and higher with corresponding rotor diameters of 60 m (200 ft) and larger, the number of rotor revolutions per minute (rpm) has dropped and the advantages of the downwind rotor have become less definite—hence, the trend toward greater use of rigid three-bladed upwind rotors. Currently two different types of upwind rotor configurations are favored: fixed-blade-pitch, constant-speed rotors; and variable-pitch, variable-speed rotors. In the fixed type, blade stall is used to passively limit the maximum power output; whereas in the variable type, maximum output is controlled by actively varying the pitch of the

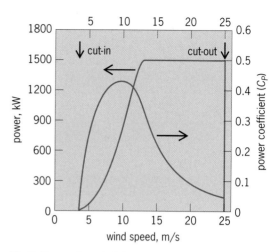

Fig. 2. Power characteristics of a typical 1.5-MW, 65-m-diameter (213-ft), variable-speed wind turbine.

blades in addition to letting the rpm of the rotor vary with wind speed. **Figure 1** shows a horizontal-axis wind turbine with upwind rotor whose blade pitch angle is actively controlled by an electric or hydraulic mechanism.

Performance. Figure 2 shows a power curve for one of the newer horizontal-axis wind turbines with a cut-in wind speed of 3.5 m/s (11–5 ft/s), a rated power of 1.5 MW reached at a wind speed of 13.0 m/s (42.7 ft/s), and a cut-out wind speed of 25 m/s (82 ft/s). This wind turbine has a three-bladed, upwind, variable-speed, pitch-regulated rotor with a diameter of 65 m (213 ft). It has been shown that, based on linear momentum theory, the maximum power that can be extracted from the wind is 16/27, or 60%, of the available wind power. For the wind turbine in Fig. 2 the efficiency or power coefficient, C_P, is determined by dividing the power output by the available wind power, $\rho V_w^3 A/2$. Here ρ is the air density (1.225 kg/m³; 0.002377 lb-s²/ft⁴), V_w is the unobstructed wind speed, and A is the swept area of the rotor (3318 m²; 35,715 ft²). This power coefficient based on the electric power output includes gearbox and generator losses. The maximum power coefficient of the 1.5-MW machine evaluated here is 43%, reached at a wind speed of 10 m/s (33 ft/s). Although this is well below the theoretical aerodynamic maximum efficiency of 60%, it is significantly higher than the efficiency of earlier wind turbines.

Rotor blade designs. A key element of the high efficiency and the relatively high power levels at low wind speeds is the rotor blade section shape. In the past, these section shapes were limited to airfoils designed for aircraft and rotorcraft and published in the open literature. More recently, airfoils specifically designed for wind turbines have become available. Airfoil aerodynamic performance characteristics that require special attention include lift-to-drag ratio, maximum lift coefficient, and sensitivity of lift to surface fouling. The lift-to-drag ratio indicates how much drag the airfoil generates for a given amount

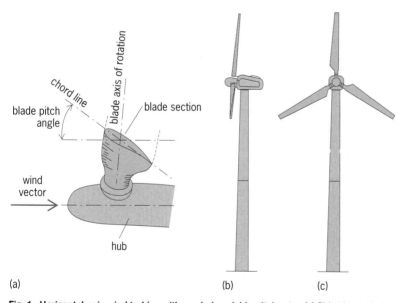

Fig. 1. Horizontal-axis wind turbine with upwind, variable-pitch rotor. (a) Side view of hub of rotor, showing base of one blade. (b) Side view. (c) Front view.

of lift, with low drag levels being especially critical at low wind speeds. The maximum lift coefficient indicates the lift level at which blade stall occurs. Reduced sensitivity to surface fouling is especially decisive for stall-regulated turbines because a loss in maximum lift due to contamination (such as insects or dust) results in a loss in power output and hence revenue.

An important parameter in the analysis of the aerodynamic performance characteristics of rotor section shapes is the Reynolds number, $Re = \rho V_r c/\mu$, where V_r is the vector sum of the wind speed and the rotational speed, c is the section chord length, and μ is the absolute viscosity of air. Typical Reynolds numbers range from approximately 0.5 million for blade sections on smaller wind turbines to 6 million for sections in the tip region on large wind turbines. In **Fig. 3** the performance characteristics of four airfoils, including one turbine-specific airfoil (S809), are compared for a typical Reynolds number of 2.0 million. The S809 was designed for stall-regulated wind turbines and hence has a low (restrained) maximum lift coefficient of approximately unity. The other three airfoils have much higher maximum lift coefficients, ranging from 1.5 for the NACA63-415 to 2.0 for the LS(1)-0417Mod.

Comparison of the S809 and NACA63-415 performance characteristics demonstrates a downside to the philosophy of passive stall control design, with a loss in lift-to-drag ratio in the lift range preceding stall for the S809 ($0.75 < c_l < 1$, where c_l is the sectional lift coefficient). Because of this performance difference, NACA63-xxx-type airfoils are sometimes preferred. However, the high-lift characteristics of the latter are less than desirable for stall-regulated turbines because of their so-called double stall behavior with two distinct maximum lift levels depending on the inflow conditions. This trade-off between lift-to-drag ratio and satisfactory maximum lift behavior illustrates one dilemma that designers face when selecting blade section shapes for stall-controlled machines. For pitch-controlled variable-speed machines, a high maximum lift coefficient is actually beneficial, because it avoids stall at high loading conditions and leads to smaller blade chords and lower blade weight and cost. The pitch control and variable-speed capability allow these turbines to operate at maximum efficiency over a range of wind speeds, providing further improvements in energy capture.

Optimum size. The push toward larger wind turbines is largely fueled by the need to reduce the cost of electricity. The cost of electricity for wind ranges $0.04–0.06/kWh for state-of-the-art wind facilities. The goal is to drive the cost down to $0.025/kWh in order to make wind energy competitive with the cost of electricity of natural-gas-fired, combined-cycle facilities. The cost of electricity of a wind facility comprises capital cost, and operating and maintenance cost. Capital cost includes all expenditures for planning, equipment purchase, construction, and installation. The operating and maintenance cost includes

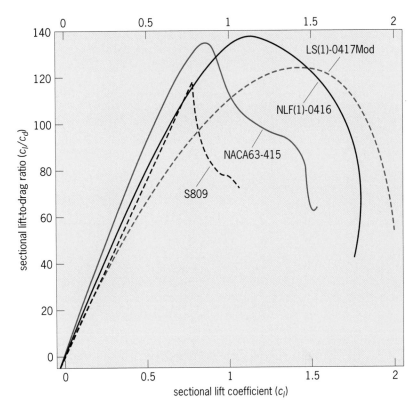

Fig. 3. Aerodynamic performance characteristics of four representative blade section shapes at Re = 2.0 million and natural transition (clean surface) conditions.

all expenditures to monitor, control, and maintain electric power output; and it is sensitive to the complexity and durability of the turbine equipment, the size and number of turbines in the wind facility, ease of maintenance, and cost of replacement parts. Currently, the operating and maintenance cost of a modern wind plant is approximately $0.01/kWh with major contributions coming from unscheduled maintenance. Because power capture is proportional to the rotor area and the wind speed cubed, it is clear why there is a trend toward megawatt turbines with large rotors installed on tall towers that minimize losses in wind speed due to the proximity of the ground (**Fig. 4**). However, large turbines require specialized equipment for turbine installation, rotor and drivetrain overhaul, and blade cleaning and adjustments. This drives up capital cost as well as operating and maintenance cost and raises the question of what turbine size is optimum. At this point it is unclear what the optimum size is, but it is undoubtedly site-dependent. For instance, turbines in the California hills with their complex wind patterns will likely be different from offshore turbines in northwestern Europe.

Prospects. Increasing fossil-fuel prices, fuel-shortage concerns, and rising levels of air pollutants and greenhouse gases have created a strong interest in the generation of electricity using "free," nonpolluting wind energy. A rapid growth in wind energy activity, including the development and installation of highly efficient turbines, can be seen worldwide,

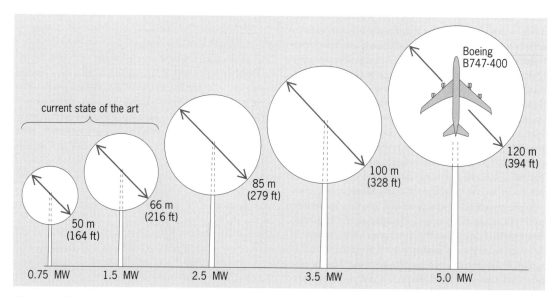

Fig. 4. Trend in size and rated power output of current and future wind turbines. (*After U.S. Department of Energy Wind Program Subcontractor Review Meeting, July 17–19, 2000*)

confirming that wind power is here to stay. However, more work is needed to make it less costly and more reliable.

For background information *see* AIRFOIL; ELECTRIC POWER GENERATION; ENERGY SOURCES; WIND; WIND POWER in the McGraw-Hill Encyclopedia of Science & Technology.   C. P. van Dam

Bibliography. A. C. Hansen and C. P. Butterfield, Aerodynamics of horizontal-axis wind turbines, *Annu. Rev. Fluid Mech.*, 25:115–149, 1993; S. Heier, *Grid Integration of Wind Energy Conversion Systems*, Wiley, Chichester, 1998; D. A. Spera (ed.), *Wind Turbine Technology*, ASME Press, New York, 1994; S. Wagner, R. Bareiss, and G. Guidati, *Wind Turbine Noise*, Springer, Berlin, 1996; J. F. Walker and N. Jenkins, *Wind Energy Technology*, Wiley, Chichester, 1997.

Wing drop

The phenomenon known as wing drop is experienced by an air vehicle during maneuvers at moderate to high angles of attack (AoA). It is caused by an abrupt reduction of lift from one side of the vehicle prior to the other side. This sudden asymmetric loss of lift creates a rolling moment, which causes loss of control of the airplane's roll attitude. Though typically an abrupt and momentary loss of control, these disturbances may be quite severe (up to 60–90°), resulting in abandonment of the task being performed in order to regain control. In operations at low altitudes, even a momentary loss of roll control could result in the loss of the aircraft.

Background. Since the advent of fighter aircraft that can operate at transonic speeds and angles of attack beyond which the airflow remains attached to the wings, there have been numerous examples of aircraft experiencing uncommanded lateral activity. This type of motion has been variously characterized as heavy wing, wing rock, or wing drop, depending on the particular manifestation in each aircraft. A recent encounter of wing drop was experienced during the Engineering & Manufacturing Development (EMD) phase of the U.S. Navy F/A-18E/F program. At transonic Mach numbers, the airplane initially exhibited abrupt roll-offs of 40–60°, passing through 8–10° angles of attack. Though clearly unsatisfactory for mission purposes, this characteristic also precluded performance of many other types of testing. This led to a concentrated effort by the Navy and its contractors to identify the aerodynamic cause of this characteristic and to determine a low-cost, low-impact remedy.

Immediate dramatic improvement was achieved by a modification of the aircraft control laws to provide a more aggressive flap schedule in the relevant flight regime, thereby permitting other test disciplines to advance. The final production solution incorporated a porous fairing over the wing-fold mechanism. Those minor alterations were sufficient to suppress any adverse rolling moment, but did little to provide a fundamental understanding of the flow physics causing the wing drop phenomenon.

At the direction of the Secretary of Defense, a special blue ribbon panel recommended that a national effort be undertaken to determine the underlying causes of the abrupt flow separation. Subsequently, a joint Navy/NASA Abrupt Wing Stall (AWS) program was formed with collaboration from industry and universities to identify figures of merit (FOM) which would indicate wing drop onset, and to provide design guidelines to prevent its occurrence in future air vehicles. The AWS program's approach to investigate the causes of wing drop proceeds on three fronts: numerical analysis through the use of computational fluid dynamics (CFD); experimental analysis through the use of static and dynamic wind tunnel testing of subscale models; and flight testing full-scale aircraft.

Physics. When an aircraft moves through the air, it produces lift by accelerating the flow over the top of the wing. At small angle of attack, the airflow remains attached to the wing, separating at the trailing edge and creating a small wake region. Increasing the angles of attack causes the airstream to move faster over the upper surface and moves the point at which the airflow separates from the wing forward. Further increases in angle of attack move the separation point forward and create a larger wake region until a maximum lift is produced. Beyond this angle of attack, the flow over the top of the wing is mostly separated and lift is reduced. This is known as wing stall. If this occurs very suddenly, asymmetries between the left and right wings can produce large rolling moments and, therefore, wing drop.

As the air moves over the wing, a small area of reduced-velocity flow close to the surface is created due to viscous effects. This region which divides the friction-dominated flow near the surface from the inertia-dominated air farther away is called the boundary layer. At a constant angle of attack, increasing the airspeed of the vehicle causes the air to move faster over both the upper and lower surfaces of the wing. But, since lift is produced by accelerating the upper-surface flow in relation to the lower-surface flow, shock waves formed by achieving supersonic speeds over the wing may be of sufficient strength to cause the boundary layer to separate from the wing and induce a loss of lift. This is referred to as shock-induced separation.

Aircraft maneuvering at transonic airspeeds encounter both of these phenomena. In the event that the separation occurs abruptly, it is unlikely that the separation will occur simultaneously on the two wings, introducing an undesirable rolling moment. The identification, mitigation, and hopefully, prediction and elimination of these adverse flight vehicle characteristics are the engineering challenges being addressed by the AWS program participants.

Computational fluid dynamics. Several computational codes have been employed to identify and understand the physical aspects of the abrupt wing stall. These codes solve the three-dimensional, compressible, Navier-Stokes equations that govern fluid flow. Both structured and unstructured grids are being employed on aircraft such as the F/A-18E and AV-8B to numerically solve these equations using high-speed computers. The grids created around the aircraft are difficult to construct, requiring several weeks or months to produce, and then additional weeks to compute the solutions and analyze the results. Typical structured grids necessary for the accurate solution of this complex flow consist of approximately 15 million grid points. Comparable unstructured grids contain about half as many points but require twice the computational memory to solve the equations. A further challenge is the application of turbulence models to describe the flow field in the boundary layer.

Wind tunnel testing. The experimental efforts of the AWS program are designed to (1) provide an understanding of the Abrupt Wing Stall phenomenon in general; (2) determine useful figures of merit from static and dynamic tunnel tests that can be used to predict Abrupt Wing Stall in flight; and (3) provide data for validation of computational fluid dynamics results. Thus far, the experimental program has involved three wind tunnel experiments—two static tests in the NASA-Langley 16-ft (5-m) transonic tunnel and one dynamic test in the Langley transonic dynamics tunnel (TDT). The objectives of the 16-ft transonic tunnel tests were to increase understanding of the stall progression over the F/A-18E and to assess potential figures of merit for static wind tunnel testing. The objective of the transonic dynamics tunnel test was to correlate the static figures of merit with the results of the dynamic test.

The tests in the Langley transonic tunnel were the first transonic tests of any aircraft conducted to provide an understanding of the Abrupt Wing Stall phenomena that utilized instrumented wings. Two left-side wings and one right-side wing of an 8% scale model of the F/A-18E were utilized. The first left-side wing included 66 static pressure ports and 20 dynamic pressure transducers on its upper surface, two wing tip accelerometers, and one wing-root-bending gauge. The right wing had 54 upper-surface static pressure transducers, two wing-tip accelerometers, and one wing-root bending gauge. The first test focused on the flow unsteadiness. It was found that when the model was tested in the region of Abrupt Wing Stall, dynamic movement was apparent. For example, the location of the shock wave moved forward and aft about 40% of the local chord length of the wing. Analysis determined that the shock movement was being caused by the unsteady nature of the flow and not by the dynamic response of the model and support structure being tested.

The second test in the Langley transonic tunnel utilized pressure-sensitive paint (PSP) as a means of establishing the pressure distributions over the wing in the angle-of-attack region where Abrupt Wing Stall is encountered. This technology was found to be extremely helpful in illustrating the flow physics over the wings. For example, when rolling moments were recorded on the model balance, an asymmetry in the extent of separation on the left and right wings was observed. Qualitative details in the flow field such as oblique shock waves, regions of recompression in front of forward-facing surfaces, and regions of high velocity were apparent and quantifiable from pressure visualizations. Regions of separation could be inferred by looking for regions of constant pressure over the middle and aft portions of the wing. Most importantly, the pressure-sensitive paint confirmed earlier computational fluid dynamics work, which pointed to the importance of the leading-edge wing extension or snag in the rapid, forward progression of the separation region.

The two transonic tunnel tests have been used to address the second major objective of the experimental program of determining useful figures of merit. On the basis of what had been learned previously, it was expected that rolling moment and its level of unsteadiness could be important indicators of asymmetric stall occurring between the left and right sides of the vehicle. It was also observed that simply looking at time-averaged values of rolling moment for an angle-of-attack range at constant heading does not always correlate with the presence or absence of Abrupt Wing Stall in flight. This absence of an asymmetric, time-averaged rolling moment can be due to model dynamics about the roll axis. While these dynamics result in large, fluctuating values of rolling moment, the time-averaged value of rolling moment may be minimal. This is why it is important to measure the level of unsteadiness. In fact, abrupt increases in the levels of unsteadiness in the rolling moment have been found to correlate with the onset of Abrupt Wing Stall in flight. In addition, it was found that the wing-root bending gauges gave very important complementary data. The wing root bending data more clearly showed areas in which the outboard portion of the wing would stall and lose lift as the model angle of attack increased through the Abrupt Wing Stall region of interest. This lift loss was much more obvious than that recorded by the model balance, which includes the integrated lift from the same wing panels but also includes lift from the wing leading-edge extension, the fuselage, and the empennage. It is highly recommended that future models include wing bending gauge instrumentation.

The dynamic test in the Langley transonic dynamics tunnel was an extension of the work in the transonic tunnel and included taking static force and moment data, similar to what was done during the transonic tunnel test, but also included taking free-to-roll data at the same transonic speeds. This test used a new, lightweight 9% scaled model of the F/A-18E. The dramatic results of this test included model activity that very much resembled both wing rock and, at times, wing drop. The Navy test pilot who discovered the Super Hornet's wing drop commented that these free-to-roll tests behaved identically to wing drop events that he had experienced in flight with the preproduction configuration. The data from this dynamic test are still under analysis, but it is hoped that this dynamic data set can help establish the utility of the static testing figures of merit.

Finally, the experimental data are being shared with the computational fluid dynamics members of the AWS program and are helping to guide the validation of the various codes and the different possible turbulence models.

Flight testing. Tests flights of a NASA F/A-18C with revised control laws are being planned to investigate the effect of wing flap deflections in relation to the Abrupt Wing Stall problem. Both left and right sides of the wing will be tufted and cameras mounted in the vertical tails to visualize the airflow over the wings during transonic maneuvering. Future flights of other aircraft are also being considered.

For background information see AERODYNAMICS; AERONAUTICAL ENGINEERING; AIRCRAFT DESIGN; AIRCRAFT TESTING; AIRPLANE; COMPUTATIONAL FLUID DYNAMICS; SUPERCRITICAL WING; TRANSONIC FLIGHT; WIND TUNNEL; WING; WING STRUCTURE in the McGraw-Hill Encyclopedia of Science & Technology.   Shawn H. Woodson

Bibliography. J. D. Anderson, *Modern Compressible Flow with Historical Perspective*, McGraw-Hill, New York, 1982; J. R. Chambers, *Independent Consultant Activities in Support of the NASA/DoD Abrupt Wing Stall (AWS) Program, Task 2— Historical Review of Wing Drop*, Ball Aerospace Systems Division Report, 1998; H. Schlichting, *Boundary-Layer Theory*, McGraw-Hill, New York, 1979.

X-ray astronomy

The *Chandra X-ray Observatory (CXO)*, the third of NASA's four great observatories (the other three being the Hubble Space Telescope, the *Compton Gamma-Ray Observatory*, and the *Space Infrared Telescope Facility*), was launched aboard the space shuttle *Columbia* just past midnight on July 23, 1999. Intended to be the x-ray observatory with the best spatial and energy resolution, *Chandra* was built on the legacy of an array of international missions, most recently *Einstein* (1978–1981), *ROSAT* (1990–1998), and *ASCA* (1993–2000). The *CXO* is designed to work with complementary x-ray telescopes currently in orbit, including *BeppoSax* and *XMM-Newton*, as well as with telescopes working at other wavelengths, in orbit and on the ground. Astronomers study x-rays to better understand the upper atmospheres of stars; the interaction of relativistic electrons with magnetic fields in the space around black holes; the plasma between the galaxies; and objects at the edge of the visible universe. Because x-rays cannot travel far through the atmosphere, their use in ground-based astronomical studies is limited. X-ray astronomy must be done from space.

CXO design. X-ray photons have much more energy (shorter wavelength and higher frequency) than photons in the optical portion of the electromagnetic spectrum. Because of their high energy, x-ray photons do not reflect off mirrors the way that optical photons do with conventional telescopes, such as the Hubble Space Telescope. Instead, the mirrors of the *CXO* are arranged in consecutive concentric rings, and x-rays are deflected from one mirror to another, much like a stone skipping across the water, then on to the focus. This is called a Wolter type II mirror. This mirror assembly, the greatest design achievement of the *CXO*, can focus x-rays over 10 times more sharply than any previous telescope.

At the focus of the 50-ft-long (15-m) telescope are two detectors, the Advanced CCD Imaging Spectrometer (ACIS) and the High Resolution Camera (HRC). The ACIS is composed of 10 charge-coupled devices (CCDs), which when struck by an x-ray can register the x-ray's energy, or color. The ACIS also measures the location and arrival time of each photon. The HRC uses a different imaging technology, called a microchannel plate, that allows it to measure x-ray arrival times and locations more accurately than ACIS, but it lacks the energy resolution of ACIS. While ACIS acts as a color camera, the HRC is black and white. Each instrument sends its information to a ground station in Cambridge, Massachusetts, where the image is reconstructed.

Located between the mirrors and the cameras are transmission gratings, another of the notable technological achievements of the *CXO*. The gratings resemble large circular screens and can be swung in and out of the optical pathway. The wires are made of gold, and the space between them is very precisely measured. X-rays are deflected by the gratings in a particular pattern. The lower the energy of the x-ray, the greater the deflection. This "dispersive spectroscopy" gives the *CXO* better x-ray energy resolution than has been seen before and allows both cameras to measure chemistry, density, and temperatures in addition to producing exquisite visual images.

Launch, ascent, and initial testing. After reaching low Earth orbit aboard *Columbia*, the *CXO* was released from the shuttle. It then used another rocket and its own thrusters to achieve a final orbit that carries it more than a third of the way to the Moon every 64 hours.

Once in its final orbit, *Chandra*'s telescope mirror covers were opened and the *CXO* observed its first target, a quasar called PKS 0637−752. This source was expected to be a single point of light, suitable for focusing the telescope's optics. But instead of focusing down to a single point, the telescope detected a point with a line off to one side. For 6 hours, astronomers were concerned that there was damage to the telescope. Then it was realized that a radio-wave image of the quasar had a jet in the exact same location as this line. The *CXO* had made its first discovery: an x-ray jet near the edge of the visible universe, driven by a supermassive black hole some 200,000 light-years away from the jet. Scientists are now trying to understand how the black hole can exert such control over particles so far away.

In spite of the surprising discovery, focus was rapidly achieved. The *CXO* then turned to its first demonstration target. The target was the brightest x-ray source in the constellation of Cassiopeia, the remains of a supernova, hundreds of years old, known simply as Cas A. The first demonstration image not only showed the remnant in much sharper focus than ever before, but also showed how the various elements in it—oxygen, neon, magnesium, sulfur, silicon, and iron—were mixed but not homogenized. The arrangement of the elements in the remnant is

an echo of the "shell burning" core of the star that Cas A once was, before it became a supernova. Observers also saw for the first time the central neutron star, the core of what was once a massive star.

Science discoveries. The *CXO* has been in science operations nearly continuously since October 1999. The following are samples of discoveries it has made during that time.

One of the largest of the *CXO*'s projects is the Chandra deep-field survey. These are regions of space that overlap with the Hubble deep fields but are much larger in area. The goal of this experiment is to look back to almost the beginning of time, to help explain when and how the first galaxies formed. The superb resolution of *Chandra* has allowed astronomers to resolve a background sea of x-rays into individual sources that are probably the precursors to modern galaxies. Also near the edge of the observable universe are elusive gamma-ray bursters (GRBs). These are small points of light that, for a moment, outshine the entire universe. Just 5 months after launch, the *CXO* became the first telescope to obtain an x-ray spectrum of a GRB. The observations strengthen the hypothesis that GRBs are the result of a "hypernova," a gigantic star collapsing on itself under its own weight. *See* GAMMA-RAY ASTRONOMY; GAMMA-RAY BURSTS.

Chandra is also studying the gas among somewhat closer galaxies. Galaxies are found in clusters, of which thousands are known. The space between the galaxies in clusters is filled with very thin, very hot gas or plasma called the intergalactic medium. While this plasma has been known for about two decades, with *Chandra* we now see that the gas between the galaxies interacts with the galaxies much like the atmosphere on Earth interacts with landmasses to produce weather. Within these galaxies *Chandra* looks for black holes thought to be at the centers of active galaxies and quasars. Often astronomers are finding jets of material racing away from the black hole at nearly the speed of light.

Nearby galaxies, such as the Magellanic Clouds, contain many supernova remnants similar to Cas A of the Milky Way Galaxy. Supernova remnants are of great interest because they are a milestone in the stellar life cycle. They mark the death of a star, and are the last witness to its interior structure. Supernova remnants also mark the place where heavy elements are reintroduced to interstellar space, from which they will become the next generation of stars. Supernova remnants have demonstrated wide variety, some with very ringlike structures and some more mottled. Not only has remnant around supernova 1987A been imaged, but multiple images have shown the expansion of the explosion into the interstellar gas.

It is within our own galaxy that the resolution of the *CXO* has had its most remarkable results. Planetary Nebulae such as The Cat's Eye (**Fig. 1**) appear to be shaped by underlying hot gas. The first short image of the rapidly rotating pulsar at the center of the Crab Nebula shows what appears to be a whirlpool of

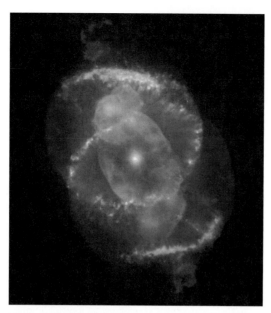

Fig. 1. Combination of Hubble and Chandra observations of the Cat's Eye Nebula. Recent observations by the *CXO* demonstrates that pressure from the expanding, very hot, x-ray emitting gas is responsible for the nebula's feline shape. (*X-ray: NASA/UIUC/Y.Chu et al.; Optical: NASA/HST; Composite: Z.G. Levay*)

material steadily falling onto the pulsar. Meanwhile, jets of material are sent away via the poles in order to conserve angular momentum (**Fig. 2**). While this result is a striking confirmation of theory, the details of the processes will take years to understand.

Among the brightest x-ray phenomena in our galaxy are x-ray binaries. In these systems, a compact

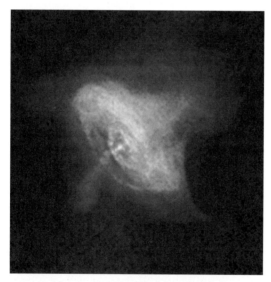

Fig. 2. X-rays from the Crab Nebula are produced by high-energy particles spiraling around magnetic field lines in the nebula. The explosion which formed it was seen on Earth in A.D. 1054. At the center of the nebula is a rapidly spinning pulsar. The image shows the central pulsar surrounded by tilted rings of high-energy particles. Perpendicular to the rings, jetlike structures produced by high-energy particles blast away from the pulsar to conserve angular momentum. The diameter of the inner ring in the image is about one light-year. (*NASA/CXC/SAO*)

star such as a neutron star or black hole has a normal star like the Sun as a companion. The strong gravity of the compact star literally rips the surface off the normal star. The material is then cannibalized by the compact star. *Chandra* has shown that the process is not complete, and that much of the material torn from the normal star is ejected into interstellar space where it becomes material for the next generation of stars.

The *CXO* is also being used to study stellar clusters. In clusters of young stars, like the Orion Nebula, x-ray emission is used as a signature of youth and cluster membership. This allows researchers to make a better census of the stars in this and other clusters. Since most stars in the Galaxy formed in such clusters, the census provides insight into the constituents of the Galaxy itself. Further, many young stars are in the process of forming planetary systems. Astronomers are interested in the effect that stellar x-rays have on the young planetary system. Similar energetics were probably at work when the Sun and Earth formed.

Individual stars do not escape the *CXO*'s study. Their upper atmospheres are common targets. These studies directly reveal the densities and pressures in the coronae above these stars. From this scientists infer the interior structure of stars. The *CXO* has also examined comets and planets within the solar system. Even 10 years ago, x-rays from such cold bodies were considered unlikely. Today such observations, while difficult, are common. They have proven that x-rays from comets are created through a process called charge exchange. In this process, highly charged carbon, oxygen, and nitrogen ions coming from the Sun collide with the upper atmosphere of the solid body. Scientists can now study the interaction of planets with the most tenuous part of the Sun's upper atmosphere, the solar wind.

The results given here are only the beginning of a potentially long career for the *CXO*. The observations already are raising many questions among scientists. Meanwhile, astronomers and engineers have designed the next large x-ray observatory, with launch tentatively scheduled for the end of this decade. The new observatory is designed as a group of four x-ray telescopes working together as a precision team dubbed Constellation-X.

For background information *see* BINARY STAR; CHARGE-COUPLED DEVICES; CRAB NEBULA; GALAXY, EXTERNAL; GAMMA-RAY ASTRONOMY; INTERSTELLAR MATTER; NEBULA; ORION NEBULA; PULSAR; QUASAR; SOLAR WIND; STAR CLUSTERS; SUPERNOVA; X-RAY ASTRONOMY; X-RAY TELESCOPE in the McGraw-Hill Encyclopedia of Science & Technology.

Scott J. Wolk

Bibliography. B. Aschenbach, H. M. Hahn, and J. E. Trumper, *The Invisible Sky ROSAT and the Age of X-ray Astronomy*, Springer-Verlag, 1998; P. A. Charles and F. D. Seward, *Exploring the X-ray Universe*, Cambridge University Press, 1995; R. Naeye, *Signals from Space: The Chandra X-ray Observatory*,

Turnstone Publishing, 2000; W. Tucker and K. Tucker, *Revealing the Universe: The Making of the Chandra X-ray Observatory*, Harvard University Press, 2001.

XML (Extensible Markup Language)

The Internet is now used to publish much primary (first disclosed) scientific information, particularly data. Protein sequences, genes, meteorological data, planetary information, and crystallographic data are now routinely deposited and managed on the Web. Top-quality data can be found almost instantly. In many disciplines, it is now a requirement of primary publication that data are electronically available. This has been made possible largely because of Hypertext Markup Language (HTML), which is used to create documents incorporating text, graphics, sound, video, and hyperlinks.

Origin. Markup is the process of adding information to a document that is not part of its content but provides information about the structure or elements. Markup languages are not new. In 1986, the Standard Generalized Markup Language (SGML) became an International Standards Organization (ISO) standard. SGML has been widely used in the publishing industry (this article was formatted using SGML) and those sectors where technical manuals are a key requirement, but it is large, complex, often expensive, and not widely used by the general public. HTML, based on SGML, made markup languages widely accessible, but it also has limitations. Readers and publishers have been frustrated by getting files in formats they cannot read without (commercial) software, and garbles are common.

HTML cannot manage complex objects and is not suited for input to machines. A major challenge comes from e-commerce, which has an essential need for a robust, machine-readable information protocol, so the World Wide Web Consortium (W3C) has developed Extensible Markup Language (XML). XML is rapidly becoming the standard information protocol for all commercial software such as office tools, messaging, and distributed databases.

Something simpler than SGML was needed, and XML is essentially "SGML-lite," with all SGML's power but with the rarely used and obsolete parts removed. In this case, smaller is better because, unlike SGML, XML is easy to implement and understand. XML was originally conceived by the SGML community as supporting the transmission of complex documents over the Internet. A surprising benefit is that it is probably used even more for data than documents and for providing the "glue" that allows information to flow freely between machines—it has been called the "digital dial tone."

Document structuring. XML is not a super-HTML but a set of rules for writing a markup language. For example, this article could be represented in XML as

```
<book id="MGHYB2002">
   <title>McGraw-Hill YearBook of Science and
Technology</title>
   <date status="publication" convention=
"ISO8601">2000-09</date>
   <isbn status="not known"/>
   <article id="murr02">
      <title>XML</title>
      <subtitle>eXtensible Markup Language
</subtitle>
      <author firstname="Peter" lastname=
"Murray-Rust"/>
      <paragraph>. . .</paragraph>
   </article>
   <article>. . .</article>
</book>
```

This looks very like HTML except that the element-Names ("tagNames"), which describe the document structure, are different. Unlike HTML, all tags must be balanced (that is, have a start and end tag or be empty as in <isbn/>), and attribute values must be quoted (for example, id="murr02"). These conditions make well-formed XML easy to author and to parse. This structured document implicitly describes a book element (information component), with, for example, title and isbn child elements (an element within another element) and many article child elements. The article elements themselves have title, subtitle, author, and many paragraph children. This hierarchical tree structure is easy to process, and there are a lot of high-quality XML tools now available, many being open source (freely distributed).

Most English-speaking humans would understand the sample document above. Many other humans, and all machines, could make no sense of it. XML allows specific disciplines to define rules or vocabularies [such as DocumentTypeDefinitions (DTDs) or XMLSchemas], often represented by FooML (Foo is used to represent a random word such as Math in MathML). The World Wide Web Consortium and the American Mathematical Society (AMS) have created MathML. In similar style, Chemical Markup Language (CML) was developed. Many other disciplines in science, technology, and medicine (astronomy, meteorology, bioscience, materials science, and so on) now have such languages. Open Financial Exchange (OFX) and Interactive Financial Exchange (IFX) were produced by financial consortia, and new markup languages (MLs) are posted almost daily.

These languages represent the ontology of the discipline (what the fundamental concepts are; how to represent them, describe them, and process them). This is a challenging task because there is not a single classification system for scientific information. Communication in some communities (such as crystallography and mathematics) is well coordinated by organizations (for example, the International Union of Crystallography and the American Mathematical Society), but in others such as chemistry the field

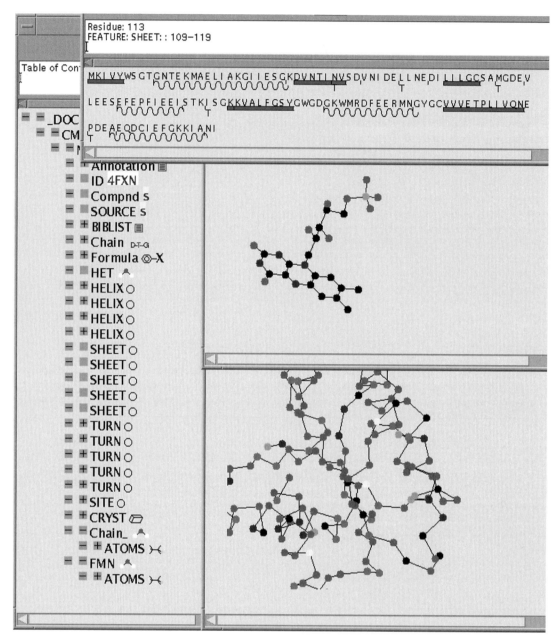

Fig. 1. CML rendering of a molecule in a scientific document.

is fragmented by incompatible, vendor-specific file formats with poor ontologies. Without coordination, software cannot be interoperable (compatible with other platforms such as HTML and SGML).

Some markup languages are generic and used in many disciplines, and the World Wide Web Consortium has activities in such areas. These include mathematics (MathML), graphics (SVG, Scalable Vector Graphics), multimedia (music, audio, streaming video, and slides, in SMIL, Synchronized Multimedia Integration Language), hypermedia ("linking," XLINK), and metadata (RDF, Resource Description Framework). The Organization for the Advancement of Structured Information Standards (OASIS) has formalized the specification for tabular data (CALS, or Continuous Acquisition and Life-Cycle). Many scientific publications (in the widest sense) could be supported with a small number of these horizontal languages. Among the immediate benefits of doing so are robustness of information exchange; a much larger user community than a single discipline; high-quality, usually free software; access to training, books, and so on; and interoperability with other domains.

Current electronic data often omit or garble scientific units and can result in errors such as the loss of the NASA *Mars Climate Observer* spacecraft because of the failed translation of English units into metric units in a segment of ground-based, navigation-related mission software. Published numeric data should have units (including "dimensionless") chosen from a single central repository from an appropriate standards organization, or from agreed domains (such as medical). Frank Olken and John

McCarthy (Lawrence Berkeley National Laboratory) are working on encoding scientific units in XML.

Graphs are also a challenge. Rainfall data, drug response, infrared spectra, and stock prices could all be handled by a generic graphical markup language, but the problem of scalar typed data will need to be solved first (XML schemas).

Other markup languages are domain-specific—for example, Astronomical Instrument Markup Language, AIML, or materials science markup language, MatML, created by Ed Begley at the National Institute of Standards and Technology (NIST). Any organization or community can create its own language, and the world is developing an "ontological marketplace." The value of any such language depends on factors such as usefulness; simplicity; interoperability with other disciplines; availability of software tools; formal (such as ∗.org or ∗.gov) or de facto acceptance by a community of critical size; and vendor marketing and support in areas such as e-commerce.

It takes time (sometimes years) to create a good markup language since it represents the consensus ontology of a community. A new markup language is likely to be successful if the community already has created data dictionaries, standards bodies, and peer-reviewed publications. The major information technology (IT) vendors have realized that single-vendor approaches are counterproductive and now collaborate under the World Wide Web Consortium. In smaller domains, it is even more important that vendors collaborate with the community.

Publishing. XML is not just a language, but also a set of related technologies and a philosophy of "open data" publishing. There are over 20 XML-related activities on the World Wide Web Consortium pages, with support for formatting/styling, searching, indexing, metadata, and so on (see **table**). One key development was the avoidance of name collisions (two entities having the same name) through the use of namespaces (attribute names). A typical example is:

```
<xsl:stylesheet
    xmlns:xsl="http://www.w3.org/1999/XSL/
Transform"
    xmlns="http://www.w3.org/TR/1999/
REC-html401-19991224"
    xmlns:cml="http://www.xmlcml.org">
<xsl:template match="/">
  <HTML>
    <cml:molecule>
      <xsl:apply-templates./>
    </cml:molecule>
  </HTML>
</xsl:stylesheet>
```

This links (arbitrary) namespace prefixes such as xsl and cml to unique strings (because they use domain names). In this way all names in the document, such as cml:molecule, are guaranteed unique. The example also shows a default namespace—the HTML element is implicitly linked to the HTML 4.01 namespace.

Markup Languages will revolutionize scientific publishing. The next generation of Web browsers, office products, and spreadsheets will be based on XML, and authors will create XML documents by default. Any discipline (such as mathematics or chemistry) which can supply high-quality XML-based authoring tools supports authors in creating structured documents. It then becomes possible (with the Extensible Stylesheet Language, XSL) to selectively filter or transform documents. A typical query (with generic tools) is "extract all <CML:molecule>s in this primary publication which also have <CML:crystallography> data." It is not necessary to know what other data are in the paper; the XSL process simply ignores the data. This molecular information might be abstracted into a database, rendered in a CML-aware molecular viewer (such as JMOL), input into a computer program, or used to initiate an online purchase or a robotic chemical synthesis.

XML development activities		
Component	Approach*	Implementation
Text	XHTML(+)	Many
Formatting	XSL(+)	Several
Images (pixel maps)	GIF, PNG(+), JPEG(+)	Many
Structured graphics	SVG($)	Several
Tables	XHTML(+), CALS(+)	Many
Graphs	?	?
Numeric/typed data	XML Schema($)	Many in near future
Units	?(+)	?
Metadata	RDF(+), XMI/ISO11179	?
Links (hypermedia)	Xlink($)	In near future
Terminology	VHG, MARTIF, ISO12620	Generic tools
Citations	Several	Generic
Mathematics	MathML(+)	Some
Molecular sciences	CML, BioML, BSML	Some
Multimedia	SMIL(+)	Some

*Firm "standards" are denoted by +, drafts by $, unsatisfied or unknown requirements by ?
Since The XML field is changing very fast, readers should scan W3C and OASIS pages which are updated daily and fully comprehensive.

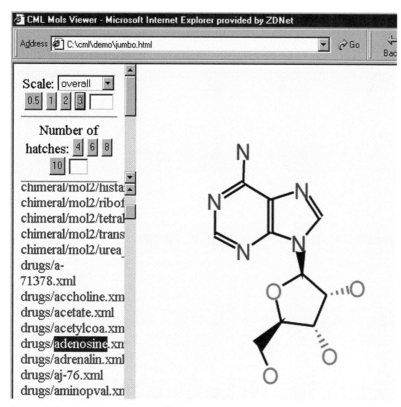

Fig. 2. CML molecules rendered by SVG within a Web browser. Three markup languages (XHTML, SVG, and CML) are interoperating in this diagram.

A typical multicomponent scientific document is shown in **Fig. 1**, which is part of a protein molecule published in the Protein DataBank (PDB). Proteins, small molecules, sequences, and numeric data are all accessible.

Everything in the table of contents (Fig. 1) can be thought of as an information object, such as the Citations (BIBLIST) and text (not shown). Thus CRYST not only holds the primary crystallographic data but also can calculate derived data, including cell volume, reciprocal cell, and orthogonalization matrix.

The World Wide Web Consortium's Scalable Vector Graphics (SVG) supports 2D graphics as a markup language. The graphics are therefore high-resolution objects rather than bitmaps and can be scaled, can be manipulated, or can hold nongraphics information. In addition, SVG can be used within a browser (**Fig. 2**).

Initiatives in primary e-publishing, including theses, will lead to automatic capture of huge amounts of reusable data. If the scientific community collaborates in the markup languages, the tools and the public availability of these data, along with appropriate management and review, will provide a revolutionary benefit in many disciplines.

For background information *see* MULTIMEDIA TECHNOLOGY; SOFTWARE in the McGraw-Hill Encyclopedia of Science & Technology.

Peter Murray-Rust

Yeast systematics

Yeasts are predominantly unicellular fungi that reproduce asexually by budding or fission and do not form sexual stages within or upon a fruiting body. They can reproduce sexually by means of spores (ascospores or basidiospores, belonging to the fungal phyla *Ascomycota* and *Basidiomycota*, respectively), or they can be asexual. There are currently over 600 recognized yeast species. Yeast identification—the assignment of a particular strain to a species and/or genus—is usually accomplished by means of diagnostic morphological and physiological characters in a process known as conventional identification. There is a general agreement that this conventional identification can lead to errors, especially when dealing with new biotypes (variants of a species that possess slightly different capabilities but do not justify the description of a new species). However, the use of molecular characters (genotypic traits) has helped yeast taxonomists with identification.

DNA sequencing. Ideally, all taxonomic categories, including species, should be monophyletic (evolved from a single interbreeding population), and any identification system should be based on the phylogenetic (evolutionary) history of the organism under study. This history is difficult to uncover through phenotypic means because the morphological and physiological traits used for identification are very likely to have evolved several independent times during an organism's evolution. Although it can only be inferred that a certain group of organisms has evolved from a single ancestor, molecular traits, such as DNA sequences, are more likely to show fidelity to phylogenetic history.

Several molecular methods have been used in yeast taxonomy, but nucleic acid sequencing seems to be the most promising. The genome of organisms consists of regions that evolve at different rates, each one being suitable for resolving a different level of relatedness among organisms. In a general matter, higher evolutionary rates are more suitable for comparing taxa that are closely related, such as species and varieties, while regions that evolve more slowly can be used for differentiating genera and families. Selecting the right region to be sequenced is imperative.

There is a general preference toward using ribosomal DNA (rDNA) for phylogenetic inferences because of its universality, the presence of multiple copies that evolve in concert (any alteration in its sequence is corrected or spread among the other copies, resulting in homogeneous copies of the same molecule), and the belief that it is homologous, having originated only once in evolutionary history. The use of homologous molecules for inferring the phylogenetic history of a group of organisms is of ultimate importance because if we want to discover if a certain group of organisms has a common ancestor, we need to use molecules that have a common ancestor.

Each rDNA copy contains rRNA structural genes (5S, 5.8S, 18S, and 26S rDNA—the S value is the sedimentation rate of the RNA during ultracentrifugation in water) alternating with spacer regions. The evolutionary rate varies among these regions; therefore it is necessary to choose the region that has an evolutionary rate compatible with the level of relationship (species/genus) to be studied. Spacer regions, especially the nontranscribed ones, are less prone to evolutionary constraints, evolving more rapidly than structural genes. Even within structural genes, the rate of evolution may vary among different regions.

18S rDNA. The 5S rDNA was the first ribosomal gene to be sequenced in order to resolve relationships among organisms, but its short length and relatively high evolutionary rate preclude its use in comparisons of large numbers of taxa. Few studies have explored the use of 5.8S rDNA for yeast taxonomy. In order to find regions with greater resolving power, systematists started sequencing the 18S rDNA. This molecule is large enough to allow good resolution and contains regions with different evolutionary rates, rendering it suitable to resolve both close and distant relationships, that is, from species to higher levels. There is an extensive database of complete and partial 18S rDNA sequences for ascomycetous and basidiomycetous yeasts, making the sequencing of this gene an important tool for yeast identification and phylogeny.

26S rDNA. Sequencing of the 26S rDNA molecule was performed at the same time as 18S rDNA in the search for regions that were informative enough to deal with closer relationships, mainly at the species level. Because this molecule is so large, sequencing for yeast identification is always partial. Several regions with different evolutionary rates were tested within the 26S rDNA. These regions were chosen because they were previously known to be variable, but no one knew at which level of relationship they could be used. Presently the most influential of DNA regions for yeast identification and phylogeny is the D1/D2 region of the 26S rDNA. This is a region of approximately 600 base pairs (bp), which proved to be valuable for species differentiation in yeasts, and there is an extensive database for yeasts of ascomycetous and basidiomycetous affinities.

Spacer regions. The search for increasing resolution led to the study of the spacer regions of the rDNA. Because these are noncoding regions, they are less subject to evolutionary constraints due to functional loss. The intergenic spacer (IGS), previously known as the nontranscribed spacer (NTS), localized between the rDNA gene clusters, has been studied mostly by restriction analysis, although some partial sequencings are available. This region has proven to be valuable as a population marker and a source of species-specific probes because it is highly variable. The external transcribed spacers (ETS), located at the flanks of each rDNA copy, have rarely been used in yeast taxonomy, whereas the internal transcribed spacers (ITS) are being used not only for yeasts but also for other fungi and plants. The internal transcribed spacers are located between the 18S rDNA and the 26S rDNA, flanking the 5.8S rDNA. One major disadvantage of ITS sequencing is the difficulty of aligning the sequences satisfactorily due to the high rate of divergence. This limits ITS sequencing to the study of very close relatives such as sister species (species that have been separated in recent times) and varieties. The principal application is the design of molecular probes for taxonomic identification.

Bootstrap method. It should be kept in mind that a phylogenetic tree is just an inference of the evolutionary history of an organism and should not be regarded as the whole truth. Trees based on sequencing of different genes or regions may have different topologies, and errors can result during the analysis of the data. One statistical way of assessing the reliability of an estimated tree is the bootstrap method, in which a data set is randomly sampled with replacement from the original set, and this subset is used for constructing a new phylogenetic tree. This process is repeated many times, and the proportion of replications in which a given cluster (branch) appears is computed. If the proportion is higher than 0.95, it is considered to be statistically significant. This is one of the most critical problems during the analysis of the phylogenetic trees. Branches with bootstrap values over 0.95 are not so easy to obtain, thus limiting confidence on the topology of phylogenetic trees whose interior branches have lower bootstrap values. Even when bootstrap values are high, it is advisable that the phylogenetic tree is corroborated by other evidence, such as phenotypic traits or another tree based on an unlinked gene.

Impact of sequencing on yeast taxonomy. The use of phylogenetic guidelines has impacted yeast taxonomy greatly during the past decade, principally in regard to the establishment of monophyletic groups for composing each taxonomic category. An example is the study concerning yeasts of the genus *Saccharomyces*. As these yeasts are of considerable industrial significance, there is a great interest in their taxonomy. Relatively recent sequencing studies have established that this genus is not monophyletic, being interdispersed with members of other yeast genera, although the members of the *Saccharomyces sensu stricto* group (the group which includes *S. cerevisiae*) form a unique cluster. As a consequence of such studies, it seems reasonable to question the current delineation of the genus *Saccharomyces*. Since the delineation of all yeast genera is currently based on phenotypic rather than phylogenetic traits, this kind of question can be extended to other yeast genera as well.

Although sequencing is not a prerequisite for yeast species description, there is a growing trend to include this kind of analysis in the works that describe new species. This approach has some advantages. One is the high confidence with which subject organisms can be assigned to the same taxa; another is the ability to place an organism in relation to known taxa, allowing an overall view of its phylogenetic relationships. The effort of performing sequencing studies for describing new yeast species is facilitating comparison among the work of yeasts systematists throughout the world.

For background information *see* CHEMOTAXONOMY; DEOXYRIBONUCLEIC ACID (DNA); FUNGAL GENETICS; FUNGI; NUCLEIC ACID; PROTEINS, EVOLUTION OF; RIBONUCLEIC ACID (RNA); YEAST in the McGraw-Hill Encyclopedia of Science & Technology.

Patricia Valente

Bibliography. J. W. Fell et al., Biodiversity and systematics of basidiomycetous yeasts as determined by large-subunit rDNA D1/D2 domain sequence analysis, *Int. J. Syst. Evol. Microbiol.*, 50:1351–1371, 2000; C. P. Kurtzman and P. A. Blanz, Ribosomal RNA/DNA sequence comparisons for assessing phylogenetic relationships, in C. P. Kurtzman and J. W. Fell (eds.), *The Yeasts: A Taxonomic Study*, pp. 69–74, Elsevier Science, Amsterdam, 1998; C. P. Kurtzman and C. J. Robnett, Identification and phylogeny of ascomycetous yeasts from analysis of nuclear large subunit (26S) ribosomal DNA partial sequences, *Antonie van Leeuwenhoek*, 73:331–371, 1998; P. Valente, J. P. Ramos, and O. Leoncini, Sequencing as a tool in yeast molecular taxonomy, *Can. J. Microbiol.*, 45:949–958, 1999.

Contributors

The affiliation of each Yearbook contributor is given, followed by the title of his or her article. An article title with the notation "coauthored" indicates that two or more authors jointly prepared an article or section.

A

Abu-Omar, Dr. Mahdi M. *Department of Chemistry and Biochemistry, University of California, Los Angeles, CA.* GREEN CHEMISTRY.

Adewumi, Dr. Michael A. *Pennsylvania State University, University Park, PA.* PIPELINE DESIGN.

Allen, Prof. Heather C. *Department of Chemistry, Ohio State University, Columbus, OH.* INTERFACIAL ANALYSIS.

Almirall, Dr. José R. *International Forensic Research Institute and Department of Chemistry, Florida International University, Miami, FL.* FORENSIC CHEMISTRY—coauthored.

Andersen, Dr. Bjarne. *ALSTOM T & D Power Electronic Systems, Ltd., Stafford, England.* DIRECT-CURRENT TRANSMISSION—coauthored.

Anderson, Dr. Steven W. *Associate Professor of Geology and Chair of The Science Department, Black Hills State University, Spearfish, SD.* LAVA FLOW.

Arndt, Prof. Markus. *Institute for Experimental Physics, University of Vienna, Vienna, Austria.* WAVE-PARTICLE DUALITY—coauthored.

Arnone, Dr. Don. *Toshiba Research Europe Limited, Cambridge, England.* TERAHERTZ TECHNOLOGY.

B

Banks, Dr. John. *Interdisciplinary Arts and Sciences, University of Washington at Tacoma, Tacoma, WA.* AGRICULTURAL ECOLOGY.

Barbara, Dr. Paul F. *Department of Chemistry and Biochemistry, University of Texas, Austin, TX.* SOLVATED ELECTRON—coauthored.

Barnes, Prof. Ted. *Physics Division, Oak Ridge National Lab, Oak Ridge, TN, and Department of Physics and Astronomy, University of Tennessee, Knoxville, TN.* TOP QUARK—coauthored.

Batterman, Dr. Scott. *Consultants Associates, Inc., Cherry Hill, NJ.* FORENSIC ENGINEERING—coauthored.

Batterman, Dr. Steven. *Consultants Associates, Inc., Cherry Hill, NJ.* FORENSIC ENGINEERING—coauthored.

Bayliss, Dr. Sue C. *Departments of Chemistry and Physics, De Montfort University, The Gateway, Leicester, U.K.* BIOCOMPATIBLE SEMICONDUCTOR—coauthored.

Beard, Dr. Chris. *Division of Earth Sciences, Carnegie Museum of Natural History, Pittsburgh, PA.* ANTHROPOID ORIGINS.

Becker, Prof. Karsten. *Institute of Medical Microbiology, University of Münster, Münster, Germany.* FUNGAL GENOTYPING.

Berden, Dr. Giel. *FOM-Institute for Plasma Physics, Rijnhuizen, Nieuwegein, The Netherlands.* MOLECULAR COOLING—coauthored.

Bergles, Prof. Arthur E. *Centerville, MA.* HEAT TRANSFER—coauthored.

Bethlem, Dr. Hendrick L. *FOM-Institute for Plasma Physics, Rijnhuizen, Nieuwegein, and Department of Molecular and Laser Physics, University of Nijmegen, Nijmegen, The Netherlands.* MOLECULAR COOLING—coauthored.

Beuselinck, Dr. Paul R. *USDA-ARS, Plant Genetics Research Unit, University of Missouri, Columbia, MO.* LOTUS.

Bjornbeberg, Dr. D. L. *Soil Scientist, Kimberly, ID.* SOIL EROSION REDUCTION—coauthored.

Black, Prof. JT. *Auburn, University, Auburn, AL.* LEAN MANUFACTURING.

Blaner, Dr. William S. *Department of Medicine, Columbia University, Presbyterian Hospital, New York, NY.* RETINOID RECEPTORS.

Bock, Dr. Jay L. *Department of Pathology, State University of New York at Stony Brook, University Medical Center, Stony Brook, NY.* IMMUNOASSAYS.

Bornhop, Prof. Darryl. *Department of Chemistry, Texas Tech University, Lubbock, TX.* CANCER DETECTION—coauthored.

Bostock, Dr. Michael. *Department of Earth and Ocean Sciences, University of British Columbia, Vancouver, Canada.* CONTINENTAL ROOTS—coauthored.

Boyd, Prof. Richard N. *Department of Physics, Smith Lab, Ohio State University, Columbus, OH.* NUCLEAR ASTROPHYSICS.

Brown, Dr. Michael. *Department of Geology, University of Maryland, College Park, MD.* METAMORPHISM (GEOLOGY).

Bryant, Dr. Harold. *Royal Saskatchewan Museum, Saskatchewan, Canada.* PHYLOGENETIC BRACKETING.

Buckberry, Dr. Lorraine D. *Departments of Chemistry and Physics, De Montfort University, The Gateway, Leicester, U.K.* BIOCOMPATIBLE SEMICONDUCTOR—coauthored.

C

Calderone, Dr. Richard. *Department of Microbiology, Georgetown University School of Medicine, Washington, DC.* CANDIDA.

Cantino, Dr. Philip D. *Department of Environmental and Plant Biology, Ohio University, Athens, OH.* NOMENCLATURE, PHYLOGENETIC.

Caseas, Raphael N. *Department of Chemistry, University of California, Berkeley, CA.* SPECTROSCOPY (CRLAS)—coauthored.

Cohen, Dr. Avis. *Department of Biology and Neuroscience and Cognitive Science, University of Maryland, College Park, MD.* NEUROMORPHIC ENGINEERING.

Collerson, Dr. Kenneth D. *Advanced Centre for Queensland University Isotope Research Excellence, University of Queensland, Brisbane, Australia.* ULTRADEEP XENOLITHS.

Collett, Dr. Thomas S. *School of Biological Sciences, University of Sussex, Brighton, U.K.* HONEYBEE INTELLIGENCE.

Collinson, Prof. Maryanne M. *Department of Chemistry, Kansas State University, Manhattan, KS.* SOL-GEL SENSORS.

Compton, Prof. Robert N. *Departments of Chemistry and Physics, University of Tennessee, Knoxville, TN.* OPTICAL ACTIVITY—coauthored.

Cowan, Dr. Thomas E. *General Atomics—Photonics Division, San Diego, CA.* LASER-INDUCED NUCLEAR REACTIONS.

Crawford, Dr. David L. *International Dark-Sky Association, Tucson, AZ.* LIGHT POLLUTION.

Cummings, Prof. Peter T. *Department of Chemical Engineering, University of Tennessee, Knoxville, TN.* MOLECULAR RHEOLOGY—coauthored.

D

Day, Dr. Simon. *Benfield Greig Hazard Research Centre, Department of Geological Sciences, University College, London, U.K.* SUBOCEANIC LANDSLIDES—coauthored.

Delefosse, Thomas. *Department of Cell and Molecular Biology, Northwestern University Medical School, Chicago, IL.* ANCIENT DNA—coauthored.

Desnoyers, Prof. Jacques E. *INRS-Énergie et Matériaux, Varennes, Québec, Canada.* COLLOID.

Di Fiore, Dr. Anthony. *Department of Anthropology, New York University, New York, NY.* PRIMATE SOCIAL ORGANIZATION.

Dillon, Dr. Robin L. *Pamplin College of Business, Virginia Tech, Falls Church, VA.* PROGRAMMATIC RISK ANALYSIS.

Dogru, Dr. Ali H. *Saudi Arabian Oil Company, Dhahran, Saudi Arabia.* PARALLEL PROCESSING (PETROLEUM ENGINEERING).

Donahoo, Dr. William T. *Division of Endocrinology, Metabolism and Diabetes, University of Colorado Health Sciences Center, Denver, CO.* LEPTIN—coauthored.

Dudka, Dr. S. *Department of Crop and Soil Sciences, University of Georgia, Athens, GA.* TRACE ELEMENTS—coauthored.

E

Eckel, Dr. Robert H. *Division of Endocrinology, Metabolism and Diabetes, University of Colorado Health Sciences Center, Denver, CO.* LEPTIN—coauthored.

Engeda, Dr. Abraham. *Turbomachinery Lab—Mechanical Engineering Department, Michigan State University, East Lansing, MI.* PUMP FLOW.

F

Filippelli, Prof. Gabriel. *Department of Geology and Center for Earth and Environmental Sciences, Indiana University-Purdue University-Indianapolis, Indianapolis, IN.* ECOSYSTEM DEVELOPMENT.

FitzGerald, Dr. Garrett A. *Center for Experimental Therapeutics, University of Pennsylvania, Philadelphia, PA.* ISOPROSTANES—coauthored.

Forsburg, Dr. Susan L. *Molecular Biology and Virology Laboratory, The Salk Institute, La Jolla, CA.* FISSION YEAST.

Franzluebbers, Dr. Alan J. *Soil Ecologist, USDA—Agricultural Research Service, Watkinsville, GA.* ENDOPHYTE GRASSES.

Frisk, Dr. George V. *Applied Ocean Physics and Engineering Department, Woods Hole Oceanographic Institute, Woods Hole, MA.* UNDERWATER SOUND.

Furton, Dr. Kenneth G. *International Forensic Research Institute and Department of Chemistry, Florida International University, Miami, FL.* FORENSIC CHEMISTRY—coauthored.

G

Garnero, Dr. Ed J. *Department of Geology, Arizona State University, Tempe, AZ.* ULTRALOW VELOCITY ZONES (SEISMOLOGY).

Gehrels, Dr. Neil. *NASA Goddard Space Flight Center, Greenbelt, MD.* GAMMA-RAY ASTRONOMY.

Gilleo, Dr. Ken. *Cookson Electronics, Foxboro, MA.* PHOTONICS.

Giuliani, Gaston. *Centre de Recherches Pétrographiques et Géochimiques, Vandoeuvrelès-Nancy, France.* EMERALD.

Goff, Dr. M. Lee. *Department of Entomology, University of Hawaii at Manoa, Honolulu, HI.* FORENSIC ENTOMOLOGY.

Goldberg, Dr. Alfred L. *Department of Cell Biology, Harvard Medical School, Boston, MA.* PROTEASOME—coauthored.

Graham, Don. *Carboloy, Inc., Warren, MI.* DRY CUTTING.

Graham, Dr. Laurie A. *Department of Biochemistry, Queen's University, Kingston, Ontario, Canada.* ANTIFREEZE PROTEIN.

Greaves, Dr. Jane. *Joint Astronomy Centre, Hilo, HI.* STARBURST GALAXY.

Griffin, Prof. John. *Department of Chemistry, Texas Tech University, Lubbock, TX.* CANCER DETECTION—coauthored.

Grimmond, Dr. C. S. B. *Indiana University, Bloomington, IN.* HYDROMETEOROLOGY, URBAN—coauthored.

Gurnis, Dr. Michael. *Seismological Laboratory, California Institute of Technology, Pasadena, CA.* AFRICAN SUPERPLUME.

H

Halpin, Dr. Daniel W. *School of Civil Engineering, Purdue University, West Lafayette, IN.* DESIGN AND BUILD.

Hamilton, John. *Gemini Observatory, Hilo, HI.* GEMINI OBSERVATORY.

Hanagud, Dr. Sathya V. *School of Aerospace Engineering, Georgia Institute of Technology, Atlanta, GA.* SMART STRUCTURES.

Henry, Prof. Eugene W. *Department of Computer Science and Engineering, University of Notre Dame, Notre Dame, IN.* COMPUTER DESIGN.

Huffman, Dr. Paul R. *National Institute of Standards and Technology, Gaithersburg, MD.* ULTRACOLD NEUTRONS.

J

Janeczek, Prof. Janusz. *Faculty of Earth Sciences, University of Silesia, Sosnowiec, Poland.* NATURAL FISSION REACTORS.

Jardetzky, Dr. Theodore. *Department of Biochemistry, Northwestern University, Evanston, IL.* ALLERGY.

Jeffery, Dr. William R. *Department of Biology, University of Maryland, College Park, MD.* EYE DEVELOPMENT.

Jeng-Leng, Dr. Liow. *Department of Mechanical Engineering, James Cook University, Townsville, Queensland, Australia.* SPLASHING DROPS—coauthored.

Jongma, Dr. Rienk T. *FOM-Institute for Plasma Physics Rijnhuizen, Nieuwegein, and Department of Molecular and Laser Physics, University of Nijmegen, Nijmegen, The Netherlands.* MOLECULAR COOLING—coauthored.

K

Kazazian, Dr. Haig H. *Department of Genetics, University of Pennsylvania, Philadelphia, PA.* GENOME INSTABILITY—coauthored.

Knapp, Prof. Alan K. *Division of Biology, Kansas State University, Manhattan, KS.* TALLGRASS PRAIRIE.

Kopylova, Dr. Maya. *Department of Earth and Ocean Sciences, University of British Columbia, Vancouver, Canada.* CONTINENTAL ROOTS—coauthored.

Krechmer, Ken R. *Action Consulting, Palo Alto, CA.* COMMUNICATIONS STANDARDS.

L

Laroussi, Dr. Mounir. *ODU Applied Research Center, Newport News, VA.* COLD PLASMA.

Larson, Dr. Eric B. *Medical Director, University of Washington Medical Center, Seattle, WA.* DEMENTIA—coauthored.

Lecker, Dr. Stewart. *Department of Cell Biology, Harvard Medical School, Boston, MA.* PROTEASOME—coauthored.

Lee, Dr. Minkyu. *Bell Labs-Lucent Technologies, Murray Hill, NJ.* TEXT-TO-SPEECH SYSTEMS.

Lee, Dr. Young C. *The Mitre Corporation, McLean, VA.* INTEGRITY (NAVIGATION).

Levin, Dr. Phillip S. *Northwest Fisheries Science Center, National Marine Fisheries Service, Seattle, WA.* FISHERIES ECOLOGY.

Levitus, Dr. Sydney. *NODC/NOAA, Silver Spring, MD.* OCEAN WARMING.

Lithgow, Dr. Gordon J. *School of Biological Sciences, University of Manchester, Manchester, U.K.* AGING (GENETICS).

Liu, Prof. Clark. *Department of Civil Engineering, University of Hawaii at Manoa, Honolulu, HI.* WASTEWATER REUSE.

Lu, Prof. J. C. *School of Industrial and Systems Engineering, Georgia Institute of Technology, Atlanta, GA.* INFORMATION ENGINEERING (INDUSTRIAL).

Lu, Dr. Zheng-Tian. *Physics Division, Argonne National Laboratory, Argonne, IL.* TRACE-ISOTOPE ANALYSIS.

Lymar, Dr. Sergei, V. *Chemistry Department, Brookhaven National Lab, Upton, NY.* PEROXYNITRITE CHEMISTRY.

M

MacCracken, Dr. Michael C. *Office of the U.S. Global Change Research Program, Washington, DC.* CLIMATE CHANGE IMPACTS.

Mallapragada, Prof. Surya K. *Department of Chemical Engineering, Iowa State University, Ames, IA.* MICROFABRICATION (NERVE REPAIR).

Margulies, Benson. *Vice President, Basis Technology Corp. Cambridge, MA.* POLYGLOT PROGRAMMING.

Mark Welch, Dr. David B. *Department of Molecular and Cellular Biology, Harvard University, MA.* ROTIFERA.

Marrero, Prof. Thomas. *Department of Chemical Engineering, University of Missouri, Columbia, MO.* GRANULAR MIXING AND SEGREGATION.

Marvier, Dr. Michelle. *Department of Biology and Environmental Studies Institute, Santa Clara University, Santa Clara, CA.* GENETICALLY ENGINEERED CROPS; TROPICAL ECOLOGY.

Masliah, Dr. Eliezer. *Departments of Neuroscience and Pathology, University of California San Diego School of Medicine, La Jolla, CA.* DEGENERATIVE NEURAL DISEASES (ANIMAL MODELS).

Mata-Toledo, Dr. Ramon. *Computer Science Department, James Madison University, Harrisonburg, VA.* PETRI NETS.

McCabe, Prof. Clare. *Department of Chemical Engineering, Colorado School of Mines, Golden, CO.* MOLECULAR RHEOLOGY—coauthored.

McCleverty, Dr. Jon A. *School of Chemistry, University of Bristol, Bristol, U.K.* MAGNETIC INTERACTIONS, LIGAND-MEDIATED—coauthored.

McCurry, Dr. Susan M. *Psychosocial and Community Health, University of Washington, Seattle, WA.* DEMENTIA—coauthored.

McDonald, Dr. Hayes. *Department of Molecular Biotechnology, University of Washington, Seattle, WA.* PROTEOMICS—coauthored.

McGrew, Dr. William. *Department of Sociology, Gerontology and Anthropology, Miami University, Oxford, OH.* TOOL EVOLUTION.

Meijer, Dr. Gerard. *FOM-Institute for Plasma Physics Rijnhuizen, Nieuwegein, and Department of Molecular and Laser Physics, University of Nijmegen, Nijmegen, The Netherlands.* MOLECULAR COOLING—coauthored.

Menzel, Prof. E. Roland. *Director, Center for Forensic Studies, Texas Tech University, Lubbock, TX.* FORENSIC PHYSICS.

Meyn, Dr. Raymond E., Jr. *Department of Experimental Radiation Oncology, The University of Texas M.D. Anderson Cancer Center, Houston, TX.* GENE THERAPY (CANCER).

Milster, Dr. Tom D. *Optical Data Storage Center, Optical Sciences Center, University of Arizona, Tucson, AZ.* OPTICAL DATA STORAGE—COAUTHORED.

Mora, Dr. Silvia. *Department of Physiology and Biophysics, University of Iowa, Iowa City, IA.* GLUCOSE TRANSPORTER—coauthored.

Murray-Rust, Prof. Peter. *Virtual School of Molecular Sciences, School of Pharmaceutical Sciences, University of Nottingham, Nottingham, U.K.* XML (EXTENSIBLE MARKUP LANGUAGE).

Murty, Prof. Katta G. *Department of Industrial and Operations Engineering, University of Michigan, Ann Arbor, MI.* SUPPLY CHAIN MANAGEMENT.

N

Nairz, Prof. Olaf. *Institute of Experimental Physics, University of Vienna, Vienna, Austria.* WAVE-PARTICLE DUALITY—coauthored.

Neuberger, Prof. Herbert. *Department of Physics and Astronomy, Rutgers University, Piscataway, NJ.* LATTICE MODELS.

O

Ohgushi, Dr. Hajime. *Department of Orthopedics, Nara Medical University, Kashibara City, Japan.* BIOCERAMICS.

Ohno, Dr. Yuzo. *Laboratory for Electronic Intelligent Systems, Research Institute of Electrical Communication, Tohoku University, Sendai, Japan.* MAGNETOELECTRONICS—coauthored.

O'Leary, Dr. Maureen A. *Department of Anatomical Sciences, State University of New York, Stony Brook, NY.* EVOLUTIONARY RELATIONSHIPS.

Ostertag, Dr. Eric M. *University of Pennsylvania School of Medicine, Department of Genetics,* GENOME INSTABILITY—coauthored.

Ostrowski, Dr. Elizabeth A. *Center for Microbial Ecology, Michigan State University, East Lansing, MI.* DISEASE.

P

Pagel, Dr. Mark. *School of Animal and Microbial Sciences, University of Reading, Reading, U.K.* TRAIT RECONSTRUCTION (EVOLUTIONARY BIOLOGY).

Pagni, Prof. R. M. *Department of Chemistry, University of Tennessee, Knoxville, TN.* OPTICAL ACTIVITY—coauthored.

Parker, Dr. Kevin J. *University of Rochester, UR-SEAS, Rochester, NY.* SONOELASTOGRAPHY—coauthored.

Pate-Cornell, Dr. M. Elisabeth. *Chair, Department of Management Sciences and Engineering, Stanford University, Stanford, CA.* PROGRAMMATIC RISK ANALYSIS—coauthored.

Pessin, Dr. Jeffrey. *Department of Physiology and Biophysics, University of Iowa College of Medicine, Iowa City, IA.* GLUCOSE TRANSPORTER—coauthored.

Peterson, Capt. Benjamin B. *U.S. Coast Guard Academy, Waterford, CT.* DIFFERENTIAL GPS.

Peterson, Dr. Ulrich. *Harry C. Dudley Professor of Economic Geology, Emeritus, Harvard University, Cambridge, MA.* HYDROTHERMAL ORE DEPOSITS.

Pfeifer, Prof. Peter. *Department of Physics and Astronomy, University of Missouri, Columbia, MO; Center for Nonlinear Studies, Los Alamos National Laboratory, Los Alamos, NM.* QUANTUM COMPUTATION.

Piran, Prof. Tsvi. *Racah Institute for Physics, The Hebrew University, Jerusalum, Israel.* GAMMA-RAY BURSTS.

R

Rasbury, Prof. E. Troy. *Department of Geosciences, State University of New York, Stony Brook, NY.* GEOISOTOPES.

Remcho, Prof. Vincent T. *Department of Chemistry, Oregon State University, Corvallis, OR.* CAPILLARY ELECTROCHROMATOGRAPHY.

Rescigno, Dr. Thomas N. *Physics Directorate, Lawrence Livermore Lab, Livermore, CA.* THREE-BODY COULOMB PROBLEM.

Reyes-Garcia, Dr. Carlos Alberto. *Laboratorio de Investigación de Tecnologias Inteligentes, Instituto Tecnologico de Apizaco, Tlaxcala, Mexico.* SOFT COMPUTING.

Reynolds, Dr. Charles F., III. *Professor of Psychiatry and Director, Intervention Research Center for Late-Life Mood*

Disorders, Department of Psychiatry, University of Pittsburgh School of Medicine, Pittsburgh, PA. DEPRESSION.

Roessett, Prof. José. *Professor of Civil and Ocean Engineering, Senior Chair of Engineering in Offshore Technology, Texas A & M University, College Station, TX.* ENGINEERING EDUCATION—coauthored.

Rogers, Dr. John A. *Lucent Bell Laboratories, Murray Hill, NJ.* IMPULSIVE STIMULATED THERMAL SCATTERING (ISTS).

Rogoff, Mortimer. *Navigational Electronic Charts System Association, Washington, DC.* ELECTRONIC CHART.

Rokach, Dr. Joshua. *Claude Pepper Institute and Department of Chemistry, Florida Institute of Technology, Melbourne, FL.* ISOPROSTANES—coauthored.

Rosenberg, Dr. Susan M. *Department of Molecular and Human Genetics, Baylor College of Medicine, Houston, TX.* MUTATION.

Rothschild, Dr. Lynn J. *Ecosystem Science and Technology Branch, NASA Ames Research Center, Moffett Field, CA.* ASTROBIOLOGY.

Rudman, Dr. Murray. *Commonwealth Scientific and Industrial Organisation, Australia.* SPLASHING DROPS—coauthored.

Rutkowski, Prof. Gregory. *Department of Chemical Engineering, University of Louisville, Louisville, KY.* MICROFABRICATION (NERVE REPAIR).

S

Sabo, Dr. John. *National Center for Ecological Analysis and Synthesis, Santa Barbara, CA.* FOOD WEB THEORY.

Sage, Dr. Andrew P. *Founding Dean Emeritus and First American Bank Professor, University Professor, School of Information Technology and Engineering, George Mason University, Fairfax, VA.* COMPLEX ADAPTIVE SYSTEMS.

Saric, Dr. Tomo. *Department of Cell Biology, Harvard Medical School, Boston, MA.* PROTEASOME—coauthored.

Saykally, Prof. Richard. *Department of Chemistry, University of California, Berkeley, CA.* SPECTROSCOPY (CRLAS)—coauthored.

Scawthorn, Dr. Charles. *EQE International, Oakland, CA.* EARTHQUAKE ENGINEERING.

Schmidt, Dr. Georg. *Physikalisches Institut der Universität Würzburg, Würzburg, Germany.* MAGNETOELECTRONICS—coauthored.

Schneiter, Dr. Albert A. *Department of Plant Sciences, North Dakota State University, Fargo, ND.* SUNFLOWER.

Schoonhoven, Laurie. *School of Forest Resources, The Pennsylvania State University, University Park, PA.* SUSTAINABLE FOREST MANAGEMENT—COAUTHORED.

Shalhoub, Dr. Victoria. *Department of Pharmacology/Pathology, Amgen, Inc., Thousand Oaks, CA.* OSTEOPROTEGERIN.

Shannon, Dr. Robert V. *House Ear Institute, Los Angeles, CA.* SPEECH PERCEPTION.

Sharum, Dr. Bernard J. *Fredericksburg, VA.* SYSTEMS ARCHITECTURE.

Shelton, Dr. Ian. *Subaru Telescope, Hilo, HI.* SUBARU TELESCOPE.

Sincerbox, Dr. Glenn. *Optical Data Storage Center, Optical Sciences Center, University of Arizona, Tucson, AZ.* OPTICAL DATA STORAGE—coauthored.

Singer, Dr. Michael J. *Department of Land, Air and Water Resources, University of California, Davis, CA.* SOIL QUALITY—coauthored.

Sojka, Dr. Robert. E. *Soil Scientist, Kimberly, ID.* SOIL EROSION REDUCTION—coauthored; SOIL QUALITY—coauthored.

Son, Dr. Dong Hee. *Department of Chemistry and Biochemistry, University of Texas, Austin, TX.* SOLVATED ELECTRON—coauthored.

Souch, Dr. C. *Indiana University, Indianapolis, IN.* HYDROMETEOROLOGY, URBAN—coauthored.

Spence, Dr. Thomas W. *Geosciences, NSF, Arlington, VA.* EARTH'S SURFACE.

Sprigg, Dr. William A. *Deputy Director, Institute for the Study of Planet Earth, University of Arizona, Tucson, AZ.* SUN-CLIMATE CONNECTIONS.

Squire, Dr. Larry R. *Veterans Affairs Medical Center, San Diego, CA.* FUNCTIONAL MAGNETIC RESONANCE IMAGING (fMRI)—coauthored.

Stark, Dr. Craig E. L. *Veterans Affairs Medical Center, San Diego, CA.* FUNCTIONAL MAGNETIC RESONANCE IMAGING (fMRI)—coauthored.

Stephanishen, Dr. Peter R. *Department of Ocean Engineering, University of Rhode Island, Narrangansett, RI.* SOUND PROPAGATION.

Sullivan, Dr. Gregory M. *Department of Psychiatry, Columbia University, New York, NY.* PANIC DISORDER.

Summers, Dr. David. *Director, Waterjet Lab, Department of Mining Engineering, University of Missouri, Rolla, MO.* WATERJET (MINING).

Sumner, Prof. Malcolm E. *Department of Crop and Soil Sciences, University of Georgia, Athens, GA.* TRACE ELEMENTS—coauthored.

T

Tao, Prof. Rongjia. *Department of Physics, Temple University, Philadelphia, PA.* SUPERCONDUCTING BALLS.

Taylor, Lawrence S. *University of Rochester, Rochester, NY.* SONOELASTOGRAPHY—coauthored.

Thompson, Dr. Anne J. B. *Petrascience Consultants, Inc. Vancouver, Canada.* PROSPECTING.

Triantafyllou, Dr. Michael S. *Department of Ocean Engineering, Massachusetts Institute of Technology, Cambridge, MA.* FISH HYDRODYNAMICS.

Tsvelik, Prof. Alexei. *Department of Physics, University of Oxford, Oxford, U.K.* STRIPE PHASES.

Turner, Dr. Jean. *Department of Physics and Astronomy, University of California, Los Angeles, CA.* STAR FORMATION.

V

Vagenas, Prof. Nick. *Laurentian University Mining Automation Laboratory, Sudbury, Ontario, Canada.* MINE MECHANIZATION.

Valente, Dr. Patrícia. *Centro Federal de Eduçāo Tecnológica de Química de Nilópolis—Unidad Rio de Janeiro, Rio de Janeiro, Brazil.* YEAST SYSTEMATICS.

van Dam, Prof. C. P. *Department of Mechanical and Aeronautical Engineering, University of California, Davis, CA.* WIND TURBINES.

Varma, Prof. Rajender S. *National Risk Management Research Laboratory, U.S. Environmental Protection Agency, Cincinnati, OH.* MICROWAVE ORGANIC SYNTHESIS.

W

Wade, David C. *Reactor Analysis Division, Argonne National Lab, Argonne, IL.* SUSTAINABLE NUCLEAR ENERGY.

Ward, Michael D. *School of Chemistry, University of Bristol, Bristol, U.K.* MAGNETIC INTERACTIONS, LIGAND-MEDIATED—coauthored.

Ward, Dr. Steven N. *IGPP, University of California, Santa Cruz, CA.* SUBOCEANIC LANDSLIDES—coauthored.

Washburn, Dr. Michael, P. *Director, Program on Forest Certification/Liaison to USDA Forest Service, Global Institute for Sustainable Forest Management, Yale School of Forestry and Environmental Studies, New Haven, CT.* SUSTAINABLE FOREST MANAGEMENT—coauthored.

Wayner, Dr. Peter C., Jr. *Isermann Department of Chemical Engineering, Rensselaer Polytechnic Institute, Troy, NY.* HEAT TRANSFER—coauthored.

Weingärtner, Dr. Hermann. *Physikalische Chemie II, Ruhr-University, Bochum, Germany.* SUPERCRITICAL FLUID TECHNOLOGY.

White, Prof. Mary Anne. *Killam Research Professor in Materials Science, Department of Chemistry, Dalhousie University, Halifax, Nova Scotia, Canada.* HEAT STORAGE SYSTEMS.

Wilbur, Dr. Paul. *Engineering Research Center, Colorado State University, Fort Collins, CO.* ELECTRIC PROPULSION.

Wilczek, Prof. Frank. *Department of Physics, Massachusetts Institute of Technology, Cambridge, MA.* QUARK-GLUON MATTER.

Wilson, Dr. Ian D. *Department of Drug Metabolism and Pharmacokinetics, AstraZeneca Pharmaceuticals, Cheshire, U.K.* HPLC-COMBINED SPECTROSCOPIC TECHNIQUES.

Winograd, Prof. Nicholas. *Department of Chemistry, Pennsylvania State University, University Park, PA.* CHEMICAL IMAGING.

Wolk, Dr. Scott J. *Harvard-Smithsonian Center for Astrophysics, Cambridge, MA.* X-RAY ASTRONOMY.

Wolkow, Dr Robert A. *National Research Council of Canada, Ottawa, Ontario, Canada.* HYBRID SILICON-MOLECULAR DEVICES.

Wong, Dr. K. T. G. *ALSTOM T & D Power Electronic Systems, Ltd., Stafford, England.* DIRECT-CURRENT TRANSMISSION—coauthored.

Woodson, Dr. Shawn H. *Naval Air Systems Command, Patuxent River, MD.* WING DROP.

Wyatt, Dr. Philip. *Chief, Department of Genetics, North York General Hospital, Ontario, Canada.* MATERNAL SERUM SCREENING.

Y

Yao, Prof. James T. P. *Professor of Civil Engineering, Texas A & M University, College Station, TX.* ENGINEERING EDUCATION—coauthored.

Yates, Dr. John R., III. *Department of Cell Biology, The Scripps Research Institute, La Jolla, CA.* PROTEOMICS—coauthored.

Yeomans, Dr. Donald C. *Jet Propulsion Laboratory, Pasadena, CA.* ASTEROID.

Yoder, Dr. Anne. *Department of Cell and Molecular Biology, Northwestern University Medical School, Chicago, IL.* ANCIENT DNA—coauthored.

Young, Dr. Grant M. *Department of Earth Sciences, University of Western Ontario, Ontario, Canada.* PROTEROZOIC.

Z

Zeilinger, Prof. Anton. *Institute for Experimental Physics, University of Vienna, Vienna, Austria.* WAVE-PARTICLE DUALITY—coauthored.

Zolensky, Michael. *NASA Johnson Space Center, Houston, TX.* METEORIC INCLUSIONS.

Index

Asterisks indicate page references to article titles.

T